P9-ELU-200

FOUNDATIONS OF

ASTRONOMY

1999 EDITION

MICHAEL A. SEEDS

JOSEPH R. GRUNDY OBSERVATORY

FRANKLIN AND MARSHALL COLLEGE

WADSWORTH PUBLISHING COMPANY

I(T)P® An International Thomson Publishing Company

Belmont, CA • Albany, NY • Boston • Cincinnati • Johannesburg • London • Madrid
Melbourne • Mexico City • New York • Pacific Grove, CA • Scottsdale, AZ • Singapore • Tokyo • Toronto

Science Publisher	GARY CARLSON
Assistant Editor	MARIE CARIGMA-SAMBILAY
Editorial Assistant	LARISA LIEBERMAN
Marketing Manager	CHRISTINE HENRY
Project Editor	JOHN WALKER
Print Buyer	STACEY J. WEINBERGER
Permissions Editor	SUSAN WALTERS
Production	MARY DOUGLAS, ROGUE VALLEY PUBLICATIONS
Text Design	CLOYCE JORDAN WALL
Art Coordinator	MYRNA ENGLER
Copy Editor	MARY ROYBAL
Illustrator	PRECISION GRAPHICS, ALEX TESHIN & ASSOCIATES
Cover Design	CUTTRISS & HAMBLETON
Cover Image	HUBBLE TELESCOPE AND DEEP FIELD, JAN.1996 (BOTTOM) ARE FROM NASA AND REPRODUCED WITH PERMISSION FROM AURA/STScL. SPIRAL GALAXY (TOP) COURTESY OF NASA
Compositor	MONOTYPE COMPOSITION COMPANY, INC.
Printer	TRANSCONTINENTAL PRINTING, INC.

COPYRIGHT © 1999 by Wadsworth Publishing Company

A Division of International Thomson Publishing Inc.

I⟨T⟩P The ITP logo is a registered trademark under license.

Printed in Canada

3 4 5 6 7 8 9 10

For more information, contact Wadsworth Publishing Company, 10 Davis Drive, Belmont, California 94002, or electronically at http://www.wadsworth.com

International Thomson Publishing Europe
Berkshire House
168-173 High Holborn
London, WC1V 7AA, United Kingdom

International Thomson Editores
Seneca, 53
Colonia Polanco
11560 México D.F. México

Nelson ITP Australia
102 Dodds Street
South Melbourne
Victoria 3205, Australia

International Thomson Publishing Asia
60 Albert Street
#15-01 Albert Complex
Singapore 189969

Nelson Canada
1120 Birchmount Road
Scarborough, Ontario
Canada M1K 5G4

International Thomson Publishing Japan
Hirakawa-cho Kyowa Building, 3F
2-2-1 Hirakawa-cho, Chiyoda-ku
Tokyo 102 Japan

International Thomson Publishing Southern Africa
Building 18, Constantia Square
138 Sixteenth Road, P.O. Box 2459
Halfway House, 1685 South Africa

All rights reserved. No part of this work covered by the copyright hereon may be reproduced or used in any form or by any means—graphic, electronic, or mechanical, including photocopying, recording, taping, or information storage and retrieval systems—without the written permission of the publisher.

Library of Congress Cataloging-in-Publication Data

Seeds, Michael A.
    Foundations of astronomy/Michael A. Seeds.—1999 ed.
        p. cm.
    Includes bibliographical references and index.
    ISBN 0-534-54765-6
    1. Astronomy. I. Title.
QB43.2.S43   1998
520—dc21   98-7479
                                                                               CIP

Contents

A Note

FROM MICHAEL SEEDS TO THE STUDENT

EACH OF US FACES THE ULTIMATE question that has intrigued humans since the beginning of awareness. Great minds like Plato, Beethoven, and Hemingway have tried to give their personal answers, but it continues to perplex every one of us. The ultimate question is simply, "What are we?"

Certainly part of the answer goes beyond the physical universe to include philosophical and cultural issues. We define ourselves by what we create, what we worship, what we admire, and what we demand of each other. But a major part of the answer is embedded in the physical universe, and for this we turn to astronomy. No other science speaks so clearly about our place in the universe. We cannot hope to understand what we are until we understand where we are in a universe filled with worlds, stars, and galaxies.

Introductory astronomy is the ideal science course to help students understand what they are, not because astronomy is a better science, but because introductory astronomy is different from most other science courses. There are only about 10,000 astronomers in the world, not enough to fill a football stadium. So the primary goal of astronomy courses is not to attract large numbers of students to the profession. Also, the principles of astronomy are not needed for any particular career, as is the case with chemistry. That means the basics of astronomy are not necessary for future physicians, engineers, lawyers, or business leaders. Thus, in introductory astronomy classrooms, the goals are quite different from those of other sciences.

GOALS

Teachers of introductory astronomy express their goals in different ways, but the words all boil down to two things. First, we want you, our students, to understand where you are in the cosmos. The universe is elegant and beautiful, and we want you to enjoy it and admire it as we do. But more important, we want you to understand astronomically where you are. Only then can you face the bigger question of what you are.

Our second goal necessarily follows our first goal. It is not enough to tell our students what we know. We must also tell you how humanity has learned to understand the physical universe. That is, we want you to understand science as a way of learning and knowing about nature. Not only must you understand the methods of science before you can accept the validity of scientific knowledge, but you must understand the nature of science if you are to function in the twenty-first century.

> *"By the end of my course, I want my students to know a few basic things: that they live in a very big universe that is described by a small set of rules; that those rules are knowable; that the human race has found a way to figure out the rules."*

The most important concept in an introductory astronomy course is the process of science. The old three-step scientific method—observe, hypothesize, experiment—is not a good description. It makes science sound like a sausage grinder that stuffs facts into theories. The real process of science is much more complex, more human, more creative, and more exciting.

Some students see astronomy as the dreary application of a few laws of physics. In fact, astronomy is tremendously exciting because those laws of physics are tools that help us discover the secrets of the stars. The central concepts of astronomy show how physical laws prescribe natural processes. Thus we see how Newtonian gravitation prescribes orbital motion, how nuclear fusion prescribes stellar structure, how the physics of condensation from a vapor prescribes planetary compositions. Our goal is not to teach physics, but to show you how physics prescribes the universe. In learning these concepts, you will see physics, not as a set of laws to be memorized, but as powerful tools that help us understand the universe.

1999 REVISION

Astronomy is exciting because it is always changing—with new photographs, new discoveries, and, more important, new understanding. The pages that follow contain the latest advances in astronomy, but they also contain a carefully refined guide to help you understand the universe.

The Hubble Space Telescope is revolutionizing astronomy with new images and spectra. Those photos and discoveries are distributed throughout this new 1999 edition. Other giant telescopes such as the Keck telescopes, new instruments such as infrared cameras, and new spacecraft such as Soho and Galileo contribute to the wealth of images and discoveries included here.

The real excitement of astronomy goes beyond pretty pictures. The real excitement lies in a new under-

WINDOW ON SCIENCE 20-2

Skeptical Storytellers

"Scientists are just a bunch of skeptics who don't believe in anything." That is a common complaint about scientists, but it misinterprets the fundamental characteristic of the scientist. Yes, scientists treat new discoveries with skepticism, but scientists do hold strong beliefs about how nature works, and their skepticism is the tool they use to test their beliefs. Ultimately, scientists try to tell stories about how nature works, and if those stories are to be correct, then every idea must be tested over and over.

When the discovery of a planet orbiting the star 51 Pegasi was first announced, astronomers were skeptical—not because they didn't expect other stars to have planets and not because they thought the observations were wrong, but because that is the way science works. Every observation is tested, and every

discovery must be confirmed. Only an idea that survives many tests begins to be accepted as a scientific truth, what a scientist would call a law of nature.

This makes scientists seem like irritable skeptics to nonscientists, but among scientists it is not bad manners to ask, "Really, how do you know that?" or "Why do you think that?" or "What evidence do you have?"

The goal of science is to tell stories about how nature works, stories that are sometimes called theories or hypotheses. A story by John Steinbeck can be a brilliant work of art invented to help us understand some aspect of human nature, and another story, such as a TV script, can be little more than entertaining chewing gum for the mind. But a scientist tries to tell a story that is different in one critical way—it is

the truth. To create such stories about nature, scientists cannot be artistic and create facts to fit their story, and they cannot invent a plot just to be entertaining. Every single link in the scientist's story must be based on evidence and clear, logical steps. Of course, evidence can be misunderstood, and logical slipups happen all too often. To test every aspect of the story, the scientist must be continuously skeptical. That is the only way to discover those scientific stories that help us understand how nature works.

Skepticism is not a refusal to hold beliefs. Scientists often believe sincerely in their theories and hypotheses once they have been tested over and over. Rather, skepticism is the tool scientists use to find those natural principles worthy of belief. ∎

standing of nature, and this edition includes our newest insights into the age of the universe, the history of Europa, the death of sunlike stars, colliding galaxies, solar neutrinos, dark matter, and more.

This 1999 edition has also provided a chance to refine the approach that has proved so successful in previous editions. It is carefully focused on helping students understand two concepts: nature on the astronomical scale and the process of science. This dual focus lies at the heart of this book, and this revision strengthens that emphasis.

It is critical that students know how science helps us understand nature, and, with that in mind, I have totally revised Chapter 16, The Milky Way Galaxy, to show you how we have learned about the star system we live in. Furthermore, I have fully revised Chapter 24, Jupiter and Saturn, to show you not just what we know about these two worlds, but how we know it. When you recognize the interplay of evidence and hypothesis, you understand something much more important than a mere astronomical technique. You understand a way of knowing about the world we live in. Many of the arguments in the book have been revised to focus on that reliance on evidence.

In revising this book, I have created some new tools to help students, and I have retained and revised other tools from previous editions. The principal tools in the book are:

- Windows on Science
- Guideposts
- Critical Inquiries
- Critical Inquiries for the Web
- Review Questions
- Discussion Questions
- Mathematical Problems
- Data Files

While you read the text and learn about the main concepts of modern astronomy, the Windows on Science provide a parallel commentary on how all of science

THE SKY

GUIDEPOST

Our goal in studying astronomy is to learn about ourselves. We search for an answer to the question "What are we?" The quick answer is that we are thinking creatures living on a planet that circles a star we call the sun. In this chapter, we begin trying to understand that answer. What does it mean to live on a planet? The preceding chapter gave us a quick overview of the universe, and chapters later in the book will discuss details. This chapter and the next help us understand what the universe looks like seen from the surface of our spinning planet.

We will see in Chapter 4 how difficult it has been for humanity to understand what we see in the sky every day. In fact, we will discover that science was born when people tried to understand the appearance of the sky. ∎

The Southern Cross I saw every night abeam.
The sun every morning came up astern;
every evening it went down ahead. I wished
for no other compass to guide me, for these
were true.

CAPTAIN JOSHUA SLOCUM

Sailing Alone Around the World

works. The Windows point out where we are using statistical evidence, where we are reasoning by analogy, where we are building a scientific model in place of a scientific hypothesis, and so on. The Windows allow you to momentarily step aside and observe *how* you are thinking instead of concentrating on *what* you are thinking.

The Guidepost at the beginning of each chapter helps you see the organization of the book. While the chapter introduction previews the astronomical content and sets the objectives of the chapter, the Guidepost connects the chapter with preceding and following chapters to provide an overarching organizational guide.

To encourage you to read carefully and test your understanding of the material, each section of a chapter ends with a review question called a Critical Inquiry. A short answer follows to show you how a scientific argument uses observations, evidence, theories, and natural laws. The Critical Inquiry then poses a further question that gives you the opportunity to construct your own argument on a related issue.

Each chapter ends with a selection of Internet Critical Inquiries for the Web, exercises for you to explore using the World Wide Web. Some of these will lead you to new discoveries, and others will help you review concepts explored within the chapter.

Two kinds of questions end each chapter. Review Questions help you test your understanding of the material and consolidate the scientific arguments you have read about. These questions are followed by a few Discussion Questions that are designed to go beyond the text and stimulate you individually or your entire class to think critically and creatively about scientific questions.

Many students feel so uncomfortable with mathematics that they are unable to see the beauty of nature. Too often the math is isolated from the main story of astronomy as if the math were a bowl of anchovies that could be added to taste. I want students to understand what science is, which means they must see mathematics as an indispensable part of science. At the same time, I want students who are less comfortable with math to be able to follow the arguments. To serve all students well, I have included necessary math in the text, not isolated in boxes, to assure students that the mathematical aspects of science are not peripheral but integral to our understanding of the universe. At the same time, I have not made arguments that depend entirely on mathematical

reasoning, and thus a less mathematical student can follow the story. My experience over the years is that students are more mathematically capable than instructors sometimes expect, more capable than they will admit, and more capable than even they expect. Thus, I have included a set of Problems that hinge on mathematical calculations or reasoning at the end of each chapter.

CRITICAL INQUIRY

Why can't we use ground-based telescopes to measure the parallax of stars farther than 100 pc from the earth?

A telescope on the earth's s the earth's atmosphere, an by the turbulent air) spread blob. At the best observato the star images are usually arc. To measure the parall the positions of these fuzzy many measurements and a the accuracy of such a mea of arc. If a star has a parall will make errors of roughly we try to measure it. In oth to measure it very accurate of 0.01 second of arc woul by 0.01, or 100 pc, so that which we are unable to me

Of course, we could p around the earth. How wo limit the accuracy then?

■ *Questions*

1. What factors resist the contraction of a cloud of interstellar matter?

2. Explain four different ways a giant molecular cloud can be triggered to contract.

3. What evidence do we have that (a) star formation is a continuing process? (b) protostars really exist? (c) the Orion region is actively forming stars?

4. How does a contracting protostar convert gravitational energy into thermal energy?

5. How does the geometry of bipolar flows and Herbig–Haro objects support our hypothesis that protostars are surrounded by rotating disks?

■ *Discussion Questions*

1. Ancient astronomers, philosophers, and poets assumed that the stars were eternal and unchanging. Is there any observation they could have made or any line of reasoning that could have led them to conclude that stars don't live forever?

2. How does hydrostatic equilibrium relate to hot-air ballooning?

■ *Critical Inquiries for the Web*

1. In 1997, the Hubble Space Telescope was outfitted with an infrared sensitive instrument called NICMOS. Search the Internet for information about recent observations of star-forming regions with this instrument. Choose a particular object and summarize how the NICMOS observations support or enhance our understanding of the process of star formation.

2. If neutrinos are so elusive, how do astronomers go about detecting them? Use an Internet search engine to browse for information on solar neutrino detectors that are currently in operation, under construction, or proposed. Determine similarities and differences between these detectors in terms of method of detection and energy range of detectable neutrinos.

Go to the Wadsworth Astronomy Resource Center (www.wadsworth.com/astronomy) for critical thinking exercises, articles, and additional readings from InfoTrac College Edition, Wadsworth's online student library.

REMEMBERING AND UNDERSTANDING

My task in teaching astronomy is not to teach facts to be remembered but to teach concepts to be understood. It is easy to teach and test facts, and most students feel safe memorizing and recalling facts when needed. But the two goals in teaching astronomy require going beyond facts. For example, a book might report that Jupiter is the most massive planet, but the real insight lies in the process that made Jupiter massive. Until we understand the concept of the solar nebula hypothesis and the processes by which the inner and outer planets grew, the large mass of Jupiter is just an isolated fact. Once we understand the concept, the mass of Jupiter fits into a bigger picture. It not only makes sense, but it is easy to remember without memorization.

This book innovates not by omitting facts but by using facts the way scientists use them, as keys to understanding. I have focused discussions to show you how astronomers use facts as clues to new hypotheses and as evidence to test existing hypotheses. Of course, certain facts should be in any astronomy book in case, for example, you need to look up the radius of Saturn. All textbooks serve, to some extent, as reference books. I have collected those basic descriptive facts into Data Files in appropriate chapters, which not only get the factual data out of the discussion of concepts but also help you form a quick impression of the principal objects of study.

There are no Oscars for college courses, but teachers of introductory astronomy can take heart that they would be big winners if there were. Introductory astronomy focuses on the things that, as nonscience students,

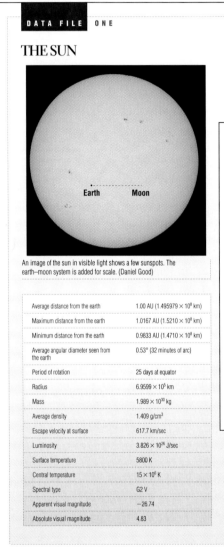

DATA FILE ONE

THE SUN

An image of the sun in visible light shows a few sunspots. The earth–moon system is added for scale. (Daniel Good)

Average distance from the earth	1.00 AU (1.495979×10^8 km)
Maximum distance from the earth	1.0167 AU (1.5210×10^8 km)
Minimum distance from the earth	0.9833 AU (1.4710×10^8 km)
Average angular diameter seen from the earth	0.53° (32 minutes of arc)
Period of rotation	25 days at equator
Radius	6.9599×10^5 km
Mass	1.989×10^{30} kg
Average density	1.409 g/cm³
Escape velocity at surface	617.7 km/sec
Luminosity	3.826×10^{26} J/sec
Surface temperature	5800 K
Central temperature	15×10^6 K
Spectral type	G2 V
Apparent visual magnitude	−26.74
Absolute visual magnitude	4.83

▪ Problems

1. In the model shown in Figure 13-1, how much of the sun's mass is hotter than 12,000,000 K?

2. What is the life expectancy of a 16-solar-mass star? of a 50-solar-mass star?

3. How massive could a star be and still survive for 5 billion years?

4. If the sun expanded to a radius 100 times its present radius, what would its density be? (HINT: The volume of a sphere is $\frac{4}{3}\pi R^3$.)

5. If a giant star 100 times the diameter of the sun were 1 pc from us, what would its angular diameter be? (HINT: Use the small-angle formula, in Chapter 3.)

6. What fraction of the volume of a 5-solar-mass giant star is occupied by its helium core? (HINTS: See Figure 13-8. The volume of a sphere is $\frac{4}{3}\pi R^3$.)

7. If the stars at the turnoff point in a star cluster have masses of about 4 solar masses, how old is the cluster?

8. If an open cluster contains 500 stars and is 25 pc in diameter, what is the average distance between the stars? (HINTS: What share of the volume of the cluster surrounds the average star? The volume of a sphere is $\frac{4}{3}\pi R^3$.)

9. Repeat Problem 8 for a typical globular cluster containing a million stars in a sphere 25 pc in diameter.

10. If a Cepheid variable star has a period of pulsation of 2 days and its period increases by 1 second, how late will it be in reaching maximum light after 1 year? after 10 years? (HINT: How many cycles will it complete in a year?)

you should understand to function in the modern world and to appreciate your lives in this beautiful universe. That is why introductory astronomy is the ideal introductory science course. It focuses not on the bare facts, but on an understanding of what we are, where we are, and how we know. Every page of this 1999 edition reflects that ideal.

Mike Seeds
m_seeds@acad.fandm.edu
http://www.fandm.edu/Departments/Astronomy/Astronomy.html

ABOUT THE AUTHOR

Mike Seeds is the John W. Wetzel Professor of Astronomy at Franklin and Marshall College as well as Director of the College's Joseph R. Grundy Observatory. His research interests focus on peculiar variable stars and the automation of astronomical photometry. He is the Principal Astronomer in charge of the Phoenix 10, the first fully robotic telescope on Mt. Hopkins in Arizona. In 1989 he received the Christian R. and Mary F. Lindback Award for Distinguished Teaching. In addition to writing textbooks, Seeds pursues an active interest in astronomical research, creates educational tools for use in computer-smart classrooms, and continues to develop his upper level courses in Archaeoastronomy and in the History of Astronomy. He has also published educational software for preliterate toddlers! Seeds was Senior Consultant in the creation of the 26-episode telecourse, *Universe: The Infinite Frontier*. Seeds is author of *Horizons: Exploring the Universe*, 5th edition and *Astronomy: The Solar System and Beyond*, for Wadsworth, as well as (with Joseph R. Holzinger) *Laboratory Exercises in Astronomy* (Prentice-Hall).

http://www.wadsworth.com/earthnet.html

THREE EASY WAYS TO REQUEST A COMPLIMENTARY REVIEW COPY

1. E-mail review@wadsworth
2. Fax 1-800-522-4923 (on school letterhead, include course title, number of students, and decision date)
3. Mail information in item two above to address below, Attn: Academic Resource Center

Wadsworth Publishing Company
I(T)P® *an International Thomson Publishing Company*

10 Davis Drive
Belmont, CA 94002

Acknowledgments

This book began life 5 years ago with a major rethinking of the goals of teaching introductory astronomy. In that time a small army have contributed, and I deeply appreciate the assistance and cheerful good will of so many astronomy people. I am glad to be part of such a welcoming community.

My thanks go to my own students and to the teachers and students around the world who have responded so enthusiastically to *Foundations of Astronomy*. Their questions, comments, and suggestions have guided every aspect of this new edition. I especially thank the many reviewers whose detailed analysis of sample chapters have refined this book and made it a teacher's tool: William N. Anderson, Phoenix College; Wallace Arthur, Fairleigh Dickinson University; Timothy Barker, Wheaton College; Henry E. Bass, University of Mississippi; William P. Bidelman, Case Western Reserve University; Michel Breger, University of Texas at Austin; Michael F. Capobianco, St. John's University; Neil F. Comins, University of Maine at Orono; Peter S. Conti, University of Colorado; John J. Cowan, University of Oklahoma; Russell J. Dubisch, Siena College; Robert J. Dukes, Jr., The College of Charleston; T. Stephen Eastmond, Rancho Santiago College; I. X. Finegold, Drexel University; Jack K. Fletcher, Eastern Kentucky University; Marjorie Harrison, Sam Houston State University; Thomas G. Harrison, North Texas State University; Paul W. Hodge, University of Washington; Terry Jay Jones, University of Minnesota; Thomas O. Krause, Towson State University; Nathan Krumm, University of Cincinnati; Robert Leacock, University of Florida; Michael E. Mickelson, Denison University; Leonard Muldawer, Temple University; John Peslak, Jr., Hardin-Simmons University; Lawrence Pinsky, University of Houston; C. W. Price, Millersville University; Robert Quigley, Western Washington University; W. L. Sanders, New Mexico State University; Gary D. Schmidt, University of Arizona; Robert F. Sears, Austin Peay State University; Don Speed, Phoenix College; Michael L. Stewart, San Antonio College; Yervant Terzian, Cornell University; Leslie J. Tomley, San Jose State University; Raymond E. White, University of Arizona; Anne G. Young, Rochester Institute of Technology. Manuscript reviewers for the fourth edition were: Wayne Christiansen, University of North Carolina, Chapel Hill; David Friend, University of Montana; Donald Foster, Wichita State University; Donald Hayes, Pima Community College; David Hufnagle, Johnson County Community College; William Keel, University of Alabama; Jim Klavetter, California State University, Sacramento; Robert Leacock, University of Florida, Gainesville; Art Litka, Seminole Community College; David Loebakka, University of Tennessee, Martin; Michael LoPresto, Henry Ford Community College; Loris Magnani, University of Georgia, Athens; John Mathis, University of Wisconsin, Madison; Gary Mechler, Pima Community College; Mark Nook, St. Cloud State University; Michael J. Ruiz, University of North Carolina, Asheville; Paul Rybski, University of Wisconsin, Whitewater; Sheldon Schafer, Bradley University; Horace Smith, Michigan State University; Donald Terndrup, Ohio State University; Bruce Twarog, University of Kansas; George Wolf, Southwest Missouri State University.

Manuscript reviewers for this book are: Gene Byrd, University of Alabama; R. Kent Clark, University of South Alabama; Geoffry Clayton, Louisiana State University; Stephen Gottesman, University of Florida; Bob Hamilton, Solano College; Bruce Hanna, Old Dominion University; Russell Harkay, Keene State College; Gary Mechler, Pima Community College, West Campus; Harrison Prosper, Florida State University; Peter Shull, Oklahoma State University; Jim Smeltzer, NW Missouri State University; Peter Strine, Bloomsburg University; and David Weinberg, Ohio State University, Columbus.

The many people listed in the illustration credits were very kind in providing photographs and diagrams. Special recognition goes to the following, who were always ready to help locate unusual images and data for the book. Sabine Alrieau, SERENDIP; David Anderson, Southern Methodist University; Edmund Bertschinger, MIT; Adam Burrows, Steward Observatory; Coral Cooksley, Anglo-Australian Observatory; Don Davis, Studio City California; Diane Dutkevitch, Northwestern University; James Gelb, MIT; Margaret Geller, Center for Astrophysics; Cheryl Gundy, Space Telescope Science Institute; Patrick Hartigan, Rice University; Floyd Herbert, Lunar and Planetary Lab; R. Scott Hudson, JPL; Shaun Hughes, Royal Greenwich Observatory; George Jacoby, National Optical Astronomy Observatory; Russell Kempton, New England Meteoric Services; James Kerr, Atmospheric Environment Service, Canada; Deborah Levine, IPAC; Gordon MacAlpine, University of Michigan; David Malin, Anglo-Australian Observatory; Larry Marschall, Gettysburg College; David Miller, Halls Creek, Australia; Steven Ostro, JPL; Andrew Perala, W. M. Kack Observatory; Carolyn C. Porso, University of Arizona; Charles Prosser, Center for Astrophysics; Pat Rawlings, SAIC, Inc.; Roger Ressmeyer, Starlight Collection; Lincoln Richman, Science; Vera Rubin, Department of Terrestrial Magnetism; Rudolph E. Schid, Center for Astrophysics; Damon Simonelli, Center for Radiophysics and Space Research; Mike Skrutskie, University of Massachusetts; Patricia Smiley, NRAO; Steven Snowden, Goddard Space Flight Center; Hyron Spinrad, University of California, Berkeley; John Stauffer, Center for Astrophysics; Lisa Vazquez, Corbis; Ray Villard, Space Telescope Science Institute; Richard Wainscoat, Institute for Astronomy; Richard E. White, Smith College.

My appreciation also goes to the following institutions for their assistance in providing figures: The Anglo-Australian Observatory; The Astrophysical Journal; Ames Research Center, Ball Laboratories; Brookhaven National Laboratories; California Association for Research in Astronomy; Celestron International; European Southern Observatory; U.S. Geological Survey; The Granger Collection; High Altitude Observatory; Hughes Aircraft; Jet Propulsion Laboratory; Johnson Space Center, Lick Observatory; Lowell Observatory; Lunar and Planetary Laboratory; Martin Marietta Aerospace Corporation; McDonald Observatory; Mt. Wilson and Las Campanas Observatories; National Aeronautics and Space Administration; National Optical Astronomy Observatories; National Radio Astronomy Observatory; Palomar Observatory; Pennsylvania Department of Transportation; Starlight Collection; TRW, Inc.; Yerkes Observatory.

I thank my partner at Franklin and Marshall College, Dana Backman, for his cheerful assistance with questions about the origin of stars and planets.

This book, which is derived from my larger volume, *Foundations of Astronomy*, is the result of Gary Carlson's helping me understand the multidimensional way the book is used by teachers and students. I thank him and my previous editors, Jennie Burger and Kim Leistner for their patience, enthusiasm, and expertise. The Wadsworth team is a talented group of professionals and a great group of friends. I have enjoyed working with them. They include Marie Carigma-Sambilay, Christine Henry, Joseph Jodar, Tami Strang, Mary Roybal, John Walker, Cloyce Jordan Wall, and Stacey Weinberger. I would especially like to thank Mary Douglas and Myrna Engler of Rogue Valley Publications for their patience with my various failings. Most of all, I want to thank my wife Janet and my daughter Kate for their patience with the fourth member of the family, "the book."

Mike Seeds

Chapter One

THE SCALE OF THE COSMOS

GUIDEPOST

How can we study something so big it includes everything, including us?

How do we start thinking about the universe? Perhaps the best way is to grab

a quick impression as we zoom from things our own size up to the largest

things in the entire universe.

That cosmic zoom, the subject of this

chapter, gives us our first glimpse of

the subject of the rest of this book—

the universe. In the next chapter, we

will return to Earth to think about the

appearance of the sky, and subsequent

chapters will discuss other aspects of

the universe around us.

As we study the universe, we must also observe the process by which we learn.

That process, science, gives us a powerful way to understand not only the uni-

verse but also ourselves. ■

The longest journey begins with a single step.

CONFUCIUS

We are going on a voyage out to the end of the universe. Marco Polo journeyed east, Columbus west, but we will travel away from our home on Earth, out past the moon, sun, and other planets, past the stars we see in the sky, and past billions more that we cannot see without the aid of the largest telescopes. We will journey through great whirlpools of stars to the most distant galaxies visible from Earth—and then we will continue on, carried only by experience and imagination—looking for the structure of the universe itself.

Besides journeying through space, we will also travel in time. We will explore the past to see the sun and planets form and search for the formation of the first stars and the origin of the universe. We will also explore forward in time to watch the sun die and the earth wither. Our imagination will become a scientific time machine searching for the ultimate end of the universe.

Though we may find an end to the universe, a time when it will cease to exist, we will not discover an edge. It is possible that our universe is infinite and extends in all directions without limit. Such vastness dwarfs our human dimensions, but not our intelligence or imagination.

Astronomy—this imagined voyage—is more than the study of stars and planets. It is the study of the universe in which we exist. Our personal lives are confined to a small corner of a small planet circling a small sun drifting through the universe, but astronomy can take us out of ourselves and thus help us understand what we are.

Our study of astronomy introduces us to sizes, distances, and times far beyond our common experience. The comparisons in this chapter are designed to help us grasp their meaning.

How big is a star? The answer—roughly 1 million miles in diameter—is meaningless. Such a large number tells us nothing. How can we humans, only 5 or 6 feet tall, hope to understand the vastness of the universe? The secret lies in the single word *scale*.

To understand the universe, we must understand the relative scale of planets, stars, galaxies, and the universe as a whole. Only when we can relate our own body size to the astronomical universe around us can we begin to understand nature on the grandest scale.

To illustrate the scale of astronomical bodies—to fit ourselves into the universe—we will journey from a campus scene to the limits of the cosmos in 12 steps. Each successive picture in this chapter will show a region of the universe that is 100 times wider than the preceding picture. That is, each step will widen our view by a factor of 100.

Figure 1-1 shows a region about 52 feet across. It is occupied by a human being, a sidewalk, and a few trees—all objects whose size we can understand. Only 12 steps separate this scene from the universe as a whole.

In Figure 1-2, we increase our field of view by a factor of 100 and see an area 1 mile across. The area of the preceding photograph is shown by the arrow. Individual

FIGURE 1-1
(Michael A. Seeds)

FIGURE 1-2
(Pennsylvania Department of Tourism, Bureau of Design)

FIGURE 1-3
(NASA infrared photograph)

and erodes the mountains, washing material down the rivers and into the sea. The mountains and valleys that we know are only temporary features; they are constantly changing. As we explore the universe, we will see that it, like the earth's surface, is evolving. Change is the norm in the universe.

Green foliage of different kinds appears as shades of red in this infrared photo, a reminder that our human eyes see only a narrow range of colors. As we explore the universe, we will learn to use a wide range of "colors,"

people, trees, and sidewalks vanish, but now we can see a college campus and the surrounding streets and houses.

These dimensions are familiar. We have been in houses, crossed streets, and walked or run a mile. We have personal experience with such dimensions, and we can relate them to the scale of our bodies.

Although we have begun our adventure using feet and miles, we should use the metric system of units because it makes arithmetic simpler (see Appendix A). To convert 1 mile to inches, we must multiply 5280 × 12 = 63,360 in. To convert 1 kilometer to centimeters, we simply multiply 1000 meters by 100 to get 100,000 cm.

One mile equals 1.609 km so our photograph is about 1.6 km across. (See Appendix A for other conversion factors.) Only 11 more steps of 100 separate us from the largest dimensions in the universe. The next photograph will span 160 km.

In Figure 1-3, our view spans 160 km, about 100 miles. The college campus is now invisible, and we see few signs of human activity. Cities are small blotches, and farmlands tiny rectangles. The suburbs of Philadelphia are visible at the lower right. At this scale, we see the natural features of the earth's surface. The Allegheny Mountains of southern Pennsylvania cross the image in the upper left, and the Susquehanna River flows southeast into Chesapeake Bay. A few puffs of clouds dot the area.

These features remind us that we live on the surface of a changing planet. Forces in the earth's crust pushed the mountain ranges up into parallel folds, like a rug wrinkled on a polished floor. The clouds remind us that the earth's atmosphere is rich in water, which falls as rain

FIGURE 1-4
(NASA)

from X rays to radio waves, to reveal sights invisible to our unaided eyes.

The next step in our journey (Figure 1-4) shows our entire planet. Earth is 12,756 km in diameter and rotates on its axis once a day.

This image shows most of the daylight side of the planet, with the sunset line at the extreme right. The rotation of the earth carries us eastward across the daylight side, and as we cross the sunset line into darkness, we say the sun has set. Thus, the rotation of our planet causes the cycle of day and night.

We know that the earth's interior is made mostly of iron and nickel and that its crust is mostly silicate rocks. Only a thin layer of water makes up the oceans, and the atmosphere is only a few hundred miles deep. On the scale of this photograph, the depth of the atmosphere on which our lives depend is less than the thickness of a piece of thread.

Our **solar system** consists of the sun, its family of nine planets, and some smaller bodies such as moons and comets. Within our solar system, our earth is the only planet with the air and liquid water that make life possible. Only recently (1996) have astronomers found planets orbiting other stars, and those planets seem quite unlike the earth. Other earthlike planets probably exist, but they are too small and too distant to detect. So far as we know, we are the only life in the universe.

Again we enlarge our field of view by a factor of 100 (Figure 1-5), and we see a region 1,600,000 km wide. Earth is the small white dot in the center, and the moon, only one-fourth the earth's diameter, is an even smaller dot along its orbit 380,000 km from the earth.

These numbers are so large that it is inconvenient to write them out. Astronomy is the science of big numbers, and we will use numbers much larger than these to discuss the depths of the universe. Rather than write out these numbers, it is convenient to write them in **scientific notation.** This is nothing more than a simple way to write big numbers without writing a great many zeros.

In scientific notation, we would write 380,000 as 3.8×10^5. The 5 tells us to move the decimal point five places to the right. The 3.8 then becomes 380,000.

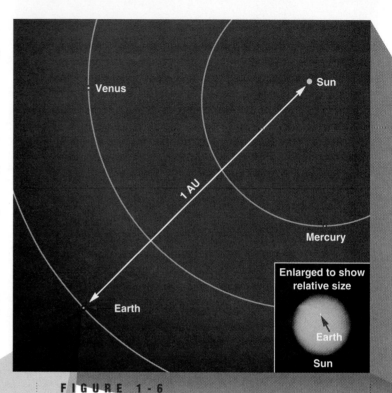

FIGURE 1-6

In the same way, we would write 1,600,000 as 1.6×10^6. Notice that we could also write this as 16×10^5 or 0.16×10^7. All represent the same quantity.

We can also use scientific notation to write very small numbers. If you are not familiar with scientific notation, consult the Appendix. The universe is too big to discuss without using scientific notation.

When we once again enlarge our field of view by a factor of 100 (Figure 1-6), Earth, the moon, and the moon's orbit all lie in the small red box at lower left. But now we see the sun and two other planets.

Venus is about the size of Earth, and Mercury is a bit larger than our moon. On this diagram, they are both too small to be seen as anything but tiny dots. The sun is 109 times larger in diameter than Earth (inset), but it too is nothing more than a dot on this diagram.

This figure spans 1.6×10^8 km. One way to deal with such large numbers is to define new units. Astronomers use the average distance from Earth to the sun as a unit of distance called the **astronomical unit (AU).** Using this unit, we can say that the average distance from Venus to the sun is about 0.7 AU. The average distance from Mercury to the sun is about 0.39 AU.

The orbits of the planets are not perfect circles, and this is particularly apparent for Mercury. Its orbit carries it as close to the sun as 0.307 AU and as far away as 0.467 AU. Earth's orbit is more circular, and its distance from the sun varies by only 1.7 percent.

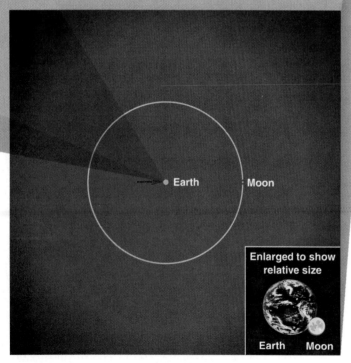

FIGURE 1-5

Our first field of view was only 52 feet (about 16 m) in width. After only six steps of enlarging our field of view by a factor of 100 at each step, we now see the entire solar system (Figure 1-7). Our field of view is 1 trillion (10^{12}) times wider than in our first view.

The details of the preceding figure are now lost in the tiny square at the center of this diagram. We see only the brighter, more widely separated objects as we back away. The sun, Mercury, Venus, and Earth lie so close together that we cannot separate them at this scale. Mars, the next outward planet, lies only 1.5 AU from the sun.

In contrast, the outer planets Jupiter, Saturn, Uranus, Neptune, and Pluto are so far from the sun that they are easy to place in this diagram. These planets are cold worlds far from the sun's warmth. Light from the sun takes over 4 hours to reach Neptune, which is slightly

When we again enlarge our field of view by a factor of 100 (Figure 1-8), our solar system vanishes. The sun is only a point of light, but all the planets and their orbits are now crowded into the small square at the center. None of the sun's family of planets are visible. They are too small and reflect too little light to be visible so near the brilliance of the sun.

Nor are any stars visible except for the sun. The sun is a fairly typical star, and it seems to be located in a fairly average neighborhood in the universe. Although there are many billions of stars like the sun, none are close enough to be visible in this diagram of only 11,000 AU in diameter. The stars are typically separated by distances about ten times larger than this diagram. We will see stars in our next field of view, but, except for the sun at the center, this diagram is empty.

It is difficult to image the isolation of the stars. If the sun were represented by a golf ball in New York City, the nearest star would be another golf ball in Chicago. Except for the widely scattered stars and a few atoms of gas drifting between the stars, the universe is nearly empty.

In Figure 1-9, our field of view has now expanded to a diameter of a bit over 1 million AU. The sun is located at the center, and we see a few of the nearest stars.

These stars are so distant that it is not reasonable to give their distances in astronomical units. We must define a new unit of distance, the light-year. One **light-year (ly)** is the distance that light travels in 1 year, roughly 10^{13} km or 63,000 AU. The nearest star to the sun is Proxima Centauri at a distance of 4.2 ly. Light from Proxima Centauri takes 4.2 years to reach Earth. The diameter of our field of view is now 17 ly.

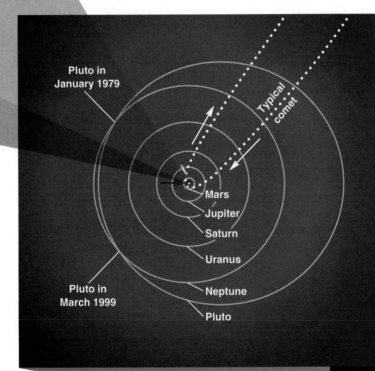

Pluto in
January 1979

Typical comet

Mars

Jupiter

Saturn

Uranus

Neptune

Pluto

Pluto in
March 1999

FIGURE 1-8

FIGURE 1-7

over 30 AU distant. In contrast, sunlight reaches Earth in only 8 minutes.

Notice that Pluto's orbit is so elliptical that Pluto can come closer to the sun than Neptune. In fact, Pluto is now closer to the sun than Neptune and will not reclaim its crown as the most distant planet in our solar system until March 14, 1999, when its orbit carries it back outside the orbit of Neptune.

FIGURE 1-9

Our sun formed from such a cloud about 5 billion years ago. We will see more star formation in our next view.

If we expanded our field of view by a factor of 100 (Figure 1-11), we would see our own **Milky Way Galaxy.** Of course, no one can journey far enough into space to photograph our galaxy, so this photo shows a similar galaxy indicating a representative location for the sun.

Our sun and the neighboring stars of the previous figure would be lost among the 100 billion stars of the galaxy. Most of the stars are smaller and fainter than our sun, but some are larger and more luminous. Most are cooler, but a few are much hotter. Why some stars are larger, more luminous, or hotter than others is one of the mysteries of the universe that we will explore.

Although these stars are roughly the same size as the sun, they are so far away that we cannot see them as anything but points of light. Even with the largest telescopes on Earth, we still see only points of light when we look at stars. In Figure 1-9, the sizes of the dots represent not the size of the stars but their brightness. This is the custom in astronomical diagrams, and it is also how star images are recorded on photographs. Bright stars make larger spots on a photograph than faint stars. Thus, the size of a star image in a photograph tells us not how big the star is but only how bright the star looks.

In Figure 1-10, we expand our field of view by another factor of 100, and the sun and its neighboring stars vanish into the background of thousands of stars. The field of view is now 1700 ly in diameter.

Of course, no one has ever journeyed thousands of light-years from the sun to photograph the solar neighborhood, so we use a representative photo of the sky. The sun is a relatively faint star, so we could no longer locate it on such a photo.

We notice a tendency for stars to occur in clusters. A loose cluster of stars lies in the lower left quarter of the photograph. We will discover that many stars are born in clusters and that both old and young clusters exist in the sky. Star clusters are forming right now.

What we do *not* see is critically important. We do not see the thin gas that fills the spaces between the stars. Although those clouds of gas are thinner than the best vacuum on earth, it is those clouds that give birth to new stars.

FIGURE 1-10

This box ▪ represents the relative size of the previous frame. (NOAO)

Typical of our galaxy are the graceful **spiral arms** marked by clusters of bright stars and clouds of gas. We will discover that stars are born in great clouds of gas and dust passing through these spiral arms.

Our galaxy is over 75,000 ly in diameter, and until the 1920s astronomers thought it was the entire universe—an island universe of stars in an otherwise empty vastness. Now we know that our galaxy is not unique. Indeed, ours is only one of many billions of galaxies scattered throughout the universe.

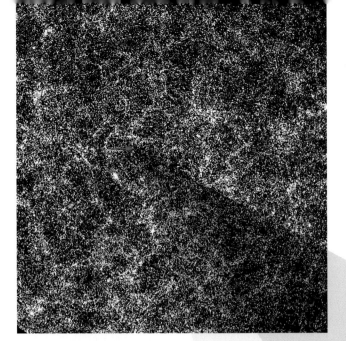

FIGURE 1-13
This box ▪ represents the relative size of the previous frame. (Detail from galaxy map from M. Seldner, B. L. Siebers, E. J. Groth, and P. J. E. Peebles, *Astronomical Journal* 82 [1977])

Among the galaxies we see here, a few are as large as our own galaxy, but most are smaller. A few have the beautiful spiral features we see in our galaxy, but most do not. Among more distant galaxies, we see a few galaxies twisted into peculiar shapes or wracked by violent eruptions. Although astronomers understand why stars differ from one another, it is not entirely clear why one galaxy differs from another. We will find some clues to the mystery when we compare the clusters of galaxies.

One theory holds that the centers of some galaxies contain supermassive black holes, which are capable of swallowing stars whole. Whatever the truth, the evolution of galaxies must occasionally be marked by events of titanic violence.

If we again expand our field of view (Figure 1-13), we see that our Local Group of galaxies is part of a large supercluster—a cluster of clusters. Other galaxies are not scattered at random throughout the universe but lie in clusters within larger superclusters.

To represent the universe at this scale, we use a diagram in which each dot represents the location of a single galaxy. At this scale, we see superclusters linked to form

As we expand our field of view by another factor of 100 (Figure 1-12), our galaxy appears as a tiny luminous speck surrounded by other specks. This diagram includes a region 17 million ly in diameter. Each of these dots represents a galaxy.

We see that our galaxy (arrow) is one of a small cluster of galaxies. This **Local Group** consists of roughly two dozen galaxies scattered throughout a region about 6 million ly in diameter.

FIGURE 1-11
(© Anglo-Australian Telescope Board)

FIGURE 1-12

long filaments outlining voids that seem nearly empty of galaxies. These appear to be the largest structures in the universe. Were we to expand our field of view yet another time, we would probably see a uniform sea of filaments and voids. When we puzzle over the origin of these structures, we are at the frontier of human knowledge.

Our problem in studying astronomy is to keep a proper sense of scale. Remember that each of the billions of galaxies contains billions of stars. Most of those stars probably have families of planets like our solar system, and on some of those billions of planets liquid-water oceans and a protective atmosphere may have spawned life. It is possible that some other planets in the universe are inhabited by intelligent creatures who share our curiosity and our wonder at the scale of the cosmos.

▪ Summary

Our goal in this chapter has been to preview the scale of astronomical objects. To do so, we journeyed outward from a familiar campus scene by expanding our field of view by factors of 100. Only 12 such steps took us to the largest structures in the universe.

The numbers in astronomy are so large it is not convenient to express them in the usual way. Instead, we use the metric system to simplify our calculations and scientific notation to more easily write big numbers. The metric system and scientific notation are discussed in Appendix A.

We live on the rotating planet Earth, which orbits a rather typical star we call the sun. We defined a unit of distance, the astronomical unit, to be the average distance from Earth to the sun. Of the eight other planets in our solar system, Mercury is closest to the sun, and Neptune is currently the most distant, at about 30 AU.

The sun, like most stars, is very far from its neighboring stars, and this leads us to define another unit of distance, the light-year. A light-year is the distance light travels in 1 year. The nearest star to the sun is Proxima Centauri at a distance of 4.2 ly.

As we enlarged our field of view, we discovered that the sun is only one of 100 billion stars in our galaxy and that our galaxy is only one of many billions of galaxies in the universe. Galaxies appear to be grouped together in clusters, superclusters, and filaments, the largest structures known.

As we explored, we noted that the universe is evolving. Earth's surface is evolving, and so are stars. Stars form from the gas in space, grow old, and eventually die. We do not yet understand how galaxies form or evolve.

Among the billions of stars in each of the billions of galaxies, many probably have planets, but detecting such planets is very difficult. Although astronomers can now detect planets orbiting other stars, we know very little about the nature of these planets. Yet we suppose that there must be many planets in the universe and that some are like Earth. We wonder if a few are inhabited by intelligent beings like ourselves.

▪ New Terms

solar system	Milky Way Galaxy
scientific notation	spiral arm
astronomical unit (AU)	Local Group
light-year (ly)	

▪ Questions

1. What is the largest dimension you have personal knowledge of? Have you run a mile? Hiked 10 miles? Run a marathon?

2. Why are astronomical units more convenient than miles or kilometers for measuring some astronomical distances?

3. In what ways is our planet changing?

4. What is the difference between our solar system, our galaxy, and the universe?

5. Why do all stars, except for the sun, look like points of light as seen from Earth?

6. Why are light-years—rather than miles or kilometers—more convenient units for measuring some astronomical distances?

7. Why is it difficult to see planets orbiting other stars?

8. Which is the outermost planet in our solar system? Why does that change?

9. Why can't we measure the diameters of stars from the size of the star images on photographs? What does the diameter of a star image really tell us about the star?

10. How long does it take light to cross the diameter of our solar system, of our galaxy, of the Local Group?

11. What are the largest known structures in the universe?

12. How many planets inhabited by intelligent life do you think the universe contains? Explain your answer.

▪ Problems

1. If 1 mile equals 1.609 km and the moon is 2160 miles in diameter, what is its diameter in kilometers? in meters?

2. If sunlight takes 8 minutes to reach the earth, how long does moonlight take?

3. How many suns would it take, laid edge to edge, to reach the nearest star?

4. How many kilometers are there in a light-minute? (Hint: The speed of light is 3×10^5 km/sec.)

5. How many galaxies like our own, laid edge to edge, would it take to reach the nearest large galaxy (which is 2×10^6 ly away)?

 Go to the Wadsworth Astronomy Resource Center (www.wadsworth.com/astronomy) for critical thinking exercises, articles, and additional readings from InfoTrac College Edition, Wadsworth's online student library.

Chapter Two

THE SKY

GUIDEPOST

Our goal in studying astronomy is to learn about ourselves. We search for an answer to the question "What are we?" The quick answer is that we are thinking creatures living on a planet that circles a star we call the sun. In this chapter, we begin trying to understand that answer. What does it mean to live on a planet? The preceding chapter gave us a quick overview of the universe, and chapters later in the book will discuss details. This chapter and the next help us understand what the universe looks like seen from the surface of our spinning planet.

We will see in Chapter 4 how difficult it has been for humanity to understand what we see in the sky every day. In fact, we will discover that science was born when people tried to understand the appearance of the sky. ∎

The Southern Cross I saw every night abeam.
The sun every morning came up astern;
every evening it went down ahead. I wished
for no other compass to guide me, for these
were true.

CAPTAIN JOSHUA SLOCUM

Sailing Alone Around the World

The night sky is the rest of the universe as seen from our planet. When we look at the stars, we look out through a layer of air only a few hundred kilometers thick. Beyond that, space is nearly empty. The nearest star to the earth is the sun, and the next nearest star beyond the sun is over 4 ly away, so distant that its light takes more than 4 years to reach us. The average bright star in our sky is about 300 ly away. Beyond the stars of our galaxy lie other galaxies millions or even billions of light-years away.

In the previous chapter, we took a quick journey through the universe to preview its scale. Now we are ready to repeat that journey more slowly. We want to understand how the universe works, and, at the same time, how humanity has discovered the nature of the universe. That is, as we study the stars and galaxies that fill our universe, we also want to understand how science empowers us to discover how nature works.

We begin by trying to understand what we see in the sky. Throughout this chapter, remember that we live on a moving planet. The earth spins on its axis once a day and revolves around the sun once a year. Because we feel stationary on the solid earth, the sky seems to spin around us. Our task is to try to understand the appearance and motions of the objects in the sky, but that task is complicated by the earth's motion. We are like observers riding the tilt-a-whirl and trying to sketch a map of the fairground from the spinning car.

This chapter will not answer all our questions about the universe and the nature of science. It merely sets the scene for the challenging explorations to follow, explorations that will help us understand what we are.

2-1 THE STARS

Constellations

Gazing at the night sky on a clear, moonless evening, we can see thousands of stars scattered in random groups. Some of these groups have been named and are called **constellations** (Figure 2-1). To learn to locate a few constellations, consult the star charts in Appendix B. Choose the chart for the appropriate month, and begin with the most prominent constellations.

Although we identify these patterns by name, we must remember that the stars in a constellation are usually not physically associated with one another. Some may be many times farther away than others (Figure 2-2). The only thing they have in common is that they lie in approximately the same direction as seen from earth.

Half of today's 88 constellations were named in ancient times. The oldest, including Taurus and Leo, seem to have originated in Mesopotamia more than 5000 years ago, and others were added by Babylonian, Egyptian, and Greek astronomers. In some cases, these star patterns had mystical or religious significance, but some appear to have been designed as practical navigation aids

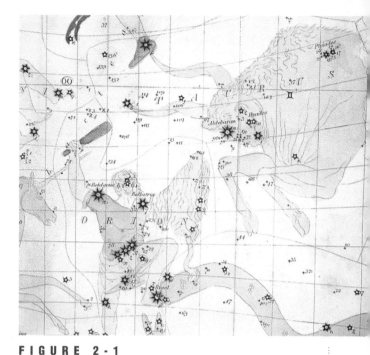

FIGURE 2-1

The constellations Orion and Taurus represent figures from Greek mythology. (From Duncan Bradford, *Wonders of the Heavens,* Boston: John B. Russell, 1837)

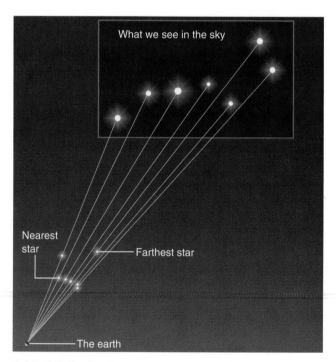

FIGURE 2-2

The stars we see in the Big Dipper—the brighter stars of the constellation Ursa Major, the Great Bear—are not at the same distance from the earth. In fact, the most distant star is not the faintest in the Big Dipper, and the nearest star is not the brightest. We see the stars in a group in the sky because they lie in the same general direction as seen from the earth. The size of the star dots in the star chart represents the apparent brightness of the stars.

a

b

FIGURE 2-3

To the ancients, the star Alpheratz was one of the stars in the great square of Pegasus and was also the eye of Andromeda. (a) In this later engraving from 1837, dotted lines mark the constellation boundaries and show that Alpheratz is part of Andromeda. (From Duncan Bradford, *Wonders of the Heavens,* Boston: John B. Russell, 1837) (b) Modern constellation boundaries are precisely defined by international agreement.

for sailors. Still others honored great heroes such as Orion and Hercules. Of course, the star patterns don't look like Orion or Hercules, but the Washington Monument does not look like George Washington. The constellations were meant to symbolize their subjects, not represent them.

Different cultures grouped stars and named constellations differently. The constellation we know as Orion was known as Al Jabbar, the giant, to the Syrians, as the White Tiger to the Chinese, and as Praja-pati in the form of a stag in ancient India. The Pawnee Indians knew the constellation Scorpius as two groupings. The long tail of the scorpion was the Snake, and the two bright stars at the tip of the scorpion's tail were the Two Swimming Ducks.

The ancient constellations were thought of as loose groupings of stars. A star could even belong to more than one constellation, as in the case of Alpheratz in the constellations Andromeda and Pegasus (Figure 2-3). As a convenience, modern astronomers have given each constellation definite boundaries as defined by the International Astronomical Union in 1928. Thus, a constellation represents not just a group of stars but a region of the sky, and any star within the region is a part of the constellation. This convention places Alpheratz in the constellation Andromeda (Figure 2-3b).

Not all regions of the sky are parts of the ancient constellations. Some regions contain no bright stars, and the extreme southern sky is never visible from the latitudes of the Mediterranean and Middle East, where classical astronomy arose. Consequently, the ancient astronomers never made these regions parts of constellations. Beginning during the Renaissance, as astronomy became more sophisticated and explorers sailed the southern seas, astronomers named 44 additional constellations to fill the empty spaces. These modern constellations include Telescopium (the telescope), Microscopium (the microscope), and Antlia (the air pump).

Modern astronomers still use the names of the constellations as a convenient way to refer to regions of the sky.

a

b

FIGURE 2-4

(a) The brighter stars in each constellation are assigned Greek letters in approximate order of brightness. In Orion, β is brighter than α and κ is brighter than η. (b) A long-exposure photograph reveals the many faint stars that lie within Orion's constellation boundaries. These stars are members of the constellation, but they do not have Greek-letter designations. Note the differences in the colors of the stars. Blue stars are hot and red stars cool. (William Hartmann)

The Names of the Stars

The brightest stars were named thousands of years ago, and these names, along with the names of the constellations, found their way into the first catalogs of stars. We take our constellation names from Greek versions translated into Latin—the language of science from the fall of Rome to the 19th century—but most star names come from ancient Arabic. Names such as Sirius (the Scorched One), Capella (the Little She Goat), and Aldebaran (the Follower of the Pleiades) are beautiful additions to the mythology of the sky.

Giving the stars individual names is not very helpful, because we see thousands of stars, and these names do not help us locate the star in the sky. In which constellation is Antares, for example? In 1603, Bavarian lawyer Johann Bayer published an atlas of the sky called *Uranometria* in which he assigned lowercase Greek letters to the brighter stars of each constellation in approximate order of brightness. Astronomers have used those Greek letters ever since. (See inside back cover for a listing of the Greek alphabet.) Thus, the brightest star is usually designated α (alpha), the second-brightest β (beta), and so on (Figure 2-4a). To identify a star in this way, we give the Greek letter followed by the possessive form of the constellation name, such as α Scorpii (for Antares). Now

we know that Antares is in the constellation Scorpius and that it is probably the brightest star in that constellation.

This method of identifying a star's brightness, however, is only approximate. In some constellations, the Greek letters were not assigned in exact order of brightness (Figure 2-4b). To be precise, we must have an accurate way of referring to the brightness of stars, and for that we must consult one of the first great astronomers.

The Brightness of Stars

Astronomers measure the brightness of stars using the magnitude scale, a system that had its start over two millennia ago when the Greek astronomer Hipparchus (160–127 BC) (Figure 2-5) divided the stars into six classes. The brightest he called first-class stars, and those that were fainter second-class. His scale continued down to sixth-class stars, the faintest visible to the human eye. Thus, the larger the magnitude number, the fainter the star. This makes sense if we think of the brightest stars as first-class stars and the faintest stars visible as sixth-class stars.

FIGURE 2-5

Hipparchus (2nd century BC) was the first great observational astronomer. Among other things, he constructed a catalog listing 1080 of the brightest stars by position and dividing them into six brightness classes now known as magnitudes. He is honored here on a Greek stamp, which also shows one of his observing instruments.

Hipparchus used his brightness scale in a great catalog of the stars, and successive generations of astronomers have continued to use his system. Later astronomers had to revise the ancient magnitude system to include very faint and very bright stars (Figure 2-6). Telescopes reveal many stars fainter than those the eye can detect, so the magnitude scale was extended to include these faint stars. The Hubble Space Telescope, currently the most sensitive astronomical telescope, can detect stars as faint as about +28 magnitude. In contrast, the brightest stars are actually brighter than the first brightness class in Hipparchus's system. Thus, modern astronomers extended the magnitude system into negative numbers to account for brighter objects. For instance, Vega (alpha Lyrae) is almost zero magnitude at 0.04, and Sirius, the brightest star in the sky, has a magnitude of −1.42. We can even place the sun on this scale at −26.5 and the moon at −12.5.

These numbers are known as **apparent visual magnitudes (m_v)**, and they describe how the stars look to human eyes seen from the earth. A star that is a million times more luminous than the sun might appear very faint if it is far away, and a star that is much less luminous might look bright if it is nearby. Look at the brightness and distance of the stars in Figure 2-2. In Chapter 9, we will develop a magnitude scale that tells us how bright the stars really are. Apparent visual magnitudes tell us only how bright the stars appear in our sky.

Magnitude and Intensity

Although astronomers have adopted the ancient Greek magnitude scale, measurements of stellar brightness require that the scale be given mathematical precision. Our estimates of brightness are subjective and depend on things such as the physiology of the human eye and the psychology of perception. For purposes of measurement, we should use the more precise term *intensity*—a measure of the light energy from a star that hits 1 square meter in 1 second. If two stars have intensities I_A and I_B, we can compare them by writing the ratio of their intensities, I_A/I_B.

The human eye senses the brightness of objects by comparing the ratios of their intensities. The magnitude system that Hipparchus devised is based on a constant intensity ratio of about 2.5 for each magnitude. That is, if two stars differ by 1 magnitude, then they have an intensity ratio of about 2.5. If they differ by 2 magnitudes, their ratio is 2.5 × 2.5, and so on.

When 19th-century astronomers began measuring starlight, they realized they needed to define a mathematical magnitude system that was precise, but for convenience they wanted a system that agreed at least roughly with that of Hipparchus. The stars that Hipparchus classified as first and sixth differ by 5 magnitudes and have an intensity ratio of almost exactly 100, so the modern system of magnitudes specifies that a magnitude difference of 5 magnitudes corresponds to an intensity ratio of 100. That means that 1 magnitude corresponds to an intensity ratio of precisely 2.512, the fifth root of 100. That is, $100 = (2.512)^5$.

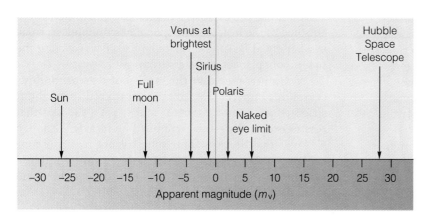

FIGURE 2-6

The scale of apparent visual magnitudes extends into negative numbers to represent the brighter objects.

Now we can compare the light we receive from two stars. The light from a first-magnitude star is 2.512 times more intense than that from a second-magnitude star. The light from a third-magnitude star is $(2.512)^2$ times more intense than the light from a fifth-magnitude star. In general, the intensity ratio equals 2.512 raised to the power of the magnitude difference. That is:

$$I_A/I_B = (2.512)^{(m_B - m_A)}$$

For instance, if two stars differ by 6.32 magnitudes, we can calculate the ratio of their intensities as $2.512^{6.32}$. A pocket calculator tells us that the ratio is 337.

This magnitude system has some advantages. It compresses a tremendous range of intensity into a small range of magnitudes (Table 2-1) and makes it possible for modern astronomers to compare their measurements with all the recorded measurements of the past, right back to the first star catalog assembled by Hipparchus over two thousand years ago.

CRITICAL **INQUIRY**

The nearest star in the Big Dipper is ζ (zeta) Ursa Majoris, with apparent magnitude of 2.40. The most distant is α Ursa Majoris with apparent magnitude 1.79. What does that tell us?

The smaller the magnitude number, the brighter the star, so we can conclude immediately that α looks brighter in the sky than ζ. We can go further with our analysis by noting that the difference in magnitude tells us the intensity ratio. In this case the magnitude difference is 2.4 − 1.79, which is 0.61 magnitudes, and the ratio of the brightnesses is $(2.512)^{0.61}$. A calculator tells us that the intensity ratio is 1.75. So light from α is 1.75 times more intense than light from ζ. We might say that α looks 1.75 times brighter than ζ, but our judgment of brightness is subjective and depends on the psychology of vision. We would be right to say α looks about 1.75 times brighter, but for precision, we should refer to the intensity of the light and not to the subjective brightness.

These two numbers can tell us even more. The nearest star, ζ, is one of the fainter stars in the Big Dipper. We might expect nearer stars to be brighter. The most distant star, α, is very nearly the brightest star in the Big Dipper. (ε Ursa Majoris is 0.03 magnitudes brighter than α.) We would expect the most distant star to be the faintest. This must mean that stars don't all emit the same amount of light. Clearly, α must be emitting much more light than ζ to be so bright even though it is farther away. We discuss this in more detail in Chapter 9.

The sun is very close to us and so looks very bright. Compare the intensity of sunlight (apparent magnitude −26.5) with the intensity of light from the brightest star, Sirius (apparent magnitude −1.42). ∎

TABLE **2-1**

Magnitude and Intensity

Magnitude Difference	Intensity Ratio
0	1
1	2.5
2	6.3
3	16
4	40
5	100
6	250
7	630
8	1600
9	4000
10	10,000
⋮	⋮
15	1,000,000
20	100,000,000
25	10,000,000,000
⋮	⋮

In this section, we have thought of the sky as if it were static and unchanging. We have reviewed the constellations, traced the origin of star names, and seen how modern astronomers refer to the apparent brightness of the stars. Now we can look at the sky as a whole and notice its motion.

2-2 THE CELESTIAL SPHERE

A Model of the Sky

Many ancient astronomers, including Hipparchus, thought of the sky as a great, hollow, crystalline sphere surrounding the earth. The stars, they imagined, were attached to the inside of the sphere like thumbtacks stuck in a ceiling. The sphere rotated once a day, carrying the sun, moon, planets, and stars from east to west.

We know now that the sky is not a great, hollow, crystalline sphere. The stars are scattered throughout space at different distances, and it isn't the sky that rotates once a day—the earth turns on its axis. Although we know that the crystalline sphere is not real, it is convenient as a model of the sky and is used by modern astronomers when they think about the locations and motions of celestial bodies.

Scientific Models

A scientific **model** is a mental conception of how something works. We all use models. For example, we might have a model in our minds of how a car works and use this model to make practical decisions about how to start the car on a cold morning. Our model doesn't have to be right to be useful. We may be totally wrong about how the engine works, but our model will probably be useful as long as we don't extend it too far. Of course, if we decided to rebuild our own carburetor, we might discover that our model is no longer adequate for our needs.

A scientific model need not be right, but it must be useful. That is, it must allow us to make useful predictions about how nature works. Scientists use models as mental crutches to help them think about nature. A chemist, for example, thinks of a molecule as little balls linked together with rods. Real molecules are much more complex than this model, but it is almost impossible to think about chemistry without using such a model to visualize molecular structure.

The astronomer's model of the celestial sphere is very helpful, and we can use it to think about the sun rising in the east and setting in the west. We can imagine the way the stars move across the sky, and we can predict the motion of the sky as a whole. Of course, the model is wrong, but as a mental aid to visualizing the motions in the sky, it is very useful within its limitations.

Some scientific models can be systems of mathematical equations expressed in computer programs that mimic the behavior of complex processes—an exploding star, for example. Our imaginations are not capable of numerical precision; such models act as mathematical crutches to help us "imagine" complicated processes with numerical precision.

Scientific models can range from general aids to visualization to mathematical equations that mimic the behavior of complex systems. In every case, the model helps us think about nature. It doesn't have to be true, but as long as we don't press a model beyond its limitations, it can be tremendously useful. In a sense, scientists are not so much searching for ultimate truths as they are trying to build better and better models of how nature works. ■

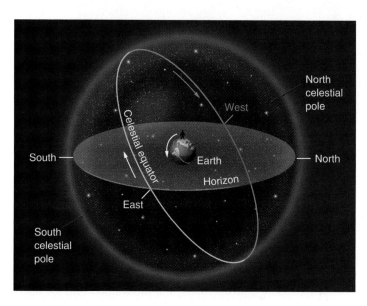

FIGURE 2-7
The modern celestial sphere models the appearance of the sky. The poles mark the pivots, and the equator divides the sky into a northern half and a southern half. An observer on the earth can see objects in the sky above the horizon, and the eastward rotation of the earth makes objects in the sky appear to rise along the eastern horizon and set along the western horizon. This figure is drawn for an observer at a latitude typical of the United States.

Our model of the sky is called the **celestial sphere** (Window on Science 2-1), an imaginary sphere of very large radius surrounding the earth and to which the stars, planets, sun, and moon seem to be attached (Figure 2-7). This sphere must have a large radius so that no part of the earth is significantly closer to a given star than any other part. Then it does not matter where on the earth we go. The sky always looks like a great sphere centered on our location.

When we think of the celestial sphere, we imagine a great sphere surrounding the earth, but, of course, we can see only half the sky at any one time. The half of the celestial sphere above our horizon is visible, and the half below our horizon is invisible. Thus, we can draw a circle around the sky and label it "horizon" (Figure 2-7), and we can even label the directions—north, south, east, and west—along our horizon. People living in different parts of the earth (Australia, for instance) will have horizons dramatically different from ours because they can see different parts of the celestial sphere.

If we watch the late afternoon sky for a few hours, we can notice movement. As the rotation of the earth carries us eastward, the sun appears to move westward and eventually sets. As darkness falls, we can see the stars, and in an hour or so it becomes obvious that the eastward

FIGURE 2-8

(a) A time exposure of a few hours (looking northeast) shows stars as streaks due to the rotation of the earth. (National Optical Astronomy Observatory) (b) From the middle latitudes of the United States, about 40°N, we find the stars of the northern constellations circling the north celestial pole. (c) In the eastern sky, stars rise at an angle to the horizon. (d) In the south, the stars circle the south celestial pole, which is invisible below the southern horizon. Compare with Figure 2-7.

rotation of the earth is making the sky appear to rotate westward (Figure 2-8). As some constellations set in the west, others rise in the east.

Reference Marks on the Sky

The pivots about which the sky seems to rotate are called the celestial poles. The **north celestial pole** is the point on the sky directly above the earth's North Pole, and the **south celestial pole** is the point directly above the earth's South Pole. As the earth rotates eastward, the sky appears to rotate westward, and stars located near the celestial poles seem to follow small circles around the celestial poles (Figure 2-8b).

Another important reference mark on the sky is the **celestial equator,** an imaginary line around the sky directly above the earth's equator (Figure 2-7). Seen from mid-northern latitudes such as the United States,

the celestial equator runs from the east point on the horizon up across the southern sky and down to the west point. Stars near the celestial equator rise along the eastern horizon parallel to the celestial equator (Figure 2-8c).

The location of the celestial poles and equator in our sky depends on our latitude. Imagine that we are standing in the ice and snow around the North Pole (Figure 2-9a). The north celestial pole is directly overhead, and the celestial equator runs along our horizon. If we begin walking southward, the north celestial pole begins to move away from the zenith, and the celestial equator rises above the horizon in front of us (Figure 2-9b). As we walk farther southward, the angle between the north celestial pole and the northern horizon always equals our latitude (Figure 2-9c). This relationship makes it simple for navigators in the earth's Northern Hemi-

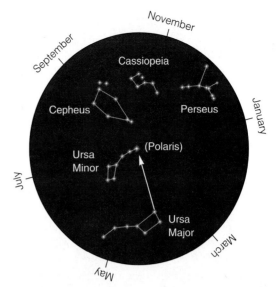

FIGURE 2-10
The northern constellations seen from mid-northern latitudes. To use the chart, face north soon after sunset and hold the chart in front of you with the current month at the top.

FIGURE 2-9
At the earth's North Pole (a), we see the north celestial pole directly overhead, and the celestial equator circles the horizon. As we journey southward (b), the angle between the north celestial pole and the northern horizon (L) always equals our latitude (c). At the earth's equator (d), we see the celestial poles on our horizon, and the celestial equator passes through our zenith. From southern latitudes (e), we see the south celestial pole above our horizon.

sphere to find their latitude—they need only measure this angle.

On our journey southward, we notice the celestial equator running from the east point on our horizon across the southern sky. As we cross the earth's equator, the celestial equator passes through our zenith, and the celestial poles lie on the northern and southern horizons (Figure 2-9d). Now as we walk farther southward, the north celestial pole sinks below our northern horizon, and the south celestial pole rises above our southern horizon.

From mid-northern latitudes we see the north celestial pole above our northern horizon. Currently, the star Polaris happens to lie very near the north celestial pole, and thus it hardly moves as the sky rotates. (The Pawnee Indian name for this star means "Star That Does Not Walk Around.") Figure 2-10 shows the position of Polaris near the north celestial pole. At any time of the night, in any season of the year, from anywhere in the earth's Northern Hemisphere, Polaris always stands above the northern horizon and is consequently known as the North Star. Later we will see that other stars have occupied this location in the past.

Because it lies below our horizon, the south celestial pole is never visible from the United States. As seen from our latitude, the constellations near the south celestial pole never rise, and the constellations near the north celestial pole never set. These constellations are known as **circumpolar constellations.** The farther north you live, the more circumpolar constellations you see. For most of us, Ursa Major, containing the Big Dipper, is a north circumpolar constellation. In fact, we can always find the North Star by first finding the Big Dipper and

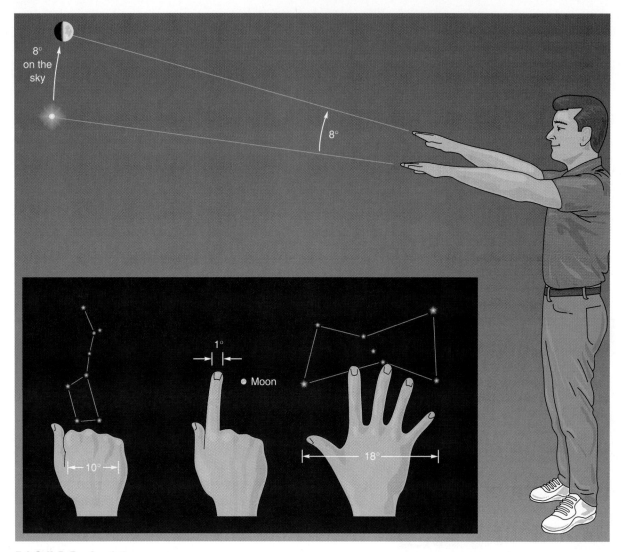

FIGURE 2-11

The angular separation between two objects is the angle your arms would make if you pointed to the two objects. Your hand held at arm's length makes a convenient measuring tool. Your fist is about 10° across, and your index finger is about 1° wide. Your spread fingers span about 18°.

then following the pointer stars in the dipper toward Polaris (Figure 2-10).

Angles on the Sky

Astronomers often use angles to describe separations across the sky. They might say, for instance, that the moon is 8° north of a certain star, meaning that if we point one arm at the moon and the other arm at the star, the angle between our arms will be 8° (Figure 2-11).

We measure angles in degrees, minutes of arc, and seconds of arc. There are 360° in a circle and 90° in a right angle. Each degree is divided into 60 **minutes of arc** (sometimes abbreviated 60′). If you view a 25¢ piece from the length of a football field, it has an angular diameter of about 1 minute of arc. Each minute of arc is divided into 60 **seconds of arc** (sometimes abbreviated 60″). The dot on this letter i seen from the length of a football field is about 1 second of arc in diameter.

When astronomers speak of angles, they often use the phrase "angles on the sky" as if the sky were a great plaster ceiling and the moon and stars were spots painted on the plaster. In this way, astronomers are using the celestial sphere as a model to help them think about angles. The moon may be 8° north of a certain star in our model, but we know that the star is hundreds of light-years away in space and the moon is nearby. The true distance between them is immense, but if we imagine them painted on the celestial sphere, we can think of their angular separation as an angle painted on the celestial sphere (Figure 2-11). Thus, we can discuss the angular separation between two objects even when we don't know their true distance from us or from each other.

a

b

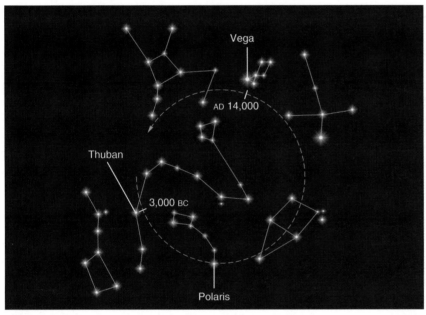

c

FIGURE 2-12

(a) The weight of a spinning gyroscope tends to make it fall over, and as a result it precesses in a conical motion about a vertical line. (b) The gravity of the sun and moon tends to twist the earth's axis upright in its orbit, and as a result it precesses. (c) Because of precession, the location of the north celestial pole describes a circle among the stars. Five thousand years ago, it was near the star Thuban. It is now near Polaris, and in 12,000 years it will be within 5° of Vega.

We can also talk about the **angular diameter** of an object in the sky—the angle our arms would make if we pointed at the left and right edges of the object as we see it in the sky. The moon and sun are each about 0.5° in angular diameter. You can use your hand as a convenient tool for estimating angles on the sky. Held at arm's length, your fist is about 10° in diameter, your index finger is about 1° in diameter, and your spread fingers span about 18° (Figure 2-11).

Angles on the sky, combined with the reference marks we discussed in the preceding section, make possible a highly useful coordinate system on the sky. Just as latitude and longitude specify the locations of points on the earth, the system of celestial coordinates can specify the locations of points on the sky. But we must be careful. The reference marks on the sky, the anchors for these coordinates, are defined by the earth's rotation, and the earth is wobbling like a gyroscope.

Precession

Over two thousand years ago, Hipparchus compared a few of his star positions with those made nearly two centuries before and realized that the celestial poles and equator were slowly moving across the sky. Later astronomers understood that this motion is caused by the toplike motion of the earth.

If you have ever played with a gyroscope, you have seen how the spinning mass resists any change in the direction of its axis of rotation. The more massive the gyroscope and the more rapidly it spins, the more difficult it is to change the direction of its axis of rotation. But you may recall that the axis of even the most rapidly spinning gyroscope does not remain absolutely fixed. A spinning gyroscope wobbles in a conical motion called **precession** (Figure 2-12a). The gyroscope precesses because of the interaction of its weight and its rotation.

The earth's gravity pulling on the gyroscope (its weight) tends to make the gyroscope tip over, and this combines with its rapid rotation to make its axis sweep around in a conical motion about a vertical line (Figure 2-12a).

The earth behaves like a giant gyroscope. Its large mass and rapid rotation keep its axis of rotation pointing near the star Polaris. If the earth were a perfect sphere, its axis of rotation would remain fixed, but the earth, because of its rotation, has a bulge around its middle. The moon's gravity pulls slightly more on the near side of the bulge than on the far side and thus twists the earth's axis of rotation, tending to set it upright in its orbit. The sun, though farther away than the moon, has much more mass, and it too twists the earth's axis of rotation. The combination of these forces and the earth's rotation causes the earth's axis to precess in a conical motion, taking about 26,000 years for one cycle (Figure 2-12b).

Because the celestial poles and the equator are defined by the earth's rotation, precession changes these reference marks. We see no change at all from night to night or year to year, but precise measurements reveal the motion of the poles and the equator.

Over centuries, precession has dramatic effects. For example, it makes the celestial poles move across the sky. Egyptian records show that 4800 years ago the north celestial pole was near the star Thuban (α Draconis). The pole is now approaching Polaris and will be closest to it about AD 2100. In about 12,000 years, the pole will have moved away from Polaris and will be within 5° of Vega (α Lyrae) (Figure 2-12c).

CRITICAL INQUIRY

Does everyone see the same circumpolar constellations?

Here we can use the celestial sphere as a convenient model, as shown in Figure 2-9. Which constellations are circumpolar depends on where we live. If we live on the earth's equator, we see all the constellations rising and setting, and there are no circumpolar constellations at all. If we live at the earth's North Pole, all the constellations north of the celestial equator never set, and all the constellations south of the celestial equator never rise. In that case, every constellation is circumpolar! (Check the star chart at the end of this book and try to decide if Orion would be north or south circumpolar if you lived at the earth's North Pole.)

At intermediate latitudes, the circumpolar regions are caps whose angular radius equals the latitude. If we live in Iceland, the caps are very large, and if we live in Egypt, near the equator, the caps are much smaller. Constellations that lie inside these caps are called circumpolar constellations. Check the star chart to locate Ursa Major. People in Canada count it as a cir-

cumpolar constellation, but people in Mexico see most of this constellation slip below the horizon.

People who lived at different times had different circumpolar constellations. Because of precession, the celestial poles are moving among the stars, so the constellations within the circumpolar caps are changing. For example, the ancient Egyptians saw slightly different north circumpolar constellations than we do. Likewise, our descendants thousands of years in the future will see other circumpolar constellations. From where you live, will Ursa Major be circumpolar in 12,000 years, when Vega is the north star? ■

The motions of the celestial sphere may seem to have little to do with our lives, but at the end of this chapter we will see how precession may be one of the causes of the ice ages. Before we can think about ice ages, however, we must consider the orbital motion of the earth around the sun and the resulting apparent motion of the sun.

2-3 THE MOTION OF THE SUN

Everything in the sky is moving. The sun, the moon, the planets, and even the stars move along their various orbits. Because the stars are so distant, their motion is not obvious to us even over decades, but the sun, moon, and planets are closer and move noticeably against the background of more distant stars. In this section, we discuss the motion of the sun.

The Ecliptic

The daily rising and setting of the sun, its diurnal motion, are a reflection of the eastward rotation* of the earth. That is, the sun appears to rise and set because the earth rotates. But the sun has a second motion, a slow eastward drift against the background of stars, that is a reflection of the motion of the earth along its orbit.

To see how the motion of the earth could cause an apparent motion of the sun, imagine that we are riding on a raft, drifting smoothly around an island. As we begin, we see the island against a background of more distant islands, but as we drift we see the island from different directions against different backgrounds. If we did not know that our raft was moving, we would imagine that the island was moving around us. This is precisely what happens when we observe the sun from the moving earth. The earth moves so smoothly along its orbit that we feel motionless, and it appears as if the sun moves around the sky.

---

*Astronomers distinguish between the words *revolve* and *rotate*. The earth revolves around the sun but rotates on its axis.

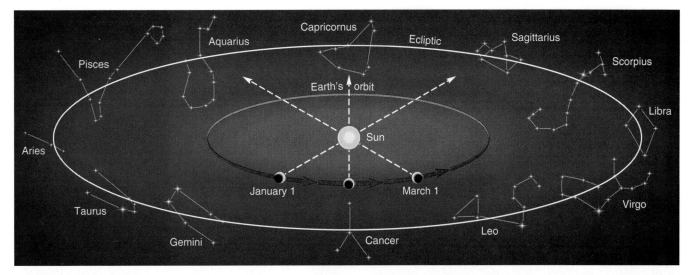

FIGURE 2-13

As the earth moves around its orbit, we see the sun in front of different constellations. In January the sun is in front of Sagittarius, but by March it has moved along the ecliptic into Aquarius. The ecliptic is the projection of the earth's orbit onto the celestial sphere.

In January we see the sun in the direction of the constellation Sagittarius (Figure 2-13). We can't see the stars of the constellation, of course, because the sun is too bright, but we can observe that the sun is located in that part of the sky merely by noting the time of sunset and the constellations in the evening sky. As the earth moves through space, we observe the sun from a different part of Earth's orbit, and the sun appears to be in Capricornus in February. Thus, as the earth moves along its orbit, the sun seems to move eastward through the constellations, taking a year to circle the sky one time.

The apparent path of the sun around the sky is called the **ecliptic.** Because the apparent motion of the sun is due to the orbital motion of the earth, it is easy to see that the ecliptic is the projection of the earth's orbit on the sky. If the celestial sphere were a great screen illuminated by the sun at the center, the shadow cast by the earth's orbit would be the ecliptic (Figure 2-13).

Because of the rotation of the earth, this slow eastward motion of the sun is not easy to visualize. Since the earth spins on its axis once each day, we see the sun, stars, moon, and planets rise in the east and set in the west. While this daily motion is taking place, the sun is moving slowly eastward along the ecliptic about 1° per day, which is about twice its own angular diameter.

An additional complication is that the earth's axis is not perpendicular to the plane of its orbit. Its axis of rotation is tipped 23.5° from the perpendicular. The spinning earth, like a spinning top, holds its axis fixed in space as it moves around the sun (Figure 2-14). (Precession alters this only very slowly.)

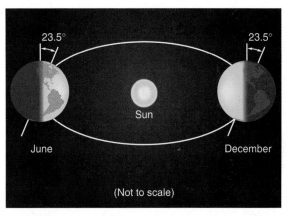

FIGURE 2-14

The cycle of the seasons occurs because Earth's axis is inclined 23.5° from the perpendicular to its orbit. Although precession moves the axis very slowly, the axis is essentially fixed in space for periods as short as human lifetimes. As Earth circles the sun, its axis remains pointing in the same direction in space. On one side of the sun (left), Earth's Northern Hemisphere is inclined toward the sun, and northern latitudes experience summer. Six months later, Earth is on the other side of the sun (right), the Northern Hemisphere is inclined away, and northern latitudes experience winter. The seasons are reversed in southern latitudes.

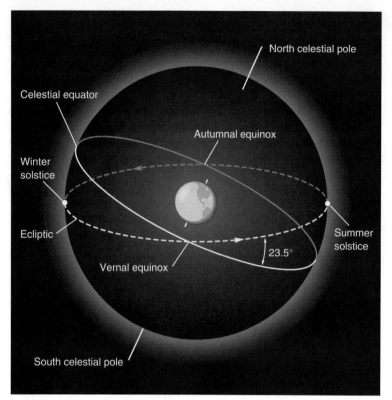

FIGURE 2-15

The ecliptic (dashed line), the sun's apparent path through the sky, crosses the celestial equator at the equinoxes. The solstices mark the most northerly and most southerly points.

Recall that the celestial equator is the projection of the earth's equator and that the ecliptic is the projection of the earth's orbit. Because the earth is tipped 23.5°, its equator is tipped 23.5° from the plane of the earth's orbit. When we project this onto the sky, we find that the ecliptic and the celestial equator meet at an angle of 23.5° (Figure 2-15).

This angle is very important to us who live on Earth. Because the ecliptic is tipped with respect to the celestial equator, Earth passes through a yearly cycle of seasons.

The Seasons

The seasonal temperature depends on the amount of heat we receive from the sun. For the temperature of North America to remain constant, there must be a balance between the amount of heat we gain and the amount we radiate into space. If we receive more heat than we lose, we get warmer; if we lose more than we gain, we get cooler.

The motion of the sun around the ecliptic tips the heat balance one way in summer and the opposite way in winter. Because the ecliptic is inclined with respect to the celestial equator, the sun spends half the year in the northern celestial hemisphere and half the year in the southern celestial hemisphere (Figure 2-15b). When

the sun is in the northern celestial hemisphere, the northern half of the earth receives more direct sunlight—and therefore more heat—than the southern half. This makes North America, Europe, and Asia warmer.

The seasons are reversed in the southern half of the earth (see Figure 2-14). While the sun is in the northern celestial hemisphere warming North America, South America becomes cooler. New Zealand has warm weather on New Year's Day and cold weather in July.

Four locations along the ecliptic help define the beginnings of the seasons (Figure 2-15b). The **vernal equinox** is the place in the sky where the sun crosses the celestial equator moving northward. *Vernal*, like "verdant" and "Vermont," comes from the Latin word for green. *Equinox* comes from the Latin word for "equal" and refers to the fact that we have equal amounts of daylight and darkness when the sun is at the equinox. The **summer solstice** is the point on the ecliptic where the sun is farthest north. *Solstice* comes from Latin words that mean "sun" and "stationary" and refers to the fact that the sun pauses there in its northward travel before beginning to move south. The **autumnal equinox** is the point where the sun crosses the equator moving southward, and the **winter solstice** is the place on the sky where the sun is farthest south (Window on Science 2-2).

According to our modern calendar, the seasons begin at the moment the sun crosses these four places on the ecliptic. The sun crosses the vernal equinox about March 21, and we say spring has begun. Summer begins about June 22 when the sun reaches the summer solstice, and autumn begins about September 22 when the sun crosses the autumnal equinox. Winter begins about December 22 when the sun reaches the winter solstice. (These dates vary slightly because of leap year.)

To see how we can get more heat from the summer sun, think about the path the sun takes across the sky between sunrise and sunset. Figure 2-16 shows these paths when the sun is at the summer solstice and at the winter solstice as seen by a person living at latitude 40°, an average latitude for most of the United States. Notice in Figure 2-16a that at the summer solstice the sun rises in the northeast, moves high across the sky, and sets in the northwest. But at the winter solstice, Figure 2-16b, the sun rises in the southeast, moves low across the sky, and sets in the southwest. Two features of these paths tip the heat balance.

First, the summer sun is above the horizon for more hours of each day than the winter sun. Summer days are long, and winter days are short. Because the sun is above our horizon longer in summer, we receive more energy each day.

Second, the sun stands high in the sky at noon on a summer day. It shines almost straight down, as shown by our small shadows (Figure 2-16a). On a winter day, how-

Naming Versus Understanding

One of the fascinations of science is that it can reveal things we normally do not sense, such as the daily motion of the sun among the stars. Science can even take our imagination into realms beyond our experience, from the inside of an atom to the inside of a star. Because these experiences are so unfamiliar, we need a vocabulary of technical terms just to talk about them, and that leads to a common confusion between naming things and understanding things.

The first step in understanding something is naming the parts, but it is only the beginning. Real understanding comes from examining the relationships in science. Cause-and-effect is one relationship, and one way we try to understand nature is by understanding causes. Geologists, for example, don't just name different kinds of earthquakes; they try to understand what causes them. Process—a sequence of events that leads to some result—is another relationship, and scientists struggle to understand natural processes. For example, some psychologists try to understand the step-by-step process that allows a newborn infant to develop three-dimensional vision. Theory and evidence comprise perhaps the key relationship in science. We use the vocabulary of science to show what known facts give us confidence that certain theories are correct. Thus, biologists studying the colors of flowers go beyond memorizing the names of the colors. They search for evidence to test their theory that the colors of flowers have evolved to attract certain kinds of insects for efficient pollination. Naming the colors is only the beginning.

Because the vocabulary of science is unfamiliar, we often confuse knowing scientific terms with understanding science. Of course, we must know the words, but real understanding comes from being able to use the vocabulary to discuss the way nature works. We must be able to tell the stories that scientists call theories and hypotheses and cite the factual evidence that makes us think those stories are true. That is real understanding.

In folklore, naming a thing gives us power over it—recall the story of Rumpelstiltskin. In science, naming things is only the beginning. Our goal lies in understanding, not just naming. ∎

---

ever, the noon sun is low in the southern sky. The ground gains less heat from the winter sun because the sunlight strikes the ground at an angle and spreads out (Figure 2-17). These two effects work to tip the heat balance and produce the seasons.

Notice that the seasons are not caused by changes in the distance from the earth to the sun. The earth's orbit is slightly elliptical, but the total variation in the earth–sun distance is only about 2 percent. The earth is actually at **perihelion** (closest to the sun) in the first week of January, when it is winter in the earth's Northern Hemisphere. The earth passes **aphelion** (farthest from the sun) in early July, when it is summer in the Northern Hemisphere. Thus, spring begins not because the earth is drawing closer to the sun but because the sun is moving into the northern half of the sky.

Of course, the weather does not turn warm the instant spring begins. The ground, air, and oceans are still cool from winter, and they take a while to warm up. Likewise, in the fall the earth slowly releases the heat it has stored through the summer. Because of this thermal lag, the average daily temperatures lag behind the solstices by about 1 month. Although the sun crosses the summer solstice about June 22, the hottest months in the Northern Hemisphere are July and August. The coldest months are January and February, even though the sun passes the winter solstice earlier, about December 22.

In ancient times, the solstices and equinoxes were celebrated with rituals and festivals. Shakespeare's play *A Midsummer Night's Dream* describes the enchantment of the summer solstice night. (In Shakespeare's time, the

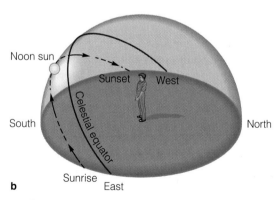

FIGURE 2-16

The seasons for middle northern latitudes typical of the United States. (a) The summer sun rises in the northeast and sets in the northwest. It is above the horizon for more than 12 hours and stands high in the sky at noon. (b) The winter sun rises in the southeast and sets in the southwest. It is above the horizon for less than 12 hours and never stands high in the sky, even at noon. Notice the figure's shadows.

a   b

FIGURE 2-17
(a) At noon on the day of the summer solstice, the sun shines from nearly overhead at the average latitude of the United States. Like the light from a flashlight shining nearly straight downward, the summer sunlight is not spread out very much. (b) On the day of the winter solstice, sunlight strikes the ground at a steep angle and spreads out. Thus, the ground receives less energy per square meter from the winter sun than from the summer sun.

equinoxes and solstices were taken to mark the midpoint of the seasons.) Many North American Indians marked the summer solstice with ceremonies and dances. Early church officials placed Christmas in late December to coincide with an earlier pagan celebration of the winter solstice.

CRITICAL INQUIRY

If Earth had a significantly elliptical orbit, how would its seasons be different?

Suppose Earth had an elliptical orbit so that perihelion occurred in July and aphelion in January. At perihelion, Earth would be closer to the sun, and the entire surface of Earth would be a bit warmer. If that happened in July, it would be summer in the Northern Hemisphere and winter in the Southern Hemisphere, and both would be warmer then they now are. It could be a dreadfully hot summer in Canada, and it might not snow at all in southern Argentina. Six months later, aphelion would occur in January, which means Earth would be farther from the sun. Winter in northern latitudes would be frigid, and summers in Argentina would be cool.

Of course, this doesn't happen. Earth's orbit is nearly circular, and the seasons are caused not by a variation in the distance of Earth from the sun but by the inclination of Earth in its orbit.

Nevertheless, Earth's orbit is slightly elliptical. Earth passes perihelion about January 4th and aphelion about July 4th. Although Earth's oceans tend to store heat and reduce the importance of this effect, this very slight variation in distance does affect the seasons. Does it make your winters warmer or cooler? ■

The sun is not the only body that moves along the ecliptic. The moon and planets maintain constant traffic along the sun's highway.

2-4 THE MOTION OF THE PLANETS

The ecliptic is important to us because it is the path of the sun around the sky, and that motion gives rise to the seasons. In a later chapter, we will consider the moon's cycle around the ecliptic, but here it is worthwhile to mention the motion of the planets. Not only are most of the planets visible to the naked eye, but their motion is the basis for one of the longest-living superstitions in human history—that our fate is written in the stars.

The Moving Planets

Most of the planets of our solar system are visible to the unaided eye, though they produce no light of their own. We see them by reflected sunlight. Mercury, Venus, Mars, Jupiter, and Saturn are all visible to the naked eye, but Uranus is usually too faint to be seen (magnitude 5.6 at its brightest) and Neptune is never bright enough. Pluto is even fainter, and we need a large telescope to find it. Although Uranus, Neptune, and Pluto are usually too faint to see, their motions are the same as those of the other planets.

All the planets of our solar system move in nearly circular orbits around the sun. If we were looking down on the solar system from the north celestial pole, we would see the planets moving in the same counterclockwise direction around their orbits (Chapter 1). The farther they are from the sun, the more slowly the planets move.

When we look for planets in the sky, we always find them near the ecliptic, because their orbits lie in nearly the same plane as the orbit of Earth. As they orbit the sun, they appear to move generally eastward along the ecliptic. In fact, the word *planet* comes from a Greek word meaning "wanderer." Mars moves completely around the ecliptic in slightly less than 2 years, but Saturn, being farther from the sun, takes nearly 30 years.

As seen from Earth, Venus and Mercury can never move far from the sun because their orbits are inside Earth's orbit. They sometimes appear near the western horizon just after sunset or near the eastern horizon just before sunrise. Venus is easier to locate because its larger orbit carries it higher above the horizon than Mercury (Figure 2-18). Mercury's orbit is so small that it can never get farther than about 28° from the sun. Consequently, it is usually hard to see against the sun's glare and is often hidden in the clouds and haze near the horizon. At certain times when it is farthest from the sun, however, Mercury shines brightly and can be located near the horizon in the evening or morning sky. (See the inside back cover for the best times to observe Venus and Mercury.)

By tradition, any planet visible in the evening sky is called an **evening star,** although planets are not stars. Any planet visible in the sky shortly before sunrise is a **morning star.** Perhaps the most beautiful is Venus, which can become as bright as minus fourth magnitude. As Venus moves around its orbit, it can dominate the western sky each evening for about half the year, but eventually its orbit carries it back toward the sun and it is lost in the haze near the horizon. In a few weeks it reappears in the dawn sky as a brilliant morning star.

Astrology

Seen from the earth, the planets move gradually eastward along the ecliptic, but they don't follow the ecliptic exactly. Also, they each travel at their own pace and seem to speed up and slow down at various times. To the ancients, this complex motion reflected the moods of the sky gods, and astrology was born.

Ancient astrologers defined a **zodiac,** a band 18° wide centered on the ecliptic as the highway the planets follow. They divided this band into 12 segments named for one of the constellations on the ecliptic—the signs of the zodiac. A **horoscope** shows the location of the sun, moon, and planets among the zodiacal signs with respect to the horizon at the moment of a person's birth.

Astrology buffs argue that a person's personality, life history, and fate are revealed in their horoscope, but the evidence contradicts this belief. Astrology has been tested many times over the centuries, and it just doesn't work. Believers, however, don't give up on it. Thus, astrology is a superstition that depends on faith and not a science that depends on evidence (Window on Science 2-3).

One reason astronomers find astrology irritating is that it has no link to the physical world. For example, precession has moved the constellations so that they no longer match the zodiacal signs. Whatever sign you were "born under," the sun was probably in the previous zodiacal constellation. In fact, if you were born on or between November 30th and December 17th, the sun was passing through a corner of the nonzodiacal constellation Ophi-

a

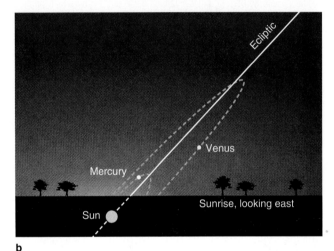

b

FIGURE 2-18

The orbits of Mercury and Venus lie inside the orbit of Earth, so they are always near the sun in the sky. They are sometimes visible in the western sky just after sunset (a) or in the eastern sky just before sunrise (b). Because the orbit of Venus is larger, it is easier to locate than Mercury.

uchus, and you have no official zodiacal sign.* Furthermore, there is no mechanism by which the planets could influence us. The gravitational field of a doctor who is delivering a baby is many times more powerful than the gravitational field of the planets.

Astrology makes sense only when we think of the world as the ancients did. They believed in multiple sky gods whose moods altered events on earth. They believed in mystical influences between natural events and human events. The ancients did not understand natural forces such as gravity as the mechanisms that cause things to happen, and thus they did not believe in cause and effect as we do. If a house burned, they might conclude that it caught fire because it was cursed and not because someone was careless with an oil lamp.

*The author of this book was born on December 14th and thus has no astrological sign. An astronomer friend claims that the author must therefore have no personality.

Astrology and Pseudoscience

Astronomers have a low opinion of astrology not so much because it is groundless but because it pretends to be a science. It is a pseudoscience, from the prefix *pseudo,* meaning false. There are many examples of pseudoscience, and it is illuminating to consider the difference between pseudoscience and science.

A pseudoscience is a set of beliefs that appear to be based on scientific ideas but that fail to obey the most basic rules of science. For example, some years ago a fad arose in which people placed objects under pyramids made of paper, plastic, wire, and so on. The claim was that the pyramidal shape would focus cosmic forces on anything inside and so preserve fruit, sharpen razor blades, and do other miraculous things. Many books promoted this idea, but simple experiments showed that any shape would protect a piece of fruit from airborne spores and allow it to dry without rotting. Likewise, any shape would allow oxidation to improve the cutting edge of a razor blade. In

short, experimental evidence contradicted the claim. Nevertheless, supporters of the theory declined to abandon or revise their claims. Thus, the fad of pyramid power was a pseudoscience.

One characteristic of a pseudoscience is that it appeals to our needs and desires. Thus, some pseudoscientific claims are self-fulfilling. For example, some people bought pyramidal tents to put over their beds and thus improve their rest. While there is no logical mechanism by which such a tent could affect a sleeper, because people wanted and expected the claim to be true they slept more soundly. Many pseudoscientific claims involve medical cures, ranging from copper bracelets and crystals to focus spiritual power to astonishingly expensive and illegal treatments for cancer. Logic is a stranger to pseudoscience, but human fears and needs are not.

Astrology is a pseudoscience. Over the centuries, astrology has been tested repeatedly,

and no correlation has been found. But it survives, and its supporters disregard any evidence that it doesn't work. Like all pseudosciences, astrology is not open to revision in the face of contradictory evidence. Furthermore, astrology fulfills our human need to believe that there is order and meaning to our lives. It may comfort us to believe that our sweetheart has rejected us because of the motions of the planets rather than to admit that we behaved badly on our last date. Comfort aside, astrology is a poor basis for life decisions.

Human nature and human needs probably ensure that pseudoscientific beliefs will continue to plague us like emotional viruses propagating from person to person. But if we recognize them for what they are, we can more easily guide our lives by rational principles, and not by giving credit for our successes and blame for our failings to the stars. ∎

Modern science left astrology behind centuries ago, but it survives as a fascinating part of human history—an early attempt to understand the meaning of the sky.

CRITICAL INQUIRY

Planets like Mars can sometimes be seen rising in the east as the sun sets. Why is the same not true for Mercury or Venus?

Mars has an orbit outside the orbit of Earth, and that means it can reach a location opposite the sun in the sky. As the sun sets in the west, Mars rises in the east. But Mercury and Venus follow orbits that are smaller than the orbit of Earth, so they can never reach a point opposite the sun in the sky. As they follow their orbits around the sun, we on Earth see them gradually swing out on the east side of the sun, reach a maximum distance from the sun, swing back toward the sun, move out to the west side of the sun, reach a maximum distance west, and then move back toward the sun again. Since they never get very far from the sun in the sky, they can never be seen rising in the east as the sun sets.

The planetary motions we see in the sky are produced by the orbital motions of the planets around the sun. What would we see if the planets followed orbits that were all in exactly the same plane as the orbit of Earth? ∎

Modern astronomers have a low tolerance for astrological superstition, yet the motions of the heavenly bodies do affect our lives. The motion of the sun produces the seasons, and, as we will see in Chapter 3, the moon governs the tides. In addition, small changes in Earth's orbit may partially control global climate.

2-5 ASTRONOMICAL INFLUENCES ON EARTH'S CLIMATE

Weather is what happens today; climate is the average of what happens over decades and centuries. We know that the earth has gone through past episodes, called ice ages, when the worldwide climate was cooler and dryer and thick layers of ice covered northern latitudes. The earliest known ice age occurred about 570 million years ago, and the next about 280 million years ago. The most recent ice age began only about 3 million years ago and is still going on. We are living in one of the periodic episodes when the glaciers melt and the earth grows slightly warmer. The current warm period seems to have begun about 20,000 years ago.

Ice ages seem to occur with a period of roughly 250 million years, and cycles of glaciation within ice ages occur with a period of about 40,000 years. Many scien-

Evidence

Science is based on evidence. Every theory and conclusion must be supported by evidence obtained from experiments or from observation. If a theory is supported by many pieces of evidence but is clearly contradicted by a single experiment or observation, scientists quickly abandon it. For a theory to be true, there can be no contradictory evidence.

Of course, scientists argue about the significance of particular evidence and often disagree on the interpretation of evidence. Some observations may seem significant at first glance, but upon closer examination we may find that the procedure was flawed and so the piece of evidence is not important. Or we might conclude

that the observational fact is correct but is being misinterpreted. The observation may not mean what it seems. Some of the most famous disagreements in science, such as those surrounding Galileo and Darwin, have arisen over the interpretation of well-established factual evidence.

Furthermore, scientists are not allowed to be selective in considering evidence. A lawyer in court can call a certain witness and intentionally fail to ask a critical question that would reveal evidence harmful to the lawyer's case. But a scientist may not ignore any known evidence. The difference in their methods is revealing. The lawyer is attempting to prove only

one side of the case and rightly may ignore contradictory evidence. The scientist, however, is searching for the truth and so must test any theory against all available evidence. In a sense, the scientist, in dealing with evidence, must act as both the prosecution and the defense.

As you read about any science, look for the evidence in the form of measurements or observations. Every theory or conclusion should have supporting evidence. If you can find and understand the evidence, the science will make sense. All scientists, from astronomers to zoologists, demand evidence. You should, too. ∎

---

tists now believe that these cyclic changes have an astronomical origin.

The Hypothesis

Sometimes a theory or hypothesis is proposed long before scientists can find the critical evidence to support it. That situation happened in 1920 when Yugoslavian meteorologist Milutin Milankovitch proposed what became known as the **Milankovitch hypothesis**—that changes in the shape of Earth's orbit, in precession, and in inclination affect Earth's climate and trigger ice ages. We will examine each of these three motions in turn.

First, astronomers know that the elliptical shape of Earth's orbit varies slightly over a period of about 100,000 years. At present, Earth's orbit carries it about 2 percent closer to the sun during Northern-Hemisphere winters and about 2 percent farther away in Northern-Hemisphere summers. This makes the northern climate warmer, and that is critical—most of the land mass where ice can accumulate is in the Northern Hemisphere. If Earth's orbit became more elliptical, Milankovitch suggested, northern winters might be warm enough to prevent the accumulation of snow and ice from forming glaciers, and Earth's climate would warm.

A second factor is also at work. Precession causes Earth's axis to wobble with a period of about 26,000 years, and that changes the location of the seasons around Earth's orbit. Northern winters now occur when Earth is 2 percent closer to the sun, but in 13,000 years northern winters will occur on the other side of Earth's orbit where Earth is farther from the sun. Northern winters will be colder and glaciers may grow.

The third factor is the inclination of Earth's equator to its orbit. Currently at 23.5°, this angle varies from 22°

to 24° with a period of roughly 41,000 years. If the inclination is large, seasons are more severe.

In 1920, Milankovitch proposed that these three factors cycled against each other to produce complex periodic variations in Earth's climate and the advance and retreat of glaciers (Figure 2-19). But no evidence was available to test the theory in 1920, and scientists treated it with skepticism. Many thought it was laughable.

The Evidence

By the middle 1970s, Earth scientists could collect the data that Milankovitch needed. Oceanographers could drill deep into the seafloor and collect samples, and geologists could determine the age of the samples from the natural radioactive atoms they contained. From all this, scientists constructed a history of ocean temperatures that convincingly matched the predictions of the Milankovitch hypothesis (Figure 2-20).

The evidence seemed very strong, and by the 1980s, the Milankovitch hypothesis was widely discussed as the leading hypothesis. But science follows a mostly unstated set of rules which holds that a hypothesis must be tested over and over against all available evidence (Window on Science 2-4). In 1988, scientists discovered contradictory evidence.

While scuba diving in a water-filled crack in Nevada called Devil's Hole, scientists drilled out samples of calcite, a mineral that contains oxygen atoms. For 500,000 years, layers of calcite have built up in Devil's Hole, recording in their oxygen atoms the temperature of the atmosphere when rain fell there. Finding the ages of the mineral samples was difficult, but the results seemed to show that the previous ice age ended thousands of years too early to have been caused by Earth's motions.

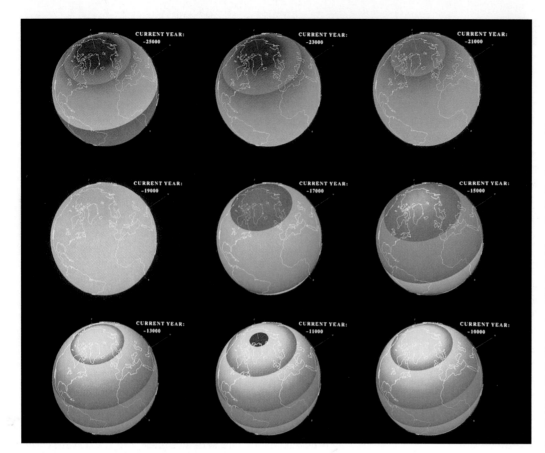

FIGURE 2-19
Color-coded globes show the importance of the Milanko-vitch effect. These globes show Earth at the summer solstice from 25,000 years ago (upper left), during the last glaciation, to only 10,000 years ago, after the end of the glaciation. Red, yellow, and green regions receive more solar energy than average; blue and violet regions receive less. Changes in Earth's rotational axis and orbital shape affect the total amount of solar energy received and thus alter the climate. (Courtesy Arizona State University, Computer Science and Geography Departments)

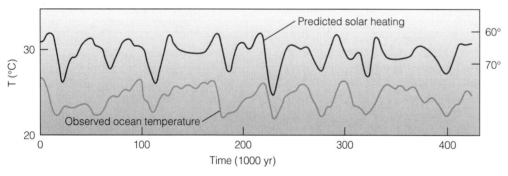

FIGURE 2-20
The Milankovitch theory predicts periodic changes in solar heating (shown here as the equivalent summer latitude of the sun). Over the last 400,000 years, these changes seem to have varied in step with ocean temperatures measured from fossils in sediment layers. (Adapted from Cesare Emiliani)

These contradictory findings are irritating because we naturally prefer certainty, but such circumstances are common in science. The disagreement between ocean floor samples and Devil's Hole samples triggered a scramble to understand the problem. Were the ages of one or the other set of samples wrong? Were the ancient temperatures wrong? Or were scientists misunderstanding the significance of the evidence?

In 1997, a new study of the ages of the samples confirmed that those from the ocean floor are correctly dated. This seems to give scientists renewed confidence in the Milankovitch hypothesis. But the same study found that the ages of the Devil's Hole samples are also correct. Many now believe the temperatures at Devil's Hole tell us about local climate changes in the region that became the southwestern United States. The ocean floor samples, which agree with the Milankovitch hypothesis, seem to tell us about global climate. Thus Milutin Milankovitch's hypothesis, first proposed in 1920, is still being tested as we try to understand the world we live on.

CRITICAL INQUIRY

How do precession and the shape of Earth's orbit interact to affect the earth's climate?

One technique that is useful in the critical analysis of an idea is exaggeration. If we exaggerate the variation in the shape of Earth's orbit, we can see dramatically the influence of precession. At present, Earth reaches perihelion during winter in the Northern Hemisphere and aphelion during summer. The variation in distance is only about 2 percent, and that difference doesn't cause much change in the severity of the seasons. But if Earth's orbit were much more elliptical, then winter in the Northern Hemisphere would be much warmer, and summer would be much cooler.

Now we can see the importance of precession. As Earth's axis precesses, it points gradually in different directions, and the seasons occur at different places in earth's orbit. In 13,000 years, northern winter will occur at aphelion, and if Earth's orbit were highly elliptical, northern winter would be terrible. Similarly, summer would occur at perihelion, and the heat would be awful. Such extremes might deposit large amounts of ice in the winter but then melt it away in the hot summer, thus preventing the accumulation of glaciers.

Continue this analysis by exaggeration. What affect would precession have if Earth's orbit were circular? ■

This chapter is about the sky, but it has told us a great deal about the earth. We have discovered that the orbital motion of the earth produces the apparent motion of the sun along the ecliptic, causes the seasons, and may even cause the cycle of the ice ages. As is often the case, astronomy helps us understand what we are and where we are. In this case, our study of the sky has helped us understand what it means to live on a planet. However, we have omitted one of the most dramatic objects in the sky—the moon. We will repair that omission in the next chapter. ■

■ Summary

Astronomers divide the sky into 88 areas called constellations. Although the constellations originated in Greek mythology, the names are Latin. Even the modern constellations, added to fill in the areas between the ancient figures, have Latin names. The names of stars usually come from ancient Arabic, though modern astronomers often refer to a star by constellation and Greek letters assigned according to brightness within each constellation.

The magnitude system is the astronomers' brightness scale. First-magnitude stars are brighter than second-magnitude stars, which are brighter than third-magnitude stars, and so on. The magnitude we see when we look at a star in the sky is its apparent visual magnitude.

The celestial sphere is a model of the sky, carrying the celestial objects around the earth. Because the earth rotates eastward, the celestial sphere appears to rotate westward on its axis. The northern and southern celestial poles are the pivots on which the sky appears to rotate. The celestial equator, an imaginary line around the sky above the earth's equator, divides the sky in half.

Because the earth orbits the sun, the sun appears to move eastward around the sky following the ecliptic. Because the

ecliptic is tipped 23.5° to the celestial equator, the sun spends half the year in the northern celestial hemisphere and half the year in the southern celestial hemisphere, producing the seasons. The seasons are reversed south of the earth's equator. That is, while the Northern Hemisphere is experiencing a warm season, the Southern Hemisphere is experiencing a cold season.

Of the nine planets in our solar system, Mercury, Venus, Mars, Jupiter, and Saturn are visible to the naked eye. Their orbital motion around the sun carries them along the zodiac, a band 18° wide centered on the ecliptic. The positions of the sun, the moon, and these five planets form the basis of astrology, an ancient superstition that originated in Babylonia about 1000 BC.

The motion of the earth may change in ways that can affect the climate. Changes in orbital shape, in precession, and in axial tilt can alter the planet's heat balance and may be responsible for the ice ages and glacial periods.

■ New Terms

constellation	ecliptic
apparent visual magnitude (m_v)	vernal equinox
model	summer solstice
celestial sphere	autumnal equinox
north and south celestial poles	winter solstice
celestial equator	perihelion
circumpolar constellation	aphelion
minute of arc	morning and evening stars
second of arc	zodiac
angular diameter	horoscope
precession	Milankovitch hypothesis

■ Questions

1. Why are most modern constellations composed of faint stars or located in the southern sky?

2. What does a star's Greek-letter designation tell us that its ancient Arabic name does not?

3. From your knowledge of star names and constellations, which of the following stars in each group is the brighter and which is the fainter? Explain your answers.

   a. α Ursae Majoris; θ Ursae Majoris
   b. λ Scorpii; β Pegasus
   c. β Telescopium; β Orionis

4. Give two reasons why the magnitude scale might be confusing.

5. Why do modern astronomers continue to use the celestial sphere when they know that stars are not all at the same distance?

6. How do we define the celestial poles and the celestial equator?

7. From what locations on the earth is the north celestial pole not visible? the south celestial pole? the celestial equator?

8. If the earth did not turn on its axis, could we still define an ecliptic? Why or why not?

9. Give two reasons why winter days are colder than summer days.

10. How do the seasons in the earth's Southern Hemisphere differ from those in the Northern Hemisphere?

11. Why don't the planets move exactly along the ecliptic?

12. Why can we be sure the astrological predictions printed in newspapers and magazines are not based on true calculations of the positions of celestial bodies in a horoscope?

13. Why should the eccentricity of the earth's orbit make winter in the earth's Northern Hemisphere different from winter in the Southern Hemisphere?

14. How might small changes in the inclination of the earth's axis to the plane of its orbit affect the growth of glaciers?

■ Discussion Questions

1. Have you thought of the sky as a ceiling? as a dome overhead? as a sphere around the earth? as a limitless void?

2. How would the seasons be different if the earth were inclined 90° instead of 23.5°? 0° instead of 23.5°?

■ Problems

1. If one star is 6.3 times brighter than another star, how many magnitudes brighter is it?

2. If one star is 40 times brighter than another star, how many magnitudes brighter is it? *4*

3. If two stars differ by 7 magnitudes, what is their intensity ratio?

4. If two stars differ by 8.6 magnitudes, what is their intensity ratio? *2800*

5. If star A is third magnitude and star B is fifth magnitude, which is brighter and by what factor?

6. If star A is magnitude 4 and star B is magnitude 9.6, which is brighter and by what factor?
A by 170

7. By what factor is the sun brighter than the full moon? (HINT: See Figure 2-6.)

8. What is the angular distance from the north celestial pole to the summer solstice? to the winter solstice?

9. As seen from your latitude, what is the angle between the north celestial pole and the northern horizon? between the southern horizon and the noon sun at the summer solstice?

10. Draw a diagram like that in Figure 2-16 to show the path of the sun across the sky at the time of the vernal equinox.

■ Critical Inquiries for the Web

1. Can you see the figure of a hunter in the constellation Orion? Orion's central location on the celestial sphere means that most cultures have noticed this familiar pattern of stars. Search the Internet for information on the mythologies associated with this star pattern. What stories do you find? Do the stories have common threads?

2. Would the stars of a familiar constellation such as Pegasus or Orion look the same if we lived on planets orbiting other stars? Find a table of distance data for the bright stars in a familiar constellation and construct a diagram that visualizes the positions of these stars in space. (See Figure 2-2.)

3. Is astrology a science or a pseudoscience? The Internet is full of astrological Web sites. Read the Windows on Science essays for this chapter and examine several astrology-related Web pages. Look for evidence that a scientific approach is being applied. What points can you make to indicate that the information presented has not been generated through scientific processes?

 Go to the Wadsworth Astronomy Resource Center (www.wadsworth.com/astronomy) for critical thinking exercises, articles, and additional readings from InfoTrac College Edition, Wadsworth's online student library.

Chapter Three

LUNAR PHASES, TIDES, AND ECLIPSES

GUIDEPOST

In the preceding chapter, we saw how the sun dominates our sky and determines the seasons. The moon is not as bright as the sun, but the moon passes through dramatic phases and occasionally participates in eclipses. The sun dominates the daytime sky, but the moon rules the night.

As we try to understand the appearance and motions of the moon in the sky, we discover that what we see is a product of light and shadow. To understand the appearance of the universe, we must understand light. Later chapters will show that much of astronomy hinges on the behavior of light.

In the next chapter, we will see how Renaissance astronomers found a new way to describe the appearance of the sky and the motions of the sun, moon, and planets. ∎

Even a man who is pure in heart

And says his prayers by night

May become a wolf when the wolfbane

   blooms

And the moon shines full and bright.

Proverb from old Wolfman movies

a

b

FIGURE 3-1
(a) A 12th-century Mayan symbol believed to represent a solar eclipse. The black-and-white sun symbol hangs from a rectangular sky symbol, and a voracious serpent approaches from below.
(b) Chinese representation of a solar eclipse as a dragon flying in front of the sun. (From the collection of Yerkes Observatory) (c) A wall carving from the ruins of a temple at Vijayanagara in southern India. It symbolizes a solar eclipse as two snakes approaching the disk of the sun. (T. Scott Smith)

c

The moon has tremendous fascination for us. It is the brightest object in the night sky, and its monthly cycle of phases has served lovers and predators since the first humans left the forests to live in the grasslands. Human culture is filled with traditions and superstitions connected with the moon.

Some people still believe that moonlight causes insanity. "Don't stare at the moon—you'll go crazy," more than one child has been warned.* The word *lunatic* comes from a time when even doctors thought that the insane were "moonstruck." A "mooncalf" is someone who has been crazy since birth, and the word is probably also related to the belief that moonlight can harm unborn children.

You have probably heard that people act less rationally when the moon is bright. Nurses, police, and teach-

*When I was very small, my grandmother told me if I gazed at the moon, I might go crazy. But it was too beautiful, and I ignored her warning. I secretly watched the moon from my window, became fascinated by the sky, and became an astronomer.

ers sometimes make this claim, but careful studies of hospital and police records show that there is no real correlation between the moon and erratic behavior. The moon is so impressive that we *expect* it to influence us.

If the moon has any effect on us, it is only because lunar phenomena are so dramatic. The most dramatic of those phenomena are eclipses, events that turn the full moon red or bring sudden darkness on a sunny day.

Many ancient cultures had special traditions concerning eclipses. The ancient Chinese, for example, greeted solar eclipses with noisemakers and arrows shot toward the heavens. The noise and arrows were intended to scare off the great dragon that was slowly devouring the sun (Figure 3-1). The ceremony never failed—the dragon always retreated.

One story tells of Hsi and Ho, two Chinese astronomers who got drunk and either failed to predict the solar eclipse of October 22, 2137 BC, or were unable to conduct the proper ceremonies to scare away the dragon. When the emperor recovered from the terror of the eclipse, he had the unfortunate astronomers beheaded.

a

Earth   Moon

b

FIGURE 3-2

(a) Seen from above the earth's North Pole, the moon's orbit carries it counterclockwise around the earth. As it follows its orbit, the moon rotates counterclockwise and keeps the same side facing the earth. A mountain on the moon that faces the earth will always face the earth.
(b) The earth and moon to scale. The moon is about 30 earth diameters from the earth.

This chapter is about lunar phenomena—phases, tides, and eclipses. Studying these events will help us understand what we see in the night sky. They will also introduce us to some of the most basic concepts of astronomy. We will discover how gravity produces tides, how light and shadow move through space, how the size and distance of an object affect what we see, and how cyclic events can be analyzed by searching for their patterns. Most of all, this chapter forces us to begin thinking of our home as a world in space.

3-1 THE PHASES OF THE MOON

Starting this evening, look for the moon in the sky. If it is a cloudy night or if the moon is in the wrong part of its orbit, you may not see it, but keep trying on successive evenings, and within a week or two you will see the moon. Then watch for the moon on following evenings, and you will see it following its orbit around the earth and cycling through its phases as it has done for billions of years.

The Motion of the Moon

When we watch the moon night after night, we notice two things about its motion. First, we see it moving eastward against the background of stars; second, we notice that the markings on its face don't change. These two observations help us understand the motion of the moon and the origin of the moon's phases.

The moon moves rapidly among the constellations. If you watch the moon for just an hour, you can see it move eastward against the background of stars by slightly more than its angular diameter. In the previous chapter, we discovered that the moon is about 0.5° in angular diameter, so it moves eastward a bit more than 0.5° per hour. In 24 hours, it moves 13°. Each night when we look at the moon, we see it about 13° eastward of its location the night before. This eastward movement is the result of the motion of the moon along its orbit around the earth.

The moon orbits around the earth at an average distance of 384,400 km, about 30 times the diameter of the earth. The orbit is slightly elliptical, so its distance from the earth can vary by about 6 percent (Figure 3-2). Seen from far above the earth's North Pole, the moon orbits counterclockwise (eastward) with a period of 27.321661 days. This is called the **sidereal period,** meaning that it is measured in relation to the stars. The moon takes 27.321661 days to circle the sky once and return to the same place among the stars.

In its eastward motion, the moon stays near the ecliptic. Its orbit is tipped by 5°9′ to the plane of Earth's orbit, so its path around the sky is inclined to the ecliptic by the same angle. It can never wander farther than 5°9′ north or south of the ecliptic. Thus, it follows the zodiac around the sky.

Whenever we look at the moon, we see the same markings on its face because it rotates on its axis to keep one side facing the earth. If it did not rotate, that is, if it remained facing the same direction with respect to the stars, then we would see different sides of the moon in different parts of its orbit. But because the moon rotates, a mountain on the moon facing the earth will always face the earth as the moon follows its orbit (Figure 3-2). Thus, we can talk about the near side of the moon (the side we see) and the back side of the moon (the side we can never see from the earth). We will discover later in this section why the moon rotates in this way.

The Phase Cycle

As we watch the moon night after night, we see its shape changing. The moon shines by reflected sunlight and as it orbits the earth, sunlight illuminates different parts of the side we can see (Figures 3-3 and 3-4). Thus, the moon's shape changes, and it passes through a cycle of phases.

The first half of the lunar cycle extends from new moon to full moon. When the moon is approximately between the earth and the sun, the side toward us is in darkness. The moon is invisible, and we refer to it as the new moon. A few days after new moon, it has move far enough along its orbit to allow the sun to illuminate a small sliver of the side toward us, and we see a thin crescent. Night by night, this crescent moon waxes (grows) until we see half of the side toward us illuminated by sunlight. We refer to this as first quarter. The moon continues to wax, becoming gibbous, or protuberant.

FIGURE 3-3

The phases of the moon are produced by the varying amounts of sunlight that reach the side of the moon facing the earth. At first quarter, for example, only half of the near side is illuminated; at full moon, all of the near side is illuminated. To locate the moon in the sky at a given time, choose the appropriate human figure. At sunset, for instance, the first-quarter moon is high in the sky, but the third-quarter moon is below the horizon. The lettered phases **a** to **g** refer to Figure 3-4.

FIGURE 3-4

The lunar phases. Compare with Figure 3-3. (UCO/Lick Observatory photos)

a

b

c

Gibbous comes from a Latin word meaning "hunch-backed" and is pronounced Gib′es—"Gib" as in Gibson and "es" as in estimate. When it is nearly opposite the sun, the side toward the earth is fully illuminated, and we see a full moon.

The second half of the lunar cycle reverses the first half. After reaching full, the moon wanes (shrinks) through gibbous phase to third quarter, then through crescent to new moon. To distinguish between the gibbous and crescent phases of the first and second half of the cycle, we refer to waning gibbous and waning crescent when the moon is shrinking, and waxing gibbous and waxing crescent when it is growing.

To summarize, let us follow the moon through one cycle of phases (Table 3-1). At new moon, the moon is nearly in line with the sun and sets in the west with the sun. Therefore, we see no moon at new moon. A few days after new moon, we see the waxing crescent above the western horizon soon after sunset, and each evening it is fatter and higher above the horizon until, about a week after new moon, it reaches first quarter and stands high in the southern sky at sunset. The first-quarter moon does not set until about midnight. In the days following first quarter, the moon waxes fatter, becoming waxing gibbous. Each evening we find it farther east among the stars, and it sets later and later. About 2 weeks after new moon, the moon reaches full, rising in the east as the sun sets in the west. The full moon is visible all night long, setting in the west at sunrise.

The waning phases of the moon may be less familiar, because the moon is not visible in the early evening sky. As it wanes through gibbous, it rises later and later. By the time it reaches third quarter, it does not rise until midnight. The waning crescent does not rise until even later, and if we want to see the thin waning crescent just before new moon, we must get up before sunrise and look for the moon above the eastern horizon (Figure 3-5).

TABLE 3-1

Times of Moonrise and Moonset

Phase	Moonrise	Moonset
New	Dawn	Sunset
First quarter	Noon	Midnight
Full	Sunset	Dawn
Third quarter	Midnight	Noon

FIGURE 3-5

When the moon is at waxing crescent phase, it is just above the western horizon soon after sunset. But when the moon is at waning crescent phase, it is above the eastern horizon just before dawn.

d

e

f

g

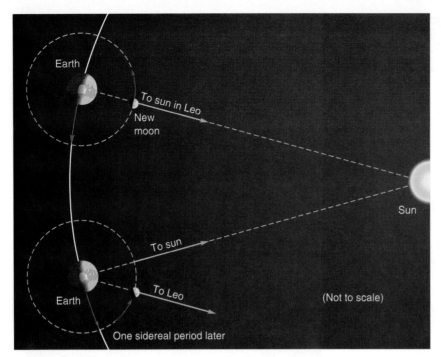

FIGURE 3-6
One sidereal period after new moon, the moon returns to the same place among the stars—the constellation Leo in this diagram—but the sun has moved eastward because of the earth's orbital motion. Thus, the moon must travel a bit over 2 more days to catch up with the sun and return to new. The period of the lunar phases (synodic period) is slightly more than 2 days longer than the moon's orbital period (sidereal period).

This cycle of phases takes longer than the moon's sidereal period. Imagine that we begin watching the moon at new moon when the sun and moon happen to be in the constellation Leo. The moon circles the sky in one sidereal period and returns to its starting point in Leo, but the sun is not there (Figure 3-6). During the 27.321661 days the moon needed to circle the sky and return to Leo, the earth moved along its orbit, and consequently the sun moved eastward along the ecliptic. One sidereal period is not enough for the moon to return to the earth–sun line. The moon must travel a bit over 2 more days to catch up with the sun and return to new moon. Thus, the lunar phases repeat with a period of 29.5305882 days. This is called the **synodic period,** the period with respect to the sun. *Synodic* comes from the Greek word *synodos,* which combines words meaning "together" and "journey."

CRITICAL **INQUIRY**

Why do we sometimes see the moon in the daytime?

The full moon rises at sunset and sets at sunrise, so it is visible in the night sky but never in the daytime sky. But at other phases, it is possible to see the moon in the daytime. For example, when the moon is a waxing gibbous moon, it rises a few hours before sunset. If you look in the right spot in the sky, you can see it in the late afternoon low in the southeast. It looks pale and washed out because of sunlight in our atmosphere, but it is quite visible once you notice it. You can also locate the waning gibbous moon in the morning sky. It sets a few hours after sunrise, so you would look for it in the southwestern sky in the morning.

If you look in the right place, you might even see the first- or third-quarter moon in the daytime sky, but you have probably never seen a crescent moon in the daytime. Why not? Where would you have to look? ■

The phases of the moon don't affect us directly—we don't really act crazier than usual at full moon. But the moon does have an important influence on the earth.

3-2 THE TIDES

Anyone who lives near the sea is familiar with a dramatic lunar effect—the ebb and flow of the tides. These periodic changes in the ocean are caused by the moon's gravity.

The Cause of the Tides

We feel the earth's gravity drawing us downward with a force we refer to as our weight, but the moon also exerts a gravitational force on us. Because the moon is less massive and more distant, its gravitational force is only about 0.0003 percent of the earth's gravitational force. That is a tiny force, but it is enough to cause tides in the earth's oceans.

Tides are produced by a *difference* between the gravitational force acting on different parts of an object. The

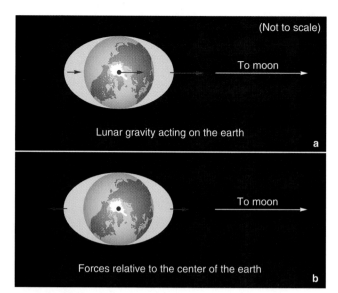

Lunar gravity acting on the earth

(Not to scale)

To moon

a

To moon

Forces relative to the center of the earth

b

FIGURE 3-7

(a) Because one side of the earth is closer to the moon than the other side, the moon's gravity does not attract all parts of the earth with the same force. (b) Relative to the center of the earth, small differences in the forces cause tidal bulges (much exaggerated here) on the sides of the earth facing toward and away from the moon.

Full moon   New moon   To sun

a

First quarter

To sun

Third quarter

b

FIGURE 3-8

(a) When the moon and the sun pull in the same direction, their tidal forces are added together, and the tidal bulges are larger (as symbolized by the dashed line). Thus, spring tides occur at new moon and full moon. (b) When the moon and sun pull at right angles, their tidal forces are not added together, and the tidal bulges are smaller. Such neap tides occur at first- and third-quarter moon.

side of the earth facing the moon is about 6400 km (4000 miles) closer to the moon than is the center of the earth, and the moon's gravity pulls more strongly on the oceans on the near side than on the center of the earth. The difference is small, only about 3 percent of the moon's total gravitational force on the earth, but it is enough to make the ocean waters flow into a bulge on the side of the earth facing the moon (Figure 3-7).

A bulge also forms on the side of the earth facing away from the moon. The far side of the earth is about 6400 km farther from the moon than is the center of the earth, and the moon's gravity pulls on it less strongly than it does on the earth's center. Thus, relative to the center of the earth, a small force makes the ocean waters on the earth's far side flow away from the moon (Figure 3-7).

We can see dramatic evidence of tidal forces if we watch the ocean shore for a few hours. Although the earth rotates on its axis, the tidal bulges remain fixed along the earth–moon line. As the turning earth carries us into a tidal bulge, the ocean water deepens, and the tide crawls up the beach. To be precise, the tide does not "come in." Rather, we are "carried into" the tidal bulge. Because there are two bulges on opposite sides of the earth, the tides rise and fall twice a day, and the times of high and low tide depend on the phase of the moon.

In reality, the tide cycle at any given location can be quite complex, depending on the latitude of the site, the shape of the shore, the north–south location of the moon, and so on. Tides in the Bay of Fundy occur twice

a day and can exceed 12 m. The northern coast of the Gulf of Mexico has only one tidal cycle of roughly 30 cm each day.

The sun, too, produces tides on the earth. The sun is roughly 27 million times more massive than the moon, but it lies almost 400 times farther from the earth. Consequently, tides on the earth caused by the sun are only about half those caused by the moon. At new moon and full moon, the moon and sun produce tidal bulges that join together to cause extreme tidal changes (Figure 3-8a); high tide is very high, and low tide is very low. Known as **spring tides,** these occur at every new and full moon. **Neap tides** occur at first- and third-quarter moon, when the moon and sun pull at right angles to each other (Figure 3-8b). Then the smaller tide caused by the sun slightly reduces the tide caused by the moon, and the resulting high and low tides are less extreme.

Though the ocean has been used as an example, you should note that tides would occur even if the earth had no oceans. The difference in the gravitational forces acting on different parts of the earth causes slight bulges in the rocky shape of the earth itself. We do not feel the mountains and plains rising and falling by a few centimeters; the changes in the fluid oceans are much more obvious.

Tidal Effects

Tidal forces can have surprising effects on both rotation and orbital motion. The friction of the earth's ocean waters against the seabeds slows the rotation of the earth, and our days are getting longer by 0.001 second per century. Some marine animals deposit layers in their shells in phase with the tides and the day–night cycle, and fossils of these shells confirm that only 400 million years ago the earth's day was 22 hours long. Tides are slowing the earth's rotation.

In addition, the earth's gravitational field exerts tidal forces on the moon, and although there are no oceans on the moon, tides do flex its rocky bulk. The resulting friction in the rock has slowed the moon's rotation. It once rotated much faster, but the tides caused by the earth have slowed the moon until it now rotates to keep one side permanently facing the earth. It is thus tidally locked to the earth. We will discover that most of the moons in the solar system are tidally locked to their planets.

Tidal forces can also affect orbital motion. For example, friction with the rotating earth drags the tidal bulges eastward out of a direct earth–moon line (Figure 3-9). These tidal bulges contain a large amount of mass, and the gravitation of the bulge near the moon pulls the moon slightly forward in its orbit. The bulge on the far side of the earth pulls the moon slightly backward in its orbit, but because it is farther from the moon, that bulge is less effective. Thus, the net effect is to drag the moon forward in its orbit. As a result, the moon's orbit is growing larger and is receding from the earth at about 3 cm per year, an effect that astronomers can measure by bouncing laser beams off reflectors left on the lunar surface by the Apollo astronauts. Thus, we have clear evidence that tides are changing the moon's orbit.

CRITICAL INQUIRY

Would the earth have tides if it had no moon?

Yes, there would be tides, but they would be much smaller. Don't forget that the sun also produces tides on the earth. The solar tides are less extreme than the lunar tides, but we would still see the advance and retreat of the oceans if the moon did not exist.

There would be an interesting difference, however, if the only tides were solar tides. The sun rises every 24 hours, but the moon moves rapidly eastward in the sky and rises, on average, every 24.8 hours. How many hours would separate high tides caused by the sun compared with high tides caused by the moon? ■

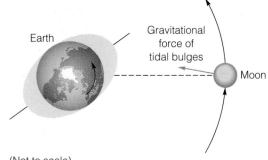

(Not to scale)

FIGURE 3-9

The rotation of the earth drags the tidal bulges ahead of the earth–moon line (exaggerated here). The gravitational attraction of these masses of water pulls the moon forward in its orbit, forcing its orbit to grow in size.

Tides and tidal forces are important in many areas of astronomy. In later chapters, we will see how tidal forces can pull gas away from stars, rip galaxies apart, and melt the interiors of satellites orbiting near massive planets. Now, however, we must consider yet another kind of lunar phenomenon—eclipses.

3-3 LUNAR ECLIPSES

A lunar eclipse occurs at full moon when the moon moves through the shadow of the earth. Because the moon shines only by reflected sunlight it gradually darkens as it enters the shadow.

The Earth's Shadow

The earth's shadow consists of two parts. The **umbra** is the region of total shadow. If we were floating in space in the umbra of the earth's shadow, we would see no portion of the sun. However, if we moved into the **penumbra,** we would be in partial shadow and would see part of the sun peeking around the edge of the earth. In the penumbra, the sunlight is dimmed but not extinguished.

We can construct a model of this by pressing a map tack into the eraser of a pencil and holding the tack between a light bulb a few feet away and a white cardboard screen (Figure 3-10). The light bulb represents the sun, and the map tack represents the earth. When we hold the screen close to the tack, we see that the umbra is nearly as large as the tack and that the penumbra is only slightly larger. However, as we move the screen away from the tack, the umbra shrinks and the penumbra expands. Beyond a certain point, the shadow has no dark

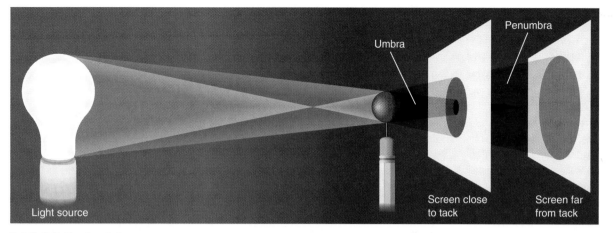

FIGURE 3-10

The shadows cast by a map tack resemble those of the earth and moon. The umbra is the region of total shadow; the penumbra is the region of partial shadow.

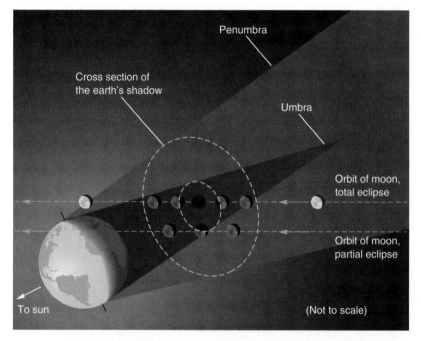

FIGURE 3-11

During a total lunar eclipse, the moon's orbit carries it through the penumbra and completely into the umbra. During a partial lunar eclipse, the moon does not completely enter the umbra. Compare the cross section of the earth's shadow with Figure 3-10.

core at all, indicating that the screen is beyond the end of the umbra.

The umbra of the earth's shadow is about 1.4 million km (860,000 miles) long and points directly away from the sun. A giant screen placed in the shadow at the average distance of the moon would reveal a dark umbra about 9000 km (5700 miles) in diameter, and the faint outer edges of the penumbra would mark a circle about 16,000 km (10,000 miles) in diameter. For comparison, the moon's diameter is only 3476 km (2160 miles). Thus, when the moon's orbit carries it through the umbra, it has plenty of room to become completely immersed in shadow.

Total Lunar Eclipses

If the orbit of the moon carries it entirely into the umbra so that no part of the moon protrudes into the partial sunlight of the penumbra, we say the eclipse is total. Figure 3-11 shows a three-dimensional drawing as seen from space, and Figure 3-12 shows a photographic image of what we would see from the earth (Window on Science 3-1).

A **total lunar eclipse** occurs gradually, although the exact timing depends on how the moon's orbit crosses the earth's shadow. Because the moon moves eastward a distance equal to its own diameter each hour, it takes about an hour for it to completely enter the outer edge of the penumbra. As it moves deeper into the penumbra,

3-D Relationships

Much of science is explained in diagrams; one particular type of diagram can be confusing. When artists draw three-dimensional diagrams on flat sheets of paper, they use lots of artistic clues that have been discovered over the centuries. Perspective, shading, color, and shadows help us see the three-dimensional figure if the drawing is familiar—a house, for example. But when a drawing shows something that is unfamiliar, as is often the case in science, the artist's clues don't work as well. When we look at drawings of a molecule, a nerve cell, layers of rock under the ocean, or the earth and its shadows, we must pay special attention to the three-dimensional nature of the figure.

When you see a three-dimensional diagram in any science book, it helps to decide what the point of view is. For example, if the diagram is a photograph, where was the camera? In Figure 3-11, the camera would have to have been in space looking back at the earth and moon. Of course, astronauts have never been that far from the earth, but we can make such a voyage in our imagination, and that helps us understand the geometry of eclipses. Other three-dimensional diagrams have other points of view, so always be sure to first determine the point of view when you look at any three-dimensional diagram. ■

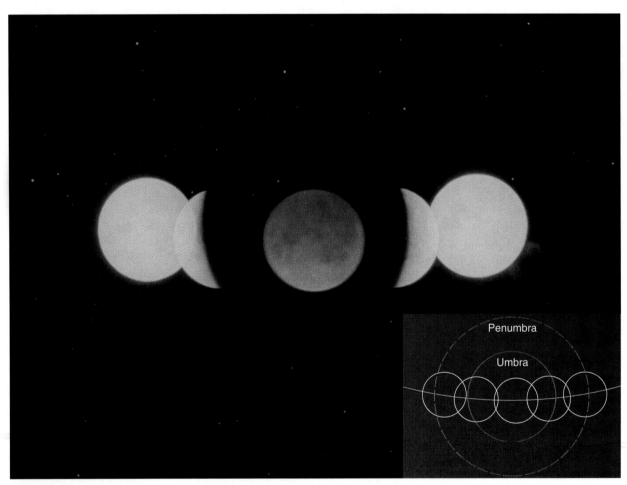

FIGURE 3-12

This multiple-exposure photo of a lunar eclipse spans 5 hours and shows the moon passing through the earth's umbra and penumbra (inset) from the right. The totally eclipsed moon (center) was 10,000 times dimmer than the full moon, so the exposure was lengthened to record the fainter image. Photographic effects make the moon's orbit appear curved here. Compare with Figure 3-11. (© 1982 by Dr. Jack B. Marling)

TABLE 3-2

Total and Partial Eclipses of the Moon, 1996 to 2011

Year	Date	Time* of Mideclipse (GMT)	Length of Totality (Hours:Min)	Length of Eclipse† (Hours:Min)
1996	Apr. 4	0:11	1:26	3:36
1996	Sept. 27	2:55	1:10	3:22
1997	Mar. 24	4:41	Partial	3:22
1997	Sept. 16	18:47	1:02	3:16
1999	July 28	11:34	Partial	2:22
2000	Jan. 21	4:45	1:16	3:22
2000	July 16	13:57	1:46	3:56
2001	Jan. 9	20:22	1:00	3:16
2001	July 5	14:57	Partial	2:38
2003	May 16	3:41	0:52	3:14
2003	Nov. 9	1:20	0:22	3:30
2004	May 4	20:32	1:16	3:22
2004	Oct. 28	3:05	1:20	3:38
2005	Oct. 17	12:04	Partial	0:56
2006	Sept. 7	18:52	Partial	1:30
2007	Mar. 3	23:22	1:14	3:40
2007	Aug. 28	10:38	1:30	3:32
2008	Feb. 21	3:27	0:50	3:24
2008	Aug. 16	21:11	Partial	3:08
2009	Dec. 31	19:24	Partial	1:00
2010	June 26	11:40	Partial	2:42
2010	Dec. 21	8:18	1:12	3:28
2011	June 15	20:13	1:40	3:38
2011	Dec. 10	14:33	0:50	3:32

*Times are Greenwich Mean Time. Subtract 5 hours for Eastern Standard Time, 6 hours for Central Standard Time, 7 hours for Mountain Standard Time, and 8 hours for Pacific Standard Time. From your time zone, lunar eclipses that occur between sunset and sunrise will be visible, and those that occur at midnight will be best placed.

†Does not include penumbral phase.

FIGURE 3-13

During a total lunar eclipse, the moon turns coppery red as sunlight refracted by the earth's atmosphere illuminates the moon in a red, sunset glow. An astronaut on the moon during such an eclipse would see this red light coming from a sunset completely encircling the earth. (Celestron International)

it grows dimmer, and about an hour after entering the penumbra it reaches the edge of the umbra. The moon takes about an hour to completely enter the umbra and become totally eclipsed (see Table 3-2).

Even when the moon is totally eclipsed, it does not disappear completely. Sunlight, bent by the earth's atmosphere, leaks into the umbra and bathes the moon in a faint glow. Because blue light is scattered by the earth's atmosphere more easily than red light, it is red light that penetrates to illuminate the moon in a coppery glow (Figure 3-13). If we were on the moon during total-

ity and looked back at the earth, we would not see any part of the sun because it would be entirely hidden behind the earth. However, we would see the earth's atmosphere illuminated from behind by the sun in a spectacular sunset completely ringing the earth. It is the red glow from this sunset that gives the totally eclipsed moon its reddish color.

How dim the totally eclipsed moon becomes depends on a number of things. If the earth's atmosphere is especially cloudy in those regions that must bend light into the umbra, the moon will be darker than usual. An unusual amount of dust in the earth's atmosphere (from volcanic eruptions, for instance) also causes a dark eclipse. Also, total lunar eclipses tend to be darkest when the moon's orbit carries it through the center of the umbra.

As the moon moves through the earth's umbral shadow, we can see that the shadow is circular. From this the Greek philosopher Aristotle (384–322 BC) concluded that the earth had to be a sphere, because only a sphere could cast a shadow that was always circular.

Depending on the geometry of the eclipse, the moon can take as long as 1 hour 40 minutes to cross the umbra and another hour to emerge into the penumbra. Still another hour passes as it emerges into full sunlight. A total eclipse of the moon, including the penumbral stage, can take almost 6 hours from start to finish.

Not all eclipses of the moon are total. If the moon only partially enters the umbra, the eclipse is termed a

partial lunar eclipse (Figure 3-11). Partial eclipses usually are not as beautiful as total eclipses because the faint coppery glow of the eclipsed moon is lost in the glare of the uneclipsed portions. If the moon does not enter the umbra at all but only passes through the penumbra, the eclipse is termed a **penumbral eclipse.** The partial dimming of the moon in the penumbra is often difficult to detect.

CRITICAL INQUIRY

Why doesn't the earth's shadow on the moon look red during a partial lunar eclipse?

During a partial lunar eclipse, part of the moon protrudes from the earth's umbral shadow into sunlight. This part of the moon is very bright compared to the fainter red light inside the earth's shadow, and the glare of the reflected sunlight makes it difficult to see the red glow. If a partial eclipse is almost total, so that only a small sliver of moon extends out of the shadow into sunlight, we can sometimes detect the red glow in the shadow.

Of course, this red glow does not happen for every planet–moon combination in the universe. Suppose a planet had a moon but no atmosphere. Would the moon glow red during a total eclipse? Why not? ■

Lunar eclipses are slow and stately. For drama and excitement, there is nothing like a solar eclipse.

3-4 SOLAR ECLIPSES

A solar eclipse occurs when the moon moves between the earth and the sun. If the moon covers the disk of the sun completely, the eclipse is a **total solar eclipse.** If the moon covers only part of the sun, the eclipse is a **partial solar eclipse.** During a particular solar eclipse, people in one place on the earth may see a total eclipse while people only a few hundred kilometers away see only a partial eclipse.

These spectacular sights are possible because we on Earth are very lucky. Our moon has the same angular diameter as our sun, so it can cover the sun almost exactly. That lucky coincidence allows us to see total solar eclipses.

The Angular Diameter of the Sun and Moon

We discussed the angular diameter of an object in Chapter 2; now we need to think carefully about how the size and distance of an object like the moon determine its angular diameter (Figure 3-14). This is the key to understanding solar eclipses.

Linear diameter is simply the distance between an object's opposite sides. We use linear diameter when we

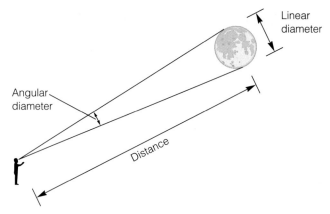

FIGURE 3-14

The small-angle formula relates angular diameter, linear diameter, and distance. Angular diameter is the angle formed by lines extending from our eye to opposite sides of the object—in this figure, the moon. Linear diameter and distance are typically measured in kilometers or meters.

order a 16-inch pizza—the pizza is 16 inches in diameter. The linear diameter of the moon is 3476 km. The angular diameter of an object is the angle formed by lines extending toward us from opposite sides of the object and meeting at our eye. Clearly, the farther away an object is, the smaller its angular diameter.

The **small-angle formula** gives us a way to figure out the angular diameter of any object, whether it is a pizza, the moon, or a galaxy. In the small-angle formula, we always express angular diameter in seconds of arc,* and we always use the same units for distance and linear diameter:

$$\frac{\text{angular diameter}}{206{,}265''} = \frac{\text{linear diameter}}{\text{distance}}$$

Of course, we can use this formula to find any one of these three quantities if we know the other two; here we are interested in finding the angular diameter of the moon.

The moon has a linear diameter of 3476 km and a distance from the earth of about 384,000 km. What is its angular diameter? The moon's linear diameter and distance are both given in the same units, so we can put them directly into the small-angle formula:

$$\frac{\text{angular diameter}}{206{,}265''} = \frac{3476 \text{ km}}{384{,}000 \text{ km}}$$

To solve for angular diameter, we multiply both sides by 206,265 and find that the angular diameter is 1870 seconds of arc. If we divide by 60, we get 31 minutes of arc or, dividing by 60 again, about 0.5°. The moon's orbit is slightly elliptical, so it can sometimes look a bit larger or smaller, but its angular diameter is always close to 0.5°.

---

*The number 206,265″ is the number of seconds of arc in a radian. When we divide by 206,265″, we convert the angle from seconds of arc to radians.

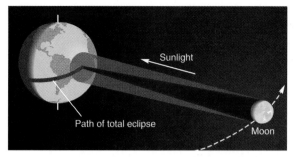

FIGURE 3-15

Observers in the path of totality see a total solar eclipse when the umbral shadow sweeps over them. Those in the penumbra see a partial eclipse.

We can repeat this calculation for the angular diameter of the sun. The sun is 1.39×10^6 km in linear diameter and 1.50×10^8 km from the earth. Using the small-angle formula, we discover that the sun has an angular diameter of 1900 seconds of arc, which is 32 minutes of arc or about 0.5°. The earth's orbit is slightly elliptical, and thus the sun can sometimes look slightly larger or smaller, but it, like the moon, is always close to 0.5° in angular diameter.

By fantastic good luck, we live on a planet with a moon that is almost exactly the same angular diameter as our sun. When the moon passes in front of the sun, it is almost exactly the right size to block the brilliant surface of the sun. Then we see the most exciting sight in astronomy—a total solar eclipse. There are few other worlds where this can happen, because the angular diameters of the sun and a satellite rarely match so closely. To see this beautiful sight, all we have to do is arrange to be in the moon's shadow when the moon crosses in front of the sun.

The Moon's Shadow

Like the earth's shadow, the moon's shadow consists of a central umbra of total shadow and a penumbra of partial shadow. What we see when the moon crosses in front of the sun depends on where we are in the moon's shadow. The moon's umbral shadow produces a spot of darkness roughly 269 km (167 miles) in diameter on the earth's surface (Figure 3-15). (The exact size of the umbral shadow depends on the location of the moon in its elliptical orbit and the angle at which the shadow strikes the earth.) If we are in this spot of total shadow, we see a total solar eclipse. If we are just outside the umbral shadow but in the penumbra, we see part of the sun peeking around the moon, and the eclipse is partial. Of course, if we are outside the penumbra, we see no eclipse at all. Because of the orbital motion of the moon, its shadow sweeps across the earth at speeds of at least 1700 km/h (1060 mph). To be sure of seeing a total solar eclipse, we must select an appropriate eclipse (Table 3-3), plan far in advance, and place ourselves in the **path of totality,** the path swept out by the umbral spot.

TABLE 3-3

Total and Annular Eclipses of the Sun, 1995 to 2009

Date	Time of Mideclipse* (GMT)	Total/ Annular (T/A)	Max. Length of Total or Annular Phase (Min:Sec)	Area of Visibility
1995 Oct. 24	5ʰ	T	2:10	Asia, Borneo, Pacific
1997 Mar. 9	1ʰ	T	2:50	Siberia
1998 Feb . 26	17ʰ	T	4:08	Pacific, N. of S. America, Atlantic
1998 Aug. 22	2ʰ	A	3:14	Sumatra, Borneo, Pacific
1999 Feb. 16	7ʰ	A	1:19	Indian Ocean, Australia
1999 Aug. 11	11ʰ	T	2:23	Atlantic, Europe, S.E. and S. Asia
2001 June 21	12ʰ	T	4:56	Atlantic, S. Africa, Madagascar
2001 Dec. 14	21ʰ	A	3:54	Pacific, Central America
2002 June 10	24ʰ	A	1:13	Pacific
2002 Dec. 4	8ʰ	T	2:04	S. Africa, Indian Ocean, Australia
2003 May 31	4ʰ	A	3:37	Iceland, Arctic
2003 Nov. 23	23ʰ	T	1:57	Antarctica
2005 Apr. 8	21ʰ	A, T	0:42	Pacific, N. of S. America
2005 Oct. 3	11ʰ	A	4:32	Atlantic, Spain, Africa
2006 Mar. 29	10ʰ	T	4:07	Atlantic, Africa, Turkey
2006 Sept. 22	12ʰ	A	7:09	N.E. of S. America, Atlantic
2008 Feb. 7	4ʰ	A	2:14	S. Pacific, Antarctica
2008 Aug. 1	10ʰ	T	2:28	N. Canada, Arctic, Siberia
2009 Jan. 26	8ʰ	A	7:56	S. Atlantic, Indian Ocean
2009 July 22	3ʰ	T	6:40	Asia, Pacific

*Times are Greenwich Mean Time. Subtract 5 hours for Eastern Standard Time, 6 hours for Central Standard Time, 7 hours for Mountain Standard Time, and 8 hours for Pacific Standard Time.

Total Solar Eclipses

A total solar eclipse begins when we first see the edge of the moon encroaching on the sun. This is also the moment when the edge of the penumbra sweeps over our location.

During the partial phase, part of the sun remains visible, and it is hazardous to look at the eclipse without protection. Dense filters and exposed film do not necessarily provide protection, because some do not block the invisible heat radiation (infrared) that can burn the retina of our eyes. This has led officials to warn the public not to look at solar eclipses and has even frightened some people into locking themselves and their children into windowless rooms during eclipses. In fact, the sun is a bit less dangerous than usual during an eclipse because part of the bright surface is covered by the moon. But an eclipse is dangerous in that it can tempt us to look at the sun directly and burn our eyes.

The safest and simplest way to observe the partial phases of a solar eclipse is to use pinhole projection. Poke a small pinhole in a sheet of cardboard. Hold the sheet with the hole in the sunlight, and allow light to pass through the hole to a second sheet of cardboard (Figure 3-16). On a day when there is no eclipse, the result is a small, round spot of light that is an image of the sun. During the partial phases of a solar eclipse, the image shows the dark silhouette of the moon obscuring part of the sun. These pinhole images of the partially eclipsed sun can also be seen in the shadows of trees as the sunlight peeks through the tiny openings between the leaves and branches. This can produce an eerie effect just before

FIGURE 3-16

A safe way to view the partial phases of a solar eclipse. Use a pinhole in a card to project an image of the sun on a second card. The greater the distance between the cards, the larger and fainter the image will be.

FIGURE 3-17

This sequence shows the first half of a total solar eclipse. The brilliant surface of the sun is gradually covered by the moon moving from the right. It is dangerous to look at the sun during the partial phases because the bright surface of the sun can burn our eyes. Once the last of the brilliant photosphere is covered, we can see the pink prominences of the chromosphere. A much longer exposure is needed to photograph the fainter corona. (Daniel Good)

a

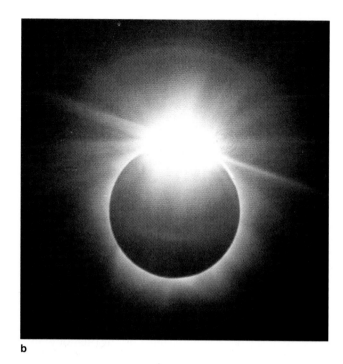

b

FIGURE 3-18

(a) During a total solar eclipse, the moon covers the photosphere, and the white corona and pink prominences are visible. Note the streamers in the corona caused by the sun's magnetic field. (Daniel Good) (b) The diamond ring effect can sometimes occur momentarily at the beginning or end of totality if a small segment of the photosphere peeks out through a valley at the edge of the lunar disk. (National Optical Astronomy Observatory)

totality as the remaining sliver of sun produces thin crescents of light on the ground under trees.

Throughout the partial phases of a solar eclipse, the moon gradually covers the bright disk of the sun (Figure 3-17). Totality begins as the last sliver of the sun's bright surface disappears behind the moon. This is the same moment when the edge of the umbra sweeps over our location. So long as any of the sun is visible, the countryside is bright, but as the last of the sun disappears, dark falls in a few seconds. Automatic streetlights come on, car drivers switch on their headlights, and birds go to roost. The darkness of totality depends on a number of factors, including the weather at the observing site, but it is usually dark enough to make it difficult to read the settings on cameras.

The totally eclipsed sun is a spectacular sight. With the moon covering the bright disk of the sun, called the **photosphere,*** we can see the sun's faint outer atmosphere, the **corona,** glowing with a pale, white light so faint we can safely look at it directly. This corona is made of low-density, hot gas, which is given a wispy appearance by the solar magnetic field (Figure 3-18a). Also visible just above the photosphere is a thin layer of bright gas called the **chromosphere.** The chromosphere is often marked by eruptions on the solar surface called **prominences** (Figure 3-18a), which glow with a clear, pink color due to the high temperature of the gases

involved. The small-angle formula tells us that a large prominence is about 3.5 times the diameter of the earth.

Totality cannot last longer than 7.5 minutes under any circumstances, and the average is only 2 to 3 minutes. Totality ends when the sun's bright surface reappears at the trailing edge of the moon. This corresponds to the moment when the trailing edge of the moon's umbra sweeps over the observer.

Just as totality begins or ends, a small part of the photosphere can peek out from behind the moon through a valley at the edge of the lunar disk. Although it is intensely bright, such a small part of the photosphere does not completely drown out the fainter corona, which forms a silvery ring of light with the brilliant spot of photosphere gleaming like a diamond (Figure 3-18b). This **diamond ring effect** is one of the most spectacular of astronomical sights, but it is not visible during every solar eclipse. Its occurrence depends on the exact orientation and motion of the moon.

Once totality is over, daylight returns quickly, and the corona and chromosphere vanish. Astronomers travel great distances to place their instruments in the path of totality to study the faint outer corona and make other measurements possible only during the few minutes of a total solar eclipse.

But not all solar eclipses are total. In more than half of all eclipses, the moon, following its slightly elliptical orbit, is too far from the earth, and its umbral shadow

---

*The photosphere, corona, chromosphere, and prominences will be discussed in detail in Chapter 8. Here the terms are used as the names of features we see during a total solar eclipse.

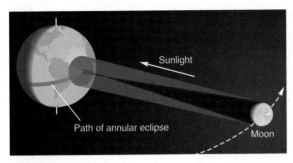

FIGURE 3-19
If the moon is near the farther part of its slightly elliptical orbit, the umbral shadow does not reach the earth, resulting in an annular eclipse.

FIGURE 3-20
An annular eclipse occurs when the moon is too far from the earth and looks too small to cover the entire disk of the sun. Here the annular eclipse of January 4, 1992 is seen as the sun and moon set among clouds in the west. The flattening of the sun and moon visible in this photograph is produced by Earth's atmosphere. (Copyright © Association of Universities for Research in Astronomy, Inc. AURA. All rights reserved)

does not reach earth's surface (Figure 3-19). In these cases, the moon's angular diameter is slightly too small to cover the sun completely. At mideclipse, when the moon is exactly centered on the sun, it is too small to cover all of the photosphere, and a bright ring, or annulus, of the sun's surface is visible around the disk of the moon, producing an **annular eclipse** (Figure 3-20). These eclipses are not nearly as interesting as total eclipses. The countryside does not get dark, and the glare from the exposed ring of photosphere hides the corona, the chromosphere, and prominences. Such an annular eclipse of the sun swept across the United States on May 10, 1994.

CRITICAL INQUIRY

If people on the earth were seeing a total solar eclipse, what would astronauts on the moon see when they looked at the earth?

Astronauts on the moon could see the earth only if they were on the side that faces the earth. Because solar eclipses always happen at new moon, the near side of the moon would be in darkness, and the far side of the moon would be in full sunlight. The astronauts would be standing in darkness, and they would be looking at the fully illuminated side of the earth. They would see a "full earth." The moon's shadow would be crossing the earth, and if the astronauts looked closely, they might be able to see the spot of darkness where the moon's umbral shadow touched the earth. It would take hours for the shadow to cross the earth.

Standing on the moon and watching the moon's umbral shadow sweep across the earth would be a cold, tedious assignment. Perhaps it would be more interesting for astronauts on the moon to watch the earth while people on the earth were seeing a total lunar eclipse. What would the astronauts see then? ■

Eclipses of the sun and moon are often dramatic and mysterious. Ancient astronomers studied them and found ways to predict the coming of an eclipse. In the section that follows, we will see how eclipses can be predicted from a basic understanding of the motion of the sun and the moon.

3-5 PREDICTING ECLIPSES

To make exact eclipse predictions, you would need to calculate the precise motions of the sun and moon, and that requires a computer and proper software. Such software is available for desktop computers, but it isn't necessary if you are satisfied with making less exact predictions. In fact, many primitive peoples, such as the builders of Stonehenge and the ancient Maya, are believed to have made eclipse predictions.

We examine eclipse prediction for three reasons. First, it is an important part of the history of science. Second, it illustrates how apparently complex phenomena can be analyzed in terms of cycles. Third, eclipse predic-

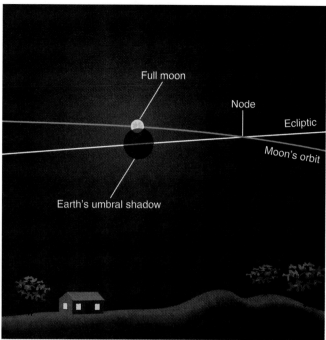

a

b

FIGURE 3-21

Eclipses can occur only near the nodes of the moon's orbit. (a) A solar eclipse occurs when the moon meets the sun near a node. (b) A lunar eclipse occurs when the sun and moon are at opposite nodes. Partial eclipses are shown here for clarity.

tion exercises our mental muscles and forces us to see the earth, the moon, and the sun as objects moving through space.

Conditions for an Eclipse

We can predict eclipses by understanding the conditions that make them possible. As we begin to think about these conditions, we must be sure we understand our point of view. (See Window on Science 3-1.) Later we will change our point of view, but to begin we will imagine that we can look up into the sky from our home on the earth and see the sun moving along the ecliptic and the moon moving along its orbit.

The orbit of the moon is tipped 5°8′43″ to the plane of the earth's orbit, so we see the moon follow a path tipped by that angle to the ecliptic. Each month, the moon crosses the ecliptic at two points called **nodes** (Figure 3-21). At one node it crosses going southward, and 2 weeks later it crosses at the other node going northward.

Eclipses can occur only when the sun is near a node. A solar eclipse is caused by the moon passing in front of the sun, and this can happen only when the sun is near a node. A lunar eclipse occurs when the moon enters the earth's shadow, but that shadow always follows the ecliptic exactly opposite the sun. Thus, the moon can enter the shadow only when the shadow is at a node, and that

means the sun must be on the opposite side of the sky at the other node.

Thus, there are two conditions for an eclipse: the sun must be crossing a node, and the moon must be crossing either the same node (solar eclipse) or the other node (lunar eclipse). Clearly, solar eclipses can occur only when the moon is new, and lunar eclipses can occur only when the moon is full.

An **eclipse season** is the period during which the sun is close enough to a node for an eclipse to occur. For solar eclipses, an eclipse season is about 32 days. Any new moon during this period will produce a solar eclipse. For lunar eclipses, the eclipse season is a bit shorter, about 22 days. Any full moon in this period will be eclipsed.

This makes eclipse prediction easy. We simply keep track of where the moon crosses the ecliptic, and when the sun is near one of these nodes we predict that the nearest new moon will cause a solar eclipse and the nearest full moon will cause a lunar eclipse. This system works fairly well, and ancient astronomers such as the Maya may have used such a system. But we can do better if we change our point of view.

The View from Space

Let us change our point of view and imagine that we are looking at the orbits of Earth and the moon from a point far away in space. We see the moon's orbit as a smaller

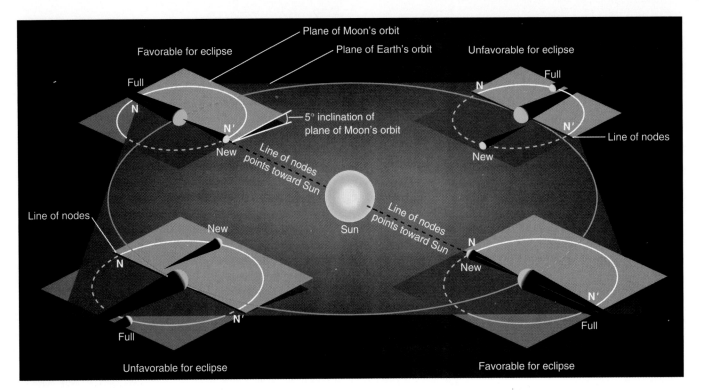

FIGURE 3-22

The moon's orbit is tipped about 5° to Earth's orbit. The nodes N and N′ are the points where the moon passes through the plane of Earth's orbit. If the line of nodes does not point at the sun, the shadows miss and there are no eclipses at new moon and full moon. At those parts of Earth's orbit where the line of nodes points toward the sun, eclipses are possible at new moon and full moon.

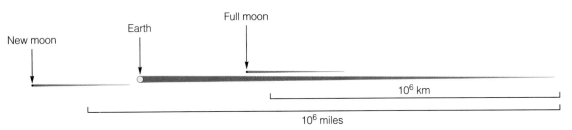

FIGURE 3-23

This scale drawing of the umbral shadows of the earth and moon shows how easy it is for the shadows to miss their mark at full moon and new moon. That is, it is easy for a full moon to occur and not enter the earth's shadow, or for a new moon to occur and not cast its shadow on the earth. The diameters of the earth and moon are exaggerated by a factor of 2 for clarity.

disk tipped at an angle to the larger disk of Earth's orbit (Figure 3-22a). As Earth orbits the sun, the moon's orbit remains fixed in direction. The nodes of the moon's orbit are the points where it passes through the plane of Earth's orbit; an eclipse season occurs each time the line connecting these nodes, the **line of nodes,** points toward the sun (Figure 3-22c).

The shadows of the earth and moon, seen from space, are very long and thin (Figure 3-23). Only at the time of an eclipse season, when the line of nodes points toward the sun, do the shadows produce eclipses.

From our point of view in space, we would see the orbit of the moon precess like a hubcap spinning on the ground. This precession is caused mostly by the gravitational influence of the sun, and it makes the line of nodes rotate once every 18.6 years. People back on the earth see the nodes slipping westward along the ecliptic 19.4° per year, and the sun takes only 346.62 days (an **eclipse year**) to return to a node. This means that, according to our calendar, the eclipse seasons begin about 19 days earlier every year (Figure 3-24).

FIGURE 3-25

The saros cycle at work. The total solar eclipse of March 7, 1970, recurred after 18 years $11\frac{1}{3}$ days over the Pacific Ocean. After another interval of 18 years $11\frac{1}{3}$ days, the same eclipse will be visible from Asia and Africa. After a similar interval, the eclipse will again be visible from the United States.

+ Total and annular solar eclipses   • Lunar eclipses

FIGURE 3-24

A calendar of eclipse seasons. Each year, the eclipse seasons begin about 19 days earlier. Any new moon or full moon that occurs during an eclipse season results in an eclipse. Not all eclipses are shown here.

The cyclic pattern of eclipses shown in Figure 3-24 makes eclipse prediction simple. We know that the eclipse seasons occur 19 days earlier each year because the moon's orbit precesses. Any new moon during an eclipse season will cross in front of the sun and cause a solar eclipse, and any full moon will enter the earth's shadow and cause a lunar eclipse.

The Saros Cycle

Ancient astronomers could predict eclipses in a crude way using the eclipse seasons, but they could have been much more accurate if they recognized that eclipses occur following certain patterns. The most important of these is the **saros cycle** (sometimes referred to simply as the saros). After one saros cycle of 18 years $11\frac{1}{3}$ days, the

pattern of eclipses repeats. In fact, *saros* comes from a Greek word that means "repetition."

The eclipses repeat because, after one saros cycle, the moon and the nodes of its orbit return to the same place with respect to the sun. One saros contains 6585.321 days, which is equal to 223 lunar months. Therefore, after one saros cycle the moon is back to the same phase it had when the cycle began. But one saros is nearly equal to 19 eclipse years. After one saros cycle, the sun has returned to the same place it occupied with respect to the nodes of the moon's orbit when the cycle began. If an eclipse occurs on a given day, then 18 years $11\frac{1}{3}$ days later the sun, the moon, and the nodes of the moon's orbit return to nearly the same relationship, and the eclipse occurs all over again.

Although the eclipse repeats almost exactly, it is not visible from the same place on earth. The saros cycle is one-third of a day longer than 18 years 11 days. When the eclipse recurs, the earth will have rotated one-third of a turn farther east, and the eclipse will occur 8 hours of longitude west of its earlier location (Figure 3-25). Thus, after three saros cycles—a period of 54 years 1 month—the same eclipse occurs in the same part of the earth.

One of the most famous predictors of eclipses was Thales of Miletus (about 640–546 BC), who supposedly learned of the saros cycle from the Chaldeans, who had discovered it. No one knows which eclipse Thales predicted, but some scholars suspect the eclipse of May 28, 585 BC. In any case, the eclipse occurred at the height of a battle between the Lydians and the Medes, and the mysterious darkness in midafternoon so startled the two Greek factions that they concluded a truce.

In fact, many historians doubt that Thales actually predicted the eclipse. It would have been very difficult to gather enough information about past eclipses, since total solar eclipses are rare. If you stay in one city, you will see a total solar eclipse about once in 360 years. Also, 585 BC is very early for the Greeks to have known of the saros cycle. The important point is not that Thales did it, but that he could have done it. If he had had records of past eclipses of the sun visible from Greece, he could have discovered that they tended to recur with a period of 54 years 1 month (three saros cycles). Indeed, he could have predicted the eclipse without ever understanding what the sun and moon were or how they moved.

CRITICAL INQUIRY

Why can't two successive full moons be totally eclipsed?

A total lunar eclipse occurs when the moon passes through the earth's shadow, and that can happen only when the sun is near one node and the moon crosses the other node. The earth's shadow always points toward the ecliptic exactly opposite the sun, so most of the time the moon will pass north or south of the earth's shadow and there will be no eclipse. Only when the shadow is near one of the nodes of the moon's orbit can the moon enter the shadow and be eclipsed.

An eclipse season for a total lunar eclipse is only 22 days long. If the moon crosses a node more than 11 days before or after the sun crosses the other node, there will be no eclipse. The moon takes 29.5 days to go from one full moon to the next, so if one full moon is totally eclipsed, the next full moon 29.5 days later will occur too late for an eclipse.

Use your knowledge of the cycles of the sun and moon to explain why the sun can be eclipsed by two successive new moons. ▪

Predicting eclipses isn't very hard, and many ancient peoples were probably familiar enough with the cycles of the sun and moon to know when eclipses were likely. We might call those first predictions early science, but science involves understanding nature, not just predicting events. In the next chapter, we will see how modern science was born out of astronomers' attempts to understand the cycles they saw in the sky. ▪

▪ Summary

Because we see the moon by reflected sunlight, its shape appears to change as it orbits the earth. The lunar phases wax from new moon to first quarter to full moon and wane from full moon to third quarter to new moon. A complete cycle of lunar phases takes 29.53 days.

The moon's gravitational field exerts tidal forces on the earth that pull the ocean waters up into two bulges, one on the side of the earth facing the moon and the other on the side away from the moon. As the rotating earth carries the continents through these bulges of deeper water, the tides ebb and flow. Friction with the seabeds slows the earth's rotation, and the gravitational force the bulges exert on the moon forces its orbit to grow larger.

A lunar eclipse occurs when the moon enters the earth's shadow. If the moon becomes completely immersed in the umbra, the central shadow, the eclipse is termed total. The moon glows coppery red due to sunlight bent as it passes through the earth's atmosphere. If the moon enters the umbral shadow only partly, the eclipse is termed partial, and the reddish glow is not visible. A penumbral eclipse occurs when the moon passes through the penumbra, the region of partial shadow. Penumbral eclipses are not very noticeable.

A solar eclipse occurs when the earth passes through the moon's shadow. To see a total solar eclipse, observers must place themselves in the path of totality, the path swept by the umbra of the moon. As the umbra sweeps over the observers, they see the bright surface of the sun, the photosphere, blotted out by the moon. Then the fainter chromosphere and corona, higher layers of the sun's atmosphere, become visible. Eruptions on the solar surface, called prominences, may be visible peeking around the edge of the moon.

The corona, chromosphere, and prominences are not visible to observers outside the path of totality. Observers in the path of the penumbra of the moon will see a partial solar eclipse, but the photosphere will never be completely hidden.

If an eclipse of the sun occurs when the moon is not in the nearer part of its orbit, the moon's umbra does not reach the earth's surface. In these cases, the eclipse is not total. The moon's angular diameter is too small to cover the sun completely. Even at maximum eclipse, therefore, a bright ring, or annulus, of the photosphere is visible around the edge of the moon. This annular eclipse is not as dramatic as a total eclipse.

Ancient astronomers could predict eclipses because they occur not randomly but in a pattern. Two conditions must be met if an eclipse is to occur. First, the moon must be on or near the ecliptic. The two points where the moon crosses the ecliptic are called the nodes of the moon's orbit. The second condition is that the sun must be at or near one of the nodes. This means that eclipses can occur only at new moon or full moon during two eclipse seasons that are about 32 days long and that occur almost 6 months apart.

Because the moon's orbit precesses, the nodes slip westward along the ecliptic, and the eclipse seasons begin 19 days earlier each year. The moon, sun, and nodes return to the same relative positions every 18 years $11\frac{1}{3}$ days in what is called the saros cycle. After the passage of a saros cycle the pattern of eclipses begins to repeat. This means that ancient astronomers could predict eclipses just by examining the date of previous eclipses. The Chaldeans discovered the saros cycle, and the ancient Greeks used it. Many other primitive cultures also used this method.

▪ New Terms

sidereal period	umbra
synodic period	penumbra
spring and neap tides	total eclipse (lunar or solar)

partial eclipse (lunar or solar)   diamond ring effect

penumbral eclipse   annular eclipse

small-angle formula   node

path of totality   eclipse season

photosphere   line of nodes

corona   eclipse year

chromosphere   saros cycle

prominences

▪ Questions

1. Which lunar phases would be visible in the sky at dawn? at midnight?

2. If you looked back at the earth from the moon, what phase would you see when the moon was full? new? a first-quarter moon? a waxing crescent?

3. Give examples to show how tides can alter the rotation and revolution of celestial bodies. (HINT: Recall from Chapter 2 the difference between rotation and revolution.)

4. Could a solar-powered spacecraft generate any electricity while passing through the earth's umbral shadow? the penumbral shadow?

5. Draw the umbral and penumbral shadows onto Figure 3-3. Explain why lunar eclipses can occur only at full moon and solar eclipses can occur only at new moon.

6. How did lunar eclipses lead Aristotle to conclude that the earth was round?

7. Why isn't the corona visible during partial or annular solar eclipses?

8. Why can't the moon be eclipsed when it is halfway between the nodes of its orbit?

9. Why aren't eclipses separated by one saros cycle visible from the same location on the earth?

10. How could Thales of Miletus have predicted the date of a solar eclipse without observing the location of the moon in the sky?

▪ Discussion Questions

1. If the moon were closer to the earth such that it had an orbital period of 24 hours, what would the tides be like?

2. How would eclipses be different if the moon's orbit were not tipped with respect to the plane of Earth's orbit?

3. Are there other planets in our solar system from whose surface we could see a lunar eclipse? a total solar eclipse?

▪ Problems

1. Identify the phases of the moon if on March 21 the moon is located at (a) the vernal equinox, (b) the autumnal equinox, (c) the summer solstice, (d) the winter solstice.

2. Identify the phases of the moon if at sunset the moon is (a) near the eastern horizon, (b) high in the southern sky, (c) in the southeastern sky, (d) in the southwestern sky.

3. About how many days must elapse between first-quarter moon and third-quarter moon?

4. How many hours would elapse between successive high tides if tides at a given location were caused only by the sun's gravity? only by the moon's gravity? Why is there a difference?

5. How many times larger than the moon is the diameter of the earth's umbral shadow at the moon's distance? (HINT: See Figure 3-11.)

6. Use the small-angle formula to calculate the angular diameter of the sun as seen from the earth.

7. During solar eclipses, large solar prominences are often seen extending 5 minutes of arc from the edge of the sun's disk. How far is this in kilometers? in earth diameters?

8. If a solar eclipse occurs on October 3, why can't there be a lunar eclipse on October 13? Why can't there be a solar eclipse on December 28?

9. A total eclipse of the sun was visible from Canada on July 10, 1972. When will this eclipse occur again? From what part of the earth will it be total?

10. When will the eclipse described in Problem 9 next be total as seen from Canada?

11. When will the eclipse seasons occur during the current year? What eclipse(s) will occur?

▪ Critical Inquiries for the Web

1. Most people see more total lunar eclipses in a lifetime than total solar eclipses. Why is this so? Compare the regions of visibility for a number of past and upcoming eclipses, and determine which future events will be visible from your area.

2. How do the tides on Earth vary with the phases of the moon? Use the Internet to explore the range of tides for a particular location through the next few weeks. Also, look up moon phase information for that same interval. How do the tidal amplitudes correlate with moon phase during that period? Consider the location of the moon in the sky during a particular date and discuss how the tidal data reflect the orientation shown in Figure 3-9.

 Go to the Wadsworth Astronomy Resource Center (www.wadsworth.com/astronomy) for critical thinking exercises, articles, and additional readings from InfoTrac College Edition, Wadsworth's online student library.

Chapter Four

THE ORIGIN OF MODERN ASTRONOMY

Oh, my dear Kepler, how I wish that we could have one hearty laugh together!

From a letter by Galileo Galilei

GUIDEPOST

The motions of the sun, moon, and planets play out a beautiful and complex dance across the heavens. Previous chapters have described that dance; this chapter describes how astronomers learned to understand the geometry of the universe by observing the motions in the sky.

In learning to interpret what they saw, Renaissance astronomers invented a new way of understanding nature, which we recognize today as modern science.

Although this chapter tells the story of heavenly motions, it omits one important idea—gravity. That is the subject of the next chapter. ∎

The history of astronomy is more like a soap opera than a situation comedy. In a simple half-hour sitcom, only a few characters carry the story, but the plot of a soap opera is so complex, so filled with characters and subplots only tenuously related to one another, that the action jumps from character to character, from subplot to subplot, with almost no connection. The history of astronomy is similarly complex, filled with brilliant people who lived at different times and worked in different parts of the world.

Two subplots twine through our story. One is the human quest to understand the place of the earth. That plot will lead us from Aristotle to Copernicus and finally to the trial of Galileo before the Inquisition. But the second subplot, the puzzle of planetary motion, is equally important to our story. That plot will involve an astronomer with a false nose and another whose mother was tried for witchcraft. We will see that the true place of the earth could be understood only when the puzzle of planetary motion was solved.

Our story goes far beyond the birth of astronomy. The quest for the place of the earth and the puzzle of planetary motion became, in the 16th and 17th centuries, the center of a storm of controversy over the best way to understand our world and ourselves. That conflict is now known as the scientific revolution. Thus, our story describes not only the birth of astronomy but also the birth of modern science.

Of course, astronomy had its beginnings long before the 16th century. It began when the first semihuman creature looked up at the moon and stars and wondered what they were. Since then, hundreds of generations of talented astronomers have lived and worked on this planet and left almost no record of their accomplishments. Only through archaeology can we get a hint of their insights and triumphs. Other cultures, such as ancient Greece, have left written records from which we can reconstruct the sophisticated astronomy of the ancient world.

4-1 THE ROOTS OF ASTRONOMY

Astronomy has its origin in that most noble of all human traits, curiosity. Just as modern children ask their parents what the stars are and why the moon changes, so did ancient humans ask themselves these questions. The answers, often couched in mythical or religious terms, reveal a great reverence for the order of the heavens.

Archaeoastronomy

The study of the astronomy of ancient peoples, **archaeoastronomy,** came to public attention in 1965 when Gerald Hawkins published a book called *Stonehenge*

Decoded. He reported that Stonehenge, the prehistoric ring of stones, was a sophisticated astronomical observatory.

Stonehenge, standing on Salisbury Plain in southern England, was built in stages from about 2800 BC to about 1075 BC, a period extending from the late Stone Age into the Bronze Age. Though the public is most familiar with the massive stones of Stonehenge (Figure 4-1), those were added late in its history. In its first stages, Stonehenge consisted of a circular ditch slightly larger in diameter than a football field, with a concentric bank just inside the ditch and a long avenue leading away toward the northeast. A massive stone, the Heelstone, stood then, as it does now, outside the ditch in the opening of the avenue.

As early as AD 1740, the English scholar W. Stukely suggested that the avenue pointed toward the rising sun at the summer solstice, but few accepted the idea. More recently, astronomers have recognized significant astronomical alignments at Stonehenge. Seen from the center of the monument, the summer-solstice sun rises behind the Heelstone. Other sight lines point toward the most northerly and most southerly risings of the moon.

The significance of these alignments has been debated. Hawkins claims that the Stone Age people who built Stonehenge were using it as a device to predict lunar eclipses. Others have had more conservative notions, but the truth may never be known. The builders of Stonehenge had no written language and left no records of their intentions. Nevertheless, the presence of solar and lunar alignments at Stonehenge and at many other Stone Age monuments dotting England and Continental Europe shows that so-called primitive peoples were paying detailed attention to the sky. The roots of astronomy lie not in sophisticated science and logic but in human curiosity and wonder.

The early inhabitants of North America were also interested in astronomy. The Big Horn Medicine Wheel,* located on a 3000-m shoulder of the Big Horn Mountains of Wyoming, was used by the Plains Indians about AD 1500–1750. This arrangement of rocks marks a 28-spoke wheel about 27 m in diameter. Piles of rocks, called cairns, mark the center and six locations on the circumference. Astronomer John Eddy has discovered that these cairns mark sight lines toward a number of important points on the eastern horizon and could have been used as a calendar to help schedule hunting, planting, harvesting, and celebrations (Figure 4-2).

Many other American Indian sites have astronomical alignments. More than three dozen medicine wheels are known, although most do not have the sophistication of the Big Horn Medicine Wheel. The Moose Mountain Wheel, for instance, seems to have been in use as early as AD 100. Alignments at some Mound Builder sites of the Midwest show that these peoples were familiar with the sky. In the Southwest, the Pueblo Indian ruins in Chaco

---

Medicine is used here to mean magical power.

a

b

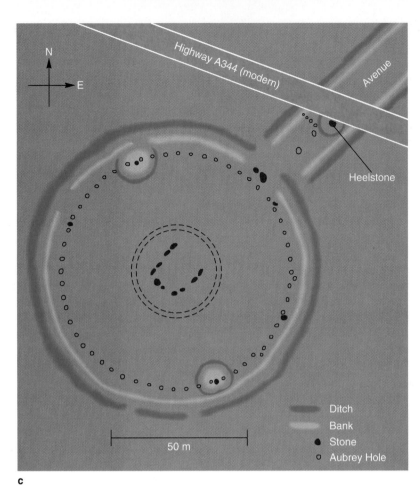

c

FIGURE 4-1

(a) A stamp issued to mark the return of Comet Halley shows the central horseshoe of upright stones at Stonehenge.
(b) On the morning of the summer solstice, an observer at the center of Stonehenge sees the sun rise (dashed line) behind the Heelstone in the center of the avenue extending to the northeast (c).

Canyon, New Mexico, among others, have alignments that point toward the rising and setting of the sun at the summer and winter solstices.

Primitive astronomy also flourished in Central and South America. The Mayan and Aztec empires built many temples aligned with the solstice rising of the sun and with the extreme points where Venus rose and set. The Caracol temple in the Yucatán is a good example (Figure 4-3). It is a circular tower containing complicated passageways and a spiral staircase (thus the name *Caracol*—"the snail's shell"). The tubelike windows at the top of the tower point toward the equinox sunset point and the most northerly and most southerly setting points of Venus. Unfortunately, only about one-third of the tower top survives, so the directions of any other windows are forever lost.

Archaeoastronomers are uncovering the remains of ancient astronomical observatories around the world. Some temples in the jungles of Southeast Asia, for instance, are believed to have astronomical alignments.

Other scholars are looking not at temples but at small artifacts from thousands of years ago. Scratches on certain bone and stone implements seem to follow a pattern and may be an attempt to keep a record of the phases of the moon (Figure 4-4). Some scientists contend that humanity's first attempts at writing were stimulated by a desire to record and predict lunar phases.

Archaeoastronomy is uncovering the earliest roots of astronomy and simultaneously revealing some of the first human efforts at systematic inquiry. The most important lesson of archaeoastronomy is that humans don't have to be technologically sophisticated to admire and study the universe.

One thing about archaeoastronomy is especially sad. Although we are learning how primitive people observed the sky, we may never know what they thought about their universe. Many had no written language. In other cases, the written record has been lost. Dozens, perhaps hundreds, of beautiful Mayan manuscripts, for instance, were burned by Spanish missionaries who believed that

F I G U R E   4 - 2

The Big Horn Medicine Wheel in Wyoming was built by Native Americans before the westward spread of European settlers. Its rock cairns appear to be aligned with the rising and setting of the summer solstice sun (red) and with the rising of three bright stars (blue). Such connections with the sky are found in many Native American structures and ceremonies. (John A. Eddy)

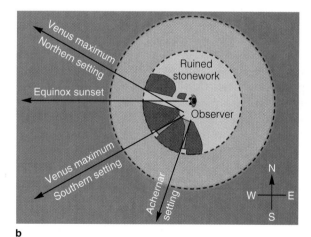

F I G U R E   4 - 3

The Caracol at the Mayan city of Chichén Itzá (a) is a 1000-year-old observatory. The remaining windows in the partially ruined tower contain sight lines that point toward the equinox sunset, the setting places of Venus at its most northerly and most southerly positions, and the setting place of the star Achernar (b). Various evidence suggests that the Caracol was directly associated with the worship of the Venus god, Quetzalcoatl. (Anthony Aveni)

FIGURE 4-4
A fragment of a 27,000-year-old mammoth tusk found at Gontzi in the Ukraine contains scribe marks on its edge, simplified in this drawing. These markings have been interpreted as a record of four cycles of lunar phases. Although controversial, such finds suggest that some of the first human attempts at recording events in written form were stimulated by astronomical phenomena.

they were the work of Satan. That is one reason why our history of astronomy really begins with the Greeks. Some of their writing has survived, and we can discover what they thought about the shape and motion of the heavens.

The Astronomy of Greece

Greek astronomy was based on the astronomy of Babylonia and Egypt, but these astronomies were heavily influenced by religion and astrology. Those ancient astronomers studied the motions of the heavens as a way of worship and divination. The Greek astronomers studied astronomy in an entirely new way—they tried to understand the universe.

This new attitude toward the heavens, a truly scientific attitude, was made possible by two early Greek philosophers. Thales of Miletus (c. 624–547 BC) lived and worked in what is now Turkey. He taught that the universe is rational and that the human mind can understand why the universe works the way it does. This view contrasts sharply with those of earlier cultures, which believed that the ultimate causes of things are mysteries beyond human understanding. To Thales and his followers, the mysteries of the universe are mysteries because they are unknown, not because they are unknowable.

The second philosopher who made the new scientific attitude possible was Pythagoras (c. 570–500 BC). He and his students noticed that many things in nature seem to be governed by geometrical or mathematical relations. Musical notes, for example, are related in a regular way to the lengths of plucked strings. This led Pythagoras to propose that all nature was underlain by musical principles, by which he meant mathematics. One result of this philosophy was the later belief that the harmony of the celestial movements produced actual music, the music of the spheres. But at a deeper level, the teachings of Pythagoras made Greek astronomers look at the universe

in a new way. Thales said that the universe could be understood, and Pythagoras said that the underlying rules were mathematical.

In trying to understand the universe, Greek astronomers did something that Babylonian astronomers had never done—they tried to construct descriptions based on geometrical forms. Anaximander (c. 611–546 BC) described a universe made up of wheels filled with fire: the sun and moon are holes in the wheels through which we see the flames. Philolaus (5th century BC) argued that the earth moves in a circular path around a central fire (not the sun), which is always hidden behind a counterearth located between the fire and the earth. This was the first theory to suppose that the earth is in motion.

Plato (428–347 BC) was not an astronomer, but his teachings influenced astronomy for 2000 years. Plato argued that the reality we see is only a distorted shadow of a perfect, ideal form. If our observations are distorted, then observation can be misleading, and the best path to truth is through pure thought on the ideal forms that underlie nature.

Plato also argued that the most perfect form was the circle and that therefore all motions in the heavens should be made up of combinations of circular motion. Since the most perfect motion is uniform motion, later astronomers tried to describe the motions of the heavens using the principle of **uniform circular motion.**

Pythagoras had taught that the earth is a sphere and that the other heavenly bodies were divine, perfect spheres moving in perfect circles. Eudoxus of Cnidus (409–356 BC), a student of Plato, combined a system of 27 nested spheres rotating at different rates about different axes with the concept of uniform circular motion to produce a mathematical description of the motions of the universe (Figure 4-5).

At the time of the Greek philosophers, it was common to refer to systems such as that of Eudoxus as descriptions of the world, where the word *world* included not only the earth but all of the heavenly spheres. The reality of these spheres was open to debate. Some thought of the spheres as nothing more than mathematical ideas that described motion in the world model, while others began to think of the spheres as real objects made of perfect celestial material. Aristotle, for example, seems to have thought of the spheres as real.

Aristotle (384–322 BC) taught and wrote on philosophy, history, politics, ethics, poetry, drama, and so on (Figure 4-6). Because of his sensitivity and insight, he became the great authority of antiquity, and astronomers for almost 2000 years cited him as their authority in adopting the Greek model of the universe.

Aristotle believed that the universe was divided into two parts—the earth, corrupt and changeable, and the heavens, perfect and immutable. Like most of his predecessors, he believed that the earth was the center of the universe, so his model was **geocentric** (earth-centered). The heavens surrounded the earth, and he added more crystalline spheres to bring the total to 56. The lowest

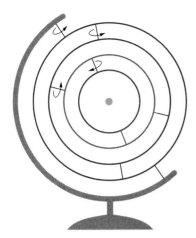

FIGURE 4-5
The spheres of Eudoxus explain the motions in the heavens by means of nested spheres rotating about various axes at different rates. The earth is located at the center. (Eudoxus used 27 spheres; Aristotle used 56.)

FIGURE 4-6
Aristotle, honored on this Greek stamp, wrote on such a wide variety of subjects and with such deep insight that he became the great authority on all matters of learning. His opinions on the nature of the earth and the sky were widely accepted for almost two millennia.

sphere, that of the moon, marked the boundary between the imperfect region of the earth and the perfection of the celestial realm above the moon.

Because he believed the earth to be immobile, he had to make these spheres whirl westward around the earth each day and move with respect to one another to produce the motions of the sun, moon, and planets. Like most other Greek philosophers, Aristotle viewed the universe as a perfect heavenly machine that was not many times larger than the earth itself (Figure 4-7).

About a century after Aristotle, the Alexandrian philosopher Aristarchus proposed a theory that the earth rotated on its axis and revolved around the sun. This theory is, of course, correct, but most of the writings of Aristarchus were lost, and his theory was not well known. Later astronomers rejected any suggestion that the earth could move, because it conflicted with Aristotle's theory and because the astronomers saw no parallax.

❯ **Parallax** (*p*) is the apparent change in the position of an object due to a change in the location of the observer. It is actually an old friend, although you may not have known its name, for you use parallax to judge the distance to things. To see how it works, close your right eye and use your thumb, held at arm's length, to cover some distant object, a building perhaps. Now look with your right eye. Your thumb seems to move to the left, uncovering the building (Figure 4-8). This apparent shift is parallax, and your brain uses it to estimate distances to objects around you.

Ancient astronomers reasoned that if the earth moved around the sun, we should be viewing the stars from different positions in the earth's orbit at different times. Then we should see parallax shifting the apparent positions of the stars. In the most extreme cases, this should distort the shapes and sizes of the constellations. That is, constellations should look largest when the earth is nearest that part of the sky. Because they did not see this parallax, they concluded that the earth did not move.

FIGURE 4-7
According to Aristotle, the earth is motionless (*Terra immobilis*) at the center of the universe. It is surrounded by spheres of water, air, and fire (*ignus*), above which lie spheres carrying the celestial bodies, beginning with the moon (*lune*) in the lowest celestial sphere. This woodcut is from C. Cornipolitanus's book *Chronographia* of 1537. (Granger Collection, New York)

| Seen by left eye | Seen by right eye |

FIGURE 4-8

To demonstrate parallax, close one eye and cover a distant object with your thumb held at arm's length. Then look with the other eye; your thumb appears to have shifted position.

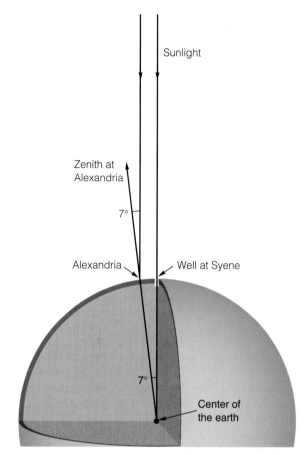

FIGURE 4-9

On the day of the summer solstice, sunlight fell to the bottom of a well at Syene, but the sun was about one-fiftieth of a circle (7°) south of the zenith at Alexandria. This told Eratosthenes that the distance from Syene to Alexandria was one-fiftieth of the circumference of the earth. Thus, he was able to calculate the radius of the earth.

Actually, they saw no parallax because the stars are much farther away than they supposed and the parallax is much too small to be visible to the naked eye.

> Aristotle had taught that the earth had to be round because it always casts a round shadow during lunar eclipses, but he could only estimate its size. About 200 BC, Eratosthenes, working in the great library in Alexandria, found a way to calculate the earth's radius. He learned from travelers that the city of Syene (Aswan) in southern Egypt contained a well into which sunlight shone vertically on the day of the summer solstice. Thus, the sun was at the zenith at Syene, but on that same day in Alexandria, he noted that the sun was one-fiftieth of the circumference of the sky (about 7°) south of the zenith. Because sunlight comes from such a great distance, its rays arrive at the earth traveling almost parallel. Thus, Eratosthenes could conclude from simple geometry (Figure 4-9) that the distance from Alexandria to Syene was one-fiftieth the circumference of the earth.

To find the circumference of the earth, Eratosthenes had to know the distance from Alexandria to Syene. Travelers told him it took 50 days to cover the distance, and he knew that a camel can travel about 100 stadia per day. Thus, the total distance was about 5000 stadia. If 5000 stadia is one-fiftieth the earth's circumference, then the earth must be 250,000 stadia around, and dividing by 2π Eratosthenes found the radius of the earth to be 40,000 stadia.

We don't know how accurate Eratosthenes was. The stadium (singular of *stadia*) had different lengths in ancient times. If we assume 6 stadia to the kilometer, then Eratosthene's result was too big by only 4 percent. If he used the Olympic stadium, his result was 14 percent too big. In any case, this was a much better measurement of the radius of the earth than Aristotle's estimate, which was only about 40 percent of the true radius.

The greatest of the ancient observers was Hipparchus (see Figure 2-5), who lived during the 2nd century BC about two centuries after Aristotle. He is usually credited with the invention of trigonometry, he compiled the first star catalog, and he discovered precession (Chapter 2). Instead of describing the motion of the sun and moon using nested spheres, as most Greek philosophers did, Hipparchus proposed that the sun and moon traveled around circles with the earth near, but not at, their centers. These off-center circles are now known as **eccentrics.** Hipparchus recognized that he could produce the same motion by having the sun, for instance, travel around a small circle that followed a larger circle around the earth. The compounded circular motion that he devised became the key element in the masterpiece of the last great astronomer of classical times, Ptolemy.

The Ptolemaic Universe

Claudius Ptolemaeus (Figure 4-10) was one of the great astronomer-mathematicians of antiquity. His nationality and birth date are unknown, but he lived and worked in the Greek settlement at Alexandria about AD 140. He ensured the survival of Aristotle's universe by fitting to it a sophisticated mathematical model.

In agreement with Aristotle, the Ptolemaic model was geocentric (earth-centered). In addition, it incorporated the Greek belief that the heavenly bodies move perfectly with uniform circular motion. But simple, circular paths centered on the earth do not account for the motions of the planets in the sky. The planets sometimes move faster and sometimes slower, and occasionally they appear to slow to a stop and move backward over a period of months, tracing a **retrograde loop** (Figure 4-11).

To describe the complicated planetary motions and yet preserve a geocentric model with uniform circular motion, Ptolemy adopted a system of wheels within wheels. The planet moves in a small circle called an **epicycle,** and the center of the epicycle moves along a larger circle around the earth called a **deferent** (Figure 4-12). By adjusting the size of the circles and the rate of their motion, Ptolemy could account for most planetary movement. But as a final adjustment, he placed the earth off center in the deferent circle and specified that the center of epicycle would appear to move at constant speed only if viewed from a point called the **equant** located on the other side of the deferent's center. Thus, by using a few dozen circles of various sizes rotating at various rates, Ptolemy's system predicted the positions of the planets (Figure 4-13).

About AD 140, Ptolemy included this work in a book now known as *Almagest*. He never knew that title; he called his work *Mathematical Syntaxis*. With his death, classical astronomy ended forever. Western civilization was slipping into the shadows of the Dark Ages, and the invading Islamic nations dominated astronomy for almost 1000 years. Arabian astronomers translated, studied, and preserved many classical manuscripts, and in Arabic Ptolemy's book became *Al Magisti* (Greatest). Beginning in the 1200s, Europeans began recovering their classical heritage through Arabic translations.* In Latin, Ptolemy's book became *Almagestum* and thus our modern *Almagest*. For 1000 years, Arab astronomers

FIGURE 4-10
The muse of astronomy guides Ptolemy (about AD 140) in his study of the heavens. (Courtesy Owen Gingerich and the Houghton Library)

*Islamic scholars knew of both Ptolemy's and Eratosthenes's estimates for the size of the earth, and they had settled on a final figure that was quite accurate. However, Columbus miscalculated the Arabic mile and thus thought the earth much smaller than it really is. That underestimation, combined with his overestimation of the eastward extent of Asia, led him to believe he could sail west to Japan. Had he known the true size of the earth, he might never have attempted his voyage.

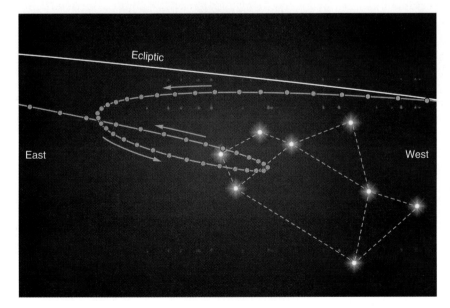

FIGURE 4-11
The motion of Mars along the ecliptic near the teapot shape of Sagittarius is shown at 4-day intervals. Though the planet moves eastward, it sometimes appears to slow to a stop and move westward in retrograde motion.

studied and preserved Ptolemy's work, but they made no significant improvement in his theory.

At first the Ptolemaic system predicted the positions of the planets well, but as centuries passed, errors accumulated. A watch that gains only 1 second a year will keep time well for many years, but the error gradually accumulates. After a century the watch will be 100 seconds fast. So, too, did the errors in the Ptolemaic system gradually accumulate as the centuries passed. Arabian and later European astronomers tried to update the system, computing new constants and sometimes adding more epicycles. In the middle of the 13th century, a team of astronomers supported by King Alfonso X of Castile

worked for 10 years revising the Ptolemaic system, publishing the result as the *Alfonsine Tables*. It was the last great adjustment of the Ptolemaic system.

CRITICAL INQUIRY

How did the astronomy of Hipparchus and Ptolemy violate the principles of the early Greek philosophers Plato and Aristotle?

Hipparchus and Ptolemy lived very late in the history of classical astronomy, and they concentrated more on the mathematical problems and less on philosophical principles. They replaced the perfect spheres of Plato with nested circles in the form of epicycles and deferents. The earth was moved slightly away from the center of the deferent, so their models of the universe were not exactly geocentric, and the epicycles moved uniformly only as seen from the equant. The celestial motions were no longer precisely uniform, and the principles of geocentrism and uniform circular motion were weakened.

The work of Hipparchus and Ptolemy led eventually to a new understanding of the heavens, but first astronomers had to abandon the principle of uniform circular motion. The great philosopher Plato had argued for uniform circular motion from what seemed the most basic of principles. What were those principles?

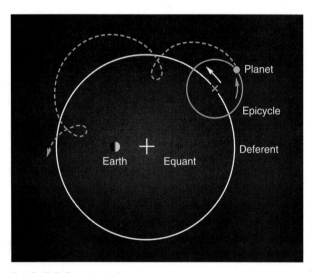

FIGURE 4-12
Ptolemy accounted for a planet's motion by placing it on a small circle (epicycle) that moved along a larger circle (deferent). Viewed from the equant, the center of the epicycle would have moved at constant speed.

Ptolemy was the last of the classical astronomers, and his work dominated astronomical thought for almost 1500 years. The collapse of the Ptolemaic model and the rise of a new model of the universe make an exciting story, but it is not just the story of an astronomical idea. It includes the invention of science as a new way of knowing and understanding what we are and where we are.

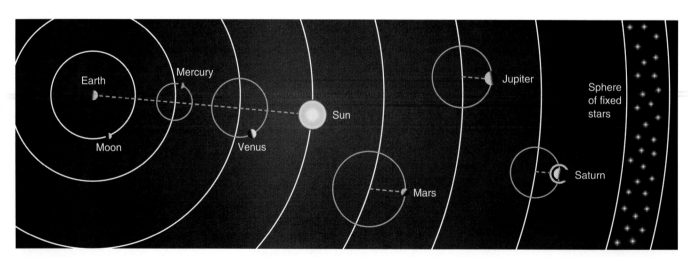

FIGURE 4-13
The Ptolemaic system was geocentric (earth-centered) and based on the uniform circular motion of epicycles. Note that the centers of the epicycles of Mercury and Venus must lie on the Earth–sun line.

4-2 THE COPERNICAN REVOLUTION

Nicolaus Copernicus (Figure 4-14) triggered an earth-shaking revision in human thought by proposing that the universe is not centered on the earth but is **heliocentric**, centered on the sun. That idea eventually led Galileo before the inquisition and changed forever how we think of our world and ourselves. This Copernican Revolution was much more than an upheaval in astronomy; it marks the birth of modern science.

Copernicus the Revolutionary

When Copernicus proposed that the universe was heliocentric, he risked controversy. According to Aristotle, the most perfect region was the heavens, and the most imperfect was the center of the earth. Thus, the classical geocentric universe matched the commonly held Christian geometry of heaven and hell, and anyone who criticized the central location of the earth challenged Christian belief and thus risked a charge of heresy.

Throughout his life, Copernicus had been associated with the church. His uncle, by whom he was raised and educated, was an important bishop in Poland, and after studying canon law and medicine in some of the finest universities in Europe, Copernicus became a canon of the church at the age of 24. He was secretary and personal physician to his powerful uncle for 15 years. When his uncle died, Copernicus moved to quarters adjoining the cathedral in Frauenburg. No doubt his long association with the church added to his reluctance to publish controversial ideas.

He first wrote about a heliocentric universe in 1507, at the age of 34, in a short pamphlet that he distributed anonymously. Over the years, Copernicus worked on his book *De Revolutionibus Orbium Coelestium* (Figure 4-15a), but he hesitated to publish it even though other astronomers knew of his theories.

This was a time of rebellion—Martin Luther was speaking harshly about fundamental church teachings, and others, scholars and scoundrels, questioned the authority of the church. Even astronomy could stir argument; moving the earth from its central place was a controversial and perhaps heretical idea. Thus, Copernicus did not permit publication of his book until 1543 when he realized he was dying.

The most important idea in the book was the placement of the sun at the center of the universe. That single innovation had an astonishing consequence—the retrograde motion of the planets was immediately explained in a straightforward way without the large epicycles that Ptolemy used. In the Copernican system, Earth moves faster along its orbit than the planets that lie further from

FIGURE 4-14

Copernicus proposed that the sun and not the earth was the center of the universe. Notice the heliocentric model on this stamp issued in 1973 to commemorate the 500th anniversary of his birth.

the sun. Consequently, Earth periodically overtakes and passes these planets, and they appear to slow and fall behind (Figure 4-16). Because the planetary orbits do not lie in precisely the same plane, a planet does not resume its eastward motion in precisely the same path it followed earlier. Consequently, it describes a loop whose shape depends on the angle between the orbital planes.

Copernicus could explain retrograde motion without epicycles, and that was impressive. But he could not do away with epicycles completely. Copernicus, a classical astronomer, had tremendous respect for the old concept of uniform circular motion. In fact, he objected strongly to Ptolemy's use of the equant. It seemed arbitrary to Copernicus, a direct violation of the elegance of Aristotle's philosophy of the heavens. Copernicus called equants "monstrous" in that they violated both geocentrism and uniform circular motion. In his model, Copernicus returned to a strong belief in uniform circular motion. Although he did not need epicycles to explain retrograde motion, Copernicus discovered that the sun, moon, and planets suffered small variations in their motions—variations that he could not explain with the concept of uniform circular motion centered on the sun. Today, we recognize those motions as typical of objects following elliptical orbits, but Copernicus held firmly to uniform circular motion, so he had to use small epicycles to produce these minor variations in the motions of the sun, moon, and planets.

Because Copernicus imposed uniform circular motion on his model, it could not accurately predict the motions of the planets. The *Prutenic Tables* (1551) were based on the Copernican model, and they were not significantly more accurate than the *Alfonsine Tables* (1251) which were based on Ptolemy's model. Both could be in error by as much as 2°, four times the angular diameter of the full moon.

The Copernican *model* was inaccurate, but the Copernican *hypothesis* that the universe is heliocentric was correct. There are probably a number of reasons why the hypothesis gradually won acceptance in spite of the inaccuracy of the epicycles and deferents. The most important factor may be the elegance of the idea. Placing the sun at the center of the universe produced a symmetry among the motions of the planets that was pleasing to the eye and to the intellect (Figure 4-15b). All of the planets moved in the same way at speeds that were simply related to their distance from the sun. Venus and Mercury moved in the same kind of orbits as did all the other planets. Thus, the model may have won support not for its accuracy, but for its elegance.

The most astonishing consequence of the Copernican hypothesis was not what it said about the sun, but what it said about Earth. By placing the sun at the center, Copernicus made the earth move along an orbit just as the other planets did. By making the earth a planet, Copernicus revolutionized humanity's view of its place in the universe and triggered a controversy that would eventually bring Galileo before the Inquisition.

Although astronomers throughout Europe read and admired *De Revolutionibus*, they did not usually accept the Copernican hypothesis. The mathematics was elegant, and the astronomical observations and calculations were of tremendous value. Yet few astronomers believed, at first, that the sun actually was the center of the planetary

a

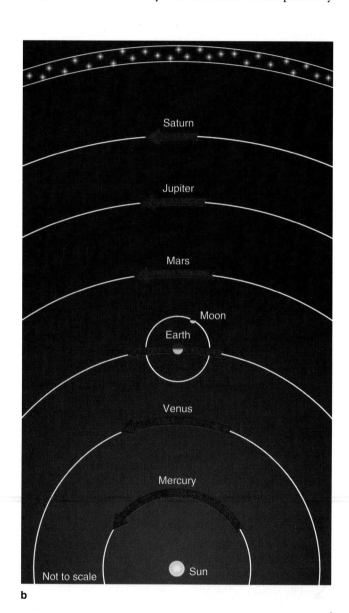

b

FIGURE 4-15

(a) The Copernican universe as reproduced in *De Revolutionibus*. The earth and all the known planets revolve in separate circular orbits about the sun (Sol) at the center. The outermost sphere carries the immobile stars of the celestial sphere. Notice the orbit of the moon around the earth. (Yerkes Observatory) (b) The model was elegant not only in its arrangement of the planets but in their motions. Orbital velocities (red arrows) decreased from Mercury, the fastest, to Saturn, the slowest. Compare the elegance of this model with the complexity of the Ptolemaic model in Figure 4-13.

Scientific Revolutions

The Copernican revolution is often cited as the perfect example of a scientific revolution. Over a few decades, astronomers abandoned a way of thinking about the universe that was almost 2000 years old and adopted a new set of ideas and assumptions. The American philosopher of science Thomas Kuhn has referred to a commonly accepted set of scientific ideas and assumptions as a scientific **paradigm.** Thus, the pre-Copernican astronomers had a geocentric paradigm that included uniform circular motion and the perfection of the heavens. That paradigm survived for many centuries until a new generation of astronomers was able to overthrow the old paradigm and establish a new

paradigm that included heliocentrism and the motion of the earth.

A scientific paradigm is powerful because it shapes our perceptions. It determines what we judge to be important questions and what we judge to be significant evidence. Thus, it is often difficult for us to recognize how our paradigms limit what we can understand. For example, the geocentric paradigm contained problems that seem obvious to us, but because astronomers before Copernicus lived and worked inside that paradigm, they had difficulty seeing the problems. Overthrowing an outdated paradigm is not easy, because we must learn to see nature in an entirely new way. Galileo, Kepler, and

Newton saw nature from a new paradigm that would have been almost incomprehensible to astronomers of earlier centuries.

We can find examples of scientific revolutions in many fields, including biology, geology, genetics, and psychology. These scientific revolutions have been difficult and controversial because they have involved the overthrow of accepted paradigms. But that is why scientific revolutions are exciting. They give us not just a new idea or a new theory, but an entirely new insight into how nature works—a new way of seeing the world. ■

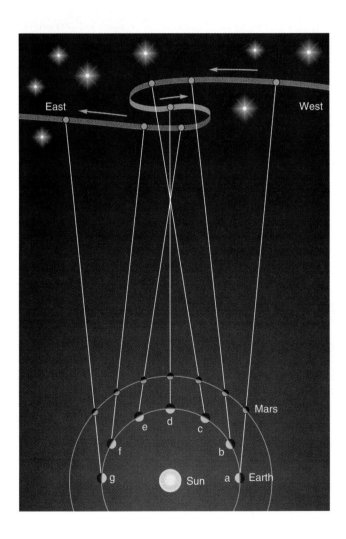

FIGURE 4-16

The Copernican explanation of retrograde motion. As Earth overtakes Mars (a–c), it appears to slow its eastward motion. As Earth passes Mars (d), it appears to move westward (retrograde). As Earth draws ahead of Mars (e–g), it resumes its eastward motion against the background stars. Compare with Figure 4-11. The positions of Earth and Mars are shown at equal intervals of 1 month.

revolution in the way astronomers thought about the place of the earth (Window on Science 4-1). In fact, we derive the word "revolutionary" from the title of Copernicus' book—as "one who overthrows established authority."

Galileo the Defender

Most people know two facts about Galileo, and both facts are wrong. We should begin our discussion by getting those facts right: Galileo did not invent the telescope, and he was not condemned by the Inquisition for believing the earth moved around the sun. Then why is Galileo so important that in 1979, almost 400 years after his trial, the Vatican reopened his case? As we discuss Galileo, we will discover that his trial concerned not just the place of the earth but a new way of finding truth, a method known today as science.

Galileo Galilei (Figure 4-17) was born in Pisa in 1564, and he studied medicine at the university there. His true love, however, was mathematics, and although he had to leave school early because of financial difficulties, he returned only four years later as a professor of mathematics. Three years later, he became professor of mathematics at the university at Padua. He remained there for 18 years.

system and that the earth moved. How the Copernican hypothesis became gradually recognized as correct has been named the Copernican revolution because it involved not just the adoption of a new idea, but a total

FIGURE 4-17
Galileo, remembered as the great defender of Copernicanism, also made important discoveries in the physics of motion. He is honored here on an Italian 2000-lira note. The reverse side shows one of his telescopes at lower right and a modern observatory above it.

FIGURE 4-18
Galileo's telescopic discoveries generated intense interest and controversy. Some critics refused to look through a telescope lest it deceive them. (Yerkes Observatory)

During this time, Galileo seems to have adopted the Copernican model, although he admitted in a 1597 letter to Kepler that he did not support Copernicanism publicly because of the criticism such a declaration would bring. It was the telescope that drove Galileo to publicly defend the heliocentric model.

Galileo did not invent the telescope. It was apparently invented around 1608 by lens makers in Holland.

Galileo, hearing descriptions in the fall of 1609, was able to build working telescopes in his workshop (Figure 4-18). Galileo was not the first person to look at the sky through a telescope, but he was the first person to observe the sky systematically and apply his observations to the theoretical problem of the day—the place of the earth.

What Galileo saw through his telescopes was so amazing he rushed a small book into print. *Sidereus*

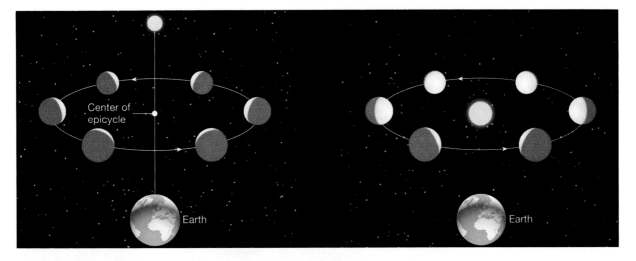

FIGURE 4-19

(a) In the Ptolemaic universe, Venus moves around an epicycle between Earth and the sun and would always appear as a crescent. (b) Galileo's telescope showed that Venus goes through a full set of phases, proving that it must orbit the sun, as in the Copernican universe.

Nuncius (*The Sidereal Messenger*) reported three major discoveries. First, the moon was not perfect. It had mountains and valleys on its surface, and Galileo used the shadows to calculate the height of the mountains. The Ptolemaic model held that the moon was perfect, but Galileo showed that it was not only imperfect, it was a world like the earth.

The second discovery reported in the book was that the Milky Way was made up of a myriad of stars too faint to see with the unaided eye. While intriguing, this discovery could not match the third discovery. Galileo's telescope revealed four new "planets" circling Jupiter, planets known today as the Galilean moons of Jupiter.

The moons of Jupiter supported the Copernican model over the Ptolemaic model. Critics of Copernicus had said Earth could not move because the moon would be left behind. But Jupiter moved yet kept its satellites, so Galileo's discovery proved that Earth, too, could move and keep its moon. Also, the Ptolemaic model included the Aristotelian belief that all heavenly motion was centered on Earth at the center of the universe. Galileo showed that Jupiter's moons revolve around Jupiter, so there could be other centers of motion.

Soon after *Sidereus Nuncius* was published, Galileo made two additional discoveries. When he observed the sun, he discovered sunspots, raising the suspicion that the sun was less than perfect. Further, by noting the movement of the spots, he concluded that the sun was a sphere and rotated on its axis. When he observed Venus, Galileo saw that it was going through phases like those of the moon. In the Ptolemaic model, Venus revolves around an epicycle located between Earth and the sun. Thus, it would always be seen as a crescent. But Galileo saw Venus go through a complete set of phases, proving that it did indeed revolve around the sun (Figure 4-19).

Sidereus Nuncius was very popular and made Galileo famous. He became chief mathematician and philosopher to the Grand Duke of Tuscany in Florence. In 1611, Galileo visited Rome and was treated with great respect. He had long, friendly discussions with the powerful Cardinal Barberini, but he also made enemies. Personally, Galileo was outspoken, forceful, and sometimes tactless. He enjoyed debate, but most of all he enjoyed being right. Thus, in lectures, debates, and letters he offended important people who questioned his telescopic discoveries.

By 1616, Galileo was the center of a storm of controversy. Some critics said he was wrong, and others said he was lying. Some refused to look through a telescope lest it mislead them, and others looked and claimed to see nothing (hardly surprising given the awkwardness of those first telescopes). Pope Paul V decided to end the disruption, so when Galileo visited Rome in 1616 Cardinal Bellarmine interviewed him privately and ordered him to cease debate. There is some controversy today about the nature of Galileo's instructions, but he did not pursue astronomy for some years after the interview. Books relevant to Copernicanism, including *De Revolutionibus*, were banned.

In 1621 Pope Paul V died, and his successor, Pope Gregory XV, died in 1623. The next pope was Galileo's friend Cardinal Barberini, who took the name Urban VIII. Galileo rushed to Rome hoping to have the prohibition of 1616 lifted, and although the new pope did not revoke the orders, he did encourage Galileo. Thus, Galileo began to write his great defense of the Copernican model, finally completing it on December 24, 1629. After some delay, the book was approved by both the local censor in Florence and the head censor of the Vatican in Rome. It was printed in February 1632.

Called *Dialogo Dei Due Massimi Sistemi* (*Dialogue Concerning the Two Chief World Systems*; Figure 4-20), the book is a debate among three friends over the Coper-

FIGURE 4-20

Aristotle, Ptolemy, and Copernicus discuss astronomy in this frontispiece from Galileo's book *Dialogue Concerning the Two Chief World Systems.* (Owen Gingerich and the Houghton Library)

nican and Ptolemaic models. Salviati, a swift-tongued defender of Copernicus, dominates the book, Sagredo is intelligent but largely uninformed. Simplicio is the dismal defender of Ptolemy. In fact, he does not seem very bright.

The publication of *Dialogo* created a storm of controversy, and it was sold out by August 1632, when the Inquisition ordered sales stopped. The book was a clear defense of Copernicus, and, either intentionally or unintentionally, Galileo exposed the pope's authority to ridicule. Urban VIII was fond of arguing that, as God was omnipotent, He could construct the universe in any form while making it appear to us to have a different form, and thus we could not deduce its true nature by mere observation. Galileo placed the pope's argument in the mouth of Simplicio, and Galileo's enemies showed the passage to the pope as an example of Galileo's disrespect. The pope thereupon ordered Galileo to face the Inquisition.

Galileo was interrogated by the Inquisition four times and was threatened with torture. He must have thought often of Giordano Bruno, tried, condemned, and burned at the stake in Rome in 1600. One of Bruno's offenses had been Copernicanism. But the trial did not center on Galileo's belief in Copernicanism. *Dialogo* had been approved by two censors. Rather, the trial centered on the instructions given Galileo in 1616. From his file in the Vatican, his accusers produced a record of the meeting between Galileo and Cardinal Bellarmine that included the statement that Galileo was "not to hold, teach, or defend in any way" the principles of Copernicus. Many historians believe that this document, which was signed neither by Galileo nor by Bellarmine nor by a legal secretary, was a forgery and that Galileo's true instructions were much less restrictive. But Bellarmine was dead, and Galileo had no defense.

The Inquisition condemned him not for heresy but for disobeying the orders given him in 1616. On June 22, 1633, at the age of 70, kneeling before the Inquisition, Galileo read a recantation admitting his errors. Tradition has it that as he rose he whispered "E pur si muove" ("Still it moves"), referring to the earth. Although he was sentenced to life imprisonment, he was actually confined at his villa for the next 10 years, perhaps through the intervention of the pope. He died there on January 8, 1642, 99 years after the death of Copernicus.

Galileo was not condemned for heresy, nor was the Inquisition interested when he tried to defend Copernicanism. He was tried and condemned on a charge we might call a technicality. Then why is his trial so important that historians have studied it for almost four centuries? Why have some of the world's greatest authors, including Bertolt Brecht, written about Galileo's trial? Why in 1979 did Pope John Paul II create a commission to reexamine the case against Galileo?

To understand the trial, we must recognize that it was the result of a conflict between two ways of understanding our universe. Since the Middle Ages, scholars had taught that the only path to true understanding was through religious faith. St. Augustine (AD 354–430) wrote "Credo ut intelligame," which can be translated as "Believe in order to understand." But Galileo and other scientists of the Renaissance used their own observations to try to understand the universe, and when their observations contradicted Scripture, they assumed their observations were correct. Galileo paraphrased Cardinal Baronius in saying, "The Bible tells us how to go to heaven, not how the heavens go."

Various passages of Scripture seemed to contradict observation. For example, Joshua is said to have commanded the sun to stand still, not the earth to stop rotating (Joshua 10:12–13). In response to such passages, Galileo argued that we should "read the book of nature"—that is, we should observe the universe with our own eyes.

The trial of Galileo was not about the place of the earth. It was not about Copernicanism. It wasn't really about the instructions Galileo received in 1616. It was about the birth of modern science as a rational way to understand our universe. The commission appointed by John Paul II in 1979, reporting its conclusions in October 1992, said of Galileo's inquisitors, "This subjective error of judgment, so clear to us today, led them to a disciplinary measure from which Galileo 'had much to suffer.'" Galileo was not found innocent so much as the Inquisition was forgiven for having charged him in the first place.

The gradual change from reliance on personal faith to scientific observation came to a climax with the trial of Galileo. Since that time, scientists have increasingly reserved Scripture and faith for personal comfort and have depended on systematic observation to describe the physical world. The final verdict of 1992 was an attempt to bring some balance to this conflict. In his remarks on the decision, Pope John Paul II said, "A tragic mutual incomprehension has been interpreted as the reflection of a fundamental opposition between science and faith. . . . this sad misunderstanding now belongs in the past." Galileo's trial is over, but it continues to echo through history as part of our struggle to understand the place of the earth in our universe.

CRITICAL INQUIRY

How were Galileo's observations of the moons of Jupiter evidence against the Ptolemaic model?

Aristotle, following Plato, argued that the universe was geocentric and that the heavens were made up of perfect spheres. The spheres rotated because that was the natural motion of spheres, and since the spheres were concentric and centered on Earth, Earth could be the only center of motion. When Galileo saw four moons revolving around Jupiter, he knew that Earth was not the only center of motion in the universe, and thus at least that one thing about the Ptolemaic model had to be wrong.

Furthermore, Ptolemaic astronomers had argued that Earth could not move or it would leave the moon behind. The fact that the moon continued to circle Earth seemed to prove that Earth was immobile. But Galileo saw moons orbit Jupiter, and everyone agreed that Jupiter moved. Thus, if Jupiter could move and keeps its moons, then Earth might move and keeps its moon.

Of all of Galileo's telescopic observations, the moons of Jupiter caused the most debate. But the craters on the moon and the phases of Venus were also critical evidence. How did they argue against the Ptolemaic model? ■

Galileo faced the Inquisition because of a conflict between two ways of knowing and understanding the world. The Church taught faith and understanding through revelation, but the scientists of the age were inventing a new way to understand nature that relied on evidence. In astronomy, evidence means observations, so our story turns now from the philosophical problem of the place of the earth to the observational problem of the motion of the planets.

4-3 THE PUZZLE OF PLANETARY MOTION

While Galileo was defending Copernicanism in Florence and Rome, two other astronomers were working in northern Europe, beyond the sway of the Inquisition. Although they, too, struggled to understand the place of the earth, they approached the problem differently. They used the most accurate observations of the positions of the planets to try to discover the rules that govern planetary motion. When that puzzle was finally solved, astronomers understood why the Ptolemaic model of the universe does not work well and how to make the Copernican universe a precise predictor of planetary motion.

FIGURE 4-21

Tycho Brahe (1546–1601), a Danish nobleman, established an observatory at Hveen and measured planetary positions with high accuracy. His artificial nose is suggested in this engraving.

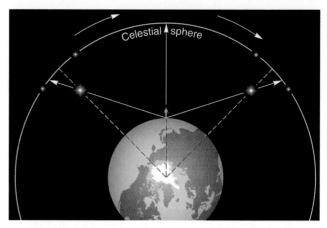

FIGURE 4-22

Daily parallax from a stationary earth. Looking down on the earth and celestial sphere from the north celestial pole, we see the celestial sphere rotating westward. The supernova is shown here closer to the earth than the stars of the celestial sphere. Because the observer is not at the earth's center, the supernova would appear to move against the background stars through the night. Tycho could not detect this daily parallax, and so he concluded that the new star was no closer to the earth than the moon and probably lay on the sphere of the stars.

Tycho the Observer

The great observational astronomer of our story is a Danish nobleman, Tycho Brahe, born December 14, 1546, only 3 years after the publication of *De Revolutionibus* (Figure 4-21). Great figures in history are often referred to by their last name, but Tycho Brahe is usually called Tycho. Were he alive today, he would no doubt object to such familiarity from his obvious inferiors. He was well known for his vanity and lordly manners.

Tycho's college days were eventful. He was officially studying law with the expectation that he would enter Danish politics, but he made it clear to his family that his real interest was astronomy and mathematics. It was also during his college days that he became involved in a duel and received a wound that disfigured his nose. For the rest of his life, he wore false noses made of gold and silver and stuck on with wax. The disfigurement probably did little to improve his disposition.

Tycho's first astronomical observations were made while he was a student. In 1563, Jupiter and Saturn passed very near each other in the sky, nearly merging into a single point on the night of August 24. Tycho found that the *Alfonsine Tables* were a full month in error and that the *Prutenic Tables* were in error by a number of days. These discrepancies dismayed Tycho and sparked his interest in the motions of the planets.

In 1572, a brilliant "new star" (now called Tycho's supernova) appeared in the sky. Such changes in the stars puzzled classically trained astronomers. Aristotle had argued that the starry sphere was perfect and unchanging, and therefore such new stars had to lie closer to the earth than the moon. Layers below the moon were thought to be less perfect and thus more changeable.

Tycho observed that the new star moved westward through the night, keeping pace with the stars, so he concluded that its position should shift slightly through the night as seen by an observer on earth (Figure 4-22). This change in position, called daily parallax, would make the star appear slightly east of its average position when it was in the eastern sky and slightly west of its average position when it was in the western sky. Tycho had great confidence in his talents for astronomy and mathematics and was sure he could detect this daily parallax, if it existed, by carefully measuring the position of the new star against background stars.

Tycho saw no daily parallax in the new star and thus concluded that it was farther away than the moon and was probably among the stars of the celestial sphere. The appearance of a new star in the supposedly unchanging heavens beyond the moon forced Tycho to question the Ptolemaic system. He summarized his results in a small book, *De Stella Nova* (*The New Star*), published in 1573.

The book attracted the attention of astronomers throughout Europe, and soon Tycho was summoned to the court of the Danish King Frederik II and offered funds to build an observatory on the island of Hveen just off the Danish coast. Tycho also received a steady source of income as landlord of a coastal district from which he collected rents. (He was not a popular landlord.) On Hveen, Tycho constructed a luxurious home with four towers specially equipped for astronomy and populated it with servants, assistants, and a dwarf to act as jester. Soon Hveen was an international center of astronomical study.

Tycho's great contribution to the birth of modern astronomy was not theoretical. In fact, his grand theory of the universe was wrong. Because he could measure no

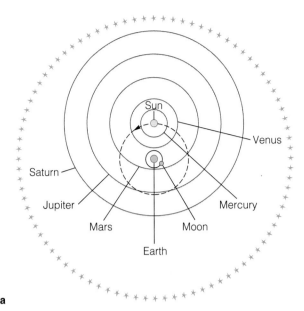

a

FIGURE 4-23

(a) Tycho Brahe's model of the universe held that the earth was fixed at the center of the starry sphere. The moon and sun circled the earth while the planets circled the sun. (b) Much of Tycho's success was due to his skill in designing large, accurate instruments. In this engraving of his mural quadrant, the figure of Tycho, his dog, and the background scene are painted on the wall within the arc of the quadrant. The observer (Tycho himself) at the extreme right peers through a sight out the loophole in the wall at the upper left to measure a star's altitude above the horizon. (Granger Collection, New York)

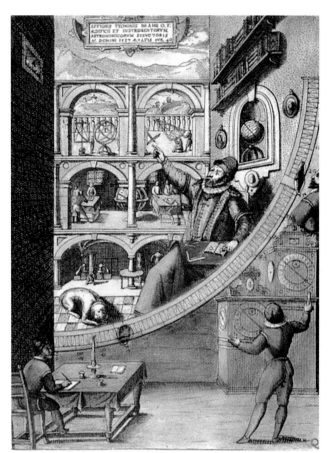

b

parallax for the stars, he concluded that the earth had to be stationary, thus rejecting the Copernican hypothesis. However, he also rejected the Ptolemaic system because of its inaccurate predictions. Instead, he devised a complex theory in which the earth is the immobile center of the universe around which the sun and moon move while the remaining planets orbit the sun (Figure 4-23a). This system is nearly the same as the Copernican system except that the earth is held fixed instead of the sun.

The true value of Tycho's work was observational. Because he was able to devise new and better observing instruments, he was able to make highly accurate observations of the positions of the stars, sun, moon, and planets. Tycho had no telescopes—they were not invented until after his death—so his observations were made by the naked eye peering along sights on his large instruments (Figure 4-23b). Despite these limitations, he measured the positions of 777 stars to better than 4 minutes of arc and measured the positions of the sun, moon, and planets almost daily for the 20 years he stayed on Hveen.

Unhappily for Tycho, King Frederik II died in 1588, and his young son took the throne. Suddenly, Tycho's temper, vanity, and noble presumptions threw him out of favor. In 1596, taking most of his instruments and books of observations, he went to Prague, the capital of Bohemia, and became imperial mathematician to the Holy Roman Emperor Rudolph II. His assignment there

FIGURE 4-24

Johannes Kepler (1571–1630) derived three laws of planetary motion from Tycho Brahe's observations of the positions of the planets. This Romanian stamp commemorates the 400th anniversary of Kepler's birth. Ironically, it contains an error—the incorrect orientation of the moon.

was to revise the *Alfonsine Tables* and publish the revision as a monument to his new patron. It would be called the *Rudolphine Tables*.

Tycho intended to base the *Rudolphine Tables* not on the Ptolemaic system but on his own Tychonic system, proving once and for all the validity of his theory. To assist him, he hired a few mathematicians and astronomers,

FIGURE 4-25
Kepler believed that the five regular solids were the spacers between the spheres containing the planetary orbits. His book *Mysterium Cosmographicum* contained this fold-out illustration of the spheres and spacers. (Owen Gingerich and the Houghton Library)

including one Johannes Kepler. Soon after, in November 1601, while dining at a nobleman's home, Tycho collapsed. Before he died 9 days later, he asked Rudolph II to make Kepler imperial mathematician. Thus, the newcomer Kepler became Tycho's replacement (though at half Tycho's salary).

Kepler the Analyst

No one could have been more different from Tycho Brahe than Johannes Kepler (Figure 4-24). He was born December 27, 1571, to a poor family in a region now included in southwestern Germany. His father was unreliable and shiftless, principally employed as a mercenary soldier fighting for whoever paid enough. He finally failed to return from a military expedition, either because he was killed or because he found circumstances more to his liking. Kepler's mother was apparently an unpleasant and unpopular woman. She was accused of witchcraft in her later years, and Kepler had to defend her in a trial

that dragged on for 3 years. She was finally acquitted but died the following year.

Kepler was the oldest of six children, and his childhood was no doubt unhappy. The family was not only poor and often lacking a father, but it was also Protestant in a predominantly Catholic region. In addition, Kepler was never healthy, even as a child, so it is surprising that he did well in the pauper's school he attended, eventually winning a scholarship to the university at Tübingen, where he studied to become a Lutheran pastor.

During his last year of study, Kepler accepted a job in Graz teaching mathematics and astronomy. Evidently, he was not a good teacher. He had few students his first year and none at all his second. His superiors put him to work teaching a few introductory courses and preparing an annual almanac that contained astronomical, astrological, and weather predictions. Through good luck, in 1595 some of his weather predictions were fulfilled and he gained a reputation as an astrologer and seer, and even in later life he earned money from his almanacs.

While still a college student, Kepler had become a believer in the Copernican theory, and at Graz he used his extensive spare time to study astronomy. By 1596, the same year Tycho left Hveen, Kepler was ready to solve the mystery of the universe. That year he published a book called *The Forerunner of Dissertations on the Universe, Containing the Mystery of the Universe*. Like nearly all scientific works, the book was in Latin, and it is now known as *Mysterium Cosmographicum*.

The book begins with a long appreciation of Copernicanism and then goes on to speculate on the spacing of the planetary orbits. Kepler, as a Copernican, knew of six planets circling the sun—Mercury, Venus, Earth, Mars, Jupiter, and Saturn. According to his model, the spheres containing the six planets were separated from one another, and their relative sizes were fixed by the five regular solids*—the cube, tetrahedron, dodecahedron, icosahedron, and octahedron (Figure 4-25). Kepler gave astrological, numerological, and musical arguments for his model (Window on Science 4-2) and so followed the tradition of Pythagorus in believing that the order of the universe is underlain by musical (meaning mathematical) principles.

In the second half of the book, Kepler tried to fit the five solids to the planetary orbits and so demonstrated that he was a talented mathematician well versed in astronomy. He sent copies to Tycho and to Galileo, and both recognized his talents.

Life was unsettled for Kepler because of the persecution of Protestants in the region, so when Tycho invited him to Prague in 1600, he went readily, anxious to work with the famous astronomer. Tycho's sudden death in 1601 left Kepler in a position to use the observations from Hveen to analyze the motions of the planets and complete the *Rudolphine Tables*. Tycho's family,

---

*A regular solid is a three-dimensional body each of whose faces is the same. For example, a cube is a regular solid each of whose faces is a square.

Hypothesis, Theory, and Law

Even scientists misuse the words *hypothesis*, *theory*, and *law*. We must try to distinguish these terms from one another because they are key elements in science.

A **hypothesis** is a single assertion or conjecture that must be tested. It could be true or false. "All Texans love chili" is a hypothesis. To know whether it is true or false, we need to test it against reality by making observations or performing experiments. Copernicus asserted that the universe was heliocentric; his assertion was a hypothesis subject to testing.

In Chapter 2, we saw that a model (Window on Science 2-1) is a description of some natural phenomenon; it can't be right or wrong. A model is not a conjecture of truth but merely a convenient way to think about a natural phenomenon. Thus, a model such as the celestial sphere cannot be a hypothesis. Copernicus used his hypothesis to build a model of the universe, and, to the extent that the model described the motion of the planets, it was useful.

A **theory** is a system of rules and principles that can be applied to a wide variety of circumstances. A theory may have begun as one or more hypotheses, but it has been tested, expanded, and generalized. Many textbooks refer to the "Copernican theory," but some experts argue that it was not complete in that it lacked a precise description of orbital motion and gravitation, features added by Kepler and Newton. Thus, it is probably more accurate to call it the Copernican hypothesis.

A **natural law** is a theory that almost everyone accepts as true. Such a theory has been tested over and over and applied to many situations. Thus, natural laws are the most fundamental principles of scientific knowledge. Kepler's laws of planetary motion and Newton's laws of gravity and motion (Chapter 5) are examples of natural laws.

Scientists work by developing a hypothesis and testing it, often by building a model based on the hypothesis. If the model is a successful description of nature, the hypothesis is more likely to be true. As a hypothesis is tested repeatedly, expanded, and applied to many circumstances, we begin to refer to it as a theory. If the theory has been so intensively tested that it is almost universally accepted as true, we refer to it as a natural law. ∎

---

recognizing that Kepler was a Copernican and guessing that he would not follow the Tychonic system in completing the *Rudolphine Tables,* sued to recover the instruments and books of observations. The legal wrangle went on for years. The family did recover the instruments Tycho had brought to Prague, but Kepler had the books, and he kept them.

Whether Kepler had any legal right to Tycho's records is debatable, but he put them to good use. He began by studying the motion of Mars, trying to deduce from the observations how the planet moves. By 1606, he had solved the puzzle of planetary motion. The orbit of Mars (and all planets) is an ellipse with the sun at one focus. Thus, he abandoned the 2000-year-old belief in circular motion. But the mystery is even more complex. The planets do not move at constant speed along their orbits—they move faster when close to the sun and slower when farther away. Kepler therefore abandoned both uniform motion and circular motion.

Kepler published his results in 1609 in a book called *Astronomia Nova (New Astronomy).* Like Copernicus's book, *Astronomia Nova* did not become an instant success. It is written in Latin for other scientists and is highly mathematical. In some ways, the book is surprisingly advanced. For instance, Kepler discusses the force that holds the planets in their orbits and comes within a paragraph of discovering the principle of mutual gravitation.

Despite the abdication of Rudolph II in 1611, Kepler continued his astronomical studies. He wrote about a supernova that had appeared in 1604 (now known as Kepler's supernova) and about comets, and he wrote a textbook about Copernican astronomy. In 1619, he published *Harmonice Mundi (The Harmony of the World),* in which he returned to the cosmic mysteries of *Mysterium Cosmographicum.* The only thing of note in *Harmonice Mundi* is his discovery that the radii of the planetary orbits are related to the planet's orbital period. That and his two previous discoveries are now recognized as Kepler's three laws of planetary motion.

Kepler's Three Laws

The puzzle of planetary motion was at least partially solved by Kepler's three laws of planetary motion, which describe how planets move around the sun. Just as the Copernican hypothesis overthrew heliocentrism, Kepler's laws overthrew the principle of uniform circular motion.

Kepler's first law states that *the orbits of the planets around the sun are ellipses with the sun at one focus.* An **ellipse** is defined as a figure drawn around two points, called the **foci** (singular, **focus**), such that the distance from one focus to any point on the ellipse back to the other focus equals a constant; that constant is the longest diameter of the ellipse. This makes it very easy to draw an ellipse with two thumbtacks and a loop of string. Press the thumbtacks into a board, loop the string around them, and place the point of a pencil in the loop, as in Figure 4-26a. If you keep the string taut as you move the pencil, the string forces the total distance from one thumbtack to the pencil to the other thumbtack to remain constant. Thus, the pencil point must follow an ellipse. The closer together the thumbtacks, the more nearly circular the ellipse.

Kepler's second law states that *a line from a planet to the sun sweeps over equal areas in equal intervals of time.*

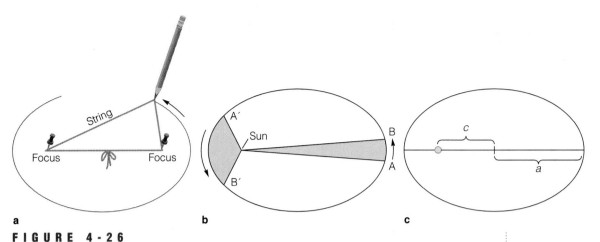

FIGURE 4-26

(a) Drawing an ellipse with two tacks and a loop of string. (b) A line from a planet to the sun sweeps over equal areas in equal intervals of time. (c) The average distance from a planet to the sun equals *a*, the semimajor axis of its orbit.

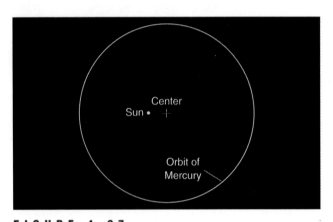

FIGURE 4-27

The orbits of the planets are nearly circular. In this scale drawing of the orbit of Mercury, the horizontal diameter and the vertical diameter of the orbit differ by less than 3 percent.

This means that when the planet is closer to the sun and thus the line connecting it to the sun is shorter, the planet must move faster if the line is to sweep over the same area. The planet in Figure 4-26b would move from point A to point B in 1 month, sweeping out the area shown. But when the planet is closer to the sun, 1 month's motion would carry it from A' to B'. Obviously, it must move faster when it is closer to the sun.

Kepler's third law states that *a planet's orbital period squared is proportional to its average distance from the sun cubed.* Because of the way a planet moves along its orbit, its average distance from the sun is equal to half the long diameter of the elliptical orbit, the so-called **semimajor axis** *a* (Figure 4-26c). If we measure this distance in astronomical units and the orbital period in years, we can summarize the third law as:

$$P_y^2 = a_{au}^3$$

The subscripts remind us that we must measure the period in years and the semimajor axis in astronomical units (AU).

We can use Kepler's third law to make simple calculations. For example, Jupiter's average distance from the sun is 5.20 AU. What is its orbital period? If *a* equals 5.20, then a^3 equals 140.6. The orbital period must be the square root of 140.6, which equals about 11.8 years.

The Significance of Kepler's Laws

The three laws of planetary motion are quite simple, but their significance lies in what they meant for the astronomy of the past and what they portend for the astronomy of the future.

Kepler's laws finally overthrew the classical principle of uniform circular motion. The path of a planet is an ellipse, not a circle, and the planet moves faster and slower as its distance from the sun varies. Of the planets known to Kepler, Mercury has the most elliptical orbit, but even it differs less than 3 percent from a perfect circle (Figure 4-27). Detecting this difference was a triumph, testimony to the precision of both Kepler's calculations and Tycho Brahe's observations.

Notice that Kepler's three laws are empirical. That is, they describe the phenomenon without explaining why it occurs. Kepler derived them from Tycho's observations, not from any fundamental assumption or theory. In fact, Kepler never knew what holds the planets in their orbits or why they continue to move around the sun.

Looking back in history from the time of Kepler, we see astronomers struggling with theory—the place of the earth and the shape of the paths of the planets. Looking forward from Kepler's time, we see a new empirical interest in the precise motion of the planets and the laws that govern their motion.

Of course, the final test of the Copernican hypothesis and Kepler's laws was their ability to predict precisely the motion of the heavenly bodies, and that required that the models be converted into tables of planetary positions. Kepler continued the slow process of creating the *Rudolphine Tables*, and at last in 1628 they were complete. He financed their printing himself, dedicating them to the memory of Tycho Brahe. In fact, Tycho's name appears in larger type on the title page than Kepler's own. This is especially surprising when we recall that the tables were based on the heliocentric models of Copernicus and the elliptical orbits of Kepler, not the geocentric system of Tycho. The reason behind Kepler's care was concern for Tycho's family, still powerful and still intent on protecting Tycho's memory. They even demanded a share of the profits and the right to review the book before publication.

The *Rudolphine Tables* was Kepler's masterpiece. The tables correctly predicted the motions of the planets, a quest begun by the ancient Greek philosophers. The *Rudolphine Tables* provided the final confirmation of the heliocentric hypothesis—the confirmation that Copernicus had sought but failed to find because he could not give up uniform circular motion.

CRITICAL INQUIRY

What were the main differences between the *Alfonsine Tables*, the *Prutenic Tables*, and the *Rudolphine Tables*?

All three of these tables predicted the motions of the heavenly bodies, but only the *Rudolphine Tables* proved accurate. The *Alfonsine Tables*, produced in Toledo around AD 1250, were based on the Ptolemaic model, so they were geocentric and used uniform circular motion; consequently, they were not very accurate. The *Prutenic Tables*, published in 1551, were based on the Copernican model and so were heliocentric. But because the *Prutenic Tables* included the classical principle of uniform circular motion, they were no more accurate than the *Alfonsine Tables*. Kepler's *Rudolphine Tables* were Copernican in that they were heliocentric, but they included Kepler's three laws of planetary motion and thus finally predicted the motions of the planets.

One of the main reasons for the success of the Copernican hypothesis was not that it was accurate but that it was elegant. Compare the motions of the planets and the explanation of retrograde motion in the Copernican model with those in the Ptolemaic and Tychonic models. How was the Copernican model more elegant? ∎

Kepler died November 15, 1630. During his life he had been an astrologer, a mystic, a numerologist, and a seer, but he became one of the world's great astronomers. His work not only overthrew the concept of uniform circular motion that had shackled the minds of astronomers for over 2000 years but also opened the way to the empirical study of planetary motion, a study that lies at the heart of modern astronomy.

4-4 MODERN ASTRONOMY

We date the origin of modern astronomy from the 99 years between the deaths of Copernicus and Galileo (1543 to 1642) because it was an age of transition. That period marked the transition between the Ptolemaic model of the universe and the Copernican with the attendant controversy over the place of the earth. But that same period also marked a transition in the nature of astronomy in particular and science in general, a transition illustrated in the resolution of the puzzle of planetary motion. The puzzle was solved by precise observation and careful computation, techniques that are the foundation of modern science.

The discoveries of Kepler and Galileo found acceptance in the 1600s because the world was in transition. Astronomy was not the only thing changing during this period. The Renaissance is commonly taken to be the period between 1300 and 1600, and thus these 99 years of astronomical history lie at the culmination of the reawakening of learning in all fields (Figure 4-28). The world was open to new ideas and new observations. Martin Luther remade religion, and other philosophers and scholars re-formed their areas of human knowledge. Had Copernicus not published his hypothesis, someone else would have suggested that the universe is heliocentric. History was ready to shed the Ptolemaic system.

In addition, this period marks the beginning of the modern scientific method. Beginning with Copernicus, Tycho, Kepler, and Galileo, scientists depended more and more on evidence, observation, and measurement. This, too, is coupled to the Renaissance and its advances in metalworking and lens making. Before our story began, no astronomer had looked through a telescope, because one could not be made. By 1642, not only telescopes but also other sensitive measuring instruments had transformed science into something new and precise. Also, the growing number of scientific societies increased the exchange of observations and hypotheses among scientists and stimulated more and better work. However, the most important advance was the application of mathematics to scientific questions. Kepler's work demonstrated the power of mathematical analysis, and as the quality of these numerical tools improved, the progress of science accelerated. Our story, therefore, is the story of the birth not just of modern astronomy but of modern science as well.

We will continue our story in the next chapter with the life of Isaac Newton. We will see how his accomplishments had their origins in the work of Galileo and Kepler. ∎

∎ Summary

Archaeoastronomy, the study of the astronomy of ancient cultures, is revealing that primitive astronomers could make

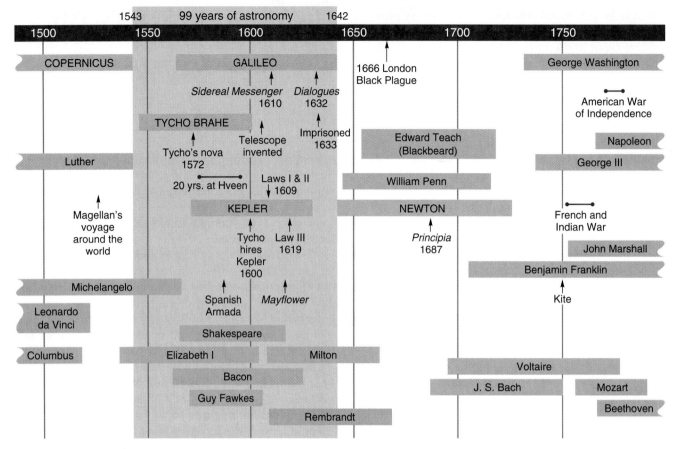

FIGURE 4-28

The 99 years between the deaths of Copernicus and Galileo marked the transition from the ancient astronomy of Ptolemy and Aristotle to the birth of modern scientific astronomy.

sophisticated astronomical observations. The Stone Age people of Britain who built Stonehenge aligned its long avenue and Heelstone to point toward the rising sun at the summer solstice. Other stones mark sight lines toward the rising moon. It seems clear that Stonehenge was a calendrical device, and some believe that its builders used it to predict eclipses.

Other archaeoastronomy sites span the world. A great many stone circles dot England and northern Europe, and related patterns, called medicine wheels, show that North American Indians also practiced astronomy. The Mound Builders of the Midwest, the Indians of the Southwest, and the Aztecs, Maya, and Incas of Central and South America also studied astronomy.

Traditional histories of astronomy usually begin with the Greeks. The Greek philosopher Aristotle held that the earth is fixed at the center of the universe, and Ptolemy based a mathematical model of the moving planets on this geocentric universe. To preserve the concept of uniform circular motion, he used epicycles, deferents, and equants for each of the planets, the sun, and the moon.

Near the end of his life, in 1543, Nicolaus Copernicus published his hypothesis that the sun is the center of the universe. Because the teachings of Aristotle had been adopted as official church doctrine, the heliocentric universe of Copernicus was considered radical.

Galileo Galilei was the first astronomer to use the newly invented telescope for a systematic study of the heavens, and

he found that the appearance of the sun, moon, and planets supported the heliocentric model. After he described his findings in a book, the church, in 1616, ordered him to stop teaching Copernicanism. After he published another book on the subject in 1632, he was tried by the Inquisition and ordered to recant. He was imprisoned in his home until his death in 1642.

In northern Europe, beyond the power of the church, Tycho Brahe rejected the Ptolemaic and Copernican systems. In Tycho's own system, the earth is fixed at the center of the universe, and the sun and moon revolve around the earth. The planets orbit the moving sun. Although his hypothesis is wrong, he was a brilliant observer and accumulated 20 years of precise measurements of the positions of the sun, moon, and planets.

At Tycho's death, one of his assistants, Johannes Kepler, became his successor. Analyzing Tycho's observations, Kepler finally discovered how the planets move. His three laws of planetary motion overthrew uniform circular motion. The first law says the orbits of the planets are elliptical rather than circular, and the second law says the planets do not move at a uniform rate but move faster when near the sun and slower when more distant. The third law relates orbital period to orbital size. These studies of the place of the earth and the nature of planetary motion led to the birth of modern astronomy.

■ New Terms

archaeoastronomy

uniform circular motion

geocentric universe

parallax (*p*)

eccentrics

retrograde loop

epicycle

deferent

equant

heliocentric universe

paradigm

hypothesis

theory

natural law

ellipse

focus

semimajor axis (*a*)

■ Questions

1. What evidence do we have that early human cultures observed astronomical phenomena?

2. Why did Plato propose that all heavenly motion was uniform and circular?

3. How do the epicycles of Mercury and Venus differ from those of Mars, Jupiter, and Saturn?

4. Why did Copernicus have to keep epicycles and deferents in his system?

5. Explain how each of Galileo's telescopic discoveries supported the Copernican hypothesis.

6. Galileo was condemned, but Kepler, also a Copernican, was not. Why not?

7. When Tycho observed the new star of 1572, he could detect no parallax. Why did that result undermine belief in the Ptolemaic system?

8. Does Tycho's model of the universe explain the phases of Venus that Galileo observed? Why or why not?

9. How do the first two of Kepler's three laws overthrow one of the basic beliefs of classical astronomy?

■ Discussion Questions

1. Suggest some reasons why Copernicus may have delayed publication of his book.

2. What parallels and contrasts do you see between the Vatican's recent reconsideration of Galileo's trial and the controversy over teaching evolution in public schools?

3. Why might Tycho have hesitated to hire Kepler? Why did Tycho appoint Kepler, a commoner and a Copernican, to be his scientific heir?

4. Tycho's description of the universe is usually called the Tychonic theory. What should it be called—a hypothesis, a model, a theory, or a law?

■ Problems

1. Draw and label a diagram of the eastern horizon from northeast to southeast, and label the rising point of the sun at the solstices and equinoxes. (See Figures 2-16 and 4-2).

2. If you lived on Mars, which planets would describe retrograde loops? Which would never be visible as crescent phases?

3. Galileo's telescope showed him that Venus has a large angular diameter (61 seconds of arc) when it is a crescent and a small angular diameter (10 seconds of arc) when it is nearly full. Use the small-angle formula to find the ratio of its maximum distance to its minimum distance. Is this ratio compatible with the Ptolemaic universe shown in Figure 4-13?

4. Galileo's telescopes were not of high quality by modern standards. He was able to see the rings of Saturn, but he never reported seeing features on Mars. Use the small-angle formula to find the angular diameter of Mars when it is closest to Earth. How does that compare with the maximum angular diameter of Saturn's rings?

5. If a planet has an average distance from the sun of 4 AU, what is its orbital period?

6. If a space probe is sent into an orbit around the sun that brings it as close as 0.5 AU and as far away as 5.5 AU, what will be its orbital period?

7. Pluto orbits the sun with a period of 247.7 years. What is its average distance from the sun?

■ Critical Inquiries for the Web

1. The trial of Galileo is an important event in the history of science. We now know, and the Church now recognizes, that Galileo's view was correct, but what were the arguments on both sides of the issue as it was unfolding? Research the Internet for documents that chronicle the trial, Galileo's observations and publications, and the position of the Church. Use this information to outline the case for and against Galileo in the context of the times in which the trial occurred.

2. Take a "virtual tour" of Stonehenge by browsing the Web for information related to this megalithic site and the possibility that it was used for astronomical purposes. (Be careful to use legitimate scientific and historical Web sites in your survey.) Summarize the various astronomical alignments evident by the layout of the site.

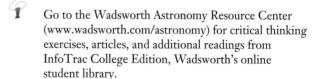

Go to the Wadsworth Astronomy Resource Center (www.wadsworth.com/astronomy) for critical thinking exercises, articles, and additional readings from InfoTrac College Edition, Wadsworth's online student library.

Chapter Five

NEWTON, EINSTEIN, AND GRAVITY

Nature and Nature's laws lay hid in night:

God said, "Let Newton be! and all was light.

ALEXANDER POPE

GUIDEPOST

Astronomers are gravity experts. All of the heavenly motions described in the preceding chapters are dominated by gravitation. Isaac Newton gets the credit for discovering gravity as a way to explain the motions of the sun, moon, and planets, but even Newton couldn't explain what gravity was.

Einstein proposed that gravity is a curvature of space, but that only pushes the mystery further away. What is curvature? we might ask.

This chapter shows how scientists build theories to explain and unify observations. Such theories can give us entirely new ways to understand nature, but no theory is an end in itself. Astronomers continue to study Einstein's theory, and they wonder if there is an even better way to understand the motions of the heavens. ■

Isaac Newton was born in Woolsthorpe, England, on December 25, 1642, and on January 4, 1643. This was not a biological anomaly but a calendrical quirk. Most of Europe, following the lead of the Catholic countries, had adopted the Gregorian calendar, but Protestant England continued to use the Julian calendar. Thus, December 25 in England was January 4 in Europe. If we take the English date, then Newton was born in the same year that Galileo Galilei died.

Newton went on to become one of the greatest scientists who ever lived (Figure 5-1), but even he admitted the debt he owed to those who had studied nature before him. He said, "If I have seen farther than other men, it is because I stood on the shoulders of giants."

One of those giants was Galileo (Figure 5-2). Although we remember Galileo as the defender of Copernicanism, he was also a trained scientist who studied the motions of falling bodies.

Newton based his own work on the discoveries of Galileo and others. During his life he studied optics, invented calculus, developed three laws of motion, and discovered the principle of mutual gravitation. Of these, the last two are the most important to us in this chapter because they make it possible to understand the orbital motion of the moon and planets.

Newton's laws of motion and gravity were astonishingly successful in describing the heavens. A friend of Newton, Edmund Halley, used Newton's laws to calculate the orbits of comets and discovered that certain comets that had been seen throughout recorded history were actually a single object returning every 75 years, now known as Comet Halley.

For over two centuries following the publication of Newton's works, astronomers used his laws to describe the universe. Then, early in this century, Albert Einstein proposed a new way to describe gravity. The new theory did not replace Newton's laws but rather showed that they were only approximately correct and could be seriously in error under special circumstances. We will see how Einstein's theories further extend our understanding of the nature of gravity. Just as Newton had stood on the shoulders of Galileo, Einstein stood on the shoulders of Newton.

5-1 GALILEO AND NEWTON

Johannes Kepler discovered three laws of planetary motion, but he never understood why the planets move along their orbits. At one place in his writings, he wonders if they are pulled along by magnetic forces emanating from the sun. At another place, he speculates that they are pushed along their orbits by angels beating their wings.

Newton refined Kepler's model of planetary motion but did not perfect it. In science, a model is an intellectual conception of how nature works (Window on

a

b

FIGURE 5-1

Isaac Newton (1642–1727) was the founder of modern physics. He made important discoveries in optics, developed three laws of motion, invented differential calculus, and discovered the law of mutual gravitation. These stamps were issued in 1987 to mark the 300th anniversary of the publication of his book *Principia*. Note that the Russian stamp uses the Gregorian calendar for the date of his birth.

FIGURE 5-2

Galileo, usually remembered as the defender of Copernicanism, was also a talented scientist. He studied the motions of falling bodies and discovered the law of inertia.

Science 2-1). No model is perfect; although Kepler's model was better than Aristotle's, it still included angels to push the planets along their orbits. Newton improved Kepler's model by expanding it into a general model of motion and gravity, but Newton's model was not perfect either. Newton never understood what gravity was. It was as mysterious as an angel pushing the moon inward toward the earth instead of forward along its orbit.

FIGURE 5-3
In Aristotle's universe, the earth is located at the center. Of the four elements, earth and water tend to move toward the center of the universe. Fire and air move up toward the sphere of stars. (From Peter Apian, *Cosmographia*, 1539)

FIGURE 5-4
Galileo found that a falling object is accelerated downward. Each second, its velocity increases by 9.8 m/sec (32 ft/sec).

To understand science, we must understand the importance of models. The scientist studies nature by either creating new models or refining old models. Yet a model can never be perfect, because it can never represent the universe in all its intricacies. Instead, a model must be a limited approximation of a single phenomenon, such as orbital motion. It is fitting that Newton's discoveries all began with Kepler's fellow Copernican, Galileo.

Galileo and Motion

Even before Galileo built his first telescope, he had begun studying the motion of freely moving bodies. After the Inquisition condemned and imprisoned him in 1633, he continued his study of motion. He seems to have realized that he would have to understand motion before he could truly understand the Copernican system. That he was eventually able to formulate principles that later led Newton to the laws of motion and the theory of gravity is a tribute to Galileo's ability to set aside authority and think for himself.

The authority of the age was Aristotle, whose ideas on motion were hopelessly confused. Aristotle said that the world is made up of four classical elements: earth, water, air, and fire. According to this idea, all earthly things—wood, rock, flesh, bone, metals, and so on—are made up of mixtures of earth and water. The motions of bodies are determined by their natural tendencies to move toward their proper places in the cosmos. Earth and water, and all things composed of them, move toward the center of the cosmos, which, in Aristotle's geocentric universe, is the center of the earth (Figure 5-3). Thus, objects fall downward because they are moving toward their proper place.* On the other hand, fire

---

*This is one reason why Aristotle had to have a geocentric universe. If the center of the earth had not also been the center of the cosmos, his theory of gravity would not have worked.

and air move upward toward the heavens, their proper place in the cosmos.

Aristotle called these motions **natural motions** to distinguish them from **violent motions** produced, for instance, when we push on an object and make it move other than toward its proper place. According to Aristotle, such motions stop as soon as the force is removed. To explain how an arrow could continue to move upward even after it had left the bowstring, he said currents in the air around the arrow carried it forward even though the bowstring was no longer exerting a force on it.

These ideas about the proper places of objects, natural and violent motion, and the necessity of a force to preserve motion were still accepted theory in Galileo's time. In fact, in 1590, when Galileo was 26, he wrote a short work called *De Motu* (*On Motion*) that deals with the proper places of objects and their natural motions.

In Galileo's time and for the two preceding millennia, scholars had commonly tried to resolve problems of science by referring to authority. To analyze the flight of a cannonball, for instance, they would turn to the writings of Aristotle and other classical philosophers and try to deduce what those philosophers would have said on the subject. This generated a great deal of discussion but little real progress. Galileo broke with this tradition and conducted his own experiments.

He began by studying the motions of falling bodies, but he quickly discovered that the velocities were so great and the times so short that he could not measure them accurately. Consequently, he turned to polished bronze balls rolling down gently sloping inclines. In that instance, the velocity is lower and the time longer. Using an ingenious water clock, he was able to measure the time the balls took to roll given distances down the incline, and he correctly recognized that these times are proportional to the times taken by falling bodies.

He found that falling bodies do not fall at constant rates, as Aristotle had said, but are accelerated. That is, they move faster with each passing second. Near the surface of the earth, a falling object will have a velocity of 9.8 m/sec (32 ft/sec) at the end of 1 second, 19.6 m/sec (64 ft/sec) after 2 seconds, 29.4 m/sec (96 ft/sec) after 3 seconds, and so on. Each passing second adds 9.8 m/sec (32 ft/sec) to the object's velocity (Figure 5-4). In modern terms, this is called the **acceleration of gravity** at the earth's surface.

Galileo also discovered that the acceleration does not depend on the weight of the object. This, too, is contrary to the teachings of Aristotle, who believed that heavy objects, containing more earth and water, fall with higher velocity. Galileo found that the acceleration of a falling body is the same whether it is heavy or light. According to some accounts, he demonstrated this by dropping balls of iron and wood from the top of the Leaning Tower of Pisa to show that they would fall together and hit the ground at the same time (Figure 5-5a). In fact, he probably didn't perform this experiment. It would not have been conclusive anyway because of air resistance. More than 300 years later, Apollo 15 astronaut David Scott,

a

Hammer   Feather

b

FIGURE 5-5

(a) According to tradition, Galileo demonstrated that the acceleration of a falling body is independent of its weight by dropping balls of iron and wood from the Leaning Tower of Pisa. In fact, air resistance would have confused the result. (b) In a historic television broadcast from the moon on August 2, 1971, David Scott dropped a hammer and a feather at the same instant. In the vacuum of the lunar surface, they fell together. (NASA)

FIGURE 5-6
Newton's three laws of motion.

standing on the airless moon, demonstrated Galileo's discovery by dropping a feather and a steel geologist's hammer. They hit the lunar surface at the same time (Figure 5-5b).

Having described natural motion, Galileo turned his attention to violent motion—that is, motion directed other than toward an object's proper place in the cosmos. He pointed out that an object rolling down an incline is accelerated toward the earth and that an object rolling up the same incline is decelerated. If the surface were perfectly horizontal and frictionless, he reasoned, there could be no acceleration or deceleration to change the object's velocity, and, in the absence of friction, the object would continue to move forever. In his own words, "any velocity once imparted to a moving body will be rigidly maintained as long as the external causes of acceleration or retardation are removed."

This is contrary to Aristotle's belief that motion can continue only if a force is present to maintain it. In fact, Galileo's statement is a perfectly valid summary of the law of inertia, which became Newton's first law of motion.

Galileo published his work on motion in 1638, 2 years after he had become entirely blind and only 4 years before his death. The book was called *Mathematical Discourses and Demonstrations Concerning Two New Sciences, Relating to Mechanics and to Local Motion*. It is known today as *Two New Sciences*.

The book is a brilliant achievement for a number of reasons. To understand motion, Galileo had to abandon the authority of the age, devise his own experiments, and draw his own conclusions. In a sense, this was the first example of experimental science. But Galileo also had to generalize his experiments to discover how nature worked. Though his apparatus was finite and plagued by friction, he was able to imagine an infinite, totally frictionless plane on which a body moves at constant velocity. In his workshop, the law of inertia was obscure, but in his imagination, it was clear and precise.

TABLE 5-1

Newton's Three Laws of Motion

I. A body continues at rest or in uniform motion in a straight line unless acted upon by some net force.

II. The acceleration of a body is inversely proportional to its mass, directly proportional to the net force, and in the same direction as the net force.

III. To every action, there is an equal and opposite reaction.

Newton and the Laws of Motion

Newton's three laws of motion (Table 5-1) are critical to our understanding of orbital motion. They apply to any moving object, from an automobile driving along a highway to galaxies colliding with each other.

The first law is really a restatement of Galileo's law of inertia. An object continues at rest or in uniform motion in a straight line unless acted upon by some force (Figure 5-6a). An astronaut drifting in space will travel at a constant rate in a straight line forever if no forces act on him or her.

Newton's first law also explains why a projectile continues to move after all forces have been removed—for instance, why an arrow continues to move after leaving the bowstring. The object continues to move because it has momentum. We can think of an object's **momentum** as a measure of the amount of motion.

Momentum depends on velocity and mass.* A low-velocity object such as a paper clip tossed across a room has little momentum, and we could easily catch it in our hand. But the same paper clip fired at the speed of a rifle

*Mathematically, momentum is the product of mass and velocity.

Mass

One of the most fundamental parameters in science is **mass,** the measure of the amount of matter in an object. A bowling ball, for example contains a large amount of mass, but a child's rubber ball contains less matter than the bowling ball, and we say it is less massive.

Mass is not the same as weight. Our weight is the force that the earth's gravity exerts on the mass of our bodies. Because gravity pulls us downward, we press against the bathroom scale and we can measure our weight. Floating in space, we would have no weight at all; a bathroom scale would be useless. But our bodies would still contain the same amount of matter, so we would still have mass.

Sports analogies illustrate the importance of mass in dramatic ways. A bowling ball, for example, must be massive in order to have a large effect on the pins it strikes. Imagine trying to knock down all the pins with a bowling ball that weighed no more than a balloon. Even in space, where the bowling ball would be weightless, a low-mass bowling ball would have little effect on the pins. On the other hand, runners want track shoes that have low mass and thus are easy to move. Imagine trying to run a 100-meter dash wearing track shoes that weighed as much as bricks. They would be very hard to move, and it would be difficult to accelerate away from the starting block. The shot put takes muscle because the shot is massive, not because it is heavy. Imagine throwing the shot in space where it would have no weight. It would still be massive, and it would take great effort to start it moving.

Mass is a unique measure of the amount of material in an object. Using the metric system (Appendix), we will measure mass in kilograms. ■

---

bullet would have tremendous momentum, and we would not dare try to catch it.

Momentum also depends on the mass of an object (Window on Science 5-1). To see how, imagine that, instead of tossing us a paper clip, someone tosses us a bowling ball. A bowling ball contains much more mass than a paper clip and therefore has much greater momentum at the same velocity.

Newton's first law explains the consequences of the conservation of momentum. When we say that momentum is conserved, we mean that it remains constant until something acts to change it. A moving object has a given amount of momentum. To change that momentum, we must exert some force on the object to change either the speed or the direction. Newton's first law and the concept of momentum came from the work of Galileo.

Newton's second law of motion discusses forces, and Galileo did not talk about forces. He spoke instead of accelerations. Newton saw that an acceleration is the result of a force acting on a mass (Figure 5-6b). Newton's second law is commonly written as:

$$F = ma$$

Once again, we must carefully define terms. An **acceleration** is a change in velocity, and a **velocity** is a directed rate of motion. By rate of motion, we mean, of course, a speed, but the word *directed* has a special meaning. Speed itself does not have any direction associated with it, but velocity does. If you drive a car in a circle at 55 mph, your speed is constant, but your velocity is changing. Thus, an object experiences an acceleration if its speed changes or if its direction of motion changes. Every automoibile has three accelerators—the gas pedal, the brake pedal, and the steering wheel. All three change the auto's velocity.

The acceleration of a body is proportional to the force applied. This is reasonable. If we push gently against a grocery cart, we do not expect a large acceleration. The second law of motion also says that the acceleration is inversely proportional to the mass of the body. This, too, is reasonable. If the cart were filled with bricks and we pushed it gently, we would expect very little result. If it were full of table tennis balls, however, it might move easily in response to a gentle push. Finally, the second law says that the resulting acceleration is in the direction of the force. This is also what we would expect. If we push on a cart that is not moving, we expect it to begin moving in the direction we push.

The second law of motion is important because it establishes a precise relationship between cause and effect (Window on Science 5-2). Objects do not just move. They accelerate due to the action of a force. Moving objects do not just stop. They decelerate due to a force. Also, moving objects don't just change direction for no reason. Any change in direction is a change in velocity and requires the presence of a force. Aristotle said that objects move because they have a tendency to move. Newton said that objects move due to a specific cause, a force.

Newton's third law of motion specifies that for every action there is an equal and opposite reaction. In other words, forces must occur in pairs directed in opposite directions. For example, if you stand on a skateboard and jump forward, the skateboard will shoot away backward. As you jump, your feet exert a force against the skateboard, which accelerates it toward the rear. But forces must occur in pairs, so the skateboard must exert an equal but opposite force on your feet that accelerates your body forward (Figure 5-6c).

Mutual Gravitation

The three laws of motion led Newton to consider the force that causes objects to fall. The first and second laws tell us that falling bodies accelerate downward because

Cause and Effect

One of the most often used and least often stated principles of science is cause and effect, and we could argue that Newton's second law of motion was the first clear statement of the principle. Ancient philosophers such as Aristotle argued that objects moved because of tendencies. They said that earth and water, and objects made of earth and water, had a natural tendency to move toward the center of the universe. This natural motion had no cause but was inherent in the nature of the objects. But Newton's second law says $F = ma$. If an object changes its motion (a in the equation), then it must be acted on by a force (F in the equation). Any effect (a) must be the result of a cause (F).

The principle of cause and effect goes far beyond motion. The principle of cause and effect gives scientists confidence that every effect has a cause. Hearing loss in certain laboratory rats, color changes in certain chemical dyes, and explosions on certain stars are all effects that must have causes. All of science is focused on understanding the causes of the effects we see. If the universe were not rational, then we could never expect to discover causes. Newton's second law of motion was arguably the first statement that the behavior of the universe depends rationally on causes. ■

---

some force must be pulling downward on them. Newton wondered what that force could be.

Newton was also aware that some force has to act on the moon. The moon follows a curved path around the earth, and motion along a curved path is accelerated motion. The second law says that a force is required in order to make the moon follow that curved path.

Newton wondered if the force that holds the moon in its orbit could be the same force that causes rocks to roll downhill—gravity. He was aware that gravity extends at least as high as the tops of mountains, but he did not know if it could extend all the way to the moon. He believed that it could, but he thought it would be weaker at greater distances. He also guessed that its strength would decrease as the square of the distance increased.

This relationship, the **inverse square law,** was familiar to Newton from his work on optics, where it applied to the intensity of light. A screen set up 1 unit from a candle flame receives a certain amount of energy on each square meter. However, if that screen is moved to a distance of 2 units, the light that originally illuminated 1 m^2 must cover 4 m^2 (Figure 5-7). Thus, the intensity of the light is inversely proportional to the square of the distance to the screen.

Newton made two assumptions that enabled him to predict the strength of the earth's gravity at the distance of the moon. He assumed that the strength of gravity follows the inverse square law and that the critical distance is not the distance from the surface of the earth but the distance from the center of the earth. Because the moon is about 60 earth radii away, the earth's gravity at the distance of the moon should be about 60^2 times less than at the earth's surface. Instead of being 9.8 m/sec^2, it should be about 0.0027 m/sec^2.

Now, Newton wondered, could this acceleration keep the moon in orbit? He knew the moon's distance and its orbital period, so he could calculate the actual acceleration needed to keep it in its curved path. The answer is 0.0027 m/sec^2. To the accuracy of Newton's data for the radius of the earth, it was exactly what his assumptions predicted. The moon is held in its orbit by gravity, and gravity obeys the inverse square law.

Newton's third law says that forces always occur in pairs, and this quickly led him to realize that gravity is mutual. If the earth pulls on the moon, then the moon must pull on the earth. Gravitation is a general property of the universe. The sun, the planets, and all their moons must also attract each other by mutual gravitation. In fact, every particle in the universe must attract every other particle, and Newtonian gravity is often called universal mutual gravitation.

Because we do not find ourselves attracting particles of mass—books, rocks, passing birds, and so on—drawn to us by our personal gravity, Newton concluded that the gravitational force depends on the mass of the object. Large masses, like the earth, have strong gravity, but smaller masses, like people, have weaker gravity. Combining this dependence on mass with the inverse square law led to the famous formula for the gravitational force between masses M and m:

$$F = -G\frac{Mm}{r^2}$$

The constant G is the gravitational constant, and r is the distance between the masses. The negative sign tells us the force is attractive, pulling the masses together and making r decrease. In plain English, Newton's law of gravitation states: The force of gravity between two masses M and m is proportional to the product of the masses and inversely proportional to the square of the distance between them.

Newton's model of gravity was a difficult idea for physicists of his time to accept because it is an example of action at a distance. The earth and moon exert forces on each other although there is no physical connection between them. Modern scientists resolve this problem by referring to gravity as a **field.** The presence of the earth produces a gravitational field directed toward the center of the earth. The strength of the field decreases according to the inverse square law. Any particle of mass in that field experiences a force that depends on the mass of the particle and the strength of the field at the particle's location. The resulting force is directed toward the center of the field.

The field is an elegant way to describe gravity, but it does not tell us what gravity is. For that we must wait until later in this chapter, when we discuss Einstein's theory of curved space-time.

CRITICAL **INQUIRY**

What do the words *universal* and *mutual* mean in the term "universal mutual gravitation"?

Newton realized that gravity was a force that drew the moon toward the earth. But his third law of motion said that forces always occur in pairs, so if the earth attracted the moon, then the moon had to attract the earth. That is, gravitation had to be *mutual* between any two objects.

When Newton looked beyond the earth–moon system, he realized that the earth's gravity must attract the sun as well as the moon, and if the earth attracted the sun, then the sun had to attract the earth. Furthermore, if the sun had gravity that acted on the earth, then the same gravity would act on all the planets and even on the distant stars. Step by step, Newton's third law of motion led him to conclude that gravitation had to apply to all masses in the universe. That is, it had to be *universal*.

Aristotle explained gravity in a totally different way. How did Aristotle account for a falling apple? Could that explanation account for a hammer falling on the surface of the moon? ■

Newton first came to understand gravity by thinking about the orbital motion of the moon. Gravitation is tremendously important in astronomy, and its most important manifestation is orbital motion.

5-2 ORBITAL MOTION

Newton's laws of motion and gravitation make it possible to understand why the planets move along their orbits. We can understand how they are held in their curved paths, and we can even discover why Kepler's laws work.

Orbiting the Earth

To illustrate the principle of orbital motion, imagine that we position a large cannon at the top of a mountain, point it horizontally, and fire it (Figure 5-8). The cannonball falls to the earth some distance from the foot of the mountain. The more gunpowder we use, the faster the ball travels and the farther from the foot of the

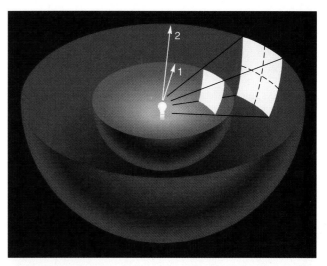

FIGURE 5-7
The inverse square law. A light source is surrounded by spheres with radii of 1 unit and 2 units. The light falling on an area of $1 m^2$ on the inner sphere spreads to illuminate an area of $4 m^2$ on the outer sphere. Thus, the brightness of the light source is inversely proportional to the square of the distance.

FIGURE 5-8
A cannon on a high mountain could put its projectile into orbit if it could achieve a high enough velocity. Newton published a similar figure in *Principia*.

mountain it falls. If we use enough powder, the ball travels so fast it never strikes the ground. The earth's gravity pulls it toward the earth's center, but the earth's surface curves away from it at the same rate at which it falls. We say it is in orbit. If the cannonball is high above the atmosphere, where there is no friction, it will continue to fall around the earth forever. Real earth satellites do fall back to earth sometimes, but that is caused by

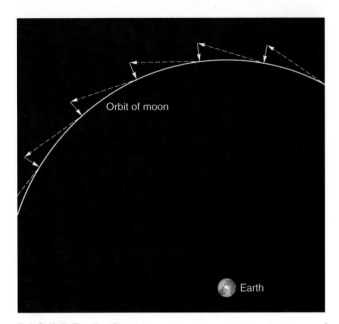

FIGURE 5-9
In the absence of forces, the moon would follow a straight-line path (dashed arrows). It follows a curved path because the earth's gravity accelerates it toward the earth (solid arrows).

friction with the tenuous upper layers of the earth's atmosphere.

The first law of motion says that an object in motion tends to stay in motion in a straight line unless acted upon by some force. Thus, the cannonball in our example travels in a curve around the earth only because the earth's gravity acts to pull it away from its straight-line motion.

Similarly, the moon would move in a straight line at constant speed (dashed arrows in Figure 5-9) were it not for gravity accelerating it toward the earth. Each second, the moon moves 1020 m (3350 ft) eastward along its orbit and falls about 1.6 mm (about $\frac{1}{16}$ inch) toward the earth. The combination of these two motions results in a closed orbit around the earth.

A space capsule orbiting the earth is not "beyond earth's gravity," to use a phrase common in old science fiction movies. Like the moon, the space capsule is accelerated toward the earth. That acceleration, combined with its lateral motion, places it in a closed orbit. Astronauts inside are weightless not because they are beyond the earth's gravity, but because they are falling at the same rate as their capsule. Rather than saying the astronauts are weightless, we should more accurately say they are in free fall.

To be precise, we should avoid saying that the moon orbits the earth. In fact, the moon and the earth orbit each other. Gravitation is mutual, and if the earth pulls on the moon, then the moon pulls on the earth. The two bodies revolve around their common **center of mass,** the balance point of the system. If the earth and moon could be connected by a massless rod and placed in a uniform gravitational field such as that near the earth's surface, the system would balance at its center of mass like a child's seesaw (Figure 5-10). As the system revolves, the moon and the earth each describe an orbit around the center of mass.

As we might expect from experience on a seesaw, the balance point, or center of mass, is closest to the more massive body. The center of mass of the earth–moon system lies only 4708 km (2926 miles) from the earth's center. This places the center of mass inside the earth. As the moon revolves around its orbit, the earth swings about the center of mass.

Orbital Velocity

If we were about to ride a rocket into orbit, we would have a critical question. How fast do we have to travel to stay in orbit? An object's **circular velocity** is the lateral velocity the object must have to remain in a circular orbit. If we assume the mass of our spaceship is small compared with the mass of the object we expect to orbit, the earth in this case, then the circular velocity is:

$$V_c = \sqrt{\frac{GM}{r}}$$

In this formula, M is the mass of the central body in kilograms, r is the radius of the orbit in meters, and G is the gravitational constant, 6.67×10^{-11} N m^2/kg^2, where N stands for a newton, a measure of force. This formula is all we need to calculate how fast an object must travel to stay in a circular orbit.

For example, how fast does the moon travel in its orbit? The mass of the earth is 5.98×10^{24} kg, and the radius of the moon's orbit is 3.84×10^8 m. Then the moon's velocity is:

$$V_c = \sqrt{\frac{6.67 \times 10^{-11} \times 5.98 \times 10^{24}}{3.84 \times 10^8}} = \sqrt{\frac{39.9 \times 10^{13}}{3.84 \times 10^8}}$$

$$= \sqrt{1.04 \times 10^6} = 1020 \text{ m/sec}$$

This calculation shows that the moon travels 1.02 km along its orbit each second. That is the circular velocity of the moon.

A satellite just above the earth's atmosphere is only about 200 km above the earth's surface, or 6578 km from the earth's center, so the earth's gravity is much stronger and the satellite must travel much faster to stay in a circular orbit. We can use the formula above to find that the circular velocity just above the earth's atmosphere is about 7790 m/sec, or 7.79 km/sec. This is about 17,400 miles per hour and shows why putting satellites into earth orbit takes such large rockets. Not only must the rocket lift the satellite above the earth's atmosphere, but

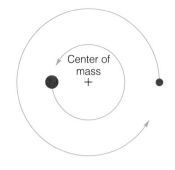

FIGURE 5-10

(a) If two bodies were connected by a massless rod and placed in a uniform gravitational field like that at the earth's surface, they would balance at their center of mass. (b) If the bodies orbit each other, the center of mass remains fixed, and the objects move around it. The center of mass of the earth–moon system lies inside the earth.

a   b

it must tip over and accelerate the satellite to circular velocity.

Geosynchronous Orbits

One orbit that is particularly useful is called a **geosynchronous orbit.** An object in such an orbit remains over the same spot on the earth (Figure 5-11). Communications satellites, for example, are often placed in a geosynchronous orbit.

The orbital velocity of an earth satellite depends on its distance from the center of the earth. If an object orbits near the earth, it has a high velocity and a small orbit, and as a result its orbital period is short, about 90 minutes. At a greater distance from the earth, the orbital velocity is lower and the orbit is larger. Thus, the satellite has a longer orbital period. At a distance of about 42,000 km (26,000 miles), the orbital period is 24 hours.

If we launch an earth satellite eastward around the earth in a geosynchronous orbit above the earth's equator, it will remain above a fixed point on the earth. The earth rotates eastward in 24 hours, and the satellite circles its orbit in 24 hours. If we aim a communications antenna at the satellite, we do not have to move the antenna in order to track the satellite. The many television dishes that dot the landscape are aimed at communications satellites orbiting 42,000 km above the earth's equator in geosynchronous orbit. Communications satellites may be one of the most commonly used applications of the orbital physics worked out by Kepler and Newton.

Open and Closed Orbits

Kepler's laws refer to elliptical orbits, which include circular orbits because a circle is a special kind of ellipse. These orbits are called **closed orbits** because they return back on themselves. If you were in a spacecraft orbiting the earth in a closed orbit, you would return to the same place in the orbit time after time.

If your spacecraft were in an elliptical orbit, we would refer to the point of closest approach as **perigee.** The point where your spacecraft was most distant we

FIGURE 5-11

An earth satellite in a geosynchronous orbit revolves eastward around the earth in an orbit with a period of 24 hours. Thus, it remains fixed above a given spot on the earth, and a dish antenna can remain pointed at the satellite 24 hours a day. Communications satellites are often put into geosynchronous orbit.

FIGURE 5-12

In an elliptical orbit around the earth, the point of closest approach is called perigee. The point on the orbit most distant from the earth is called apogee. Note that apogee and perigee lie at the ends of the major axis of the ellipse.

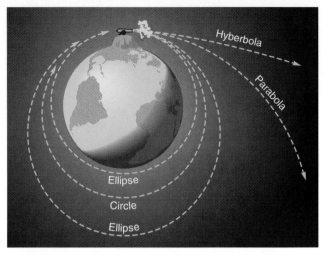

FIGURE 5-13

If an object's orbital velocity is less than escape velocity, it will follow an elliptical or circular orbit. If its velocity equals or exceeds escape velocity, it will follow a parabolic or hyperbolic orbit and escape from the earth.

would call **apogee** (Figure 5-12). Note that perigee and apogee occur on opposite sides of an elliptical orbit.

Newton's laws reveal the existence of another kind of orbit. An **open orbit** (or trajectory) leads away from the central body, never to return. If you were in a spacecraft following an open orbit around the earth, you would be leaving the earth along a path that did not return. Open orbits are also called **escape orbits.**

We can analyze these open and closed orbits by thinking again of the cannon in Figure 5-8. If the cannonball travels at the circular velocity, it will fall in a circular orbit around the earth. Of course, if the velocity is much too small, the ball will hit the earth. However, if the velocity of the ball is only slightly less than the circular velocity, the ball will follow an elliptical orbit with its highest point, apogee, at the cannon (Figure 5-13). If the velocity of the ball is slightly greater than the circular velocity, it will again follow an elliptical path, but with the lowest part of its orbit, perigee, at the cannon. All of these orbits are closed—the cannonball returns to its starting point. If we really performed this experiment, the cannonball would circle the earth in about 90 minutes, returning to its starting point and smashing the cannon to fragments.

If the cannonball's velocity equals or exceeds **escape velocity,** the velocity needed to escape from the surface of a body, the cannonball will follow an open orbit. If the velocity equals the escape velocity, the orbit is a parabola, and if the velocity exceeds the escape velocity, the orbit is a hyperbola. The mathematical definitions of the parabola and hyperbola are not important here, but in both cases the cannonball leaves the earth, never to return.

Calculating Escape Velocity

If we launch a rocket upward, it will consume its fuel in a few moments and reach its maximum speed. From that point on, it will coast upward. How fast must a rocket travel to coast away from the earth and escape? We know, of course, that no matter how far it travels, it can

never escape from earth's gravity. The effects of earth's gravity extend to infinity. It is possible, however, for a rocket to travel so fast initially that gravity can never slow it to a stop. Thus, it could leave the earth.

The escape velocity is the velocity required to escape from the surface of an astronomical body. Here we are interested in escaping from the earth or a planet; in later chapters, we will consider the escape velocity from stars, galaxies, and even a black hole.

The escape velocity, V_e, is given by a simple formula:

$$V_e = \sqrt{\frac{2GM}{r}}$$

Here G is the gravitational constant 6.67×10^{-11} N m^2/kg^2, M is the mass of the astronomical body in kilograms, and r is its radius in meters. (This formula is very similar to the formula for circular velocity.)

We can find the escape velocity from the earth by looking up its mass, 5.98×10^{24} kg, and its radius, 6.38×10^6 m. Then the escape velocity is:

$$V_e = \sqrt{\frac{2 \times 6.67 \times 10^{-11} \times 5.98 \times 10^{24}}{6.38 \times 10^6}} = \sqrt{\frac{7.98 \times 10^{14}}{6.38 \times 10^6}}$$

$$= \sqrt{1.25 \times 10^8} = 11{,}200 \text{ m/sec} = 11.2 \text{ km/sec}$$

This is equal to about 25,000 miles per hour.

Notice that the formula tells us that escape velocity depends on both mass and radius. A massive body might have a low escape velocity if it has a very large radius. We will meet such objects when we discuss giant stars. On the other hand, a rather low mass body can have a very large escape velocity if it has a very small radius, a condition we will discuss when we meet black holes.

Circular velocity and escape velocity are two aspects of Newton's laws of gravity and motion. Once Newton understood gravity and motion, he could do what Kepler had failed to do—he could explain why the planets obey Kepler's laws of planetary motion.

Kepler's Laws Reexamined

Now that we understand Newton's laws, gravity, and orbital motion, we can understand Kepler's laws of planetary motion in a new way.

Kepler's first law says that the orbits of the planets are ellipses with the sun at one focus. Kepler wondered why the planets keep moving along these orbits, and now we know the answer. They move because there is nothing to slow them down. Newton's first law says that a body in motion stays in motion unless acted on by some force. The gravity of the sun accelerates the planets inward toward the sun and holds them in their orbits, but it doesn't pull backward on the planets, so they don't slow to a stop. With no friction, they must continue to move.

The orbits of the planets are ellipses because gravity follows the inverse square law. In one of his most famous problems, Newton proved that if a planet moves in a closed orbit under the influence of an attractive force that follows the inverse square law, then the planet must follow an elliptical path.

Kepler's second law says that a planet moves faster when it is near the sun and slower when it is farther away. Once again, Newton's discoveries explain why. Earlier we saw that a body moving on a frictionless surface will continue to move in a straight line until it is acted on by some force; that is, the object has momentum. But an object set rotating on a frictionless surface will continue rotating until somethings acts to speed it up or slow it down. Such an object has **angular momentum,** a measure of the rotation of the body about some point. A planet circling the sun has a given amount of angular momentum, and with no outside influences to speed it up or slow it down, it must conserve its angular momentum. That is, its angular momentum must remain constant.

Mathematically, a planet's angular momentum is the product of its mass, velocity, and distance from the sun. This explains why a planet must speed up as it comes closer to the sun along an elliptical orbit. Its angular momentum is conserved, so as its distance from the sun decreases, its velocity must increase. In the same way, the planet's velocity must decrease as its distance from the sun increases.

This conservation of angular momentum is actually a common human experience. Skaters spinning slowly can draw their arms and legs closer to their axis of rotation and, through conservation of angular momentum, spin faster (Figure 5-14). To slow their rotation, they again extend their arms. Similarly, divers can spin rapidly in the tuck position and slow their rotation by stretching into the extended position.

FIGURE 5-14

Skaters demonstrate conservation of angular momentum when they spin faster by drawing their arms and legs closer to their axis of rotation.

Kepler's third law is also explained by a conservation law, but in this case it is the law of conservation of energy (Window on Science 5-3). A planet orbiting the sun has a specific amount of energy that depends only on its average distance from the sun. That energy can be divided between energy of motion and energy stored in the gravitational attraction between the planet and the sun. The energy of motion depends on how fast the planet moves, and the stored energy depends on the size of its orbit. The relation between these two kinds of energy is fixed by Newton's laws. That means there has to be a fixed relationship between the rate at which a planet moves around its orbit and the size of the orbit—between its orbital period P and the orbit's semimajor axis a. This is just Kepler's third law.

Newton's Version of Kepler's Third Law

The equation for circular velocity is actually a version of Kepler's third law, as we can prove with a few lines of algebra. The result is one of the most useful formulas in astronomy.

The equation for circular velocity, as we have seen, is:

$$V_c = \sqrt{\frac{GM}{r}}$$

The orbital velocity of a planet is simply the circumference of its orbit divided by the orbital period:

$$V = \frac{2\pi R}{P}$$

If we substitute this for V in the equation for circular velocity and solve for P^2, we get:

$$P^2 = \frac{4\pi^2}{GM} R^3$$

Energy

Energy is one of the most fundamental of all scientific parameters. Physicists define energy as the ability to do work, but we might paraphrase that definition as the ability to produce a change. Certainly a moving body has energy. A planet moving along its orbit, a cement truck rolling down the highway, and a golf ball sailing down the fairway all have the ability to produce a change. Imagine colliding with any of these objects! But energy need not be represented by motion. Sunlight falling on a green plant, on photographic film, or on unprotected skin can produce chemical changes, and thus light is a form of energy.

Energy can take many forms. Batteries and gasoline are examples of chemical energy, and uranium fuel rods contain nuclear energy. A tank of hot water contains thermal energy. Even a weight on a high shelf can represent stored energy. Imagine a bowling ball falling off a high shelf onto your desk. That would produce significant change.

Much of science is the study of how energy flows from one place to another place, producing changes. A biologist might study the way a nerve cell transmits energy along its length to a muscle, while a geologist might study how energy flows as heat from the earth's interior and deforms the earth's surface. In such processes, we see energy being transformed from one state to another. Sunlight (energy) is absorbed by ocean plants and stored as sugars and starches (energy). When the plant dies, it and other ocean life are buried and become oil (energy), which we pump to the surface and burn in automobile engines to produce motion (energy).

Aristotle believed that all change originated in the motion of the starry sphere and flowed down to the earth. Modern science has found a more sophisticated description of the continual change we see around us, but we are still interested in the way energy flows through the world and produces change. Energy is the pulse of the natural world. ■

Here M is just the total mass of the system in kilograms. For a planet orbiting the sun, we can use the mass of the sun for M because the mass of the planet is negligible compared to the mass of the sun. In a later chapter, we will apply this formula to two stars orbiting each other, and then the mass M will be the sum of the two masses. For a circular orbit, R equals the semimajor axis a, so this formula is a general version of Kepler's third law, $P^2 = a^3$. In Kepler's version, we use the units AU and years, but in this formula, known as Newton's version of Kepler's third law, we use units of meters, seconds, and kilograms. G, of course, is the gravitational constant.

This is a very powerful formula. Astronomers use it to find the masses of bodies by observing orbital motion. If, for example, we observe a moon orbiting a planet and we can determine the size of its orbit, R, and the orbital period, P, we can use this formula to solve for M, the total mass of the planet plus the moon. There is no other way to find masses in astronomy, and we will use this formula in later chapters to find the masses of stars, galaxies, and planets.

This discussion is a good illustration of the power of Newton's work. By carefully defining motion and gravity and by giving them mathematical expression, Newton was able to derive new truths, among them Newton's version of Kepler's third law. His work changed science and made it into something new, precise, and exciting.

Astronomy After Newton

Newton published his work in July 1687 in a book called *Philosophiae Naturalis Principia Mathematica* (*Mathematical Principles of Natural Philosophy*), now known simply as *Principia* (Figure 5-15). It is one of the most important books ever written. The principles changed astronomy, science, and the way we think about nature.

Principia changed astronomy and ushered in the age of gravitational astronomy. No longer did astronomers appeal to the whim of the gods to explain things in the heavens. No longer did they speculate on why the planets move. They now knew that the motions of the heavenly bodies are governed by simple, universal rules that describe the motions of everything from planets to falling apples. Suddenly the universe was understandable in simple terms.

Newton's laws of motion and gravity made it possible for astronomers to calculate the orbits of planets and moons. Not only could they explain how the heavenly bodies move, they could predict future motions (Window on Science 5-4). This subject, known as gravitational astronomy, dominated astronomy for almost 200 years and is still important. It included the calculation of the orbits of comets and asteroids and the theoretical prediction of the existence of two planets, Neptune and Pluto.

Principia also changed science in general. The works of Copernicus and Kepler had been mathematical, but no book before had so clearly demonstrated the power of mathematics as a language of precision. Newton's arguments were couched in geometrical terms instead of the new analytical methods developed by European mathematicians, but *Principia* was so powerful an illustration of the quantitative study of nature that scientists around the world adopted mathematics as their most powerful tool.

Also, *Principia* changed the way we think about nature. Newton showed that the rules that govern the universe are simple. Particles move according to three rules of motion and attract each other with a force called

Prediction in Science

When you read about any science, you should notice that scientific theories face in two directions. They look back into the past and explain phenomena we have previously observed. For example, Newton's laws of motion and gravity explained how the planets moved. But theories also face forward in that they enable us to make predictions about what we should find as we explore further. Thus, Newton's laws allowed astronomers to calculate the orbits of comets, predict their return, and eventually understand their origin.

Scientific predictions are important in two ways. First, if a theory leads to a prediction and scientists later discover the prediction was true, the theory is confirmed, and scientists gain confidence that it is a true description of nature. But predictions are important in science in a second way. Using an existing theory to make a prediction may lead us into an unexplored avenue of knowledge. Thus, the first theories of genetics made predictions that confirmed the genetic theory of inheritance, but those predictions also created a new understanding of how living creatures evolve.

As you read about any scientific theory, think about both what it can explain and what it can predict. ■

FIGURE 5-15

Newton, working from the discoveries of Galileo and Kepler, derived three laws of motion and the principle of mutual gravitation. He and some of his discoveries are honored on this English pound note. Notice the diagram of orbital motion in the background and the open copy of *Principia* in Newton's hands.

gravity. These motions are predictable, and that makes the universe a vast machine based on a few simple rules. It is complex only in that it contains a vast number of particles. In Newton's view, if he knew the location and motion of every particle in the universe, he could, in principle, derive the past and future of the universe in every detail. This mechanical determinism has been undermined by modern quantum mechanics, but it dominated science for more than two centuries during which scientists thought of nature as a beautiful clockwork that would be perfectly predictable if we knew how all the gears meshed.

Most of all, Newton's work broke the last bonds between science and formal philosophy. Newton did not speculate on the good or evil of gravity. He did not debate its meaning. Not more than a hundred years before, scientists would have argued over the "reality" of gravity. Newton didn't care for these debates. He wrote, "It is enough that gravity exists and suffices to explain the phenomena of the heavens."

CRITICAL INQUIRY

How do Newton's laws of motion explain the orbital motion of the moon?

If the earth and moon did not attract each other, the moon would move in a straight line in accord with Newton's first law of motion and vanish into space. Instead, gravity pulls the moon toward the center of the earth, and the moon accelerates toward the earth. This acceleration is just enough to pull the moon away from its straight-line motion and cause it to follow a curve around the earth. In fact, it is correct to say that the moon is falling, but because of its lateral motion it continuously misses the earth.

Every orbiting object is falling toward the center of its orbit but is moving laterally fast enough to compensate for the inward motion, and it follows a curved orbit. If this is true, then how can astronauts float inside spacecraft in a "weightless" state? Why might "free fall" be a more accurate term? ■

FIGURE 5-16
Einstein's greatest accomplishments, the special theory of relativity and the general theory of relativity, were developed as he analyzed the nature of motion and time.

Newton's laws of motion and gravitation are critical in astronomy not only because they describe orbital motion but also because they describe the interaction of astronomical bodies. Newton made other discoveries, but we reserve these for later chapters. Now we move forward two centuries to see how Einstein described gravity in a new and powerful way.

5-3 EINSTEIN AND RELATIVITY

In the early years of this century, Albert Einstein (1879–1955), then a clerk in the Swiss patent office (Figure 5-16), began thinking about how motion and gravity interact. He soon gained international fame by showing that Newton's laws of motion and gravity were only partially correct. The revised theory became known as the theory of relativity. As we will see, there are really two theories of relativity.

Special Relativity

Einstein began by thinking about how moving observers see events around them. His analysis led him to the first postulate of relativity, also known as the principle of relativity:

> **First postulate** (the principle of relativity)
> Observers can never detect their *uniform* motion except relative to other objects.

You may have experienced the first postulate while sitting on a train in a station. You suddenly notice that the train on the next track has begun to creep out of the station. However, after several moments you realize that it is your own train that is moving and that the other train is still motionless on its track.

Consider another example. Suppose you are floating in a spaceship in interstellar space and another spaceship comes coasting by (Figure 5-17a). You might conclude that it is moving and you are not, but someone in the other ship might be equally sure that you are moving and it is not. The principle of relativity says that there is no experiment you can perform to decide which ship is moving and which is not. This means that there is no such thing as absolute rest—all motion is relative.

Because neither you nor the people in the other spaceship could perform any experiment to detect your absolute motion through space, the laws of physics must have the same form in both spaceships. Otherwise, experiments would give different results in the two ships, and you could decide who was moving. Thus, a more general way of stating the first postulate refers to these laws of physics:

> **First postulate** (alternate version)
> The laws of physics are the same for all observers, no matter what their motion, so long as they are not *accelerated*.

The words *uniform* and *accelerated* are important. If either spaceship were to fire its rockets, then its velocity would change. The crew of that ship would know it because they would feel the acceleration pressing them into their couches. Accelerated motion, therefore, is different—we can always tell which ship is accelerating and which is not. The postulates of relativity discussed here apply only to observers in uniform motion. That is why the theory is called **special relativity.**

The first postulate fit with Einstein's conclusion that the speed of light must be constant for all observers. No matter how you move, your measurement of the speed of light has to give the same result (Figure 5-17b). This became the second postulate of special relativity:

> **Second postulate** The velocity of light is constant and will be the same for all observers independent of their motion relative to the light source.

Once Einstein had accepted the basic postulates of relativity, he was led to some startling discoveries. Newton's laws of motion and gravity worked well as long as distances were small and velocities were low. But when we begin to think of very large distances or very high velocities, Newton's laws are no longer adequate to describe what happens. Instead, we must use relativistic physics. For example, special relativity shows that the observed mass of a moving particle depends on its velocity. The higher the velocity, the greater the mass of the particle. This is not significant at low velocities, but it becomes very important as the velocity approaches the velocity of light. Such increases in mass are observed

a

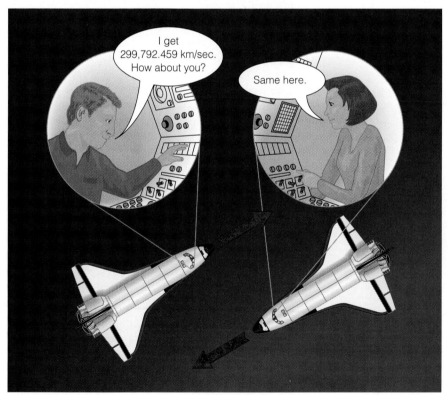

b

FIGURE 5-17

(a) The principle of relativity says that observers can never detect their uniform motion, except relative to other observers. Thus, neither of these travelers can decide who is moving and who is not. (b) If the principle of relativity is correct, the velocity of light must be a constant for all observers. If the velocity of light depended on the motion of the observer through space, then these travelers could decide who was moving and who was not.

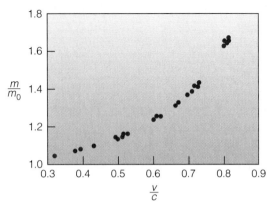

FIGURE 5-18

The observed mass of moving electrons depends on their velocity. As the ratio of their velocity to the velocity of light, *v/c*, gets larger, the mass of the electrons in terms of their mass at rest, *m/m*$_0$, increases. Such relativistic effects are quite evident in particle accelerators, which accelerate atomic particles to very high velocities.

whenever physicists accelerate atomic particles to high velocities (Figure 5-18).

This discovery led to yet another insight. The relativistic equations that describe the energy of a moving particle predict that the energy of a motionless particle is not zero. Rather, its energy at rest is $m_0 c^2$. This is of course the famous equation:

$$E = m_0 c^2$$

The c is the speed of light and the m_0 is the mass of the particle when it is at rest. (We must specify mass this way because one of the consequences of relativity is that a particle's mass depends on its velocity.) This simple formula suggests that mass and energy are related, and we will see in later chapters how nature can convert one into the other inside stars.

For example, suppose that we convert 1 kg of matter into energy. We must express the velocity of light as 3×10^8 m/sec, and our result is 9×10^{16} joules (J) (approximately equal to a 20-megaton nuclear bomb). (A **joule** is a unit of energy roughly equivalent to the energy given up when an apple falls from a table to the floor.)

a

b

Our simple calculation shows that the energy equivalent of even a small mass is very large.

Other relativistic effects include the slowing of moving clocks and the shrinkage of lengths measured in the direction of motion. A detailed discussion of the major consequences of the special theory of relativity is beyond the scope of this book. Instead, we must consider Einstein's second advance, the general theory.

The General Theory of Relativity

In 1916, Einstein published a more general version of the theory of relativity that dealt with accelerated as well as uniform motion. This **general theory of relativity** contained a new description of gravity.

Einstein began by thinking about observers in accelerated motion. Imagine an observer sitting in a spaceship (Figure 5-19). Such an observer cannot distinguish between the force of gravity and the inertial forces produced by the acceleration of the spaceship. This led Einstein to conclude that gravity and motion through space-time are related, a conclusion now known as the equivalence principle:

> **Equivalence principle** Observers cannot distinguish locally between inertial forces due to acceleration and uniform gravitational forces due to the presence of a massive body.

The importance of the general theory of relativity lies in its description of gravity. Einstein concluded that gravity, inertia, and acceleration are all associated with the way space and time are related. This relation is often referred to as curvature, and a one-line description of general relativity explains a gravitational field as a curved region of space-time:

> **Gravity according to general relativity** Mass tells space-time how to curve, and the curvature of space-time (gravity) tells mass how to accelerate.

Thus, we feel gravity because the mass of the earth causes a curvature of space-time. The mass of our bodies responds to that curvature by accelerating toward the center of the earth. According to general relativity, all masses cause curvature, and the larger the mass, the more severe the curvature.

Confirmation of the Curvature of Space-Time

Einstein's general theory of relativity has been confirmed by a number of experiments, but two are worth mention-

a

b

FIGURE 5-20

(a) Mercury's orbit precesses 5600.73 seconds of arc per century—43.11 seconds of arc per century faster than predicted by Newton's laws. (b) Even when we ignore the influences of the other planets, Mercury's orbit is not a perfect ellipse. Curved space-time near the sun distorts the orbit from an ellipse into a rosette. The advance of Mercury's perihelion is exaggerated about a million times in this figure.

ing here because they were among the first tests of the theory. One involves Mercury's orbit, and the other involves eclipses of the sun.

Johannes Kepler understood that the orbit of Mercury is elliptical, but only since 1859 have astronomers known that the long axis of the orbit sweeps around the sun in a motion called precession (Figure 5-20). The total observed precession is 5600.73 seconds of arc per century (as seen from Earth), or about 1.5° per century. This precession is produced by the gravitation of Venus, Earth, and the other planets. However, when astronomers used Newton's description of gravity, they calculated that the precession should amount to only 5557.62 seconds of arc per century. Thus, Mercury's orbit is advancing 43.11 seconds of arc per century faster than Newton's law predicted.

◄ **FIGURE 5-19**

(a) An observer in a closed spaceship on the surface of a planet feels gravity. (b) In space, with the rockets smoothly firing and accelerating the spaceship, the observer feels inertial forces that are equivalent to gravitational forces.

TABLE 5 - 2

Precession in Excess of Newtonian Physics

Planet	Observed Excess Precession (seconds of arc per century)	Relativistic Prediction (seconds of arc per century)
Mercury	43.11 ± 0.45	43.03
Venus	8.4 ± 0.48	8.6
Earth	5.0 ± 1.2	3.8
Icarus	9.8 ± 0.8	10.3

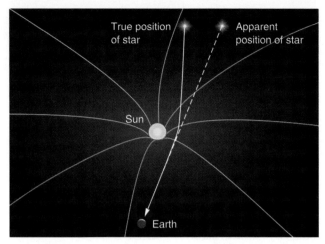

FIGURE 5-21

Like a depression in a putting green, the curved space-time near the sun deflects light from distant stars and makes them appear to lie in slightly different positions.

This is a tiny effect. Each time Mercury returns to perihelion, its closest point to the sun, it is about 29 km (18 miles) past the position predicted by Newton's laws. This is such a small distance compared with the planet's diameter of 4850 km that it could never have been detected had it not been cumulative. Each orbit, Mercury gains 29 km, and in a century it gains over 12,000 km—more than twice its own diameter. Thus, this tiny effect, called the advance of perihelion of Mercury's orbit, accumulated into a serious discrepancy in the Newtonian description of the universe.

The advance of perihelion of Mercury's orbit was one of the first problems to which Einstein applied the principles of general relativity. First he calculated how much the sun's mass curves space-time in the region of Mercury's orbit, and then he calculated how Mercury moves through the space-time. The theory predicted that the curved space-time should cause Mercury's orbit to advance by 43.03 seconds of arc per century, well within the observational accuracy of the excess.

Einstein was elated with this result, and he would be even happier with modern studies that have shown that Mercury, Venus, Earth, and even Icarus, an asteroid that comes close to the sun, have orbits observed to be slipping forward due to the curvature of space-time near the sun (Table 5-2).

This same effect has been detected in pairs of stars that orbit each other. In some cases, the advance of perihelion agrees with general relativity; in many cases, the sizes and masses of the stars are not well enough known for us to be certain of the theoretical rate of advance we should expect. But in a few cases, the stars' orbits appear to be changing faster than predicted. This may be a critical test for Einstein's theory. (This shows how science continues to test theories over and over even after they are widely accepted.)

A second test of the curvature of space-time was directly related to the motion of light through the curved space-time near the sun. The equations of general relativity predicted that light would be deflected by curved space-time just as a rolling golf ball is deflected by undu-

lations in a putting green (Figure 5-21). Einstein predicted that starlight grazing the sun's surface would be deflected by 1.75 seconds of arc. Starlight passing near the sun is normally lost in the sun's glare, but during a total solar eclipse stars beyond the sun could be seen. As soon as Einstein published his theory, astronomers rushed to observe such stars and thus test the curvature of space-time.

The first solar eclipse following Einstein's announcement in 1916 was June 8, 1918. It was cloudy. The next occurred on May 29, 1919, only months after the end of World War I, and was visible from Africa and South America. British teams went to both Brazil and Príncipe, an island off the coast of Africa. First, they photographed that part of the sky where the sun would be located during the eclipse and measured the positions of the stars on the plates. Then during the eclipse they photographed the same star field with the eclipsed sun located in the middle. After measuring the plates, they found slight changes in the positions of the stars. During the eclipse, the positions of the stars on the plates were shifted outward, away from the sun (Figure 5-22). If a star had been located at the edge of the solar disk, it would have been shifted outward by about 1.8 seconds of arc. This represents good agreement with the theory's prediction.

This test has been repeated at many total solar eclipses since 1919, with similar results. The most accurate results were obtained in 1973 when a Texas–Princeton team measured a deflection of 1.66 ± 0.18 seconds of arc—good agreement with Einstein's theory.

The general theory of relativity is critically important in modern astronomy. We will discuss it again when we meet black holes, distant galaxies, and the big bang universe. The theory revolutionized modern physics by

a

b

FIGURE 5-22

(a) Schematic drawing of the deflection of starlight by the sun's gravity. Dots show the true positions of the stars as photographed months before. Lines point toward the positions of the stars during the eclipse. (b) Actual data from the eclipse of 1922. Random errors of observation cause some scatter in the data, but in general the stars appear to move away from the sun by 1.77 seconds of arc at the edge of the sun's disk. The deflection of stars is magnified by a factor of 2300 in both (a) and (b).

providing a theory of gravity based on the geometry of curved space-time. Thus, Galileo's inertia and Newton's mutual gravitation are shown to be fundamental properties of space and time.

CRITICAL INQUIRY

What does the equivalence principle tell us?

The equivalence principle says that there is no observation we can make inside a closed spaceship to distinguish between uniform acceleration and gravitation. Of course, we could open a window and look outside, but then we would no longer be in a closed spaceship. As long as we make no outside observations, we can't tell whether our spaceship is firing its rockets and accelerating through space or resting on the surface of a planet where gravity gives us weight.

Einstein took the equivalence principle to mean that gravity and acceleration through space-time are somehow related. The general theory of relativity gives that relationship mathematical form and shows that gravity is really a disturbance in space-time that physicists refer to as curvature. Thus, we say "mass tells space-time how to curve, and space-time tells mass how to move." The equivalence principle led Einstein to an explanation for gravity.

But what about the second postulate of special relativity? Why does it have to be true if the first postulate is true? And what does the second postulate tell us about the nature of uniform motion? ■

Our discussion of the origin of astronomy began with the builders of Stonehenge and reaches the modern day with Einstein's general theory of relativity. Now that we have seen where astronomy came from, we are ready to see how it helps us understand the nature of the universe. Our first question should be "How do astronomers get information?" The answer involves the astronomer's most basic tool, the telescope; that is the subject of the next chapter. ■

■ *Summary*

Galileo took the first step toward understanding motion and gravity when he began to study falling bodies. He found that a falling object is accelerated; that is, it falls faster and faster with each passing second. The rate at which it accelerates, termed the acceleration of gravity, is 9.8 m/sec² (32 ft/sec²) at the earth's surface and does not depend on the weight of the object. According to tradition, Galileo demonstrated this by dropping balls of iron and wood from the Leaning Tower of Pisa to show that they would fall together. Finally, Galileo stated the law of inertia. In the absence of friction, a moving body on a horizontal plane will continue moving forever.

Newton adopted Galileo's law of inertia as his first law of motion. The second law of motion establishes the relationship between the force acting on a body, its mass, and the resulting acceleration. The third law says that forces occur in pairs acting in opposite directions.

Newton also developed an explanation for the accelerations that Galileo had discovered—gravity. By considering the motion of the moon, Newton was able to show that objects attract each other with a gravitational force that is proportional to the product of their masses and inversely proportional to the square of the distance between them.

If we understand Newton's laws of motion and gravity, we can better understand orbital motion. An object in space near the earth would move along a straight line and quickly leave the earth were it not for the earth's gravity accelerating the object toward the earth's center and forcing it to follow a curved path, an orbit. If there is no friction, the object will fall around its orbit forever.

Newton's laws also illuminate the meaning of Kepler's three laws of planetary motion. The planets follow elliptical orbits because gravity follows the inverse square law. The planets move faster when closer to the sun and slower when farther away because they conserve angular momentum. The same law makes ice skaters spin faster when they draw their arms and legs nearer their body. The planet's orbital period squared is proportional to their orbital radius cubed because the moving planets conserve energy.

In fact, Newtonian gravity and motion show that an ellipse is only one of a number of orbits that a body can follow. The circle and ellipse are closed orbits that return to their starting points. If an object moves at circular velocity, V_c, it will follow a circular orbit. If its velocity equals or exceeds the escape velocity, V_e, it will follow a parabola or hyperbola. These orbits are termed open because the object never returns to its starting place.

Newton's laws changed astronomy and our view of nature. They made it possible for astronomers to predict the motions of the heavenly bodies using the analytic power of mathematics. Newton's laws also show that the apparent complexity of the universe is based on a few simple principles, or natural laws.

Einstein published two theories that extended Newton's laws of motion and gravity. The special theory of relativity, published in 1905, applies to observers in uniform motion. The theory holds that the speed of light is a constant for all observers and that mass and energy are related by the expression $E = m_0c^2$.

The general theory of relativity, published in 1916, holds that a gravitational field is a curvature of space-time caused by the presence of a mass. Thus, the mass of the earth curves space-time, and the mass of our bodies responds to that curvature by accelerating toward the center of the earth. This curvature of space-time was confirmed by the slow advance in perihelion of the orbit of Mercury and by the deflection of starlight observed during a 1919 total solar eclipse.

▪ New Terms

natural motion	velocity
violent motion	inverse square law
acceleration of gravity	field
momentum	center of mass
mass	circular velocity
acceleration	geosynchronous orbit
closed orbit	angular momentum
perigee	special relativity
apogee	joule (J)
open (escape) orbit	general theory of relativity
escape velocity	

▪ Questions

1. Why wouldn't Aristotle's explanation of gravity work if the earth was not the center of the universe?

2. According to the principles of Aristotle, what part of the motion of a baseball pitched across the home plate is natural motion? What part is violent motion?

3. If we drop a feather and a steel hammer at the same moment, they should hit the ground at the same instant. Why doesn't this work on the earth, and why does it work on the moon?

4. What is the difference between mass and weight? between speed and velocity?

5. Why did Newton conclude that some force had to pull the moon toward the earth?

6. Why did Newton conclude that gravity had to be mutual and universal?

7. How does the concept of a field explain action at a distance? Name another kind of field also associated with action at a distance.

8. Why can't a spacecraft go "beyond the earth's gravity"?

9. What is the center of mass of the earth–moon system? Where is it?

10. How do planets orbiting the sun and skaters conserve angular momentum?

11. Why is the period of an open orbit undefined?

12. How does the first postulate of special relativity imply the second?

13. When we ride a fast elevator upward, we feel slightly heavier as the trip begins and slightly lighter as the trip ends. How is this phenomenon related to the equivalence principle?

14. From your knowledge of general relativity, would you expect radio waves from distant galaxies to be deflected as they pass near the sun? Why or why not?

▪ Discussion Questions

1. How did Galileo idealize his inclines to conclude that an object in motion stays in motion until it is acted on by some force?

2. Give an example from everyday life to illustrate each of Newton's laws.

▪ Problems

1. Compared with the strength of the earth's gravity at its surface, how much weaker is gravity at a distance of 10 earth radii from the center of the earth? at 20 earth radii?

2. If a lead ball falls from a high tower on earth, what will be its velocity after 2 seconds? after 4 seconds?

3. What is the circular velocity of an earth satellite 1000 km above the earth's surface? (HINT: The radius of the earth is 6380 km.)

4. Calculate the circular velocity of an earth satellite orbiting 36,000 km above the earth's surface. What is its orbital period?

5. Describe the orbit followed by the slowest cannonball in Figure 5-8 on the assumption that the cannonball could pass freely through the earth. (Newton got this problem wrong the first time he tried to solve it.)

6. If you visited an asteroid 30 km in radius with a mass of 4×10^{17} kg, what would be the circular velocity at its surface? A major league fastball travels 90 mph. Could a good pitcher throw a baseball into orbit around the asteroid?

7. What would be the escape velocity at the surface of the asteriod in Problem 6? Could a major league pitcher throw a baseball off of the asteroid?

▪ *Critical Inquiries to the Web*

1. Einstein's general theory of relativity predicts the curvature of space-time, but here on Earth we have little opportunity to observe such effects. Find an astronomical situation in which space-time curvature is evident from our observations, and describe the effect of the curvature on what we see when we view these objects.

2. Communications satellites are obvious uses of the geosynchronous orbit, but can you think of other uses for such orbits? Find an Internet site that uses or displays information gleaned from geosynchronous orbit that provides a useful service.

Go to the Wadsworth Astronomy Resource Center (www.wadsworth.com/astronomy) for critical thinking exercises, articles, and additional readings from InfoTrac College Edition, Wadsworth's online student library.

Chapter Six

LIGHT AND TELESCOPES

GUIDEPOST

Previous chapters have described the sky as it appears to our unaided eyes, but modern astronomers turn powerful telescopes on the sky. Chapter 6 introduces us to the modern astronomical telescope and its delicate instruments.

The study of the universe is so challenging, astronomers cannot ignore any

He burned his house down for the

fire insurance

And spent the proceeds on

a telescope.

ROBERT FROST

The Star-Splitter

source of information; that is why they use the entire spectrum, from gamma rays to radio waves. This chapter shows how critical it is for astronomers to understand the nature of light.

In each of the 22 chapters that follow, we will study the universe using information gathered by the telescopes and instruments described in this chapter. ∎

What do fleas living on rats have to do with modern astronomy? That may sound like the beginning of a bad joke, but it is actually related to the subject of this chapter. We will examine the tools that modern astronomers use, and those tools are connected by an interesting sequence of events to rats and their fleas.

The most horrible disease in history—the black plague—was spread by flea bites, and the fleas lived on the rats that infested the cities. Plague broke out in London in 1665, and, although no one knew how the disease spread, people who lived in the country were less likely to get the plague than city dwellers. All who could left the cities. When the plague reached Cambridge, the colleges were closed, and both students and faculty fled to the English countryside. One who fled was the young Isaac Newton. From the summer of 1665 to 1667, he spent most of his time in his mother's cottage in the small village of Woolsthorpe. While he was there, he conducted an experiment that changed the history of science.

Boring a hole in a shutter, he admitted a thin beam of sunlight into his darkened room. A glass prism placed in the beam threw a rainbow of color—a spectrum—across the wall. When he used a second prism to recombine the colors, they produced white light. From this and other experiments conducted in his bedroom, he concluded that white light was made up of a mixture of all the colors of the rainbow.

When the plague abated, Newton returned to the university, where he began experimenting with telescopes. He discovered that telescopes made of lenses produced colored fringes around bright objects in the field of view because the glass lenses broke the light into colors just as his prism broke up the sunlight. To solve the problem, Newton designed and built a telescope containing a mirror instead of a lens. Although his first model hardly exceeded 1 inch in diameter, when Newton presented it to the Royal Society in 1671, it established his reputation as a scientist.

For a century after Newton's first telescope, astronomers did little with such devices, but as instrument makers grew more skilled, large telescopes became the principal tool of the astronomer. Telescopes are important in astronomy because they gather light and concentrate it for study. The larger the telescope, the more light it gathers. Thus astronomers are still striving to build bigger telescopes to gather more light from the objects in the sky. Like Newton's original telescope, almost all modern telescopes use mirrors rather than lenses to avoid spreading the light into its component colors.

6-1 RADIATION: INFORMATION FROM SPACE

Astronomers are in the light business. Almost everything we know about the universe, we learn by analyzing the light gathered by telescopes. Thus, to understand astronomy, we must understand light.

Electromagnetic Radiation

Light is merely one form of radiation called **electromagnetic radiation** because it is associated with changing electric and magnetic fields that travel through space and transfer energy from one place to another. When light enters our eye, the fluctuating electric and magnetic fields carry energy that stimulates nerve endings, and we see what we call light.

The oscillating electric and magnetic fields that constitute electromagnetic radiation move through space at about 300,000 km/sec (186,000 m/sec). This speed is commonly referred to as the speed of light c, but it is in fact the speed of all such radiation in a vacuum.

Electromagnetic radiation is a wave phenomenon; that is, it is associated with a periodically repeating disturbance, or wave. We are familiar with waves in water. If we disturb a quiet pool of water, waves spread across the surface. Imagine that we use a meter stick to measure the distance between the successive peaks of a wave. This distance is the **wavelength,** usually represented by the Greek letter lambda (λ). If we were measuring ripples in a pond, we might find that the wavelength is a few centimeters, whereas the wavelength of ocean waves might be a hundred meters or more. There is no restriction on the wavelength of electromagnetic radiation. Wavelengths can range from smaller than the diameter of an atom to larger than that of the earth.

Because all electromagnetic radiation travels at the speed of light, wavelength is related to **frequency,** the number of cycles that pass in one second. Short-wavelength radiation has a high frequency; long-wavelength radiation has a low frequency. To understand this, imagine watching an electromagnetic wave race past us while we count its peaks (Figure 6-1). If the wavelength is short, we will count many peaks in 1 second; if the wavelength is long, we will count few peaks per second.

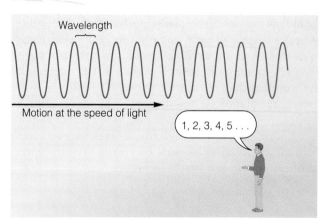

FIGURE 6-1

All electromagnetic waves travel at the speed of light. The wavelength is the distance between successive peaks. The frequency of the wave is the number of peaks that pass us in 1 second.

The dials on radios are marked in frequency, but they could just as easily be marked in wavelength. The relation between wavelength and frequency is a simple one:

$$\lambda = \frac{c}{f}$$

That is, the wavelength equals the speed of light c divided by the frequency f. Notice that the larger (higher) the frequency, the smaller (shorter) the wavelength. In most cases, astronomers use wavelength rather than frequency.

Radio waves can have wavelengths from a few millimeters for microwaves to kilometers. In contrast, the wavelength of light is so short that we must use more convenient units. In this book, we will use **nanometers (nm)** because this unit is consistent with the International System of units. One nanometers is 10^{-9} meter, and visible light has wavelengths that range from about 400 nm to about 700 nm. Another unit that astronomers commonly use, and a unit that you will see in many references on astronomy, is the **Angstrom (Å).** One Angstrom is 10^{-10} meter, and visible light has wavelengths between 4000 Å and 7000 Å.

You may find radio astronomers describing wavelengths in centimeters or millimeters, and infrared astronomers often refer to wavelengths in micrometers (or microns). One micrometer (μm) is 10^{-6} meter. Whatever unit is used to describe the wavelength, we must keep in mind that all electromagnetic radiation is the same phenomenon.

What exactly is electromagnetic radiation? Although we have been discussing its wavelength, it is incorrect, or at least incomplete, to say that electromagnetic radiation is a wave. It sometimes has the properties of a wave and sometimes has the properties of a particle. For instance, the beautiful colors in a soap bubble arise from the wave nature of light. On the other hand, when light strikes the photoelectric cell in a camera's light meter, it behaves like a stream of particles carrying specific amounts of energy. Throughout his life, Newton believed that light was made up of particles, but we now recognize that light can behave as both particle and wave. Our model of light is thus more complete than Newton's. We will refer to "a particle of light" as a **photon,** and we can recognize its dual nature by thinking of it as a bundle of waves.

The amount of energy a photon carries depends on its wavelength. The shorter the wavelength, the more energy the photon carries; the longer the wavelength, the less energy it contains. This is easy to remember because short wavelengths have high frequencies, and we expect rapid fluctuations to be more energetic. We can express this relationship in a simple formula:

$$E = \frac{hc}{\lambda}$$

Here h is Planck's constant (6.6262×10^{-34} joule sec), c is the speed of light (3×10^8 m/sec), and λ is the wavelength in meters. A photon of visible light carries a very small amount of energy, but a photon with a very short wavelength can carry much more.

The Electromagnetic Spectrum

A spectrum is an array of electromagnetic radiation in order of wavelength. We are most familiar with the spectrum of visible light, which we see in rainbows, for instance, but the visible spectrum is merely a small segment of the much larger electromagnetic spectrum (Figure 6-2).

The average wavelength of visible light is about 0.00005 cm. We could put 50 light waves end to end across the thickness of a sheet of household plastic wrap. Measured in nanometers, the wavelength of visible light ranges from about 400–700 nm. Just as we sense the wavelength of sound as pitch, we sense the wavelength of light as color. Light near the short-wavelength end of the visible spectrum (400 nm) looks violet to our eyes, and light near the long-wavelength end (700 nm) looks red (Figure 6-2).

Beyond the red end of the visible spectrum lies **infrared radiation,** where wavelengths range from 700 nm to about 0.1 cm. Our eyes are not sensitive to this radiation, but our skin senses it as heat. A "heat lamp" is just a bulb that gives off principally infrared radiation.

Beyond the infrared part of the electromagnetic spectrum lie radio waves. Microwaves have wavelengths of a millimeter to a few centimeters and are used for radar and long-distance telephone communication. Longer wavelengths are used for UHF and VHF television transmissions. FM, military, governmental, and ham radio signals have wavelengths up to a few meters, and AM radio waves can have wavelengths of kilometers.

The distinction between the wavelength ranges is not sharp. Long-wavelength infrared radiation and the shortest microwave radio waves are the same. Similarly, there is no clear division between the short-wavelength infrared and the long-wavelength part of the visible spectrum. It is all electromagnetic radiation.

At wavelengths shorter than violet, we find **ultraviolet radiation,** with wavelengths ranging from 400 nm down to about 10 nm. At even shorter wavelengths lie X rays and gamma rays. Again, the boundaries between these wavelength ranges are not clearly defined.

X rays and gamma rays can be dangerous, and even ultraviolet photons have enough energy to do us harm. Small doses produce a suntan, sunburn, and skin cancers. Contrast this to the lower-energy infrared photons. Individually they have too little energy to affect skin pigment, a fact that explains why you can't get a tan from a heat lamp. Only by concentrating many low-energy photons in a small area, as in a microwave oven, can we transfer significant amounts of energy.

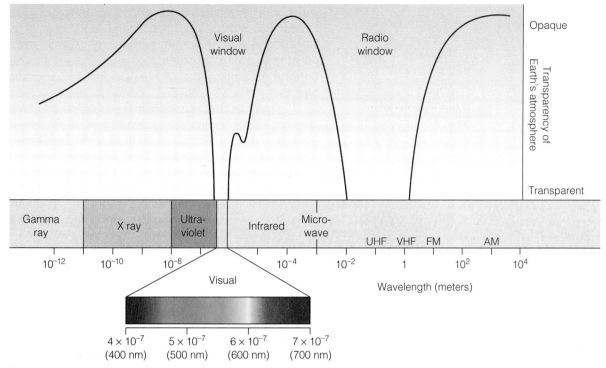

FIGURE 6-2

The electromagnetic spectrum includes all wavelengths of electromagnetic radiation. The earth's atmosphere is relatively opaque at most wavelengths. Visual and radio windows allow light and some radio waves to reach the earth's surface.

We are interested in electromagnetic radiation because it brings us clues to the nature of stars, planets, and other celestial objects. However, only a small part of this radiation can get through the earth's atmosphere. Only visible light and some radio waves can reach the surface of the earth; other wavelengths are absorbed. The highest parts of the atmosphere absorb X rays, gamma rays, and some radio waves, and a layer of ozone (O_3) at an altitude of about 30 km absorbs ultraviolet radiation. In addition, water vapor in the lower atmosphere absorbs infrared radiation. The wavelength regions in which our atmosphere is transparent are called **atmospheric windows** (Figure 6-2). If we wish to study the sky from the earth's surface, we must look out through one of these windows.

CRITICAL INQUIRY

Why can't astronomers on the earth's surface observe the stars at ultraviolet and X-ray wavelengths?

The earth's atmosphere is transparent to visible light, and starlight can reach the earth's surface. That means that when we look upward on a clear, dark night, the starlight can enter our eyes and we can see the stars. But the earth's atmosphere is not transparent at ultraviolet and X-ray wavelengths. Most ultraviolet radiation is

absorbed in the ozone layer, and nearly all X-ray photons are absorbed by the highest layers of the earth's atmosphere. If our eyes worked only at ultraviolet and X-ray wavelengths, we would live in a dark world; when we looked upward at the stars, we would see nothing but the opaque atmosphere above us.

Suppose our eyes were sensitive only to radiation at infrared or radio wavelengths. Would we be able to see stars? Before you answer, study the atmospheric windows shown in Figure 6-2. ■

Having explored the nature of electromagnetic radiation and the electromagnetic spectrum, we can now study the tools astronomers use to analyze radiation.

6-2 ASTRONOMICAL TELESCOPES

The telescope is the symbol of the astronomer. Later in this chapter, we will see how astronomers are using special telescopes to observe radio waves and X rays and how some telescopes are venturing into space. Here, however,

FIGURE 6-3

(a) A refracting telescope uses a primary lens to focus starlight into an image that is magnified by a lens called an eyepiece. The primary lens has a long focal length, and the eyepiece has a short focal length. (b) A reflecting telescope uses a primary mirror to focus the light by reflection. A small secondary mirror reflects the starlight back down through a hole in the middle of the primary lens to the eyepiece.

we consider the traditional earth-based astronomical telescope that gathers visible light.

Astronomers have used two kinds of telescopes (Figure 6-3). Most early astronomical telescopes, beginning with those made by Galileo, were **refracting telescopes** using a lens to bend the light to a focus. The newer **reflecting telescopes** use a concave mirror to reflect the light to a focus. Reflecting telescopes have some strong advantages.

Refracting Telescopes

The main element in a refracting telescope is a lens, a piece of glass carefully shaped so that light striking it is refracted, or bent, into an image. The **focal length** of a lens is the distance from the lens to the point where it focuses parallel rays of light. If the surface of the lens is strongly curved, the light is brought to a focus close to the lens, and we say it has a short focal length. If the glass is less strongly curved, the lens has a longer focal length, and the image is formed farther from the lens.

The image in an astronomical telescope is inverted (Figure 6-4). This is true for all such optical systems,

whether they use a lens or a mirror. Even your eye forms an inverted image. The image could be reinverted by an extra lens, but this is an unnecessary expense and a possible source of distortion. Consequently, most astronomical telescopes, like microscopes, produce inverted images.

To build a refracting telescope, we need a lens of relatively long focal length to form an image of the object we wish to view. This lens is often called the **objective lens** because it is closest to the object. To view the image, we add a lens of short focal length called an **eyepiece** to enlarge the image and make it easy to see (Figure 6-3a). Thus, the eyepiece acts as a magnifier. By changing eyepieces, we can easily change the magnification of the telescope.

Refracting telescopes suffer from a serious optical distortion (aberration) that limits what we can see through them. When light is refracted through glass, shorter wavelengths bend more than longer wavelengths, and blue light comes to a focus closer to the lens than does red light (Figure 6-5a). If we focus the eyepiece on the blue image, the red light is out of focus, producing a red blur around the image. If we focus on the red image, the blue light blurs. This color separation is called **chromatic aberration.**

A telescope designer can partially correct for this aberration by replacing the single objective lens with one made of two lenses ground from different kinds of glass. Such lenses, called **achromatic lenses,** can be designed to bring any two colors to the same focus (Figure 6-5b). Because our eyes are most sensitive to red and yellow light, we might bring these two colors to the same focus, but blue and violet would still be out of focus, producing a hazy blue fringe around bright objects.

Refracting telescopes were popular through the 19th century, but they are no longer economical for professional astronomy. A large achromatic lens is very expensive because it contains four matched optical surfaces and must be made of high-quality glass. Refractors can't be made larger than about 1 m in diameter because such large lenses sag under their own weight. Also, large refractors have very long telescope tubes that require large observatory domes. Modern telescopes, like Newton's first telescope, focus light with mirrors.

Reflecting Telescopes

In a reflecting telescope, a concave mirror, the **objective mirror,** focuses the starlight into an image that can be viewed with an eyepiece (Figure 6-3b). Objective mirrors are usually made of special kinds of glass or quartz covered with a thin layer of aluminum to act as a reflecting surface.

It is relatively easy to grind an astronomical mirror to form a concave surface that is spherical; that is, a cross section is a segment of a circle. But such mirrors suffer from **spherical aberration** in that they cannot bring all light rays to the same focus. To correct such a mirror, the center must be ground a few millionths of a meter

FIGURE 6-4
(a) To see how a lens can focus light, we trace three light rays from the flame and base of a candle through a lens, where they are refracted to form an inverted image. (b) A mirror forms an image by reflection from a concave surface. Notice that the light is reflected from the aluminized front surface of the mirror and does not enter the glass. Thus, mirrors do not produce chromatic aberration.

deeper to make the shape parabolic; that is, a cross section of the mirror is a parabola. Then all the rays of light from a star are brought to a single, sharp focus.

The objective mirror forms an image at the location called the **prime focus** at the upper end of the telescope tube (Figure 6-6). Because it is usually inconvenient to view the image there, a smaller **secondary mirror** reflects the light to a more accessible location. In one popular arrangement, the secondary mirror reflects the light back down the telescope tube through a hole in the center of the objective mirror. This kind of telescope is called a **Cassegrain telescope** (Figure 6-7).

Other focal arrangements are also common (Figure 6-6). The largest telescopes usually have prime focus cages where the astronomer can ride inside the telescope tube to observe very faint objects. The **Newtonian focus,** named after Newton's first telescope, is common for smaller telescopes but awkward for large instruments. The **coudé focus** (from the French word for "elbow") bends the light path and sends it to a separate observing room, where very large or very delicate instruments such as spectrographs can be used. Most large telescopes can be used in more than one focal position, but **Schmidt cameras** are especially designed for wide-field photography and can't be changed to other arrangements. The **Schmidt–Cassegrain telescope** combines the Schmidt's thin correcting lens with the Cassegrain's perforated mirror and is very popular for small telescopes. Many amateur astronomers own Schmidt–Cassegrain telescopes.

Nearly all recently built telescopes are reflectors. Because the light does not enter the glass, there is no

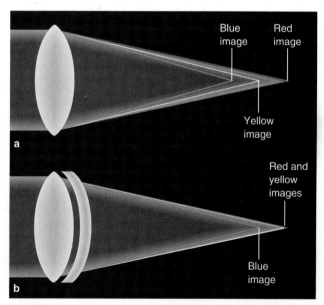

FIGURE 6-5
(a) A normal lens suffers from chromatic aberration because short wavelengths bend more than long wavelengths. (b) An achromatic lens, made in two parts, can bring any two colors to the same focus, but other colors remain slightly out of focus.

chromatic aberration, the glass need not be of perfect optical quality, and the mirror can be supported over its back to reduce sagging. Also, reflecting telescopes tend to be shorter and thus require smaller mountings and smaller observatory buildings.

FIGURE 6-6

Many large telescopes can be used at a number of focal positions. Schmidt cameras, however, are used only for photography and can't be changed. The Schmidt–Cassegrain arrangement is popular for small telescopes.

a

b

FIGURE 6-7

(a) An astronomer, C. R. Lynds, adjusts an instrument mounted on the Cassegrain observing cage beneath the objective mirror of the 4-m Mayall Telescope at Kitt Peak National Observatory in Arizona. (b) During observations, astronomers run the telescope and the instruments mounted on it remotely from a nearby control room. (© Association of Universities for Research in Astronomy, Inc., Kitt Peak National Observatory)

Telescope Mountings

A telescope mounting can be more expensive than the telescope itself. The mounting must support the optics, protect against vibration, move accurately to designated objects, and then compensate for the earth's rotation. Because the earth rotates eastward, the telescope mounting must contain a **sidereal drive** (literally, a star drive) that can move the telescope smoothly westward to follow the object being studied. To simplify this motion, most telescope mountings are **equatorial mountings** with one axis of rotation, the **polar axis,** parallel to the earth's axis. Rotation around the polar axis moves the telescope parallel to the celestial equator (Figure 6-8).

Improvements in computer technology are making telescope mountings and observatories much less expensive. Instead of building a large, awkward equatorial mounting inside a large, heavy dome, astronomers now take advantage of the short focal lengths typical of reflecting telescopes. Such telescopes are so short they can be mounted like a cannon, that is, the mounting can move the telescope both in altitude (perpendicular to the horizon) and in azimuth (parallel to the horizon). Such **alt-azimuth mountings** are compact, but they must move simultaneously about both axes at varying rates in order to track stars in different parts of the sky. This complicated motion is made possible by computers that guide the telescope.

Such computer-controlled alt-azimuth mountings are stronger and smaller and do not require buildings as large as those required by equatorial mountings. In fact, many of the newest telescopes are designed so that the telescope is part of the observatory building. Like the turret of a tank, the entire building rotates as the telescope moves.

Further advances in optics and in computer technology are changing the way astronomers build telescopes.

New-Generation Telescopes

For most of this century, astronomers faced a serious limitation in the size of telescopes. A large telescope mirror sags under its own weight and cannot focus light accurately. To reduce this sagging, astronomers made telescope mirrors very thick, but that produced two further problems—weight and cost.

Supporting such a heavy mirror required a massive telescope tube and mounting. The two largest conventional telescopes in the world are the 6-m (236 in.) reflector in the former Soviet Union and the 5-m (200-in.) Hale Telescope on Mount Palomar (Figure 6-9). The 5-m mirror alone weighs 14.5 tons, and its mounting weighs about 530 tons.

Grinding a conventional telescope mirror to shape is expensive because it requires the slow removal of large amounts of glass. The 5-m mirror, for instance, was begun in 1934 and finished in 1948, and 5 tons of glass were ground away.

FIGURE 6-8

Westward motion around the polar axis of an equatorial mounting counters the eastward rotation of the earth and keeps the telescope pointed at a given star.

a

b

FIGURE 6-9

Until recent decades, telescope mirrors were single, thick pieces of glass that required massive telescope mountings and large observatory domes. (a) The world's largest traditional telescope is the 6-m reflector built by the former U.S.S.R. (b) The 5-m (200-in.) Hale Telescope on Mount Palomar contains a mirror weighing 14.5 tons. The telescope dome weighs 1000 tons.

FIGURE 6-10

The problem and two solutions: Conventional telescope mirrors (a) would sag out of shape if they were not very thick. To reduce the weight and build bigger mirrors, astronomers have begun making mirrors of segments (b) or of relatively thin disks (c). The mirrors' shape is maintained by computer-controlled thrusters (red) positioned under the mirrors.

Astronomers are developing new techniques to make large mirrors that weigh and cost less. For example, a revolving oven under the football stands at the University of Arizona is now producing preshaped mirrors. The oven turns like a merry-go-round, and the molten glass flows outward in the mold to form a concave upper surface. Once cooled, the spun-cast mirrors can be ground quickly to final shape. The oven can cast preshaped mirrors up to 8 meters in diameter.

An 8-m diameter mirror of conventional thickness would be too heavy to support in a telescope, so astronomers have devised ways to make the mirrors thinner (Figure 6-10). One technique is to make the mirror in segments. Small segments are less expensive and sag less under their own weight. The largest optical telescope in the world, the 10-m Keck Telescope in Hawaii, uses 36 hexagonal mirror segments held in alignment by computer-controlled thrusters to form a single mirror (Figure 6-11a). This telescope has been so successful that a companion 10-m telescope has been built nearby. Another giant telescope with a segmented mirror is the Hobby–Eberly Telescope (HET), built in Texas by a consortium of five major universities (Figure 6-11b). This new telescope uses 91 hexagonal segments to make a mirror 11 m in diameter. The telescope will be used exclusively to record the spectra of stars.

Another way to reduce the weight of telescope mirrors is to make them thin. Thin mirrors sag so easily under their own weight they are called floppy mirrors, but a computer can control their shape in what astronomers call **active optics.** The New Technology Telescope at the European Southern Observatory in Chile contains a 3.58-m (141-in.) mirror that is only 24 cm (10 in.) thick. With a computer controlling the mirror shape, the telescope has produced very high quality images.

The Multiple Mirror Telescope (MMT) was originally built using six round mirrors, each 1.8 m (72 in.) in diameter, on one mounting (Figure 6-11c). The mirrors could combine their light to produce the equivalent of a 4.5-m (176-in.) telescope. The success of active optics and floppy mirrors has led astronomers to replace the six mirrors with a single thin mirror 6.5 m (256 in.) in diameter.

Segmented or floppy, thin telescope mirrors have yet another advantage—they cool quickly. During the day, a telescope mirror warms to air temperature, but after sunset the air cools quickly. A massive telescope mirror cannot cool rapidly and uniformly, and because some parts of the mirror cool faster than other parts, the material expands and contracts, producing strains in the mirror that distort the image. Thin telescope mirrors cool more quickly and more uniformly as night falls and thus produce better images.

The success of active optics, segmented mirrors, and floppy mirrors has stimulated the construction of a number of large telescopes. The European Southern Observatory is building four 8.2-m telescopes on a mountaintop in Chile. The four telescopes will be able to observe either individually or as a single telescope with an equivalent diameter of 16.4 m. U.S. astronomers and international collaborators are building the Gemini telescopes, a pair of 8-m telescopes located in Hawaii and Chile. From those two sites, the Gemini telescopes will be able to scan the entire sky. Japanese astronomers are building the Subaru (Pleiades) Telescope, which will contain an 8.3-m mirror only 23 cm (9 in.) thick.

Still more giant telescopes are being planned and built thanks to new advances in optics and computers. All of these giant reflecting telescopes are designed to give human eyes the power to explore the secrets of the sky.

The Powers of a Telescope

Whether refractor or reflector, a telescope aids our eyes in three ways: light-gathering power, resolving power, and magnifying power.

Light-gathering power refers to the ability of a telescope to collect light. Nearly all the interesting

a

b

c

FIGURE 6-11

Segmented mirrors can be assembled to create very large diameter telescopes. (a) The two Keck Telescopes in Hawaii contain mirrors made up of hexagonal segments with a total diameter of 10 m. Find the human figure at the left. (Matt Weinberg/Santa Barbara Research Corp.) (b) The hexagonal mirror segments in the keck I Telescope are clearly visible in this photo. The human figure is crouching in the Cassegrain opening at the center of the mirror. Each segment is 1.8 m (70 inches) in diameter and is supported and aligned by computer-controlled actuators (© Russ Underwood/W. M. Keck Observatory) (c) The 91 hexagonal mirrors of the Hobby-Eberly Telescope in Texas reflect the structure of the interior of the telescope dome. Spanning 11 meters, the segmented mirror is specially designed to feed starlight into either of two state-of-the-art spectrographs. (Photo courtesy Dr. Thomas G. Barnes III, McDonald Observatory, The University of Texas at Austin)

objects in the sky are faint, so we need a telescope that can gather large amounts of light to produce a bright image. Catching light in a telescope is like catching rain in a bucket—the bigger the bucket, the more it catches (Figure 6-12). This is the main reason why astronomers use large telescopes.

Light-gathering power is proportional to the area of the telescope objective. A lens or mirror with a large area gathers a large amount of light. Because the area of a circle is proportional to the square of its diameter, the light-gathering power of a telescope is proportional to the square of the diameter of its objective. We compare two telescopes by comparing the squares of their diameters. That is, the ratio of the light-gathering powers of two

telescopes A and B is equal to the ratio of their diameters squared:

$$\frac{\text{LGP}_A}{\text{LGP}_B} = \left(\frac{D_A}{D_B}\right)^2$$

For example, suppose we compare a telescope 24 cm in diameter with a telescope 4 cm in diameter. The ratio of the diameters is 24/4, or 6, but the larger telescope does not gather 6 times as much light. Light-gathering power increases as the ratio of diameters squared, so it gathers 36 times more light than the smaller telescope.

This example shows the importance of diameter in astronomical telescopes. Even a small increase in diameter produces a large increase in light-gathering power and allows astronomers to study much fainter objects.

Resolving power, the second telescopic power, refers to the ability of a telescope to reveal fine detail. Whenever light is focused to form an image, a small, blurred fringe surrounds the image (Figure 6-13). Because this **diffraction fringe** surrounds every point of light in the image, we cannot see fine detail. There is nothing we can do to eliminate diffraction fringes; they are produced by the wave nature of light as it passes through the telescope. If we use a large-diameter telescope, however, the fringes are smaller and we can see smaller details. Thus, the larger the telescope, the better its resolving power. Resolving power also depends on wavelength. Telescopes that observe at longer wavelengths have larger fringes and poorer resolving power.

For optical telescopes, we estimate the resolving power by calculating the angular distance between two stars that are just barely visible through the telescope as two separate images. The resolving power, α, in seconds of arc, equals 11.6 divided by the diameter of the telescope in centimeters:

$$\alpha = \frac{11.6}{D}$$

For example, the resolving power of a 25-cm telescope is 11.6 divided by 25, or 0.46 second of arc. No matter how perfect the telescope optics, this is the smallest detail we can see through that telescope.

In addition to resolving power, two other factors—lens quality and atmospheric conditions—limit the detail we can see through a telescope. A telescope must contain high-quality optics to achieve its full potential resolving power. Even a large telescope shows us little detail if its optics are marred with imperfections. Also, when we look through a telescope, we are looking through miles of turbulent air in the earth's atmosphere, which makes the image dance and blur, a condition called **seeing.** On a night when the atmosphere is unsteady and the images are blurred, the seeing is bad. Even under good seeing conditions, the detail visible through a large telescope is limited, not by its diffraction fringes, but by the air through which the observer must look. A telescope performs better on a high mountaintop where the air is thin and steady, but even there the earth's atmosphere limits the detail the best telescopes can reveal to about 0.5 second of arc (Window on Science 6-1).

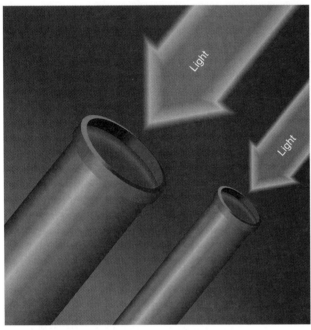

FIGURE 6-12
Gathering light is like catching rain in a bucket. A large-diameter telescope gathers more light and has a brighter image than a smaller telescope of the same focal length.

FIGURE 6-13
(a) The stars are so far away, we should see nothing but a point of light. However, because of the wave nature of light, every star image is a tiny disk surrounded by diffraction fringes (much magnified in this computer model).
(b) When two stars lie close to each other, their diffraction fringes overlap, and the stars become impossible to detect separately. (Computer graphics by M. A. Seeds)

The Resolution of a Measurement

Have you ever seen a movie in which the hero magnifies a newspaper photo and reads some tiny detail? It isn't really possible, because newspaper photos are made up of tiny dots of ink, and no detail smaller than a single dot will be visible no matter how much you magnify the photo. In fact, all images are made up of elements of some sort, and that means there is a limit to the amount of detail you can see in an image. In an astronomical image, the resolution is often set by seeing. It is foolish to attempt to see a detail in the image that is smaller than the resolution.

This limitation is true of all measurements in science. A zoologist might be trying to measure the length of a live snake, or a sociologist might be trying to measure the attitudes of people toward drunk driving, but both face limits to the resolution of their measurements. The zoologist might specify that the snake was 43.28932 cm long, and the sociologist might say that 98.2491 percent of people oppose drunk driving, but a critic might point out that it isn't possible to make these measurements that accurately. The resolution of the techniques does not justify the accuracy implied.

Science is based on measurement, and whenever we make a measurement we should ask ourselves how accurate that measurement can be. The accuracy of the measurement is limited by the resolution of the measurement technique, just as the amount of detail in a photograph is limited by the resolution of the photo. ∎

A technique called **adaptive optics** uses a high-speed computer to monitor atmospheric distortion and adjust the optics to partially compensate for seeing. This can produce dramatically sharper images. Of course, the ultimate solution to the problem of atmospheric seeing is to send the telescope above the earth's atmosphere. We will discuss space telescopes later in this chapter.

The third and least important power of a telescope is **magnifying power,** the ability to make the image bigger. Because the amount of detail we can see is limited by the seeing conditions and the resolving power, very high magnification does not necessarily show us more detail. Also, we can change the magnification by changing the eyepiece, but we cannot alter the telescope's light-gathering power or resolving power.

We calculate the magnification of a telescope by dividing the focal length of the objective by the focal length of the eyepiece:

$$M = \frac{F_o}{F_e}$$

For example, if a telescope has an objective with a focal length of 80 cm and we use an eyepiece whose focal length is 0.5 cm, the magnification is 80/0.5, or 160 times.

If you visit a department store to shop for a telescope, you will probably find telescopes described according to magnification. One might be labeled an "80-power telescope" and another a "40-power telescope." However, the magnifying power really tells you little about the telescopes. Astronomers identify telescopes by diameter, because that determines both light-gathering power and resolving power.

CRITICAL INQUIRY

Why do astronomers build observatories at the tops of mountains?

Astronomers have joked that the hardest part of building a new observatory is constructing the road to the top of the mountain. It certainly isn't easy to build a large, delicate telescope at the top of a high mountain, but it is worth the effort. A telescope on top of a high mountain is above the thickest part of the earth's atmosphere. There is less air to dim the light, and there is less water vapor to absorb infrared radiation. Even more important, the thin air on a mountaintop causes less disturbance to the image, and thus the seeing is better. A large telescope on the earth's surface has a resolving power much better than the distortion caused by the earth's atmosphere. Thus, it is limited by seeing, not by its own diffraction. It really is worth the trouble to build telescopes atop high mountains.

Astronomers not only build telescopes on mountaintops; they also build gigantic telescopes many meters in diameter. What are the problems and advantages in building such giant telescopes? ∎

Astronomers sometimes refer to a telescope that produces distorted images as a "light bucket." In a sense, all astronomical telescopes are just light buckets, because the light they focus into images tells us very little until it is recorded and analyzed by special instruments attached to the telescopes.

6-3 SPECIAL INSTRUMENTS

Looking through a telescope doesn't tell us much. To use an astronomical telescope to learn about stars, we must be able to analyze the light the telescope gathers. Special instruments attached to the telescope make that possible.

Imaging Systems

The original imaging device in astronomy was the photographic plate. It could record faint objects in long exposures and could be stored for later analysis. But photographic plates have been almost entirely replaced in astronomy by electronic imaging systems.

Low-light television cameras were the first electronic replacements for the photographic plate, but the newest and most common system used in astronomy is a **charge-coupled device (CCD)** (Figure 6-14a)—typically a quarter-million microscopic, light-sensitive diodes in an array about the size of a postage stamp. These devices can be used like a small photographic plate, but they have dramatic advantages. They can detect both bright and faint objects in a single exposure, are much

more sensitive than photographic plates, and can be read directly into computer memory for later analysis.

However an image is recorded, astronomers often reproduce the image in exaggerated ways to bring out subtle details. If an image of a faint object is reproduced as a negative, the sky is white and the stars are dark (Figure 6-14b). This makes the faint parts of the image easier to see. With a computer, astronomers can easily manipulate an image to produce **false-color** images (Figure 6-14c). The colors in such images are merely codes to intensity and are not related to the true colors of the objects. To measure the true colors, we need a photometer.

The Photometer

A **photometer** is nothing more than a sensitive light meter that can measure the brightness and color of stars. Such a photometer contains a sensitive detector that

a

b

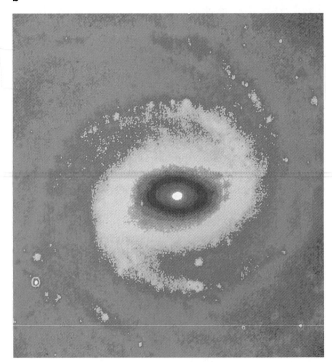

c

FIGURE 6-14

(a) This large CCD chip contains over 4 million light-sensitive detectors in an array 55.3 mm square. (Copyright ©1986 Tektronix, Inc. All rights reserved. Reprinted by permission of Tektronix, Inc.) (b) An electronic image of a giant galaxy (arrow) so distant that light took 10 billion years to reach us. This image is reproduced as a negative. The sky is light, and the stars and galaxy are black. (Hyron Spinrad) (c) This false-color image of the galaxy NGC 1232 displays different levels of brightness in different colors. The original CCD image was manipulated by computer. (California Association for Research in Astronomy) produces an electric current when struck by light. The telescope focuses starlight onto the detector, and the resulting electric current is proportional to the amount of light. Because a CCD chip contains an array of light-sensitive devices, astronomers can use a CCD image to make photometric measurements.

Whether the photometer contains a single detector or a CCD chip, astronomers can measure the color of stars by passing the light through filters. A filter transmits only those wavelengths in a certain range—for instance, blue light between 400 nm and 480 nm. The difference between the brightness of the star through a red and a blue filter tells us the color of the star. (We will examine this measurement further in Chapter 7.)

We also use a photometer to make measurements in the near ultraviolet or near infrared. With the proper detector and filters, the photometer measures the brightness at wavelengths we cannot see. Of course, if we go too far into the ultraviolet or infrared, the earth's atmosphere is not transparent, and we cannot make such observations from the earth's surface.

The Spectrograph

A **spectrograph** is a device that separates starlight according to wavelength to produce a spectrum. We can see how this works by reproducing Newton's original experiment. In place of a hole in a shutter, we might use a narrow slit to produce a thin beam of light. When that beam passes through a prism, the angle through which it is bent depends on wavelength—violet bends most and red least—so the light leaving the prism is spread into a spectrum (Figure 6-15). A typical prism spectrograph contains more than one prism to spread the light further and lenses to guide the light into the prism and to focus the light onto a photographic plate.

Nearly all modern spectrographs use a grating in place of a prism. A **grating** is a piece of glass with thousands of microscopic parallel lines scribed onto its surface. Different wavelengths of light reflect from the grating at slightly different angles, so white light is spread into a spectrum. Modern grating spectrographs can be designed using all reflection optics so that the light never

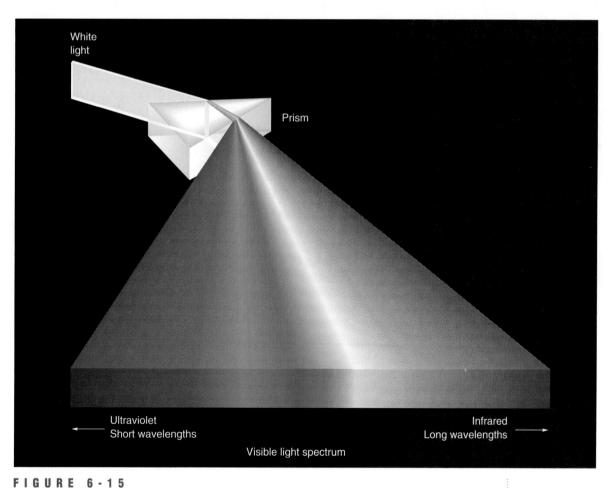

White light

Prism

Ultraviolet
Short wavelengths

Infrared
Long wavelengths

Visible light spectrum

FIGURE 6-15

A prism bends light by an angle that depends on the wavelength of the light. Short wavelengths bend most, and long wavelengths least. Thus, white light passing through a prism is spread into a spectrum.

FIGURE 6-16

In this grating spectrograph (on a Cassegrain telescope), starlight enters the spectrograph from above, and the light of a single star is isolated by the slit. A mirror directs the light onto the grating, which breaks the light into a spectrum. The camera optics focus the light onto a CCD chip, where the spectrum is recorded by a computer. Note that no lenses are used in this design in order to avoid chromatic aberration, which would distort the spectrum.

What is the difference between light going through prisms in a spectrograph and light passing through lenses in a refracting telescope?

A refracting telescope producing chromatic aberration and a prism dispersing light into a spectrum are two examples of the same thing, but one is bad and one is good. When light passes through the curved surfaces of a lens, different wavelengths are bent by slightly different amounts, and the different colors of light come to focus at different focal lengths. This produces the color fringes in an image called chromatic aberration, and that's bad. But the surfaces of a prism are made to be precisely flat, so all of the light enters the prism at the same angle, and any given wavelength is bent by the same amount wherever it meets the prism. Thus, white light is dispersed into a spectrum. We could call the dispersion of light by a prism "controlled chromatic aberration," and that's good.

If, however, we build a spectrograph that uses lenses to focus the spectrum on a photographic plate or on a CCD detector, we will get chromatic aberration in the lens. Different colors will be focused at different distances from the lens, and we may not be able to focus the blue, yellow, and red parts of the spectrum at the same time. How could we design a spectrograph to avoid this chromatic aberration? ∎

So far, our discussion has been limited to visual wavelengths. Now it is time to consider the rest of the electromagnetic spectrum.

6-4 RADIO TELESCOPES

All the telescopes and instruments we have discussed look out through the visible light window in the earth's atmosphere, but there is another window running from a wavelength of 1 cm to about 1 m (see Figure 6-2). By building the proper kinds of instruments, we can study the universe through this radio window.

Operation of a Radio Telescope

A radio telescope usually consists of four parts: a dish reflector, an antenna, an amplifier, and a recorder (Figure 6-17a). The dish reflector gathers incoming radio waves and focuses them on the antenna, which absorbs the energy. The signal is amplified, and its intensity is recorded, usually under the direction of a computer.

Like the mirror of an optical telescope, the dish reflector must be of large diameter in order to gather as much radio energy as possible, but because radio photons have such long wavelengths, the surface of the dish

passes through glass. This assures that no chromatic aberration can distort the spectrum (Figure 6-16).

Some spectrographs can be attached to the telescope; others are so large they must be placed in a special room at the coudé focus. Specialized spectrographs can record spectra in the ultraviolet or infrared. Whatever the size and type, a spectrograph is the astronomer's most powerful instrument. An astronomer recently remarked, "We don't know anything till we get a spectrum," and that comment is only a slight exaggeration.

a

FIGURE 6-17

In most radio telescopes, a dish reflector concentrates the radio signal on the antenna. The signal is then amplified and recorded. For all but the shortest radio waves, wire mesh is an adequate reflector (above). (Courtesy Seth Shostak/SETI Institute.)

need not be as smooth as a mirror. In fact, wire mesh is a very good reflector of all but the shortest radio waves (Figure 6-17b).

The dish also acts to exclude all radio energy except that coming from a small region of the sky. By scanning from spot to spot, the radio astronomer can construct a map showing the intensity of the radio energy coming from different regions of the object being studied. These maps are usually drawn as colorful contour maps, which mark areas of constant radio intensity (Figure 6-18). Of course, radio energy has no color, so such maps are false-color images.

Because they focus photons of much longer wavelength, radio telescopes have much worse resolving power than optical telescopes. A dish 30 m in diameter receiving radiation at a wavelength of 21 cm has a resolving power of about 0.5°. Such a radio telescope would be

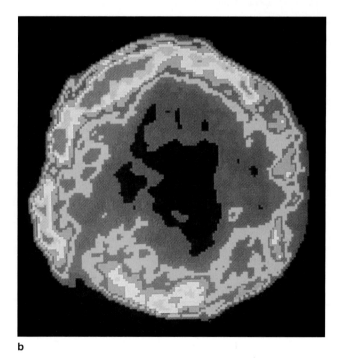

b

FIGURE 6-18

(a) A contour map of a baseball stadium shows regions of similar admission prices. The most expensive seats are those behind home plate.
(b) A false-color-image radio map of Tycho's supernova remnant, the expanding shell of gas produced by the explosion of a star in 1572. The radio contour map has been color-coded to show intensity. Red is the strongest radio intensity, and violet is the weakest. (National Radio Astronomy Observatory—Associated Universities, Inc.)

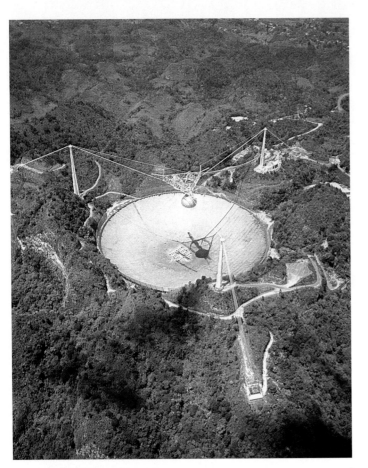

FIGURE 6-19
The largest radio telescope in the world is the 300-m (1000-ft) dish suspended in a valley in Arecibo, Puerto Rico. The antenna hangs above the dish on cables stretching from towers. The Arecibo Observatory is part of the National Astronomy and Ionosphere Center, which is operated by Cornell University under contract with the National Science Foundation. (David Parker/Science Photo Library)

Radio astronomers have built radio interferometers of various sizes. The Very Large Array (VLA) is a Y-shaped pattern of 27 radio dishes spread across the New Mexico desert (Figure 6-20b). The combination has the resolving power of a radio telescope 36 km (22 mi) in diameter and can produce radio maps ten times as detailed as the best photographs taken from the earth. (See Chapter 18 for numerous examples.)

Very long baseline interferometry (VLBI) links telescopes on opposite sides of the world to achieve very high resolution. This technique has been used with various existing radio telescopes; the Very Long Baseline Array (VLBA) is a specially built network of ten matched 25-m radio dishes spread from Hawaii to the Virgin Islands. With an effective diameter of 8000 km, the VLBA has a resolution of 0.0002 second of arc—500 times better than the Hubble Space Telescope. It is the equivalent of being able to read a newspaper 600 miles away.

VLBI requires large, fast computers to combine signals from different telescopes, so European radio astronomers have created a computing center at Dwingeloo Observatory in the Netherlands that will allow them to combine existing radio telescopes in Europe, North America, and Australia into a planet-wide interferometer.

Advantages of a Radio Telescope

The radio telescope has three advantages over an optical telescope. First, it can detect cold clouds of gas in space. These gas clouds are important because they give birth to stars, but the gas is so cold that it emits no visible light. In Chapter 7, we will see how cold hydrogen can emit photons with a wavelength of 21 cm and how other gases such as carbon monoxide can emit other wavelengths.

A radio telescope can also detect very hot gas trapped in a magnetic field even though such gas almost never radiates visible light. This is a rather common phenomenon in certain kinds of nebulae produced by exploding stars and peculiar galaxies.

The third advantage of radio telescopes is that they can see through the dust clouds that block our view at visible wavelengths. Because radio photons have longer wavelengths, they are not affected by the microscopic particles of dust that float in space. But those dust specks scatter visual-wavelength photons and make the dust clouds opaque. Radio signals from far across the galaxy pass unhindered through the dust, giving us an unobscured view.

The disadvantage of radio telescopes is that they are very susceptible to radio noise. Radio and TV transmitters, noisy electric motors, and even the spark plugs in a car can overload the delicate radio ears of a radio telescope. Many radio telescopes are hidden in valleys far from cities to avoid radio noise.

unable to show us any details in the sky smaller than the angular diameter of the moon. The larger the diameter of the dish, the better the resolving power (and the stronger the signal), so radio telescopes tend to be very large. The largest radio dish in the world is 300 m (1000 ft) in diameter. It is built into a mountain valley in Arecibo, Puerto Rico (Figure 6-19), and the operators depend on the rotation of the earth to point the telescope at different regions of the sky.

The Radio Interferometer

To improve the resolving power of their telescopes, radio astronomers have linked radio telescopes together to form **radio interferometers**—two or more radio telescopes that combine their signals to simulate one big telescope (Figure 6-20a). The system has the resolving power of a telescope whose diameter is equal to the separation between the outer edges of the two smaller telescopes.

Diameter of equivalent
radio telescope

Signal comparator

Recording device

a

FIGURE 6-20
(a) A radio interferometer consists of two or more radio telescopes whose signals are combined to give the resolving power of a much larger instrument. (b) The Very Large Array (VLA) radio interferometer uses up to 27 dishes along the 20-km arms of a Y to simulate a radio telescope 36 km in diameter. (National Radio Astronomy Observatory, operated by Associated Universities, Inc., under a contract with the National Science Foundation)

b

CRITICAL INQUIRY

Why do optical astronomers build big telescopes, while radio astronomers build groups of widely separated smaller telescopes?

Optical astronomers build large telescopes to maximize light-gathering power, but the problem for radio telescopes is resolving power. Because radio waves are so much longer than light waves, a single radio telescope can't see details in the sky much smaller than the moon. By linking radio telescopes miles apart, radio astronomers build a radio interferometer that can simulate a radio telescope miles in diameter and thus increase the resolving power.

The difference between the wavelengths of light and radio waves makes a big difference in building the best telescopes. But why don't radio astronomers want to build their telescopes on mountaintops as optical astronomers do? ∎

Our atmosphere causes trouble for the earth's astronomers in two ways. It distorts images, and it absorbs many wavelengths. The only way to avoid these limitations completely is to send telescopes above the atmosphere, into space.

6-5 SPACE ASTRONOMY

Celestial bodies emit radiation at many different wavelengths, but the earth's atmosphere absorbs most such photons. From the earth's surface, we can observe at visual wavelengths, in the near infrared, in the near ultra-violet, and at some radio wavelengths. But if we wish to observe at other wavelengths, we must get above the earth's atmosphere, and that means we must put telescopes into space.

Infrared Astronomy

Hot objects emit light, but cooler objects may emit so little light that they are undetectable at visible wavelengths. Many such objects, however, are warm enough to emit infrared radiation. Thus, we can use an infrared telescope to study very cool stars, planets, comets, asteroids, and dust in space. Another advantage is that infrared radiation has a long enough wavelength to penetrate dust clouds, so we can use infrared telescopes to study young stars hidden deep inside clouds of dust.

FIGURE 6-21
Comet Hale–Bopp hangs over the 3-meter NASA Infrared Telescope atop Mauna Kea. The air at high altitudes is steady and so dry that it is transparent to shorter infrared photons. Infrared astronomers often observe with the lights on in the telescope dome. Their instruments are usually insensitive to visible light. (Courtesy William Keel)

Infrared observations are hampered by the earth's atmosphere. Some infrared radiation penetrates partially open atmospheric windows scattered from 1.2 μm (microns) to 40 μm. In this wavelength range, called the near infrared, most of the radiation is absorbed by water vapor in the earth's atmosphere, so it is an advantage to place telescopes on mountains where the air is thin and dry. Two major infrared telescopes, a 3.0-m (118-in.) and a 3.8-m (150-in.), observe from 13,600-ft Mauna Kea in Hawaii. At this altitude, they are above most of the water vapor in the atmosphere (Figure 6-21).

The far infrared includes wavelengths longer than 40 μm, but such photons are strongly absorbed by the atmosphere. To observe in the far infrared, telescopes must venture high in the atmosphere. Remotely operated infrared telescopes suspended under balloons have reached altitudes as high as 41 km (25 mi).

Another solution is to fly the infrared telescope to high altitudes in an airplane. NASA has modified a Lockheed C-141 jet transport to carry a 91-cm infrared telescope and a crew of a dozen astronomers to altitudes of 40,000 ft to get above 99 percent of the water vapor in the earth's atmosphere. The rings of Uranus, for example, were discovered by astronomers flying in this aircraft.

The ultimate solution to the problem of atmospheric absorption is to place the telescope in orbit above the atmosphere. The Infrared Astronomical Satellite (IRAS) (Figure 6-22) carried a 56-cm (22-in.) telescope cooled to nearly absolute zero by liquid helium. It observed from an orbit 900 km high through most of 1983 before its helium coolant was exhausted. Among its many discoveries, IRAS found a disk of cold matter around the bright star Vega—the first clear evidence of planetlike material orbiting a star other than our sun.

IRAS, like all infrared telescopes, had to be cooled to very low temperatures. Infrared radiation is heat, and if the telescope is warm, it will emit many times more infrared than that coming from a distant object. Imagine trying to look at a dim, moonlit scene through binoculars that were glowing brightly. In the near infrared, only the detector, the element on which the infrared radiation is focused, must be cooled. To observe in the far infrared, however, the entire telescope must be cooled.

European astronomers are developing the Infrared Space Observatory (ISO). It is similar in size to IRAS, but its detectors are much more sensitive. Also, NASA is now planning the Space Infrared Telescope Facility (SIRTF). Once launched, SIRTF will become a major infrared observatory in space.

Ultraviolet Astronomy

Astronomers observe at ultraviolet wavelengths for two main reasons. First, many hot objects emit much more ultraviolet radiation than visible light. Not only hot stars but also hot regions in the atmospheres of cooler stars are strong sources of ultraviolet. Second, many atoms—including hydrogen, the most common atom—have strong spectral lines in the ultraviolet part of the spectrum. Thus, the ultraviolet region of the spectrum is packed with information about a wide range of celestial bodies.

Ultraviolet radiation beyond a wavelength of 300 nm is completely absorbed by the ozone layer extending from 20 km to above 40 km in our atmosphere. To get above this layer, ultraviolet telescopes must go into space.

The first far-ultraviolet observations of a celestial body were made in 1946 when a captured German V-2 rocket lifted instruments to an altitude of 100 km and

FIGURE 6-22

The advantage of infrared observations is illustrated in this mosaic of images of the constellations Vela and Puppis obtained by the IRAS satellite (inset). Three images made at different infrared wavelengths are color-coded and combined. Stars, the hottest objects in the photo, appear as blue dots, while clouds of interstellar dust heated by starlight are color-coded red or yellow. These clouds are colder than 100 K (−280°F) and are undetectable at visible wavelengths. (NASA)

recorded the spectrum of the sun. Since then, many rockets have carried ultraviolet instruments on short trips into space, but extended studies are possible only from satellites.

One of the most successful ultraviolet observatories was the International Ultraviolet Explorer (IUE) (Figure 6-23). Launched in January 1978, it carried a 45-cm (18-in.) telescope with attached spectrographs. Spectra were transmitted to a control room on the earth for analysis. Although the IUE was expected to operate for only a few years, it proved so useful that NASA engineers and scientists designed new operating procedures each time a component failed, and the IUE gathered data into the mid-1990s before it was finally retired to make way for the operation of new orbiting telescopes.

Like a sleek new-model IUE, the Extreme Ultraviolet Explorer (EUVE) was launched into earth orbit on June 7, 1992. Observing at wavelengths between 7 and 76 nm, even shorter than the range of the IUE, the EUVE has explored astronomical spectra near the wavelength range we call X rays.

X-Ray Astronomy

Beyond the far ultraviolet, at wavelengths from 10 nm to 0.01 nm, lie the X rays. These very high energy photons can be produced only by violent, high-energy events. We detect X rays coming from exploding stars, from colliding galaxies, and from matter smashing into the surface of a neutron star or falling into a black hole—phenomena we will examine in greater detail in later chapters.

HISTORY OF SPACE EXPLORATION 1957-1989

ULTRAVIOLET EXPLORER

1978 USA, UK, EUROPEAN SPACE AGENCY

SIERRA LEONE   LE15

FIGURE 6-23

The International Ultraviolet Explorer, commemorated on this stamp from Sierra Leone, was launched in January 1978 with an expected life of 3 years. It operated into the mid-1990s as an ultraviolet observatory in space run remotely by astronomers working in control rooms on the earth.

Although early X-ray observations of the sky were made from balloons and small rockets in the 1960s, the age of X-ray astronomy did not really begin until 1970, when an X-ray telescope named Uhuru (Swahili for "freedom") was put into orbit. Uhuru detected nearly 170 separate sources of celestial X rays. Later, the three High Energy Astronomy Observatory (HEAO) satellites,

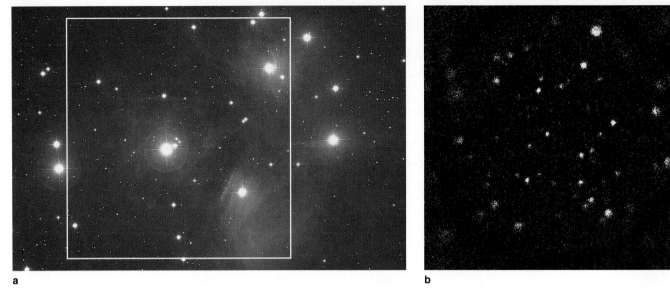

FIGURE 6-24

The value of observing at nonvisible wavelengths is shown by this pair of images. (a) A visible image of the Pleiades star cluster shows the brightest hot stars. (The Observatories of the Carnegie Institution of Washington) (b) An X-ray image of the boxed area in part (a) reveals many low-luminosity stars producing X rays in their outer layers. Such stars cannot be distinguished using visible images. (Courtesy John Stauffer and Charles Prosser)

carrying more sensitive and more sophisticated equipment, pushed the total to many hundreds. The second HEAO satellite, named the Einstein Observatory, used special optics to produce X-ray images. Similar X-ray optics aboard Skylab, an orbiting space station, produced X-ray images of the sun.

The European Space Agency launched an X-ray observatory (Exosat) in May 1983, and it was very successful in studying sites of high-energy violence in the universe. Unfortunately, Exosat failed in April 1986. The versatile U.S.–German–British satellite Rosat was launched June 1, 1990, and after completing an X-ray survey of the sky, it began observing specific objects of interest (Figure 6-24).

Specialized X-ray telescopes have been put in orbit by the Soviet Union and by Japan, but the world's astronomers need a large-diameter X-ray telescope in order to detect faint objects. The Advanced X-Ray Astrophysics Facility (AXAF) is a proposed NASA satellite that was originally planned to carry a 1.2-m (47-in.) telescope observing from 6 nm to 0.15 nm. Budget cuts first delayed the start of construction and then forced major changes in the design of the telescope. It may be some years before a large X-ray observatory reaches orbit.

Gamma-Ray Telescopes

Gamma rays have wavelengths even shorter than X rays, and that means they have even higher energies. They are not understood very well, but they appear to be produced by the hottest and most violent objects in the universe:

exploding stars, erupting galaxies, neutron stars, and black holes.

We don't understand gamma rays well because they are so difficult to detect. First, they cannot be focused or detected as X rays can. Thus, gamma-ray telescopes are more complex and can see less detail than similar X-ray telescopes. Also, gamma rays are such high-energy photons that natural processes produce only a few. Gamma-ray telescopes must count gamma rays one at a time. For example, a gamma-ray telescope observed an intense source for a month and counted only 3000 gamma rays—one every 15 minutes.

In addition, gamma rays are almost totally absorbed by our atmosphere, so gamma-ray telescopes must observe from orbit. NASA has operated three important gamma-ray telescopes aboard satellites, including one on the Small Astronomy Satellite 2 (SAS 2). The Soviet Union has orbited two, the most recent of which was aboard the satellite COS B, launched in 1975 by the European Space Agency. It surveyed the sky and detected gamma rays from clouds of gas in space, from the center of our galaxy, and from the remains of exploded stars.

NASA launched the 17-ton Gamma Ray Observatory in April 1991. At 10 to 50 times the sensitivity of any previous gamma-ray telescope, it first mapped the entire sky and then observed selected targets.

The Hubble Space Telescope

Astronomers have dreamed of having a giant telescope in orbit since the 1920s. Not only could they observe at nonvisible wavelengths, they could avoid the turbulence

a

b

FIGURE 6-25

(a) The Hubble Space Telescope is the largest orbiting telescope ever launched. It was carried into orbit by the space shuttle in 1990. (b) In December 1993, astronauts made five space walks to install corrective optics in the telescope and upgrade equipment, and a second visit by astronauts in early 1997 installed new instruments. For scale, locate the astronaut at the end of the manipulator arm at top center and the astronaut inside the base of the telescope. After these visits the telescope was released back into Earth orbit, where it continues to be a powerful observatory in space. (NASA)

in the Earth's atmosphere. That dream came true on April 24, 1990, when the Hubble Space Telescope (HST) was released from a space shuttle into orbit (Figure 6-25). Named after Edwin Hubble, the man who discovered the expansion of the universe (Chapter 17), and carrying a mirror 2.4-m (96-inch) in diameter, it is the largest orbiting telescope ever built.

An error in its mirror caused some image distortion at first, but in 1993 astronauts visited HST and fitted it with corrective optics. The telescope has performed beyond expectations ever since.

Astronomers on Earth can control the telescope and direct the light its main mirror gathers to a number of cameras and spectrographs. Although larger telescopes exist on Earth, HST orbits above Earth's atmosphere, so it is limited only by diffraction in its optics and the sensitivity of its detectors. It can resolve details 10 times smaller and detect objects 50 times fainter than can any telescope on earth.

The telescope is operated from the Space Telescope Science Institute at Johns Hopkins University where astronomers plan the observing schedule and engineers monitor telescope operations. Instructions are transmitted to HST, and after it completes its assigned observations, it transmits its data to the institute where it is analyzed and distributed to astronomers. Because it is a national facility, anyone can propose observing projects

for the telescope, but competition is fierce and only the most worthy projects win approval.

It is impossible here to list the discoveries that HST has made; they will be found all through this book. But we can note that it has observed a wide range of objects from nearby planets to the most distant galaxies with precision impossible from Earth-based telescopes (Figure 6-26). HST is a general purpose astronomical telescope. Astronauts visited the telescope in early 1997 to install new equipment, and another upgrade spacewalk is planned in the near future. Such upgrades will keep HST at the frontier of astronomy for years to come.

Cosmic Rays

All of the radiation we have discussed in this chapter has been electromagnetic radiation. **Cosmic rays,** however, are not really rays; they are subatomic particles traveling at tremendous velocities that strike our atmosphere from space. Almost no cosmic rays reach the ground, but they do smash into gas atoms in the upper atmosphere, and fragments of these collisions shower down on us day and night over our entire lives. These secondary cosmic rays are passing through you as you read this sentence.

Some cosmic-ray research can be done from high mountains or high-flying aircraft, but to study cosmic rays in detail, detectors must go into space. A number of

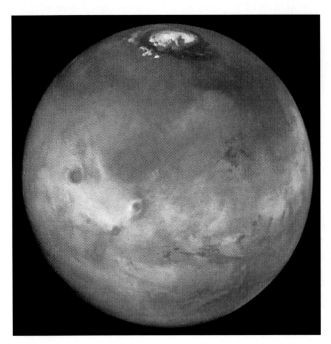

FIGURE 6-26
Observing from above Earth's atmosphere, the Hubble Space Telescope has achieved results that exceed its designers' expectations. This image of Mars shows details only seen before from spacecraft orbiting the planet. The polar cap is carbon dioxide ice, and the spots at center left are giant extinct volcanoes extending up through thin clouds. Similar spectacular images and spectra are allowing astronomers to explore a wide range of objects from nearby planets to the most distant visible galaxies. (NASA)

cosmic-ray detectors have been carried into orbit, but this area of astronomical research is just beginning to bear fruit.

We can't be sure where cosmic rays come from. Since they are atomic particles with electric charges, they are deflected by the magnetic fields spread through our galaxy, and that means we can't tell where they are coming from. The space between the stars is a glowing fog of cosmic rays. Some lower-energy cosmic rays come from the sun, but many cosmic rays are probably produced by the violent explosions of dying stars. At present, cosmic rays largely remain a mystery. We will discuss them again in future chapters.

CRITICAL INQUIRY

Why can infrared astronomers observe from high mountaintops, while X-ray astronomers must observe from space?

Infrared radiation is absorbed by water vapor in the earth's atmosphere. If we built our infrared telescope on top of a high mountain, we would be above most of the water vapor in the atmosphere, and we could collect some infrared radiation from the stars. The longer-wavelength infrared radiation is absorbed much higher in the atmosphere, so we couldn't observe it from our mountaintop. Similarly, X rays are absorbed in the uppermost layers of the atmosphere, and we would not be able to find any mountain high enough to get an X-ray telescope above those absorbing layers. To observe the stars at X-ray wavelengths, we would need to put our telescope in space, above the earth's atmosphere.

X-ray and far infrared telescopes must observe from space, but the Hubble Space Telescope observes in the visual wavelength range. Then why must the Hubble Space Telescope observe from orbit? ∎

The tools of the astronomer are designed to gather radiation from the sky and extract information. Perhaps no tool is as important as the spectrograph, because no form of observation is as loaded with information as a spectrum. In the next chapter, we will see how we can harvest the information in a star's spectrum. ∎

∎ Summary

Electromagnetic radiation is an electric and magnetic disturbance that transports energy at the speed of light. The electromagnetic spectrum includes radio waves, infrared radiation, visible light, ultraviolet radiation, X rays, and gamma rays.

We can think of "a particle of light," a photon, as a bundle of waves that sometimes acts as a particle and sometimes as a wave. The energy a photon carries depends on its wavelength. The wavelength of visible light, usually measured in nanometers (10^{-9} m), ranges from 400 nm to 700 nm. Infrared and radio photons have longer wavelengths and carry less energy. Ultraviolet, X-ray, and gamma-ray photons have shorter wavelengths and carry more energy.

Astronomical telescopes are of two types, refractor and reflector. A refractor uses a lens to bend the light and focus it into an image. Because of chromatic aberration, refracting telescopes cannot bring all colors to the same focus, resulting in color fringes around the images. An achromatic lens partially corrects for this, but such lenses are expensive and cannot be made larger than about 1 m in diameter.

Reflecting telescopes use a mirror to focus the light and are less expensive than refracting telescopes of the same diameter. In addition, reflecting telescopes do not suffer from chromatic aberration. Thus, most recently built telescopes are reflectors.

The largest traditional telescopes in the world are the 6-m telescope in the former Soviet Union and the 5-m telescope in California. They have thick, solid mirrors of great weight. New telescopes use thin mirrors controlled by computers or mirrors made up of segments. The Keck Telescope in Hawaii uses segments to make up a mirror 10 m in diameter.

The powers of a telescope are light-gathering power, resolving power, and magnifying power. The first two of these depend on the telescope's diameter; thus, astronomical telescopes often have large diameters.

Special instruments attached to a telescope analyze the light it gathers. The photographic plate records vast amounts of detail for later analysis, but electronic imaging systems such as charge-coupled devices have largely replaced photographic plates. CCDs are more sensitive, record both faint and bright objects, and can be read directly into computer memory. A photometer measures the brightness and color of the light entering a telescope. Spectrographs, using prisms or gratings, break the starlight into a spectrum, which then can be photographed or electronically recorded.

To observe radio signals from celestial objects, we need a radio telescope, which usually consists of a dish reflector, an antenna, an amplifier, and a recorder. Such an instrument measures the intensity of radio signals over the sky and constructs radio maps. The poor resolution of the radio telescope can be improved by combining it with another radio telescope to make a radio interferometer. The three principal advantages of radio telescopes are that they can detect the very cold gas clouds in space, they can detect regions of very hot gas produced by exploding stars or erupting galaxies, and they can look through the dust clouds that block our view at optical wavelengths.

Observations at some wavelengths in the near infrared are possible from high mountaintops or from high-flying aircraft. At these altitudes, the air is thin and dry, and the infrared radiation can reach the telescope. For instance, infrared telescopes can operate at 4150 m (13,600 ft) atop the volcano Mauna Kea in Hawaii.

At other infrared wavelengths, the telescope must observe from above the earth's atmosphere—from orbit. The Infrared Astronomy Satellite observed from its orbit about 900 km high. It carried a 56-cm telescope and was cooled by liquid helium.

To observe at the short ultraviolet wavelengths, the telescope must be above the ozone layer in the earth's atmosphere. That means it must be in orbit. The International Ultraviolet Explorer was an ultraviolet astronomy satellite operated jointly by U.S. and European astronomers. X-ray and gamma-ray telescopes must also be placed in orbit above the absorbing layers of the earth's atmosphere.

The largest orbiting telescope is the Hubble Space Telescope, launched in the spring of 1990.

▪ New Terms

electromagnetic radiation	Cassegrain telescope
wavelength	Newtonian focus
frequency	coudé focus
nanometer (nm)	Schmidt camera
Angstrom (Å)	Schmidt–Cassegrain telescope
photon	sidereal drive
infrared radiation	equatorial mounting
ultraviolet radiation	polar axis
atmospheric window	alt-azimuth mounting
refracting telescope	active optics
reflecting telescope	light-gathering power
focal length	resolving power
objective lens	diffraction fringe
eyepiece	seeing
chromatic aberration	adaptive optics
achromatic lens	magnifying power
objective mirror	charge-coupled device (CCD)
spherical aberration	false-color image
prime focus	photometer
secondary mirror	spectrograph
grating	very long baseline interferometry (VLBI)
radio interferometer	cosmic rays

▪ Questions

1. Why do nocturnal animals usually have large pupils in their eyes? How is that related to astronomical telescopes?

2. Astronomers have not constructed a single major refracting telescope during this century. Why not?

3. How have computers and new technology changed the design of astronomical telescopes and their mountings?

4. Small telescopes are often advertised as "200 power" or "magnifies 200 times." As someone knowledgeable about astronomical telescopes, how would you improve such advertisements?

5. An astronomer recently said, "Some people think I should give up photographic plates." Why might she change to something else?

6. What purpose do the colors in a false-color image or false-color radio map serve?

7. Why can we observe in the infrared from high mountains and aircraft but must go into space to observe in the far ultraviolet?

8. The moon has no atmosphere at all. At what wavelengths could we observe if we had an observatory on the lunar surface?

▪ Discussion Questions

1. Why does the wavelength response of the human eye match so well the visual window of the earth's atmosphere?

2. Basic research in chemistry, physics, biology, and so on is supported in part by industry. How is astronomy different? Who funds the major observatories?

3. Some astronomers hope to place radio telescopes in very large orbits around the earth and use them with earth-based radio telescopes. Why would this be an advantage?

▪ Problems

1. The thickness of the plastic in plastic bags is about 0.001 mm. How many wavelengths of red light is this?

2. Measure the actual wavelength of the wave in Figure 6-1. In what portion of the electromagnetic spectrum would it belong?

3. Compare the light-gathering powers of the 5-m telescope and a 0.5-m telescope.

4. How does the light-gathering power of the largest telescope in the world compare with that of the human eye? (HINT: Assume that the pupil of your eye can open to about 0.8 cm.)

5. What is the resolving power of a 25-cm telescope? What do two stars 1.5 seconds of arc apart look like through this telescope?

6. Most of Galileo's telescopes were only about 2 cm in diameter. Should he have been able to resolve the two stars mentioned in Problem 5?

7. How does the resolving power of the 5-m telescope compare with that of the Hubble Space Telescope? Why will the HST outperform the 5-m telescope?

8. If we build a telescope with a focal length of 1.3 m, what focal length should the eyepiece have to give a magnification of 100 times?

9. Astronauts observing from a space station need a telescope with a light-gathering power 15,000 times that of the human eye, capable of resolving detail as small as 0.1 second of arc, and having a magnifying power of 250. Design a telescope to meet their needs. Could you test your design by observing stars from the earth?

10. A spy satellite orbiting 400 km above the earth is supposedly capable of counting individual people in a crowd. What minimum-diameter telescope must the satellite carry? (HINT: Use the small-angle formula.)

▪ *Critical Inquiries for the Web*

1. How do professional astronomers go about making observations at major astronomical facilities? Visit several observatory Web sites to determine the process an astronomer would go through to secure observing time and to make observations at the facility.

2. NASA is in the process of completing a fleet of 4 space-based "Great Observatories." (The Hubble Space Telescope is one. What are the others?) Examine the current state of these missions by visiting their home pages on the Internet. What advantages would these facilities have over ground-based observatories?

Go to the Wadsworth Astronomy Resource Center (www.wadsworth.com/astronomy) for critical thinking exercises, articles, and additional readings from InfoTrac College Edition, Wadsworth's online student library.

Chapter Seven

STARLIGHT AND ATOMS

GUIDEPOST

Some chapters in textbooks do little more than present facts. The chapters in this book attempt to present organized understanding. But this chapter is special. It presents a tool. The interaction of light with matter gives astronomers clues about the nature of the heavens, but the clues are meaningless unless astronomers understand how atoms leave their traces on starlight. Thus, we dedicate an entire chapter to understanding how atoms interact with light.

This chapter marks a transition in the way we look at nature. Earlier chapters described what we see with our eyes and explained those observations using models and theories. With this chapter, we turn to modern astrophysics, the application of modern physics to the study of the sky. Now we can search out secrets of the stars that lie beyond the grasp of our eyes.

If this chapter presents us with a tool, then we should use it immediately. The next chapter will apply our new tool to understanding the sun. ∎

Awake!

for Morning in the Bowl of Night Has flung

the Stone that puts the Stars to Flight:

And Lo! the Hunter of the East has caught

The Sultan's Turret in a Noose of Light.

THE RUBÁIYÁT OF OMAR KHAYYÁM

Trans. Edward FitzGerald

FIGURE 7-1

The Eagle Nebula is a cloud of gas and dust associated with a star cluster that must have formed recently from the nebula. Light from the star cluster heats and excites the atoms of the nebula to glow like a giant neon sign. Only by studying the light from the nebula can astronomers deduce its composition, temperature, density, age, and so on. (National Optical Astronomy Observatories)

Sunlight is nothing but bright starlight—the sun is about 10 billion times brighter than the next brightest star, but that is only because it is so close. The sun is, in fact, an ordinary star whose light differs in no detail from the light of hundreds of millions of similar stars scattered across our galaxy. If we want to understand what stars are, we have only to look at the sun.

However close the sun may seem in comparison with other stars, it is far beyond our reach. The sun is 93 million miles away—a distance so great that a commercial jet could not travel that far in 5 years. Thus, we cannot probe and sample the sun. No laboratory jar on Earth holds a sample labeled "sun stuff," and no instrument has ever descended into the sun to tell us the temperature, pressure, composition, and so on. As with all stars, the only information we can obtain about the sun comes to

us hidden in light. Whatever we want to know about stars, we must catch in a "noose of light."

The stars are so far away that it would not be surprising if earthbound humans knew almost nothing about them. But we can learn about the stars by analyzing the light we receive (Figure 7-1). We begin by considering the bulk properties of that light—how much red and how much blue are present in a star's light. This will tell us the approximate temperature of the star, but to obtain more detail we must examine the starlight more closely.

If we spread out starlight into a spectrum, we can see dark lines that contain information about the gases that make up the star. To analyze those lines and learn the secrets of the stars, we must know how atoms interact with light. We begin our study of atoms with the hydrogen atom, the most common and simplest atom in the universe.

Temperature and Heat

Temperature and heat are two fundamental physical properties that are closely related. It is critical, however, for us to distinguish between them. Temperature is an intensity, and heat is an amount.

Temperature is a measure of the average motion of the atoms and molecules that make up an object. In a hot object, the particles vibrate at higher speeds than in a cool object. If you have your temperature taken, it will probably be 98.6°F, an indication that the atoms and molecules in your body are vibrating at a normal pace. If you measure the temperature of a month-old baby, the thermometer should register the same temperature, showing that the atoms and molecules in the baby's body are moving at the same average velocity as the atoms and molecules in your body. Thus, temperature is a measure of the intensity of the motion.

Heat is a measure of the total amount of thermal energy in a body. Although you and the infant have the same temperature, you contain much more mass than the infant, so you must contain much more heat. Thus, the thermal energy in your body and that in the baby's body have the same intensity (temperature) but different amounts (heat).

Temperature is the intensity of the thermal energy, and heat is the amount of energy. We might explore this idea by using sound as an analogy. A pocket radio might have enough power to fill a small room with loud music, but it could never fill a large auditorium with music at the same loudness. Loudness is a measure of the *intensity* of the sound; the *quantity* of sound is the total amount of sound energy in the room or auditorium.

When you think of temperature and heat, remember the difference between intensity and amount. ∎

7-1 STARLIGHT

If you look at the stars in the constellation Orion, you will notice that they are not all the same color. Betelgeuse, in the upper left corner, is quite red; Rigel, in the lower right corner, is blue. These differences in color arise from the way the stars produce light and give us our first clue to the temperatures of stars.

Temperature and Heat

Temperature is one of the defining characteristics of a star. That is, if we know a star's temperature and a few other properties such as size, we can understand the star. But if we don't know a star's temperature, we can know almost nothing about it.

A gas is made up of particles—atoms and molecules—that are in constant motion, colliding with one another millions of times a second. The **temperature** of a gas is a measure of the average velocity of the particles. If a gas is hot, the particles are moving very rapidly; if it is cool, the particles are moving more slowly. **Heat,** on the other hand, is a measure of the total energy of motion stored in the moving gas particles. The particles in a liquid or a solid are not as free to move as particles in a gas, but they do vibrate, so it makes sense to extend the definition of temperature to include liquids and solids. Hot coffee is hot because its atoms and molecules are in rapid motion. Notice the distinction between temperature and heat. A cup of coffee and a giant vat of coffee can have the same temperature, but the vat of coffee obviously contains more heat (Window on Science 7-1).

We will measure the temperature of stars on the **Kelvin temperature scale.** The Kelvin scale sets its zero point at **absolute zero,** the temperature at which the particles of a gas have no remaining motion that we can extract as heat. Absolute zero is −273.2°C and −459.7°F (see Appendix A). We will see that the temperatures of the stars, ranging from 2000 K to 40,000 K or more, can be deduced from starlight.

The Origin of Starlight

The starlight we see comes from gases in the outer surface of the star—the photosphere. The gases deep inside the star also emit light, but it is absorbed before it can escape, and the low-density gas above the photosphere is too thin to emit significant amounts of light. Thus, the photosphere of a star is that layer of gases dense enough to emit significant amounts of light but thin enough to allow the photons to escape. (Recall that we met the photosphere of the sun in Chapter 3.)

To see how the photosphere of a star can produce light, we must consider two things: how photons are produced and how the temperature of a material is related to motion among its atoms.

First, a photon can be produced by a changing electric field. An **electron** is a negatively charged subatomic particle, and if we disturb the motion of an electron, the sudden change in the electric field around it can cause the emission of a photon. For example, if you run a comb through your hair while standing near an AM radio, you can produce popping noises on the radio. The moving comb disturbs the electrons in both the comb and your hair, building static electricity. The sudden sparks of static electricity produce electromagnetic waves that the radio picks up as noise. This illustrates an important principle: When we change the motion of an electron, we generate an electromagnetic wave.

Second, recall that temperature is a measure of the average energy of motion among the particles in a material. In the gases of a hot star, the atoms move faster, on average, than the atoms in a cool star.

With these two ideas in mind, we can understand how hot objects emit photons. All of the particles in the object are in rapid motion, colliding with one another over and over, and when an electron is involved in a collision, the sudden change in its motion can produce a photon. Thus, we can expect a hot object to emit electromagnetic radiation. Such radiation is called **black body radiation.** Although only a perfect radiator can emit perfect black body radiation, most hot objects behave roughly like black bodies. The hot filament in an incandescent light bulb is a good example.

In the heated filament, gentle collisions produce low-energy photons with long wavelengths, and violent collisions produce high-energy photons with short wavelengths. If we graph the energy emitted at different wavelengths, we get a curve like those shown in Figure 7-2. The curve shows that gentle collisions and violent collisons are rare. Most collisions are intermediate in violence, producing photons of intermediate wavelength.

Now we can understand how the photosphere of a star produces light. The light is black body radiation produced by the rapidly moving atoms in the gas.

One characteristic of black body radiation is that hot stars emit more radiation per second from 1 square meter of their surface than do cooler stars. In a hot gas, the atoms move more rapidly and collide more often, so they emit more photons per second. In Figure 7-2, we see that a difference of only 1000 K makes a big difference in the total amount of radiation emitted.

Another characteristic is that hot stars look bluer than red stars. Because the gases at the surface of a hot star are hot, collisions tend to be more violent, and the star tends to radiate higher-energy photons. Higher-energy photons have shorter wavelengths, so the hot star will tend to emit a larger fraction of blue light. In the gas at the surface of a cool star, collisions tend to be less violent, so the emitted photons tend to have lower energies and thus longer wavelengths. Cool stars look red.

In Figure 7-2, notice that the hottest object emits almost twice as much blue light as red light, but the object only 1000 K cooler emits roughly equal amounts of blue and red. The coolest object in the figure emits only about half as much blue as red.

The two characteristics of black body radiation described here can be expressed as two laws of radiation and given simple mathematical form, as shown in the next section.

Two Radiation Laws

Black body radiation can be described by two simple laws that will help us understand the nature of starlight. One law is related to energy and one to color.

As we saw in the previous section, a hot surface emits more black body radiation than a cool surface. That is, it emits more energy. Recall from Chapter 5 that we measure energy in units called joules (J); 1 joule is about the energy of an apple falling from a table to the floor. The total radiation given off by 1 square meter of the object in joules per second equals a constant number, represented by σ, times the temperature raised to the fourth power.* This relationship is often called the Stefan–Boltzmann law:

$$E = \sigma T^4 \; (\text{J/sec/m}^2)$$

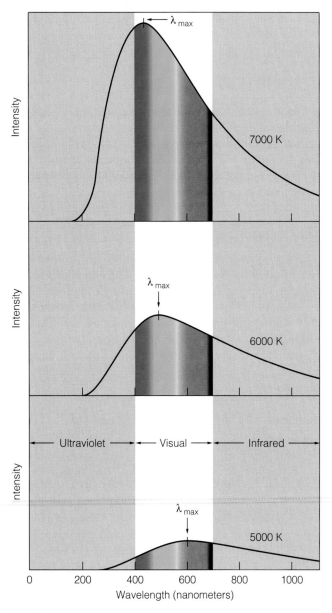

FIGURE 7-2

Black body radiation from three objects of nearly the same temperature demonstrates that a hotter object emits more short-wavelength radiation than a cooler object. The wavelength of maximum intensity, λ_{max}, shifts to longer wavelengths (as shown by its shift to the right) as temperature falls.

*For the sake of completeness, we should note that the constant σ equals 5.67×10^{-8} J/m²sec degree⁴.

How does this help us understand stars? Suppose a star the same size as the sun had a surface temperature that was twice as hot as the sun's surface. Then each square meter of that star would radiate not twice as much energy but 2^4 or 16 times as much energy. From this law we see that a small difference in temperature can produce a very large difference in the amount of energy produced.

The second radiation law is related to the color of stars. In the previous section, we saw that hot stars look blue and cool stars look red. Wien's law tells us that the wavelength at which a star radiates the most energy, the **wavelength of maximum (λ_{max}),** depends only on the star's temperature:

$$\lambda_{max} = \frac{3,000,000}{T}$$

That is, the wavelength of maximum radiation in nanometers equals 3 million divided by the temperature on the Kelvin scale.

This is a powerful tool in astronomy because it means we can relate the temperature of a star and its wavelength of maximum. For example, we might find a star that has a surface temperature of 3000 K. Then its wavelength of maximum would be 3,000,000/3000, or 1000 nm, in the near infrared. Later we will meet objects much hotter than most stars; such objects radiate most of their energy at very short wavelengths. The hottest stars, for instance, radiate most of their energy in the ultraviolet.

These two laws of radiation are powerful tools in astronomy, but astronomers also use a simpler method to estimate the temperature of a star. They measure its color directly.

The Color Index

Hot stars look blue, but if we want to know the temperature of the star's surface, we must have a way of measuring color. In astronomy, the **color index** is a measure of color.

To measure the color index of a star, we need a photometer and a set of standard filters. The most commonly used filters are the blue (B) and visual (V). Each filter isolates a specific part of the spectrum. The B filter, for example, will not allow any light through except those photons with wavelengths between about 400 nm and 480 nm. The V filter is transparent to photons with wavelengths between about 500 nm and 600 nm, roughly approximating the sensitivity of the human eye.

If we measure the magnitude of a star through each of these filters, the difference between the two magnitudes is a number related to the color of the star. This is commonly written B−V and is referred to as the B-V color index.

For example, a hot star radiates much more blue light than red and will be brighter through the B filter (Figure 7-2). That means that the B magnitude will be smaller and the B-V color index will be negative. The bluest stars have a B-V color index of about −0.4 and have surface temperatures of about 50,000 K. Red stars radiate much more red light than blue, so they will be fainter through the blue filter. Then B-V for a red star will be positive. The reddest stars have a B-V color index of about 2 and surface temperatures of about 2000 K.

CRITICAL INQUIRY

Why are hot stars bluer than cool stars?

Temperature is a measure of the average velocity of the randomly moving particles of a material. If a star is hot, then its atoms must be bouncing around at high velocities. When they interact with electrons and cause the emission of photons, most of the interactions are quite violent, and most of the photons that are emitted are very energetic. If photons have high energies, they have short wavelengths, so the gas in a hot star tends to emit larger numbers of blue photons than red photons, and the star looks blue.

In a cool star, the atoms are not moving as rapidly, collisions are less violent, and the average photon emitted is not as energetic. Lower-energy photons have longer wavelengths and look red, so cooler stars emit a larger proportion of red photons.

This interesting bit of physics explains why hot stars look blue and cool stars look red. But the surface of a hot star emits more energy than that of a cool star. How is this related to the agitation of the atoms? ■

The colors of the stars are beautiful, but now we know that their color reveals their temperature. To get even more information from starlight, we must understand the structure of atoms and how atoms interact with light.

7-2 ATOMS

The atoms in the surface layers of stars leave their marks on the light the stars emit. By understanding what atoms are and how they interact with light, we can decode the spectra of the stars. We begin this section by constructing a working model of an atom.

A Model Atom

In Chapter 2, we devised a model of the sky, the celestial sphere, to help us think about the nature and motion of the heavens. In the case of the atom, we again need a model.

Our model of the atom consists of a small central **nucleus** surrounded by a cloud of orbiting electrons. The nucleus has a diameter of about 0.0000016 nm, and the cloud of electrons has a diameter of about 0.1–0.5 nm. (Recall from Chapter 6 that 1 nm is 10^{-9} m.) Household plastic wrap is about 100,000 atoms thick. This makes the atom seem very small, but the nucleus is 100,000 times smaller.

Atoms are so small that any scale model we might make would have to be magnified tremendously. Suppose that we could make a hydrogen atom bigger by a factor of 10^{12} (1 million million). Only then would it be big enough to examine.

The nucleus of a hydrogen atom is a proton whose diameter is about 0.0000016 nm, or 1.6×10^{-13} cm. Multiplying by a factor of 10^{12} magnifies it to 0.16 cm, about the size of a grape seed. The electron cloud* has a diameter of about 0.4 nm, or 4×10^{-8} cm. When we magnify the atom by 10^{12}, this becomes 400 m, or about $4\frac{1}{2}$ football fields laid end to end (Figure 7-3). When you imagine a grape seed in the midst of $4\frac{1}{2}$ football fields, orbited by one magnified electron still too small to be visible, you can see that an atom is mostly empty space.

The mass of a hydrogen atom is very small—about 2×10^{-24} g. For comparison, a paper clip has a mass of about 1 g. Individual atoms have such small masses that only by assembling vast numbers can nature build massive objects such as the stars. The sun, for instance, contains about 10^{57} atoms.

The nucleus of our model atom contains nearly all of the mass. The mass of the proton is about 1836 times the mass of the electron. The electrons in an atom never represent more than about 0.05 percent of the atom's mass.

The nucleus of a typical atom is more complicated than that of hydrogen in that it contains two different kinds of particles, protons and neutrons. **Protons** carry a positive electrical charge, and **neutrons** have no charge. Consequently, an atomic nucleus, made of protons and neutrons, has a net positive charge.

The electrons surrounding the nucleus carry a charge equal to but opposite that on the nucleus. In a neutral atom, the number of electrons equals the number of protons. Thus, the positive charge on each proton is balanced by the negative charge on an electron, and the atom is electrically neutral.

There are over a hundred kinds of atoms, called chemical elements. The kind of element an atom represents depends only on the number of protons in its nucleus. For example, carbon has six protons and six neutrons in its nucleus. Adding a proton produces nitrogen, and subtracting a proton produces boron.

However, we can change the number of neutrons in an atom's nucleus without changing the atom significantly. For instance, if we add a neutron to the carbon nucleus, we still have carbon, but it is slightly heavier than the more common form of carbon. Atoms that have the same number of protons but a different number of neutrons are called **isotopes.**

*For a representative diameter, we take the size of the atom's second orbit.

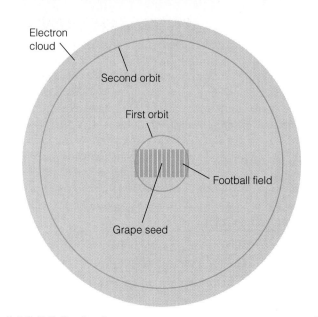

FIGURE 7-3
Magnifying a hydrogen atom by 10^{12} makes the nucleus the size of a grape seed and the diameter of the outer electron cloud about $4\frac{1}{2}$ times longer than a football field. The electron itself is still too small to see.

Hydrogen has a single proton as its nucleus, but a rare isotope of hydrogen, deuterium, has a proton and a neutron in its nucleus (Figure 7-4). Carbon has two stable isotopes. One form contains six protons and six neutrons, making a total of twelve particles, and is thus called carbon-12. Carbon-13 has six protons and seven neutrons in its nucleus. Figure 7-4 shows schematically the nuclei of these isotopes.

Protons and neutrons are bound tightly into the nucleus, but electrons are held loosely in the electron cloud. Running a comb through your hair creates a static charge by removing a few electrons from their atoms. This process is called **ionization,** and the atom that has lost one or more electrons is an **ion.** The neutral carbon atom, with six protons and six neutrons in its nucleus, has six electrons, which balance the positive charge of the nucleus. If we ionize the atom by removing one or more electrons, the atom is left with a net positive charge. Under some circumstances, an atom may capture one or more extra electrons, giving it a net negative charge. Such a negatively charged atom is also an ion.

Atoms form bonds with each other by exchanging or sharing electrons. Two or more atoms bonded together form a **molecule.** Few atoms can form chemical bonds in stars. The high temperatures produce such violent collisions between atoms that most molecules would quickly break up. Only in the coolest stars are the collisions gentle enough to permit chemical bonds. We will see later that the presence of molecules such as titanium oxide (TiO) in a star is a clue that the star is cool. In later chapters, we will also see that molecules can form in cool gas clouds in space and in the atmospheres of planets.

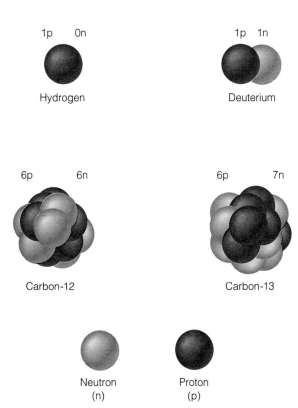

FIGURE 7-4
Some common isotopes. A rare isotope of hydrogen, deuterium, contains a proton and a neutron in its nucleus. Two isotopes of carbon are carbon-12 and carbon-13.

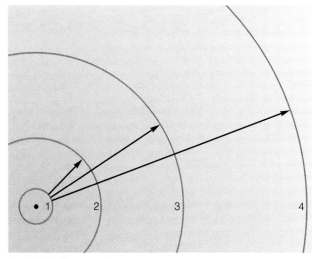

FIGURE 7-5
The electron in a hydrogen atom can occupy any permitted energy level but not energy levels in between. If an atom with its electron in the lowest energy level can absorb the correct energy, it can move its electron to a higher energy level.

Electron Shells

So far we have described the electron cloud only in a general way, but the specific way electrons behave within a cloud is very important in astronomy.

The electrons are bound to the atom by the attraction between their negative charge and the nucleus's positive charge. If we wish to ionize the atom, we need a certain amount of energy to pull an electron away from its nucleus. This energy is the electron's **binding energy,** the energy that holds it to the atom.

The rules that govern the motion of an electron within an atom are the same rules that govern our lives, but, because the atom is very small, peculiar effects show up. Called **quantum mechanics,** these rules specify that an electron can have only certain amounts of binding energy. Physicists refer to these possible amounts of binding energy as **energy levels** (Figure 7-5). An electron with a certain amount of binding energy is said to be "in" the corresponding energy level. In our model atom, we think of the different energy levels as different-size orbits within the electron cloud (Window on Science 7-2).

These energy levels are like steps in a staircase: you can stand on the first step or the second step but not in between. Similarly, the electron can be in any of the permitted energy levels but cannot occupy any level in between. That is, the electron can have any of the permitted energies but no other energy.

The arrangement of permitted energy levels depends mainly on the charge on the nucleus and the number of electrons in the atom. Thus, each kind of element has its own pattern of energy levels. Isotopes of the same element have nearly the same pattern because they have the same number of protons. However, ionized atoms have energy-level patterns that differ from their un-ionized forms. Thus, the arrangement of energy levels differs for every kind of atom and ion.

CRITICAL **INQUIRY**

How many hydrogen atoms would it take to cross the head of a pin?

This is not a frivolous question. In answering it, we will discover how small atoms really are, and we will see how powerful physics can be as a way to understand nature. First, we will assume that the head of a pin is about 1 mm in diameter. That is 0.001 m. The size of a hydrogen atom is represented by the diameter of the electron cloud, and we will assume that the electron in our atom is in the second orbit. Then the diameter of the electron cloud is about 0.4 nm. Since 1 nm equals 10^{-9} m, we multiply and discover that 0.4 nm equals 4×10^{-10} m. To find out how many atoms would stretch 0.001 m, we divide the diameter of the pinhead by the diameter of an atom. That is, we divide 0.001 m by 4×10^{-10} m, and we get 2.5×10^6. It would thus take 2.5 million hydrogen atoms lined up side by side to cross the head of a pin.

Quantum Mechanics

Quantum mechanics is the set of rules that describe how atoms and subatomic particles behave. When we think about large objects such as stars, planets, aircraft carriers, and hummingbirds, we don't have to think about quantum mechanics, but on the atomic scale, particles behave in ways that seem unfamiliar to us.

One of the principles of quantum mechanics specifies that we can not know simultaneously the exact location and motion of a particle. This is why physicists refer to the electrons in an atom as if they were a cloud of negative charge surrounding the nucleus. Since we can't know the position and motion of the electron, we can't really describe it as a small particle following an orbit. We can use that image as a model to help our imaginations, but the reality is much more interesting, and describing the electrons as a charge cloud gives us a better and more sophisticated model of an atom.

This raises some serious questions about reality. Is an electron really a particle at all? Quantum mechanics describes particles as waves, and waves as particles. If we can't know simultaneously the position and motion of a specific particle, how can we know how it will react to a collision with a photon or another particle? The answer is that we can't know, and that seems to violate the principle of cause and effect (Window on Science 5-2).

Needless to say, we can't explore these discrepancies here. Scientists and philosophers of science continue to struggle with the meaning of reality on the quantum-mechanical level. Here we should note that the reality we see on the scale of stars and hummingbirds is only part of nature. We have constructed some models to help us think about nature on the scale of atoms, but the truth is much more interesting and much more exciting than anything we see on larger scales. Although we use models of atoms to study stars, there is still much to learn about the atoms themselves. ∎

This shows how tiny an atom is and also how powerful basic physics is. A bit of arithmetic gives us a view of nature beyond the capability of our eyes. Now use another bit of arithmetic to calculate how many hydrogen atoms you would need to add up to the mass of a paper clip (1 g). ∎

The energy levels in atoms are familiar territory to astronomers because the electrons in those levels can interact with light. Such interactions fill starlight with clues to the nature of the stars.

7-3 THE INTERACTION OF LIGHT AND MATTER

If light did not interact with matter, you would not be able to see these words. In fact, you would not exist, because, among other problems, photosynthesis would be impossible and there would be no grass, wheat, bread, beef, cheeseburgers, or any other kind of food. The interaction of light and matter makes our lives possible, and it also makes it possible for us to understand our universe.

Astronomers are experts on the interaction of light and matter, so we must review in detail how the atoms in a star interact with light. We begin with the hydrogen atom because it is both the simplest atom and the most common. Roughly 90 percent of all atoms in the universe are hydrogen. Once we understand the energy levels of hydrogen atoms, understanding other atoms will require only a simple elaboration.

The Excitation of Atoms

The hydrogen atom in Figure 7-5 has its electron in the lowest permitted energy level, where it is tightly bound to the atom. We can move the electron to a higher level by supplying some energy. This is like moving a flowerpot from a low shelf to a high shelf; the higher we move the pot, the more energy we must expend. The amount of energy needed to move the electron from one level to a higher level is just the energy difference between the levels.

If we move the electron from a low energy level to a higher level, we say the atom is **excited.** That is, we have added energy to the atom in moving its electron. If the electron falls back to a lower energy level, the energy is returned.

At atom can become excited if it collides with another atom. During the collision, some of the energy of motion of the atoms can be absorbed by one or both of the atoms, leaving them in an excited state. This is very common in a hot, dense gas where the atoms move rapidly and collide often.

An atom can also become excited if it absorbs a photon. Only a photon with exactly the right energy can move the electron from one energy level to another. Too much or too little energy, and the photon cannot be absorbed. Because the energy of a photon depends on its

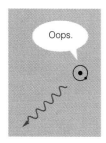

FIGURE 7-6

An atom can absorb a photon only if the photon has the correct energy. The excited atom is unstable and within a fraction of a second returns the electron to a lower energy level, reradiating the photon in a random direction.

wavelength, only photons of certain wavelengths can be absorbed by a given kind of atom. The atom in Figure 7-5, for example, can absorb photons of three different wavelengths, moving the electron up to any of three permitted energy levels. Photons of any other wavelength have either too much or too little energy to be absorbed.

Atoms, like humans, cannot remain excited forever. The excited atom is unstable and must eventually (usually within 10^{-6} to 10^{-9} second) give up the energy it has absorbed, and the electron returns to the lowest energy level. Because the electrons eventually tumble down to this bottom level, it is known as the **ground state.**

When the electron drops from a higher energy level to a lower level, it must give up the excess energy—usually by emitting a photon. Study the sequence of events in Figure 7-6 to see how an atom can absorb and emit photons of certain wavelengths.

Because only certain energy levels are permitted in an atom, only certain energy differences can occur. Each type of atom or ion has its unique set of energy levels, so each atom or ion absorbs and emits photons with a unique set of wavelengths. Thus, we can identify the elements in a gas by studying the characteristic wavelengths of light absorbed or emitted.

This process of excitation and emission produces a common sight. The gas in a neon sign glows because a high voltage forces electrons to flow through the gas, exciting the atoms by collisions. Almost as soon as an atom is excited, its electron drops back to lower energy levels, emitting the surplus energy as a photon of a certain wavelength. Neon is a popular gas for such signs because it emits a large number of photons of red, orange, and yellow wavelengths, which we see as a rich reddish orange. So-called neon signs of other colors contain other gases or mixtures of gases selected to produce those colors.

The Formation of a Spectrum

To see how an astronomical object can produce a spectrum, imagine a cloud of hydrogen floating in space with an incandescent light bulb glowing inside it. The bulb glows because its filament is hot, producing black body radiation. Thus, it emits photons of all wavelengths, and a spectrum of its light would reveal an uninterrupted band of color called a **continuous spectrum.**

However, the light from this bulb must pass through the hydrogen gas before it can reach our telescope (Figure 7-7). Most of the photons will pass through the gas unaffected because they have wavelengths that the hydrogen atoms cannot absorb, but a few photons will have the right wavelengths. These photons cannot pass through the gas because they are absorbed by the first atom they meet. The atom is excited for a fraction of a second, and the electron then drops back to a lower orbit and a new photon is emitted. The original photon was traveling through the gas toward our telescope, but the new photon is emitted in some random direction. Very few of these new photons leave the cloud in the direction of our telescope, so the light that finally enters the telescope has very few photons at the wavelengths the atoms can absorb. When we form a spectrum from this light, photons of these wavelengths are missing, and the spectrum has dark lines at the positions these photons would have occupied. These dark lines are called **absorption lines** because the atoms absorbed the photons. A spectrum containing absorption lines is an **absorption spectrum** (also called a **dark line spectrum**).

What happens to the photons that were absorbed? They bounce from atom to atom, being absorbed and emitted over and over until they escape from the cloud. If, instead of aiming our telescope at the bulb, we swing it to one side so that no light from the bulb enters the telescope, we can photograph a spectrum of the light emitted by the gas atoms (Figure 7-8). In that case, the only photons entering the telescope are photons that were absorbed and reemitted. A spectrum of this light is almost entirely dark except for the wavelengths corresponding to the photons the gas can absorb and reemit. Thus, we will see a spectrum containing only bright lines on a dark background. These bright lines are called **emission lines,** and a spectrum with emission lines is an **emission spectrum** (also called a **bright line spectrum**). The spectrum of a neon sign, for instance, is

FIGURE 7-7

Photons of the proper wavelengths can be absorbed by the gas atoms and reemitted in random directions. Because most of these particular photons do not reach the telescope, the spectrum is dark at the wavelengths of the missing photons.

FIGURE 7-8

Pointing the telescope away from the bulb, we can receive only those photons the atoms can absorb and reemit, producing emission lines in the spectrum.

TABLE 7-1

Kirchhoff's Laws

Law I: The Continuous Spectrum

A solid, liquid, or dense gas excited to emit light will radiate at all wavelengths and thus produce a continuous spectrum.

Law II: The Emission Spectrum

A low-density gas excited to emit light will do so at specific wavelengths and thus produce an emission spectrum.

Law III: The Absorption Spectrum

If light comprising a continuous spectrum is allowed to pass through a cool, low-density gas, the resulting spectrum will have dark lines at certain wavelengths. That is, it will be an absorption spectrum.

an emission spectrum. Also, the bluish purple color of mercury-vapor streetlights and the pinkish orange color of sodium-vapor streetlights are produced by the emission lines of those elements.

These three types of spectra (Figure 7-9) are summarized in **Kirchhoff's laws** (Table 7-1), which give us our first tool in the analysis of astronomical spectra. If the spectrum of an astronomical object contains emission lines, Kirchhoff's second law tells us that the object probably contains an excited, low-density gas. On the other hand, if an object's spectrum contains absorption lines, the third law tells us that we are probably looking at an object in which light is passing through a gas.

A star produces a spectrum in much the same way as the light bulb and gas cloud produce a spectrum in Figure 7-7. The outer layers of the star, which make up the photosphere, are relatively dense gas and emit a continuous spectrum much as does the filament in the light bulb. Above these layers lie the thinner gases of the star's atmosphere; as the light travels upward through these layers, photons of certain wavelengths are absorbed by atoms and so never reach us. The spectrum of a star is an absorption spectrum whose dark lines indicate which wavelengths the atoms absorbed (Figure 7-10).

The absorption lines in stellar spectra provide a windfall of data about the star's surface layers. By study-

400 nm   500 nm   600 nm   700 nm

Continuous
spectrum

Absorption
spectrum
of hydrogen

Emission
spectrum
of hydrogen

H δ   H γ   H β   H α
(410) (434) (486) (656)

FIGURE 7-9

The three types of spectra. A continuous spectrum (top) contains no bright or dark lines, but an absorption spectrum (middle) is interrupted by dark absorption lines. An emission spectrum (bottom) is dark except at certain wavelengths were emission lines occur. Note that the lines in the absorption spectrum of hydrogen have the same wavelengths as the lines in the emission spectrum of hydrogen.

Absorption spectrum

FIGURE 7-10

A star produces an absorption spectrum because its atmosphere absorbs certain wavelengths in the spectrum.

ing the spectral lines, we can identify the elements in the stellar atmosphere and find the temperature of the atoms. To understand how to get this information, we need to look carefully at the way the hydrogen atom produces lines in a star's spectrum.

The Hydrogen Spectrum

As you must have gathered by now, each element has its own spectrum, as unique as a human fingerprint, and can be recognized by its spectrum across trillions of miles. To see how hydrogen produces its spectrum, we must draw a scale diagram of its permitted energy levels with the radius of the level proportional to its energy (Figure 7-11). Then we can examine the way hydrogen atoms interact with light.

A **transition** occurs in an atom when an electron changes energy levels. In our diagram of a hydrogen atom, we represent transitions by arrows pointing from one level to another. If the arrow points upward, the

atom must absorb energy, and if the arrow points downward, the atom must emit energy.

If the transition results in the absorption or emission of a photon, the length of the arrow tells us its energy. Long arrows represent large amounts of energy and thus short-wavelength photons. Short arrows represent smaller amounts of energy and longer-wavelength photons.

We can divide the possible transitions in a hydrogen atom into groups, called series, according to their lowest energy level. Those arrows whose lower ends rest on the ground state represent the **Lyman series;** those resting on the second energy level, the **Balmer series;** and those resting on the third, the **Paschen series.** In principle, each series contains an infinite number of transitions, and there are an infinite number of series. Figure 7-11 shows only the first few transitions in the first few series.

The Lyman series transitions involve large energies, as shown by the long arrows in Figure 7-11. These energetic transitions produce lines in the ultraviolet part of the spectrum, where they are invisible to the human eye.

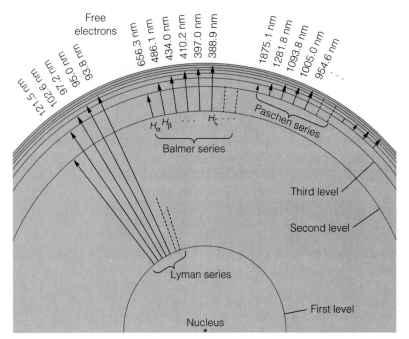

Free electrons

121.5 nm
102.6 nm
97.2 nm
95.0 nm
93.8 nm
. . .

656.3 nm
486.1 nm
434.0 nm
410.2 nm
397.0 nm
388.9 nm

1875.1 nm
1281.8 nm
1093.8 nm
1005.0 nm
954.6 nm
. . .

H_α H_β . . . H_ζ . . .

Balmer series

Paschen series

Third level

Second level

Lyman series

First level

Nucleus

FIGURE 7-11

The levels in this diagram are spaced to represent the energy the hydrogen atom's electron can have. The transitions, drawn as arrows, can be grouped into series according to the lowest level. This drawing shows only a few of the infinity of transitions and series possible. Note that the arrows point upward and thus represent transitions that would absorb energy.

FIGURE 7-12

The Lagoon Nebula in Sagittarius is a cloud of gas and dust about 60 ly in diameter. Its gases are excited by the ultraviolet radiation of the hot, young stars within, and it glows in the pink color produced by the mixture of the red, blue, and violet Balmer lines. (National Optical Astronomy Observatories)

Balmer lines

365 nm   370 nm   375 nm

FIGURE 7-13

The Balmer lines photographed in the near ultraviolet. (The Observatories of the Carnegie Institution of Washington)

The very low energy Paschen transitions produce spectral lines in the infrared.

The first few Balmer lines are the only hydrogen lines in the visible spectrum (Figure 7-11). The Balmer lines are labeled by Greek letters in order of decreasing wavelength. H_α is a red line, H_β is blue, and H_γ is violet. These lines blend together to produce the purple-red color characteristic of glowing clouds of hydrogen (Figure 7-12). The remaining Balmer lines are in the near ultraviolet and are invisible to the eye, but they can be photographed easily (Figure 7-13).

CRITICAL INQUIRY

What spectrum would we see if we observed molten iron?

Molten iron is a dense liquid, and the atoms and molecules collide so often they emit all wavelengths and we would see a continuous spectrum. That is Kirchhoff's first law. But in order to see the molten iron, we would have to look through the hot vapors rising from it. Photons on their way to our spectrograph would pass through these gases, and atoms in the gases would absorb certain wavelengths. So what we would really see would be the continuous spectrum of the molten iron with weak absorption lines caused by the gases above the iron. This is what Kirchhoff's third law describes.

But suppose we had a very sensitive spectrograph that could look at the hot gases above the molten iron from the side so as to avoid looking directly at the molten iron. What kind of spectrum would the hot gases emit? ∎

Whatever kind of spectrum astronomers look at, the most common spectral lines are the Balmer lines of hydrogen, the only hydrogen lines we can study from the earth's surface. In the next section, we will see how Balmer lines can tell us a star's temperature.

7-4 STELLAR SPECTRA

In later chapters, we use spectra to study galaxies and planets, but we begin here by studying the spectra of stars. Such spectra are the easiest to understand, and the nature of stars is central to our study of all celestial objects.

The Balmer Thermometer

We can use the Balmer absorption lines as a thermometer to find the temperatures of stars. Earlier we saw how to estimate temperature from color, but the strengths of the Balmer lines in a star's spectrum give a much more accurate estimate of the star's temperature.

The Balmer thermometer works because the Balmer absorption lines are produced only by atoms whose elec-

trons are in the second energy level (Figure 7-11). If the star is cool, there are few violent collisons between atoms to excite the electrons, and most atoms have their electron in the ground state. If most electrons are in the ground state, they can't absorb photons in the Balmer series. As a result, we should expect to find weak Balmer absorption lines in the spectra of cool stars.

In hot stars, on the other hand, there are many violent collisions between atoms, exciting electrons to high energy levels or knocking the electron clear out of some atoms; that is, some atoms are ionized. Thus, few atoms have their electron in the second orbit to form Balmer absorption lines, and we should expect hot stars, like cool stars, to have weak Balmer absorption lines.

At some intermediate temperature, the collisions are just right to excite large numbers of electrons into the second energy level. With many atoms excited to the second level, the gas absorbs Balmer-wavelength photons strongly and thus produces strong Balmer lines.

To summarize, the strength of the Balmer lines depends on the temperature of the star's surface layers. Both hot and cool stars have weak Balmer lines, but medium-temperature stars have strong Balmer lines.

Theoretical calculations can predict just how strong the Balmer lines should be for stars of various temperatures. The details of these calculations are not important to us, but the results are. Figure 7-14a shows the strength of the Balmer lines for various stellar temperatures. We could use this as a temperature indicator except that the curve gives us two answers. A star with Balmer lines of a certain strength might have either of two temperatures, one high and one low. We must examine other spectral lines to choose the correct temperature.

We have seen how the strength of the Balmer lines depends on temperature. The same process affects the spectral lines of other elements, but the temperature at which they reach maximum strength differs for each element (Figure 7-14b). If we add these elements to our graph, we get a handy tool for finding the stars' temperatures (Figure 7-14c).

If we photograph the spectra of many stars and arrange the photographic plates in order of decreasing temperature (Figure 7-15), we can see how spectral features depend on temperature. If we see a star whose spectrum contains medium-strength Balmer lines and strong helium lines, we can conclude that it has a temperature of about 16,000 K. If a star has weak hydrogen lines and strong lines of ionized calcium, we can conclude that it has a temperature of about 5000 K, similar to that of the sun. If we see broad bands of absorption from molecules of titanium oxide, as in the bottom spectrum in Figure 7-15, we know immediately that we are observing a very cool star. Compare Figure 7-15 with Figure 7-14c.

From a careful comparison of the strength of different kinds of stellar features, astronomers can determine the temperatures of stars. The hottest normal stars have surface temperatures of 40,000 K, and the coolest have temperatures of 2000 K. Compare these temperatures with the surface temperature of the sun, about 5800 K.

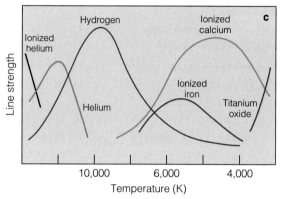

FIGURE 7-14

(a) The strength of the Balmer lines in a stellar spectrum depends on the temperature of the star. A star with medium-strength Balmer lines could have one of two possible temperatures. (b) The curve for lines of once-ionized calcium reaches maximum strength at a different temperature. (c) A graph of the principal features in stellar spectra can be used as the basis for a spectral classification system based on temperature. Given the strengths of the different features in a star's spectrum, the graph identifies its spectral type and temperature uniquely.

Spectral Classification

We have seen that the strengths of spectral lines depend on the surface temperature of the star. From this we can predict that all stars of a given temperature should have similar spectra. If we learn to recognize the pattern of spectral lines produced by a 6000-K star, for instance, we need not use Figure 7-14c every time we see that kind of spectrum. In other words, we can save time by classifying stellar spectra rather than by analyzing each one individually.

The first widely used classification system was devised by astronomers at Harvard during the 1890s and 1900s. One of them, Annie J. Cannon, personally inspected and classified the spectra of over 250,000 stars. The spectra were first classified in groups labeled A through Q, but some groups were later dropped, merged with others, or reordered. The final classification includes the seven **spectral classes,** or **types,** still used today: O, B, A, F, G, K, M.*

This sequence of spectral types, called the **spectral sequence,** is important because it is a temperature sequence. The O stars are the hottest, the B stars next hottest, and so on. The temperature continues to decrease down to the M stars, the coolest of all.

We can classify a star by examining features in its spectrum, as described in Table 7-2. For example, if it

has weak Balmer lines and lines of ionized helium, it must be an O star. This table is based on the same information used in Figure 7-14c.

Annie J. Cannon worked with tiny spectra on photographic plates that she studied with a magnifier. Modern spectrographs usually record spectra with CCD cameras, and it is natural to represent such spectra as graphs of intensity versus wavelength (Figure 7-16). In such a graph, a dark absorption line becomes a sharp dip in the curve. From such spectra, it is easy to measure the strengths of spectral features with great precision. Note

TABLE 7-2

Spectral Classes

Spectral Class	Approximate Temperature (K)	Hydrogen Balmer Lines	Other Spectral Features	Naked-Eye Example
O	40,000	Weak	Ionized helium	Meissa (O8)
B	20,000	Medium	Neutral helium	Achernar (B3)
A	10,000	Strong	Ionized calcium weak	Sirius (A1)
F	7,500	Medium	Ionized calcium weak	Canopus (F0)
G	5,500	Weak	Ionized calcium medium	Sun (G2)
K	4,500	Very weak	Ionized calcium strong	Arcturus (K2)
M	3,000	Very weak	TiO strong	Betelgeuse (M2)

*Generations of astronomers have remembered the spectral sequence using the mnemonic "Oh, Be A Fine Girl, Kiss Me." Nationwide contests to find a less sexist mnemonic have failed to displace this traditional sentence.

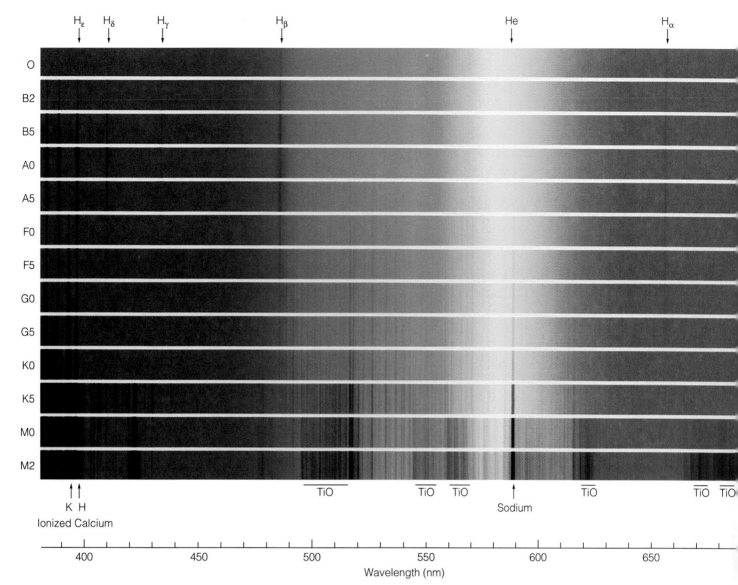

FIGURE 7-15

Stellar spectra from class O to class M2 show how spectral lines vary with stellar temperature. In the spectra of hot stars we see weak Balmer lines and helium lines. The Balmer lines are strongest about A0, but are very weak in cool stars. The two ultraviolet lines of ionized calcium are strong in cooler stars, while sodium and titanium oxide (TiO) bands are strong in the spectra of the coolest stars. Note that these synthetic spectra have been extended past the short wavelength limit for the human eye, about 420 nm, to show the H and K lines of calcium. (Courtesy Roger Bell and Michael Briley)

that the spectra in Figure 7-16 show the same dependence on temperature as the photographic spectra in Figure 7-15 and the graph in Figure 7-14c.

The spectra shown in Figure 7-15 illustrate how spectral lines change from class to class. Study the Balmer line H_β in the center of the spectra. In the spectrum of the O5 star, H_β is not very strong. As we run our eye down the spectra, H_β becomes stronger, reaching maximum strength about A0, and then becomes weaker in the cooler stars. The hotter O5 star has weak Balmer lines because most of the hydrogen atoms are excited to energy levels above the second and thus cannot absorb Balmer photons. The cooler stars (F through M) have weak Balmer lines because most of the hydrogen atoms are not excited out of the ground state.

The spectral lines of other atoms also change from class to class. Helium is visible only in the spectra of the hottest classes, and titanium oxide bands only in the coolest. The two lines of ionized calcium, labeled H and K, increase in strength from A to K and then decrease from K to M. Because the strength of these spectral features depends on temperature, it requires only a few minutes to compare a star's spectrum with Table 7-2 or Figure 7-15 and determine its temperature.

To be more precise, we can divide each spectral class into ten subclasses. For example, spectral class A consists of the subclasses A0, A1, A2, . . . , A8, A9. Next come F0, F1, F2, and so on. Thus, A5 lies halfway between A0 and F0. This finer division, of course, demands that we look carefully at a spectrum, but it is worth the effort, for the

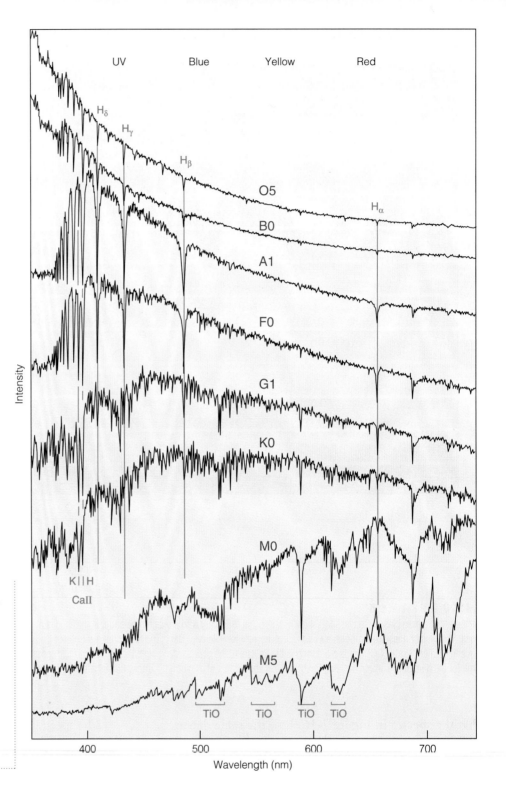

FIGURE 7-16

Modern digital spectra show how stellar spectra depend on spectral class. Here spectra are represented by graphs of intensity versus wavelength, and dark absorption lines appear as sharp dips in intensity. Hydrogen Balmer lines are strongest at about A0, while lines caused by CaII are strong in K stars. Bands produced by TiO molecules are strong in the coolest stars. Compare these spectra with Figures 7-14c and 7-15. (Courtesy NOAO, G. Jacoby, D. Hunter, and C. Christian)

subclasses give us a star's temperature within an accuracy of about 5 percent. The sun, for example, is not just a G star but a G2 star. That means its temperature is about 5800 K. Often, astronomers round this to 6000 K for convenience in approximate calculations.

Stellar spectra tell us the temperature of stars, but spectra contain other kinds of information as well. A bit of simple physics can extract a wide range of information from the spectrum of a star, such as its composition, motion, and atmospheric density.

The Composition of the Stars

Many branches of science and industry use spectrographs to analyze everything from chocolate candy to steel pipe, and astronomers, too, can use spectra to discover the chemical composition of the stars. The story of how astronomers first discovered the composition of the stars is worth telling, not only because it is the story of an important American astronomer who never received proper credit, but also because the story illustrates the special problems astronomers have to solve in order to find composition from spectra. The story begins in England.

As a child in England, Cecilia Payne (1900–1979) excelled in classics, languages, mathematics, and literature, but her first love was astronomy. After finishing Newnham College in Cambridge, she left England, sensing that there were no opportunities in England for a woman of science. In 1922, Payne arrived at Harvard, where she eventually earned her Ph.D., although the degree was awarded by Radcliffe because Harvard did not then admit women.

In her thesis, Payne attempted to relate the strength of the absorption lines in stellar spectra to the physical conditions in the atmospheres of the stars. This was not easy because, as we have seen in this chapter, a given spectral line can be weak because the atom is rare or because the temperature is too high or too low for that atom to be able to absorb efficiently. If we see sodium lines in a star's spectrum, we can be sure that the star contains sodium atoms, but if we see no sodium lines, we must consider the possibility that the star is too hot or too cool for sodium to produce spectral lines.

Payne's problem was to untangle these two factors and find the true temperatures of the stars and the true abundance of the atoms in their atmospheres. Recent advances in atomic physics gave her the theoretical tools she needed. About the time Payne left Newnham College, Indian physicist Meghnad Saha published his work on the ionization of atoms. Drawing from such theoretical work, Payne was able to show that over 90 percent of the atoms in stars (including the sun) were hydrogen and most of the rest helium (Table 7-3). The heavier atoms seemed more abundant only because they are better at absorbing photons at the temperatures of stars.

At the time, astronomers found it hard to believe that hydrogen and helium were so abundant in stars. After all, hydrogen lines are not strong in most stars, and helium lines are almost invisible in all but the hottest stars. Rather, nearly all astronomers assumed that the stars had roughly the same composition as the surface of the earth; that is, they believed that the stars were composed mainly of heavier atoms such as carbon, silicon, iron, aluminum, and so on. Even the most eminent astronomers dismissed Payne's result as illusory. Faced with this pressure and realizing the limited opportunities available to women in science in the 1920s, Payne could not press her discovery.

It was 1929 before astronomers generally understood the importance of temperature on measurements of composition derived from stellar spectra. At that point, astronomers recognized that stars are mostly hydrogen and helium, but Payne received no credit.

Payne worked for many years as a staff astronomer at the Harvard College Observatory with no formal position on the faculty. She married Russian astronomer Sergei Gaposchkin in 1934 and was afterward known as Cecilia Payne-Gaposchkin. In 1956, when Harvard accepted women to its faculty, she was appointed a full professor and chair of the Harvard astronomy department.

TABLE 7-3

The Most Abundant Elements in the Sun

Element	Percentage by Number of Atoms	Percentage by Mass
Hydrogen	92.0	73.4
Helium	7.8	25.0
Carbon	0.03	0.3
Nitrogen	0.008	0.1
Oxygen	0.06	0.8
Neon	0.008	0.1
Magnesium	0.002	0.05
Silicon	0.003	0.07
Sulfur	0.002	0.04
Iron	0.004	0.2

SOURCE: Adapted from C. W. Allen, *Astrophysical Quantities*, London: Athlone Press, 1976.

Cecilia Payne-Gaposchkin's work on the chemical composition of the stars illustrates the importance of fully understanding the interaction between light and matter. Only a detailed understanding of the physics could lead her to the correct composition. As we turn our attention to other information that can be derived from stellar spectra, we again discover the importance of understanding how light and atoms interact.

Radial Velocity

The spectra of stars contain information not only about temperature and composition but also about motion. By understanding how moving atoms emit and absorb light, we can discover how the stars are moving through space.

The wavelengths of spectra lines are determined by the structure of atoms, but small differences in the precise wavelengths of spectral lines are caused by the motion of the object that emitted the light. This change in the observed wavelength due to the relative motion of the source and the observer is called the **Doppler effect.** Although it affects both light and sound, we are most familiar with the Doppler effect on the sounds of moving objects. The sound of a passing car, for example, has slightly shorter wavelengths as it approaches us and slightly longer wavelengths after it passes us. We hear this as a drop in the pitch of the sound as the car rushes past. The faster the car travels, the larger the drop in pitch, and for very high velocity objects such as low-flying airplanes, the drop in pitch is dramatic.

Blue
shift

Red
shift

a

FIGURE 7-17

(a) Successive clangs of the engine bell (marked by dots) occur closer to the observer ahead, decreasing the distance the sound must travel. Thus, the observer hears the bell ring more often than it really does. The observer behind the train hears the bell ring less often. This is an example of the Doppler effect. (b) The upper spectrum of Arcturus was taken when the earth's orbital motion carried it toward the star. The lower spectrum was taken 6 months later when the earth was receding from Arcturus. The difference in the wavelengths of the lines is due to the Doppler shift. (The Observatories of the Carnegie Institution of Washington)

—Approaching

—Receding

← Red   Blue →

b

To understand how the Doppler shift works, it is helpful to think about a railroad locomotive with a bell. Imagine standing on a railroad track as a train approaches with the engine bell ringing once each second. (Figure 7-17). When the bell rings, the sound travels ahead of the engine to reach your ears. One second later the bell rings again, but not at the same place. During that 1 second, the engine moved closer to you, so the bell is closer at its second ringing. Now the sound has a shorter distance to travel and reaches your ears a little sooner than it would have if the engine had not moved. The third time the bell rings, it is even closer. By timing the ringing of the bell, you would observe that the bell seemed to be ringing more often than once each second, all because the engine was approaching.

Standing behind the engine would produce the opposite effect. You would find that each successive ring took place farther away from you, and the rings would sound as if they were more than 1 second apart. These apparent changes in the rate of the ringing bell are an example of the Doppler effect.

We can think of the peaks of the electromagnetic waves leaving a star as a series of clangs from a bell. If a star is moving toward us, we see the peaks of the light waves closer together than expected, making the wavelengths slightly shorter than they would have been if the star were not moving. If the star is going away from us, the peaks of the light waves are slightly farther apart and the wavelengths are longer. Thus, the lines in a star's spectrum suffer a small **blue shift** if the star is approaching and a small **red shift** if it is receding, although the

changes in wavelength are generally much too small to actually affect the color.

For convenience, this example has assumed that the earth is stationary and the star is moving, but the Doppler effect depends only on relative motion. Thus, we cannot say that either the earth or the star is stationary—only that there is relative motion between them. In addition, the Doppler effect depends only on the **radial velocity,** that part of the velocity directed away from or toward the earth (Figure 7-18). The Doppler effect cannot reveal relative motion to the right or left—the **transverse velocity.**

How much the spectral lines change depends on the radial velocity. This can be expressed as a simple ratio relating the radial velocity V_r divided by the speed of light c to the change in wavelength $\Delta\lambda$ divided by the unshifted wavelength λ_0:

$$\frac{V_r}{c} = \frac{\Delta\lambda}{\lambda_0}$$

This expression is quite accurate for the low radial velocities of stars, but we will need a better version later when we discuss objects moving with very high velocities.

For example, we might observe a line in a star's spectrum with a wavelength of 600.1 nm. Laboratory measurements show that the line should have a wavelength of 600 nm. That is, its unshifted wavelength is 600 nm. What is the star's radial velocity? First we note that the change in wavelength is 0.1 nm:

$$\frac{V_r}{c} = \frac{0.1}{600} = 0.000167$$

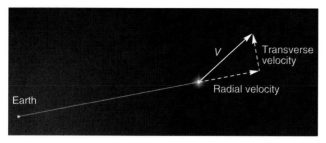

FIGURE 7-18
The radial velocity of a star is the part of the star's velocity *V* that is directed away from or toward the earth. The Doppler effect can tell us a star's radial velocity but cannot tell us the transverse velocity—the part of the velocity perpendicular to the radial direction.

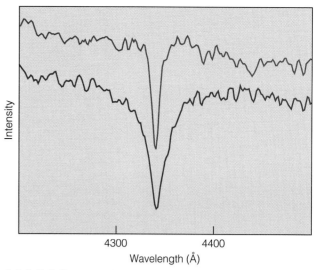

FIGURE 7-19
A line profile is a graph that shows the intensity of starlight as a function of wavelength across a spectral line. In these two examples, the spectra show the same spectral line from the spectra of two A1 stars. Thus, both stars have the same temperature. Differences in the density of the gas in the atmospheres of the stars cause the differences in the widths of the spectral lines. Such differences in the width of the spectral lines are easily measured from such line profiles. (Courtesy NOAO, G. Jacoby, D. Hunter, and C. Christian)

Multiplying by the speed of light, 3×10^5 km/sec, gives the radial velocity, 50 km/sec. Because the wavelength is shifted to the red (lengthened), the star must be receding from us.

You may be quite familiar with this method of speed measurement. Police radar uses the Doppler shift in reflected radio waves to determine the velocity of approaching cars.

Understanding the Doppler shift leads us to a final illustration of the information hidden in stellar spectra. Even the shapes of the spectral lines can tell us secrets about the stars.

The Shapes of Spectral Lines

When astronomers refer to the shape of a spectral line, they mean the variation of intensity across the line. An absorption line, for instance, is darkest in the center and grows brighter to each side. This shape is represented by a **line profile,** a graph of brightness as a function of wavelength across a spectral line (Figure 7-19).

The exact shape of a line profile can tell us a great deal about a star, but the most important characteristic is the width of the line. Spectral lines are not perfectly narrow; if they were, we could not see them. They have a natural width because nature allows an atom some leeway in the energy it may absorb or emit. The quantum mechanics behind this effect is beyond the scope of our discussion, but the result is very simple. In the absence of all other effects, spectral lines have a natural width of about 0.001 to 0.00001 nm—very narrow indeed.

The natural widths of spectral lines are not important in most branches of astronomy because other effects smear out the lines and make them much broader. For example, if a star spins rapidly, the Doppler effect will broaden the spectral lines. As the star rotates, one side will recede from us, and the other side will approach. Light from the receding side will be red-shifted, and light from the approaching side will be blue-shifted, so any spectral lines will be broadened.

Another important process is called **Doppler broadening.** To consider this process, let us imagine that we photograph the spectrum of a jar full of hydrogen atoms (Figure 7-20). Because the gas has some thermal energy (it is not at absolute zero), the gas atoms are in motion. Some will be coming toward our spectrograph, and some will be receding. Most, of course, will not be traveling very fast, but some will be moving very quickly. The photons emitted by the atoms approaching us will have slightly shorter wavelengths because of the Doppler effect, and photons emitted by atoms receding from us will have slightly longer wavelengths. Thus, the Doppler shifts due to the motions of the individual atoms will smear the spectral line out and make it broader. We have described the Doppler broadening of an emission line, but the effect is the same for absorption lines.

The extent of Doppler broadening depends on the temperature of the gas. If the gas is cold, the atoms travel at low velocities, and the Doppler shifts are small (Figure 7-20a). If the gas is hot, however, the atoms travel faster, Doppler shifts are larger, and the lines will be wider (Figure 7-20b).

Another form of broadening, **collisional broadening,** is caused by collisions between atoms, and thus it

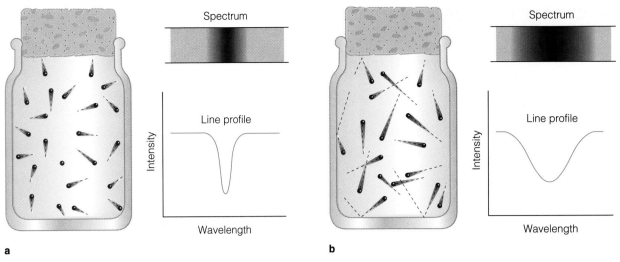

FIGURE 7-20

Doppler broadening. The atoms of a gas are in constant motion. Photons emitted by atoms moving toward the observer will have slightly shorter wavelengths, and those emitted by atoms moving away will have slightly longer wavelengths. This broadens the spectral line. If the gas is cool (a), the atoms do not move very fast, the Doppler shifts are small, and the line is narrow. If the gas is hot (b), the atoms move faster, the Doppler shifts are larger, and the line is broader.

depends on the density of the gas (Window on Science 7-3). Densities in astronomy cover an enormous range, from one atom per cubic centimeter in space to millions of tons of atoms per cubic centimeter inside dead stars. Clearly, we should expect such densities to affect the way atoms collide with one another and thus how they absorb and emit photons.

Collisional broadening spreads out spectral lines when the atoms absorb or emit photons while they are colliding with other atoms, ions, or electrons. The collisions disturb the energy levels in the atoms, making it possible for the atoms to absorb a slightly wider range of wavelengths. Thus, the spectral lines are wider. Because atoms in a dense gas collide more often than atoms in a low-density gas, collisional broadening depends on the density of the gas, but temperature is also a factor. Atoms in a hot gas travel faster and collide more often and more violently than atoms in a cool gas. Once again, the physics of the interaction of light and matter gives us a tool to understand starlight. In later chapters, we will see how collisional broadening affects the widths of spectral lines that are formed in normal stars like the sun, in giant stars, and in gas in interstellar space.

CRITICAL INQUIRY

If helium is much more common in the sun than calcium, why are helium lines weak and calcium lines strong in the visible solar spectrum?

To answer this question, we must ride with a few photons as they leave the sun. Photons that can be absorbed by calcium ions have a difficult time because, although they encounter few calcium ions, almost every calcium ion is at just the right energy level to absorb those photons. Few of those photons make it out of the sun, and we see dark spaces where they would be in the solar spectrum—the lines of ionized calcium. At the temperatures in the solar atmosphere, calcium is excited to just the right energy level to absorb photons, but helium is much harder to excite. At solar temperatures, helium is mostly in the ground state, and the visible-wavelength photons that it could absorb, although they will encounter many helium atoms, almost never encounter a helium atom excited to the right energy level to absorb the photon. Those photons escape from the sun, and we don't see strong helium lines in the solar spectrum.

This explains why astronomers in the 1920s found it hard to believe Cecilia Payne-Gaposchkin's claim that 10 percent of the atoms in the sun were helium. We have to think carefully about how atoms interact with light before we can interpret spectra. Use what you have learned to explain why TiO bands appear in the spectra of only the coolest stars. ■

This chapter has described some of astronomy's most powerful techniques. By understanding how light interacts with matter, how spectra are produced, and how the spectral lines acquire their shapes, astronomers can discover the composition, temperature, density, and motions of gas anywhere in the universe. All they need is a glimmer of light to spread out into a spectrum.

In the next chapter, we will apply these tools to the sun, our own star. In subsequent chapters, we will explore farther into the universe, analyzing the spectra of other stars, nebulae, and distant galaxies. ■

Density

One of the fundamental parameters in science is **density,** the measure of the amount of matter in a given volume. Density is expressed as mass per volume, such as grams per cubic centimeter. The density of water, for example, is about 1 g/cm³.

To get a feel for density, imagine holding a brick in one hand and a similar-size block of styrofoam in the other hand. We can easily tell that the brick contains more matter than the styrofoam block even though both are the same size. The brick weighs more than the styrofoam, but it isn't really the weight that we should consider. Rather, we should think about the mass of the two objects. In space, where they have no weight, the brick and the styrofoam would still have mass, and we could tell by moving them

about that the brick contains more mass than the styrofoam. For example, imagine tapping each object gently against your ear. The massive brick would be easy to distinguish from the low-mass styrofoam block even in weightlessness.

When we think of density, we divide mass by volume, and our minds make that comparison at an instinctive level. We can sense the density of an object just by handling it. Gift shops sometimes sell imitation rocks made of styrofoam as humorous gifts. "Rocks" made of styrofoam seem odd when we handle them because our brain expects rocks to be dense.

Density is a common parameter in science because it is a general property of materials. Lead, for example, has a density of about 7 g/cm³, and rock has a density of 3–4 g/cm³.

Water and ice have densities of about 1 g/cm³. If we knew that a small moon had a density of 1.5 g/cm³, we could immediately draw some conclusions about what kinds of materials it might be made of—ice and a little rock, but not much lead. The density of an object is a basic clue to its composition.

Density also determines how materials behave. The low-density gases in a nebula behave in one way, but the same gases under high density inside a star behave in a different way. Thus, astronomers must consider the density of materials when they describe astronomical objects. ∎

▪ Summary

Stars emit black body radiation from the dense gases of their photospheres, and as that radiation passes through the less dense gas of the star's atmosphere, the atoms absorb photons of certain wavelengths to produce dark lines in the spectrum.

A heated solid, liquid, or dense gas emits black body radiation that contains all wavelengths and thus produces a continuous spectrum. Black body radiation is most intense at the wavelength of maximum, λ_{max}, which depends on the temperature of the radiating body. Hot objects emit mostly short-wavelength radiation, whereas cool objects emit mostly long-wavelength radiation. This effect gives us clues to the temperatures of stars—hot stars appear blue, and cool stars red.

A low-density gas excited to emit radiation will produce an emission (or bright line) spectrum. Light from a source of a continuous spectrum passing through a low-density gas will produce an absorption (or dark line) spectrum. These three kinds of spectra are described by Kirchhoff's laws.

The lines in spectra are produced by the electrons that surround the nucleus of the atom. The electrons may occupy only certain permitted energy levels, and photons may be absorbed or emitted when electrons move from one energy level to another. Because each kind of atom and ion has a different set of energy levels, each kind of atom and ion can absorb or emit only photons of a certain wavelength. The hydrogen atom can produce the Lyman series lines in the ultraviolet, the Balmer series in the visual and near ultraviolet, the Paschen series in the infrared, and many more.

In cool stars, the Balmer lines are weak because most atoms are not excited out of the ground state. In hot stars, the Balmer lines are weak because most atoms are excited to higher orbits or ionized. Only at medium temperatures are the Balmer lines strong. We can use this effect as a thermometer for determining the temperature of a star. In its simplest form,

this amounts to classifying the stars' spectra in the spectral sequence O, B, A, F, G, K, M.

Stellar spectra can tell us the chemical compositions of the stars, but we must be careful to consider the temperature of the star in our analysis. In general, 90 percent of the atoms in a star are hydrogen.

When a source of radiation is approaching us, we observe shorter wavelengths, and when it is receding, we observe longer wavelengths. This Doppler effect makes it possible for the astronomer to measure a star's radial velocity, that part of its velocity directed toward or away from the earth.

The widths of spectral lines are affected by a number of processes. With no outside influences, a spectral line has a natural width that is very small. Doppler broadening, due to the thermal motion of the gas atoms, depends on the temperature of the gas. Collisional broadening occurs when the atoms emitting or absorbing photons collide with other atoms, ions, or electrons. The collisions disturb the atomic energy levels and slightly alter the wavelengths the atoms can absorb or emit. In a dense, hot gas, where the gas atoms collide often, collisional broadening is important.

■ New Terms

temperature
heat
Kelvin temperature scale
absolute zero
electron
black body radiation
wavelength of maximum (λ_{max})
color index
nucleus
proton
neutron
isotope
ionization
ion
molecule
binding energy
quantum mechanics
energy level
excited atom
ground state
continuous spectrum
absorption line
absorption spectrum
   (dark line spectrum)
emission line
emission spectrum
   (bright line spectrum)
Kirchhoff's laws
transition
Lyman, Balmer, and
   Paschen series
spectral class, or type
spectral sequence
Doppler effect
blue and red shifts
radial velocity (V_r)
transverse velocity
line profile
Doppler broadening
collisional broadening
density

■ Questions

1. Why can a good blacksmith judge the temperature of a piece of heated iron by its color?

2. Why do hot stars look bluer than cool stars?

3. Use black body radiation to explain why you could hold a cigarette between your lips and light it with a match, but you couldn't hold the cigarette in your lips while you light it by sticking the tip into a bonfire. The match and bonfire have about the same temperature. (This is just one of the hazards of smoking.)

4. What is the difference between a neutral atom, an ion, and an excited atom?

5. How do the energy levels in an atom determine which wavelength photons it can absorb or emit?

6. What kind of spectrum would you expect to record if you observed molten lava? In practice, to view molten lava you must look through gases boiling out of the lava. What kind of spectrum might you see in that case?

7. Why do the strengths of the Balmer lines depend on the temperature of the star?

8. Why does a stellar spectrum tell us about the surface layers but not the deeper layers of the star?

9. Explain the similarities between Figure 7-15, Figure 7-16, and Table 7-2.

10. Why do we not see TiO bands in the spectra of hot stars?

11. How could Kirchhoff have identified sodium in the solar spectrum in the 1850s, long before physicists understood how atoms interact with light?

12. In observing a star's spectrum, we note that the lines of a certain element are not present. Are we safe in concluding that the star does not contain this element? Why or why not?

13. If a star moves exactly perpendicular to a line connecting it to the earth, will the Doppler effect change its spectrum? Why or why not?

14. Why would we expect a star that rotates very rapidly to have broad spectral lines? What else could broaden the lines?

■ Discussion Questions

1. In what ways is our model of an atom a scientific model? How can we use it when it is not a completely correct description of an atom?

2. Can you think of classification systems we commonly use to simplify what would otherwise be very complex measurements? Consider foods, movies, cars, grades, clothes, and so on.

■ Problems

1. Human body temperature is about 310 K (98.6°F). At what wavelength do humans radiate the most energy? What kind of radiation do we emit?

2. If a star has a surface temperature of 20,000 K, at what wavelength will it radiate the most energy?

3. Infrared observations of a star show that it is most intense at a wavelength of 2000 nm. What is the temperature of the star's surface?

4. If we double the temperature of a black body, by what factor will the total energy radiated per second per square meter increase?

5. If one star has a temperature of 6000 K and another star has a temperature of 7000 K, how much more energy per second will the hotter star radiate from each square meter of its surface?

6. Transition A produces light with a wavelength of 500 nm. Transition B involves twice as much energy as A. What wavelength light does it produce?

7. Determine the temperatures of the following stars based on their spectra. Use Figure 7-14c.

   a. medium-strength Balmer lines, strong helium lines
   b. medium-strength Balmer lines, weak ionized calcium lines
   c. TiO bands very strong
   d. very weak Balmer lines, strong ionized calcium lines

8. To which spectral classes do the stars in Problem 7 belong?

9. In a laboratory, the Balmer beta line has a wavelength of 486.1 nm. If the line appears in a star's spectrum at 486.3 nm, what is the star's radial velocity? Is it approaching or receding?

10. The highest-velocity stars an astronomer might observe have velocities of about 400 km/sec. What change in wavelength would this cause in the Balmer gamma line? (HINT: Wavelengths are given in Figure 7-11).

▪ Critical Inquiries for the Web

1. To what extent is it possible to tell the spectral type of a star based on observation of its color? Search for color index data on the stars in Table A-9 (Appendix A) to see how B-V indices correlate with spectral type. Observational follow-up: Is it possible to rate star colors in such a way as to determine spectral type through visual observation? (Search the Web for indications that others have tried this.)

2. Is it possible to tell the direction of motion of the sun in space by observing radial velocities of other stars? Search the Internet for stellar radial velocity data for stars in a variety of directions on the celestial sphere, and look for trends that might indicate our direction in space.

 Go to the Wadsworth Astronomy Resource Center (www.wadsworth.com/astronomy) for critical thinking exercises, articles, and additional readings from InfoTrac College Edition, Wadsworth's online student library.

Chapter Eight

THE SUN — OUR STAR

GUIDEPOST

The preceding chapter described how atoms interact with light to produce spectra. In this chapter, we see how we can learn a great deal about the surface of the sun by applying what we know about atoms and spectra.

This chapter gives us our first close look at scientists at work, and we dis-

All cannot live on the piazza, but everyone

may enjoy the sun.

Italian Proverb

cover that much of science consists of confirmation of previous hypotheses and the gradual consolidation of our understanding.

Most important, this chapter gives us our first look at a real star. The chapters that follow will concentrate on the billions of stars in the heavens, but this

chapter shows us that each of them is both complex and beautiful; each is a sun.

A wit once remarked that solar astronomers would know a lot more about the sun if it were farther away. This contains a grain of truth; the sun is just a humdrum star, and there are billions like it in the sky, but the sun is the only one close enough to show surface detail. Solar astronomers can see so much detail in the swirling currents of gas and arching bridges of magnetic force that present theories seem inadequate to describe it. Yet the sun is not a complicated body. It is just a star.

In their general properties, stars are very simple. They are great balls of hot gas held together by their own gravity. Their gravity would make them collapse into small, dense bodies were they not so hot. The tremendously hot gas inside stars has such a high pressure that the stars would surely explode were it not for their own confining gravity. Thus, stars, like soap bubbles, are simple structures balanced between opposing forces that individually would destroy them.

We know that stars are hot inside because nuclear reactions near their centers generate energy, but we postpone a discussion of these nuclear reactions until later chapters. Here we concentrate on what we see on the surface of the sun. The spots, eruptions, and storms on the face of the sun are important because they tell us what an average star is like. By understanding some of the complex processes we see on the sun, we may better understand more distant stars.

Another reason that this solar activity is important is that it affects the earth. The sun's atmosphere of very thin gas reaches out past the orbit of the earth, and thus any change in the sun can have a direct effect on the earth. Also, we get nearly all our energy from the sun— oil and coal are merely stored sunlight—and our pleasant climate is maintained by energy from the sun. Should the sun's energy output vary by even a small amount, life on the earth might vanish.

8-1 THE SOLAR ATMOSPHERE

With a radius 109.1 times the earth's (109.1 $R_\oplus$),* the sun is gaseous throughout (Data File One). When we look at the sun, we see only the surface layers, what astronomers call the atmosphere. These layers extend from the visible surface out to about 5×10^6 km (7 $R_\odot$). If we include the low-density gas flowing away from the sun, the atmosphere extends to envelop the earth.

Below the Surface

The solar atmosphere consists of the outermost layers of the sun, and the structure of those layers is determined by processes that go on deep below the visible surface. In this chapter, we concentrate on the observable nature of the sun, not its deep interior. We will discuss the sun's

*In astronomy the symbols $\odot$ and $\oplus$ represent the sun and the earth, respectively.

THE SUN

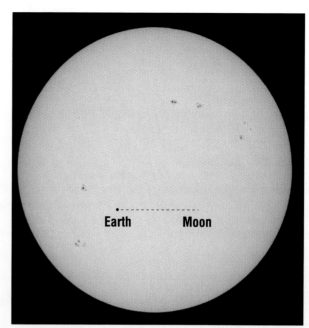

An image of the sun in visible light shows a few sunspots. The earth–moon system is added for scale. (Daniel Good)

Average distance from the earth	1.00 AU (1.495979×10^8 km)
Maximum distance from the earth	1.0167 AU (1.5210×10^8 km)
Minimum distance from the earth	0.9833 AU (1.4710×10^8 km)
Average angular diameter seen from the earth	0.53° (32 minutes of arc)
Period of rotation	25 days at equator
Radius	6.9599×10^5 km
Mass	1.989×10^{30} kg
Average density	1.409 g/cm³
Escape velocity at surface	617.7 km/sec
Luminosity	3.826×10^{26} J/sec
Surface temperature	5800 K
Central temperature	15×10^6 K
Spectral type	G2 V
Apparent visual magnitude	−26.74
Absolute visual magnitude	4.83

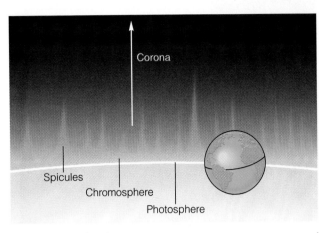

FIGURE 8-1

The layers in the solar atmosphere. A cross section at the edge of the sun shows the relative thickness of the photosphere and the chromosphere. The earth is shown for scale. Spicules are flamelike extensions of the chromosphere that reach up into the corona. On this scale, the corona extends far beyond the top of this figure.

interior in Chapters 12, 13, and 14, when we explore the structure, evolution, and death of stars. Here it is sufficient to summarize the energy flow in the sun.

All of the energy in the sun is generated by nuclear reactions that occur in the central core, a region that contains just over 10 percent of the sun's mass. That energy flows outward through the sun and heats its surface, which radiates that energy as light and heat. The flow of energy through the solar surface creates all the features we see.

As we observe the solar surface in this chapter, we must be alert for evidence of energy flow from the interior. We will find such evidence in the form of hot and cool regions, gas motions, magnetic fields, and so on. All stars make their energy near their centers, so all stars are dominated by the outward flow of energy. The features we see on the surface of the sun are characteristic of the features on the surfaces of other stars.

The Photosphere

The visible surface of the sun, the photosphere, is a layer of gas only about 500 km deep from which we receive most of the sun's light. Below the photosphere, the gas is denser and hotter and therefore radiates more light. However, that light cannot escape from the sun because of the outer layers of gas. Thus, we cannot detect light directly from these deeper layers. Above the photosphere, the gas is less dense and thus unable to radiate much light. The photosphere is the layer in the sun's atmosphere that is dense enough to emit plenty of light but of low enough density to allow the light to escape.

The photosphere is actually a very narrow layer. If the sun were to shrink to the size of a bowling ball, the photosphere would be no thicker than a layer of tissue paper wrapped without wrinkles around the ball (Figure 8-1).

One reason the photosphere is so shallow is related to the hydrogen atom. Because the temperature of the photosphere is sufficient to ionize some atoms, there are a large number of free electrons in the gas. Neutral hydrogen atoms can add an extra electron and become an H^- (H-minus) ion, but this extra electron is held so loosely that almost any photon has energy enough to free it. In the process, of course, the photon is absorbed. Thus, the H^- ions are very good absorbers of photons and make the gas of the photosphere very opaque. Light from below cannot escape easily, and we see a well-defined surface—the thin photosphere.

Most of the light we see comes from a region of the photosphere with a temperature of about 6000 K, roughly the temperature of a hot welding arc. We receive smaller amounts of light from deeper regions of the photosphere where the temperature is about 8000 K and from higher regions where the temperature is about 4000 K. The light we see coming from the sun is therefore a mixture of photons emitted by gases at various temperatures.

Although the photosphere appears to be substantial, it is really a very-low-density gas. The density in the middle of the photosphere is only 0.1 percent that of air at sea level. This is a rather good vacuum. To find gases as dense as the air we breathe, we would have to descend about 48,000 km below the photosphere—about 7 percent of the way to the sun's center. If someone could invent a fantastically efficient insulation, we could fly a spaceship right through the photosphere.

Good photographs of the photosphere show that it is not uniform but rather is mottled by a pattern of bright cells called **granulation** (Figure 8-2a). Each granule is about 1000 km in diameter—about the size of Texas—and is separated from its neighbors by a dark boundary. A granule lasts for about 20 minutes before it dissipates or merges with neighboring granules. Measurements of Doppler shifts show that the centers of granules are rising gas slightly hotter than the sinking gas in the dark boundary regions.

Solar astronomers believe the granules are the tops of rising currents below the photosphere (Figure 8-2b). Rising currents of hot gas reach the photosphere and heat it, making it slightly brighter above the rising gas. As the gas cools, it sinks, and the region above the sinking gas is slightly darker because it is slightly cooler.

Recent spectroscopic studies of the solar surface have revealed another kind of granulation. **Supergranules** are regions about 30,000 km in diameter (about 2.3 times the diameter of the earth) and include about 300 granules. These supergranules are regions of very slowly rising currents that last a day or two. They may be the surface traces of larger currents of gas deeper under the photosphere.

The edge, or **limb**, of the solar disk is dimmer than the center (see Figure in Data File One). This **limb darkening** is caused by the absorption of light in the photosphere. When we look at the center of the solar disk, we are looking directly down into the sun, and we see deep,

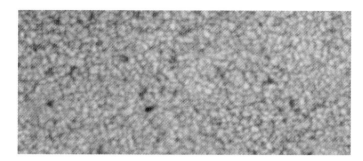

a

FIGURE 8-2

(a) A visible-light photo of the solar surface shows granulation. Each granule is about the size of Texas and lasts about 20 minutes before it is replaced by new granules. (b) In this model of granulation, heat flows upward just below the surface as rising currents of hot gas and sinking currents of cool gas. The rising currents heat the solar surface in small regions that we see as granules. (National Optical Astronomy Observatories)

b

FIGURE 8-3

The flash spectrum of the chromosphere shows emission lines of high ionization. The lines are curved because the thin cresent of chromosphere acts as the slit of the spectrograph. (The Observatories of the Carnegie Institution of Washington)

hot, bright layers in the photosphere. But when we look near the limb of the solar disk, we are looking at a steep angle and cannot see as deeply. The photons we see come from shallower, cooler, dimmer layers in the photosphere. Limb darkening proves that the temperature in the photosphere decreases with height, as we would expect if energy is flowing up from below.

The Chromosphere

Above the photosphere lies a nearly invisible layer of gas about 10,000 km thick (Figure 8-1). It is about 1000 times fainter than the photosphere, so we can see it only during a total solar eclipse when the moon covers the

brilliant photosphere. Then, for a few seconds, the chromosphere flashes into view as a thin line of pink just above the photosphere. The term *chromosphere* comes from the Greek work *chroma*, meaning "color."

If astronomers observe the spectrum of the sun during a total solar eclipse, they see the usual absorption spectrum as long as any part of the photosphere is visible. When the moon covers the photosphere, however, the chromosphere is visible for a few seconds extending beyond the edge of the moon in a thin crescent, and the solar absorption spectrum suddenly flashes into an emission spectrum produced by the hot, low-density gas of the chromosphere (Figure 8-3). The lines of this **flash spectrum** curve because the spectrograph does not contain a slit to isolate the light but merely uses the narrow crescent of chromosphere. The lines in the emission spectrum are generally the same as those in the solar absorption spectrum, with the addition of the lines of unionized helium, which are normally invisible in the sun's spectrum.

As the moon covers more and more of the chromosphere, the lines in the flash spectrum change in peculiar ways. Let's analyze these changes in detail. The analysis will not only reveal the structure of the sun's atmosphere but also illustrate how we can extract information from a spectrum if we know how atoms interact with light.

The presence of emission lines in the flash spectrum is itself important. Recall from Figures 7-7 and 7-8 our example of a gas cloud surrounding a light bulb. There we saw an absorption spectrum when we looked at the bulb through the gas and an emission spectrum when we looked at the gas alone. In the case of the sun, the photosphere plays the role of the glowing bulb, and the chromosphere plays the part of the gas cloud. The atoms in the chromosphere absorb photons as they leave the photosphere below. This forms the absorption lines we see when we look at the photosphere. But these same atoms emit photons in random directions. When the moon blots out the brighter photosphere, we can see the fainter light emitted by the atoms of the chromosphere. Thus, we see emission lines where we saw absorption lines before. This accounts for the beautiful color of the chromosphere—it is glowing like a giant neon sign (see Figure 3-18a).

When the flash spectrum first appears, we receive light from the top, middle, and bottom of the chromosphere. Because the bottom is brightest, it dominates the spectrum. In the spectrum of this lowest layer, we see Balmer emission lines plus lines of neutral helium. For helium to produce even weak emission lines, the temperature must be at least 10,000 K, and neutral helium lines don't become strong until the temperature is nearly 50,000 K. But at a higher temperature, the hydrogen atoms would become ionized, so the presence of Balmer lines in the spectrum assures us that the lowest layers of the chromosphere cannot be extremely hot.

When the moon conceals these lowest layers, we can see the light emitted by the middle layers dominating the flash spectrum. The Balmer lines fade, and lines of ionized helium appear along with lines of ionized iron and titanium. To ionize even a few helium atoms, the temperature must be at least 20,000 K, and a detailed analysis of the strengths of the emission lines assures solar astronomers that the middle chromosphere is much hotter than the lower chromosphere.

When the moon moves on and covers all but the top of the chromosphere, we see weak lines of very highly ionized atoms such as calcium, iron, and strontium. One line, for example, is produced by iron atoms that have lost 13 electrons. The temperature must be very high indeed to produce such extreme ionization.

Notice the similarity between our analysis of the flash spectrum and our classification of stars in the previous chapter. In each case, we used the excitation of different elements to judge temperature.

The flash spectrum, analyzed in this way, can tell us the temperature at each layer in the sun's atmosphere (Figure 8-4a). At the photosphere, the temperature is about 6000 K. It decreases slightly as we go upward (as shown by limb darkening), reaching a minimum of about 4000 K just above the photosphere. Above that, in the chromosphere, the temperature increases rapidly to 1,000,000 K at a height of 10,000 km, the beginning of the corona.

The flash spectrum can even tell us how dense the sun's atmosphere is. The emission lines fade away not only because the temperature increases with height but also because the density of the gas decreases. Near the photosphere the gas is rather dense, only about 10^4 times thinner than the air we breathe, but at the top of the chromosphere it is nearly a vacuum, about 10^{13} times thinner than air.

Although the chromosphere is not visible to the naked eye outside of solar eclipses, it can be photographed if special filters are used to admit only those photons easily absorbed by certain atoms and ions. Such photographs are called **filtergrams.** An H-alpha filtergram (Figure 8-4b) for instance, is formed by photons with wavelengths in the Balmer alpha line of hydrogen. Because hydrogen can absorb these photons so readily, they cannot have come from deep in the chromosphere. Photons with these wavelengths could only have escaped from the upper layers of the chromosphere. Thus, the H-alpha filtergram shows detail in the uppermost layers of the chromosphere. By tuning the filter to the wavelengths of photons slightly less likely to be absorbed, the solar astronomer can photograph different depths in the solar atmosphere.

Filtergrams of the chromosphere reveal **spicules**—flamelike structures 100–1000 km in diameter extending up to 12,000 km above the photosphere and lasting from 5 to 15 minutes (Figures 8-1 and 8-4b). These spicules appear to be cool regions (about 10,000 K) extending up into the much hotter corona (about 500,000 K). Seen at the edge of the solar disk, the spicules blend together and look like flames covering a burning prairie, but filtergrams of spicules located near the center of the solar disk show that they spring up around the edges of supergranules like weeds around flagstones. Some astronomers have suggested that the spicules are channels through which energy flows from below the photosphere into the corona.

The Corona

The sun's atmosphere extending above the chromosphere is termed the corona, after the Greek word for "crown." Although these outermost layers of the sun are far removed from the sun's surface, they are closely coupled with events in the chromosphere.

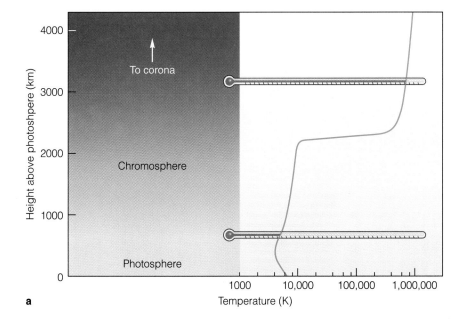

a

FIGURE 8-4

The chromosphere. (a) If we could place thermometers in the sun's atmosphere, we would discover that the temperture increases from 6000 K at the photosphere to 1,000,000 K at the top of the chromosphere. (b) An H-alpha filtergram shows that the chromosphere is not uniform. Spicules appear, looking like black grass blades, at the edges of supergranules over twice the diameter of the earth. (© 1971 National Optical Astronomy Observatories, NSO) (c) An H-alpha filtergram of the entire disk of the sun reveals details of the structure of the chromosphere. (NOAO)

b

c

The corona is visible to the naked eye only during total solar eclipses when the moon covers the bright photosphere (see Figure 3-18). Then the corona shines with a milky glow not quite as bright as the full moon. Eclipse photographs taken from the ground can trace the corona out to a distance of about 10 solar radii, and photographs taken from high-flying balloons or aircraft can trace the corona out to 30 solar radii.

By using a special telescope called a **coronagraph,** earthbound astronomers can see the corona at times when there are no eclipses. The coronagraph uses a disk to cover the brilliant photosphere and light baffles to reduce scattered light in the telescope. Also, coronagraphs are generally placed on high mountains to reduce light scattered from the earth's atmosphere. Good coronagraphs can detect the corona out to about 1.3 solar radii (Figure 8-5).

The spectrum of the corona consists of a continuous spectrum with superimposed emission lines. The continuous spectrum is produced by sunlight scattered from dust and free electrons in the corona. Because of the very high temperatures (about 1,000,000 K), the electrons travel at high velocities, and the resulting Doppler shifts in photons scattered from electrons smear out any absorption lines in the sunlight to produce a continuous spectrum. The emission lines are produced by very low density, highly ionized gases. In the lower corona, atoms such as ionized oxygen emit photons, but in the outer corona, the atoms are more highly ionized. Emission from these ions is clear evidence of low density and high temperature.

The temperature in the corona rises as we travel outward. Just above the chromosphere, in the region called the transition region, the temperature increases 500,000 K in only 300 km. In the lower corona, the temperature is about 500,000 K, and in the outer corona it may be as high as 3,500,000 K. The density of this gas must be very low indeed, or it would emit a great deal of light. In fact, at the base of the corona, the gas is 100 billion times less dense than the air we breathe, and the outer corona is even more tenuous.

How can the corona be hotter than the photosphere? This question has puzzled solar astronomers for many years. If we could safely visit the sun, we might hear (and feel) an extremely low-pitched rumble caused by the circulation of gas just below the photosphere; most solar astronomers first assumed that these sound waves were the source of heat in the corona. As the sound waves traveled upward through the photosphere and chromosphere, the falling density of the gas would convert them into **shock waves** (the astrophysical equivalent of sonic booms). Energy from the shock waves would agitate the gas atoms, thus raising the temperature.

In the 1970s, astronomers found reason to doubt the sound-wave hypothesis. Theoretical calculations suggested that the shock waves would dissipate in the lower chromosphere and never reach the corona. Observations of the sun made from space at ultraviolet and X-ray wavelengths showed no shock waves in the lower corona.

Ultraviolet observations of other stars, however, showed that many stars have emission lines in their spectra, lines which could come only from coronae. Even the K and M main-sequence stars, which have very gentle circulation below their photospheres and should not generate shock waves, have hot coronae. Something other than shock waves has to be carrying energy outward into the corona.

Studies of the sun's magnetic field show that it could carry energy from below the photosphere into the corona. As we will see later in this chapter, the solar magnetic field is very complicated and can store large amounts of energy. Apparently, it can participate in the outward energy flow in the sun and heat the corona.

The hot gases of the corona blow away from the sun as the **solar wind**—the moving outer extension of the corona. It contains mostly ionized hydrogen (protons with free electrons) but also heavier elements and a trapped magnetic field. Containing only a few particles per cubic centimeter, it blows past the earth at about 400 km/sec and interacts with the earth's magnetic field in complex ways (see Chapter 21).

Because of the gas carried away by the solar wind, the sun is slowly losing mass. This is a minor loss for the sun, amounting to some 10^7 tons per year, only about 10^{-14} of a solar mass per year. Other stars, however, at other stages in their lives, can lose mass rapidly.

Do other stars have chromospheres, coronae, and stellar winds like the sun's? Ultraviolet spectra taken by orbiting space telescopes such as the IUE (see Figure 6-23) suggest that the answer is yes. The spectra of many stars contain emission lines in the far ultraviolet that

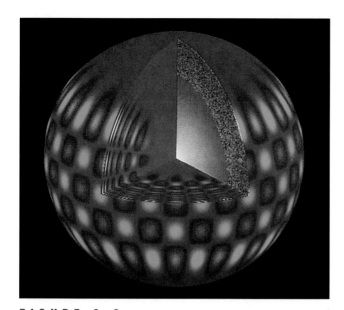

FIGURE 8-6

Helioseismology is the study of the modes of vibration of the sun. This computer-generated image shows one of nearly 10 million possible modes of oscillation. Red regions are receding, and blue approaching. Comparison of observations with such models reveals details about the sun's interior. (National Optical Astronomy Observatories)

could have been formed only in the low-density, high-temperature gases of the upper chromosphere and lower corona. Thus, the sun, for all its complexity, seems to be a normal star.

Helioseismology

Although the sun's interior is forever hidden from our eyes, we are beginning to explore it through **helioseismology,** the study of the way the sun vibrates.

Just as geologists can study the earth's interior by observing how sound waves produced by earthquakes are reflected and transmitted by the layers of the earth's interior, so too can helioseismologists explore the sun's interior. We can't put detectors on the sun, but we can measure the motions of the sun's surface using the Doppler effect. As the vibrations pass through the sun's interior, the solar surface we see—the photosphere—moves up and down in a complicated pattern (Figure 8-6). The Doppler shifts caused by the moving surface, although very small, can tell solar astronomers what frequency vibrations are present. Some frequencies penetrate deeper than others, and conditions in the sun's interior layers can weaken or strengthen vibrations. By observing which of the many million modes of vibration actually occur, solar astronomers can determine the temperature, density, pressure, composition, and motion of the sun's internal layers.

Helioseismology sounds almost magical, but we can understand it better if we think of a duck pond. If we stood at the shore of a duck pond and looked down at the water, we would see ripples arriving from all parts of the pond. Since every duck on the pond contributes to the ripples, we could, in principle, study the ripples near the shore and draw a map showing the position and velocity of every duck on the pond. Of course, it would be difficult to untangle all the different ripples, but all of the information would be there, lapping on the rocks at our feet.

Just as we could map the ducks on a pond, helioseismologists can study the vibrations in the surface of the sun and deduce the characteristics of the sun's interior. Because there are so many possible modes of vibration, they need large masses of data in order to separate the frequencies. The Global Oscillation Network Group (GONG) has set up small telescopes around the world to observe the sun nonstop for up to 3 years. Needless to say, supercomputers are needed to analyze the data and map the sun's interior. This exciting form of solar astronomy is just beginning to produce results.

CRITICAL INQUIRY

How deeply into the sun can we see?

This is a simple question, but it has a very interesting answer. When we look into the layers of the sun, our sight does not really penetrate into the sun. Rather, our eyes record photons that have escaped from the sun and traveled outward through the layers of the sun's atmosphere. If we observe at a wavelength at the center of a dark absorption line, then the photosphere and lower chromosphere are opaque, photons can't escape to our eyes, and the only photons we can see come from the upper chromosphere. What we see are the details of the upper chromosphere—a filtergram. On the other hand, if we observe at a wavelength that is not easily absorbed (a wavelength between spectral lines), the atmosphere is more transparent, and photons from deep inside the photosphere can escape to our eyes. There is a limit, however, set by the H^- ion, a hydrogen atom with an extra electron. Below a certain point, there is so much of this ion that the sun's atmosphere is opaque for almost all wavelengths, few photons can escape, and we can't see deeper.

By choosing the proper wavelength, solar astronomers can observe to different depths in the sun's photosphere. But the corona is so thin and the gas below the photosphere so dense that this method doesn't work in these regions. How can we observe the corona and the deeper layers of the sun? ■

So far we have thought of the sun as a static, unchanging ball of gas with energy flowing outward from the interior and through the atmospheric layers. In fact, the sun is a highly variable body whose appearance is constantly changing. It is now time to think of the active sun.

a

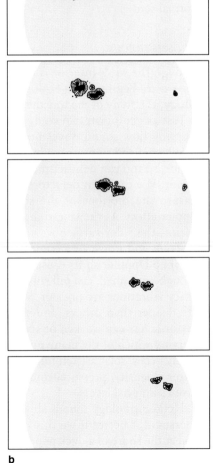

b   **c**

FIGURE 8-7

(a) A sunspot slightly larger than the earth appears dark against the bright photosphere because it is slightly cooler than the photosphere. (b) The slit of a spectrograph placed across a sunspot admits light from the photosphere, where the magnetic field is weak, and light from the spot, where the field is strong. The resulting spectrum (c) shows a single line in the regions outside the sunspot, but the Zeeman effect splits the line into three components (arrow) inside the spot. This allows the measurement of the magnetic fields inside sunspots. (National Optical Astronomy Observatories)

FIGURE 8-8

Looking through a telescope at the sun is dangerous, but you can always view the sun safely with a small telescope by projecting its image on a white screen (a). If you sketch the location and structure of sunspots on successive days (b), you will see the rotation of the sun and gradual changes in the size and structure of sunspots.

a   **b**

8-2 SOLAR ACTIVITY

The secret of solar activity seems to be magnetic fields. As we explore all aspects of solar activity, from giant eruptions and titanic explosions to dark, quiescent spots, we see the influence of the sun's complex magnetic field.

Sunspots

A **sunspot** is a cool, dark area of the solar surface (Figure 8-7a). The center of the spot, the umbra, is darker than the outer border, the penumbra. The average spot is about twice the diameter of the earth and may last for a week or so. Sunspots tend to form in groups, and a large group may contain up to 100 individual spots and last as long as 2 months or more.

Observing the sun with the unprotected eye can be very dangerous, because the sun is so bright it can burn the retina of the eye; *looking through a telescope at the sun could burn your eye in an instant.* Some small telescopes come with filters, but these should be used with great care, because a badly made or damaged filter could allow the sun to burn your eye. Nevertheless, there are safe ways to observe the sun and see sunspots (Figure 8-8a).

a

b

c

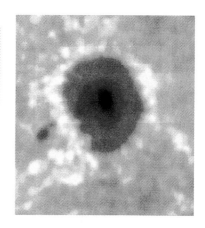
d

FIGURE 8-9

Active regions on the sun. (a) A white-light image of the sun shows the location of sunspot groups. (b) A magnetic map constructed using the Zeeman effect shows the location of strong magnetic fields on the same day as the image in (a). A radio image of the sun (c) and an X-ray image (d), made on different days, reveal a similar pattern of high-temperature coronal gas trapped in the magnetic fields above sunspot groups. (a and b, National Optical Astronomy Observatories; c, National Radio Astronomy Observatory, operated by Associated Universities, Inc., under contract with the National Science Foundation; d, IBM Research and Smithsonian Astrophysical Observatory)

In fact, sunspots are sometimes large enough to be visible to the naked eye when the sun is dimmed at sunset; the Chinese observed sunspots in this way as early as the 5th century BC. Galileo used his telescope in 1610 to observe sunspots and to study the rotation of the sun (Figure 8-8b).

Sunspots look dark because they are cooler than the photosphere. The center of a large sunspot is about 4240 K. The temperature of the photosphere is about 5800 K, so the cooler spot looks dark in contrast. In fact, a sunspot emits quite a bit of radiation. If the sun were magically removed and only an average-size sunspot were left behind, it would glow a brilliant orange-red and would be brighter than the full moon.

A clue to the origin of sunspots appeared in 1908 when American astronomer George Ellery Hale discovered magnetic fields in sunspots. A magnetic field affects the permitted energy levels in an atom. With no magnetic field present, a particular atom might absorb photons of a certain wavelength, producing a spectral line. If the atom is located in a magnetic field, however, the energy levels are split into multiple levels, and the single spectral line could appear as three or more spectral lines at slightly different wavelengths. This is known as the **Zeeman effect.** The separation of the spectral lines depends on the strength of the magnetic field, so Hale could measure the strength of magnetic fields on the sun by looking for this effect. He found that the field in a sunspot is about 1000 times stronger than the sun's average field (Figure 8-9).

This suggests that the powerful magnetic field causes a sunspot by inhibiting circulation. Ionized gas is made up of electrically charged particles, and such particles

FIGURE 8-10

This infrared image of a sunspot shows a brightening of the photosphere around the sunspot, apparently caused by energy flowing outward from the interior and unable to emerge through the strongly magnetic sunspot. (Dan Gezari, NASA/Goddard)

cannot move freely in a magnetic field. Astronomers often say the magnetic field is "frozen into" the ionized gas. This simply means that the gas and magnetic field are locked together. Rising currents of hot gas just under the photosphere might be slowed by the magnetic fields in sunspots, causing a decrease in the temperature and producing a dark spot. This hypothesis received strong support when an infrared image of a sunspot revealed the photosphere around the sunspot to be brighter than elsewhere on the sun (Figure 8-10). This brightening appears to be energy flowing upward from the sun's

FIGURE 8-11

Spotted stars: (a) An analysis of Doppler shifts in the spectrum of the rotating star HR 1099 allowed the construction of this map showing the location of dark spots on its surface. (b) A similar study of Gamma-2 Arietis reveals spots deficient (green) and overabundant (red) in silicon. (Steven Vogt, Artie Hatzes, and Dan Penrod, Lick Observatory–University of California, Santa Cruz)

a

b

a

b

FIGURE 8-12

(a) The number of sunspots varies in an 11-year period. (b) Magnetograms of the sun show the magnetic field over the sun's disk at sunspot minimum (left) and at sunspot maximum (right). (National Optical Astronomy Observatories)

interior. The energy was unable to emerge through the sunspot, so it was deflected to emerge from the photosphere around the sunspot.

Do other stars have sunspots, or rather "starspots," on their surfaces? This is a difficult question, because, except for the sun, the stars are so far away that no surface detail is visible. Some stars, however, vary in brightness in ways that suggest they are mottled by randomly placed, dark spots. As the star rotates, its total brightness changes slightly, depending on the number of spots facing in our direction. This has been suggested as an explanation for the variation of the RS Canum Venaticorum stars, whose spots may cover as much as 25 percent of the surface.

Also, some stars show spectral features that suggest the presence of magnetic fields and starspots. Ultraviolet observations reveal stars whose spectra contain emission lines commonly produced by the regions around spots on the sun. This suggests that these stars, too, have spots. One team of astronomers has used the Doppler shifts in the spectrum of the rotating star HR 1099 to construct a map showing the distribution of dark spots on its surface (Figure 8-11). Such results suggest that the sunspots we see on our sun are not unusual.

The Sunspot Cycle

The total number of sunspots visible on our sun is not constant. In 1843, the German amateur astronomer Heinrich Schwabe noticed that the number of sunspots varies in a period of about 11 years. This is now known as the sunspot cycle (Figure 8-12). At sunspot maximum, there are often as many as 100 spots visible at any one time, but at sunspot minimum there are only a few small spots. The most recent sunspot maximum occurred in 1990.

At the beginning of each sunspot cycle, the spots begin to appear in the sun's middle latitudes about 35° above and below the sun's equator. As the cycle proceeds, the spots appear at lower latitudes until, near the end of

FIGURE 8-13

The Maunder butterfly diagram is a plot of the latitude on the sun where sunspots first appear. Early in a sunspot cycle, the spots appear at higher latitudes. Later in the cycle, they appear nearer the sun's equator.

a

b

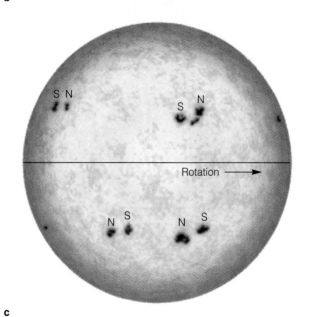

c

FIGURE 8-14

An H-alpha filtergram of a sunspot group (a) shows that structures in the chromosphere over the spots follow a pattern much like iron filings sprinkled over a bar magnet (b). (a, © Association of Universities for Research in Astronomy, Inc., Sacramento Peak Observatory; b, Grundy Observatory) The polarity of sunspot groups in the sun's southern hemisphere is the reverse of that in the northern hemisphere (c).

the cycle, they are appearing within 5° of the sun's equator. If we plot the latitude of the appearance of sunspots versus time, the diagram takes on the shape of butterfly wings (Figure 8-13). Such diagrams are known as **Maunder butterfly diagrams,** named after E. Walter Maunder of the Royal Greenwich Observatory, who first published the diagram in 1922.

The Magnetic Cycle

The solar magnetic field is apparently generated by a process called the **dynamo effect.** The interior gases of the sun are ionized, making them a good electrical conductor—much better than the copper wire that power companies use in dynamos on the earth. The outward flow of energy stirs some parts of the sun with rising and falling gas, and that circulation, coupled with the sun's rotation, is capable of creating a magnetic field. The dynamo effect in the sun is not well understood; in Chapter 21, we will see that this dynamo effect also operates inside the earth and produces the earth's magnetic field.

While the dynamo effect generates the solar magnetic field, other processes cause the field to cycle from strong to weak and back to strong. The most obvious consequence of this magnetic cycle is the sunspot cycle, and the characteristics of sunspots give us clues to how the solar magnetic cycle arises. Alternate sunspot cycles feature spots of reversed polarity. For example, if we used the Zeeman effect to study the magnetic fields in sunspots, we might find that every sunspot pair is made up of a magnetic pair (Figure 8-14a, b). One spot would be a magnetic north pole and one a south pole. We might also discover that the leading spot of each pair in the northern hemisphere is a north pole and that each trailing spot

a

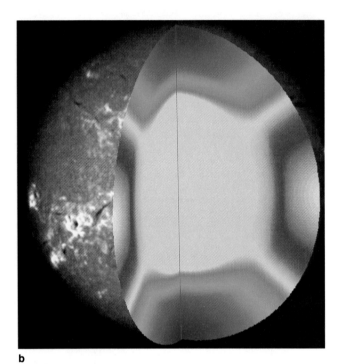

b

FIGURE 8-15

(a) The differential rotation of the sun's surface has been known for many years. The equator rotates faster than higher latitudes. (b) Helioseismology shows how the interior rotates with the surface. In this cutaway diagram, the equator rotates once in 25 days, but regions shaded red rotate more slowly. The rotation period near the poles, shaded blue, is 35 days. (Kenneth Libbrecht, California Institute of Technology)

FIGURE 8-16

The Babcock model suggests that the differential rotation of the sun winds up the magnetic field. When the field becomes tangled and bursts through the surface, it forms sunspot pairs. For the sake of clarity, only one line in the sun's magnetic field is shown here.

Confirmation and Consolidation

While many textbooks describe science as the process of testing hypotheses by observation and experiment, you should not think that every astronomer approaches the telescope with the expectation of making an observation that will disprove long-held beliefs and trigger a revolution in science. Then what is the daily grind of science really about?

First, many observations and experiments merely confirm already tested hypotheses. The biologist knows that all worker bees in a hive are sisters, but a careful study of the DNA from different workers further confirms that hypothesis. By repeatedly confirming a hypothesis, scientists build confidence in the hypothesis and

may be able to extend it to a wider application. Of course, there is always the chance that a new observation or experiment will disprove the hypothesis, but that is usually very unlikely. Much of the daily grind of science is confirmation.

Another aspect of routine science is consolidation, the linking of a hypothesis to other well-studied phenomena. Chemists may understand certain kinds of carbon molecules shaped like rings, but by repeated study they find a carbon molecule shaped like a hollow sphere. To consolidate their findings, they must show that the chemical bonding in the two molecules follows the same rules and that the molecules have certain properties in common. No hypoth-

esis is overthrown, but the chemists consolidate their knowledge and understand carbon molecules better.

The Babcock model of the solar magnetic cycle is an astronomical example of the scientific process. Solar astronomers know that the model explains some solar features but has shortcomings. Although most astronomers don't expect to discard the entire model, they work through confirmation and consolidation to better understand how the solar magnetic cycle works and how it is related to cycles in other stars. ∎

---

is a south pole. South of the sun's equator, we would find the polarities reversed (Figure 8-14c). Even more mysterious, if we watched for an entire sunspot cycle, we would discover that the overall polarity reverses from cycle to cycle.

Although this magnetic cycle is not fully understood, a model proposed in 1961 by Mount Wilson astronomer Horace Babcock explains the cycle as an interaction between the sun's magnetic field and the sun's rotation.

The earth, being solid, rotates as a rigid body, but the sun is a sphere of gas, and some parts of it rotate faster than other parts. The sun's equator rotates once in about 25 days, whereas the higher latitudes take up to 29 days to rotate once. This means that objects near the equator pull ahead of objects farther north or south in **differential rotation** (Figure 8-15).

A magnetic field is elastic and can be stretched and twisted. The differential rotation of the sun drags the equatorial parts of the magnetic field around the sun. The field winds up, like a loop of elastic thread on a spool. Turbulence below the photosphere further twists the field into ropelike tubes of magnetic field, which tend to float upward. Where these magnetic tubes burst through the sun's surface, sunspot pairs occur (Figure 8-16).

Babcock's model explains the butterfly diagram. The magnetic field becomes tangled first at higher latitudes and later at lower latitudes. Thus, the sunspots appear at higher latitudes early in the cycle and at progressively lower latitudes as the magnetic field becomes more tightly wound.

The Babcock model even explains the reversal of the sun's magnetic field from cycle to cycle. When the magnetic field becomes severely tangled, it breaks and reorders itself into a simpler pattern, and the differential

rotation begins winding it up again. This marks the beginning of a new cycle, but because of the way the field is reordered, it is reversed, and the new cycle begins with magnetic north replaced by magnetic south.

We have seen that the period of the sunspot cycle is 11 years, but the overall magnetic cycle in the sun repeats every 22 years. For 11 years the sun's magnetic field maintains its orientation, and the sun goes through a sunspot cycle. Then the magnetic field reverses, and the sun goes through another sunspot cycle. Thus, the overall period of the Babcock cycle in the sun is 22 years.

Notice that Babcock's idea is a model of the sun's magnetic field. Because no model can perfectly represent nature, we should ask not whether the model is correct but rather how well it explains the sun's magnetic field. In fact, the Babcock model is very successful at explaining the sun's magnetic cycles (Window on Science 8-1).

The activity on the sun is driven by the outward flow of energy from its center, but the activity is controlled by the magnetic cycle. Before we go on to consider other aspects of solar activity, let us examine whether evidence exists to suggest that other stars go through similar cycles of magnetic activity.

Magnetic Cycles on Other Stars

Because we believe that the sun is a representative star, we should expect other stars to have similar cycles of starspots. We can't see individual spots from the earth, of course, but certain features in stellar spectra are associated with magnetic fields. Regions of strong magnetic fields on the solar surface emit strongly at the central wavelengths of the H and K lines of ionized calcium. This calcium emission appears in the spectra of other sunlike stars and tells us that these stars, too, have strong magnetic fields on their surfaces. These stars presumably have starspots as well.

In 1966, astronomers began measuring the strength of this H and K emission in the spectra of stars. The stars to be observed were selected to be similar to the sun. With temperatures ranging from 1000 K hotter than the sun to 3000 K cooler, these stars were considered most likely to have sunlike magnetic activity on their surfaces.

The observations show that the strength of the emissions in the spectra of these stars varies from year to year. The H and K emission averaged over the sun's disk varies with the sunspot cycle, and similar periodic variations can be seen in the spectra of the stars studied (Figure 8-17). The star 107 Piscium, for instance, appears to have a starspot cycle lasting 9 years. Thus, we can be sure that stars like the sun do have magnetic fields and are subject to magnetic cycles.

The rotation periods of these stars are also apparent from the observations. If a star has a major region of magnetic activity and rotates every 20 days, we see the emission appear for 10 days and then disappear for 10 days as the rotation of the star carries the region to the far side of the star.

These observations confirm our belief that the sun is an average sort of star, that is, not peculiar. Most other stars like our sun have magnetic fields and starspots and go through magnetic cycles. More important, these studies are helping us understand our sun better.

Prominences and Flares

Although the connection between the magnetic cycle in the sun and the sunspot cycle is not clearly understood, it does seem clear that the sun's magnetic field governs the production of sunspots. It also controls other forms of solar activity such as prominences and flares.

Prominences are visible during total solar eclipses as red protrusions at the edge of the solar disk (see Figure 3-18b). The red color is the same as the red color of the chromosphere and comes from the emission lines of hydrogen.

Prominences seem to be controlled by magnetic fields. Many are arch-shaped, looking much like the patterns made by iron filings sprinkled over a magnet (Figures 8-14b and 8-18). In filtergrams of the sun, prominences show as dark filaments that wind through the magnetically active regions around sunspots. Eruptive prominences burst out of these complex magnetic fields and may shoot upward 500,000 km in a few hours. Quiescent prominences may develop as graceful arches over sunspot groups and can last weeks or even months. Eruptive or quiescent, prominences are clearly ionized gases trapped in the twisted magnetic fields of active regions.

FIGURE 8-17

The average amount of emission in the H and K lines of calcium (mean H-K flux) is related to magnetic activity. In the sun, this emission is stronger when sunspot activity is higher. The sunlike star 107 Piscium appears to have a magnetic cycle, while the star Tau Ceti does not. (Adapted from data by Baliunas and Saar)

Flares are much more violent than prominences (Figure 8-19). A **flare** is an eruption on the solar surface that rises to maximum in a few minutes and decays in an hour or less. During that time, it emits vast amounts of X-ray, ultraviolet, and visible radiation and streams of high-energy protons and electrons. A large flare can release 10^{25} J, the equivalent of 2 billion megatons of TNT.

Solar flares seem clearly linked to the magnetic field. They almost always occur near sunspot groups and may recur over and over at the same place. A large sunspot group may experience 100 small flares a day, although only one flare a year may be bright enough to be seen in visible light. Various theories suggest that flares occur when sharp twists in the magnetic field store up great quantities of energy and then release it all at once.

Solar flares can have important effects on the earth. X-ray and ultraviolet radiation reaches the earth in only 8 minutes and increases the ionization in the earth's upper atmosphere. This alters the reflection of shortwave radio signals and can even absorb them completely, thereby interfering with communications. Flares can eject high-energy particles at one-third the speed of light, but most particles ejected from flares have lower velocities and reach the earth hours or days after the flare as gusts in the solar wind. These gusts have long been blamed for disturbances in the earth's magnetic field, but new research shows that the true origin of such disturbances lies higher in the sun's atmosphere, in the corona.

a

b

c

FIGURE 8-18

(a and b) These H-alpha filtergrams show prominences, chromospheric eruptions of hot gas trapped in strong magnetic fields as shown by their looped structure. (Copyright © Association of Universities for Research in Astronomy, Inc. AURA, All rights reserved) (c) Photographed from space in the ultraviolet, a loop prominence rises far out into the corona. (NASA Skylab)

a

b

FIGURE 8-19

(a) An H-alpha filtergram of a solar flare shows a surge of heated gas being ejected from the surface. (National Optical Astronomy Observatories) (b) Events on the sun can affect the outer corona, as in this photograph of a solar eclipse made from a jet 11 km (36,000 ft) above the Indian Ocean. Computer processing reveals twisted streamers at right extending beyond 20 $R_{\odot}$. (C. Keller, Los Alamos National Laboratory)

FIGURE 8-20

Auroras, commonly seen at high latitudes, sometimes look like widespread glowing clouds or great curtains of light. (a) Both are visible in this photograph of auroras in southern Australia. (Courtesy David Miller) (b) This ultraviolet image of the earth was recorded by the Dynamics Explorer 1 satellite. An oval ring of auroras encircles the earth's north magnetic pole and shows the influence of the earth's magnetic field. (L. A. Frank, University of Iowa–NASA)

Coronal Activity

The corona of the sun also takes part in the solar activity cycle. The general experience of eclipse viewers is that during sunspot minimum the corona is small and slightly flattened, and at sunspot maximum the corona is larger and roughly circular. This has led to a model of the corona as a spherical cloud of thin gas extending outward from the chromosphere.

However, observing eclipses from the earth's surface is not the best way to study the corona. Scientists from Los Alamos National Laboratory have mounted equipment in an Air Force NC-135 and flown at nearly 12,200 m (40,000 ft) to photograph the eclipsed sun from above 80 percent of the earth's atmosphere. They could detect the corona out to 20 solar radii (Figure 8-19b). Skylab, an American space station, was launched in 1973 and occupied successively by three astronaut crews who made extensive observations of the corona in the X-ray and ultraviolet parts of the spectrum. Since then, other orbiting X-ray telescopes have added to our understanding of the corona.

The results of such studies show that the corona is not the uniform halo of gas that earlier astronomers imagined. It is actually composed of streamers shaped by the solar magnetic field (see Figure 8-5). The corona looks uniform during some eclipses because we see the streamers in projection at the sun's edge. Photographs made by the Los Alamos team can trace coronal streamers out to 20 solar radii.

Furthermore, recent observations suggest that the outer corona may contain a disklike concentration of streamers whose orientation depends on the state of the solar cycle. At sunspot minimum, the disk lies in the plane of the solar equator, but as the sunspot cycle progresses, the magnetic field of the outer corona and the disk component become more inclined. By sunspot maximum, the disk is believed to encircle the sun from pole to pole, and eventually it flips over completely to reverse the magnetic polarity (after one 11-year cycle). The distribution of coronal streamers we see from the earth is affected by when solar eclipses occur in the sunspot cycle and the orientation of the outer corona at the moment of the eclipse. If the above hypothesis is confirmed, then the corona is much more complex than the simple model of a spherical distribution of streamers.

Whatever the overall shape of the corona, telescopes reveal gradual changes in the coronal streamers that can be interrupted by sudden releases of magnetic energy. These releases eject billion-ton blobs of hot gas into the solar wind. If one of these ejections strikes the earth, it can trigger a magnetic storm as the earth's magnetic field is distorted and powerful electric currents run through the earth. These currents can be as much as a million megawatts, and the interaction with the upper atmosphere can excite the atoms of the atmosphere to glow in displays called **auroras** at altitudes of 100–400 km (Figure 8-20). Because these coronal mass ejections often occur above flares, astronomers have for many years blamed the flares for the earthly disruptions. The culprit, however, seems to lie far above the flares out among the coronal streamers.

The streamers seem to draw their bulbous shapes from loops of magnetic fields that extend above the solar surface, where the charged particles trapped in the fields are able to escape in long, thin streams. In some regions of the solar surface, however, the magnetic fields do not loop back, and the particles stream away from the sun unimpeded. These regions appear in X-ray images of the corona as cooler, lower-density regions called **coronal holes** (Figure 8-21). Although there are permanent coronal holes at the sun's north and south poles, the distribution of coronal holes over the rest of the sun depends on the solar activity cycle, further evidence that it is a magnetic phenomenon. Coronal activity is important because it affects the solar wind and thus affects the earth. Many solar astronomers believe that the solar wind is composed of particles streaming away from coronal holes. Understanding the solar wind, therefore, seems to depend on understanding coronal activity.

The solar activity we see in the photosphere, chromosphere, and corona is complex and beautiful, and some solar activities are important for what they tell us about the sun and stars. A few are important because they can affect the earth. But one aspect of solar activity is critically important to life on Earth—the constancy of the sun.

The Solar Constant

If the sun's energy output varied by even a small amount, life on Earth might end. The continued existence of our civilization and our species depends on the constancy of our sun, but we know very little about the variation of the sun's energy output.

The energy production of the sun can be measured by adding up all of the energy falling on 1 square meter of the earth's surface during 1 second. Of course, some correction for the absorption of the earth's atmosphere is necessary, and we must count all wavelengths from X rays to radio waves. The result, which is called the **solar constant,** amounts to about 1360 joules per square meter per second. A change in the solar constant of only 1 percent could change the average temperature of the earth by 1–2°C (about 1.8–3.6°F). For comparison, during the last ice age the average temperature on the earth was about 5°C cooler than it is now.

Some of the best measurements of the solar constant were made by instruments aboard the Solar Maximum Mission satellite. These have shown variations in the energy received from the sun of about 0.01 percent that lasted for days or weeks (Figure 8-22). Superimposed on that random variation is a long-term decrease of about 0.018 percent per year that has been confirmed by observations made by sounding rockets and balloons and by the NIMBUS 7 satellite. This long-term decrease may

FIGURE 8-21

An X-ray image of the sun reveals intensely hot regions in the solar atmosphere, linked by magnetic fields. Bright regions lie above active regions on the solar surface and reveal magnetic fields that loop back to the solar surface and confine hot gas. Dark regions are coronal holes, where the magnetic field does not loop back to the surface and thus does not restrict the motion of hot gas. (NASA and ISAS)

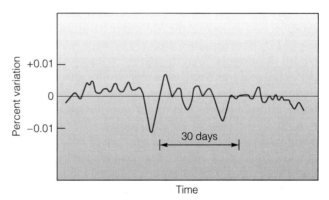

FIGURE 8-22

Variations in the amount of energy received from the sun were detected by the Solar Maximum Mission satellite. Variations of roughly 0.01 percent are caused by the passage of sunspots across the disk. Longer-term variations of similar amplitude could alter the earth's climate.

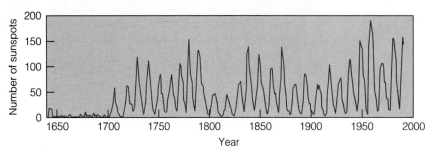

FIGURE 8-23

Counts of sunspots gathered from historical sources show that there were very few sunspots seen between 1645 and 1715. Called the Maunder Minimum, this period of low solar activity suggests that the sun's cycle of magnetic activity may not be continuous.

be related to a cycle of activity on the sun with a period longer than the 22-year magnetic cycle.

Small, random fluctuations will not affect our climate, but a long-term decrease over a decade or more could cause worldwide cooling. History contains some evidence that the solar constant may have varied in the past. The "little ice age" was a period of unusually cool weather in Europe and America that lasted from about 1430 to 1850. The average temperature worldwide was about 1 K cooler than it is now. Although many things affect climate, this cooling could have been caused by a decrease in the solar constant of only 1 percent.

This period of cool weather included an era of few sunspots now known as the **Maunder Minimum*** (Figure 8-23). Between 1645 and 1715, there is almost no record of sunspots, even though the telescope had been perfected and was being actively used by astronomers. Reports of total solar eclipses during this period make no mention of the corona or chromosphere, and there is almost no record of auroral displays. The Maunder Minimum seems to have been a period of reduced solar activity. Solar astronomer John Eddy believes that the historical record shows traces of at least 11 similar periods extending back to 3000 BC. If that is true, our sun is not a perfectly stable star, and the survival of humanity may depend on our ability to adapt to changes in the solar "constant."

CRITICAL INQUIRY

What kind of activity would the sun have if it didn't rotate differentially?

This is a really difficult question because we can see only one star close up and thus have no other examples. Nevertheless, we can make an educated guess by thinking about the Babcock model. If the sun didn't rotate differentially, with its equator traveling faster than

*Ironically, the Maunder Minimum coincides with the reign of Louis XIV of France, the "Sun King."

higher latitudes, then the magnetic field would not get wound up, and it would not get twisted and rise through the surface to form sunspot pairs. Of course, the magnetic field would not get progressively more wound up, so there would also be no magnetic cycle. Loops of magnetic field might not rise above the surface to trap ionized gases and form prominences. Furthermore, twists and kinks might never form in the field, and thus there would be no flares. In fact, with no differential rotation to stir up the magnetic field, the chromosphere and corona might not be heated to such high temperatures.

This is very speculative, but it suggests some interesting ideas about other stars. Most stars probably rotate differentially, so we should expect many stars to have activity on their surfaces much as the sun does. What observational evidence do we have for this expectation? ■

The sun is beautiful and complex, with great eruptions and spots sweeping across its surface like magnetic weather. All of this activity is driven by the steady flow of energy outward from the sun's center through its photosphere and into space. Other stars generate energy and radiate it into space, so we can conclude that many of those stars must have surface activity similar to that on our sun. But other stars are so far away we can't see their surfaces. Through the largest telescopes, they look like nothing more than points of light. How can we learn about other stars? We begin to answer that question in the next chapter. ■

■ *Summary*

The atmosphere of the sun consists of three layers: photosphere, chromosphere, and corona. The photosphere, or visible surface, is a thin layer of low-density gas from which visible photons most easily escape. It is marked by granulation, a pattern produced by circulation below the photosphere.

The chromosphere is most easily visible during total solar eclipses when it flashes into view for a few seconds and produces the flash spectrum. This spectrum shows that the chromosphere is hot, ionized gas. Filtergrams taken in the light emitted by specific atoms show that the chromosphere is filled with large jets called spicules.

The corona is the sun's outermost atmospheric layer. It is composed of very-low-density, very hot gas extending at least 20 solar radii from the sun. Although it can be studied with a special telescope called a coronagraph, its outer layers are visible only during total solar eclipses. Its high temperature—up to 3,500,000 K—is believed to be maintained by interaction with the solar magnetic field. The outer parts of the corona merge with the solar wind, a breeze of low-density ionized gas streaming away from the sun. Thus, the earth, which is bathed in the solar wind, is orbiting within the sun's outer atmosphere.

Sunspots are the most prominent example of solar activity. A sunspot appears to have a dark center, called the umbra, and a slightly lighter border, called the penumbra. A sunspot seems dark because it is slightly cooler than the rest of the photosphere. The average sunspot is about twice the size of the earth and contains magnetic fields about 1000 times stronger than the sun's average field. Sunspots are thought to form because the magnetic field inhibits convection. The average number of sunspots visible varies with a period of about 11 years and appears to be related to the solar magnetic cycle.

Alternate sunspot cycles have reversed magnetic polarity, and this has been explained by the Babcock model of the magnetic cycle. In this model, the differential rotation of the sun winds up the magnetic field. Tangles in the field rise to the surface and cause sunspot pairs. When the field becomes strongly tangled, it reorders itself into a simpler but reversed field, and the cycle starts over.

Prominences and flares are other examples of solar activity. Prominences occur in the chromosphere; their arch shape show that they are formed of ionized gas trapped in the magnetic field. Flares, too, seem to be related to the magnetic field. They are sudden eruptions of X-ray, ultraviolet, and visible radiation and high-energy particles that occur among the twisted magnetic fields around sunspot groups

Activity in the corona is also guided by the magnetic field. The corona seems to be composed of streamers of thin, hot gas escaping from the magnetic field. In some regions of the corona, the magnetic field does not loop back to the sun, and the gas escapes unimpeded. These regions are called coronal holes and are believed to be the source of the solar wind.

Because the sun is so active, it is not surprising that its total energy output varies. Such variations could have important consequences for the earth. The solar constant, a measure of the total energy coming from the sun, has been measured by instruments in spacecraft and found to vary by about 0.01 percent. Such short-term variations are not dangerous. However, the observations also show that the sun is currently fading by about 0.018 percent per year. This is probably a cyclic phenomenon, but if it continues for a decade or more it could affect climate. Historical records show that from 1645 to 1715 there were almost no sunspots. This Maunder Minimum occurred during a period of cool weather in Europe and America called the little ice age, which suggests that the solar constant may change occasionally and alter the earth's climate.

Although solar activity is complex, there is no reason to believe that it is unique. Most stars are similar to the sun and probably have magnetic fields and differential rotation. The activity we see on the sun is presumably typical of most stars.

■ New Terms

granulation	sunspot
supergranule	Zeeman effect
limb	Maunder butterfly diagram
limb darkening	dynamo effect
flash spectrum	differential rotation
filtergram	prominence
spicule	flare
coronagraph	aurora
shock wave	coronal hole
solar wind	solar constant
helioseismology	Maunder Minimum

■ Questions

1. Why can't we see deeper into the sun than the photosphere?

2. What does granulation tell us about the sun?

3. What kinds of spectra do the photosphere, chromosphere, and corona produce? Why?

4. Why is the chromosphere pink?

5. Why does the flash spectrum change as the moon moves across the chromosphere? What do those changes tell us?

6. How can a filtergram reveal details in layers above the photosphere?

7. Why does the flash spectrum contain spectral lines in emission that are not present as absorption lines in the spectrum of the photosphere?

8. Why are sunspots dark?

9. Why do we think the sunspot cycle is controlled by the magnetic cycle of the sun?

10. How can flares on the sun affect the earth?

11. In a best-selling novel of a few years ago, astronauts on the moon were killed by high-energy protons from a solar flare that reached the moon 8 minutes after the flare occurred. What is wrong with this plot device?

12. What evidence do we have that other stars have stellar winds, chromospheres, coronas, starspots, and magnetic cycles?

■ Discussion Questions

1. What energy sources on the earth cannot be thought of as stored sunlight?

2. What would the spectrum of an auroral display look like? Why?

■ Problems

1. The radius of the sun is 0.7 million km. What percentage of the radius is taken up by the chromosphere?

2. The smallest detail visible with ground-based solar telescopes is about 1 second of arc. How large a region does this represent on the sun? (HINT: Use the small-angle formula.)

3. What is the angular diameter of a star like the sun located 5 ly from the earth? Is the Hubble Space telescope able to detect detail on the surface of such a star?

4. If a sunspot has a temperature of 4200 K and the solar surface has a temperature of 5800 K, how many times brighter is the surface compared to the sunspot? (HINT: Use the Stefan–Boltzmann law, Chapter 7.

5. A solar flare can release 10^{25} J. How many megatons of TNT would be equivalent? (HINT: a 1-megaton bomb produces about 4×10^{15} J.)

6. The United States consumes about 2.5×10^{19} J of energy in all forms in a year. How many years could we run the United States on the energy released by the solar flare in Problem 5?

7. Neglecting energy absorbed or reflected by our atmosphere, the solar energy hitting 1 square meter of Earth's surface is 1360 J/sec (the solar constant). How long does it take a baseball diamond (90 ft on a side) to receive 1 megaton of solar energy? (HINT: See Problem 5.)

▪ Critical Inquiries for the Web

1. Do disturbances in one layer of the solar atmosphere produce effects in other layers? We have seen that filtergrams are useful in identifying the layers of the solar atmosphere and the structures within them. Visit a Web site that provides daily solar images. Then, choose today's date (or one near it) and examine the sun in several wavelengths to explore the relationships among disturbances in various layers.

2. How can images like those discussed in the previous question be used to estimate the rotation period of the sun? Can these data be used to demonstrate that the sun rotates differentially? Examine images at a particular wavelength over a period of two weeks. From these observations, estimate the period of rotation of the sun.

 Go to the Wadsworth Astronomy Resource Center (www.wadsworth.com/astronomy) for critical thinking exercises, articles, and additional readings from InfoTrac College Edition, Wadsworth's online student library.

Chapter Nine

MEASURING STARS

GUIDEPOST

Science is based on measurement, but measurement in astronomy is very difficult. Even with the powerful modern telescopes described in Chapter 6, it is impossible to measure directly simple parameters such as the diameter of a star. This chapter shows how we can use the simple observations that are possible, combined with the basic laws of physics, to discover the properties of stars.

With this chapter, we leave our sun behind and begin our study of the billions of stars that dot the sky. In a sense, the star is the basic building block of the universe. If we hope to understand what the universe is, what our sun is, what our earth is, and what we are, we must understand the stars.

The next chapter will show how we can make yet another measurement—the mass of the stars. Armed with these basic stellar data, we will be ready to trace the life stories of the stars from birth to death. ■

If it can't be expressed in figures, it is not science; it is opinion.

ROBERT HEINLEIN

The Notebooks of Lazarus Long

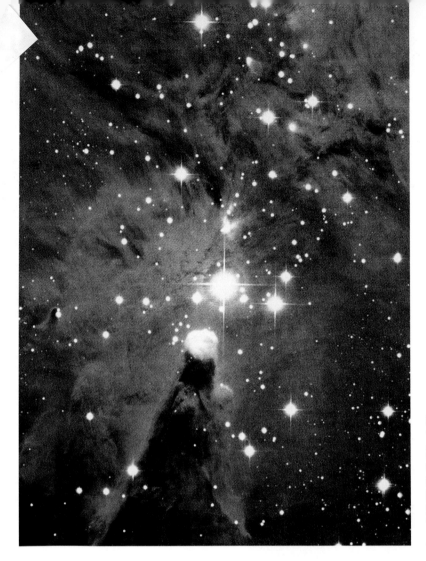

FIGURE 9-1

The central question of modern astronomy is manifest in this photo of the Cone Nebula, where clouds of gas and dust are actively forming new stars. How do stars form, evolve, and die? To answer that question, we must know the luminosities, diameters, and masses of the stars—their fundamental properties. (Copyright Anglo-Australian Telescope Board)

The stars are unimaginably remote. The nearest star is our sun, only 1.5×10^8 km (93 million miles) away, so close that light takes only 8 minutes to reach the earth. The next nearest star, α Centauri, is nearly 300,000 times farther away, about 4 ly. Recall from Chapter 1 that a light-year is the distance light travels in 1 year—about 9.5×10^{12} km (5.8 trillion miles).

Despite their great distances, the stars are the key to astronomy. The universe is filled with stars, and if we are to understand the universe, we must discover how stars are born, live, and die. We begin our study in this chapter by gathering data about the intrinsic properties of the stars (Figure 9-1). In later chapters, we will use these data to deduce the life cycles of different kinds of stars.

Unfortunately, determining a star's inherent properties is quite difficult. When we look at a star through a telescope, we see only a point of light that tells us nothing about the star's energy production, temperature, diameter, or mass. Because we cannot visit stars, we can observe only from the earth and must unravel the properties of stars through the analysis of starlight. One of the reasons astronomy is interesting is that it contains so many such puzzles, each demanding a different method of solution.

To simplify our task, this chapter concentrates on two intrinsic stellar properties. Our goals will be to find out how much energy stars emit and how large they are. In the next chapter, we will discuss the masses of stars. These three characteristics, combined with temperature—an intrinsic stellar property discussed in Chapter 7—will give us an overview of the nature of stars and provide us with the data we need to consider the lives of stars.

Although we begin this chapter with two goals firmly in mind, we immediately meet a short detour. To find out how much energy a star emits, we must know how far away it is. If at night we see bright lights approaching on the highway, we cannot tell whether they are the intrinsically bright headlights of a distant truck or the intrinsically faint lights of a nearby bicycle (Figure 9-2). Only when we know the distance to the lights can we judge their intrinsic brightness. In the same way, to find the intrinsic brightness of a star and thus the amount of energy it emits, we must know its distance. Our short detour will provide us with a method of measuring stellar distances.

Once we know the distance to a star, finding its total energy output and estimating its diameter are simple tasks. When we do this for many stars, we discover that

FIGURE 9-2
On a dark night, we see lights approaching us, and, with no further clues, we cannot tell a low-power light nearby from a more powerful but more distant light. In the darkness the headlight of a nearby bicycle might look as bright as the headlight of a distant truck. Similarly, we cannot judge the distance to a star from its brightness.

there are different kinds of stars, and we will conclude this chapter by considering the frequency of stellar types—that is, we will try to decide which kinds of stars are common and which are rare. These family relations among the stellar types are an important clue to the life stories of the stars.

9-1 MEASURING THE DISTANCES TO STARS

Distance is both the most important and the most difficult measurement in astronomy, and astronomers have found many different ways to estimate the distance to stars. Yet all those ways depend on a direct geometrical method that is much like the method surveyors use to measure the distance across a river they cannot cross. We begin by reviewing this method and then apply it to stars.

The Surveyor's Method

To measure the distance across a river, a team of surveyors begins by driving two stakes into the ground. The distance between the stakes is the baseline of the measurement. The surveyors then choose a landmark on the opposite side of the river, a tree perhaps, thus establishing a large triangle marked by the stakes and the tree. Using surveyors' instruments, they sight the tree from the two ends of the baseline and measure the two angles on their side of the river (Figure 9-3).

Knowing two angles and the length of the side between them (the baseline), the surveyors can find the distance across the river by using trigonometry or by constructing a scale drawing. For example, if the baseline was 50 m and the angles were 66° and 72° they could draw a line 50 mm long to represent the baseline. Using a protractor, they could construct angles of 66° and 72° at each end of the baseline and then extend the two sides until they met at C, the location of the tree. Measuring the height of the triangle in the drawing, they would find it was 65 mm high and thus conclude that the distance across the river to the tree was 65 m.

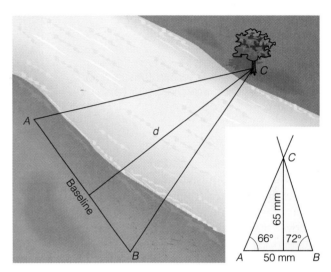

FIGURE 9-3
Surveyors can find the distance *d* across the river by measuring the baseline and the angles A and B and then constructing a scale drawing of the triangle.

The more distant an object is, the longer the baseline we must use to measure the distance to the object. We could use a baseline 50 m long to find the distance across a river, but to measure the distance to a mountain on the horizon, we might need a baseline 5 km long.

The Astronomer's Method

No baseline we could draw on the earth would be long enough to allow us to measure the distance to the stars. Instead, we must use the diameter of the earth's orbit—the longest baseline possible at present. We can use this baseline because the earth carries us from one side of its orbit to the other as it circles the sun.

If we took a photograph of a nearby star and then waited 6 months, Earth would have moved halfway around its orbit. We could then take another photograph of the star at a point in space 2 AU (astronomical units) from the point where the first photograph was taken. Thus, our baseline would equal the diameter of Earth's orbit, or 2 AU.

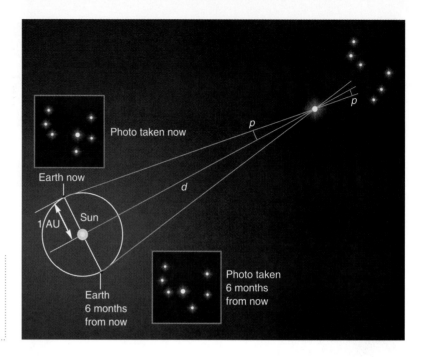

FIGURE 9-4
We can measure a nearby star's parallax by photographing it from two points along Earth's orbit, A and B. Half of the star's total change in position from one photograph to the other is its parallax, p.

We would then have two photographs of the same part of the sky taken from slightly different locations in space. If we examined the two photographs, we would discover that the nearby star was not in exactly the same place as seen against the background of more distant stars (Figure 9-4). This apparent shift in the position of the star is called parallax. We saw in Chapter 4 that parallax is an everyday experience. Our thumb, held at arm's length, appears to shift position against a distant background when we look with first one eye and then the other (Figure 4-8).

The larger the distance to a star, the smaller the star's parallax. The distant stars in the background of our photographs are so distant they hardly shift at all, and we can use them as references against which to measure the parallax of nearby stars.

The quantity that astronomers call the **stellar parallax** (p) is half of the total shift of the star (Figure 9-4). Once this angle and the baseline of the observations are known, astronomers can calculate the distance to the star. Astronomers measure parallax, and surveyors measure the angles at the ends of the baseline, but both measurements tell us the same thing—the shape of the triangle and thus the distance to the object in question.

The distance to a star with parallax p is given by the simple formula

$$d = \frac{1}{p}$$

where the parallax is measured in seconds of arc and the distance is measured in a unit of distance invented by astronomers, the **parsec (pc),*** defined as the distance to

*The parsec is used throughout astronomy because it simplifies the calculation of distance. However, there are instances when the light-year is also convenient. Consequently, the chapters that follow use either parsecs or light-years as convenience and custom dictate.

a star with a parallax of 1 second of arc. One parsec turns out to equal 206,265 AU, or 3.26 ly. This makes it very easy to calculate the distance to a star given the parallax. For example, the star Altair has a parallax of 0.20 second of arc. Then the distance to Altair in parsecs is 1 divided by 0.20, which equals 5 pc. To convert to light-years, we multiply 5 pc by 3.26 and discover that Altair is 16.3 ly away.

Measuring the small angle p is very difficult. The nearest star to the sun, α Centauri, has a parallax of only 0.76 second of arc, and more distant stars have even smaller parallaxes. To see how small these angles are, hold a piece of paper edgewise at arm's length. The thickness of the paper covers an angle of about 30 seconds of arc. You can see that the parallax of a star, smaller than 1 second of arc, must be very difficult to measure accurately.

The blurring of the earth's atmosphere limits the accuracy of parallax measurements made from earthbound telescopes. The seeing (Chapter 6) at the best observatories is limited to about 0.5 second of arc, and even the average of many parallax measurements cannot determine parallaxes smaller than about 0.01 second of arc. This means that ground-based telescopes can't measure the parallax of a star if it is farther away than about 100 pc. Fewer than 1000 stars have had their parallaxes measured to within a few percent.

From above the earth's turbulent atmosphere, spacecraft can image stars sharply and measure parallaxes with high precision. The Hipparcos satellite was launched by the European Space Agency in 1989, and although a rocket malfunction trapped it in a low orbit, it has observed the position, brightness, and color of every star in the sky brighter than ninth magnitude. The reduction of these data has been a monstrous computing task, but it is giving astronomers accurate parallaxes of over 100,000

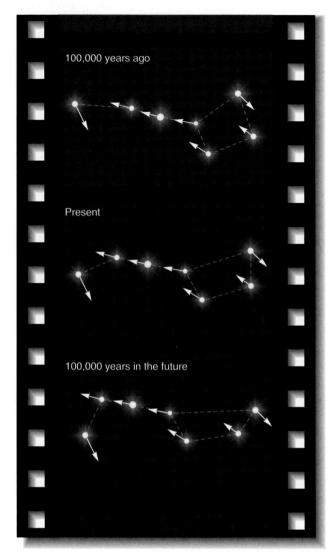

FIGURE 9-5

Proper motion slowly alters the positions of the stars in the sky. The Big Dipper looked quite different 100,000 years ago, and the proper motions of the stars (arrows) continue to change its shape.

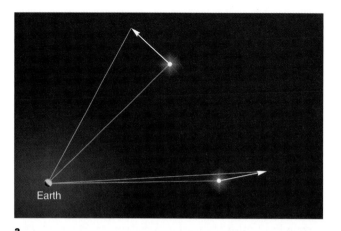

FIGURE 9-6

Proper motion: (a) Two stars with exactly the same velocity through space may have different proper motions because one is moving perpendicular to the line of sight and the other is moving nearly parallel to the line of sight. (b) Proper motion also depends on distance. If two stars have identical velocities through space, both perpendicular to the line of sight, the nearer star will have a larger proper motion.

stars good to 0.002 second of arc. That includes stars out to a distance of 500 pc. These data will allow astronomers to define the properties of stars much more accurately than can be done now using ground-based parallaxes.

Parallax observations tell us the distance to the nearer stars, but how can we find out which stars are nearby? The answer is to watch how stars move through space as the years pass.

Proper Motion

All the stars in the sky, including our sun, are moving along orbits around the center of our galaxy. We don't notice that motion over periods of years, but over the centuries it can significantly distort the shape of constellations (Figure 9-5).

If we photograph a small area of the sky on two dates separated by 10 years or more, we can notice that some of the stars in the photograph have moved very slightly against the background stars. This motion, expressed in units of seconds of arc per year, is the **proper motion** of the stars. For example, the star Altair (α Aquilae) has a proper motion of 0.662 second of arc per year, but Albireo (β Cygni) has a proper motion of only 0.002 second of arc per year.

Why should one star have a larger proper motion than another star? For one thing, a star might be moving almost directly toward or away from us, and thus its position on the sky would change very slowly. That is, its transverse velocity (Chapter 7) could be quite low, resulting in a small proper motion (Figure 9-6a).

Another reason why a star might have a small proper motion is that it could be quite far away from us. Then even a large transverse velocity would not produce a large proper motion (Figure 9-6b). Thus Albireo, at a distance of 116 pc, has a smaller proper motion, and Altair, at a distance of only 4.9 pc, has a larger proper motion.

We can use proper motion to look for nearby stars. If we see a star with a small (or zero) proper motion, it is probably a distant star. We cannot be sure, because it could be a nearby star traveling almost directly toward or away from us. But on the average, stars with small proper motions are distant. You have probably seen this effect if you watch birds. Distant geese move slowly across the sky, but a nearby bird flits quickly across our field of view. Similarly, stars with large proper motions are usually nearby stars. In fact, stars with proper motions above a certain limit must be nearby. A distant star traveling fast enough to have a proper motion above this limit would have such a high velocity that it would have escaped from the galaxy long ago.

Thus, proper motions can give us statistical clues to distance. We can locate stars that are probably nearby—good candidates for parallax measurements—by looking for stars with large proper motions. Other statistical techniques use proper motions to find the average distance to selected groups of stars.

CRITICAL INQUIRY

Why can't we use ground-based telescopes to measure the parallax of stars farther than 100 pc from the earth?

A telescope on the earth's surface must look up through the earth's atmosphere, and seeing (distortion caused by the turbulent air) spreads the star image into a fuzzy blob. At the best observatories on the highest mountains, the star images are usually no smaller than 0.5 second of arc. To measure the parallax of a star, we must measure the positions of these fuzzy blobs, and even if we make many measurements and average the results, the limit on the accuracy of such a measurement is about 0.01 second of arc. If a star has a parallax of 0.01 second of arc, we will make errors of roughly plus or minus 100% when we try to measure it. In other words, we won't be able to measure it very accurately at all. A star with a parallax of 0.01 second of arc would have a distance of 1 divided by 0.01, or 100 pc, so that is the practical limit beyond which we are unable to measure a star's parallax.

Of course, we could put a telescope into orbit around the earth. How would that help? What would limit the accuracy then? ■

Having found a way to locate nearby stars and measure their distances, we are ready to discuss the first of the three stellar properties—brightness. Our goal is to find out how much energy stars emit.

FIGURE 9-7
The lights of Pittsburgh glitter across the river. Without knowing the distance to one of these lights, we cannot estimate its intrinsic brightness. Similarly, we must know a star's distance before we can find its intrinsic brightness. (Michael A. Seeds)

9-2 INTRINSIC BRIGHTNESS

If we view a streetlight from nearby, it may seem quite bright, but if we view it from a hilltop miles away, it appears faint. Its apparent brightness depends on its distance, but its intrinsic brightness, the amount of light it emits, is independent of distance. When we look at stars, we face the same problem we might face trying to judge the brightness of city lights viewed from a distant hilltop (Figure 9-7). We can judge apparent brightness easily, but unless we know the distances to individual points of light, we cannot determine their intrinsic brightnesses. We could not, for instance, tell distant streetlights from dimmer, but nearer, light bulbs. Once an astronomer determines the distance to a star, however, it is simple to calculate its intrinsic brightness from its apparent brightness and its distance.

We will use two terms to refer to a star's intrinsic brightness. One, absolute visual magnitude, is common in astronomy because its use simplifies calculations involving distance. A second term, luminosity, refers directly to the amount of energy the star emits in 1 second.

Absolute Visual Magnitude

Judging the intrinsic brightness of stars would be easier if they were all at the same distance. Although astronomers can't move stars about and line them up at some standard distance, they can calculate how bright a star of known distance would appear at any other distance. Using this method, they refer to the intrinsic brightness of a star as its **absolute visual magnitude** (M_v)—the apparent visual magnitude the star would have if it were 10 pc away.

The symbol for absolute visual magnitude is an uppercase M with a subscript v. The symbol for apparent visual magnitude is a lowercase m with a subscript v. The subscript tells us that the visual magnitude system is based only on the wavelengths of light we can see. Other magnitude systems are based on other parts of the electromagnetic spectrum such as the infrared, ultraviolet, and so on. Yet another magnitude system refers to the total energy emitted at all wavelengths. We will limit our discussion to visual magnitudes.

The intrinsically brightest stars known have absolute magnitudes of about −8, and the faintest about +19. The nearest star to the sun, α Centauri, is only 1.4 pc away, and its apparent magnitude is 0.0, indicating that it looks bright in the sky. However, its absolute magnitude is 4.39, telling us it is not intrinsically very bright. Because we know the distance to the sun and can measure its apparent magnitude, we can find its absolute magnitude—about 4.78. If the sun were only 10 pc away from us, it would look no brighter than the faintest star in the handle of the Little Dipper.

Calculating Absolute Visual Magnitude

How can astronomers find the absolute visual magnitude of a star? This question leads us to one of the most common formulas in astronomy, a formula that relates a star's magnitude to its distance.

The **magnitude-distance formula** relates the apparent magnitude m_v, the absolute magnitude M_v, and the distance d in parsecs:

$$m_v - M_v = -5 + 5 \log_{10}(d)$$

If we know any two of the parameters in this formula, we can easily calculate the third. If we are interested in knowing how we might find the absolute magnitude of a star, then we must consider an example in which we know the distance and apparent magnitude of a star. Suppose a star has a distance of 50 pc and an apparent magnitude of 4.5. A pocket calculator tells us that the log of 50 is 1.70, and −5 + 5 × 1.70 equals 3.5, so we know that the absolute magnitude is 3.5 magnitudes brighter than the apparent magnitude. The absolute magnitude is thus 1.0, since 4.5 minus 3.5 is 1.0. (Remember that smaller numbers mean brighter magnitudes.) If this star were 10 pc away, it would be a first-magnitude star.

Astronomers also use the magnitude-distance formula to calculate the distance to a star if the apparent and absolute magnitudes are known. For that purpose, it is handy to rewrite the formula in the following form:

$$d = 10^{\frac{m_v - M_v + 5}{5}}$$

If we knew that a star had an apparent magnitude of 7 and an absolute magnitude of 2, then $m_v - M_v$ is 5 magnitudes, and the distance would be 10^2 or 100 parsecs.

The magnitude difference $m_v - M_v$ is known as the **distance modulus,** a measure of how far away the star

TABLE 9-1 Distance Moduli	
$m_v - M_v$	d (pc)
0	10
1	16
2	25
3	40
4	63
5	100
6	160
7	250
8	400
9	630
10	1000
⋮	⋮
15	10,000
⋮	⋮
20	100,000
⋮	⋮

is. The larger the distance modulus, the more distant the star. We can use the magnitude-distance formula to construct a table of distance and distance modulus (Table 9-1).

The magnitude-distance formula may seem awkward at first, but a pocket calculator makes it easy to use. It is important because it performs a critical function in astronomy: it allows us to convert observations of distance and apparent magnitude into absolute magnitude, a measure of the true brightness of the star. Once we know the absolute magnitude, we can go one step further and figure out the total amount of energy a star is radiating into space.

Luminosity

The **luminosity (L)** of a star is the total amount of energy the star radiates in 1 second—not just visible light, but all wavelengths. To find a star's luminosity, we begin with its absolute visual magnitude, make a small correction, and compare the star with the sun.

The correction we must make adjusts for the radiation emitted at wavelengths we cannot see. Absolute visual magnitude includes only visible light. The absolute magnitudes of hot stars and cool stars will underestimate their total luminosities because those stars radiate significant amounts of radiation in the ultraviolet or infrared parts of the spectrum. We can correct for the missing radiation because the amount of missing energy depends only on the star's temperature. For hot and cool stars, the correction can be large, but for medium-temperature stars like the sun, the correction is small. Adding the proper correction to the absolute visual magnitude changes it into the **absolute bolometric magnitude**— the absolute magnitude the star would have if we could see all wavelengths.

Once we know a star's absolute bolometric magnitude, we can find its luminosity by comparing it with the sun. The absolute bolometric magnitude of the sun is +4.7. For every magnitude a star is brighter than 4.7, it is 2.512 times more luminous than the sun. (Recall from Chapter 2 that a difference of 1 magnitude corresponds to an intensity ratio of 2.512.) Thus, a star with an absolute bolometric magnitude of 2.7 is 2 magnitudes brighter than the sun and 6.3 times more luminous (6.3 is approximately 2.512×2.512).

Arcturus, for example, has an absolute bolometric magnitude of −0.3. That makes it 5 magnitudes brighter than the sun. A difference of 5 magnitudes is defined to be a factor of 100 in brightness, so the luminosity of Arcturus is 100 times the sun's luminosity, or 100 $L_\odot$.

The symbol $L_\odot$ represents the luminosity of the sun, a number we can calculate in a direct way. Because we know how much solar energy hits 1 square meter in 1 second just above the earth's atmosphere (the solar constant defined in the previous chapter) and how far it is from the earth to the sun, it is a simple matter to calculate how much energy the sun must radiate in all directions to provide the earth with the energy it receives per second (see Problem 9 at the end of this chapter). We find that the luminosity of the sun is about 4×10^{26} joules/sec.

Let's review: If we can measure the parallax of a star, we can find its distance, calculate its absolute visual magnitude, correct for the light we can't see to find the absolute bolometric magnitude, and then find the luminosity in terms of the sun. If we want to know the luminosity in joules per second, we need only multiply by the sun's luminosity.

CRITICAL INQUIRY

Why do hot stars have an absolute bolometric magnitude that is quite different from their absolute visual magnitude?

Recall that the absolute visual magnitude, M_v, is the magnitude we would see if the star were 10 pc away. But our eyes can't see all wavelengths, so a very hot star, which radiates much of its energy in the ultraviolet, would not look very bright to our eyes. The absolute bolometric magnitude includes all wavelengths, so hot stars have absolute bolometric magnitudes much brighter (smaller numbers) than their absolute visual

magnitudes because of the ultraviolet energy they radiate. If we want to calculate the luminosities of hot stars, we must use absolute bolometric magnitudes. Otherwise, we'll leave out the ultraviolet radiation they emit, and the luminosities we calculate will be too small.

Stars like the sun radiate most of their energy at visible wavelengths, and we can see that energy, so there isn't a big difference between the absolute bolometric and absolute visual magnitudes for these medium-temperature stars. But what about the cool stars? Explain their absolute bolometric magnitudes. ■

Although we had to make a detour in order to find the distances to the stars, we have reached our first goal. We have found a way to discover the energy emitted by stars. The range of stellar luminosities is very large. The most luminous stars have luminosities of about $10^5 L_\odot$, and the least luminous are roughly $10^{-4} L_\odot$. Knowing the luminosities of stars helps us toward our second goal, finding their diameters.

9-3 THE DIAMETERS OF STARS

One fundamental property of stars is their diameter. Are they all the same size, or are some larger and some smaller? We know little about stars until we know their diameters. We certainly can't see their diameters through a telescope; the stars appear much too small for us to resolve their disks and measure their diameters. But there is a way to find out how big stars really are. If we know their temperature and luminosity, we can find their diameter. That relationship will introduce us to the most important diagram in astronomy, where we will discover the family relations among the stars.

Luminosity, Radius, and Temperature

The luminosity and temperature of a star can tell us its diameter if we understand the two factors that affect a star's luminosity: surface area and temperature. For example, you can eat dinner by candlelight because the candle flame has a small surface area, and although it is very hot, it cannot radiate much heat; it has a low luminosity. However, if the candle flame were 12 ft tall, it would have a very large surface area from which to radiate, and although it would be no hotter than a normal candle flame, its luminosity would drive you from the table.

In a similar way, a hot star may not be very luminous if it has a small surface area. It could be highly luminous if it were larger, and even a cool star could be very luminous if it were very large and so had a large surface area from which to radiate. Thus, we see that both temperature and surface area help determine the luminosity of a star.

Stars are spheres, and the surface area of a sphere is $4\pi R^2$. If we measure the radius in meters, this is the number of square meters on the surface of a star. Each square meter radiates like a black body, and the total energy given off each second is σT^4. The luminosity of the star is the surface area multiplied by the energy radiated per square meter:

$$L = 4\pi R^2 \sigma T^4$$

If we divide by the same quantities for the sun, we can cancel out the constants and get a simple formula for the luminosity of a star in terms of its radius and temperature:

$$\frac{L}{L_\odot} = \left(\frac{R}{R_\odot}\right)^2 \left(\frac{T}{T_\odot}\right)^4$$

Here the symbol $\odot$ stands for the sun, and the formula tells us that the luminosity in terms of the sun equals the radius in terms of the sun squared times the temperature in terms of the sun raised to the fourth power.

Suppose a star is 10 times the sun's radius but only half as hot. How luminous would it be?

$$\frac{L}{L_\odot} = \left(\frac{10}{1}\right)^2 \left(\frac{1}{2}\right)^4 = \frac{100}{1}\,\frac{1}{16} = 6.25$$

The formula tells us that it would be 6.25 times the sun's luminosity.

How can we use this to tell us the diameters of the stars? If we see a cool star that is very luminous, we know it must be very large, and if we see a hot star that is not very luminous, we know it must be very small. The formula allows us to calculate these sizes. For instance, suppose that a star is 40 times the luminosity of the sun and twice as hot. If we put these numbers into our formula, we get:

$$\frac{40}{1} = \left(\frac{R}{R_\odot}\right)^2 \left(\frac{2}{1}\right)^4$$

Solving for the radius, we get:

$$\left(\frac{R}{R_\odot}\right)^2 = \frac{40}{2^4} = \frac{40}{16} = 2.5$$

So the radius is:

$$\frac{R}{R_\odot} = \sqrt{2.5} = 1.58$$

The star is 58 percent larger in radius than the sun.

Because a star's luminosity depends on its surface area and its temperature, we can sort the stars by diameter and thus discover which are large and which are small if we can sort them by luminosity and temperature. Astronomers use a special diagram for that sorting.

The H–R Diagram

The **Hertzsprung–Russell diagram,** named after its discoverers, Ejnar Hertzsprung and Henry Norris Russell, is a graph that separates the effects of temperature and

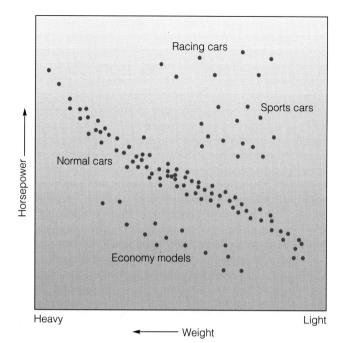

FIGURE 9-8
We could analyze automobiles by plotting their horsepower versus their weight and thus reveal relationships between various models. Most would lie somewhere along the main sequence of "normal" cars.

surface area on stellar luminosities and enables us to sort the stars according to their diameters. The **H–R diagram** (as it is often called) is the most important diagram in astronomy. It appears in this book 28 times. Referring to it sometimes as "the diagram," astronomers use it as a graphic way of thinking about the different kinds of stars. The diagram is a powerful tool because it sorts the stars by their diameters.

To see how a simple diagram can reveal hidden relationships among similar objects—to see how the H–R diagram can sort the stars into family groups according to size—let us begin with a similar diagram we might use to sort automobiles.

We could plot a diagram such as that in Figure 9-8 to show horsepower versus weight for various makes of cars. We would find that, in general, the more a car weighs, the more horsepower it has. Most cars would fall somewhere along the sequence of cars running from heavy, high-powered cars to lightweight, low-powered models. We could call this the main sequence of cars. But some would have much more horsepower than normal for their weight—the sports or racing models—and the economy models would have less power than normal for cars of the same weight. Just as this diagram would help us understand the different kinds of autos, the H–R diagram helps us understand the different kinds of stars.

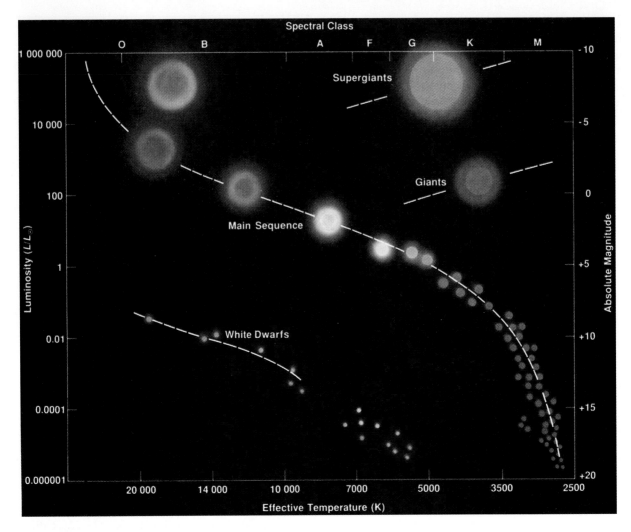

FIGURE 9-9

An H–R diagram. Roughly 90 percent of all stars are main-sequence stars, including the sun—a G2 star with an absolute magnitude of about +5. Star diameters are not to scale. (William K. Hartmann)

The H–R diagram relates the intrinsic brightness of stars to their surface temperatures (Figure 9-9). We can plot either absolute magnitude or luminosity on the vertical axis of the graph, because both refer to intrinsic brightness. Remember from Chapter 7 that spectral type is related to temperature, so we can plot either spectral type or temperature on the horizontal axis. Technically, only graphs of absolute magnitude versus spectral type are H–R diagrams. However, we will refer to plots of luminosity versus either spectral type or surface temperature by the generic term *H–R diagram.*

A point on an H–R diagram shows a star's luminosity and surface temperature. Points near the top of the diagram represent very luminous stars, and points near the bottom represent very faint stars (Window on Science 9-1). Points on the left represent hot stars, and points on the right represent cool stars. Notice that the location of a star in the H–R diagram has nothing to do with its location in space. Two stars near each other in the H–R diagram have similar properties but need not

be near each other in space. Also, as a star ages, its luminosity and surface temperature change, and the point that represents it moves in the H–R diagram, but this has nothing to do with the star's actual motion through space.

Giants, Supergiants, and Dwarfs

The **main sequence** is the region of the H–R diagram running from upper left to lower right in Figure 9-9, which includes roughly 90 percent of all stars. These are the "ordinary" stars. As we might expect, the hot main-sequence stars are brighter than the cool main-sequence stars. The sun is a medium-temperature main-sequence star.

Just as sports cars do not fit in with the normal cars in Figure 9-8, some stars do not fit in with the main-sequence stars in Figure 9-9. The **giant stars** lie at the upper right of the H–R diagram. These stars are cool, radiating little energy per square meter. Nevertheless,

Exponential Scales

Graphs are important in all branches of science, and many graphs have scales that are exponential (related to the power of some base, such as 10^2, 10^3, 10^4, and so on). Because we are accustomed to thinking about graphs with linear scales, exponential scales can be misleading. H–R diagrams are good examples of graphs with exponential scales, but such scales are common in all the sciences, including chemistry, biology, and geology. They also appear in economics, government, and other fields that deal with quantitative data.

In an exponential scale, each step (tick mark) along the axis corresponds to a constant factor. For example, suppose we plotted a diagram in which the vertical axis showed weight on an exponential scale. At the very bottom of the graph, we could plot the weight of a cat, and above that our own weight. Near the top of the graph, we could plot the weight of a bull elephant, and just above that the weight of a whale. In fact, just above that we could plot the weight of all the whales on the earth plus all the elephants, people, and cats. An exponential scale compresses the high numbers and spreads out the low numbers. Thus, our weight on the graph would seem much greater than that of a cat and surprisingly close to that of an elephant. Exponential scales are very helpful for compressing a wide range of numbers into a single graph, but they can be misleading.

For an astronomical example, notice that the vertical axis of the H–R diagram in Figure 9-9 has tick marks separated by a constant factor of 100 in stellar luminosity. If a star is just one tick mark higher than another star, it is *100 times* more luminous.

To see how misleading this can be, try drawing an H–R diagram with a linear scale on the vertical axis. Put zero luminosity at the bottom, and make each tick mark equal to an increase of 100,000 $L_\odot$, with 1,000,000 $L_\odot$ at the top. You will discover that it is difficult to plot all of the different kinds of stars, but their relative luminosities will be much more clearly illustrated. ■

a

b

they are highly luminous because they have enormous surface areas; hence the name *giant stars*. In fact, we can estimate the size of these giants with a simple calculation. Notice from the H–R diagram that they are about 100 times more luminous than the sun although they have about the same surface temperature. Thus, they must have about 100 times more surface area than the sun, indicating that their diameters must be about 10 times larger than the sun's (Figure 9-10).

Near the top of the H–R diagram, we find a few stars called **supergiants.** These exceptionally luminous stars are 10 to 1000 times the sun's diameter. The stars Betelgeuse and Rigel (α and β Orionis) are both supergiants. The largest known supergiant may be μ Cephei, an M2 star whose radius is believed to be about 3700 $R_\odot$ If μ Cephei were to replace the sun at the center of our solar

FIGURE 9-10

(a) This H–R diagram shows the relative sizes of stars. Giant stars are 10 to 100 times larger than the sun, and white dwarfs are about the size of the earth. (The dots representing white dwarfs here are much too large.) Supergiants are too large for this diagram. (b) To visualize the size of the largest stars, imagine that the sun is the size of one of your eyeballs. Then the largest supergiants would be the size of a hot air balloon.

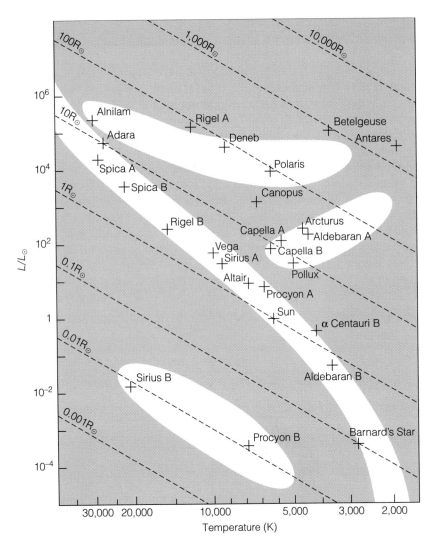

FIGURE 9-11

An H–R diagram drawn with luminosity versus surface temperatures. Diagonal lines are lines of constant radius. White areas mark the approximate location of supergiants, giants, main-sequence stars, and white dwarfs. Note that giant stars are 10 to 1000 times larger than the sun and that white dwarfs are about 100 times smaller than the sun—that is, about the size of the earth. (Individual stars that orbit each other are designated by A and B, as in Spica A and Spica B.)

system, it would engulf Mercury, Venus, Earth, Mars, Jupiter, Saturn and nearly reach the orbit of Uranus.

In the lower left of the H–R diagram are the "economy models," stars that are very faint although they are hot. Clearly, such stars must be small. These are the **white dwarf stars,** dying stars that have collapsed to about the size of the earth and are slowly cooling off (Figure 9-11).

Luminosity Classification

We can tell from a star's spectrum what kind of star it is. Recall from Chapter 7 that collisional broadening can make spectral lines wider when the density of a gas is relatively high and the atoms collide often.

Main-sequence stars are relatively small and have dense atmospheres in which the gas atoms collide often and distort their electron energy levels. Thus, lines in the spectra of main-sequence stars are broad. On the other hand, giant stars are larger, their atmospheres are less dense, and the atoms disturb one another relatively little.

Collisional broadening is not as much of a factor in the atmospheres of giant stars, and their spectra contain sharp lines. As we might expect, the lines in the spectra of supergiants are even sharper.

Thus, we can look at a star's spectrum and classify its luminosity (Figure 9-12). We can tell whether it is a supergiant, a bright or ordinary giant, a subgiant, or a main-sequence star. Although these are the **luminosity classes,** the names refer to the sizes of the stars because size is the dominating factor in determining luminosity. Supergiants, for example, are very bright because they are very large.

The luminosity classes are represented by the roman numerals I through V, as shown in the following list:

- Ia Bright supergiant
- Ib Supergiant
- II Bright giant
- III Giant
- IV Subgiant
- V Main-sequence star

Using letters for subclasses, we can distinguish between the bright supergiants (Ia) such as Rigel (β Orionis) and the regular supergiants (Ib) such as Polaris, the North

Star. The star Adara (ε Canis Majoris) is a bright giant (II), Capella (α Aurigae) is a giant (III), and Altair (α Aquilae) is a subgiant (IV). The sun is a main-sequence star (V). The luminosity class usually appears after the spectral type, as in G2 V for the sun. White dwarf stars don't enter into this classification because their spectra are quite unlike those of other stars.

We can, as in Figure 9-13, plot the locations of the luminosity classes on the H–R diagram. Remember that these are rather broad classifications. A star of luminosity class III may lie slightly above or below the line labeled III. The lines are only approximate.

The luminosity classification is subtle and not very accurate, but it is an important tool in modern astronomy. As we will see in the next section, it gives us a way to find the distances to stars.

FIGURE 9-13
The approximate location of the luminosity classes on the H–R diagram.

Spectroscopic Parallax

We can measure the parallax of nearby stars, but most stars are too distant to have measurable parallaxes. We can find the distances to these stars if we can photograph their spectra and determine their luminosity classes. A process called **spectroscopic parallax** allows us to estimate the distance to a star from its spectral type and luminosity class. Spectroscopic parallax does not actually involve measuring parallax, but it does tell us the distance to the star.

The method of spectroscopic parallax depends on the H–R diagram. If we photograph the spectrum of a star, we can determine its spectral class, which tells us its horizontal location in the H–R diagram. We can also determine its luminosity class by looking at the widths of its spectral lines, and that tells us the star's vertical location in the diagram. Once we plot the point that represents the star in the H–R diagram, we can read off its absolute magnitude. As we have seen earlier in this chapter, we can find the distance to a star by comparing its apparent and absolute magnitudes.

For example, Spica is classified B1 V, and its apparent magnitude is +1. We can plot this star in an H–R diagram such as that in Figure 9-13, where we would find that it should have an absolute magnitude of about −3. Therefore, its distance modulus is 1 minus (−3), or 4, and the distance (from Table 9-1) is about 63 pc.

CRITICAL INQUIRY

What evidence do we have that giant stars really are bigger than the sun?

We can find stars that have the same spectral type as the sun, but they are clearly more luminous than the sun. Capella, for example, is a G star with an absolute magnitude of 0. If it is a G star, then it must have about the same temperature as the sun, but its absolute magnitude is five magnitudes brighter than the sun's. A magnitude difference of five magnitudes corresponds to an intensity ratio of 100, so Capella must be about 100 times more luminous than the sun. If it has the same surface temperature as the sun but is 100 times more luminous, then it must have a surface area that is 100 times bigger than the sun's. Since the surface area of a sphere is proportional to the square of the radius, Capella must be 10 times larger in radius. That is clear observational evidence that Capella is a giant star.

In Figure 9-11, we see that Procyon B is a white dwarf only slightly warmer than the sun but about 10,000 times less luminous. Use the logic above to prove that the white dwarfs must be small stars. ■

The two goals of this chapter have been achieved; we know how to find the luminosities and diameters of stars. We have discovered that there are four different kinds of stars—main-sequence stars, giants, supergiants, and white dwarfs—and that these types differ in their intrinsic properties, luminosity and diameter. But this discovery raises an entirely new question. How common are these different kinds of stars? To find out, we can use the tools we have developed to take a survey among the stars.

9-4 THE FREQUENCY OF STELLAR TYPES

Astronomers are gathering clues to the birth, life, and death of the stars, and one important clue is the frequency with which different kinds of stars occur. By surveying a large number of stars, we discover that some types are rare and some are common. To understand the astronomer's surveying techniques, we begin by considering a survey of people.

Surveying a Representative Sample

Suppose you took a survey in your neighborhood to find out how many people have gray eyes. If you knew the area of your neighborhood in square miles, you could then say, "In my neighborhood, x people per square mile have gray eyes." Next you might think of extending your conclusion to the country as a whole: "In America, x people per square mile have gray eyes." Of course, if you did not live in an average neighborhood, your result would be wrong, but if you had sampled a truly representative neighborhood, your survey would give valid results about the entire population.

We can do the same thing with the stars. We can ask how many stars of each spectral type are whirling through each million cubic parsecs of space. We can't count every star in our galaxy, but we can take a survey in the region of space near the sun. Because we think the solar neighborhood is a fairly average sort of place, we can use this local survey to reach conclusions about the entire population of stars.

Units of Stellar Density

The result of such a survey of stars is called a **stellar density function,** a description of the abundance of different types of stars in space. A simple form of the stellar density function appears in Figure 9-14, giving the abundance of the stars of each spectral type in terms of the number of each type we would expect to find in 1 million cubic parsecs.

Clearly, the most common kind of star is an M star. Every million cubic parsecs of space contains 50,000 M

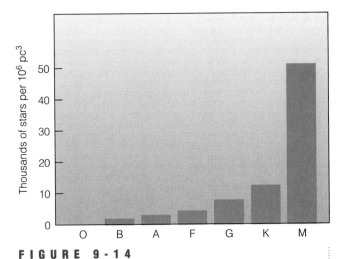

FIGURE 9-14
The stellar density function shows that M stars are the most common stars in space and that O stars are very rare.

stars, but only about 2000 B stars. One of the questions we must resolve in the next few chapters is Why are M stars so common?

But notice that the stellar density function does not tell us whether these M stars are main-sequence stars, giants, or supergiants. It lumps all M stars into one category. To distinguish among the luminosity classes, we must consider both spectral type and absolute magnitude when we take our survey. Unfortunately, certain problems stand in our way.

Three Problems in Counting Stars

The astronomer could carry out these surveys in the neighborhood of the sun by counting all the stars within a given distance. A sphere of radius 62 pc contains 1 million cubic parsecs. Thus, if we could count all the M stars, for example, within 62 pc of the sun, we would have the frequency of M stars per million cubic parsecs.

But this survey of the stars poses three problems. First, to determine which stars are within 62 pc of the sun, we must measure their distances. However, stars near the outer edge of a sphere 62 pc in radius have small parallaxes that are difficult to measure accurately, and the method of spectroscopic parallax is not accurate enough for this purpose. We could count stars in a smaller sphere, but some stars are so rare that we might not find any in such a small volume of space.

A second problem for the stellar surveyor is the intrinsic faintness of stars such as M stars and white dwarfs. These are so faint that they are very hard to see if they are only a few dozen parsecs away. For example, a white dwarf 62 pc away is over 1500 times fainter than the faintest star visible to the naked eye—very hard to find indeed.

FIGURE 9-15

In this histogram, bars rise from an H–R diagram to represent the number of stars of given spectral type and absolute magnitude found in 1 million cubic parsecs. The main-sequence M stars are the most common stars, followed closely by the white dwarfs. The upper-main-sequence stars, giants, and supergiants are so rare their bars are not visible in this diagram.

The third problem with these surveys is that the hottest stars are rare. There are no O stars at all within 62 pc of the sun. We must extend our survey to great distances before we find any of these hot, luminous stars. At such distances, the parallaxes are too small to be measured, and we have to find distances in other, less accurate ways, such as spectroscopic parallax.

The Frequency of Stellar Types

Despite these difficulties, astronomers have been able to piece together the stellar density function of the stars in our galaxy. Using statistical methods, they can now tell us not only the abundance of stars of a given spectral type but also how many stars of that type are likely to be giants, how many supergiants, and how many white dwarfs. This detailed version of the stellar density function is shown in Figure 9-15.

Notice how common the main-sequence stars are. There are about 50,000 in every million cubic parsecs. Notice also that M giants are very rare; there are only about 20 per million cubic parsecs. The white dwarfs and the main-sequence M stars are the most common kinds of star (Figure 9-16). Fortunately for us, these stars are very faint, for if they were as bright as supergiants, the sky would be filled with their glare and we would barely

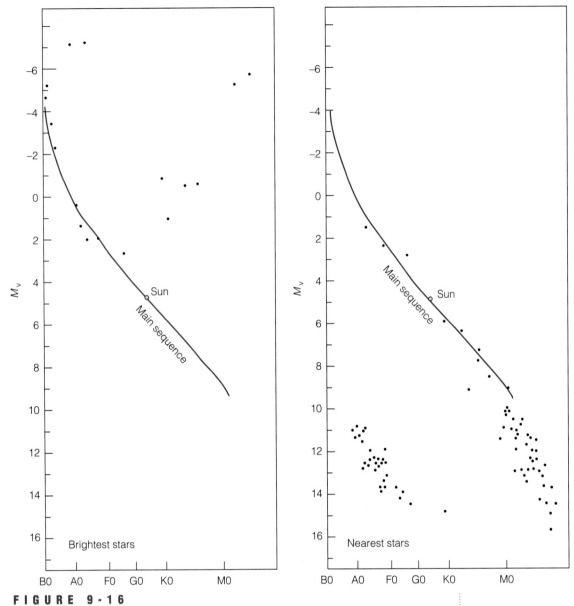

FIGURE 9-16

Representative H–R diagrams of the brightest stars in the sky and the nearest stars in space show that the most common kind of star is a lower main-sequence M star or a white dwarf. Of the bright stars we see in the sky, many are giants and supergiants, and none are lower main-sequence M stars or white dwarfs.

see anything else. As it is, these most common stars are so faint they are hard to find even with large telescopes.

The most luminous stars are very rare. On the average, there is only 0.03 main-sequence O star per million cubic parsecs. That is, we would have to search through about 30 million cubic parsecs to find an O star. Put yet another way, only one star in 4 million is an O star. The giants are slightly more abundant than this. There are a few hundred giants and subgiants in every million cubic parsecs, but the supergiants are very rare. There is about 0.07 supergiant per million cubic parsecs. That is one supergiant in every 14 million cubic parsecs. Luckily, these very rare kinds of stars are also very luminous, so

we can see them from great distances. If they were as faint as the sun, we might never know they existed.

In this chapter, we set out to measure the basic properties of stars. Once we found the distance to the stars, we were able to find their luminosities and their diameters —rather mundane data. But we have now discovered a puzzling situation. The largest and most luminous stars are so rare we might joke that they hardly exist, and the average stars are such dinky, low-mass things they are hard to see even when they are near us in space. Why does nature make stars in this peculiar way? To answer that question, we must explore the birth, life, and death of stars. We begin that quest in the next chapter, where we find a way to measure the masses of the stars. ∎

▪ Summary

We can measure the distance to the nearer stars by observing their parallaxes. The more distant stars are so far away that their parallaxes are unmeasurably small. To find the distances to these stars, we must use spectroscopic parallax. Stellar distances are commonly expressed in light-years or parsecs. One light-year is the distance light travels in 1 year, about 9.5×10^{12} km or about 5.8 trillion miles. One parsec is about 206,265 AU—the distance to an imaginary star whose parallax is 1 second of arc. One parsec equals 3.26 ly.

One way to locate nearby stars is to look for stars with large proper motions. Proper motion is the angular motion of a star across the sky and is usually expressed in seconds of arc per year. Nearby stars are likely to have larger proper motions than more distant stars.

Once we know the distance to a star, we can find its intrinsic brightness, expressed as its absolute visual magnitude. We can add a correction to account for the light at wavelengths we cannot see and thus transform absolute visual magnitude into absolute bolometric magnitude. From this we can find the luminosity of the star, the total energy radiated in 1 second. Luminosity is often expressed in terms of the sun, as in 10 $L_{\odot}$.

The H–R diagram is a graph in which stars are plotted according to their intrinsic brightness and their surface temperature. In the diagram, roughly 90 percent of all stars fall on the main sequence, the more massive being hotter, larger, and more luminous. The giants and supergiants are much larger and lie above the main sequence; they are more luminous than main-sequence stars of the same temperature. The white dwarfs are hot stars, but they fall below the main sequence because they are so small.

Because the atmospheres of giant and supergiant stars have low density, collisional broadening does not widen the lines in their spectra very much. Thus, giants and supergiants have sharper spectral lines than do main-sequence stars. In fact, it is possible to assign stars to luminosity classes by the widths of their spectral lines. Class V stars are main-sequence stars with broad spectral lines. Giant stars (III) have sharper lines, and supergiants (I) have extremely sharp spectral lines.

A survey in the neighborhood of the sun shows us that the most common stars are the lower main-sequence stars. The hot stars of the upper main sequence are very rare. Giants and supergiants are also rare, but white dwarfs are quite common, although they are faint and hard to find.

▪ New Terms

stellar parallax (p)	main sequence
parsec (pc)	giant stars
proper motion	supergiants
absolute visual magnitude (M_v)	white dwarf stars
magnitude-distance formula	luminosity class
distance modulus ($m_v - M_v$)	spectroscopic parallax
luminosity (L)	stellar density function
absolute bolometric magnitude	
H–R (Hertzsprung–Russell) diagram	

▪ Questions

1. Why are parallax measurements limited to the nearest stars?

2. How would having an observatory on Mars help astronomers measure parallax?

3. How would having an observatory in orbit above the earth's atmosphere help astronomers measure parallax?

4. For which stars does absolute visual magnitude differ least from absolute bolometric magnitude? Why?

5. How can a cool star be more luminous than a hot star? Give some examples.

6. How can we be certain that the giant stars are actually larger than the sun?

7. Describe the steps in using the methods of spectroscopic parallax. Do we really measure a parallax?

8. Give the approximate radii and intrinsic brightnesses of stars in the following classes: G2 V, G2 III, G2 Ia.

▪ Discussion Questions

1. If someone asked you to compile a list of the nearest stars to the sun based on your own observations, what measurements would you make, and how would you analyze them to detect nearby stars?

2. The sun is sometimes described as an average star. What is the average star really like?

▪ Problems

1. If a star has a parallax of 0.050 second of arc, what is its distance in parsecs? in light-years? in AU?

2. If a star has a parallax of 0.016 second of arc and an apparent magnitude of 6, how far away is it, and what is its absolute magnitude?

3. Complete the following table:

m_v	M_v	d (pc)	p (sec of arc)
—	7	10	—
11	—	1000	—
—	−2	—	0.025
4	—	—	0.040

4. If a main-sequence star has a luminosity of 400 $L_{\odot}$, what is its spectral type? (HINT: See Figure 9-9).

5. If a star has an apparent magnitude equal to its absolute magnitude, how far away is it in parsecs? in light-years?

6. If a star has an absolute bolometric magnitude that is eight magnitudes brighter than the sun, what is the star's luminosity?

7. If a star has an absolute bolometric magnitude that is one magnitude fainter than the sun, what is the star's luminosity?

8. An O8 V star has an apparent magnitude of +1. Use the method of spectroscopic parallax to find the distance to the star. Why might this distance be inaccurate?

9. Find the luminosity of the sun given the radius of the earth's orbit and the solar constant (Chapter 8). Make your calculation in two steps. First, use $4\pi R^2$ to calculate the surface area in square meters of a sphere surrounding the sun with a radius of 1 AU. Second, multiply by the solar constant to find the total solar energy passing through the sphere in 1 second. That is the luminosity of the sun. Compare your result with that in Data File One.

10. In the following table, which star is brightest in apparent magnitude? most luminous in absolute magnitude? largest? farthest away?

Star	Spectral Type	m_v
a	G2 V	5
b	B1 V	8
c	G2 Ib	10
d	M5 III	19
e	White dwarf	15

▪ Critical Inquiries for the Web

1. The Hipparcos mission (Section 9.1) has improved our knowledge of the distances to stars manyfold. Determine the celestial coordinates of the stars listed next.

Star	RA	Dec	Star	RA	Dec
Vega	279	39.8	Betelgeuse	89	07.4
Arcturus	219	19.0	Aldebaran	69	16.5
Pollux	116	28.0	Spica	201	−11.1

Visit the Hipparcos Web site, search the catalog for these stars, and determine the trigonometric parallax for each. To do this, enter the coordinates given above in the Hipparcos sky plot applet and access the data by clicking on the stars and the "get information" button. Note that Hipparcos data files list parallax angle in milliarcseconds (thousandths of arcseconds). Use the parallax formula (p. 70) to determine each star's distance in parsecs and compare the results with the distances given in Table A-9? Why are the revised distances to some stars dramatically different, yet others see only slight corrections?

2. Hertzsprung–Russell diagrams allow astronomers to graphically represent the distribution of stellar properties. Examine H–R diagrams at various Web sites and determine which types of stars are most (and least) numerous in our galaxy.

 Go to the Wadsworth Astronomy Resource Center (www.wadsworth.com/astronomy) for critical thinking exercises, articles, and additional readings from InfoTrac College Edition, Wadsworth's online student library.

Chapter Ten

BINARY STARS

GUIDEPOST

Astronomers find two aspects of nature fascinating: light and gravity. Both are critical to understanding the universe. No aspect of astronomy expresses the interplay of light and gravity better than binary stars—elegant puzzles made up of luminous stars circling each other in a dance of gravity.

Puzzles aside, we study binary stars in this chapter because they give us our only method for finding the masses of stars. Those data, plus the luminosities and diameters from the preceding chapter, are the basis for our discussion of the birth and death of stars in the following chapters.

In addition, this chapter illustrates a basic principle of science. Scientists are not satisfied with just knowing a fact—a star's mass, for example. To think critically and clearly, they must know where the fact came from. We must do the same. Only when we know the source of basic data can we judge their reliability and think critically about science or anything else. ∎

LQMZISF LQMZISF SMLLSF � LOH

KCQ M QCZAFH QKOL TCN OHF

ND OBCPF LKF QCHSA ꓙC KMGK

SMIF O AMOECZA MZ LKF ꓙIT*

*This cryptogram contains two messages. One is easy to find, but the other is hidden deeper.

FIGURE 10-1

Castor in the constellation Gemini was the first binary star to be recognized as such. William Herschel announced its orbital motion in 1803. Since he began observing it, astronomers have watched it move through about half an orbit. Its period is about 400 years, and the stars are separated by about 90 AU.

Everybody loves hidden messages. They can be cryptograms like the one here or secret messages in invisible ink. We can discover meaning in ancient hieroglyphics or systems of cosmology in ancient monuments like Stonehenge. Perhaps that is one reason why astronomers find **binary stars**—pairs of stars that orbit around each other—so fascinating. Properly decoded, their message tells us something we can learn in no other way: the masses of the stars.

If we look at a star through a telescope, there is nothing to tell us the star's mass. There is no clue buried in its luminosity, temperature, color, or spectral type. Only by studying pairs of stars that orbit each other can we find the masses of the stars. The orbits are governed by the gravity of the stars, and the gravity depends on the amount of mass in the stars. Thus, the orbits contain clues to the masses.

The behavior of certain binary stars also contains clues to the diameters of the stars, and this information, combined with the masses, will give us important hints about the formation and evolution of stars. When we look at the masses of stars, we will discover a pattern concealed in the H–R diagram, and when we look at the density of matter in stars, we will find a different pattern. We would never unravel the puzzle of stellar evolution without this information from binary stars.

More than half of all stars are binary systems, and all such systems are basically the same—two stars orbiting around each other. What they look like depends on how close the stars are to each other, what their masses are, and how far they are from the earth. In this chapter, we will discover a number of different kinds of binaries that can be observed and analyzed in different ways. But remember that in every case the secret is hidden in the orbital motion.

10-1 VISUAL BINARIES

Through even a small telescope, we can see hundreds of **double stars**—close pairs of stars. Some of these are **visual binaries,** pairs of stars that are physically associated with each other. That is, they orbit around each other. Rarely are double stars composed of a nearby star and a much more distant star that only seem to be associated. Clearly, these **optical doubles** do not orbit each other, but although they are no longer important in modern astronomy, they sparked the discovery of binary stars almost two centuries ago.

Starting about 1782, William Herschel began looking for optical doubles. He hoped to detect the parallax of the nearby star by comparing it with the more distant star. Although he cataloged hundreds of pairs of stars, none were optical doubles. By 1803, he had accumulated enough measurements of the pair of stars known as Castor to show that they were in orbit around each other (Figure 10-1). This was the first orbital motion detected

a

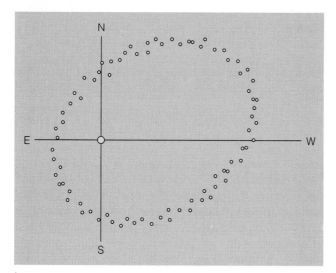

b

FIGURE 10-2

(a) To observe a visual binary, we measure the separation between the stars in seconds of arc, and the position angle of the fainter star with respect to the brighter star. Position angle is always measured from the north toward the east. (b) Many years of such measurements can be graphed to reveal the orbital motion of the fainter star around the brighter star. This is called the apparent relative orbit.

outside the solar system, and it marked the beginning of the observation of binary stars.

Observing Visual Binaries

In a visual binary, the stars usually appear only a few seconds of arc apart, and they orbit around each other with a period that is usually a few decades long. Because the star images appear so close to each other, it is difficult to measure their positions precisely. Nevertheless, because their periods are so long, astronomers can make many measurements over the years and average the results. Some of the errors of measurement average out.

FIGURE 10-3

Just as the circular dial of a clock (a) tipped at an angle is distorted into a ellipse (b), so the elliptical orbit of a binary star tipped at an angle is distorted into a different ellipse. To find the mass of a binary system, astronomers must observe the apparent relative orbit (d) and mathematically untip it to find the true relative orbit (c).

If we were to begin observing a visual binary, we would want to measure the separation between the stars in seconds of arc and the position of the stars with respect to a north–south line. We could do this by drawing a line from the brighter star to the fainter star and measuring the angle this line makes with a north–south line. This is called the **position angle** and is always measured from the north toward the east to the fainter star (Figure 10-2).

Note from Figure 10-2 that east is to the left in diagrams of the sky. On a map of the earth, we place north at the top and east to the right, but sky charts are made to be held overhead. If you face south and hold Figure 10-2 overhead, you will see that east must be to the left in such diagrams.

If we were studying a visual binary, making the observations might take many years, but that is only the first step. We would next plot the data to produce a chart of the orbit (Figure 10-2b). Of course, both stars are moving around the center of mass of the system (Chapter 5). However, because of the way we measure separations and position angles, the bright star appears to be stationary in our data. Our graph would show the motion of the fainter star relative to the brighter star, and thus we call it the **apparent relative orbit.**

The orbit shown in Figure 10-2b is an apparent relative orbit. The word *apparent* refers to the fact that we see the orbit tipped at some unknown angle (Figure 10-3). To find the **true relative orbit,** we must untip the apparent relative orbit. The shape of the apparent relative orbit contains clues to the shape of the true orbit.

a

b

FIGURE 10-4

(a) The orbit of the visual binary Gamma Leonis has a very long period. Attempts to analyze the partial orbit have given results for the period that range from 400 to 620 years. (b) The visual binary A88 has such a small separation that it is very difficult to measure accurately, although its period is only 22 years. (From data by Robert G. Aitken)

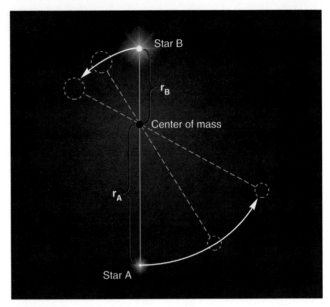

FIGURE 10-5

As the stars in a binary system revolve around their orbits, the line connecting them always passes through the center of mass. The more massive star is closer to the center of mass, and the ratio of the masses, M_A/M_B, is equal to the inverse ratio of the distances of the stars from the center of mass, r_B/r_A.

ries, and binaries with orbits only partially observed cannot be analyzed accurately (Figure 10-4a). On the other hand, if the orbital period were short enough for us to observe conveniently, the stars would probably be so close together that we couldn't observe their separations with any accuracy (Figure 10-4b).

The Masses of Visual Binaries

We can find the masses of the two stars in a visual binary system by using some simple physics. First, we need to find the ratio of the masses—how many times more massive one star is than the other. Second, we need to find the sum of the masses. Once we know the ratio and the sum, we can find the individual masses.

The ratio of the masses in a visual binary is easy to discover if we watch both stars move around the center of mass. We note that the line connecting the stars always passes through the center of mass of the system, so we can quickly locate the center of mass (Figure 10-5). We also note that the most massive star is closest to the center of mass, and the least massive is farthest away. In fact, the ratio of the masses M_A/M_B is just the inverse of the ratio of the distances from the center of mass, r_B/r_A. We can use a ruler on a diagram such as Figure 10-5 to determine the ratio of the masses. Star A is about twice as far from the center of mass as star B. This tells us that star B is twice the mass of star A, or, to express it another way, the mass ratio is 2.

For example, the center of the apparent orbit must be the same as the center of the true orbit; tipping an orbit at an angle distorts its shape but does not move the center. We also know that the two stars obey Kepler's laws, so the bright star must be at one focus of the true orbit. Then a line through the bright star and the center of the apparent orbit must be the major axis of the true ellipse. Various methods exist for solving the puzzle and discovering the shape of the true orbit. Once we know the true relative orbit, we are almost ready to calculate the masses of the stars.

Unfortunately, not all visual binaries can be observed well enough to analyze. If the stars were far enough apart so we could measure their separations accurately, they would be so far apart in space that they would have very long orbital periods, often thousands of years. Humans have not been accumulating observations long enough to plot the apparent orbits of such long-period visual bina-

Chains of Inference

Scientists can rarely observe the things they really want to know, so they must construct chains of inference. We can't observe the mass of stars directly, so we must find a way to use what we can observe—orbital period and angular separation—and step-by-step figure out the parameters we need to reach our ultimate goal. In this case, our goal is the masses of the two stars. We follow the chain of inference from the observable parameters to the unobservable quantities we want to know.

Chains of inference can be mathematical, as when geologists use earthquakes to calculate the temperature of the earth's interior. Since

there is no way to drill a hole to the center of the earth and lower a thermometer, the internal temperature is not directly observable. Nevertheless, the speed of vibrations from distant earthquakes depends on the temperature of the rock they pass through. Geologists can't measure the speed of the vibrations deep inside the earth, but they can measure the delay in the arrival time at different locations on the surface, and that allows them to work their way back to the speed and finally the temperature.

Chains of inference can also be nonmathematical. Biologists studying the migration of whales can't follow whales for years at a time,

but they can observe feeding and mating in different locations, take into consideration food sources, ocean currents, and water temperatures, and construct a chain of inference that leads back to the seasonal migration pattern for whales.

Almost all sciences use chains of inference. If you can link the observable parameters step-by-step to the final conclusions, you have gained a strong insight into the nature of science. ■

Before we can find the total mass, we must know the average separation between the stars in AU. The average distance between the two stars equals the semimajor axis of the true relative orbit, but we know this distance as an angle in seconds of arc, not in AU. To convert an angular measurement into a linear measurement, we need to know the distance to the star. Then we can use the small-angle formula (Chapter 3) to find the average separation in AU.

Once we know the period and the separation in AU, we can find the sum of the two masses. We can use Newton's version of Kepler's third law (Chapter 5), but that formula requires the separation in meters and the period in seconds. It is easier if we measure the period in years and the separation in AU. Then the total mass in solar masses is given by a simple formula:

$$M_A + M_B = \frac{a^3}{P^2}$$

This formula says that the sum of the two masses M_A and M_B equals the cube of the average separation in AU divided by the square of the orbital period in years.

Of course, we recognize that this formula is closely related to Kepler's third law of planetary motion. In fact, if we write the formula above for any planet in our solar system, with the mass of the sun as 1 and the mass of the planet as negligible, we get Kepler's third law: $P^2 = a^3$.

We can use this simple formula to find the total mass of a binary system. For example, suppose we observe a binary system with an average separation of 16 AU and an orbital period of 32 years. Then the sum of the two masses is just $16^3/32^2$, which is 4 solar masses.

If we know the ratio and sum of the masses, we can figure out the individual masses. For example, if we know that $M_A + M_B = 4$ solar masses and that $M_A/M_B = 3$, we could solve the two equations simultaneously to find the

two masses. It is often easier to use a calculator and try a few guesses. In this case, we would quickly find that M_A is 3 solar masses and M_B is 1 solar mass.

Finding the masses of visual binary star systems requires a number of steps to get from what we can observe, the orbital period and angular separation, to what we really want to know, the individual masses (Window on Science 10-1). Nonetheless, studying visual binaries is well worth the effort, because these systems give us a direct way to measure the mass of individual stars. Only a few dozen visual binaries have given us good masses. In many visual binaries, the stars are too close together to allow us to observe them precisely. In other visual binaries, the stars are far apart but the period is too long, and not enough observations have been accumulated to allow us to determine the shape of the orbit. In still other binaries, one of the stars is too faint to see; these binaries are especially difficult to analyze, as we will see in the next section.

Astrometric Binaries

A binary is called an **astrometric binary** when one of the stars is too faint to be seen. In that case, we can tell that the system is a binary only by noticing the motion of the visible star around the center of mass. This requires highly precise measurements of the position of the star.

All stars move through space, and over periods of years or decades we can detect the motion of many of these stars as their proper motion (Chapter 9). Single stars follow paths that are, over periods as short as decades or centuries, straight lines, but binary stars follow more complicated paths.

The center of mass of a binary system follows a straight line, but the stars themselves follow wavy paths

FIGURE 10-6

In an astrometric binary, one star is too faint to be visible. The binary nature of the system can be recognized by the visible star's wavy path across the sky due to its motion around the center of mass of the system.

as they orbit the center of mass (Figure 10-6). Because one of the stars is usually too faint to be visible, we recognize an astrometric binary by the proper motion of the visible star as it travels around the moving center of mass of the system.

It is interesting to note that this may be one way to search for planets orbiting nearby stars. A massive planet like Jupiter is much less massive than a star, but it could affect the motion of its star. Such a variation in the motion of a nearby star may eventually give us direct evidence that other stars have planets just like the sun.

Famous Visual Binaries

Sirius, the brightest star in the sky, is an important example of a visual binary. In 1844, German astronomer Friedrich Wilhelm Bessel (1784–1846) discovered that Sirius was an astrometric binary. That is, Sirius does not follow a straight path across the sky but moves along a wavy path (Figure 10-7a). In 1862, while testing a new telescope lens, American telescope maker Alvan Clark discovered the companion to Sirius (Figure 10-7b). Dubbed Sirius B, it is about 9 magnitudes fainter than Sirius A and is never farther away than 11.5 seconds of arc. Nevertheless, careful observations over the years have shown that both stars are moving around the center of mass as it moves through space (Figure 10-8). Once both stars were seen, Sirius could be classified a visual binary.

Observations of the masses of Sirius A and Sirius B show that Sirius A is a fairly normal star of about 2.35 solar masses. Sirius B has a mass of 0.98 solar mass but is not much larger than the earth. It is one of the first white dwarfs ever discovered, and its high density is characteristic of such objects.

Cygnus (the Swan) contains a number of interesting binary stars (Figure 10-9). Viewed through a small telescope, Albireo (β Cygni) is a beautiful sight, appearing as a golden yellow K3 star and a sapphire blue B8 star. For years, astronomers thought Albireo was a binary with a

a

b

FIGURE 10-7

(a) The proper motion of the bright star Sirius (solid line) does not follow a straight line. The wavy motion is caused by the gravitational attraction of a fainter star, Sirius B, that orbits the brighter star, Sirius A. (The path of Sirius B is shown by the dotted line.) The presence of the companion, a white dwarf, was predicted long before it was first seen in 1862. (b) A modern photograph shows the companion just visible against the glare of Sirius A. (UCO/Lick Observatory Photo)

FIGURE 10-8
Subtracting the average proper motion of Sirius A and Sirius B from the diagram in Figure 10-7a reveals the orbital motion of the two stars around each other. The intersection of the lines connecting the stars marks the center of mass. The more massive star has the smaller orbit. (Diameter of star images is not to scale.)

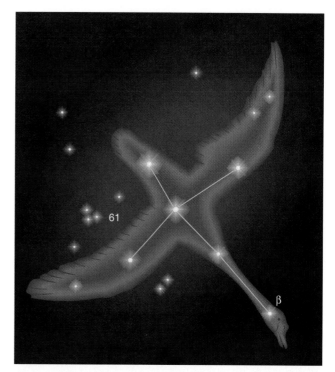

FIGURE 10-9
The constellation Cygnus, the Swan, contains many interesting binary stars. Albireo (β Cygni) consists of a yellow and a blue star and is a beautiful sight through a small telescope. Long thought to be a long-period binary, it is now recognized as a chance alignment of stars at different distances. The star 61 Cygni, a true visual binary, is one of the closest star systems to the sun and the first star to have its parallax measured (see Figure 10-10). The brighter stars of Cygnus make up the Northern Cross.

period as long as 100,000 years, but it shows no orbital motion, and a recent study suggests that the stars lie at different distances from the earth and thus are not a real binary system. Another visual binary in Cygnus is 61 Cygni (Figure 10-10). Its period is 653 years, which is too long for convenient analysis, but it is interesting in a number of ways. It is the 11th closest star (3.4 pc), and in 1838 it became the first star to have its parallax measured. Also, spectroscopic analysis suggests that one of its stars may have a planetlike companion with about eight times the mass of Jupiter.

CRITICAL INQUIRY

Why have so few visual binary star systems been fully analyzed to reveal individual masses?

Although many visual binary star systems are known, most can't be fully analyzed. In many of the binary systems, the stars are far apart and the orbital periods are very long, perhaps hundreds or thousands of years. Astronomers on the earth haven't been observing these binaries long enough to have seen them complete one orbit, so the shapes of their orbits are not well known and they can't be precisely analyzed. There are short-period

a

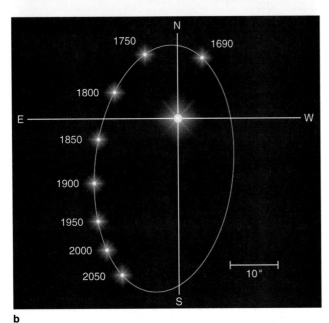

b

FIGURE 10-10

The visual binary 61 Cygni (a), photographed about 1960, and its relative apparent orbit (b). (UCO/Lick Observatory Photo)

visual binary systems, but the stars in many of those systems are so close to each other that the images blur together in the telescope. Such images can't be measured accurately, and the shapes of the orbits are hard to determine. Again, if the shapes of the orbits are not well known, the system can't be analyzed.

The only visual binary systems that can be precisely analyzed are those in which the stars are neither too far apart nor too close together. Of course, if we could observe some of these troublesome binary star systems from earth orbit, they would be easier to analyze. Which ones could we study profitably with an orbiting telescope? Why? ■

Binary stars are very common, but many pairs are so close to each other that they cannot be resolved individually. Then other methods must be used to find their orbital period and separation. Though the methods differ, the goal is the same—discovering the masses of the stars.

10-2 SPECTROSCOPIC BINARIES

If the stars of a binary system are too close together to be visible separately, the telescope shows us a single point of light. Only by taking a spectrum, which is formed by light from both stars, can we sometimes tell that there are two stars, not one, present. Such a system is called a **spectroscopic binary.**

Double-Line Spectroscopic Binaries

If the stars of a spectroscopic binary have about the same brightness, we see spectral lines from both stars. Such a binary is called a **double-line spectroscopic binary.**

Because the stars in a spectroscopic binary orbit each other, they alternately approach and recede from us, as shown in Figure 10-11. As one star comes toward us, its spectral lines are Doppler-shifted toward the blue. The other star is moving away from us, and its spectral lines are shifted toward the red. Half an orbit later, the star that was approaching is receding. As we watch the spectrum of the binary system, we see the spectral lines split into two parts that move apart and then move together as the stars follow their orbits (Figures 10-11 and 10-12).

If we convert these Doppler shifts into radial velocities, we can plot them in a graph called a **radial velocity curve.** For a double-line spectroscopic binary, this graph shows the radial velocity of both stars. The curves in Figure 10-13 are drawn for a hypothetical binary in which the orbits are circular. If the orbits are elliptical, the curves are more complicated, but properly analyzed the radial velocity curves can tell us the eccentricity of the elliptical orbits.

Except for the complication of elliptical orbits, analyzing a double-line spectroscopic binary seems straightforward. The radial velocity curve in Figure 10-13 shows that the two stars are moving about their common center of mass and that the center of mass has a radial velocity of its own. We would expect this, of course; it would be unusual to find a star system that is not moving with respect to our solar system. From the curves, we can tell that the velocity of star A never differs from the velocity of the center of mass by more than 40 km/sec and that the velocity of star B never differs by more than 20 km/sec. If we are seeing the orbits edge-on, the orbital velocity of star A with respect to star B must be 60 km/sec.

We can find the mass if we can find the radius of the orbit, and we can do that because we know the orbital velocity and the period. Multiplying the velocity times

FIGURE 10-11

As the stars of a spectroscopic binary revolve around their common center of mass, they alternately approach and recede from the earth. The Doppler shifts cause their spectral lines to move back and forth across each other.

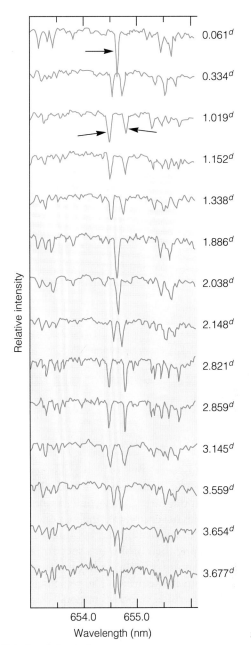

FIGURE 10-12

Fourteen spectra of the spectroscopic binary HD 80715 show how the single line of iron (arrow in first spectrum) is split into two components (arrows in third spectrum) by the orbital motion of the stars. The laboratory wavelength of the line is 654.315 nm. Times of each spectrum are given in days from an arbitrary starting time. (Adapted from data courtesy of Samuel C. Barden and Harold L. Nations)

the period, we find the total distance around the orbit, its circumference. In our idealized example shown in Figure 10-13, we assumed a circular orbit, so it is easy to find the radius of the orbit from its circumference. The radius of the orbit is just the circumference divided by 2π. Once we know the radius and the period, we can find the mass.

We can even find the individual masses. The velocity of star A relative to the center of mass is twice that of star B, so star B must be twice as massive as star A.

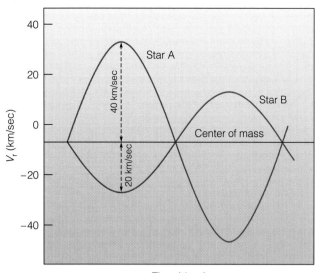

FIGURE 10-13

The radial velocity curve for a hypothetical spectroscopic binary with circular orbits. The radial velocity of the center of mass, about −7 km/sec, shows that the binary is moving toward our solar system while the stars circle the center of mass. From the amplitudes of the curves, 40 km/sec and 20 km/sec, we can conclude that star B is twice as massive as star A.

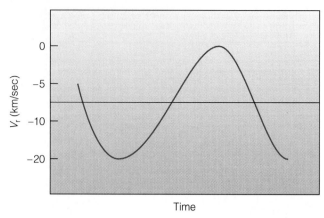

FIGURE 10-14

The radial velocity curve for a single-line spectroscopic binary shows the motion of only one star.

This sounds easy, but we have skipped over one important fact. We don't know that the orbits are edge-on as seen from the earth. They could be inclined at any angle, and we can never find that inclination. We can find the inclination of a visual binary, because we can see the stars moving along their orbits. In a spectroscopic binary, however, we cannot see the individual stars, find the inclination, or untip the orbits. The velocities we observe are not the true orbital velocities but only the part of that velocity directed radially toward or away from the earth. Because we cannot find the inclination, we cannot correct these radial velocities to their true orbital velocities. Thus, we cannot find the true masses. All we can find from a spectroscopic binary is a lower limit to the masses.

Single-Line Spectroscopic Binaries

If one of the stars in a spectroscopic binary is too faint, we will not be able to see its spectral features in the spectrum of the system. Such a binary is called a **single-line spectroscopic binary** because only a single set of lines is visible in its spectrum.

If we measure the wavelengths of the visible lines, we find them shifting first to the red and then to the blue as the brighter star orbits the center of mass. We could convert these Doppler shifts into radial velocities and plot them to produce a radial velocity curve for the system, but the curve would show the motion of only one star (Figure 10-14). We cannot find the orbital velocity of one star around the other, and so we cannot even find the lower limit to the masses of the individual stars. All

we can find is the **mass function,** a number related to the mass of the invisible star, the total mass of the system, and the inclination. There is no way to unscramble these quantities.

The Use of Spectroscopic Binaries

It might seem that spectroscopic binaries are of no use, because neither double-line nor single-line systems can tell us stellar masses. However, statistical analysis of spectroscopic binaries can give useful results.

For instance, if we had data from a dozen spectroscopic binaries containing A stars, we might assume that the orbits were randomly tipped and average the data. Some orbits would be edge-on and some nearly face-on. If we combined the lower limits set by the different systems, they could give us a good estimate for the masses of A stars.

This kind of statistical analysis is important because visual binaries that can be analyzed are very rare, and in some cases the only way to get data on stellar masses is to turn to spectroscopic binaries. Also, spectroscopic binaries are very common, so large numbers of them can be analyzed.

In fact, spectroscopic binaries are so common that many of the familiar stars in the sky are spectroscopic binaries. Capella (α Aurigae), for instance, is a spectroscopic binary with a period of 104 days. A small telescope shows that Mizar, the star at the bend of the handle of the Big Dipper, is a visual binary (Figure 10-15). Spectroscopic observations show that both of the stars in the visual binary are themselves spectroscopic binaries, making Mizar a "double double star." Near Mizar is Alcor, a fainter star just visible to the naked eye. It, too, is a spectroscopic binary. Castor in Gemini is a visual binary (Figure 10-1) in which both of the visible stars are themselves spectroscopic binaries. In addition, a fainter third star much farther from the center of gravity of the system is also a spectroscopic binary, so Castor is really a six-star system made up of three spectroscopic binaries.

a

b

FIGURE 10-15
At the bend of the handle of the Big Dipper lies Mizar, a visual binary (a). Mizar, its companion (visible through small telescopes), and the nearby star Alcor are all spectroscopic binaries. Two spectra of Mizar (b) recorded at different times show how a spectral line is separated into two components by the orbital motion of the stars. (Compare with Figure 10-12.) (The Observatories of the Carnegie Institution of Washington)

CRITICAL INQUIRY

Why do spectroscopic binaries usually have short periods?

The orbital period of a binary star system depends on the separation. If the stars are very far apart, they will have a long period, and we are likely to see the system as two separate stars—a visual binary system—and not as a spectroscopic binary. If the stars in a binary system are close enough together to blend into a single point of light in our telescope, then they will most likely have a short period. Consequently, those binaries we are most likely to see as spectroscopic binaries are those with short periods.

Spectroscopic binaries are very common. Suppose you had a large telescope and a spectrograph. Design a research program that would give you the average mass of an F star. What observations would you make, what kinds of spectroscopic binaries would you choose to observe, and how would you analyze the data? ∎

Spectroscopic binaries are common, but they do not give good results individually. Visual binaries give good results when they can be analyzed, but few have short enough periods and large enough separations. Yet another kind of binary star can give us information about stellar masses and also tell us the diameters of the stars.

10-3 ECLIPSING BINARIES

Rare among binary stars are those with orbits inclined so that the stars cross in front of each other. Seen from the earth, the two stars are not resolvable; that is, they look like a single point of light. However, when one star crosses in front of the other, part of the light is eclipsed, and we recognize the system as an **eclipsing binary**.

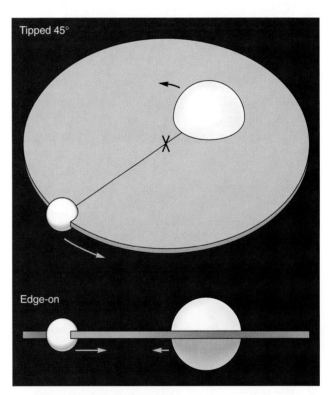

FIGURE 10-16
Imagine a model of a binary system with balls for stars and a disk of cardboard for the plane of the orbits. Only if we view the system edge-on do we see the stars cross in front of each other. These eclipsing binary systems are rare.

Masses and Diameters

Imagine a model of a binary star system in which a cardboard disk represents the orbital plane, as in Figure 10-16. If the orbits are seen edge-on from the earth, the two stars cross in front of each other. The small star

FIGURE 10-17

As the stars in an eclipsing binary cross in front of each other, the total brightness of the system changes. The light curve is a graph of magnitude versus time. In this example, a small, hot star orbits a giant, cool star.

crosses in front of the large star, and then, half an orbit later, the large star crosses in front of the small star. Each time one star crosses in front of the other, the total brightness of the system decreases in an eclipse (Figure 10-17).

There are two eclipses during each orbital revolution. The deeper eclipse is called **primary minimum,** and the shallower is called **secondary minimum.** If we observe such a system and plot the changing magnitude versus time, we get a graph called a **light curve,** as shown for the eclipsing binary in Figure 10-17.

This graph can tell us a great deal about the stars. For instance, we can find the ratio of the temperatures of the stars. During both eclipses, the same amount of area is hidden from our view. First, the small star hides an area of the large star that is equal to its own cross section. Then, when the small star is eclipsed, the same amount of area is hidden. Any difference in the amount of light lost during the two eclipses must arise from the differences in the temperature of the two stars, not from differences in the amount of area hidden.

For example, study the system shown in Figure 10-17. When the small star crosses in front of the larger, cooler star, the total brightness of the system declines slightly because a small area of the cooler star is hidden from view. When the small star is eclipsed behind the larger star, the same amount of area is hidden, but it is much hotter than that hidden before, so the eclipse results in a much deeper decline in total brightness.

We can find the masses of the stars in an eclipsing system if we can get spectra showing the Doppler shifts of the two stars. We can't analyze the Doppler shifts alone to find masses because we don't know the inclination, but the light curve can tell us how the orbits are tipped. We know we are seeing the orbits nearly edge-on or we would not see eclipses at all, and the shape of the eclipses can tell us if the orbits are tipped slightly from exactly edge-on. If we know the inclination, we can use the radial velocity curve to get the orbital velocities and analyze the system like a double-line spectroscopic binary. In this case, because we know the inclination, we can find the true masses of the stars.

Eclipsing binaries are especially important because their light curves enable us to measure the diameters of the stars. From the light curve, we can tell how long it took for the small star to cross the large star. Multiplying this time interval by the orbital velocity of the small star gives us the diameter of the larger star. We could also determine the diameter of the small star by noting how long it took to disappear behind the edge of the large star. For example, if it took 300 seconds for the small star to disappear while traveling 500 km/sec relative to the large star, it must be 150,000 km in diameter.

Of course, there are complications. Although many eclipsing binary star systems have circular orbits due to the influences of tides between the stars, some have elliptical orbits, and these affect the shape of the light curves. In some cases, the orbits are tipped slightly, so the stars do not cross directly in front of each other, and neither

FIGURE 10-18

In this idealized binary, the orbit is tipped slightly, and the stars do not cross directly in front of each other. The eclipses are only partial (a and d), because neither star ever disappears completely. (Compare with the light curve for Algol shown in Figure 10-20.)

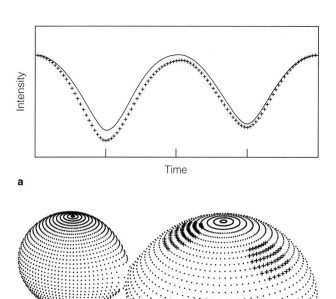

FIGURE 10-19

The light curve of an eclipsing binary star can contain large amounts of information. (a) The light curve of VW Cephei (lower curve) is typical of two stars that are so close together their gravity has distorted their shapes and they actually touch each other. Slight distortions in the light curve are caused by dark spots, much like sunspots, near the pole of the larger star. The upper curve shows what the light curve would look like if there were no spots. (b) Through the detailed analysis of such light curves, astronomers can construct a model showing the size and shape of the stars and the location of the dark spots. (Graphics created with Binary Maker 2.0.)

star is completely eclipsed (Figure 10-18). These are called partial eclipses. In some cases, the stars are so close together that they distort each other from perfect spheres, and the rotation of these nonspherical stars causes confusing variations in the brightness of the system (Figure 10-19). It is even possible for the stars to be so close that the hotter star heats up one side of the cooler star and thus further confuses the light variation.

Some of these complications can be accounted for, and some can make a system so complex that it cannot be analyzed. Those systems that can be solved give us important information not only about stellar masses but also about stellar diameters.

Algol and Other Eclipsing Binaries

Algol (β Persei) is one of the best-known eclipsing binaries because its eclipses are visible to the naked eye. Normally about magnitude 2.1, its brightness drops to magnitude 3.4 in eclipses lasting about 10 hours

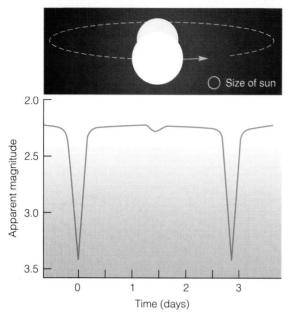

FIGURE 10-20

Algol, the famous eclipsing binary in Perseus, consists of a B8 V star orbiting a smaller star only slightly hotter than the sun. The partial eclipses, which are visible to the naked eye, are deepest when part of the hot star is hidden and are shallow when part of the cooler star is hidden (as shown).

FIGURE 10-21

The eclipsing binary Algol is the star on the demon's forehead in this 1837 engraving of Perseus and the head of the gorgon Medusa. *Algol* comes from the Arabic for "the demon's head." (From Duncan Bradford, *The Wonders of the Heavens,* Boston: John B. Russell, 1837)

(Figure 10-20). This is a decrease of about 68 percent and is easily visible to the naked eye.

The eclipses were first reported in 1669 by Geminiano Montanari, a mathematics professor at the universities of Bologna and Padua, but the star's variation was not explained until 1783. In that year, English astronomer John Goodricke realized that the variations of Algol are periodic, occurring every $68\frac{3}{4}$ hours, and he offered two possible explanations. It might be a rotating star with dark spots on its surface, or it might be a star orbited by a dark companion that periodically passes between it and the earth. Of course, the latter explanation has proved correct, and the dark companion is merely the fainter of the two stars.

Because the variation of Algol is so easy to observe, some have speculated that it was known three millennia ago when the oldest constellations originated. *Algol* comes from the Arabic for "the demon's head," and it is associated in constellation mythology with the severed head of Medusa, the sight of whose serpentine locks turned mortals to stone (Figure 10-21). Indeed in some accounts, Algol is the winking eye of the demon.

A number of other bright stars are eclipsing binaries. The star β Lyrae is a peculiar eclipsing binary with a period of 12.9 days. Its eclipses are about 1 magnitude deep, and its light curve is complicated by the distorted shape of the stars and by glowing clouds of gas floating between the two stars. Another star, ε Aurigae, has an orbital period of 27.06 years, and its eclipses last about 700 days. Its light curve is so distorted that astronomers cannot agree on the characteristics of the stars.

Compare the very long period of ε Aurigae with the 12-hour period of the faint star cataloged as +16°516. The stars that compose it orbit so quickly that its eclipses last only about 45 minutes, and it drops from maximum to minimum brightness in only 48 seconds.

CRITICAL INQUIRY

When we look at the light curve for an eclipsing binary with total eclipses, how can we tell which star is hotter?

If we assume that the two stars in an eclipsing binary are not the same size, then we can refer to them as the larger star and the smaller star. When the smaller star moves behind the larger star, we lose the light coming from the total area of the smaller star. When the smaller star moves in front of the larger star, it blocks off light from the same amount of area on the larger star. In both cases, the same amount of area (the same number of square meters) is hidden from our sight. Then the amount of light lost during an eclipse depends only on the temperature, because that is what determines how much a single square meter can radiate per second. When the surface of the hot star is hidden, the brightness will fall dramatically, but when the surface of the cooler star is hidden, the brightness will not fall as

Basic Scientific Data

In a simple sense, science is the process by which we look at data and search for relationships that tell us how nature works. But that means that science sometimes requires large amounts of data. For example, astronomers need to know the masses and luminosities of many stars before they can begin looking for relationships.

Compiling basic data is one common form of scientific study. This work may not seem very exciting to an outsider, but scientists often love their work not so much because they want to know nature's secrets but because they love the process of studying nature. Using a microscope or a telescope is fun. Gathering plants in a rain forest or geological samples from a cliff face can be tremendously exciting and satisfying. Sometimes the process of science is what is most rewarding, and that can lead scientists to gather significant amounts of information.

Solving a single binary star to find the masses of the stars does not tell an astronomer a great deal about nature, but solving a binary star is like solving a puzzle. It is fun, and it is satisfying. Over the years, many astronomers have added their results to the growing data file on stellar masses. We can now analyze that data to search for relationships between the masses of stars and other parameters, such as diameter and luminosity.

The history of science is filled with hard-working scientists who compiled large amounts of data that later scientists used to make important discoveries. For example, Tycho Brahe spent 20 years at his island observatory recording the positions of the stars and planets (Chapter 4). He must have loved his work to spend 20 years at it on his windy island, but he did not live to use his data. It was his successor, Johannes Kepler, who used Brahe's data to discover the laws of planetary motion.

Whatever science you study, you will encounter accumulations of measurements and observations that have been compiled over the years, including everything from the hardness of rocks to the attention span of infants. Determining these basic scientific data is as much a part of science as is testing a hypothesis. ∎

---

much. Thus, we can look at the light curve and point to the deeper of the two eclipses and say, "That is where the hotter star is behind the cooler star."

We haven't talked about the diameters of the stars yet. How could we look at the light curve of an eclipsing binary with total eclipses and find the ratio of the diameters? ∎

The diverse properties of eclipsing binaries make them fascinating puzzles in their own right, but they are most valuable when they are decoded to reveal the true masses and diameters of stars. In the next section, we assemble such data from all types of binaries and look for relationships among the stars.

10-4 STELLAR MASS AND DENSITY

Our entire discussion of binary stars has been aimed at a single goal: we want to know the masses of stars. Binary stars are entertaining puzzles, and many astronomers specialize in their study. But entertaining or not, binaries are critical to our discussion because they give us a way to find stellar masses.

Knowing the mass and diameter of a single star is of little use. But when we assemble data for many stars, we begin to see two patterns: a simple and universal connection between the mass of a star and its luminosity, and a division of the stars into groups according to their density. These two patterns give us clues that will eventually lead us to an understanding of how stars are born, age, and die (Window on Science 10-2).

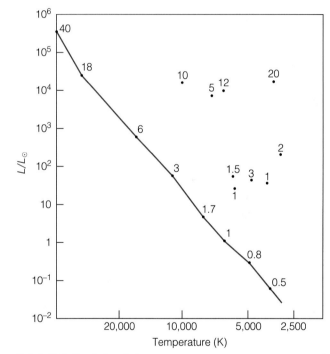

FIGURE 10-22

Stellar masses plotted in the H–R diagram. Main-sequence stars are ordered. That is, the more massive a star is, the more luminous it is. But giants and supergiants are not ordered. In the giant and supergiant region of the diagram, different masses are jumbled together with no pattern.

Stellar Masses

If we began plotting stars in the H–R diagram and writing their mass beside each dot, we would discover that the supergiants tend to be a bit more massive than the sun but that there is no pattern to their masses (Figure 10-22). That is, massive stars and lower-mass stars are all

FIGURE 10-23

The mass–luminosity relation shows that the more massive a main-sequence star is, the more luminous it is. The open circles represent white dwarfs, which do not obey the relation. The straight line represents the equation $L = M^{3.5}$.

jumbled together in the supergiant region. The same is true for the giants. Some stars more massive than the sun and some about the same mass as the sun are all jumbled together in the giant region.

When we begin to plot main-sequence stars, however, we find a pattern. The farther up the main sequence a star is located, the more massive it is. Stars near the top of the main sequence, the O and B stars, are quite massive. The F and G stars in the middle of the main sequence are about the mass of the sun, and the M stars at the lower end are much less massive than the sun. This ordering of main-sequence stars by their mass is a clue to how stars work.

Another way to represent this relation is to plot the luminosities of stars against their masses (Figure 10-23). With few exceptions, we discover that the luminosity depends only on the mass of the star. Astronomers refer to this as the **mass–luminosity relation**, and it is one of the most fundamental of all observational facts about stars. The mass–luminosity relation will eventually tell us how stars work.

The mass–luminosity relation can be summarized in a simple formula:

$$L = M^{3.5}$$

That is, a star's luminosity (in terms of the sun's luminosity) equals its mass (in solar masses) raised to the 3.5 power. For example, a star of 4 solar masses has a luminosity of approximately $4^{3.5}$, or $4 \times 4 \times 4 \times \sqrt{4}$. This equals 64×2, or 128. So a 4-solar-mass star will have a luminosity about 128 times the luminosity of the sun. This is only an approximate equation, as shown by the straight line in Figure 10-23.

Notice how large the range in luminosity is. The range of masses extends from about 0.08 solar mass to about 50 solar masses—a factor of 600. But the range of luminosities extends from about 10^{-6} to about 10^{6} solar luminosities—a factor of 10^{12}. Clearly, a small difference in mass causes a large difference in luminosity.

Stellar Densities

The data we get from binary stars can be used in another way. We can combine them with stellar diameters to find the average densities of the stars. All we have to do is divide the mass of a star by the volume it occupies to find its average density. We can be sure that stars are denser near their centers and less dense near their surfaces, but average densities are sufficient to reveal another important pattern in the H–R diagram.

The average densities of main-sequence stars are about like that of the sun. Some stars are a bit denser, and some are a bit less dense, but in general main-sequence stars are about as dense as water.

Binary stars tell us that giant stars and supergiants are not tremendously massive stars, yet we know from their positions in the H–R diagram that they must be very large. If we divide their ordinary masses by their gigantic volumes, we find that they have very low densities. Giant stars have densities ranging from 0.1 to 0.01 g/cm³. The enormous supergiants have still lower densities, ranging from 0.001 to 0.000001 g/cm³. This is thinner than the air we breathe.

On the other hand, binary stars such as Sirius show us that the white dwarfs have masses of about 1 solar mass. Yet they have to be small stars not much larger than the earth. This means that the average density of a white dwarf is about 2,000,000 g/cm³ or more. On the earth, a cubic centimeter would weigh as much as a limousine.

Density divides stars into three groups. Most stars are main-sequence stars with densities like the sun's. Giants and supergiants are very-low-density stars, and white dwarfs are high-density objects. This sorting of the stars into groups by density is a tremendously important clue to the internal structure of stars.

CRITICAL INQUIRY

What observable parameters must astronomers know in order to plot a graph of luminosity versus mass?

Since astronomers can't observe the luminosity or mass of a star directly, they must use a chain of inference to connect the observational properties of stars to luminosity and mass. We have seen that luminosity can be found from the apparent magnitude and the distance to a star. We also know that mass can be found from observations of the orbital period and separation of binary stars. Of course, we need to find the distance to the binary star system, perhaps by spectroscopic parallax (Chapter 9), in order to find the separation in AU. A simple graph of the mass–luminosity relation is the result of the combination of many different observable properties of stars leading step-by-step to their luminosity and mass.

What observable parameters would we have to add to this list in order to find the densities of stars? ■

This and the preceding chapter have given us the power to describe stars by their luminosity, diameter, and mass. We know which kinds of stars are common and which are rare, and we can divide the stars into categories based on their density. Our next task is to understand how stars are born, evolve, and die. Before we discuss star formation, however, we must consider the clouds of gas and dust from which stars form. This matter between the stars is the subject of the next chapter. ■

■ Summary

The only way to get information about the masses of stars is to find them in binary systems—that is, in systems of two stars orbiting each other. The size and period of the orbits depend on the masses of the stars. If we can find the average separation of the two stars in AU and the orbital period in years, the total mass in solar masses is a^3/P^2.

The individual masses can be found by studying the individual motions of the stars. The more massive star will be located closer to the center of mass of the system, and the less massive star will be farther away. Thus, the ratio of the distances of the stars from the center of mass tells us the ratio of their masses. If we know the sum of the masses and their ratio, we can find the individual masses.

Visual binaries are those systems in which both stars are visible. They often have periods of tens or hundreds of years, though only the shorter-period visual binaries can be analyzed. The advantage to studying visual binaries is that we can see the actual shape of the orbit and thus can discover the angle at which it is tipped. This means the true masses of the stars can be found.

In an astrometric binary, only one of the stars is bright enough to see, and we can recognize it as a binary star system only by watching the star wobble back and forth as it orbits the center of mass. These binaries do not yield stellar masses. Sirius was originally discovered to be an astrometric binary, though its faint, white-dwarf companion was later detected visually.

If the two stars in a binary system are close together, we may not be able to see them individually from the earth. The system may look like a single point of light. However, we may be able to study the orbital motion of the system by measuring the Doppler shifts of the spectral lines of the two stars. If lines are visible from both stars, the system is called a double-line spectroscopic binary. If only one set of lines is visible, it is called a single-line spectroscopic binary. In both cases, we cannot calculate true masses because we cannot find the inclination of the orbit. Thus, the results of spectroscopic binaries are most useful in statistical studies, where the effects of orbital inclination can be averaged out.

If the two stars cross in front of each other, part of the light is periodically eclipsed. We can recognize the binary nature of the star from its light curve. The shape of the light curve can tell us the surface brightness of the two stars, their diameters, and the inclination of the orbit. If we can find the inclination, we can find the masses from the radial velocity curve.

These data on masses and diameters reveal important patterns in the H–R diagram. The giants and supergiants consist of stars of various masses jumbled together in no apparent order. However, the stars on the main sequence are ordered according to mass. The most massive stars lie on the upper main sequence, and the least massive on the lower main sequence. This relationship is reflected in the mass–luminosity relation.

If we divide the mass of a star by the volume it occupies, we get its average density. These average densities divide the stars in the H–R diagram into three groups. The giants and supergiants are very-low-density stars—some are thinner than air. Main-sequence stars are all about as dense as water. White dwarfs are all about as massive as the sun but about the size of the earth, so they are very dense—about 2,000,000 g/cm^3.

These patterns hidden in the H–R diagram are important clues to how stars work. We will follow this trail through the next chapters.

■ New Terms

binary stars

double stars

visual binary

optical double

position angle

apparent relative orbit

true relative orbit

astrometric binary

spectroscopic binary

double-line spectroscopic binary

radial velocity curve

single-line spectroscopic binary

mass function

eclipsing binary

primary minimum

secondary minimum

light curve

mass–luminosity relation

■ Questions

1. Why do we have to know the distance to a visual binary before we can compute the masses of the stars?

2. How do we find the ratio of the masses in a visual binary system? in a spectroscopic binary system?

3. Explain how you can be certain that the orbit shown in Figure 10-2b is not a true relative orbit.

4. Why can't we find the masses of the stars in a spectroscopic binary system?

5. How do astrometric binaries resemble single-line spectroscopic binaries?

6. Why can we find the masses of the stars in an eclipsing binary but not in a spectroscopic binary?

7. How could we find the diameters of the stars in an eclipsing binary (assuming that the orbit is exactly edge-on)?

8. What would the light curve in Figure 10-17 look like if the two stars had the same temperature?

9. What is the mass–luminosity relation?

10. How does average density divide stars into three categories?

▪ Discussion Questions

1. Why didn't William Herschel detect parallax in his study of double stars?

2. Sometimes binary stars are so close together that the hot star heats up the near side of the cool star. How would that change the light curve of an eclipsing binary?

▪ Problems

1. Draw in the center and the major axis of the true orbit in Figure 10-2b.

2. What is the total mass of a visual binary system if its average separation is 8 AU and its period is 20 years?

3. What is the mass ratio of the stars in Figure 10-5? in Figure 10-8?

4. Assume that the stars in Figure 10-5 have a separation of 18 AU and a period of 30 years. What is the total mass? What is the mass ratio? What are their individual masses?

5. Measure the orbit of 61 Cygni in Figure 10-10, and compute the semimajor axis in astronomical units. The distance to the system is 3.4 pc. (Disregard the slight distortion of the orbit caused by its inclination.)

6. Use the result of Problem 5 and the given period and orbit of 61 Cygni to compute the total mass of the stars.

7. If the period of the spectroscopic binary in Figure 10-13 is 67 days and the orbit is edge-on, what is the total mass of the system? What are the masses of the two stars?

8. Measure the wavelengths of the iron lines in Figure 10-12, and plot a radial velocity curve for this system. What is the period? Assuming that the orbit is circular and edge-on, what is the total mass? What are the individual masses?

9. If the eclipsing binary in Figure 10-17 has a period of 32 days, an orbital velocity of 153 km/sec, and an orbit that is nearly edge-on, what is the circumference of the orbit? the radius of the orbit? the mass of the system?

10. If the orbital velocity of the eclipsing binary in Figure 10-17 is 153 km/sec and the smaller star becomes completely eclipsed in $2\frac{1}{2}$ hours, what is its diameter?

11. What is the luminosity of a 4-solar-mass star? of a 9-solar-mass star? of a 7-solar-mass star?

▪ Critical Inquiries for the Web

1. What if Algol were oriented in such a way that its stars did not eclipse each other as they do now when seen from Earth? Use the Internet to find information about Algol and binary stars in general and discuss how astronomers might still determine that Algol is a multiple star system.

2. It takes many years of observation to plot orbit diagrams for visual binary systems. Search the Internet to find plotted orbits for several visual binaries, and find three stars that have gone through significant change in orientation during your lifetime. Would you expect that these stars lie close to our sun or at large distances from our sun (or does it matter)?

 Go to the Wadsworth Astronomy Resource Center (www.wadsworth.com/astronomy) for critical thinking exercises, articles, and additional readings from InfoTrac College Edition, Wadsworth's online student library.

Chapter Eleven

THE INTERSTELLAR MEDIUM

GUIDEPOST

In a discussion of bread baking, we might begin with a chapter on wheat and flour. In our discussion of the birth and death of stars, the theme of the next five chapters, we begin with a chapter about the gas and dust between the stars. It is the flour from which nature bakes stars.

This chapter clearly illustrates how astronomers use the interaction of light and matter to learn about nature on the astronomical scale. That tool, which we developed in Chapter 7, "Starlight and Atoms," is powerfully employed here, especially when we include observations at many different wavelengths.

We also see in this chapter the interplay of observation and theory. Neither is useful alone, but together they are a powerful method for studying nature, a method generally known as science. ∎

when he shall die,

Take him and cut him out in little stars,

And he will make the face of heaven so fine

That all the world will be in love with night,

And pay no worship to the garish sun.

SHAKESPEARE

Romeo and Juliet III,ii,21

Juliet loved Romeo so much she compared him to the beauty of the stars, but had she known what was between the stars, she might have compared him to that instead. True, the gas and dust between the stars are mostly dark and cold, but where they are illuminated by stars they create beautiful nebulae, and where they are densest they give birth to beautiful stars. If there is beauty in vast extent and sweeping power, then the **interstellar medium,** the gas and dust between the stars, could steal worship from the garish stars.

We are interested in the interstellar medium because it is the matter from which stars are born. We will discover both large and small clouds of gas and dust, and both hot and cold gas, and we will find that the largest and densest clouds can contract under the influence of their own gravity and create new stars. Clearly, before we can understand how stars are born, we must understand what fills the spaces between the stars.

Another reason we are interested in studying the interstellar medium is that it further illustrates the importance of the interaction of light and matter. Our study of starlight and atoms in Chapter 7 will help us understand the vast but nearly invisible clouds of gas and dust between the stars.

In Chapter 1, we saw that we live in a disk-shaped galaxy, the Milky Way (Figure 1-11). The interstellar medium is confined mostly within a few hundred parsecs of the plane of the disk of the galaxy; if we journeyed out of the galaxy, we would find conditions to be quite different. In our study of the interstellar medium, we confine ourselves to the disk of our galaxy in the neighborhood of the sun—probably a typical part of interstellar space.

To detect the interstellar medium, we must use all the tools of the astronomer. Observations at radio, infrared, visual, ultraviolet, and X-ray wavelengths reveal the properties of the interstellar medium. We will analyze images and spectra, consider the nature of atoms, and utilize the physics of light in order to unlock the secrets hidden in the material between the stars.

11-1 VISIBLE-WAVELENGTH OBSERVATIONS

A quick glance at the night sky does not reveal any matter between the stars, but evidence of its existence is clear when we examine photographs and spectra. In fact, if we look carefully, one important piece of evidence is visible to the naked eye in a familiar constellation.

Nebulae

On a cold, clear winter night, Orion hangs high in the southern sky, a large constellation composed of brilliant stars. If you look carefully at Orion's sword, you will see that one of the stars is a hazy cloud (Figure 2-4). A small telescope reveals even more such clouds of gas and dust. Astronomers refer to these clouds as **nebulae** (singular, nebula), from the Latin word for mist or cloud.

The **emission nebulae** glow with the light emitted by excited atoms and ions. A spectrum of an emission nebula contains many emission lines, and Kirchhoff's second law (Chapter 7) tells us that the light must have been emitted by a low-density gas. Emission lines of hydrogen are strong, and the red, blue, and violet Balmer lines blend together to give the nebulae a characteristic pink-red color (Figure 11-1). Other lines are produced by helium, oxygen, nitrogen, and so on. From an analysis of the strengths of the various spectral lines, astronomers conclude that the gas has a temperature of about 10,000 K and a density ranging from 100 to 1000 atoms/cm^3. This is equivalent to the best vacuums produced in laboratories on the earth.

Emission nebulae glow in much the same way that a "neon" sign glows. The atoms in the sign are excited by electric current flowing through the gas, but the gas in an emission nebula is excited by ultraviolet light from a nearby star, often a star embedded in the nebula. Most of

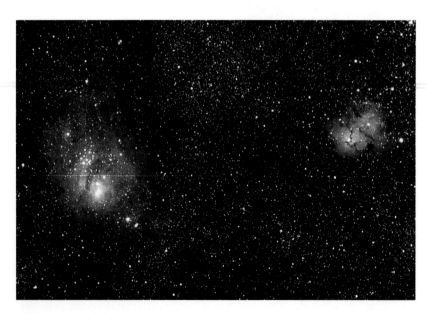

FIGURE 11-1

The Lagoon Nebula at left is an emission nebula. The hydrogen gas is ionized by hot blue stars inside the nebula. The red, blue, and violet Balmer lines of hydrogen blend together to produce the pink color characteristic of ionized hydrogen. The Trifid nebula at right contains a region of ionized hydrogen as shown by the pink color, but it also contains a reflection nebula, the blue cloud. (Daniel Good)

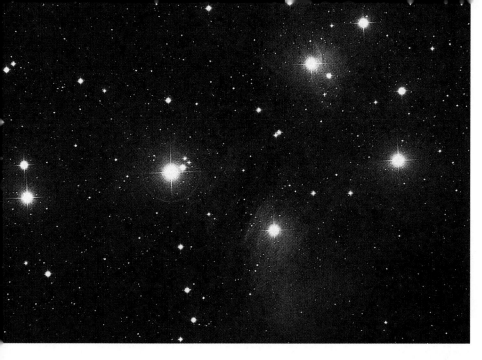

FIGURE 11-2

The Pleiades star cluster, just visible to the naked eye in Taurus, is surrounded by a faint nebula caused by the reflection of starlight from dust in the nebula. In this case, the stars do not seem to have formed from this gas and dust but are merely passing through the nebula. (Palomar Obs/Caltech)

the gas is hydrogen, and ionizing hydrogen takes quite a bit of energy. Only photons with wavelengths shorter than 91.2 nm have enough energy, and only the hottest stars can produce significant numbers of such photons. Emission nebulae are usually observed only around stars hotter than 25,000 K, which is about spectral type B1.

Because emission nebulae are clouds of ionized hydrogen, they are often called **HII regions,** following the convention of naming ionized gas with roman numerals to indicate the state of ionization. HI is neutral hydrogen, and HII is ionized hydrogen. Whenever we see the pink glow of an HII region, we know we are looking at gas ionized by one or more hot stars.

Most of the gas in an emission nebula is ionized by photons. A hydrogen atom in such a nebula is not able to keep its electron very long before an ultraviolet photon strikes it and the electron escapes. Of course, there are lots of electrons flying around, so the atom can capture a new electron; as that electron cascades down through the atomic energy levels, photons of different wavelengths are emitted. Those are the photons we see coming from an emission nebula.

Certain lines in the spectra of emission nebulae are called **forbidden lines** because they are almost never seen in excited gas on the earth. Two good examples are the strong green lines at 495.9 nm and 500.7 nm produced by oxygen atoms that have lost two electrons. (Following our convention for naming ions, twice-ionized oxygen is OIII.) The oxygen ions can become excited by collision with a rapidly moving ion or electron, and the atom can emit various-wavelength photons as its electron cascades back down to lower energy levels. However, transitions between certain energy levels are so unlikely they are called "forbidden." If an electron enters such an energy level, it will remain there for a relatively long time before it can decay further and emit the appropriate photon. For a normal transition, the electron might wait 10^{-8} to 10^{-7} second. If it enters one of these metastable levels, the electron can wait as long as an hour before it is able to decay and emit the proper photon.

This explains why we don't see these forbidden lines in laboratories on the earth. The atoms in a dense gas collide with each other so often that there isn't time for an electron in a metastable level to decay and emit a photon. Such electrons get excited back up to higher levels before they can drop downward and emit a photon. In an emission nebula, the gas has a very low density, and an atom could go for an hour or more between collisions, giving an electron in a metastable level time to decay to a lower level and emit a photon at a supposedly forbidden wavelength. This is a dramatic example of how astronomers can use knowledge of atomic physics to understand astronomical objects.

Whenever we see the pink clouds of gas glowing near hot bright stars, we know we are seeing part of the interstellar medium excited by nearby hot stars. But there are other kinds of nebulae that reveal still more about the interstellar medium.

Reflection Nebulae

The visible spectrum of a **reflection nebula** is mostly the reflected spectrum of the stars that illuminate the nebula. This tells us that the nebula is not made up of excited gas. Gas is almost certainly present, but the nebula glows because of dust specks that scatter the light from the star.

One of the best-known reflection nebulae surrounds the stars in the Pleiades star cluster (Figure 11-2). Although these stars are visible to the naked eye and the star cluster is bright and beautiful through a small telescope, the faint light of the reflection nebula is evident only on time-exposure photographs. Then it shows up as blue wisps of nebulosity around the brightest stars.

a

b

FIGURE 11-3

(a) Reflection nebulae are blue because light from stars is scattered by dust in the nebula. The slight pink color near the center of this nebula is caused by ionized hydrogen. (Anglo-Australian Telescope Board)
(b) Light entering the earth's atmosphere is scattered by molecules, and shorter-wavelength photons (blue) are scattered more often than longer-wavelength photons (red, orange, and yellow). Thus, blue photons are scattered throughout the atmosphere, and the sky looks blue.

The Pleiades star cluster is embedded in a cloud of gas and dust, but it is not an emission nebula because none of the Pleiades stars is hot enough to ionize a significant amount of gas. To produce an HII region, a star must be hotter than about B1, and the hottest star in the Pleiades is Merope, a B3 star. Rather, the light is scattered by dust, and the Pleiades nebulosity is a reflection nebula.

Reflection nebulae tend to be blue for the same reason the daytime sky is blue (Figure 11-3), and that bit of physics tells us something interesting about the dust. When a photon encounters a particle, such as an air molecule, that is small compared to the wavelength of the photon, its scattering depends strongly on the wavelength. Blue photons, having shorter wavelengths, are scattered more than yellow and red photons, so our atmosphere is filled with blue photons bouncing around in all directions. Whichever direction we look in the sky, the blue photons pour into our eyes, and the sky looks blue. Sunlight looks a bit less blue because some of the blue light was removed by the scattering.

When we look at a reflection nebula, we see more blue photons than yellow or red, and the nebula looks blue. This appearance is caused by the scattering of starlight from the dust in the nebula, and it proves that the dust particles must be very small. If the dust particles were the size of olives, for example, they would not scatter blue light better than red light, and the nebula would not look blue. The blue color of the reflection nebulae tells us that the dust in the interstellar medium must be mostly microscopic particles. In fact, the dust particles range in diameter from 0.003 mm down to 1 nm or so, in the range of the wavelength of light.

Dark Clouds

We see emission and reflection nebulae because of the light they send us, but we see **dark nebulae** because these dense clouds of gas and dust block the light of distant stars. Such clouds are visible even to the naked eye as dark regions along the Milky Way, for example, the Coalsack in the southern sky and the Northern Coalsack in Cygnus (Figure 11-4).

Some dark nebulae are more or less round, but many are twisted and distorted. This suggests that, even where there are no hot stars to ionize the gas or bright stars to illuminate it as a reflection nebula, there are breezes and currents pushing through the interstellar medium. We will find more evidence of such effects later.

The smallest of the dark nebulae are the **Bok globules,*** small, dark clouds less than 1 pc in diameter containing 10 to 100 solar masses. Bok globules are seen silhouetted against bright nebulae (Figure 11-5). Astronomers have suspected for decades that at least some Bok globules are contracting to form stars. Infrared observations confirm that some contain warm centers, as they would if they were contracting.

*Named after astronomer Bart Bok.

a

b

FIGURE 11-4

(a) The constellation Cygnus, the swan, flies southward along the Milky Way, with dark clouds of dust blocking our view of distant stars in some regions. The Northern Coalsack, near the tail of the swan, is one of the darkest dust clouds in the region. (Compare with Figure 10-9.) (b) The star cluster NGC 6520 in Sagittarius lies near the dark cloud Barnard 86. The cloud is visible because it contains thick dust that blocks our view of more distant stars. (Anglo-Australian Telescope Board)

In figure (a): Northern Coalsack

Extinction and Reddening

One way we know that dust is present in the interstellar medium is that it makes distant stars appear fainter than they would if space were perfectly transparent. This phenomenon is called **interstellar extinction,** and in the neighborhood of the sun it amounts to about 2 magnitude per thousand parsecs. That is, if a star lies 1000 pc from the earth, it will look about 2 magnitude fainter than it would if space were perfectly transparent. If it were 2000 pc away, it would look about 4 magnitudes dimmer, and so on. This is a dramatic effect, and it shows that the interstellar medium is not confined to a few nebulae scattered here and there. The spaces between the stars are far from empty.

Another way we can detect the presence of dust is through the effect it has on the colors of stars. An O star should be blue, but some stars with the spectrum of an O star seem much redder than they should be. Termed **interstellar reddening,** this effect is caused by dust particles scattering light. As we saw in the case of the

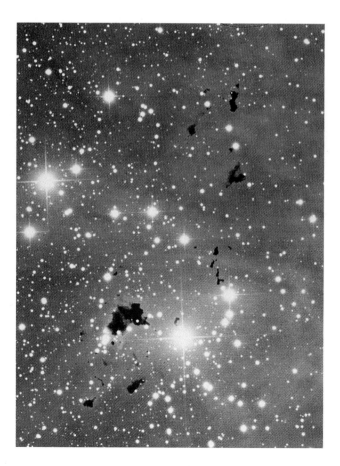

FIGURE 11-5 ▶

Bok globules are small, dense, dusty clouds that are visible silhouetted against more distant nebulosity. Typical globules are only a parsec or so in diameter and are linked to star formation. These globules are located in the nebula IC 2944. (Anglo-Australian Telescope Board)

FIGURE 11-6

Seen through a dust cloud, a star appears redder because the blue photons, having shorter wavelengths, are more likely to be scattered by the dust grains.

FIGURE 11-7

Interstellar extinction, the dimming of starlight by dust between the stars, depends strongly on wavelength. Infrared radiation is only slightly affected, but ultraviolet light is strongly scattered. The strong extinction at about 220 nm is caused by a certain form of carbon dust in the interstellar medium.

a

b

FIGURE 11-8

Interstellar absorption lines can be recognized in two ways. (a) The B0 supergiant ε Orionis is much too hot to show spectral lines of once-ionized calcium (CaII), yet this short segment of its spectrum reveals narrow, multiple lines of CaII, which must have been produced in the interstellar medium. (The Observatories of the Carnegie Institution of Washington) (b) Spectral lines produced in the atmospheres of stars are much broader than the spectral lines produced in the interstellar medium. (Adapted from a diagram by Binnendijk) In both (a) and (b), the multiple interstellar lines are produced by separate interstellar clouds with slightly different radial velocities.

reflection nebulae, the dust particles are small, with diameters roughly equal to the wavelength of light, and they scatter blue light better than red light. Thus, the light from a distant star has lost some of its blue photons because of scattering, and thus the star looks redder (Figure 11-6).

As we discussed earlier, this scattering of blue light is what makes the sky blue, but it is also what makes distant city lights look yellow. If you view the lights of a city at night from a high-flying aircraft or a distant mountaintop, the lights will look yellow. As you descend toward the city, the lights will become bluer. The light from the city is reddened by microscopic particles in the air. If the particles are especially dense, we call them smog.

Astronomers can measure the amount of reddening by comparing two stars of the same spectral type, one of which is dimmed more than the other. The more obscured star will look redder. If we plot the difference in brightness between the two stars as a function of wavelength, we get a curve that shows the reddening. That is, it shows how the starlight is dimmed as a function of wavelength (Figure 11-7). In general, the light is dimmed in proportion to the reciprocal of the wavelength, a pattern typical of scattering from small dust particles. Laboratory measurements show that the high extinction at about 220 nm is caused by a form of carbon, so we must conclude that some of the dust particles are carbon. Other evidence suggests that some grains contain silicates and metals and may have coatings of carbon-based molecules.

Interstellar Absorption Lines

If we look at the spectra of distant stars, we can see dramatic evidence of an interstellar medium. Of course, we see spectral lines produced by the gas in the atmospheres of the stars, but we also see sharp spectral lines produced by the gas in the interstellar medium (Figure 11-8).

Pressure

Pressure is one of the most fundamental parameters in science. Doctors measure blood pressure, and astronomers measure gas pressure, but both are the same thing—a force per unit area.

Pressure is expressed in the units of force per unit area. When we inflate the tires on a car, we use the unit pounds per square inch. A typical pressure might be 34 lb/in.^{2}. It is important to note that this is not the total force pushing out on the inside of the tire but only the force exerted on a single square inch. When you stand, your weight exerts a force on the floor, and the pressure under your shoes is your weight divided by the surface area of your shoes' soles. A typical pressure might be only 4 lb/in.^{2}. If you step on someone's toe, that is the pressure you exert. Of course, if you were

wearing ice skates, your weight would be spread over a much smaller area, the area of the bottom of the blade, and you might exert a pressure of 150 lb/in.^{2} or more. We must be careful not to step on someone's toe while we are wearing ice skates. The pressure would be dangerously high.

Astronomers are most commonly interested in the behavior of matter when it is a gas, and pressure in a gas arises when atoms or molecules collide. Consider, for example, how the gas molecules colliding with the inside of a balloon exert an outward force on the rubber and keep the balloon inflated. If the gas is hot, the atoms or molecules move rapidly, and the resulting pressure is higher than for a cooler gas. If the gas is dense, there will be many gas particles colliding with the inside of the bal-

loon, and the pressure will be higher than for a lower-density gas. Thus, pressure depends on both the temperature and the density of the gas.

Notice that pressure and density are related, but they are not at all the same thing. Density is a measure of the amount of matter in a given volume, and pressure is a measure of the force that matter exerts on its surroundings. A very-low-density gas and a very-high-density gas might have the same pressure if they had different temperatures.

In daily life, we think of pressure when we inflate an automobile tire, but pressure is common in nearly all of the sciences. Astronomers must consider pressure in thinking about the gas inside stars and the thin gas between the stars. ∎

These **interstellar absorption lines** give us a new way to study the gas between the stars.

We can recognize interstellar absorption lines in two ways: by their ionization and by their width. Often, we see spectral lines in a star's spectrum that just don't belong there. They are the wrong kind of line, the wrong ionization state. For example, if we look at the spectrum of a very hot star such as an O star, we would expect to see no lines of once-ionized calcium (CaII) because that ion cannot exist in the atmosphere of such a hot star. But many O-star spectra contain lines of CaII, so we must conclude that these lines were produced not in the star but in the interstellar medium.

In addition, the widths of the interstellar lines give away their identity. The spectral lines in a stellar spectrum are broad, and even in the spectrum of a giant or supergiant, pressure broadening (Chapter 7) smears the lines to a relatively broad profile. The interstellar lines, on the other hand, are exceedingly sharp. This tells us that the interstellar matter is cold and has a low density. If it were hot, Doppler broadening would smear out the lines due to the motions of individual atoms. If the gas were dense, collisional broadening would produce wider lines. The exceedingly narrow widths of the interstellar absorption lines are typical of cold, low-density gas.

Another revealing characteristic of the interstellar lines is that they are often split into two or more components. These multiple components have slightly different wavelengths and appear to have been produced when the light from the star passed through different clouds of gas on its way to the earth. Because the clouds of gas have slightly different radial velocities, they produce absorption lines with slightly Doppler-shifted wavelengths.

Astronomers disagree as to the structure of the interstellar medium, and the boundaries and characteristics of clouds are ill-defined. Nevertheless, astronomers find it convenient to categorize clouds into a few main varieties. Studies of interstellar absorption lines reveal clouds of neutral gas (and presumably dust) with densities of 10 to a few hundred atoms/cm^3. Because these clouds are not ionized, they are called **HI clouds.** These clouds must be 50 to 150 pc in diameter and have masses of a few solar masses. The gas temperature is only about 100 K. Starlight seems to pass through six to ten of these clouds for every 1000 pc near the plane of our galaxy. While it is easy to think of these clouds as more or less spherical blobs of gas, observations show that they are usually twisted into long filaments, flattened into thin sheets, or tangled into chaotic shapes—further evidence that the interstellar medium is not static and motionless.

Between these HI clouds of neutral gas lies a hot **intercloud medium** with a temperature of a few thousand K and a density of only about 0.1 atom/cm^3. This intercloud medium is ionized by the starlight from distant stars, and it must be in equilibrium with the HI clouds. The pressure of a gas depends on its density and its temperature (Window on Science 11-1). A gas can exert a high pressure either by being very hot or by being very dense. In the interstellar medium, the HI clouds are cool but dense, while the intercloud medium is not very dense but is quite hot. Thus, the two components have about the same pressure, they are in equilibrium, and the HI clouds float through the intercloud medium like ice cubes floating through lemonade.

How can the intercloud medium be ionized when it is not close to hot stars? To understand how, imagine

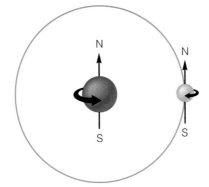

Same spins
Magnetic fields reversed

Opposite spins
Magnetic fields the same

FIGURE 11-9
The proton (red) and electron (blue) in a hydrogen atom spin and so produce small magnetic fields. If they spin in the same direction, the magnetic fields are reversed. If they spin in opposite directions, the fields are the same. Because the fields repel or attract, the electron is slightly more tightly bound to the proton in one case and slightly less tightly bound in the other case.

that we are atoms floating in the intercloud medium. Ultraviolet photons from distant stars are not common, but they do whiz by now and then. Soon we become ionized by absorbing one of these photons and losing an electron. In a denser gas, we would quickly find another electron, capture it, and become neutral again, but the interstellar medium has such a low density that we must wait a very long time to find an electron. Thus, the atoms of the intercloud medium spend almost all their time in an ionized state because of the low density.

Studies of interstellar absorption lines give us further clues about the composition of the interstellar gas. It is much like the composition of the sun. Hydrogen is most abundant, with helium second. Light elements such as carbon, nitrogen, and oxygen are common, but elements such as iron, calcium, and titanium are less abundant than they are in the sun. Most likely, these elements are missing from the gas because they have condensed to form the dust.

CRITICAL INQUIRY

What evidence do we have that there is an interstellar medium?

Everything in science is based on evidence, so this is a good question, and we should answer it by listing observations. First, there is the simple fact that we can see certain parts of the interstellar medium—the nebulae. Emission, reflection, and dark nebulae show that interstellar space is not totally empty. Further, we see interstellar extinction and reddening. The more distant stars look fainter than they should, and they also look redder. This not only shows that all of interstellar space is filled by some thin material, it also shows that some of the material is in the form of tiny dust specks. Gas atoms would not be very effective at reddening starlight, so the interstellar medium must contain dust as well as gas.

Perhaps this is enough evidence to convince ourselves that the spaces between the stars are not empty, but there is more. How do interstellar absorption lines give us information about the nature of the interstellar gas? ■

Observations at visible wavelengths can give us important information about the interstellar medium, but in order to paint a complete picture of the matter between the stars, we must observe at other wavelengths.

11-2 RADIO, INFRARED, ULTRAVIOLET, AND X-RAY OBSERVATIONS

The physics of atoms and light has no better illustration than in the study of the interstellar medium. The kind of information we need determines which property of the atoms we probe, and that determines the wavelengths we use.

21-cm Observations

As mentioned in Chapter 6, radio telescopes are important because they can detect the 21-cm-wavelength radiation emitted by clouds of cool hydrogen in space. This radiation is emitted when a hydrogen atom's electron changes its energy by changing the direction of its spin.

This **21-cm radiation** was predicted in the mid-1940s by H. C. van de Hulst. (It was not actually detected until 1951.) We can understand how such a theoretical prediction was made if we think of the structure of a hydrogen atom. It consists of a proton and an electron, both of which must spin like tiny tops. Spinning a charged top is similar to making electricity flow through a coil of wire. It produces a magnetic field. If a proton and an electron spin in the same direction, their magnetic fields are reversed because they carry opposite charges.

We have all played with small magnets and noticed that they repel each other in one orientation and attract each other in the opposite. In the same way, the small magnetic fields produced by the spinning proton and electron can repel or attract each other. In one orienta-

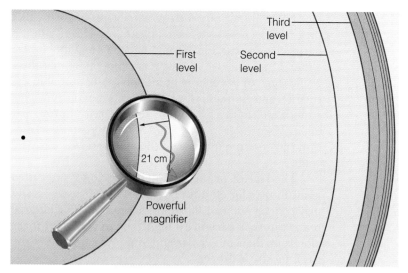

FIGURE 11-10

If we represent the energy levels in a hydrogen atom with radii proportional to their energy, the ground state seems to be a single level. However, the interaction of the magnetic fields of the proton and electron causes the ground state to split into two very close energy levels. In our diagram, we would need a powerful magnifier to distinguish the two energy levels that make up the ground state. A transition from the upper to the lower level emits a photon with a wavelength of 21 cm. This is the source of the 21-cm radiation from neutral hydrogen.

tion, the electron is slightly less tightly bound, and in the other orientation it is slightly more tightly bound (Figure 11-9).

Because of these magnetic fields, the ground state of the hydrogen atom is really two energy levels of very slightly different energy (Figure 11-10). If the electron is in the higher orbit, it can spontaneously flip over and spin the other way, dropping to the lower orbit and emitting a photon. The two energy levels are so close together that the photon emitted in the transition must have a very low energy—corresponding, in fact, to a wavelength of 21 cm.

Only very cold, low-density clouds of atomic hydrogen will emit 21-cm radiation. Once a hydrogen atom is excited into the upper of the two levels, it will, on average, stay there for 11 million years before spontaneously dropping to the lower level and emitting a 21-cm photon. The atoms of hot, dense hydrogen in stars collide much too often, disturbing the electron before it can radiate a 21-cm photon. But atoms in the interstellar medium collide much less often, so a few do manage to radiate 21-cm photons. Also, hydrogen atoms linked into molecules cannot radiate 21-cm photons because the energy levels for a molecule are different from those for an atom.

Extensive mapping of the sky at a wavelength of 21 cm has revealed that hydrogen is everywhere. Clouds of cold hydrogen fill the plane of our galaxy and are denser in the galaxy's spiral arms than between them. Since each cloud of hydrogen follows its own orbit around the galaxy, the separate clouds are often distinguishable in radio observations because of their differing Doppler shifts (Figure 11-11).

Molecules in Space

Radio telescopes can also detect radiation from various molecules in the interstellar medium. A molecule can store energy in a number of different ways. For example,

FIGURE 11-11

These 21-cm radio observations were made along the Milky Way in the direction of Cygnus and Sagittarius. Slightly different wavelengths due to Doppler shifts are plotted here as different radial velocities. The 21-cm line is made up of many peaks, each produced by a separate cloud of neutral hydrogen with its own radial velocity. (Adapted from observations by Burton)

it can rotate at different rates, or the atoms in a molecule can vibrate as if they were linked together by small springs. If a molecule suffers a collision or absorbs a photon, it can be excited to vibrate and rotate in some higher energy state. Quickly, however, it will return to a lower energy state and radiate the excess energy as a photon. Because these energy levels are closely spaced, the emitted photons typically have low energies, and we detect them in the radio or far infrared part of the electromagnetic spectrum. Just as neutral hydrogen radiates at a specific wavelength of 21 cm, many natural molecules radiate at their unique wavelengths.

TABLE 11-1

Selected Molecules Detected in the Interstellar Medium

H_2	molecular hydrogen	H_2S	hydrogen sulfide
C_2	diatomic carbon	N_2O	nitrous oxide
CN	cyanogen	H_2CO	formaldehyde
CO	carbon monoxide	C_2H_2	acetylene
NO	nitric oxide	NH_3	ammonia
OH	hydroxyl	HCO_2H	formic acid
NaCl	common table salt	CH_4	methane
HCN	hydrogen cyanide	CH_3OH	methyl alcohol
H_2O	water	CH_3CH_2OH	ethyl alcohol

Unfortunately for astronomers, molecules of hydrogen (H_2) do not emit radio-wavelength photons efficiently, so vast clouds of hydrogen dense enough to form molecules can't be detected by radio emission from molecular hydrogen. But other molecules can form in tiny amounts in these clouds, and many of them are good emitters of radio energy. Nearly 100 different molecules have been detected (Table 11-1). Some are quite complex, and it is not clear how they form. Most astronomers believe that the atoms meet and bond to form molecules on the surfaces of dust grains. Some of these molecules have not yet been synthesized on Earth, but others are common, such as N_2O (nitrous oxide), also known as laughing gas. Ethyl alcohol, which some humans drink, has also been detected. Although this is a very rare molecule compared to molecular hydrogen, an interstellar cloud can contain ethyl alcohol in amounts equivalent to 10^{28} fifths of whisky (about 100 earth masses).

One of the most important of the interstellar molecules may be carbon monoxide (CO). It is one of the poisonous gases that comes out of the tail pipes of cars, but it is also a very good emitter of radio energy at a wavelength of 2.6 mm. An interstellar cloud may contain only 1 CO molecule for every 10,000 molecules of hydrogen, but the CO can be detected while the molecular hydrogen cannot. Thus, radio astronomers can map the interstellar clouds by searching for the CO radio emission; where they detect CO, they can be certain that molecular hydrogen is common.

These molecules are very fragile, and a high-energy photon such as an ultraviolet photon has enough energy to break the molecule into fragments. Thus, the molecules cannot exist outside the dense clouds. Only deep inside the densest clouds, where dust absorbs and scatters the short-wavelength, high-energy photons, can the molecules survive. The very fact that these molecules are detected tells us that some of the clouds must be very dense.

The molecules are such good radiators of energy that they cool the clouds to low temperatures. Heat is present in the cloud as motion among the atoms and molecules. When a molecule collides with an atom or another molecule, some of the energy of motion can be stored in the rotation and vibration of a molecule. When the molecule emits that energy as a radio or infrared photon, the energy escapes from the cloud. Thus, molecular radiation can cool the interior of the cloud and keep it very cold.

The largest of these cool, dense clouds are called **giant molecular clouds.** They are 15 to 60 pc across and may contain 100 to 1,000,000 solar masses. The internal temperature is a frigid 10 K. Although we detect these clouds by their molecular emission, remember that the gas is mostly hydrogen.

Giant molecular clouds are the nests of star birth. Deep inside these great clouds, gravity can pull the matter inward and create new stars. That is a story we will discuss in detail in the next chapter. For now, we must study the dirty part of the interstellar medium—the dust. To do this, we must trade our radio telescope for an infrared telescope.

Infrared Radiation from Dust

The dust in the interstellar medium makes up only about 1 percent of the mass, and at a temperature of 100 K ($-143°C$) or less it is very cold. Nevertheless, it is easy to detect at infrared wavelengths. To see how such cold dust can radiate a lot of energy, consider a simple experiment with the dust in an imaginary giant molecular cloud.

For the sake of quick calculation, let us assume that our giant molecular cloud has a mass of about 10^5 solar masses. Only 1 percent of that mass is dust, so all of the dust in the cloud amounts to a mass of 10^3 solar masses. Imagine that we could collect all of the dust into a single sphere. It would be only about 10 times the diameter of the sun, and its surface area would be about 100 times that of the sun. But suppose we left the dust as separate specks, each 0.0005 mm in diameter, a typical size. Then the cloud would contain about 10^{43} dust specks, and the total surface area of the dust would be about 10^{29} times that of the sun. When matter is finely divided into dust, it has a very large surface area, and even though it is much colder than the sun, its vast surface area can radiate tremendous amounts of infrared radiation. The dust in the interstellar medium can emit tremendous amounts of energy.

In 1983, the Infrared Astronomy Satellite mapped the sky at far infrared wavelengths and found the galaxy filled with infrared radiation from dust. Most of this dust is confined to the region near the plane of our galaxy, and, as we would expect, the dust is thickest where the gas is thickest. Much of the dust is distributed in wispy clouds that became known as the **infrared cirrus** because of their overall resemblance to cirrus clouds in the earth's atmosphere. The infrared cirrus consists of dusty clouds of interstellar matter with temperatures of about 30 K

FIGURE 11-12
The Infrared Astronomy Satellite observed the sky in the far infrared and discovered that much of the sky is filled with the infrared cirrus, a wispy distribution of dust slightly warmed by starlight. The cirrus is visible only in the infrared. (NASA/IPAC, courtesy Deborah Levine)

FIGURE 11-13
The Cygnus superbubble, invisible behind the dust clouds of the constellation Cygnus, is dramatically revealed in this section of the X-ray maps produced by the ROSAT All Sky Survey. The outer boundary of the bubble is visible as the yellow arch of gas clouds (dashed line) at lower left. Filled with high-temperature gas and spanning 450 pc, the bubble has been inflated by many supernova explosions and by gas streaming away from hot stars. The strong X-ray source at bottom center is the Cygnus Loop, a bubble blown by a single supernova. This single bubble is only about 20 pc in diameter; it looks large because it is much closer than the superbubble. (Courtesy Steve Snowden and Max-Planck-Institute for Extraterrestrial Physics, Germany)

(Figure 11-12). Recent studies of CO emission from molecular clouds show that at least some of the infrared cirrus is associated with molecular clouds within a few hundred parsecs of the sun. This suggests that the gas and dust in the molecular clouds is not uniform but rather very patchy.

Infrared studies, combined with observations at visible and radio wavelengths, allow us to construct a model of an interstellar dust grain. Some are made of carbon, while others contain silicates and metals. Some may have coatings of carbon-based molecules. Clearly, even the individual dust grains can be very complex.

Infrared observations can help us understand the interstellar medium, but X-ray and ultraviolet observations complete the picture by allowing us to detect a part of the gas between the stars that is otherwise invisible.

X Rays from the Interstellar Medium

The interstellar medium seems very cold, so we would hardly expect to detect X rays. These high-energy photons are commonly produced by high temperatures. Nevertheless, X-ray telescopes can help us understand the interstellar medium.

X-ray telescopes above the earth's atmosphere have detected X rays from a part of the interstellar medium with a very high temperature. This gas has been called the **coronal gas** because it has temperatures of 10^6 K or higher, as does the sun's corona. Such hot gas would exert a very high pressure were it not of such low density. With a density of 0.0004 to 0.003 atoms/cm³, the coronal gas is the least dense part of the interstellar medium.

The coronal gas appears to originate in supernova explosions, which blast large amounts of very hot gas outward in expanding shells. As these shells grow larger and merge with neighboring shells, they may create a network of tunnels filled with hot coronal gas that interlace the cooler interstellar medium. Gas flowing away from very hot, young stars may add to the coronal gas. It is not clear how much of the volume of the interstellar medium is occupied by this coronal gas. Estimates range from 20 percent to 80 percent.

The Cygnus superbubble appears to be related to the coronal gas. Located in Cygnus, it is a very large shell of hot gas about 450 pc in diameter (Figure 11-13). The energy needed to create such a shell is equivalent to hundreds of supernova explosions. The bubble may have developed as a large cluster of stars was born and grew

Separating Facts from Theories

The fundamental work of science is testing theories by comparing them with facts. As we think about science, we need to distinguish clearly between facts and theories. The facts are the evidence against which we test the theories.

Scientific facts are those observations or experimental results of which we are confident. An astronomer makes observations of stars, a botanist collects samples of related plants, and a chemist performs an experiment to measure the rate of a chemical reaction. A fact could be a precise measurement, such as the mass of a star expressed as a specific number, or it could be a simple observation, such as that a certain butterfly no longer visits a certain mountain valley. In each case, the scientist is gathering facts.

A theory, however, is a conjecture as to how nature works. If we are uncertain of the theory,

we might call it a hypothesis. In any case, these conjectures are not facts; they are attempts to explain how nature works. In a sense, a theory or hypothesis is a story that scientists have made up to explain how nature works in some specific case. These stories can be wonderfully detailed and ingenious, but without evidence they are nothing more than hunches.

When one of these stories is tested against the facts and confirmed, we have more confidence that the story is more or less right. The more a story is tested successfully, the more confidence we have in it. The facts represent reality, and every theory or hypothesis must be repeatedly tested against reality.

We can't test one theory against another theory. Theories are not evidence; they are conjectures. If we were allowed to test one theory against another theory, we might fall into the

trap of circular reasoning. "Elves make the flowers bloom. I know that elves exist because the flowers bloom." That is using two theories to confirm each other, and it leads us to nonsense. Only facts can be evidence.

Nor can we use a theory to deduce facts. We can use a theory to make predictions, which we can then test against facts, but the predictions themselves can never be certainties, so they can't be facts. The only way to arrive at facts is to consult nature and make direct measurements or observations.

As we study different problems in astronomy, we must carefully distinguish between facts and theories. Facts are the basic building blocks of all science, and they can come only from the careful study of nature. As Galileo would say, we must read the book of nature. ∎

---

old, and the most massive stars died in supernova explosions. Several of these large bubbles are known.

Ultraviolet Observations of the Interstellar Medium

We can divide the ultraviolet spectrum into the near ultraviolet, with wavelengths only slightly shorter than visible light, and the far ultraviolet, with much shorter wavelengths. Only a few decades ago, astronomers believed that far-ultraviolet photons could not travel far through the interstellar medium because they would be absorbed so easily by neutral hydrogen atoms. These atoms of neutral hydrogen are good absorbers of far-ultraviolet photons because the photons have enough energy to ionize the atoms. In fact, even one atom of HI per cubic centimeter would make the interstellar medium opaque to far-ultraviolet photons, and we would be unable to see more than a light-year into space with a far-ultraviolet telescope.

When the Extreme Ultraviolet Explorer (EUVE) satellite was put into earth orbit in 1992, it discovered that the interstellar medium was only partly cloudy. While some regions were filled with clouds of neutral hydrogen and so were opaque, other regions were filled with hot, ionized hydrogen that was transparent to far-ultraviolet photons. This confirms the description of the interstellar medium produced by X-ray observations.

The far-ultraviolet observations tell us that the sun is located just inside a large region—a bubble—of hot, ionized hydrogen, while only a few light-years from the sun lies a cool, neutral hydrogen cloud that is opaque to far-ultraviolet photons. The hot bubble of gas in which the sun is located appears to be linked as if by tunnels to

other hot, transparent regions in a network that presumably extends throughout our galaxy. Thus, ultraviolet observations, combined with X-ray observations, suggest that the apparently cold and empty regions between the stars are filled with a complex, evolving mixture of hot and cold gas.

CRITICAL INQUIRY

If hydrogen is the most common molecule, why do astronomers depend on the CO molecule to map molecular clouds?

Although hydrogen is the most common atom in the universe and molecular hydrogen the most common molecule, a molecule of hydrogen does not radiate in the radio part of the spectrum. Thus, radio astronomers cannot detect it. But the much less common CO (carbon monoxide) molecule is a very efficient radiator of radio energy, so radio astronomers use it as a tracer of molecular clouds. Wherever radio telescopes reveal a great cloud of CO, we can be confident that most of the gas is molecular hydrogen.

We can also be confident that the molecular clouds contain dust, because it is the dust that protects the molecules in the cloud from the ultraviolet radiation that would otherwise break down the molecules into atoms.

Dust doesn't radiate radio energy, so we must study it in the infrared. Although the dust in a molecular cloud is very cold and makes up a small percentage of the total mass, it is a very good radiator of infrared radiation. How can a small amount of cold dust radiate vast amounts of infrared radiation? ∎

TABLE 11-2

Four Components of the Interstellar Medium

Component	Temperature (K)	Density (atoms/cm³)	Gas
HI clouds	50–150	1–1000	Neutral hydrogen Other atoms ionized
Intercloud medium	10^3–10^4	0.01	Partially ionized
Coronal gas	10^5–10^6	10^{-4}–10^{-3}	Highly ionized
Molecular clouds	20–50	10^3–10^5	Molecules

FIGURE 11-14

This infrared mosaic made up of IRAS images shows the wake left behind by the Pleiades as it passes through the interstellar medium. The plus marks (+) show the locations of the brightest stars. The curving wake is created as the stars of the cluster heat the gas of the interstellar medium and it expands to produce a low-density cavity. (Courtesy Richard E. White. Image rendered by Duncan Chesley of American Image, Inc.)

From one end of the spectrum to the other, we have used every wavelength to study the interstellar medium. Now we will try to put our data in order and create a model of the gas and dust between the stars. That is the first step toward understanding how stars are born.

11-3 A MODEL OF THE INTERSTELLAR MEDIUM

When we look at bright nebulae like the Great Nebula in Orion, we see a part of the interstellar medium, but what we see is only a very special region that happens to be close enough to hot bright stars to become excited. Most of the interstellar medium is invisible to our eyes, so we must gather the evidence—the observational facts (Window on Science 11-2)—and use them to develop a model of the interstellar medium.

We can divide the interstellar medium into four basic components (Table 11-2). Our problem is to describe these components and explain how they interact and evolve. We will discover that their evolution is intimately connected to the process of star formation and death.

Four Components of the Interstellar Medium

The interstellar medium is not at all uniform. Rather, it is lumpy, and the lumps differ dramatically in temperature and density.

HI clouds are cool, with temperatures of 50 to 150 K and densities of ten to a few hundreds of atoms per cubic centimeter. These clouds are only a few parsecs in diameter and contain a few solar masses.

Between the cool HI clouds lies the warm intercloud medium, with temperatures of a few thousand Kelvin and densities of 0.01 atoms/cm³. The intercloud medium is in approximate equilibrium with the HI clouds in that the hot, low-density gas has about the same pressure as the colder, denser gas in the HI clouds.

Molecular clouds are especially dense. Molecules cannot survive if they are exposed to ultraviolet photons in starlight, so they can form only in the densest clouds, where dust absorbs and scatters ultraviolet photons. The molecular clouds can be very large, with diameters of 60 pc and masses of a million solar masses, but they are also very cold. Strong evidence suggests that stars are born when giant molecular clouds contract under the influence of their own gravity, a subject we will explore in detail in the next chapter.

Pushing through this stew of interstellar clouds are regions of coronal gas. Most of this very hot gas is probably produced in supernova explosions, although some may be gas flowing away from very hot stars. With temperatures up to a million degrees and densities as low as 10^{-4} atoms/cm³, the coronal gas occupies a large part of the interstellar medium.

Astronomers believe that the HI clouds make up about 25 percent of the interstellar mass, and the intercloud medium about 50 percent. The coronal gas contributes only 5 percent of the mass, although it seems to occupy a large part of the volume. The giant molecular clouds amount to about 25 percent of the mass. If we add up these percentages, we get slightly more than 100 percent, which illustrates the uncertainty inherent in our model of the interstellar medium.

As the stars move through the interstellar medium, they meet no resistance, but they do have a dramatic effect on the gas and dust. The Pleiades, for example, is a relatively young star cluster that is moving rapidly through space. As it moves through the interstellar medium, it is leaving behind a trail like that left by a boat in water (Figure 11-14). This wake has been detected in IRAS images and is apparently produced by the ultraviolet radiation from the stars in the cluster. None of the stars is hot enough to ionize the gas, but there are a number of relatively hot stars; the ultraviolet radiation from those stars heats, but does not quite ionize, the gas. The heated gas then expands and forms a wake to show the path of the cluster through the interstellar medium.

The Pleiades clearly illustrates the close relationship between the stars and the interstellar medium. In fact, we can now outline a cycle that links the stars to the gas and dust between them.

The Interstellar Cycle

The story of the interstellar medium is closely linked to star formation. We will see in the next chapter that stars form in giant molecular clouds. As soon as a group of stars forms, the hottest stars begin ionizing the gas to produce emission nebulae. The pressure of the starlight and the gas flowing away from the hot, young stars pushes the interstellar cloud outward and may disrupt the cloud entirely.

The composition of the interstellar dust suggests that it is formed mostly in the atmospheres of cool stars. There the temperatures are low enough for some atoms to condense into specks of solid matter, much as soot can condense in a candle flame. The pressure of the starlight can push these dust specks out of the star and thus replenish the interstellar medium. Other stars that eject mass into space, such as supernovae, probably also add to the supply of interstellar dust.

The most massive stars die quickly in supernova explosions, and those tremendously violent events blast high-temperature gas outward that further disrupts the interstellar medium. Much of the motion we see in interstellar clouds and their twisted shapes are probably produced by these supernova explosions and the hot coronal gas they produce. It seems that our galaxy produces about four supernova explosions per century (although most are not visible from the earth because of intervening dust clouds) and that these explosions keep the interstellar medium stirred and create the vast regions filled with coronal gas. In fact, the sun lies inside such a region of high-temperature, low-density gas called the **local bubble**. With a typical diameter of a few hundred parsecs, the local bubble may be a cavity in the interstellar medium inflated by a supernova explosion within the last million years or so.

While supernovae and hot stars keep the interstellar medium in motion, the natural gravitation of gas clouds and collisions between clouds may gradually build more massive clouds. In the most massive clouds, the dust protects the interior from ultraviolet photons, and molecules can form. These giant molecular clouds eventually give birth to new stars, and thus the cycle begins all over again.

The Trifid Nebula (Figure 11-15) is a dramatic illustration of this cycle. Measuring over 12 pc in diameter, the nebula is illuminated by a number of hot, young stars that have apparently formed recently from the gas. Near the stars, the gas is ionized and glows as a pink-red HII region, but farther from the stars, where the ultraviolet radiation is weaker, the gas is not ionized. Nevertheless, dust in the nebula scatters blue light, so this un-ionized part of the nebula is visible as a blue reflection nebula. Dark lanes of obscuring dust cross the face of the nebula as if to remind us again of the importance of dust in the interstellar cycle.

CRITICAL INQUIRY

How can the coronal gas occupy most of the space but represent only 5 percent of the mass?

The HI clouds of neutral hydrogen make up about a quarter of the mass of the interstellar medium, the intercloud medium makes up roughly half of the mass, and the giant molecular clouds make up roughly a quarter of the mass. But these three components occupy a relatively small part of the volume. The coronal gas is so hot, as evidenced by the X rays it emits, that it can expand to fill a large volume. Although it occupies much of the volume, the coronal gas has a very low density, only about 1 atom for every 10,000 cm^3, so its total mass is only a small fraction of the interstellar medium.

The coronal gas has the lowest density in the interstellar medium. What component has the highest density, and how does that affect the way we can observe it? ■

Our study of the interstellar medium is incomplete for two reasons. First, astronomers don't yet understand all of its components or how those components interact. Second, we can't fully understand the interstellar medium until we understand how stars are born and how they die. We begin that story in the next chapter.

■ Summary

The interstellar medium, the gas and dust between the stars, is confined near the plane of our Milky Way Galaxy. We see clear evidence of an interstellar medium when we look at nebulae such as the Great Nebula in Orion. This is an emission nebula, a cloud of gas near one or more hot stars whose ultraviolet radiation ionizes the hydrogen and makes the nebula glow like a giant neon sign. The red, blue, and violet Balmer lines blend together to produce the characteristic pink-red glow of ionized hydrogen.

A reflection nebula is produced by gas and dust illuminated by a star that is not hot enough to ionize the gas. Rather, the dust scatters the starlight to produce a reflection of the stellar absorption spectrum. Because shorter-wavelength photons scatter more than longer-wavelength photons, reflection nebulae look blue. The daytime sky looks blue for the same reason.

A dark nebula is a cloud of gas and dust that is visible because it blocks the light of distant stars. We see such nebulae

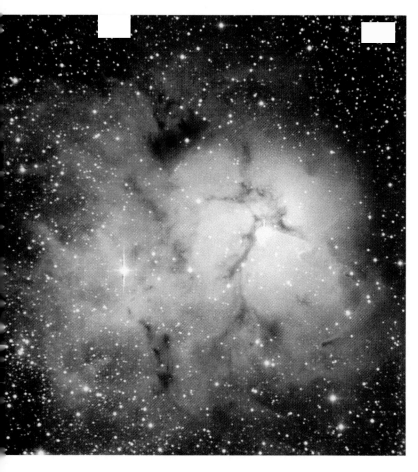

FIGURE 11-15

The Trifid Nebula in Sagittarius is a combination of an emission nebula and a reflection nebula. The newly formed hot stars near the center of the nebula ionize the hydrogen to produce the pink emission nebula. Gas farther from the stars is not ionized, but dust scatters starlight and makes that portion of the nebula visible as a blue reflection nebula. Thick dust lanes divide the nebula into three major regions and give rise to the name Trifid. (NOAO and Nigel Sharp)

as dark shapes, and the smallest, the Bok globules, are only a parsec or so in diameter.

Further evidence of an interstellar medium is the extinction, or dimming, of the light of distant stars, and interstellar reddening. Light from distant stars suffers scattering by dust particles in the interstellar medium, and blue light is scattered more than red light. This makes distant stars look redder than their spectral types suggest. The dependence of this extinction on wavelength tells us that the scattering dust particles are very small and that some of them are made of carbon.

Interstellar absorption lines in the spectra of distant stars are very narrow. The interstellar gas is cold and has a very low density, and this makes the interstellar lines much narrower than the spectral lines produced in stars. Multiple interstellar lines tell us that the light has passed through more than one interstellar cloud on its way to the earth. Radio observations at a wavelength of 21 cm also reveal multiple clouds of neutral hydrogen. These neutral clouds drift through a warmer but lower-density intercloud medium. Radio telescopes tuned to other wavelengths have detected nearly 100 different molecules in the interstellar medium, most of them found in giant molecular clouds.

The Infrared Astronomy Satellite has detected dust in the interstellar medium in the form of the infrared cirrus. X-ray and far-ultraviolet observations have detected very hot coronal gas produced by supernova explosions.

The four main components of the interstellar medium are the small neutral HI clouds, the warm intercloud medium, coronal gas, and molecular clouds. Stars are born in the dense molecular clouds, and the energy from hot stars and supernova explosions causes currents in the interstellar medium and creates the coronal gas.

▪ New Terms

interstellar medium	interstellar reddening
nebula	interstellar absorption lines
emission nebula	HI clouds
HII region	intercloud medium
forbidden line	21-cm radiation
metastable level	giant molecular clouds
reflection nebula	infrared cirrus
dark nebula	coronal gas
Bok globule	local bubble
interstellar extinction	

Questions

1. What evidence do we have that the spaces between the stars are not totally empty?

2. What evidence do we have that the interstellar medium contains both gas and dust?

3. How do the spectra of HII regions differ from the spectra of reflection nebulae? Why?

4. Why are interstellar lines so narrow? Why do some spectral lines forbidden in spectra on the earth appear in spectra of interstellar clouds and nebulae?

5. How is the blue color of a reflection nebula related to the blue color of the daytime sky?

6. Why are distant stars redder than their spectral types suggest?

7. If starlight on its way to the earth passed through a cloud of interstellar gas that was hot instead of very cold, would you expect the interstellar absorption lines to be broader or narrower than usual? Why?

8. How can the HI clouds and the intercloud medium have similar pressures when their temperatures are so different?

9. What does the shape of the 21-cm radio emission line of neutral hydrogen tell us about the interstellar medium?

10. What produces the coronal gas?

Discussion Questions

1. When we see distant streetlights through smog, they look dimmer and redder than they do normally. But when we see the same streetlights through fog or falling snow, they look dimmer but not redder. Use your knowledge of the interstellar medium to discuss the relative sizes of the particles in smog, fog, and snowstorms compared to the wavelength of light.

2. If you could see a few stars through a dark nebula, how would you expect their spectra to differ from similar stars just in front of the dark nebula?

Problems

1. A small Bok globule has a diameter of 20 seconds of arc and a distance of 1000 pc from the earth. What is the diameter of the globule in parsecs? in meters?

2. The dust in a molecular cloud has a temperature of about 50 K. At what wavelength does it emit the maximum energy? (HINT: Consider black body radiation, Chapter 7.)

3. Extinction dims starlight by about 1 magnitude per 1000 pc. What fraction of photons survives a trip of 1000 pc? (HINT: Consider the definition of the magnitude scale in Chapter 2.)

4. If the total extinction through a dark nebula is 10 magnitudes, what fraction of photons makes it through the cloud? (HINT: See Problem 3.)

5. The density of air in a child's balloon 20 cm in diameter is roughly the same as the density of air at sea level, 10^{19} particles/cm^3. To how large a diameter would you have to expand the balloon to make the gas inside the same density as the interstellar medium, about 1 particle/cm^3? (HINT: The volume of a sphere is $\frac{4}{3}\pi R^3$.)

6. If a giant molecular cloud has a diameter of 30 pc and drifts relative to neighboring clouds at 20 km/sec, how long will it take to travel its own diameter?

7. An HI cloud is 4 pc in diameter and has a density of 100 hydrogen atoms/cm^3. What is its total mass in kilograms? (HINTS: The volume of a sphere is $\frac{4}{3}\pi R^3$, and the mass of a hydrogen atom is 1.67×10^{-27} kg.)

8. Find the mass in kilograms of a giant molecular cloud that is 30 pc in diameter and has a density of 300 hydrogen molecules/cm^3. (HINT: See Problem 7.)

9. At what wavelength does the coronal gas radiate most strongly? (HINT: Consider black body radiation, Chapter 7.)

Critical Inquiries for the Web

1. What causes the varied colors in images of gaseous nebulae that grace textbooks and Web sites? Search the Web for a color image of a nebula in the Messier catalog. Describe the structure and colors you see in terms of the concepts discussed in this chapter. (Be careful to distinguish real color from false color when answering this question. See Window on Science 14-1 for more information about false-color images.)

2. An interesting highlight in the history of astronomy is the discovery of "Nebulium." Search the Internet for information on this once mysterious source of nebular spectral lines. How can astronomers today explain these lines in terms of known elements?

Go to the Wadsworth Astronomy Resource Center (www.wadsworth.com/astronomy) for critical thinking exercises, articles, and additional readings from InfoTrac College Edition, Wadsworth's online student library.

Chapter Twelve

THE FORMATION OF STARS

GUIDEPOST

Previous chapters have used the basic principles of physics as a way to deduce things about stars and the interstellar medium. In Chapter 10, for example, we used orbital motion to find the masses of the stars. All of the data we have amassed will now help us understand the life stories of the stars in this chapter and those that follow.

In this chapter, we use the laws of physics in a new way. We develop theories and models based on physics that help us understand how stars work. For instance, what stops a contracting star and gives it stability? We can understand this phenomenon because we understand some of the basic laws of physics.

Throughout this chapter and the chapters that follow, we search for evidence. What observational facts confirm or contradict our theories? That is the basis of all science, and it must be part of any critical analysis of what we know and how we know it. ■

Jim he allowed [the stars] was made, but I allowed they happened. Jim said the moon could'a laid them; well, that looked kind of reasonable, so I didn't say nothing against it, because I've seen a frog lay most as many, so of course it could be done.

MARK TWAIN

The Adventures of Huckleberry Finn

FIGURE 12-1
Where we find thick clouds of gas and dust, we usually find young stars, a correlation that suggests that stars are born from these clouds. Here, the great dust cloud known as the Horsehead Nebula (0.55 pc from nose to ear) is silhouetted against glowing gas excited by ultraviolet radiation from hot stars embedded in the dense clouds below. (NOAO and Nigel Sharp)

Stars exist because of gravity. They form because gravity makes clouds of gas contract, and they generate nuclear energy because gravity squeezes their cores to unearthly densities and temperatures. In the end, stars die because they exhaust their fuel supply and can no longer withstand the force of their own gravity.

A star can remain stable only by maintaining great pressure in its interior. Gravity tries to make it contract. However, if the internal temperature is high enough, pressure pushes outward just enough to balance gravity. Thus, a star is a battlefield where pressure and gravity struggle for dominance.

Only by generating tremendous amounts of energy can a star maintain the gravity–pressure balance. The sun, for example, generates 6×10^{13} times more energy per second than all of the coal, oil, natural gas, and nuclear power plants on the earth. Like the sun, most stars generate their power by fusion reactions that consume hydrogen and produce helium. However, we will discover that not all stars fuse their hydrogen in the same way as does the sun.

In this chapter, we will see how gravity creates stars from the thin gas of space and how nuclear reactions inside stars generate energy. We will see how the flow of that energy outward toward the surface of the star balances gravity and makes the stars stable. In the next two chapters, we will follow the life story of stars from the beginning of their stable lives to their ultimate deaths.

12-1 MAKING STARS FROM THE INTERSTELLAR MEDIUM

Stars have been forming continuously since our galaxy took shape over 10 billion years ago. We know this for two reasons. First, the sun is only about 5 billion years old, a relative newcomer compared to the older stars in our galaxy. Second, we can see hot, blue stars such as Spica (α Virginis), a B1 main-sequence star. As we will see in the next chapter, such massive stars have very short lives. In fact, a star like Spica can last only 10 million years and thus must have formed recently.

The key to understanding star formation is the correlation between young stars and clouds of gas. Where we find the youngest groups of stars, we also find large clouds of gas illuminated by the hottest and brightest of the new stars (Figure 12-1). This leads us to suspect that stars form from such clouds, much as raindrops condense from the water vapor in a thundercloud. Indeed, the giant molecular clouds discussed in the preceding chapter can give birth to entire clusters of new stars.

The central problem for our discussion of star formation is how these large, low-density, cold clouds of gas become comparatively small, high-density, hot stars. Gravity is the key.

Star Birth in Giant Molecular Clouds

The giant molecular clouds are the sites of active star formation, yet they are very unlike stars. With a typical diameter of 50 pc and a typical mass exceeding 10^5 solar masses, a giant molecular cloud is vastly larger than a star. Also, the gas in a giant molecular cloud is about 10^{20} times less dense than a star and has temperatures of only a few degrees Kelvin. These clouds can form stars because gravity can force some small regions of the clouds to contract to high density and high temperature.

Radio observations show that at least some giant molecular clouds develop dense cores that are only 0.1 pc in radius and that contain roughly 1 solar mass. A single giant molecular cloud may contain many of these dense cores and thus can give birth to star clusters containing hundreds of stars. However, both theory and observations suggest that giant molecular clouds cannot begin the formation of dense cores spontaneously. At least three factors resist the contraction of a gas cloud, and gravity must overcome those factors before star formation can begin.

First, the heat in the gas is present as motion among the atoms and molecules. Even at temperatures of 10 K, the average hydrogen molecule moves at about 0.35 km/sec (almost 800 mph). This thermal motion would make the cloud drift apart if gravity were too weak to hold it together.

The interstellar magnetic field is the second factor that gravity must overcome to make the cloud contract. Neutral atoms and molecules are unaffected by a magnetic field, but ions, having an electric charge, cannot move freely through a magnetic field. Although the gas in a molecular cloud is mostly neutral, there are some ions, and thus a magnetic field can exert a force on the gas. The magnetic field present all through our galaxy averages only about 10^{-4} times as strong as that on the earth, but it can act like an internal spring to resist the contraction of the gas cloud.

The third factor is rotation. Everything in the universe rotates to some extent. As the gas cloud begins to contract, it spins more and more rapidly as it conserves angular momentum, just as ice skaters spin faster as they pull in their arms (Figure 5-14). This rotation can become so rapid that it resists further contraction of the cloud.

The increasing rotation of a contracting cloud can sometimes cause it to fragment into two clouds, thus forming a binary star system. In some cases, the cloud may fragment a number of times to produce multiple star systems or even clusters of stars. Exactly how and why a rotating cloud fragments are poorly understood.

Given these three resistive factors, it seems surprising that any giant molecular cloud can begin producing dense cores destined to become stars, but both theory and observation suggest that a passing shock wave can compress the cloud and trigger star formation (Figure 12-2). During such a triggering event, a few regions of the large cloud can be compressed to such high densities that the resistive factors can no longer oppose gravity, and star formation begins.

At least four different processes can produce shock waves that trigger star formation. Supernova explosions (Chapter 14) can produce powerful shock waves that rush through the interstellar medium. Also, the ignition of very hot stars can ionize nearby gas and drive it away to produce a shock wave where it pushes into the colder, denser interstellar matter. A third trigger is the collision of molecular clouds. Because the clouds are large, they are likely to run into each other occasionally, and because they contain magnetic fields, they cannot pass through each other. A collision between such clouds can compress parts of the clouds and trigger star formation. The fourth trigger is the spiral pattern of our Milky Way Galaxy (see Figure 1-11). One theory suggests that the spiral arms are shock waves that travel around the galaxy like the moving hands of a clock (Chapter 16). As a cloud passes through a spiral arm, the cloud could be compressed, and star formation could begin.

FIGURE 12-2

An interstellar gas cloud has such a low density it is unlikely to form stars without an outside stimulus. Computer models show that a passing shock wave (pink line) can compress and fragment a cloud, driving some regions to high enough densities to trigger star formation (bottom). From first frame to last, this figure spans about 6 million years.

FIGURE 12-3

(a) Star formation begins in a gas cloud when dense cores begin to form as gravitation draws gas and dust into condensations. (b) The in-falling matter forms a dense, hot central object destined to become a star at the center of a larger, opaque cocoon of gas and dust. Rotation forces some of the matter to fall into a disk rotating around the central body. (c) The central object, now a protostar, and the rotating disk eventually begin to expel gas along the axis of rotation while matter continues to fall into the disk. The bipolar flows, guided and perhaps driven by magnetic fields (red), push out of the surrounding cloud of gas and dust to produce jets pointing in opposite directions.

Heating by Contraction

We have explained how clouds of interstellar gas can become dense enough to make stars, but we have not explained how the gas can become hot enough. The answer, once again, is gravity.

Once a small cloud of gas begins to contract, gravity draws the atoms toward the center, and the atoms gather speed as they fall. In fact, astronomers refer to this early stage in the formation of a star as **free-fall contraction.**

Whereas the atoms may have had low velocities to start with, by the time they have fallen most of the way to the center of the cloud they are traveling at high velocities. We have defined heat as the agitation of the particles in a gas, so this increase in velocity is a step toward heating the gas. But we can't say that the gas is hot simply because all of the atoms are moving rapidly. The air in the cabin of a jet airplane is traveling rapidly, but it isn't hot, because all of the atoms are moving in generally the same direction with the plane. To convert the high velocity of the in-falling atoms into heat, the motion must become randomized, and that happens when the atoms begin to collide with one another as they fall into the central region of the cloud. The jumbled, random motion of the atoms is heat, and the temperature of the gas increases.

This is an important principle in astronomy. Whenever a cloud of gas contracts, gravitational energy is converted into thermal energy, and the gas grows hotter. Whenever a gas cloud expands, thermal energy is converted into gravitational energy, and the gas cools. This principle applies not only to clouds of interstellar gas, but also to contracting and expanding stars, as we will see in the following chapters.

Our study of gas clouds has shown us how nature can begin the contraction of dense cores in giant molecular clouds and how contraction can generate heat. Now we will construct a detailed story of the transformation from gas cloud to star.

Protostars

To follow the story of star formation farther, we must concentrate on a single fragment of a collapsing cloud as it contracts, heats up, and begins to behave like a star. Although the term **protostar** is used rather loosely by astronomers, we will define it here to be a prestellar object that is hot enough to radiate infrared radiation but not hot enough to generate energy by nuclear fusion.

A protostar begins life as a contracting cloud of gas and dust falling inward under the influence of gravity (Figure 12-3a). As the cloud contracts, it develops a higher-density region at the center and a low-density envelope. Mass continues to flow inward from the outer parts of the cloud. That is, the cloud contracts from the inside out, with the protostar taking shape deep inside an enveloping cloud of cold, dusty gas. These clouds have been called **cocoons** because they hide the forming protostar from our view as it takes shape.

As the contracting cloud adds to the mass of the protostar, some of the gas settles into a rapidly spinning disk around the forming star (Figure 12-3b), much like a blob of pizza dough spreading into a thin disk when it is spun into the air. The contracting, rotating gas cloud naturally settles into a disk, and this disk is important to us for two reasons. First, such disks form planets. The earth may have formed in such a disk around the proto-sun

FIGURE 12-4

The Cone Nebula (shown in Figure 9-1) is a region of star formation, but the visible wavelength image at left shows no trace of newborn stars in the small white box. The infrared image at right reveals a luminous, massive star and 6 nearby sun-like stars hidden in the dust and gas. The smaller stars lie only 0.04 to 0.08 ly from their luminous neighbor and were probably formed when energy from the massive star compressed the surrounding gas. (Rodger Thompson, Marcia Rieke, and Glenn Schneider, University of Arizona, and NASA)

4.6 billion years ago. The evidence seems very strong that planetary systems, including our solar system, form in such disks around stars.

The second reason why disks around protostars are important to us is that they produce an astonishing phenomenon that is both beautiful and revealing. As the protostar grows hotter and the disk grows denser, a strong wind begins to blow outward. The cause of this wind is not well understood, but it probably involves the magnetic field being drawn inward by matter falling into the disk. In any case, the wind is blocked by the thick disk around the protostar and can escape only along the axis of rotation in two jets pointing in opposite directions (Figure 12-3c). While matter continues to flow into the disk and then into the protostar, these **bipolar flows** punch out of the dark cocoon nebula like two searchlights announcing the presence of a new star.

If we could see a protostar without its cocoon, it would be a very luminous and cool red star in the upper right part of the H–R diagram, but the dust cocoon absorbs almost all the visible radiation. The warmed dust reradiates the energy as infrared radiation, so we must search for protostars with infrared telescopes (Figure 12-4). When the protostar becomes hot enough, it can drive away the surrounding cocoon of gas and dust and become visible. The location in the H–R diagram where protostars first emerge from their cocoons and become visible is called the **birth line.** Once a star crosses the birth line, it continues to contract toward the main sequence (Figure 12-5).

How long this contraction from gas to main-sequence star takes depends on the mass of the star. More massive stars have stronger gravity and contract rapidly (Figure 12-5). The sun took about 30 million years to reach the main sequence, but a 30-solar-mass star takes only 30,000 years. A 0.2-solar-mass star needs about 1 billion years to reach the main sequence.

The theory of star formation takes us into an unearthly realm filled with unfamiliar processes and objects. We might think it was nothing more than a fairy tale if observations did not support the theory.

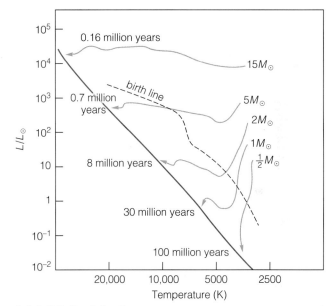

FIGURE 12-5

The more massive a protostar is, the faster it contracts. A 1 $M_\odot$ star requires 30 million years to reach the main sequence. The dashed line is the birth line, where contracting stars first become visible. (Illustration design by author)

Observations of Star Formation

What evidence do we have that stars form as theory predicts? Can we find real examples of protostars and newborn stars? Our search requires observations at visible, infrared, and radio wavelengths.

T Tauri stars, named after variable star T in Taurus, are detectable at visible wavelengths and seem to be stars that have just emerged from their cocoons. We know that they are between 100,000 and 100 million years old because they occur in some young star clusters such as NGC 2264. This cluster of stars has formed so recently that the less massive stars, contracting more

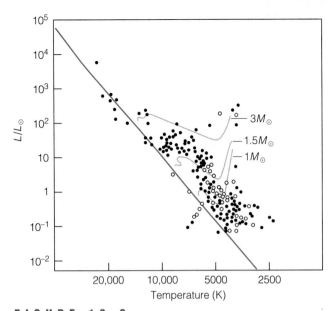

FIGURE 12-6

An H–R diagram of the star cluster NGC 2264 reveals many low-mass stars that have not yet reached the main sequence. Many of these stars are T Tauri stars (open circles). The cluster is only a few million years old. (Adapted from a diagram by M. Walker)

slowly than the massive stars, have not yet reached the main sequence (Figure 12-6). The temperature and luminosity of the T Tauri stars place them to the right of the main sequence under the birth line, just where we would expect to find newborn stars.

Other characteristics support our belief that T Tauri stars are just clearing away the clouds from which they formed. Some T Tauri stars are strong infrared sources, and the infrared radiation must be coming from dust clouds near the stars. Their spectra contain Doppler-shifted spectral lines that show that the stars are surrounded by expanding clouds of gas. Also, their spectra show signs of strongly active chromospheres. As we saw when we studied the sun, chromospheric activity is a sign of magnetic fields, so the T Tauri stars must have strong magnetic fields. This, too, is a characteristic of young, rapidly spinning stars.

The constellation Orion is filled with T Tauri stars, and other groups are known. The stars in these **T associations** must have formed together as an extended cluster of stars so large it is not bound together by its own gravity. The stars will gradually move apart as they age and the association dissipates. T Tauri stars have masses ranging from about 0.75 solar mass to 3 solar masses.

a

FIGURE 12-7

(a) This computer-enhanced image reveals Herbig–Haro objects HH34S (left) and the fainter HH34N (right) excited by jets from a newly formed star (center). (Reinhard Mundt, Calar Alto 3.5-m telescope) The inset is a Hubble Space Telescope image showing beaded structure along the jet, suggesting surges in the flow of the jet. (J. Hester, Arizona State University, and NASA) (b) Herbig–Haro objects HH46 and HH47 are part of a jet of gas moving toward us from a young protostar buried inside a dark globule. (NOAO and Patrick Hartigan) (c) In this Doppler-effect image, color represents velocity, with blue regions approaching us and red wisps moving away from us. (Courtesy Patrick Hartigan, Rice University, and Jon Morse, University of Colorado) (d) A Hubble Space Telescope image reveals complex turbulence in the jet of HH47. (J. Morse/STScI, and NASA)

b

c

FIGURE 12-8

The Eagle Nebula (see Figure 7-1) contains dark clouds of gas and dust that, in this Hubble Space Telescope image, are revealed as nests of star formation. Hot stars nearby are evaporating the dust and driving away the gas to expose EGGs, evaporating gaseous globules (upper left). These small, dense clouds are evidently embryonic stars in the act of contracting under their own gravitation. At least one seems to contain a newborn star (arrow) exposed by the evaporation of the gaseous globule around it. (Jeff Hester and Paul Scowen, Arizona State University and NASA)

O associations, groups of O stars, are also known, and they appear to be similar to T associations, but the stars are more massive. They, too, must have formed recently from a single large cloud of gas. It is not clear why some gas clouds give birth to star clusters held together by their own gravity and others give birth to larger associations not bound by gravity.

Other visible objects associated with star formation are the **Herbig–Haro objects.** Named after American astronomer George Herbig and Mexican astronomer Guillermo Haro, these small, luminous clouds of gas were first thought to be protostars, but more recent results show that most are small nebulae excited to glowing by young stars nearby. Many Herbig–Haro objects vary irregularly in brightness and may change their appearance over periods of only a few years, apparently because of changes in the way they are excited by gas and radiation from forming stars. For example, the Herbig–Haro objects HH34S and HH34N are clearly related to a young star located halfway between them (Figure 12-7a). HH 46, which has brightened and changed its shape over the years, is now clearly known to be one of a number of such objects excited by gas flowing away from a young, hot star (Figure 12-7b, c, d).

The Herbig–Haro objects led astronomers to look for beams of gas and radiation flowing away from young stars, and observations at optical, infrared, and radio wavelengths have revealed many such jets. Matter in these bipolar flows can travel as fast as 100 km/sec and can carry as much as a few solar masses of gas away from a protostar. With a length of a few light-years, these jets pack a powerful punch, and where they strike surrounding gas, they can excite Herbig–Haro objects (Figure 12-7). Such powerful outflows cannot last very long, and astronomers estimate that they have lifetimes of no more than 10,000 years. Some T Tauri stars are sources of such bipolar flows, but many of the stars producing the flows are hidden deep inside dust clouds.

Normally, the dense cores of contracting gas that are the very beginnings of stars are buried deep inside nebulae, and we cannot observe them easily. But the Hubble Space Telescope found an exciting exception in the Eagle Nebula (Figure 12-8). In this nebula, a previous generation of stars includes a few that are hot and luminous, and

d

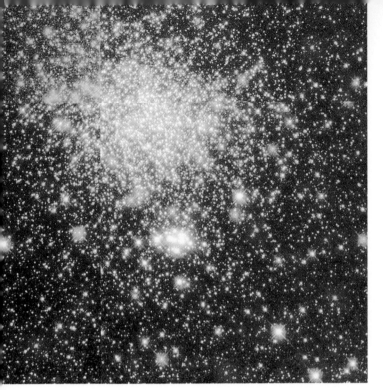

FIGURE 12-9

Generations of stars appear in the two star clusters in this Hubble Space Telescope image mosaic. The large cluster of yellow stars is about 50 million years old, but the bright white stars scattered through the image are massive young stars in a cluster only about 4 million years old. The younger cluster lies 200 ly beyond the older cluster, suggesting that star formation began 50 million years ago and created the nearer cluster. Supernova explosions in that cluster could have compressed gas clouds nearby and triggered the formation of the slightly more distant younger cluster. (R. Gilmozzi, STScI/ESA, Shawn Ewald, JPL, and NASA)

the energy from these stars is shredding and vaporizing the densest regions of the nebula. As the dust evaporates and the gas is driven away, small dense cores dubbed "EGGs," evaporating gaseous globules, are exposed. Because these dense globules protect the nebula behind them, they tend to produce features like elephant trunks as the nebula dissipates. Some of these dense cores may fail to form stars because they have been exposed too soon, but at least one seems to have hatched a new star (arrow in Figure 12-8). Thus, the Eagle Nebula and its EGGs give us an insight into the early stages of star formation.

The birth of a few hot stars can trigger the formation of still more stars. If a gas cloud produces massive stars, a few will be hot enough to ionize the gas nearby and drive it away. Where that gas impacts the remaining gas cloud, it can compress the cloud and trigger star formation. We can find places where this process has operated and produced generations of stars with older stars far from the ionized region and younger stars nearby (Figure 12-9). Like a grass fire spreading through the cloud, star formation can reignite itself if it creates massive stars. Of course, as we saw in Chapter 11, stars cooler than about B1 do not radiate many photons energetic enough to ionize hydrogen, so this process only

works for clouds forming massive stars. Lower-mass stars may form in the same clouds with the more massive stars, but evidence also suggests that some low-mass stars form singly as opposed to forming in clusters and associations.

The study of star formation is an exciting part of modern astronomy, and astronomers are using every wavelength to search for more information about the birth of stars. Although much remains to be discovered, the general outline is clear. Gravity creates stars by forcing interstellar clouds to contract and grow hotter.

CRITICAL INQUIRY

What evidence do we have that stars are forming right now?

First, we should note that some very massive stars can't live very long, so when we see such stars, such as the hot, blue stars in Orion, we know they must have formed in the last few million years. Star formation must be a continuous process, or we would not see any of these short-lived stars. But we have more direct evidence when we look at T Tauri stars, which lie above the main sequence and are often associated with gas and dust. In fact, we see entire associations of T Tauri stars as well as other associations of short-lived O stars. Many regions of gas and dust contain dark globules, which appear to be in the early stages of star formation, and small nebulae called Herbig–Haro objects, which are regions of the nebula excited by energy from nearby protostars. Of course, infrared astronomers can detect actual protostars buried deep inside the dusty nebulae from which the stars are forming.

There seems to be no doubt that star formation is an ongoing process, and we can even discuss how it happens. For example, what evidence do we have that protostars are often surrounded by disks of gas and dust? ■

Gravity pulls the gas and dust together to make new stars, but this raises a natural question. If all this matter is falling inward to make a star, why does it stop? How do contracting protostars turn themselves into stable main-sequence stars? The answer is that the contracting stars begin to generate their own energy.

12-2 THE SOURCE OF STELLAR ENERGY

Stars are born when gravity pulls matter together, but what stops the contraction? Contracting stars somehow begin to make large amounts of energy, and somehow that energy stops the contraction and stabilizes the stars.

In this section, we see how stars make energy; in the next section, we will see how contracting stars reach stability. This story will lead our imaginations into a region where our bodies can never go—the heart of a star. We begin our journey, however, at a backyard barbecue.

Chemical Energy

Most of the energy sources with which we are familiar are chemical in nature. The energy we extract from a fuel comes from the chemical bonds that link atoms together into molecules. For example, when we light the charcoal in the backyard barbecue, we burn carbon to produce carbon dioxide and release energy. We could write the reaction as follows:

$$C_2 + 2\,O_2 \rightarrow 2\,CO_2 + energy$$

Two carbon atoms linked to each other in a carbon molecule are broken apart and combine with two oxygen molecules in the air to form two molecules of carbon dioxide plus energy. The energy, which we use to heat our hot dogs, originally came from the rearrangement of the bonds that linked the atoms into molecules.

The atoms in a molecule are held to each other by one of the four forces of nature, the electrostatic force. This is the same force that holds cat hair to your sweater. Gravity, another of the four forces, is very weak in atoms and molecules and does not affect their structure. The **weak force** is involved in certain kinds of radioactive decay, and the **strong force** holds atomic nuclei together. When we burn a fuel, we extract energy from only one of these forces—the electrostatic force.

The stars make energy not through chemical processes but through nuclear processes. The strong force locks protons and neutrons into atomic nuclei, and it is much stronger than the chemical bonds that hold molecules together. When a star breaks these nuclear bonds and rearranges the particles to make new nuclei, it can generate tremendous amounts of energy. Chemical energy can cook a hot dog, but nuclear energy can fuel a star.

To review the details of these nuclear reactions, we begin by looking into the center of the sun.

Solar Energy Generation

The sun is a typical star. It continuously radiates light and heat into space, so it would cool noticeably in only a hundred thousand years if it could not make more energy to replace the energy it loses.

Not until the 1930s did astronomers realize how the sun generates its energy. The energy is produced deep in the interior of the sun, where the temperature is so high that the atoms have lost their electrons. The matter is a stormy sea of high-speed atomic nuclei and free electrons. The sun's energy is produced by fusion reactions that occur when the nuclei collide. These reactions, much like those that occur in hydrogen bombs, are called fusion reactions because they fuse nuclei together.

In the case of the sun, the reactions fuse four hydrogen nuclei to make a single helium nucleus. Because one helium nucleus has 0.7 percent less mass than four hydrogen nuclei, it seems that some mass vanishes in the process:

$$\begin{aligned}
4 \text{ hydrogen nuclei} &= 6.693 \times 10^{-27} \text{ kg} \\
\underline{1 \text{ helium nucleus}} &= \underline{6.645 \times 10^{-27} \text{ kg}} \\
\text{difference in mass} &= 0.048 \times 10^{-27} \text{ kg}
\end{aligned}$$

However, this mass does not actually vanish; it merely changes form. The equation $E = m_0 c^2$ reminds us that mass and energy are related, and under certain circumstances mass may become energy, and vice versa. Thus, the 0.048×10^{-27} kg does not vanish but merely becomes energy. To see how much, we use Einstein's equation:

$$\begin{aligned}
E &= m_0 c^2 \\
&= (0.048 \times 10^{-27} \text{ kg})(3 \times 10^8 \text{ m/sec})^2 \\
&= 0.43 \times 10^{-11} \text{ J}
\end{aligned}$$

This is a very small amount of energy, hardly enough to raise a housefly one-thousandth of an inch. Because one reaction produces such a small amount of energy, it is obvious that many reactions are necessary to supply the energy needs of a star. The sun, for example, needs 10^{38} reactions per second, transforming 5 million tons of mass into energy every second, just to stay hot enough to resist its own gravity.

Two nuclei can fuse only if they come close together, but atomic nuclei have positive charges and repel each other with an electrostatic force. To overcome this force, atomic nuclei must collide at high velocity. If the particles in a material are moving fast, we say that the material is hot. Thus, atomic fusion reactions can occur only if the gas is very hot—at least 10^7 K.

Fusion reactions in the sun also require that the gas be very dense—denser than solid lead. We know that the sun requires 10^{38} reactions per second to manufacture sufficient energy. But fusion occurs in only a small percentage of all collisions, so the sun requires many collisions between nuclei each second. Only where the gas is very dense are there enough collisions to meet the sun's energy needs.

We can symbolize the fusion reactions in the sun with a simple nuclear reaction:

$$4\,^1\text{H} \rightarrow \,^4\text{He} + energy$$

In this equation, ^1H represents a proton, the nucleus of the hydrogen atom, and ^4He represents the nucleus of a helium atom. The superscripts indicate the approximate weight of the nuclei (the number of protons plus the number of neutrons). The actual steps in the process are more complicated than this convenient summary suggests. Instead of waiting for four hydrogen nuclei to collide simultaneously, a highly unlikely event, the process can proceed step by step in a chain of reactions—the proton–proton chain.

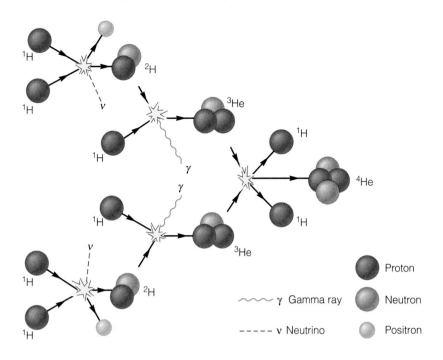

FIGURE 12-10

The proton–proton chain combines four protons (at far left) to produce one helium nucleus (at right) plus energy.

Proton — Neutron — Positron

~~~ γ Gamma ray

----- ν Neutrino

The **proton–proton chain** is a series of three nuclear reactions that builds a helium nucleus by adding together protons. This process is efficient at temperatures above 10,000,000 K. The sun, for example, manufactures over 90 percent of its energy in this way.

The three steps in the proton–proton chain entail these reactions:

$$^1H + {}^1H \rightarrow {}^2H + e^+ + \nu$$
$$^2H + {}^1H \rightarrow {}^3He + \gamma$$
$$^3He + {}^3He \rightarrow {}^4He + {}^1H + {}^1H$$

In the first step, two hydrogen nuclei (two protons) combine to form a heavy hydrogen nucleus, emitting a particle called a positron, e^+ (a positively charged electron), and a **neutrino**, ν (a subatomic particle originally thought to have zero mass and to travel at the velocity of light). In the second reaction, the heavy hydrogen nucleus absorbs another proton and, with the emission of a gamma ray, γ, becomes a lightweight helium nucleus. Finally, two light helium nuclei combine to form a common helium nucleus and two hydrogen nuclei. Because the last reaction needs two ^3He nuclei, the first and second reactions must occur twice (Figure 12-10). The net result of this chain reaction is the transformation of four hydrogen nuclei into one helium nucleus plus energy.

The energy appears in the form of gamma rays, positrons, the energy of motion of the nuclei, and neutrinos. The gamma rays are photons that are absorbed by the surrounding gas before they can travel more than a fraction of a millimeter. This heats the gas and helps maintain the pressure. The positrons produced in the first reaction combine with free electrons, and both particles vanish, converting their mass into gamma rays. Thus, the positrons also help keep the center of the star hot. In addition, when fusion produces new nuclei, they fly apart at high velocity. This energy of motion helps raise the temperature of the gas. The neutrinos, however, resemble photons except that they almost never interact with other particles. The average neutrino could pass unhindered through a lead wall 1 ly thick. Thus, the neutrinos do not help heat the gas but race out of the star at the speed of light, carrying away roughly 2 percent of the energy produced.

Theoretical studies tell us that the sun has been fusing hydrogen into helium for about 5 billion years and that it has enough hydrogen fuel to last another 5 billion years. Other stars generate their energy in similar ways. But before we extend our study to other stars, we should examine an experiment that attempts to probe the sun's central fires.

Solar Neutrinos: Evidence and Theory

One of the great conflicts in modern science arises because observations of the number of neutrinos produced inside the sun fails to match the predictions of theory. Science becomes exciting, not when evidence fits theories, but rather when it fails to fit. That warns us that there is more to learn. In this case, scientists are trying to figure out where the misunderstanding lies—in our models of the sun or in our understanding of neutrinos.

The sun produces floods of neutrinos, but because these particles almost never interact with normal matter, they race out of the sun and into space. Even at night, solar neutrinos rush through the earth as if it weren't there, through our beds, through us, and onward into space. Every second 10^{12} neutrinos pass through our bodies. Obviously we are lucky to be transparent to neutrinos, but this means that they are very hard to detect. Models of the sun combined with theories of nuclear fusion predict how many neutrinos should pass through the earth, but how do you count ghost particles?

In the late 1960s, chemist Raymond Davis, Jr., built the first solar neutrino detector. He filled a 100,000-gallon tank with the cleaning fluid perchloroethylene (C_2Cl_4). Hidden nearly a mile deep in a South Dakota gold mine to protect it from cosmic rays, the tank catches an occasional solar neutrino when it converts a chlorine atom into an argon atom (Figure 12-11). Theory predicts the tank should catch one neutrino a day, but for many years, the tank has only caught one neutrino every three days.

Other detectors have been built to test these results. The Kamiokande detector in Japan and its larger successor Super Kamiokande used water to catch neutrinos, and the Sage experiment in Russia and Gallex in Italy use the metallic element gallium. They catch only one-third to one-half the predicted number of neutrinos. So far, every experiment that can detect solar neutrinos counts fewer than theory predicts.

Can there be something wrong with the theories of stellar structure? The luminosity of the sun is well known, so we know how much energy it must produce. Nuclear physics predicts how fast hydrogen fusion must run in the sun's core to make that energy, and that predicts more neutrinos than are observed. To fit the neutrino evidence, the center of the sun should be a million degrees cooler than expected, but then the sun could not produce enough energy to maintain its present brightness. If its fusion reactions suddenly died down, it would take thousands of years to fade, but this idea strikes most astronomers as contrived. Furthermore, helioseismology (Chapter 8) tells us that the sun's center is as hot as is expected, 15.6 million degrees. There seems to be nothing wrong with theories of stellar structure.

Theorists have suggested that the sun contains particles called weakly interacting massive particles (**WIMP**s), which spread heat through a larger region of the core and allow the sun to make its energy at a lower temperature. That would create fewer neutrinos. But helioseismology tells us that the sun's center is not cooler than expected. Also, no WIMP has ever been detected in the laboratory, so most scientists think the WIMP theory is contrived.

A more promising suggestion is that neutrinos oscillate. In fact, nature allows three kinds of neutrinos (called flavors), but the sun produces only one flavor. All of the detectors can detect (or should we say "taste") only one flavor. Some nuclear particle theory holds that neutrinos may oscillate among these three flavors as they travel through dense matter. Thus, the solar neutrinos may oscillate as they rush out through the sun and emerge scrambled among three flavors. That would explain why we detect only a third of what theory predicts.

The conflict between evidence and theory may be telling us that we don't understand neutrinos as well as we thought. The standard theory of nuclear particles predicts that neutrinos have no mass and travel at the speed of light, as do photons. But if they oscillate, then neutrinos must have at least some mass, and they can't travel as fast as light. This theory provides a way to test neutrino oscillation. Laboratory measurements of the mass of

FIGURE 12-11

The solar neutrino experiment consists of 100,000 gallons of cleaning fluid held in a tank nearly a mile underground. Solar neutrinos trapped in the cleaning fluid convert chlorine atoms into argon atoms that can be counted by their radioactivity. (Brookhaven National Laboratory)

ghostly neutrinos are very difficult, but early results hint that neutrinos have some small mass. Also, observations of neutrinos arriving from a distant star that exploded show that they traveled at close to the speed of light. Thus, if the neutrino has a mass it must be very small.

Even a small mass would allow neutrinos to oscillate, and that would explain the conflict over solar neutrinos. But there is more excitement in this conflict. The universe contains so many neutrinos—over 10^8 for every normal particle—even a tiny mass for the neutrinos would make the universe so massive it would affect its structure. Thus, our theory of the entire universe (Chapter 15) may hinge on the mass of the neutrino.

Conflicts between evidence and theory signal excitement in science and force us to reexamine our most basic scientific beliefs (Window on Science 12-1). In this case, the deficit in the solar neutrinos may be telling us that there is more for us to know about the smallest atomic particles and the universe as a whole.

Hydrogen Fusion in Stars

We have studied hydrogen fusion in the sun in detail because it is typical of energy production in stars. A protostar contracts until its central temperature is hot

Scientific Faith

Scientists like to claim that every scientific belief is based on evidence, that every theory has been tested, and that the moment a theory fails a test it is discarded or revised. The truth is much more complicated than that, and the solar neutrino problem is a good illustration. If the detection of solar neutrinos contradicts the theory of stellar structure, why hasn't the theory been abandoned?

While scientists do indeed have tremendous respect for evidence, they also have faith in theories that have been tested successfully many times. If a theory has been tested and confirmed over and over, they may even begin to call it a natural law, and that means that scientists have great faith in its truth. That is, they have confidence that the theory or law is a good description of how nature works.

Nevertheless, it is not unusual for an experiment or an observation to contradict well-established theories. In many cases, the experiments and observations are simple mistakes or have not been interpreted correctly.

Scientists resist abandoning a well-tested theory even when an observation continues to contradict it. If confidence in the theory is stronger than the evidence, scientists begin testing the evidence. Can it be right? Do we understand it correctly? Of course, if the evidence cannot be impeached and it continues to contradict the theory, scientists must eventually abandon or modify the theory, no matter how many times it has previously been tested and confirmed. This ultimate reliance on evidence is the distinguishing characteristic of science.

It is human nature to hang on to the principles you have come to trust, and that confidence in well-tested scientific principles helps scientists avoid rushing to faulty judgments. For example, claims for perpetual motion machines occasionally crop up in the news, but the world's scientists don't instantly abandon the laws of energy and motion pending an analysis of the latest claim. Of course, if such a claim did prove true, the entire structure of scientific knowledge would come crashing down.

But because the known laws of energy and motion have been well tested and no perpetual motion machine has ever been successful, scientists know which way to bet. Like the keel on a ship, confidence in well-tested theories and laws keeps the scientific boat from rocking before every little breeze.

Some forms of faith must be absolute and unshakable—religious faith, for example. But scientific faith may be better described as scientific confidence, because it must be open to change. If, ultimately, a single experiment or observation conclusively contradicts our most cherished law of nature, we as scientists must abandon that law and find a new way to understand nature. We can see this scientific confidence at work in many controversies, from the origin of the human race to the meaning of IQ measurements, but one of the best examples is the solar neutrino controversy. While we struggle to understand the origin of solar neutrinos, we continue to have confidence that we do indeed understand how stars work. ∎

---

enough to ignite hydrogen fusion, which halts the contraction and converts the protostar into a stable star. Thus, hydrogen fusion is the first energy source for a newly born star.

Hydrogen fusion is also the longest-lasting source of energy in a star. Although some stars can fuse other fuels at later stages, all stars fuse hydrogen for about 90 percent of their total lifetime. Once a contracting protostar ignites hydrogen, it becomes a stable star and remains stable for a very long time.

About 90 percent of all stars (excluding white dwarfs) are now fusing hydrogen. The sun is a hydrogen-fueled star, and we can see evidence that a great many other stars are fusing hydrogen when we look at the H–R diagram. All main-sequence stars generate their energy from hydrogen. The numbers of main-sequence stars that we found in the survey of the neighborhood around the sun (Figures 9-15 and 9-16) show how common hydrogen-powered stars are.

Most hydrogen-fusing stars generate their energy in the same way the sun does—by the proton–proton chain. However, about 10 percent of the sun's energy comes from a different set of reactions. This process, the CNO cycle, is common in some main-sequence stars, but it is merely a different way of fusing hydrogen.

Much like the proton–proton chain, the **CNO (carbon–nitrogen–oxygen) cycle** is a series of nuclear reactions that produces energy through hydrogen fusion. However, the CNO cycle uses carbon as a catalyst; that

is, the carbon nucleus makes the reaction possible but is not altered in the end. Although it seems quite different from the proton–proton chain, the CNO cycle combines four hydrogen nuclei to make one helium nucleus plus energy, which we can verify by counting protons in Figure 12-12).

Because the carbon nucleus has a charge six times that of hydrogen, the electrostatic force of repulsion between a carbon nucleus and a proton is high, and much higher temperatures are necessary to force the proton into the carbon nucleus. Thus, the CNO cycle is dominant only in stars more massive than about 1.1 $M_\odot$. These stars have central temperatures hotter than 16,000,000 K. Stars less massive than this, such as the sun, are cooler at their centers and are dominated by the proton–proton chain.

CRITICAL INQUIRY

Why does nuclear fusion require that the gas be very hot?

Nuclear fusion occurs when the nuclei of atoms fuse together to form a new nucleus. Inside a star, the gas is ionized, so the electrons have been stripped off the atoms, and the nuclei are bare and have a positive charge. For hydrogen fusion, the nuclei are single protons. These atomic nuclei repel each other because of

$$^{12}C + {}^1H \rightarrow {}^{13}N +$$
$$^{13}N \rightarrow {}^{13}C + e^+ +$$
$$^{13}C + {}^1H \rightarrow {}^{14}N +$$
$$^{14}N + {}^1H \rightarrow {}^{15}O +$$
$$^{15}O \rightarrow {}^{15}N + e^+ +$$
$$^{15}N + {}^1H \rightarrow {}^4He + {}^{12}C$$

~~~~ Gamma ray
----- Neutrino
○ Positron

FIGURE 12-12

The CNO cycle uses ^{12}C as a catalyst to combine four hydrogen atoms (^1H) to make one helium atom (^4He) plus energy. The carbon atom reappears at the end of the process, ready to start the cycle over.

FIGURE 12-13

Conduction, radiation, and convection are the three modes of energy transport within a star.

their positive charges, so they must collide with each other violently in order to overcome that repulsion and get close enough together to fuse. If the atoms in a gas are moving rapidly, we say the gas has a high temperature, so nuclear fusion requires that the gas have a very high temperature. If the gas is cooler than about 10 million K, hydrogen can't fuse, because the protons don't collide violently enough to overcome the repulsion of their positive charges.

Hydrogen gas won't fuse at room temperature because it is not hot enough. But fusion requires that the gas be dense as well as hot. Why must the gas be dense in order for it to fuse? ■

Because the law of gravity and the rules of nuclear fusion determine how stars work, we can understand what the inside of a star is like. That is, we can describe the internal structure of a star.

12-3 STELLAR STRUCTURE

Gravity makes stars contract, and when the density and temperature at the centers of the stars are high enough, nuclear fusion begins making energy. Why do the stars stop contracting? Somehow, the energy generated at the center of the star flows outward to the surface, and that flow of energy stops the contraction. To understand the stable structure of a star, we must first understand how energy flows through the star and how that energy flow affects the entire star.

Energy Transport

The sun and other stars generate nuclear energy in their deep interiors, but that energy must flow outward to their surfaces to replace the energy radiated into space as light and heat. If this outward flow of energy in the sun were suddenly shut off, the sun's surface would gradually cool and dim, and the sun would begin to shrink. Thus, stars can exist only as long as energy can move from their cores to their surfaces. In the material of which stars are made, energy can move by conduction, radiation, or convection.

Conduction is the most familiar form of heat flow. If you hold the bowl of a spoon in a candle flame, the handle of the spoon grows warmer. Heat, in the form of the motion of the particles in the metal, is conducted from particle to particle up the handle, until the particles under your fingers begin to move faster and you sense heat (Figure 12-13). Thus, conduction requires close contact between the particles. Because matter in most stars is gaseous, conduction is an unimportant means of energy flow. Conduction is significant only in peculiar stars that have tremendous internal densities.

The transport of energy by radiation is another familiar experience. Put your hand beside a candle flame, and you can feel the heat. What you actually feel are infrared photons radiated by the flame (Figure 12-13). Because photons are packets of energy, your hand grows warm as it absorbs them. Radiation is the principal means of energy transport in the sun's interior. Photons are absorbed and reemitted in random directions over and over as they work their way outward. The process of absorption and reemission breaks the high-energy photons common near the sun's center into large numbers of low-energy photons in the cooler outer layers. The energy of a single high-energy photon at the center of the sun takes about 1 million years to reach the sun's surface, where it emerges as roughly 2500 photons of visible light.

The flow of energy by radiation depends on how difficult it is for the photons to move through the gas. If the gas is cool and dense, the photons are more likely to be absorbed or scattered, and thus the radiation does not penetrate the gas very well. We call such a gas opaque. In a hotter, lower-density gas, the photons can penetrate more easily; such a gas is less opaque. The **opacity** of a gas—its resistance to the flow of radiation—depends strongly on its temperature.

If the opacity of a gas is high, radiation cannot flow through it easily. Energy moving outward from the center of a star first moves through the hotter gas of the deep interior. Because the gas is so hot, it is transparent, and the energy moves as radiation. But the outer portions of the star are cooler and therefore more opaque. Like water behind a dam, energy builds up, raising the temperature until the gas begins to churn. Hot gas, being less dense, rises, and cool gas, being denser, sinks. This is convection, the third way energy can move in a star.

Convection is a common experience; the wisp of smoke rising above a candle flame travels upward in a small convection current (Figure 12-13). If you hold your hand above the flame, you can feel the rising current of hot gas. In stars, energy may be carried upward by rising currents of hot gas hundreds or thousands of miles in diameter.

Convection is important in stars because it both carries energy and mixes the gas. Convection currents flowing through the layers of star tend to homogenize the gas, giving it a uniform composition throughout the convective zone. As you might expect, this mixing affects the fuel supply of the nuclear reactions, just as the stirring of a campfire makes it burn more efficiently.

The convection in the sun stirs a zone about 200,000 km deep, just below the visible surface (Figure 12-14). We see the tops of these rising currents of hot gas as granulation in the solar photosphere (Figure 8-2). That is, we can see the upper layers of the solar convection zone. Below that, the energy moves as radiation, and the gas does not get mixed. Deep in the interior of the sun, there is no convection. Hydrogen nuclei are fusing to form helium, and nothing removes the helium ashes or brings fresh hydrogen down into the core. Thus, the sun is like a great pot of mashed potatoes burning at the bottom and stirred only slightly at the top. We will see later that most stars have interiors like the sun's.

What Supports the Sun?

From its surface to its interior, the sun is gaseous. On the earth, a puff of gas, vapor from a smokestack perhaps, dissipates rapidly, driven by the motion of the air and by the random motions of the atoms in the gas. However, the sun differs from a puff of gas in a very important characteristic—its mass. The sun is over 300,000 times more massive than the earth, and that mass produces a tremendous gravitational field that draws the sun's gases into a sphere.

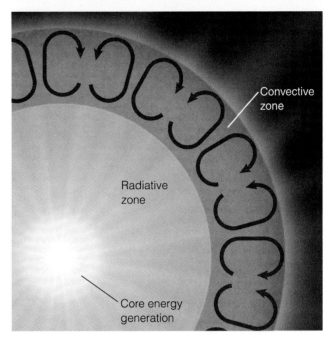

FIGURE 12-14

A cross section of the sun. Near the center, nuclear fusion reactions generate high temperatures. Energy flows outward through the radiation zone as photons. In the cooler, more opaque outer layers, the energy is carried outward by rising convection currents of hot gas.

With such a strong gravitational field, it might seem as if the gases of the sun should compress into a tiny ball, but a second force balances gravity and prevents the sun from shrinking. The gas of which the sun is made is quite hot, and because it is hot, it has a high pressure. The pressure pushes outward; indeed, without the restraint of the sun's gravity, it would blast the sun apart. Thus, the sun is balanced between two forces—gravity trying to squeeze it tighter and gas pressure trying to make it expand.

To discuss the forces inside the sun, we can imagine that the sun's interior is divided into concentric shells like those in an onion (Figure 12-15). We can then discuss the temperature, density, pressure, and so on in each shell. Keep in mind, however, that these helpful shells do not really exist. The sun and stars are not composed of separable layers. The layers are only a convenience in our discussion.

The gravity–pressure balance that supports the sun is a fundamental part of stellar structure known as the law of **hydrostatic equilibrium.** It says that, in a stable star like the sun, the weight of the material pressing downward on a layer must be balanced by the pressure of the gas in that layer. *Hydro* implies that we are discussing a fluid—the gases of the star. *Static* implies that the fluid is stable—neither expanding nor contracting.

The law of hydrostatic equilibrium can prove to us that the interior of the sun must be very hot. Near the sun's surface, there is little weight pressing down on the gas, so the pressure must be low, implying a low temper-

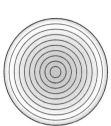

FIGURE 12-15
To discuss the structure of a star, it is helpful to divide its interior into concentric shells much like the layers in an onion. This model is, of course, only an aid to our imaginations. Stars are not really divided into separable layers.

ature. But as we go deeper into the sun, the weight becomes larger, so the pressure, and therefore the temperature, must also increase (Figure 12-16). Near the sun's surface, the pressure is only about 100 times the atmospheric pressure of the earth, and the temperature is only 6000 K. At the center of the sun, the weight pressing down equals a pressure of about 3×10^{14} times the earth's atmospheric pressure. To support that weight, the gas temperature has to be 10–15 million K.

Of course, the interior of the sun is kept hot by the nuclear reactions occurring at the core, and the outward flow of energy keeps each layer in the sun hot enough to support the weight pressing down from above. In a sense, the sun is supported by the flow of energy from its center to its surface. Turn off that energy, and gravity would gradually force the sun to collapse into its center.

Inside Stars

In Chapter 10, we discovered that the stars on the main sequence are ordered according to mass, with the upper main-sequence stars being most massive and the lower main-sequence stars being least massive. Combining this with what we know about hydrogen burning tells us that there are two kinds of main-sequence stars: upper main-sequence stars, which fuse hydrogen on the CNO cycle, and lower main-sequence stars, which fuse hydrogen on the proton–proton chain. Viewed from the outside, these stars differ only in size, temperature, and luminosity, but inside they are quite different.

The upper main-sequence stars are more massive and thus must have higher central temperatures to withstand their own gravity. These high central temperatures permit the star to fuse hydrogen on the CNO cycle, and that affects the internal structure of the star.

The CNO cycle is very temperature-sensitive. If the central temperature of the sun rose by 10 percent, energy production by the proton–proton chain would rise by

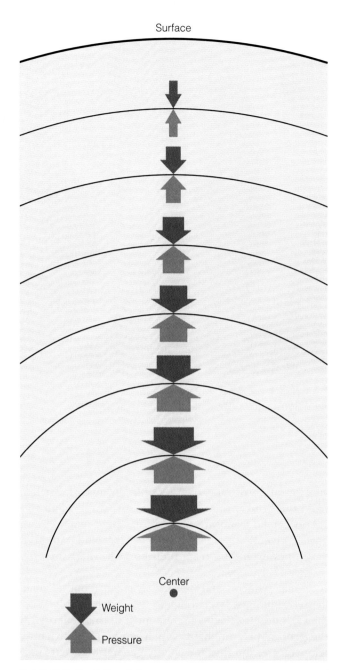

FIGURE 12-16
The law of hydrostatic equilibrium says the pressure in each layer must balance the weight on that layer. As a result, pressure and temperature must increase from the surface of a star to its center.

about 46 percent, but energy production by the CNO cycle would shoot up 350 percent. This means that the more massive stars generate almost all of their energy in a tiny region at their very centers where the temperature is highest. A 10-solar-mass star, for instance, generates 50 percent of its energy in its central 2 percent of mass.

This concentration of energy production at the very center of the star causes a "traffic jam" as the energy tries to flow away from the center. Transport of energy by radiation can't drain away the energy fast enough, and the central core of the star churns in convection as hot

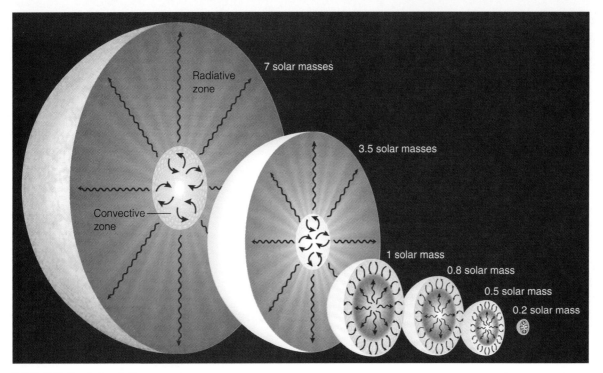

FIGURE 12-17

Inside stars. The more massive stars have small convective cores and radiative envelopes. Stars like the sun have radiative cores and convective envelopes. The lowest-mass stars are convective throughout. (Illustration design by author)

gas rises upward and cooler gas sinks downward. Farther from the center, the traffic jam is less severe, and the energy can flow outward as radiation. Thus, massive stars have convective cores at their centers and radiative envelopes extending from their cores to their surfaces (Figure 12-17).

Main-sequence stars less massive than about 1.1 solar masses cannot get hot enough to fuse much hydrogen on the CNO cycle. They generate most of their energy by the proton–proton chain, which is not as sensitive to temperature, and thus the energy generation occurs in a larger region in the star's core. The sun, for example, generates 50 percent of its energy in a region that contains 11 percent of its mass. Because the energy generation is not concentrated at the very center of the star, no traffic jam develops, and the energy flows outward as radiation. Only near the surface, where the gas is cooler and therefore more opaque, does convection stir the gas. Thus, the less massive stars on the main sequence have radiative cores and convective envelopes.

The lowest-mass stars have a slightly different kind of structure. For stars less than about 0.4 solar mass, the gas is relatively cool compared to the inside of more massive stars, and the radiation cannot flow outward easily. Thus, the entire bulk of these low-mass stars is stirred by convection.

We have answered the questions of how the energy in stars is made and how it gets from the core to the surface. We are now ready to answer the question, What makes stars stable?

The Pressure–Temperature Thermostat

Nuclear reactions in stars manufacture energy and heavy nuclei under the supervision of a built-in thermostat that keeps the reactions from erupting out of control. That thermostat is the relation between pressure and temperature discussed earlier in this chapter.

In a star, the nuclear reactions generate just enough energy to balance the inward pull of gravity. Consider what would happen if the reactions began to produce too much energy. The extra energy would raise the internal temperature of the star, and, because the pressure of the gas depends on its temperature, the pressure would also rise. The increased pressure would make the star expand. Expansion would cool the gas and lower its density, slowing the nuclear reactions. Thus, the star has a built-in regulator that keeps the nuclear reactions from going too fast.

The same thermostat also keeps the reactions from slowing. Suppose the nuclear reactions began to produce too little energy. Then the inner temperature would fall, lowering the pressure and allowing gravity to squeeze the star slightly. Compression would heat the gas and make it denser, thus increasing nuclear energy generation until the star was stable again.

The stability of a star depends on this relation between pressure and temperature. If an increase or decrease in temperature produces a corresponding change in pressure, the thermostat functions correctly, and the star is stable. We will see in the next chapter how the

FIGURE 12-18

The Great Nebula in Orion is a glowing cloud of gas and dust over 8 pc in diameter. The hydrogen glows red because it is ionized by the hottest of the stars of the Trapezium, a cluster of young, hot stars just visible at the very center of the nebula. Compare with Figure 12-21. (Daniel Good)

Stars are elegant in their simplicity. Nothing more than a cloud of gas held together by gravity and warmed by nuclear fusion, a star can achieve stability by balancing its weight through the generation of nuclear energy. But how does the star manage to make exactly the right amount of energy to support its weight? ■

We have traced the birth of stars from the first instability in the interstellar medium to the final equilibrium of nuclear fusion and gravity. We have constructed a complete story of star formation, but we should demand evidence to support our theories. One of the best places to search for evidence of star formation is in the Great Nebula in Orion. It is just part of a great storm of star birth sweeping through Orion.

12-4 THE ORION NEBULA

On a clear winter night, you can see with your naked eye the Great Nebula of Orion as a fuzzy wisp in Orion's sword. With binoculars or a small telescope, it is striking, and through a large telescope it is breathtaking. At the center lie four brilliant blue-white stars known as the Trapezium, and surrounding them are the glowing filaments of a nebula more than 8 pc across (Figure 12-18). Like a great thundercloud illuminated from within, the churning currents of gas and dust suggest immense power. A deeper significance lies hidden, figuratively and literally, behind the visible nebula, for radio and infrared astronomers have discovered a vast dark cloud just beyond the visible nebula—a cloud in which stars are now being created.

Evidence of Young Stars

We should not be surprised to find star formation in Orion. Many of the stars in the constellation are hot, blue, upper main-sequence stars that have short lifetimes. They must have formed recently. The region is also rich in T Tauri stars, which are known to be pre–main-sequence stars. The Hubble Space Telescope has found

thermostat accounts for the mass–luminosity relation. In Chapter 14, we will see what happens to a star when the thermostat breaks down completely and the nuclear fires burn unregulated.

CRITICAL INQUIRY

What would happen if the sun stopped generating energy?

Stars are supported by the outward flow of energy generated by nuclear fusion in their interiors. That energy keeps each layer of the star just hot enough for the gas pressure to support the weight of the layers above. Each layer in the star must be in hydrostatic equilibrium; that is, the inward weight must be balanced by outward pressure. If the sun stopped making energy in its interior, nothing would happen at first, but over many thousands of years the loss of energy from its surface would reduce the sun's ability to withstand its own gravity, and it would begin to contract. We wouldn't notice much for 100,000 years or so, but eventually the sun would lose its battle with gravity.

a

b

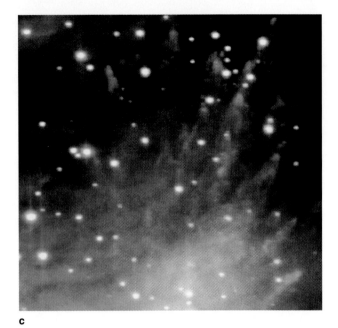

c

FIGURE 12-19

Evidence of star formation: (a) The Hubble Space Telescope photographed this jet of gas flowing away from a star. Such flows are typical of young stars surrounded by disks. (b) Other Space Telescope images reveal gas boiling away from disks of gas and dust around young stars. (c) An infrared image reveals streamers of gas flowing away from a newly luminous star, possibly the bright infrared object IRc2. (a and b, NASA; c, Anglo-Australian Telescope Board)

jets of gas streaming away from some stars (Figure 12-19a), evidence that the stars are young and surrounded by disks. Other Hubble Space Telescope images reveal that many stars in the region have these disks (Figure 12-19b). Furthermore, an infrared image of the region shows streamers of gas rushing away from a star that has recently become luminous (Figure 12-19c). All of these observations are evidence of star formation.

In fact, the glowing nebula itself is evidence of recent star formation. The nebula is an HII region excited by young stars—the Trapezium stars—that reached the main sequence no more than a million years ago. The most massive, a star of about 40 solar masses, must consume its fuel at a tremendous rate in order to support its large mass. Thus, it is hot and luminous—over 30,000 K

and over 300,000 $L_\odot$. The other stars are too cool to affect the surrounding gas very much, but the massive star is so hot it radiates large amounts of ultraviolet radiation (Figure 12-20). These photons have enough energy to ionize hydrogen near the Trapezium. Thus, the glowing gas of the nebula tells us that at least one massive star lies inside, and because such massive stars are very short-lived, it must be very young.

The Orion Nebula looks impressive, but it is nearly a vacuum, containing a mere 600 atoms/cm^3. For comparison, the interstellar medium contains about 1 atom/cm^3 and air at sea level about 10^{19} atoms/cm^3. Infrared observations show that the nebula also contains sparsely scattered dust warmed by the stars to about 70 K.

The Molecular Cloud

The significance of the Orion Nebula was established when observations at infrared and radio wavelengths revealed dense clouds of gas and dust just beyond the Great Nebula. At these wavelengths, hot stars and ionized gas are invisible, but cool, dense gas and dust can be detected. In space, ultraviolet photons break up most molecules, but hidden deep in the dark clouds, where ultraviolet photons cannot penetrate, molecules such as carbon monoxide have formed. These molecules emit

FIGURE 12-20

Photons with wavelengths shorter than 91.2 nm have enough energy to ionize hydrogen. The O6 star is the only star in the Trapezium hot enough to produce appreciable ionization.

radio photons of characteristic wavelengths that allow radio astronomers to map these molecular clouds (Figure 12-21). Dust heated by stars radiates strongly at infrared wavelengths, and thus infrared telescopes can map the location of the dust.

These clouds are significantly different from the Great Nebula. The molecular clouds contain at least 10^6 atoms/cm^3 and large amounts of dust. Because the clouds seem warmest near their centers, astronomers conclude that they are contracting.

The Infrared Clusters

Infrared observations reveal that groups of warm objects lie at the centers of the two densest regions, OMC1 and OMC2. Invisible at optical wavelengths, these objects are evidently stars in pre–main-sequence stages, wrapped deep in cocoons of dust. Two especially interesting objects lie near the center of OMC1.

The Becklin–Neugebauer object (BN object), named after its discoverers, was first thought to be a protostar, but infrared data show that the gas near its center is ionized and flowing outward. Thus, the object is probably a young B0 star that has just reached the main sequence.

The second infrared object in OMC1 was originally called the Kleinmann–Low Nebula, but further studies

FIGURE 12-21

Behind the Orion Nebula: The Hubble Space Telescope visible light image at left shows the central few light years of the Orion Nebula with the bright stars of the Trapezium at lower left. Because the dust clouds are transparent to infrared, the near infrared image at right shows the warm protostars hidden deep inside the giant molecular cloud OMC1. The brightest object is the BN object, and the twisted blue regions are hydrogen gas excited by the hottest of the newly formed stars hidden in the nebula. (NASA/STScI)

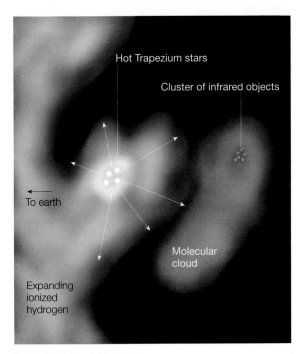

FIGURE 12-22

A side view of the Great Nebula would show the newly formed Trapezium stars heating and driving away the surrounding gas as new stars form in OMC1.

contracting protostars buried within a molecular cloud. As they approached the main sequence, their temperatures increased and their radiation drove away their cocoons. But not until the 40-solar-mass star turned on was there sufficient ultraviolet radiation to ionize the gas. The ionization transformed the cloud from a cold, dusty cocoon into a hot, transparent bubble of gas expanding away from the Trapezium. We can see evidence of this expansion in the Doppler shifts of spectral lines and in the twisted filaments of gas within the nebula.

Once the hot star ionized the gas, star formation in the Trapezium region stopped. In more distant parts of the cloud, the contraction of the gas into protostars continues. Indeed, the ionized gases pushing against the remains of the cloud may have triggered the collapse of more protostars. We now see the front of the cloud torn by the expanding ionized gases around the Trapezium cluster, while deeper within the cloud more protostars are forming (Figure 12-22).

In the next few thousand years, the familiar outline of the Great Nebula will change, and a new nebula may form as the protostars in the molecular clouds reach the main sequence and the most massive become hot enough to ionize the surrounding gas. Thus, the Great Nebula in Orion and its parent molecular cloud show us how the formation of stars continues even today.

show that it is a cluster of infrared objects, at least five of which are protostars. Radio observations show that one of these objects, IRc2, is surrounded by an expanding doughnut of dense gas and is ejecting a powerful bipolar flow along the axis of the doughnut. The object cannot sustain such mass loss for more than about 1000 years, so it must be a very young protostar. (Figure 12-19c).

Telling the Story of Star Formation in Orion

The goal of science is to use theory supported by evidence to tell stories of how nature works. Modern astronomers can now tell the story of the Orion Nebula with some confidence. The history of star formation in Orion is complex, but we can find evidence that the entire region has been a site of star formation for at least three generations of massive stars. The ages of the stars tell the story. The stars at Orion's west shoulder are about 12 million years old, while the stars of Orion's belt are about 8 million years old. The stars of the Trapezium are no older than a few million years. Apparently, star formation began near the west shoulder, and the massive stars that formed there triggered star formation near the belt. That star formation triggered the formation of the stars we see in the great nebula. Like a grass fire, star formation has swept across Orion from northwest to southeast.

We can read the story of the Great Nebula in the Trapezium stars and the molecular cloud that lies behind it. Just a few million years ago, the Trapezium stars were

CRITICAL INQUIRY

What did Orion look like to the ancient Egyptians, to the first humans, and to the dinosaurs?

The Egyptian civilization had its beginning only a few thousand years ago, and that is not very long in terms of the history of Orion. The stars we see in the constellation are hot and young, but they are a few million years old, so the Egyptians saw the same constellation we see. (They called it Osiris.) Even the Orion Nebula hasn't changed very much in a few thousand years, and the ancient Egyptians may have admired it in the dark skies along the Nile.

Our oldest human ancestors lived about 3 million years ago, about the time that the youngest stars in Orion were forming. Thus, they may have looked up and seen some of the stars we see, but other stars have formed since that time. Also, the Great Nebula is excited by the Trapezium stars, which are not more than a few million years old, so our early ancestors probably didn't see the Great Nebula.

The dinosaurs would have seen something quite different. The last of the dinosaurs died about 65 million years ago, long before the birth of the brightest stars in the constellation we see today. The dinosaurs, had they had the brains to appreciate the view, might have seen bright stars along the Milky Way, but when we think back millions of years, we must remember that the stars are moving through space and the sun is orbiting the center of our galaxy. The night sky above the dinosaurs contained totally different star patterns.

Of course, a giant molecular cloud can't continue to spawn new stars forever. What processes limit star formation in a molecular cloud and will eventually end star formation in the Orion region? ∎

The ancient Aztecs of central Mexico told the story of how the stars, known as the Four Hundred Southerners, were scattered across the sky when they lost a cosmic battle with their brother, the great war god Huitzilopochtli. Modern astronomy tells a less colorful story of the birth of stars, but the modern story is supported by evidence and leads us to ask further questions. If stars are born, then how do they die? We will begin that story in the next chapter. ∎

∎ Summary

The largest and densest clouds in the interstellar medium are the giant molecular clouds. Triggered by the compression of a passing shock wave, the densest regions of such a cloud may begin to contract under the pull of their own gravity and eventually fragment to form a cluster or association of stars.

A contracting protostar begins as a very large, cool object, but the acceleration of the atoms as they fall inward causes the temperature of the gas to rise. Rotation in the gas causes some of the material to settle into a rapidly rotating disk around the growing protostar. Although the protostar is hot and luminous, it is hidden inside the dusty remains of the cloud from which it formed and thus remains invisible until the cloud is driven away by the increasing luminosity of the protostar.

Protostars can be detected at infrared wavelengths because the longer infrared-wavelength photons can escape from the dusty clouds. Bipolar flows are ejected along the axis of rotation of the protostar and its disk. Where these jets strike the surrounding gas, they can excite small nebulae called Herbig–Haro objects. T Tauri stars appear to be very young stars just shedding their cocoons of gas and dust. Very young star clusters contain T Tauri stars, and some are born in very large, loosely bound groups called T associations.

As a protostar's center grows hot enough to fuse hydrogen into helium, the star settles onto the main sequence to begin its long, stable, hydrogen-fusion life. The sun is typical of these stars. It generates energy in its interior through the proton–proton chain, and that energy flows outward from the core as radiation and, in the outermost layers, as convection. All lower-mass stars such as the sun generate energy by the proton–proton chain, but more massive stars have hotter interiors and use the CNO cycle.

Because of the temperature sensitivity of the CNO cycle, most of the energy generated by stars more massive than about 1.1 solar masses is produced very near the star's center. The resulting traffic jam as this energy flows outward causes the core to be convective, whereas the outer part of the star transports energy by radiation. In stars less massive than about 1.1 solar masses, the energy is generated on the proton–proton chain in a larger volume of the core. Thus, there is no traffic jam, and these stars have radiative cores and convective envelopes.

In all stable stars, the nuclear reactions are regulated by the pressure–temperature thermostat. Thus, stars cannot generate less energy or more energy than that needed to support them against their own gravity.

The Orion Nebula is an example of a star-forming region. The bright stars we now see are ionizing part of the gas and forcing it to expand, and this gas is pushing into cooler regions of the cloud. Radio and infrared observations show that the compressed regions of the gas cloud are giving birth to a new generation of stars.

∎ New Terms

free-fall contraction	weak force
protostar	strong force
cocoon	proton–proton chain
bipolar flow	neutrino
birth line	WIMPs
T Tauri star	CNO (carbon–nitrogen–oxygen) cycle
T association	
O association	opacity
Herbig–Haro object	hydrostatic equilibrium

∎ Questions

1. What factors resist the contraction of a cloud of interstellar matter?

2. Explain four different ways a giant molecular cloud can be triggered to contract.

3. What evidence do we have that (a) star formation is a continuing process? (b) protostars really exist? (c) the Orion region is actively forming stars?

4. How does a contracting protostar convert gravitational energy into thermal energy?

5. How does the geometry of bipolar flows and Herbig–Haro objects support our hypothesis that protostars are surrounded by rotating disks?

6. Why can't hydrogen fusion occur in the outer layers of a star?

7. Why are solar neutrino experiments usually buried deep underground?

8. How does energy get from the core of a star, where it is generated, to the surface, where it is radiated into space?

9. Describe the principle of hydrostatic equilibrium as it relates to the internal structure of a star.

10. How does the pressure–temperature thermostat control the nuclear reactions inside stars?

∎ Discussion Questions

1. Ancient astronomers, philosophers, and poets assumed that the stars were eternal and unchanging. Is there any observation they could have made or any line of reasoning that could have led them to conclude that stars don't live forever?

2. How does hydrostatic equilibrium relate to hot-air ballooning?

Problems

1. If a giant molecular cloud has a mass of 10^{35} kg and it converts 1 percent of its mass into stars during a single encounter with a shock wave, how many stars can it make? (HINT: Assume the average mass of the stars is 1 solar mass.)

2. If a contracting protostar is five times the radius of the sun and has a temperature of only 2000 K, how luminous will it be? (HINT: See Chapter 9.)

3. The gas in a bipolar flow can travel as fast as 100 km/sec. If the length of the jet is 1 ly, how long does it take for a blob of gas to travel from the protostar to the end of the jet?

4. If a T Tauri star is the same temperature as the sun but is ten times more luminous, what is its radius? (HINT: See Chapter 9.)

5. Circle all of the ^{1}H and ^{4}He nuclei in Figure 12-10 and explain how the proton–proton chain can be summarized by $4 \, ^1$H $\rightarrow \, ^4$He + energy.

6. How much energy is produced when the sun converts 1 kg of mass into energy?

7. How much energy is produced when the sun converts 1 kg of hydrogen into helium? (HINT: How does this problem differ from Problem 6?)

8. A 1-megaton nuclear weapon produces about 4×10^{15} J of energy. How much mass must vanish when a 5-megaton weapon explodes?

9. If the Orion Nebula is 8 pc in diameter and has a density of about 600 hydrogen atoms/cm^3, what is its total mass? (HINT: The volume of a sphere is $\frac{4}{3} \pi R^3$.)

10. The hottest star in the Orion Nebula has a surface temperature of 30,000 K. At what wavelength does it radiate the most energy? (HINT: See Chapter 7.)

Critical Inquiries for the Web

1. In 1997, the Hubble Space Telescope was outfitted with an infrared sensitive instrument called NICMOS. Search the Internet for information about recent observations of star-forming regions with this instrument. Choose a particular object and summarize how the NICMOS observations support or enhance our understanding of the process of star formation.

2. If neutrinos are so elusive, how do astronomers go about detecting them? Use an Internet search engine to browse for information on solar neutrino detectors that are currently in operation, under construction, or proposed. Determine similarities and differences between these detectors in terms of method of detection and energy range of detectable neutrinos.

Go to the Wadsworth Astronomy Resource Center (www.wadsworth.com/astronomy) for critical thinking exercises, articles, and additional readings from InfoTrac College Edition, Wadsworth's online student library.

Chapter Thirteen

STELLAR EVOLUTION

GUIDEPOST

This chapter is the heart of any discussion of astronomy. Previous chapters showed how astronomers make observations with telescopes and how they analyze their observations to find the luminosity, diameter, and mass of stars. All of that aims at understanding what stars are.

This is the middle of three chapters that tell the story of stars. The preceding chapter told us how stars form, and the next chapter tells us how stars die. This chapter is the heart of the story—how stars live.

As always, we accept nothing at face value. We expect evidence to confirm theories. We expect carefully constructed models to help us understand the structure inside stars. In short, we exercise our critical faculties and analyze the story of stellar evolution rather than merely accepting it.

After this chapter, we will know how stars work, and we will be ready to study the rest of the universe, from galaxies that contain billions of stars to the planets that form around individual stars. ■

We should be unwise to trust scientific inference very far when it becomes divorced from opportunity for observational test.

Sir Arthur Eddington

The Internal Constitution of the Stars

The stars are going out. Although stars enjoy long, stable lives on the main sequence, they suffer the first pangs of stellar age as they gradually consume their hydrogen fuel. Once hydrogen is exhausted at their centers, they swell rapidly into giant stars 10 to 1000 times the size of the sun. For a short time, a few million to a billion years, they are majestic beacons visible across thousands of parsecs.

Although we say a star swells rapidly and spends a short time as a giant, these changes happen slowly compared with a human life. How can we be sure stars really do age? Stellar evolution and the formation of giant stars might be just an astronomer's daydream.

In this chapter, we see the full interplay of theory and evidence. We use the basic laws of physics, combined with our knowledge of the nature of stars, to create a theory to describe how main-sequence stars maintain their equilibrium and how such stars change when they exhaust their nuclear fuels. Theory alone, however, is never enough. At each step, we compare these theories with the evidence. We will discover that clusters of stars can make the slow evolution of stars visible even during our short lifetimes, and we will find that some evolving stars can become unstable and pulsate like beating hearts. Those pulsating stars tell us more about how stars evolve.

From beginning to end, this chapter tells the story of the evolution of stars from main sequence to giant. We will trace the passage of stars through this phase of their existence and see how they generate their energy and why they swell so large. We begin with the main-sequence stars.

13-1 MAIN-SEQUENCE STARS

If Shakespeare were alive today, he would probably have something sarcastic to say about 20th-century astronomers. "How do these stargazers pretend to know the hearts of the eternal stars?" he might ask. In fact, one of the greatest triumphs of 20th-century astronomy is the discovery that the stars are not eternal and that mere humans can indeed probe their centers. We can know the conditions at the center of a star because stars are fundamentally very simple objects. We begin by considering how we can use stellar models to understand the interior of main-sequence stars; later we will use these same models to understand what happens to stars when they grow old.

Stellar Models

The structure of a star is described by four simple laws of physics, two of which we have already discussed. In Chapter 12, we met the law of hydrostatic equilibrium, which says that the weight pressing down on each layer in a star must be balanced by the pressure pushing outward in that layer. That is, at every level in a star, the weight of the overlying layers must be supported by the pressure at that level. Of course, that means that there

TABLE 13-1

The Four Laws of Stellar Structure

1. Hydrostatic equilibrium	The weight on each layer is balanced by the pressure in that layer.
2. Energy transport	Energy moves from hot to cool by radiation, convection, or conduction.
3. Continuity of mass	Total mass equals the sum of the shell masses. No gaps are allowed.
4. Continuity of energy	Total luminosity equals the sum of the energies generated in each shell.

can be no empty layers; an empty layer could exert no pressure and would be unable to support the weight of the layers above. We also encountered the law of energy transport, which says that energy flows from hot to cool regions by radiation, convection, or conduction.

To these two laws we add two very simple laws that are laws of continuity. The **continuity of mass law** says that the total mass of the star must equal the sum of the masses of its shells. This is really just a version of the law of conservation of mass. It is equivalent to saying that the weight of a cake must be equal to the sum of the weights of its layers.

The **continuity of energy law** says that the amount of energy flowing out the top of a layer in the star must be equal to the amount of energy coming in the bottom plus whatever energy is generated within the layer, and that means that the energy leaving the surface of the star, its luminosity, must equal the sum of the energies generated in all of the layers inside the star. This is really just the law of conservation of energy—energy can neither disappear nor appear from nowhere.

These four laws of stellar structure are much more precise than the four statements summarized in Table 13-1. In fact, the four laws can be expressed as four equations (Figure 13-1). What these equations look like is not important here, but applying them to build mathematical models of stars can tell 20th-century astronomers how stars work, how they are born, and how they die. Mathematical models of stars can carry our imaginations into the very hearts of the stars, a place where our bodies and our instruments can never go (Window on Science 13-1).

A stellar model is a mathematical imitation of a real star. A computer calculates a table that shows the temperature, density, pressure, and other conditions at various levels from the surface to the center of the star. Though we can't see inside a real star, we can study the inside of a stellar model and draw conclusions about the nature of real stars. Also, we can experiment on a model star—change its composition, for instance—whereas we can't experiment on a real star.

Mathematical Models

One of the most powerful tools in science is the mathematical model, a group of equations carefully designed to mimic the behavior of the object scientists want to study. Astronomers build mathematical models of stars using only four equations, but other systems are much more complicated and may require many more equations.

Many sciences use mathematical models. Medical scientists have built mathematical models of the nerves that control the human heart, and physicists have built mathematical models of the inside of an atomic nucleus. Economists have built mathematical models of certain aspects of economic systems, such as the municipal bond market, and earth scientists have built mathematical models of the earth's atmosphere. In each case, the mathematical model allows the scientists to study something

that is too difficult to study in the real world. The model can reveal regions we cannot observe, speed up a slow process, slow down a fast process, and allow scientists to perform experiments that would be impossible in reality. Astronomers, for example, can change the abundance of certain chemical elements in a model star to see how its structure depends on its composition.

Many of these mathematical models require very large and fast computers, in some cases supercomputers. A modern computer takes only seconds to compute a stellar model, but models such as the motions of the planets in our solar system millions of years into the future require the largest and fastest computers in the world. Most astronomers are apt computer programmers, and some are experts. A few astronomers have built their own specialized computers to

calculate specific kinds of models. Thus, the mathematical model is one of the most important tools in astronomy.

As is true for any scientific model, a mathematical model is only as reliable as the assumptions that go into its creation. In Window on Science 2-1, we saw that the celestial sphere was an adequate model of the sky for some purposes but breaks down if we extend it too far. So, too, with mathematical models. We can think of a mathematical model as a numerical expression of one or more theories. While such models can be very helpful, they are always based on theory and so must be compared with the real world at every opportunity. Models must always be tested against experiment and observation. Otherwise, they might lead us astray. ■

$R/R_\odot$	T (10^6 K)	Density (g/cm^3)	$M/M_\odot$	$L/L_\odot$
1.00	0.006	0.00	1.00	1.00
0.90	0.60	0.009	0.999	1.00
0.80	1.27	0.035	0.996	1.00
0.70	1.80	0.12	0.990	1.00
0.60	2.42	0.40	0.97	1.00
0.50	3.42	1.3	0.92	1.00
0.40	4.74	4.1	0.82	1.00
0.30	6.65	13.	0.63	0.99
0.20	9.35	36.	0.34	0.91
0.10	12.65	85.	0.073	0.40
0.00	14.62	134.	0.000	0.00

$$\frac{dM}{dr} = 4\pi r^2 \rho$$

$$\frac{dL}{dr} = 4\pi r^2 \rho e$$

$$\frac{dP}{dr} = -\frac{GM}{r^2}\rho$$

$$\frac{dT}{dr} = \frac{-3}{16\pi ac}\frac{\bar{\kappa}\rho}{T^3}\frac{L}{r^2}$$

Surface

Convective zone

Radiative zone

Center

FIGURE 13-1

A stellar model is a table of numbers showing conditions inside a star. Such a table can be calculated by a computer using the four laws of stellar structure, expressed as equations. The complex relations between the equations are symbolized here by a few arrows. This model describes the sun. Each line in the table tells us the radius, temperature, and density of a shell in the sun, plus the mass the shell encloses and the energy flowing through it. (Illustration design by author)

Notice that stellar models are quantitative; that is, properties have specific numerical values. Earlier in this book, we examined models that were qualitative—our model of the sunspot cycle, for instance. Both kinds of models are useful, but a quantitative model is much more useful, reflecting the power of mathematics as a way of thinking about nature.

The model shown in Figure 13-1 represents our sun. As we scan the table from top to bottom, we descend from the surface of the sun to its center. The temperature increases rapidly as we move downward, reaching a maximum of about 15,000,000 K at the center. At this temperature, the gas is not very opaque, and the energy can flow outward as radiation. In the cooler outer layers, the gas is more opaque, and the outward-flowing energy forces these layers to churn in convection.

Stellar models also let us look into the star's past and future. In fact, we can use models as time machines to follow the evolution of stars over billions of years. To look into a star's future, for instance, we use a stellar model to determine how fast the star consumes its fuel in each shell. As the fuel is consumed, the chemical composition of the gas changes, and the amount of energy generated declines. By calculating the rate of these changes, we can predict what the star will look like at any point in the future.

The Ends of the Main Sequence

Stellar models give us a way to study the extreme ends of the main sequence, the most massive and least massive stars. We will discover, however, that theoretical models alone can lead us astray if we do not check them against observational evidence at every opportunity.

Stellar models predict that the most massive star cannot exceed a limit of about 80–120 solar masses. Such stars generate tremendous energy and have very high temperatures. These high temperatures produce floods of radiation flowing outward, and this radiation should blast gas outward from the star's surface in powerful stellar winds. One model, for example, predicts that a 60-solar-mass star would lose mass so rapidly it would waste away to 28 solar masses in less than a million years. Thus, the models predict that very massive stars will be unstable. In addition, models of contracting protostars suggest that very massive protostars tend to break in half and form binary stars instead of a single massive star.

When we compare our theory with observations, we discover that it is difficult to find truly massive stars. Most of the O and B stars we see in the sky have masses of 10–25 solar masses. Our survey of stars at the end of Chapter 9 revealed that the stars at the upper end of the main sequence are very rare, so we must search to great distances to find such stars. The most massive stars known have masses thought to be about 55 solar masses.

F I G U R E   1 3 - 2
The Hubble Space Telescope searched in the star cluster NGC 6397 for the very low mass stars that theory predicts should exist in large numbers (yellow diamonds at left). But the images revealed no such stars (right). This evidence suggests that nature makes very few stars with masses below about 0.2 solar mass. The reason for this limitation is not yet understood. (F. Paresce, STScI, ESA, and NASA)

Spectra of very massive stars reveal blue-shifted emission lines. As we saw in Chapter 7 when we discussed Kirchhoff's laws, emission lines must originate in excited, low-density gas, and the blue shifts must be caused by the Doppler effect in gas flowing toward us. Thus, these spectroscopic observations confirm our expectation that very massive stars have strong stellar winds.

The upper end of the main sequence is difficult to study because massive stars are very rare and lose mass rapidly. Both models and observations suggest that stars more massive than 60 to 100 solar masses are unstable.

The lower end of the main sequence is also difficult to study. Nature makes lots of stars with masses less than 0.5 solar masses, but these stars are so faint they are difficult to find. If a red dwarf replaced the sun, it would shine only a few times brighter than the full moon, and, just a few light-years distant, it would be extremely faint. Thus, finding and studying these stars is difficult.

Theory predicts that stars less massive than 0.08 solar masses (about 80 Jupiter masses) cannot get hot enough to ignite hydrogen fusion. Models predict that these **brown dwarfs** would be warm from their contraction but very faint because of their small size. Their low temperatures would give them a color even more red than the red dwarfs. Thus the term "brown dwarf."

Searching for brown dwarfs is especially difficult, but astronomers have good reason to expect them to exist. Nature makes lots of red dwarfs, so we might expect that many stars would form with slightly smaller masses and "miss" the lower end of the hydrogen-fusion main sequence. These objects are distinguished by their low temperature, low luminosity, and low mass. Of course, we can measure the mass of such a star only if it is a member of a binary system. Another characteristic is the spectroscopic presence of methane and lithium. Methane molecules would be broken up in the higher temperatures of a red dwarf, and lithium is present in young stars, but is quickly fused into heavier elements if a star is massive enough to ignite nuclear fusion.

Astronomers use these clues to search for brown dwarfs. The star Gliese 229B is believed to be a brown dwarf because its mass is only about 20 times the mass of Jupiter, and its spectrum shows methane. Binary star PPL15 in the Pleiades contains two good candidates; they are cool and faint, and they contain lithium. The star Kelu-1 in the constellation Hydra, is faint and very cool, and its spectrum shows both lithium and methane. In this way, astronomers search for brown dwarfs by looking for stars with the properties that theory predicts.

Brown dwarfs have such low masses that we might wonder how to tell them from planets. Evidence and theory tell us that planets form in dusty disks around stars as solid bits stick together to build larger objects. Stars, however, form when a cloud of gas becomes unstable and contracts. Thus brown dwarfs are indeed star-like, and understanding them would help us understand star formation as well as the lower end of the main sequence.

Another reason astronomers are interested in brown

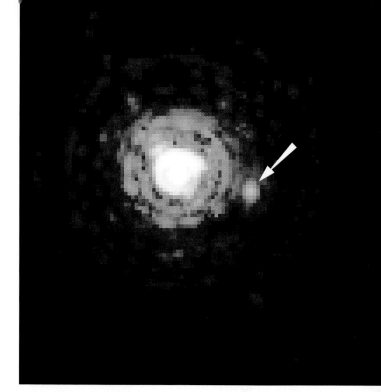

FIGURE 13-3

Gliese 623b (arrow) is estimated to have a mass of 0.1 solar mass and is consequently one of the lowest-mass stars known. Seen in the glare of its brighter companion in this Hubble Space Telescope image, the star may be a true star or a brown dwarf. Further studies of its orbital motion will be necessary to better define its mass. (C. Barbieri, University of Padua; NASA/ESA)

dwarfs is that they may represent a large amount of matter. Some theories predict that our galaxy and the universe itself contain more matter than we can see (Chapters 16 and 19). This matter makes its presence known through its gravity, but produces no light. If brown dwarfs are common, they may represent a significant amount of matter.

The search for brown dwarfs is continuing, but some results suggest that such low mass objects are not very common (Figure 13-2). Searches for these stars have revealed that there are few stars less massive than 0.2 solar masses and even fewer brown dwarfs. Nevertheless, astronomers are finding a few brown dwarfs (Figure 13-3), and they may find more as we come to understand their properties.

Observation and theory can help us understand why the main sequence has ends, but a simple question remains. How does a contracting protostar of a certain mass arrive at the right point on the main sequence. The answer to that question will tell us how stars work.

The Mass–Luminosity Relation Explained

In Chapter 10, we used binary stars to determine the mass of stars, and we discovered that the masses were ordered along the main sequence. The lowest-mass stars were the faintest, and the highest-mass stars were the most luminous. Further, we discovered a direct relationship between the mass of a star and its luminosity—the mass–luminosity relation. This is one of the most fundamental observations in astronomy, and now we can understand it by using stellar models.

The keys to the mass–luminosity relation are the law of hydrostatic equilibrium, which says that pressure must balance weight, and the pressure–temperature thermostat, which regulates energy production. We have seen that a star's internal pressure stays high because the generation of thermonuclear energy keeps its interior hot. Because more massive stars have more weight pressing down on the inner layers, their interiors must have high pressures and thus must be hot. For example, the temperature at the center of a 15-solar-mass star is about 34,000,000 K, more than twice the central temperature of the sun.

Because massive stars have hotter cores, their nuclear reactions burn more fiercely. That is, their pressure–temperature thermostat is set higher. The nuclear fuel at the center of a 15-solar-mass star fuses over 3000 times more rapidly than the fuel at the center of the sun. The rapid reactions in massive stars make them more luminous than the lower-mass stars. Thus, the mass–luminosity relation results from the requirement that a star support its weight by generating nuclear energy.

Stellar models show clearly that massive stars must be more luminous than low-mass stars. They must generate tremendous amounts of energy to balance their great weight, and so they must be very luminous. Stellar models not only help us understand the mass–luminosity relation, they also help us understand how stars age as they consume their nuclear fuels.

The Life of a Main-Sequence Star

A main-sequence star supports its weight by fusing hydrogen, but its supply of hydrogen is limited. Thus, it is inevitable that the star changes as it exhausts its hydrogen fuel. We might expect such a star to just cool off and fade away, but stellar models reveal that the star evolves in surprising ways. Even while it is on the main sequence, a star converts hydrogen to helium, thus changing its chemical composition in subtle ways that force it to change in size and temperature. Even now, the sun is slowly evolving in ways that will eventually destroy the earth.

Hydrogen fusion combines four nuclei into one. Thus, as a main-sequence star consumes its hydrogen, the total number of nuclei in its interior decreases. Each newly made helium nucleus can exert the same pressure as a hydrogen nucleus, but because the gas has fewer nuclei, its total pressure is less. This unbalances the gravity–pressure stability, and gravity squeezes the core of the star more tightly. As the core contracts, its temperature increases, and the nuclear reactions burn faster, releasing more energy and making the star more lumi-

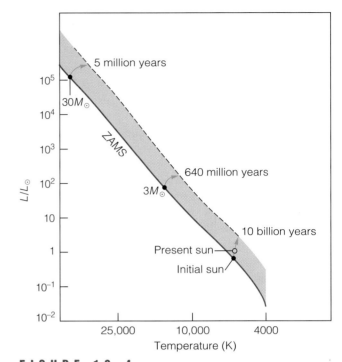

FIGURE 13-4

The main sequence is not a line but a band (blue). Stars begin their main-sequence lives on the lower edge, which is called the zero-age main sequence (ZAMS). As hydrogen fusion changes their composition, the stars slowly move across the band.

nous. This additional energy flowing outward through the envelope forces the outer layers to expand and cool, so the star becomes slightly larger, brighter, and cooler.

As a result of these gradual changes in main-sequence stars, the main sequence is not a sharp line across the H–R diagram but rather a band (blue in Figure 13-4). A star begins its stable life fusing hydrogen and falls on the lower edge of this band, the **zero-age main sequence (ZAMS)**. As it combines hydrogen nuclei to make helium nuclei, the star slowly changes. In the H–R diagram, the point that represents the star's luminosity and surface temperature moves upward and to the right, eventually reaching the upper edge of the main sequence (the dashed line in Figure 13-4) just as the star exhausts nearly all of the hydrogen in its center. Thus, we find main-sequence stars plotted throughout this band at various stages of their main-sequence lives.

These gradual changes in the sun will spell trouble for the earth. When the sun began its main-sequence life about 5 billion years ago, it was only about 60 percent of its present luminosity. This, by the way, makes it difficult to explain how the earth has remained at roughly its present temperature for at least 3 billion years. Some experts suggest that the earth's atmosphere has gradually changed and thus compensated for the increasing luminosity of the sun.

By the time the sun leaves the main sequence in 5 billion years, it will have twice its present luminosity.

TABLE 13-2

Main-Sequence Stars

Spectral Type	Mass (sun = 1)	Luminosity (sun = 1)	Approximate Years on Main Sequence
O5	40	405,000	1×10^6
B0	15	13,000	11×10^6
A0	3.5	80	440×10^6
F0	1.7	6.4	3×10^9
G0	1.1	1.4	8×10^9
K0	0.8	0.46	17×10^9
M0	0.5	0.08	56×10^9

This will raise the average temperature on the earth by at least 19°C (34°F). As this happens over the next few billion years, the polar caps will melt, the oceans will evaporate, and much of the atmosphere will vanish into space. Clearly, the future of the earth as the home of life is limited by the evolution of the sun.

Once a star leaves the main sequence, it evolves rapidly and dies. The average star spends 90 percent of its life burning hydrogen on the main sequence. This explains why 90 percent of all stars are main-sequence stars. We are most likely to see a star during that long, stable period when it is on the main sequence.

The number of years a star spends on the main sequence depends on its mass (Table 13-2). Massive stars use fuel rapidly and live short lives, but low-mass stars conserve their fuel and shine for billions of years. For example, a 25-solar-mass star will exhaust its hydrogen and die in only about 7 million years. This means that life is very unlikely to develop on planets orbiting massive stars. These stars do not live long enough for life to get started and evolve into complex creatures. We will discuss this problem in detail in Chapter 27.

Very low mass stars, the red dwarfs, use their fuel so slowly they should survive for 200–300 billion years. Because the universe seems to be only 10–20 billion years old, red dwarfs must still be in their infancy. None of them should have exhausted their hydrogen fuel yet. However, this conventional description of the life of a red dwarf is based on a model that ignores mass loss. More detailed models suggest that these low-mass stars have strong stellar winds that carry mass away from the star. If that is true, then the lifetimes of these lowest-mass stars may be much shorter than that implied by their supply of nuclear fuel.

Nature makes more low-mass stars than massive stars, but this fact is not sufficient to explain the vast numbers of low-mass stars that fill the sky. An additional factor is the stellar lifetimes. Because low-mass stars live

long lives, there are more of them in the sky than massive stars. Look at Figure 9-15 and notice how much more common the lower main-sequence stars are than the massive O and B stars. The main-sequence K and M stars are so faint they are difficult to locate, but they are very common. The O and B stars are luminous and easy to locate, but because of their fleeting lives, there are never more than a few on the main sequence at any one time.

The Life Expectancies of Stars

To understand how nature makes stars and how stars evolve, we must be able to estimate how long they can survive. In general, massive stars live short lives and lower-mass stars live long lives, but we can make more accurate estimates by using simple stellar models. In fact, we can calculate the approximate life expectancy of a star from its mass.

Because main-sequence stars consume their fuel at a constant rate, we can estimate the amount of time a star spends on the main sequence—its life expectancy T—by dividing the amount of fuel by the rate of fuel consumption. This is a common calculation. If we drive a truck that carries 20 gallons of fuel and uses 5 gallons of fuel per hour, we know the truck can run for 4 hours.

The amount of fuel a star has is proportional to its mass, and the rate at which it burns its fuel is proportional to its luminosity; thus, we could make a first estimate of the star's life expectancy by dividing mass by luminosity. A 2-solar-mass star that is 6 times more luminous than the sun should live about 2/6, or 1/3, as long as the sun. We can, however, make the calculation even easier if we remember that the mass–luminosity relation tells us that the luminosity of a star equals $M^{3.5}$. The life expectancy then is:

$$T = \frac{M}{L} = \frac{M}{M^{3.5}}, \text{ or } T = \frac{1}{M^{2.5}}$$

This means that we can estimate the life expectancy of a star by dividing 1 by the star's mass raised to the 2.5 power. If we express the mass in solar masses, the life expectancy will be in solar lifetimes.

For example, how long can a 4-solar-mass star live?

$$T = \frac{1}{4^{2.5}} = \frac{1}{4 \cdot 4 \cdot \sqrt{4}} = \frac{1}{32} \text{ solar lifetimes}$$

Detailed studies of models of the sun show that the sun, presently 5 billion years old, can last another 5 billion years. Thus, a solar lifetime is approximately 10 billion years, and a 4-solar-mass star will last about (10 billion)/32 or about 310 million years.

Our estimation of solar life expectancies is very approximate. For example, our model ignores mass loss, which may affect the life expectancies of very luminous and very faint stars. Nevertheless, it serves to illustrate a very important point. Stars that are only slightly more massive than the sun have dramatically shorter lifetimes on the main sequence.

Why are main-sequence O stars so rare and M stars so common?

We observed in Section 9-4 that space contains only 0.03 main-sequence O star per million cubic parsecs, but there are about 10,000 main-sequence M stars per million cubic parsecs (Figure 9-15). Now we can explain these observations using stellar models to estimate lifetimes. O stars live very short lifetimes. We can use our simple formula to estimate that a 20-solar-mass O star has a life expectancy of about 6 million years. But a 0.2-solar-mass M star has a life expectancy of about 60 billion years. That is, M stars live about 10,000 times longer than O stars. Assuming that nature makes equal numbers of different kinds of stars, at any given moment in the history of the universe we should expect to see about 10,000 times more M stars than O stars. How does that compare with the observed numbers of O stars and M stars given in the first sentence of this paragraph? How does this tell us which kinds of stars nature makes most often? ■

When a star finally exhausts its hydrogen fuel, it can no longer resist the pull of its own gravity. The contraction that began when it was a protostar resumes and begins the process that leads to its death. It can delay its end by fusing other fuels, but, as we will discover in Chapter 14, nothing can steal gravity's final victory.

13-2 POST–MAIN-SEQUENCE EVOLUTION

In earlier chapters, we asked three questions: Why are giant stars so large? Why are they so uncommon? And why do they have such low densities? Now we are ready to answer those questions by discussing the evolution of stars after they leave the main sequence.

Expansion into a Giant

To understand how stars evolve, we must recognize that they are not well mixed; that is, their interiors are not stirred. The centers of stars like the sun are radiative, meaning the energy moves as radiation and not as circulating currents of heated gas. The gas does not move in such stars, and thus they are not mixed at all. More massive stars have convective cores that mix the central regions (Figure 12-17), but these regions are not very large, and thus, for the most part, these stars, too, are not mixed. (The lowest-mass stars are an exception that we will examine in the next chapter.)

In this respect, stars are like a campfire that is not stirred; the ashes accumulate at the center, and the fuel

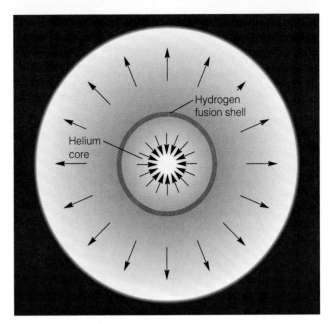

FIGURE 13-5

When a star runs out of hydrogen at its center, it ignites a hydrogen-fusing shell. The helium core contracts and heats, while the envelope expands and cools. (For a scale drawing, see Figure 13-8.)

in the outer parts never gets used. Nuclear fusion consumes hydrogen nuclei and produces helium nuclei, the "ashes," at the star's center. Nothing mixes the interior of the star, so the helium nuclei remain where they are in the center of the star, and the hydrogen in the outer parts of the star is not mixed down to the center where it can be fused.

The helium ashes that accumulate in the star's core cannot fuse into heavier elements because the temperature is too low. As a result, the helium accumulates, the hydrogen is used up, and the core becomes an inert ball of helium. As this happens, the energy production in the core falls, and the weight of the outer layers forces the core to contract.

Although the contracting helium core cannot generate nuclear energy, it does grow hotter because it converts gravitational energy into thermal energy (see Chapter 12). The rising temperature heats the unprocessed hydrogen just outside the core, hydrogen that was never before hot enough to fuse. When the temperature of the surrounding hydrogen becomes high enough, it ignites in a shell of fusing hydrogen. Like a grass fire burning outward from an exhausted campfire, the hydrogen-fusing shell burns outward, leaving helium ash behind and increasing the mass of the helium core.

At this stage in its evolution, the star overproduces energy; that is, it produces more energy than it needs to balance its own gravity. The helium core, having no nuclear energy sources, must contract, and that contraction converts gravitational energy into thermal energy—the contraction heats the helium core. Some of that heat leaks outward through the star. At the same time, the

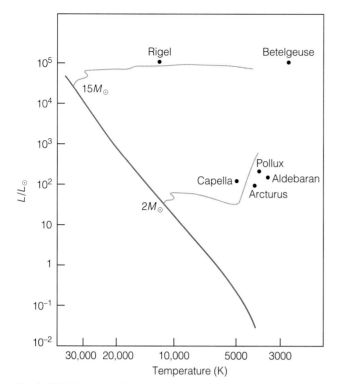

FIGURE 13-6

Many of the bright, red stars in the night sky are giants. Not long ago they were main-sequence stars, but they expanded when their cores ran out of hydrogen. Supergiants like Rigel and Betelgeuse are more massive stars from the upper main sequence.

hydrogen-fusing shell produces energy as the contracting core brings fresh hydrogen closer to the center of the star and heats it to high temperature. The result is a flood of energy flowing outward through the outer layers of the star, forcing them to expand and swelling the star into a giant (Figure 13-5).

The expansion of the envelope dramatically changes the star's location in the H–R diagram. As the outer layers of the star expand outward, the surface cools, and the point that represents the star in the H–R diagram moves quickly to the right into the red giant region. As the radius of the star continues to increase, the enlarging surface area makes the star more luminous, moving its point upward in the H–R diagram (Figure 13-6). Aldebaran (α Tauri), the glowing red eye of Taurus the bull, is such a red giant, with a diameter 25 times that of the sun but with a surface temperature only half that of the sun. Rigel (β Orionis) is a supergiant some 50 times larger in diameter than our sun. Rigel looks very blue to our eyes. With a surface temperature of 12,000 K, Rigel is not among the hottest stars, but because of the limited wavelength sensitivity of the human eye, all such stars look quite blue (see Figure 7-2).

This model explains the large diameters, low densities, and mixed masses of giants and supergiants. They are mid- and upper main-sequence stars that expanded to large size and low density as hydrogen-shell fusion

began (Figure 13-6). Most become giants, but the most massive become supergiants. Stars of different masses funnel through the giant region, which explains why giant stars have assorted masses (see Figure 10-22).

Degenerate Matter

Although the hydrogen-fusing shell can force the envelope of the star to expand, it cannot stop the contraction of the helium core. Because the core has no energy source, gravity squeezes it tighter, and it becomes very small. If we represent the helium core of a 5-solar-mass star with a quarter, the outer envelope of the star would be about the size of a baseball diamond. Yet the core would contain about 12 percent of the star's mass compressed to very high density. When gas is compressed to such extreme densities, it begins to behave in astonishing ways that can alter the evolution of the star. Thus, to follow the story of stellar evolution, we must consider the behavior of gas at extremely high densities.

Normally, the pressure in a gas depends on its temperature. The hotter the gas is, the faster its particles move and the more pressure it exerts. But the gas inside a star is ionized, so there are two kinds of particles: atomic nuclei and free electrons. If the gas is compressed to very high densities, for example, in the core of a giant star, the difference between these two kinds of particles is important.

If the density is very high, the particles of the gas are forced close together, and two laws of quantum mechanics become important. First, quantum mechanics says that the moving electrons confined in the star's core can have only certain amounts of energy, just as the electron in an atom can occupy only certain energy levels (see Chapter 7). We can think of these permitted energies as the rungs of a ladder. An electron can occupy any rung, but not the spaces between.

The second quantum-mechanical law (called the Pauli Exclusion Principle) says that two identical electrons cannot occupy the same energy level. Because electrons spin in one direction or the other, two electrons can occupy an energy level if they spin in opposite directions. That level is then completely filled, and a third electron cannot enter because, whichever way it spins, it will be identical to one or the other of the two electrons already in the level. Thus, no more than two electrons can occupy the same energy level.

A low-density gas has few electrons per cubic centimeter, so there are plenty of energy levels (Figure 13-7). If a gas becomes very dense, however, nearly all of the lower energy levels may be occupied, and the gas is termed degenerate. In such matter, a moving electron cannot slow down, because slowing down would decrease its energy and there are no open energy levels for it to drop down to. It can speed up only if it can absorb enough energy to leap to the top of the energy ladder, where there are empty energy levels.

Low-density gas (nondegenerate)   High-density gas (degenerate)

FIGURE 13-7
Electron energy levels are arranged like rungs on a ladder. In a low-density gas, many levels are open, but in a degenerate gas, all lower energy levels are filled.

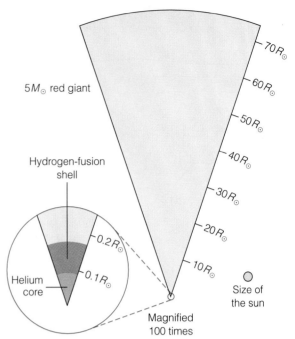

FIGURE 13-8
A plug cut from a 5-$M_\odot$ red giant would be almost all low-density envelope. The magnified image shows the hydrogen-fusing shell and the helium core. The core contains roughly 12 percent of the star's mass. (Illustration design by author)

Such a **degenerate gas** has two properties that can affect the core of a star. First, the degenerate gas resists compression. To compress the gas, we must push against the moving electrons, and changing their motion means changing their energy. That requires tremendous effort, because we must boost them to the top of the energy ladder. Thus, degenerate matter, though still a gas, takes on the consistency of hardened steel.

Second, the pressure of degenerate gas does not depend on temperature but rather on the speed of the electrons, which cannot be changed without tremendous effort. The temperature, however, depends on the motion of all the particles in the gas, both electrons and nuclei. If we add heat to the gas, most of that energy goes to speed up the motions of the nuclei, and only a few electrons can absorb enough energy to reach the empty energy levels at the top of the energy ladder. Thus, changing the temperature of the gas has almost no effect on the pressure.

These two properties of degenerate matter become important when stars end their main-sequence lives and approach their final collapse (Window on Science 13-2). Eventually, many stars collapse into white dwarfs, and we will discover that these tiny stars are made of degenerate matter. But long before that, the cores of many giant stars become so dense that they are degenerate, a situation that can produce a cosmic bomb.

Helium Fusion

Hydrogen fusion in main-sequence stars leaves behind helium ash, which cannot yet begin fusing. Helium nuclei have a positive charge twice that of a proton, so they must collide at higher velocity to overcome the repulsion between nuclei. The temperature required for

hydrogen fusion is too cool to fuse helium. As the star becomes a giant star, fusing hydrogen in a shell, the inner core of helium contracts and grows hotter. It may even become degenerate, but when it finally reaches a temperature of 100,000,000 K, it begins to fuse helium nuclei to make carbon.

We can summarize the helium-fusing process in two steps:

$$^4\text{He} + {}^4\text{He} \rightarrow {}^8\text{Be} + \gamma$$
$$^8\text{Be} + {}^4\text{He} \rightarrow {}^{12}\text{C} + \gamma$$

This process is complicated by the fact that beryllium-8 is very unstable and may break up into two helium nuclei before it can absorb another helium nucleus. Three helium nuclei can also form carbon directly, but such a triple collision is unlikely. Because a helium nucleus is called an alpha particle, astronomers often refer to helium fusion as the **triple alpha process.**

Some stars begin helium fusion gradually, but stars in a certain mass range begin helium fusion with an explosion called the **helium flash.** This explosion is caused by the density of the helium, which can reach 1,000,000 g/cm³. On the earth a teaspoon of this material would weigh as much as a large truck. At these densities, the gas is degenerate, and its pressure no longer depends on temperature. Thus, the pressure–temperature thermostat that controls the nuclear fusion reactions no longer works.

The Very Small and the Very Large

One of the most interesting lessons of science is that the behavior of very small things often determines the structure and behavior of very large things. In degenerate matter, the quantum-mechanical behavior of electrons helps determine the evolution of giant stars. Such links between the very small and the very large are common in astronomy, and we will see in later chapters how the nature of certain subatomic particles may determine the fate of the entire universe.

This is a second reason why astronomers study atoms. Recall that the first reason is that atoms interact with light and astronomers must understand that interaction in order to analyze the light. The second reason is the link between the very small and the very large. It would be impossible to understand the evolution of the largest stars if we failed to understand how degenerate electrons behave. Similarly, astronomers studying galaxies must understand how microscopic dust in space reddens starlight, and astronomers studying planets must know how molecules form crystals.

As you study any science, look for the way very small things affect very large things. Sociologists and psychologists know that the mass behavior of very large groups of people can depend on the behavior of a few key individuals. Biologists study the visible consequences of atomic bonds in molecules we call genes, and meteorologists know that tiny changes in temperature in one part of the world can affect world-wide climate weeks or months later.

Nature is the sum of many tiny parts, and science is the study of nature. Thus, one way to do science is to search out the tiny causes that determine the way nature behaves on the grandest scales. ■

When the helium ignites, it generates energy, which raises the temperature. Because the pressure–temperature thermostat is not working in the degenerate gas, the core does not respond to the higher temperature by expanding. Rather, the higher temperature forces the reactions to go faster, which makes more energy, which raises the temperature, which makes the reactions go faster, and so on. Thus, the ignition of helium fusion in a degenerate gas results in a runaway explosion so violent that for a few minutes the helium core generates more energy than an entire galaxy. At its peak, the core generates 10^{14} times more energy per second than the sun.

Although the helium flash is sudden and powerful, it does not destroy the star. In fact, if you were observing a giant star as it experienced a helium flash, you would see no outward evidence of an eruption. The helium core is quite small (Figure 13-8), and all of the energy of the explosion is absorbed by the distended envelope. Also, the helium flash is a very short lived event. In a matter of minutes, the core of the star becomes so hot it is no longer degenerate, the pressure–temperature thermostat brings the helium fusion under control, and the star proceeds to fuse helium steadily in its core. We will discuss this post–helium-flash evolution later.

Not all stars experience a helium flash. Stars less massive than about 0.4 solar mass can never get hot enough to ignite helium, and stars more massive than about 3 solar masses ignite helium before their cores become degenerate (Figure 13-9). In such stars, pressure depends on temperature, so the pressure–temperature thermostat keeps the helium fusion under control.

If the helium flash occurs only in some stars and is a very short lived event that is not visible from outside the star, why should we worry about it? The answer is that it limits the reliability of the mathematical models astronomers use to study stellar evolution. The medium-mass stars, which experience the helium flash, are those that we must consider most carefully in studying stellar evo-

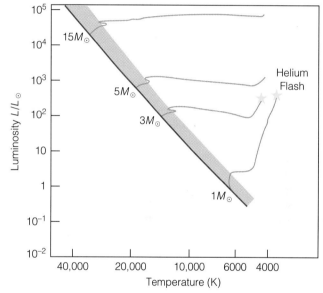

FIGURE 13-9

The helium flash occurs in stars less than about 3 solar masses when helium fusion begins in the degenerate matter in the core. With no pressure–temperature thermostat to control the fusion, it runs out of control until the core expands and the matter is no longer degenerate. More massive stars never develop degenerate cores, so they never experience helium flashes. In those stars, helium fusion begins under the control of the pressure–temperature thermostat. The zero-age main sequence is shown in red, with the band of the main sequence in blue.

lution. Massive stars are not very common; low-mass stars evolve so slowly that we cannot see much evidence of low-mass stellar evolution. Thus, studies of stellar evolution must concentrate on medium-mass stars, which do experience the helium flash. But the helium flash occurs so rapidly and so violently that computer programs cannot follow the changes in the star's internal structure in

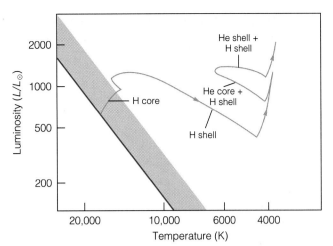

FIGURE 13-10

The evolution of a 5-$M_\odot$ star illustrates stages in energy generation. From the zero-age main sequence (red) across the band of the main sequence (blue), the star makes energy in a hydrogen-fusion core. Hydrogen-shell fusion begins as the star leaves the main sequence and continues after helium fusion ignites in the core. When the core exhausts its helium, the star makes energy with a helium-fusion shell around the core and a hydrogen-fusion shell outside the helium-fusion shell.

TABLE 13-3

Nuclear Reactions in Massive Stars

Nuclear Fuel	Nuclear Products	Minimum Ignition Temperature	Main-Sequence Mass Needed to Ignite Fusion	Duration of Fusion in a 25-$M_\odot$ Star
H	He	4×10^6 K	0.1 $M_\odot$	7×10^6 yr
He	C, O	120×10^6 K	0.4 $M_\odot$	0.5×10^6 yr
C	Ne, Na, Mg, O	600×10^6 K	4 $M_\odot$	600 yr
Ne	O, Mg	1.2×10^9 K	~8 $M_\odot$	1 yr
O	Si, S, P	1.5×10^9 K	~8 $M_\odot$	~0.5 yr
Si	Ni to Fe	2.7×10^9 K	~8 $M_\odot$	~1 day

The Fusion of Elements Heavier Than Helium

The evolution of a star following helium exhaustion is uncertain, but the general plan is clear. The inert carbon–oxygen core contracts and becomes hotter. Stars more massive than about 3 solar masses can reach temperatures of 600,000,000 K and ignite carbon. Subsequent contraction may be able to fuse oxygen, silicon, and other heavy elements (Table 13-3).

At temperatures above 600,000,000 K, carbon fuses rapidly in a complex network of reactions illustrated in Figure 13-11, where each arrow represents a different nuclear reaction. The process is complicated because nuclei can react by adding a proton, a neutron, or a helium nucleus or by combining directly with other nuclei. Unstable nuclei can decay by ejecting an electron, a positron, or a helium nucleus or by splitting into fragments. The complexity of this process makes it difficult to determine exactly how much energy will be generated and how many heavy nuclei will be produced.

Reactions at still higher temperatures can convert magnesium and silicon into yet heavier nuclei, and the star can continue to balance its high internal pressure against its own gravity. These stars become complexly structured, as shells fusing different elements lie one on top of the other like the layers in a spherical cake.

Eventually, all stars face collapse. Less massive stars cannot get hot enough to ignite the heavier nuclei. The more massive stars can ignite heavy nuclei, but these stars consume their fuels at a tremendous rate, so they run through their available fuel quickly. No matter what the star's mass, it will eventually run out of usable fuels, and gravity will win the struggle with pressure. We will explore the ultimate deaths of stars in the next chapter.

The time a star spends as a giant or supergiant is small compared with its life on the main sequence. The sun, for example, will spend about 10 billion years as a main-sequence star but only about a billion years as a giant. The more massive stars pass through the giant stage even more rapidly. Because of the short time a star spends as a giant, we are unlikely to see many such stars.

detail. To follow the evolution of medium-mass stars like the sun past the helium flash, astronomers must make assumptions about the way the helium flash affects stars' internal structure.

Post–helium-flash evolution is generally understood. As the core temperature rises, the gas ceases to be degenerate, and helium fusion proceeds under the control of the pressure–temperature thermostat. Throughout these events, the hydrogen-fusion shell continues to produce energy, but the new helium-fusion energy produced in the core makes the core expand. That expansion absorbs energy previously used to support the outer layers of the star. As a result, the outer layers contract, and the surface of the star grows slightly hotter. In the H–R diagram, the point that represents the star initially moves down toward lower luminosity and to the left toward higher temperature (Figure 13-10). Later, as the star stabilizes fusing helium in its core, its luminosity recovers, but its surface grows hotter.

Helium fusion produces carbon and oxygen as "ash," and these accumulate at the center of the star. They do not fuse because the core is not hot enough. Thus, the helium in the core is gradually converted into carbon–oxygen ashes. As this happens, the core contracts, grows hotter, and ignites a helium-fusion shell. At this stage, the star has two shells producing energy. A hydrogen-fusion shell eats its way outward, leaving behind helium ash; a helium-fusion shell eats its way outward into the helium, leaving behind carbon–oxygen ash. Unable to generate energy in its carbon–oxygen core, the star expands its outer layers and its point in the H–R diagram moves back toward the right, completing a loop (Figure 13-10).

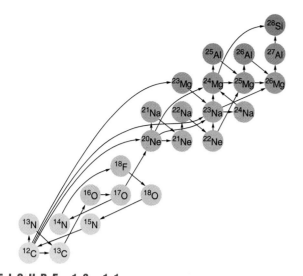

FIGURE 13-11
Carbon fusion involves many possible reactions that build numerous heavy nuclei.

This illustrates an important principle in astronomy: the shorter the time a given evolutionary stage takes, the less likely we are to see stars in that particular stage. Thus, we see a great many main-sequence stars near the sun, but few giants (Figure 9-16).

CRITICAL INQUIRY

What uncertainties limit the accuracy of our story of stellar evolution?

Mathematical models of stars apply four basic laws of physics to describe the inside of a star. Modern astronomers have used such models to study stellar evolution, a process that takes billions of years and thus is unobservable for humans. But two things limit the accuracy of the models. First, some events in the history of a star happen so fast that computers can't predict the results in a reasonable amount of time. In fact, some astronomers estimate that a computer following the evolution of a model star undergoing the helium flash would have to recompute the model in such short time steps that the computer program would take millions of years to predict events that occur inside the star in just seconds. Thus, truly rapid changes such as explosions in stars make the models uncertain.

Second, we have seen how the general properties of a giant star can depend on the subatomic properties of matter. Not all of those subatomic properties are well understood. For example, the solar neutrino problem hints that physicists might not understand exactly how neutrinos and other subatomic particles behave, and if that is true, then stellar models may be slightly different from the real stars they are supposed to represent.

Can you think of any other physical processes that have been omitted from stellar models? What features of the sun's surface suggest motions and forces that are ignored in most stellar models? ■

We cannot see the stars evolving, so we must test our theoretical story of stellar evolution against observation. To paraphrase the quotation at the beginning of this chapter, we shouldn't believe any theory until we compare it with reality—seeing is believing. Thus, in the rest of this chapter, we will search for evidence to support the theory of stellar evolution. At the same time, we will discover some interesting consequences of the fact that the stars are not eternal.

13-3 STAR CLUSTERS

The theory of stellar evolution is so complex and involves so many assumptions that astronomers would have little confidence in it were it not for H–R diagrams of clusters of stars. By observing the properties of star clusters of different ages, we can see clear evidence of the evolution of stars.

To grasp the difficulty of understanding stellar evolution, consider an analogy suggested by William Herschel. Suppose a visitor who had never seen a tree wandered through a forest for an hour looking at the falling seeds, rising sprouts, young saplings, mature trees, and fallen logs. Could such an observer understand the life cycle of the trees? Astronomers face the same problem when they try to understand the life story of the stars.

We do not live long enough to see stars evolve. We see only a momentary glimpse of the universe, a snapshot in which all the stages in the life cycle of the stars are represented. Unscrambling these stages and putting them in order is an almost impossible task. Only by looking at selected groups of stars—star clusters—can we see the pattern.

Observing Star Clusters

Star clusters are often dramatically beautiful, but they are also important sources of evidence against which we can test our theories of star formation. Like members of a very large family, the stars in a cluster share certain characteristics, such as age, composition, and distance, and that makes them ideal subjects in our search for observational confirmation of stellar evolution.

All of the stars in a cluster presumably formed from the same cloud of gas and began contracting at about the same time. Massive stars evolve faster than low-mass stars, so they reach the main sequence sooner and then evolve to become giants sooner than their low-mass siblings. Thus, the mass of a star is very important in determining how it will evolve. If we choose stars at random over the sky, we will choose some stars that formed recently and some that formed billions of years ago, some that are massive and some that have low mass. With such a mixture of stars, we will not be able to decide which characteristics of the stars are caused by different ages and which are caused by different masses. The stars in a star cluster, however, all have the same age, because they all began contracting at the same time. Then any differences we see among the stars in a cluster must be caused by different rates of evolution.

Another important factor in our study of stars is chemical composition. Because all of the stars in a cluster formed from the same cloud of gas, we might expect them to have similar chemical compositions. Although there are exceptions to this rule, it is probably true for most stars in most clusters. This means that the stars in a cluster differ from one another only because of the mass they contain, not because of differences in composition.

Clusters are also convenient laboratories for the study of stellar evolution because all of the stars lie at about the same distance from the earth. Stars on the near side of the cluster are slightly closer to us, of course, but compared with the distance to the cluster, this difference is not significant. With all of the stars at the same distance, we can easily arrange them in order of luminosity. The stars that are brightest in apparent magnitude are also the most luminous. Thus, we can study the relative luminosities of the stars without worrying about the distances to the individual stars (Figure 13-12).

We can also study the surface temperatures of the stars in a cluster. We might record the spectrum of each star and classify the stars according to spectral type, but that would be very time-consuming. Some clusters contain a thousand stars. We could use a photometer and measure the brightness of the stars through a blue and a yellow filter to find their color index (see Chapter 6). This, too, would be time-consuming. The most efficient way to measure temperature would be to record CCD images of the cluster through the proper filters. We could then instruct our computer to measure each star on the two images and compute the color index. Because color index is related to temperature, we could then deduce the surface temperatures of the stars.

This would be enough information to draw an H–R diagram for the cluster. We would plot color, spectral type, or surface temperature on the horizontal axis—all three are related to the star's temperature. If we didn't know the distance to the cluster, we could not convert apparent magnitude into absolute magnitude, but we could still plot an H–R diagram using apparent magnitudes on the vertical axis. All of the stars are at about the same distance and thus have the same distance modulus, so plotting apparent magnitude instead of absolute magnitude would not change the distribution of dots.

If we made such observations of a star cluster, we might get a diagram like that in Figure 13-13. We would discover that most of the stars are on the main sequence but that some have evolved off of the main sequence to

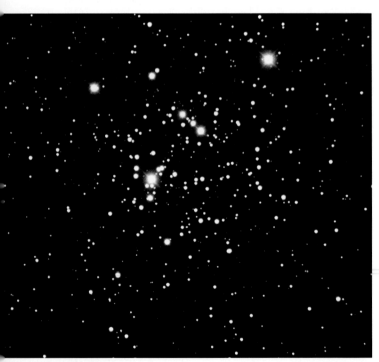

FIGURE 13-12

The Jewel Box is an open star cluster (to be defined later) containing a few hundred stars in a region about 8 pc in diameter. All of the stars in such a cluster are about the same age, have the same chemical composition, and are the same distance from the earth, so we can detect the effects of stellar evolution by observing the magnitude and temperature of the stars. (National Optical Astronomy Observatories)

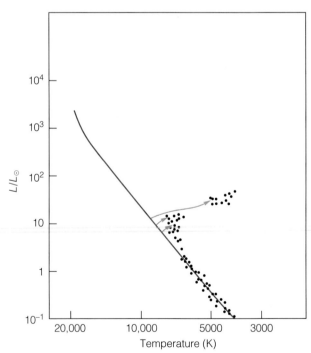

FIGURE 13-13

The H–R diagram of a hypothetical cluster of stars would reveal that the less massive stars are still on the main sequence, that some stars are just beginning to leave the main sequence, and that some have become giants.

FIGURE 13-14
A series of H–R diagrams, like frames in a film, illustrates the evolution of a cluster of stars. Massive stars approach the main sequence faster, live shorter lives, and die sooner than lower-mass stars. Compare with Figure 13-16.

become giants. The massive stars on the upper main sequence would be missing entirely because they live such short lives. Properly interpreted, such a diagram can tell us a great deal about stellar evolution.

Stellar Evolution Confirmed

To interpret H–R diagrams of star clusters, we must imagine how the diagram changes with time. Suppose we follow the evolution of a star cluster by making H–R diagrams like frames in a film (Figure 13-14). Our first frame shows the cluster only 10^6 years after it began forming, and already the most massive stars have reached the main sequence, consumed their fuel, and moved off to become supergiants. However, the medium- to low-mass stars have not yet reached the main sequence.

Because evolution is such a slow process, we cannot make the time step between frames equal, or we would fill more than 1000 pages with nearly identical diagrams. Instead, we increase the time step by a factor of 10 with each frame. Thus, the second frame shows the cluster after 10^7 years and the third after 10^8 years.

By the third frame, all massive stars have died, and stars slightly more massive than the sun are beginning to leave the main sequence. Notice that the lowest-mass stars have finally begun to fuse hydrogen. Only after 10^{10} years does a star of the sun's mass begin to swell into a giant.

These five frames were made from theoretical models of stellar evolution, but they compare quite well with H–R diagrams of real star clusters. For example, NGC 2264 is a very young cluster still embedded in the clouds of gas and dust from which it formed (Figure 13-15). At

FIGURE 13-15
The star cluster NGC 2264, with an age of only a few million years, is one of the youngest star clusters known. Its stars still lie in the nebula of dust and gas from which they formed. The cluster is rich in T Tauri stars, and its lower-mass members are still contracting toward the main sequence. (Anglo-Australian Telescope Board)

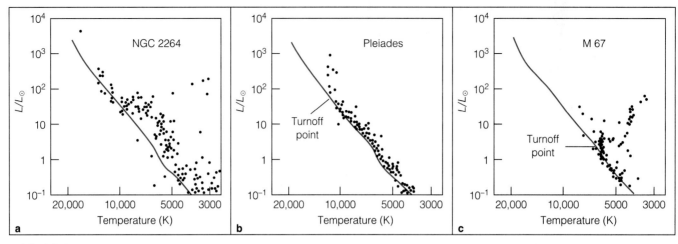

FIGURE 13-16

(a) The star cluster NGC 2264 is only a few million years old, and its lower-mass stars are still contracting toward the main sequence. (b) Most of the stars in the Pleiades star cluster (age about 76 million years) have reached the main sequence, and the most massive stars have begun to evolve toward the giant region of the H–R diagram. (c) M 67 is about 4 billion years old, and all of its more massive stars have died.

a distance of about 800 pc, it contains about 250 stars, including large numbers of T Tauri stars (see Figure 12-6). The H–R diagram of NGC 2264 (Figure 13-16a) contains many medium- and lower-mass stars still contracting toward the main sequence. The age of the cluster is only a few million years.

The Pleiades star cluster is well known, and although the stars are surrounded by a faint reflection nebula (Figure 11-2), we know that this is not the nebula from which the stars formed. The cluster is just passing through this cloud of dust and gas. The age of the Pleiades is about 76 million years, and its H–R diagram (Figure 13-16b) shows that most of its stars have reached the main sequence. The most massive stars are B stars, and they have already begun to leave the main sequence.

Compare these younger star clusters with M 67, a faint cluster of stars in the constellation Cancer the Crab. The few hundred stars in the cluster show no sign of surrounding nebulosity, nor should they. The gas and dust from which these stars formed must have dissipated billions of years ago. The H–R diagram of the cluster (Figure 13-16c) shows that all of the upper main-sequence stars have died and that only stars of about 1 solar mass or less remain. At about 4 billion years of age, this cluster is one of the older clusters known.

Notice that the older a cluster is, the fewer upper main-sequence stars remain. These massive stars evolve first, so, like a candle burning from the top down, the main sequence of a star cluster grows shorter with age.

We can estimate the age of a star cluster by noting the point at which its stars turn off the main sequence and move toward the red giant region. The masses of the stars at this **turnoff point** tell us the age of the cluster, because those stars are on the verge of exhausting their hydrogen-fusing cores. Thus, the life expectancy of the stars at the turnoff point equals the age of the cluster. (See Figure 13-16.)

Two Kinds of Clusters

Even a glance through a small telescope will convince us that there are two kinds of star clusters in the sky (Figure 13-17). The H–R diagrams of these two kinds of clusters confirm that they differ dramatically in a number of ways.

One kind of star cluster is known as an **open cluster** because its stars are uncrowded and the cluster has an open, transparent appearance. Figure 13-17a shows an open cluster. Such a cluster can contain from ten to a few thousand stars in a region about 25 pc in diameter.

The other kind of cluster is known as a **globular cluster*** because it is shaped like a globe (Figure 13-17b). Such clusters can contain 10^5 to 10^6 stars in a region 10–30 pc in diameter. The stars in a globular cluster are crowded a thousand times more densely than the stars near the sun. The average distance between the stars at the center of a globular cluster is only a few light-months. If the earth were located inside a globular cluster, we would see thousands of stars brighter than first magnitude in our sky.

The H–R diagrams of these two kinds of clusters tell us that they are quite different. Open clusters tend to be young to middle-aged. If we combine their H–R diagrams (Figure 13-18), we see a range of turnoff points demonstrating a range of ages. Nevertheless, even the oldest open clusters are only a few billion years old. In contrast, the H–R diagrams of globular clusters reveal very red turnoff points, showing that these clusters are

---

*Glob in "globular cluster" is pronounced like "glob of butter," not like "globe."

a

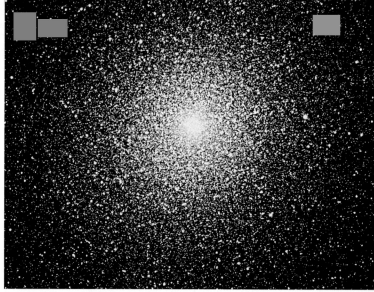

b

FIGURE 13-17

Two kinds of star clusters. (a) The open cluster NGC 3293 contains a few hundred stars. Notice the bright red giant. (b) The globular cluster 47 Tucanae contains about a million stars and is roughly spherical in shape. (Anglo-Australian Telescope Board)

very old (Figure 13-19a). Calculated ages range from 10–15 billion years or more, making them the oldest clusters of stars known. We will discuss the ages of globular clusters again in Chapter 16, because they give us clues to the age of our galaxy.

Another thing we see in the H–R diagrams of globular clusters is that their main-sequence stars are slightly fainter and bluer than the zero-age main sequence (Figure 13-19a). We can understand the reasons for this difference by applying what we have learned in this chapter about stellar structure and evolution. The spectra of globular cluster stars reveal that they are poor in elements heavier than helium, and much of the opacity of the gas in a star is caused by the heavier elements. (Recall from Chapter 12 that the opacity of a gas is its resistance to the flow of radiation.) If the globular cluster stars are composed of gas that is slightly less opaque, then the energy generated inside the globular cluster stars can flow outward more easily. As a result, the stars will be slightly smaller and hotter than the zero-age main sequence. In the H–R diagram, they will lie slightly below and to the left of the main sequence. This example illustrates how the outward flow of radiation interacting with the gas in a star supports the star and determines its structure.

Finally, we notice that globular cluster H–R diagrams have stars located along a **horizontal branch** extending toward the blue side of the diagram. Studies of stellar evolution using mathematical models of stars show that the giant stars in globular clusters contract and heat up dramatically after they begin fusing helium in their cores. In the H–R diagram, these stars loop toward the left (Figure 13-19b). Once they exhaust the helium in

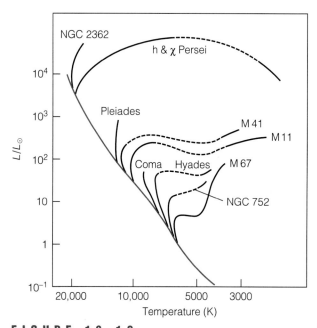

FIGURE 13-18

The combined H–R diagrams of nine open star clusters illustrate that clusters of different ages have different turnoff points.

their cores and begin fusing helium in a shell around the core, the stars expand and cool, and in the H–R diagram they move back to the giant region. Thus, the location of the stars in the H–R diagrams of globular clusters trace the evolutionary tracks followed by these low-mass stars and once again confirm that our theories of stellar evolution are, at least in general, correct.

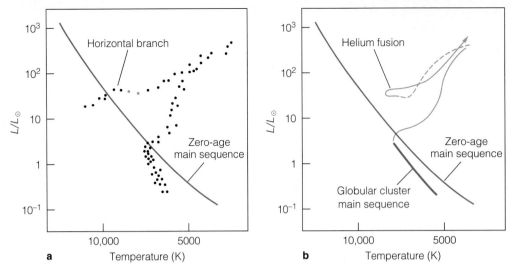

FIGURE 13-19

(a) An observational H–R diagram of a globular cluster. The main sequence has a very faint turnoff point, showing that the cluster is old, and the main-sequence stars are fainter and bluer than the zero-age main sequence. Some stars lie along the horizontal branch extending to the left. (b) A schematic evolutionary track of a low-mass star in a globular cluster shows how the evolution of the star carries it through the region of the H–R diagram occupied by globular cluster stars. During helium core fusion, the star becomes smaller and hotter, and the evolutionary path carries it to the left along the horizontal branch.

CRITICAL INQUIRY

Why do open clusters contain only a small number of giant stars but many main-sequence stars?

When we look at a star cluster, we see a snapshot of the cluster that freezes it in time. In the H–R diagram of an open cluster, we see a main sequence containing many stars, but typically we see many fewer giant stars. Those giant stars were once main-sequence stars in the cluster, but they exhausted the hydrogen fuel in their cores and expanded to become giants. A star like the sun can fuse hydrogen on the main sequence for many billions of years but can remain a giant for only a billion years or less. More massive stars evolve even faster. Thus, at any given moment, we see most of the stars in a star cluster on the main sequence and only a few, caught in the snapshot, as giant stars.

Star clusters are important in astronomy because their H–R diagrams reveal the progress of stellar evolution. How do they reveal the age of the star cluster? ■

Star clusters give us dramatic confirmation that stars evolve and also reveal some of the details of stellar structure and evolution. Certain pulsating giant stars give us further information about how stars work.

13-4 VARIABLE STARS

For millennia, poets and astronomers believed that the stars were constant and unchanging. Now we know that the stars are slowly evolving as they consume their fuels. Some stars, however, change rapidly, growing brighter and then fainter as they pulsate. Exploring how and why these stars pulsate gives us further confirmation that the stars really are evolving and a further illustration of the importance of the energy flowing outward through the stars.

A **variable star** is any star that changes its brightness in a periodic way. Variable stars include many different kinds of stars. Some variable stars are eclipsing binaries (Chapter 10), but many others are stars that expand and contract, heat and cool, and grow more luminous and less luminous because of internal characteristics. These stars are often called **intrinsic variables** to distinguish them from eclipsing binaries. Of these intrinsic variables, one particular kind is centrally important to modern astronomy even though the first star of that class was discovered centuries ago.

Cepheid and RR Lyrae Variable Stars

In 1784, the deaf and mute English astronomer John Goodricke, then 19 years old, discovered that the star δ Cephei is variable, changing its brightness by almost a magnitude over a period of 5.37 days (Figure 13-20a).

a

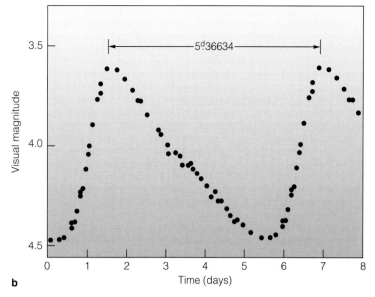

b

FIGURE 13-20

(a) Cepheid variable stars are named after the star δ in Cepheus. It varies from about magnitude 3.6 at brightest to about magnitude 4.5 at faintest. The magnitudes of nearby stars are given here for comparison. The constellation is shown as it appears on October evenings. (b) A graph of visual magnitude versus time is called a light curve. Cepheids typically have a rapid rise to maximum brightness and a slow decline.

Goodricke died at the age of 21, but his discovery ensures his place in the history of astronomy. Since 1784, hundreds of stars like δ Cephei have been found, and they are now known as **Cepheid variable stars.**

Cepheid variables are supergiant or bright giant stars of spectral type F or G. The fastest complete a cycle—bright to faint to bright—in about 2 days, whereas the slowest take as long as 60 days. A plot of a variable's magnitude versus time, a light curve, shows a typical rapid rise to maximum brightness and a slower decline (Figure 13-20b). Some Cepheids change their brightness by only 0.5 magnitude (about 10 percent), while others change by more than a magnitude.

Supergiants and bright giants are rare, and not all are Cepheids. Nevertheless, some familiar stars are Cepheids. Polaris, the North Star, is a Cepheid with a period of 3.9696 days and a very low amplitude. In fact, in the early 1990s, astronomers noticed that Polaris is pulsating less and less, and it seems possible that the pulsations may stop completely by the year 2000. Why the pulsations are declining is not known.

The **RR Lyrae variable stars** are related to the Cepheids. The RR Lyrae stars pulsate with periods of less than a day and are fainter than the faintest Cepheids.

Studies of Cepheids and RR Lyrae stars reveal two related facts of great importance. First, there is a **period–luminosity relation** that connects the period of pulsation of a Cepheid to its luminosity. The longer-period Cepheids are about 40,000 times more luminous than the sun, while the shortest-period Cepheids are only a few hundred times more luminous than the sun. Second, there are two types of Cepheids. Type I Cepheids, including δ Cephei itself, have chemical compositions like that of the sun, but Type II Cepheids and the RR Lyrae stars are poor in elements heavier than helium. These two facts can be summarized in a graph of absolute magnitude versus period (Figure 13-21). The two types of Cepheids are clearly separated, and the fainter RR Lyrae stars lie at the lower left.

Cepheid and RR Lyrae variable stars are critically important in astronomy because they can be used to find the distance to star clusters and galaxies. We will discuss this use of variable stars in later chapters; here we are interested in them as peculiar giant stars that are unstable.

Pulsating Stars

Why should a star pulsate? Why should there be a period–luminosity relation? Why are there two types of Cepheids? The answers to these questions help us understand the insides of stars. To find the answers, we start with the evolution of stars in the H–R diagram.

After a star leaves the main sequence, it fuses hydrogen in a shell, and the point that represents it in the H–R diagram moves to the right as the star becomes a cool giant. When it ignites helium in its core, the point moves back to the left. Soon, however, the helium core is exhausted, and helium fusion continues only in a shell,

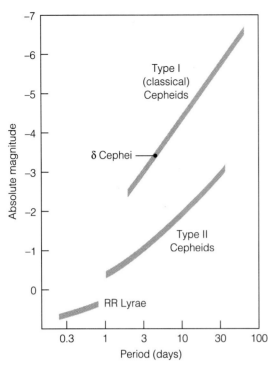

FIGURE 13-21

The Cepheid variable stars obey a period–luminosity relation: the longer-period stars are more luminous. The Cepheids are divided into two types. Type I Cepheids have chemical compositions like that of the sun, but Type II Cepheids are poor in elements heavier than helium. The RR Lyrae stars are variable stars with lower luminosity and shorter periods.

FIGURE 13-22

Both the Cepheid and RR Lyrae variable stars lie in the instability strip (dashed lines) because their variability is caused by the same mechanism. The long-period variables are much cooler red giants and have periods as long as 2 years.

forcing the star to expand again, and the point in the H–R diagram moves back to the right. Because giant stars evolve through these stages, their size and temperature change in complicated ways. Certain combinations of size and temperature make the star unstable, and it pulsates as an intrinsic variable star. In the H–R diagram, the region where size and temperature lead to pulsation is called the **instability strip** (Figure 13-22).

Why do some stars pulsate? The answer has to do with the flow of energy out of the star's interior. We have seen that a star supports its own weight by generating energy and that the flow of that energy outward prevents the star from contracting. The star is balanced between the outward flow of energy and the inward force of gravity. We could make a stable star pulsate if we squeezed it and then released it. The star would rebound outward, and it might oscillate for a few cycles, but the oscillation would eventually run down because of friction. The energy of the moving gas would be converted to heat and radiated away, and the star would return to stability. To make a star pulsate continuously, some process must drive the oscillation, just as a spring in a wind-up clock is needed to keep it ticking.

Studies using stellar models show that variable stars in the instability strip pulsate like beating hearts because of an energy-absorbing layer in their outer envelopes. This layer is the region where helium is partially ionized.

Above this layer the temperature is too low to ionize helium, and below the layer it is hot enough to ionize all of the helium. Like a spring, the helium ionization zone can absorb energy when it is compressed and release it when the zone expands. That is enough to keep the star pulsating.

We can follow this pulsation in our imaginations and see how the helium ionization zone makes a star pulsate. As the outer layers of a pulsating star expand, the ionization zone expands, and the ionized helium becomes less ionized and releases stored energy, which forces the expansion to go even faster. The surface of the star overshoots its equilibrium position—it expands too far—and eventually falls back. As the surface layers contract, the ionization zone is compressed and becomes more ionized, which absorbs energy. Robbed of some of their energy, the layers of the star can't support the weight, and they compress further. This allows the contraction to go even faster, and the in-falling layers overshoot the equilibrium point until the pressure inside the star slows them to a stop and makes them expand again. Thus, the surface of these stars is not in hydrostatic equilibrium but rather pulsates in and out.

Cepheids change their radius by 5–10 percent as they pulsate, and this motion can be observed as Doppler shifts in their spectra. When they expand, the surface layers approach, and we see a blue shift. When they contract, the surface layers recede, and we see a red shift. Although a 10 percent change in radius seems like a large change, it affects only the outer layers of the star. The center of the star is much too dense to be affected.

Now we can understand why there is an instability strip. Only stars in the instability strip have their helium ionization zone at the right depth to act as a spring and drive the pulsation. A star will pulsate if the zone lies at the right depth below the surface. In cool stars, the zone is too deep and cannot overcome the weight of the layers above it. That is, the spring is overloaded and can't expand. In hot stars, the helium ionization zone lies near the surface, and there is little weight above it to compress it. That is, the spring never gets squeezed. The stars in the instability strip have the right temperature and radius for the helium ionization zone to fall exactly where it is most effective, and thus those stars pulsate.

We can also understand why the Cepheids exhibit a period–luminosity relation. The more massive a star is, the larger it becomes when it leaves the main sequence (Figure 13-23). But the larger a star is, the more slowly it pulsates—just as large bells vibrate more slowly than small bells. Thus, pulsation period depends on size, which depends on mass, so the period depends on mass. But the mass–luminosity relation tells us that mass determines luminosity. If both period and luminosity depend on mass, there has to be a period–luminosity relation.

We also know enough about stars to understand why there are two types of Cepheid variable stars. Type I Cepheids have chemical abundances roughly like those of the sun, but Type II Cepheids are poor in elements heavier than helium, and that means the gases in Type II Cepheids are not as opaque as the gases in Type I Cepheids. The all-important energy flowing outward can escape more easily in Type II Cepheids, and they reach a slightly different equilibrium. For stars of the same period of pulsation, Type II Cepheids are less luminous (Figure 13-21).

We now understand why some stars pulsate. It is all connected to the energy flowing outward through the star. Anything that alters that flow can affect the structure of the star. Thus, variable stars give us a peek at the insides of stars.

Period Changes in Variable Stars

Some variable stars may be able to give us direct evidence that stars are evolving. A few stars have been observed to suddenly change their pulsation, and many stars suffer slow changes to their periods. These changes may arise from the evolution of the stars.

The evolution of a star may carry it through the instability strip a number of times, and each time it can become a variable star. If we could watch a star as it entered the instability strip, we could see it begin to pulsate. Similarly, if we could watch a star leave the instability strip, we could watch it stop pulsating. Such an observation would confirm our belief that the stars are evolving. Unfortunately, stars evolve so slowly that these events are surely rare.

One famous Cepheid, RU Camelopardalis, was observed to stop pulsating temporarily. Its pulsations died away in 1966, and many astronomers assumed that

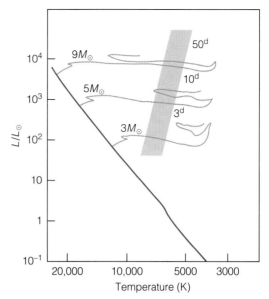

FIGURE 13-23

The instability strip (shaded) is a region of the H–R diagram in which stars are unstable and pulsate as variable stars. Massive stars cross the strip at high luminosities, and because of their large radius they pulsate with longer periods, such as 50 days. Lower-mass stars cross the instability strip at lower luminosities, and because of their smaller radius they pulsate with shorter periods, such as 3 days. This accounts for the period–luminosity relation.

it had stopped pulsating because it was leaving the instability strip. But its pulsations resumed in 1967. Polaris may be going through a similar change now, but it does not lie near the edge of the instability strip in the H–R diagram. Thus, it may not be an example of a star evolving out of the instability strip.

Clearer evidence of stellar evolution can be found in the slowly changing periods of many Cepheids. Like a clock that gains a second every day, Cepheids with slowly changing periods gradually run "fast" or "slow" by an appreciable amount. These changes may occur because the evolution of the star gradually changes its radius and thus the period of pulsation. Such changing periods of pulsation are dramatic evidence that the stars really are evolving.

Not all variable stars lie in the instability strip. The **long-period variables,** for example, are very cool giants that vary with periods ranging from 100 days to 2 years (Figure 13-22). They may change their brightness by as much as 7 magnitudes, making their variability quite dramatic. In fact, the star Omicron Ceti was the first variable star discovered, in 1596. (The first Cepheid variable was found in 1784.) Its brightness varies from about 9.3 to 3.5 magnitudes. In recognition of its dramatic variation, the star was named Mira, meaning "wonderful."

The long-period variables pulsate in part because of ionization zones deep in their distended envelopes. But they also produce large amounts of dust, which is expelled into space as they pulsate. Thus, these stars may be an important source of the interstellar dust we see between the stars. These long-period variables also suffer changes to their periods, but since the cause of their pulsation is not fully understood, the period changes remain a mystery.

The intrinsic variable stars are only partially understood. Their characteristics are determined by the interaction of their surface layers and the energy flowing outward from their interiors. As we understand more about these peculiar giant and supergiant stars, we will learn more about the behavior of stars in general.

CRITICAL INQUIRY

How do Cepheid variable stars give us evidence that stars are evolving?

If we observe Cepheids for a few years, we may notice that some pulsations are slowing down and others are speeding up. This is a very small change in the period of pulsation, but if we observe the stars for long periods of time, the change in their pulsation causes them to gain or lose just as a clock can gain or lose time over many weeks. These changes in period are caused by the expansion or contraction of the stars as they evolve. The period of pulsation depends on the size of the star, so pulsations of a star that is expanding should slow down slightly, and those of a star that is contracting should speed up slightly. When we observe these small changes in the pulsation periods of Cepheid variable stars, we are seeing direct evidence that the stars are evolving.

If we wanted to conduct this research project, what observations would we need to make and how would we analyze them to detect changes in period? Think about a clock. How would we determine whether a clock was ticking a tiny bit too fast or too slow? ■

This chapter has discussed the evolution of stars, a process that occurs so slowly we cannot expect to see much happen in a single human lifetime. Thus, the study of stellar evolution depends heavily on theory and its verification by evidence from observations of real stars. We have seen that star clusters make the evolution of stars visible in the H–R diagram and that the pulsation of the intrinsic variable stars can give us insight into how giant stars evolve through the instability strip. We have not yet considered, however, the most interesting question of stellar evolution. What happens to a star when it uses up the last of its nuclear fuel? Clearly, stars must die— the question is how. We will explore that question in the next chapter. ■

■ Summary

Almost everything we know about the internal structure of stars comes from mathematical stellar models. The models are based on four simple laws of stellar structure. Two laws say that mass and energy must be conserved and spread smoothly through the star. Another, the law of hydrostatic equilibrium, says that the star must balance the weight of its layers by its internal pressure. Yet another says that energy can flow outward only by conduction, convection, or radiation.

The mass–luminosity relation is explained by the requirement that a star support the weight of its layers by its internal pressure. The more massive a star is, the more weight it must support and the higher its internal pressure must be. To keep its pressure high, it must be hot and generate large amounts of energy. Thus, the mass of a star determines its luminosity. The massive stars are very luminous and lie along the upper main sequence. The less massive stars are fainter and lie lower on the main sequence.

How long a star can stay on the main sequence depends on its mass. The more massive a star is, the faster it uses up its hydrogen fuel. A 25-solar-mass star will exhaust its hydrogen and die in only about 7 million years, but the sun can last for 10 billion years.

When a main-sequence star exhausts its hydrogen, it does not just wink out. Its core contracts, and it begins to fuse hydrogen in a shell around its core. The outer parts of the star— its envelope—swell, and the star becomes a giant. Because of this expansion, the surface of the star cools, and it moves toward the right in the H–R diagram. The most massive stars move across the top of the diagram as supergiants.

As the core of the star continues to contract, it finally ignites helium fusion, a process that converts helium into carbon. If the core becomes degenerate before helium ignites, the pressure of the gas does not depend on its temperature, and when helium ignites, it explodes in the helium flash. Although the helium flash is violent, the star absorbs the extra energy and quickly brings the helium-fusing reactions under control.

After helium is exhausted in the core, it can fuse in a shell around the core. Then, if the star is massive enough, it can contract and ignite other fuels, such as carbon.

Confirmation of stellar evolution comes from star clusters. Because all the stars in a cluster have about the same distance, composition, and age, we can see the effects of stellar evolution in the H–R diagram of a cluster. Massive stars evolve faster than low-mass stars, so in a given cluster the most massive stars leave the main sequence first. We can judge the age of such a cluster by looking at the turnoff point, the location on the main sequence where the stars turn off to the right and become giants. The life expectancy of a star at the turnoff point equals the age of the cluster.

There are two types of star clusters. Open clusters contain 10 to 1000 stars and have an open, transparent appearance. Globular clusters contain 10^5 to 10^6 stars densely packed into a spherical shape. The open clusters tend to be young to middle-aged, but globular clusters tend to be very old. The ages of globular clusters range from 10–15 billion years or more. Also, globular clusters tend to be poor in elements heavier than helium.

A simple understanding of stellar structure explains many of the properties of pulsating stars such as the Cepheid and RR Lyrae variables. These intrinsic variable stars lie in an instability strip in the H–R diagram because they contain a layer in their envelopes that stores and releases energy as they expand and contract. Stars outside the instability strip do not pulsate, because the layer is too deep or too shallow to make the stars unstable. The Cepheids obey a period–luminosity relation because more massive stars, which are more luminous, become larger and pulsate more slowly.

Some Cepheids have periods that are slowly changing, showing that the evolution of the star is changing the star's radius and thus its period of pulsation.

New Terms

continuity of mass law
continuity of energy law
brown dwarf
zero-age main sequence (ZAMS)
degenerate gas
triple alpha process
helium flash
turnoff point
open cluster

globular cluster
horizontal branch
variable star
intrinsic variable
Cepheid variable star
RR Lyrae variable star
period–luminosity relation
instability strip
long-period variable

Questions

1. Why is there a lower limit to the mass a star can have?
2. Why is there a mass–luminosity relation?
3. Why does a star's life expectancy depend on its mass?
4. Why do we say that stars are not mixed when some have convective cores?
5. Why do expanding stars become cooler and more luminous?
6. What causes the helium flash? Why does it make it difficult for astronomers to understand the later stages of stellar evolution?
7. How do some stars avoid the helium flash?
8. Why are giant stars so low in density?
9. Why are lower-mass stars unable to ignite more massive nuclear fuels such as carbon?
10. How can we estimate the age of a star cluster?
11. How do star clusters confirm that stars evolve?
12. How do some variable stars prove that stars are evolving?

Discussion Questions

1. How do we know that the helium flash occurs if it cannot be observed? Can we accept an event as real if we can never observe it?
2. Can you think of ways that chemical differences could arise in stars in a single star cluster? Consider the mechanism that triggered their formation.

Problems

1. In the model shown in Figure 13-1, how much of the sun's mass is hotter than 12,000,000 K?

2. What is the life expectancy of a 16-solar-mass star? of a 50-solar-mass star?
3. How massive could a star be and still survive for 5 billion years?
4. If the sun expanded to a radius 100 times its present radius, what would its density be? (HINT: The volume of a sphere is $\frac{4}{3}\pi R^3$.)
5. If a giant star 100 times the diameter of the sun were 1 pc from us, what would its angular diameter be? (HINT: Use the small-angle formula, in Chapter 3.)
6. What fraction of the volume of a 5-solar-mass giant star is occupied by its helium core? (HINTS: See Figure 13-8. The volume of a sphere is $\frac{4}{3}\pi R^3$.)
7. If the stars at the turnoff point in a star cluster have masses of about 4 solar masses, how old is the cluster?
8. If an open cluster contains 500 stars and is 25 pc in diameter, what is the average distance between the stars? (HINTS: What share of the volume of the cluster surrounds the average star? The volume of a sphere is $\frac{4}{3}\pi R^3$.)
9. Repeat Problem 8 for a typical globular cluster containing a million stars in a sphere 25 pc in diameter.
10. If a Cepheid variable star has a period of pulsation of 2 days and its period increases by 1 second, how late will it be in reaching maximum light after 1 year? after 10 years? (HINT: How many cycles will it complete in a year?)

Critical Inquiries for the Web

1. Find Hertzsprung–Russell (or color-magnitude) diagrams for five star clusters. Based on the distributions of stars on the diagrams, rank the clusters in order from oldest to youngest. At what value (spectral type, temperature, or B-V color index) does each cluster have its turnoff?
2. Astronomers have been searching for brown dwarfs for years, but few candidates have been identified. Search the Web for information on efforts to locate these substellar objects. Why are they so difficult to detect? List any likely brown dwarfs found so far.

 Go to the Wadsworth Astronomy Resource Center (www.wadsworth.com/astronomy) for critical thinking exercises, articles, and additional readings from InfoTrac College Edition, Wadsworth's online student library.

Chapter Fourteen

THE DEATHS
OF STARS

GUIDEPOST

As you read this chapter, take a moment to be astonished and proud that the human race knows how stars die. Finding the mass of stars is difficult, and understanding the invisible matter between the stars takes ingenuity, but we human beings have used what we know about stars and what we know about the

physics of energy and matter to figure out how stars die.

The death of stars is important to us because our own sun will die, but it is also important because the explosive deaths of massive stars contaminate the universe with atoms heavier than helium. The atoms we are made of were made inside stars. If stars didn't die, we would not exist.

In the chapters that follow, we will see that some matter from dying stars becomes trapped in dead ends—neutron stars and black holes—but some escapes into the interstellar medium and is incorporated into new stars and the planets that circle them. ∎

Natural laws have no pity.

ROBERT HEINLEIN

The Notebooks of Lazarus Long

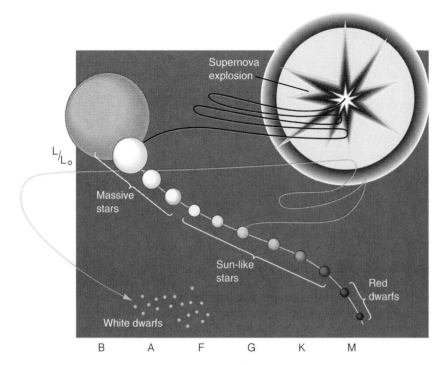

FIGURE 14-1

Three different modes of stellar death are summarized in this schematic H–R diagram. The lowest-mass stars, the red dwarfs, live much longer than the present age of the universe. Medium-mass stars like the sun become giant stars and then collapse to become white dwarfs. The most massive stars evolve rapidly to become supergiants and explode as supernovae, leaving behind either a neutron star or a black hole.

Gravity is patient—so patient it can kill stars. As we have seen, a star is a ball of gas caught in a battle between gravity trying to make it contract and nuclear fusion trying to make it expand. As long as the star can fuse nuclear fuels and keep its interior hot, it can withstand its own gravity and shine in the darkness. But no star has unlimited fuel, and when the fuel is gone, gravity wins, and the star dies.

To study the deaths of stars we will divide the main-sequence stars into three groups according to mass (Figure 14-1). We will discover that the lowest-mass stars live long, uneventful lives, whereas sunlike stars become giants before they die. The most massive stars die in violent explosions—supernovae. We will study the death throes of a specific star, the great supernova of 1987.

In Chapter 12, we learned that a protostar becomes a true star when it ignites its nuclear fuels. According to that definition, a star's life ends when its nuclear fires go out, and that is the subject of this chapter—the process by which stars die.

14-1 LOWER MAIN-SEQUENCE STARS

The stars of the lower main sequence share a common characteristic: they have relatively low masses. That means that they face similar fates as they exhaust their nuclear fuels.

When a star exhausts one nuclear fuel, its interior contracts and grows hotter until the next nuclear fuel ignites. The contracting star heats up by converting gravitational energy into thermal energy, so low-mass stars cannot get very hot, and this limits the fuels they can ignite. The lowest-mass stars, for example, cannot get hot enough to ignite helium fusion.

Structural differences divide the low-mass stars into two subgroups—very low mass stars and medium-mass stars such as the sun (see Figure 12-17). The critical difference between the two groups is the extent of interior convection. If the star is convective, fuel is constantly mixed, and the resulting evolution of the star is drastically altered.

Red Dwarfs

The stars at the low end of the main sequence are called **red dwarfs** because they are cool and small. Less massive than 0.4 solar mass, (Chapter 12), they are convective from their centers to their surfaces. The gas is constantly mixed, so hydrogen is consumed and helium accumulates uniformly through the star. Thus, the star cannot develop an inert helium core surrounded by a shell of unprocessed hydrogen, and it cannot ignite a hydrogen shell or become a giant. Rather, nuclear fusion converts hydrogen into helium, which the small star cannot fuse because it cannot get hot enough. With their low masses, such stars could live for trillions (thousands of billions) of years (Chapter 13). Recent models of a 0.1 solar mass star took 2 billion years to contract to the main sequence and 6 trillion years to become a white dwarf.

On the other hand, we cannot assume that the white dwarfs we see in the sky now were once red dwarfs. The evolution of red dwarfs is so slow that none could have exhausted their fuels yet. The universe appears to be only 10 to 15 billion years old, and the red dwarfs could last for a trillion years or more. Thus, the white dwarfs must have been produced by more massive stars.

Sunlike Stars

Stars with masses between roughly 4 solar masses and 0.4 solar mass,* including the sun, evolve in the same way. They can ignite hydrogen and helium and become giants, but they cannot get hot enough to ignite carbon, the next fuel in the sequence (see Table 13-3). When they reach that impasse, they can no longer maintain their stability, and their interiors contract while their envelopes expand.

To understand the fate of these stars, we must consider two concepts: mixing and expansion. The interiors of these sunlike stars are not well-mixed (see Figure 12-17). As we learned in Chapter 12, stars of 1.1 solar masses or less, including the sun, have no convection near their centers, so they are not mixed at all. Stars more massive than 1.1 solar masses have small zones of convection at their centers, but this mixes no more than about 12 percent of the star's mass. Thus, medium-mass stars, whether they have convective cores or not, are not mixed, and the helium accumulates in an inert helium core surrounded by unprocessed hydrogen. When this core contracts, the unprocessed hydrogen just outside the core ignites in a shell and swells the star into a giant.

As a giant, the star fuses helium in its core and then in a shell surrounding a core of carbon and oxygen. This core contracts and grows hotter, but it cannot become hot enough to ignite the carbon. Thus, the carbon–oxygen core is a dead end for these medium-mass stars.

Because no nuclear reactions can begin in the carbon–oxygen core, it cannot resist the weight pressing down on it. In addition, just outside the carbon–oxygen core, the helium-fusion shell converts helium into carbon and thus increases the mass of the carbon–oxygen core, so the core must contract. The energy released by the contracting core, plus the energy generated in the helium- and hydrogen-fusing shells, flows outward and makes the envelope of the star expand.

This forces the star to become a very large giant. Its radius may become as large as the radius of the earth's orbit, and its surface becomes as cool as 2000 K. Such a star can lose large amounts of mass from its surface.

Mass Loss from Stars

We know that stars can lose mass, because we can see mass streaming away from the sun. The solar wind is a breeze of gas flowing out of the sun's hot corona and escaping into space.

Other stars like the sun are also losing mass. Observations in the ultraviolet and X-ray parts of the spectrum have been made with space telescopes such as the International Ultraviolet Explorer (IUE) and the High Energy Astrophysical Observatories (HEAO). Such observations show strong emission lines from many main-sequence stars, emission lines that must originate in hot chromospheres and coronas like the sun's. If these stars have outer atmospheres like the sun's, they presumably have similar winds of hot gas.

This kind of mass loss does not appear to be sufficient to alter the evolution of the star. The sun, for example, loses about 0.001 solar mass per billion years. Even over its entire lifetime of 10 billion years, the sun will not lose an appreciable fraction of its mass.

However, the sun will not always be a simple main-sequence star. It will eventually become a giant star, and the spectra of some giants contain distorted spectral lines caused by a rapid flow of gas away from the star. Because the spectra of some of these stars do not show the characteristic emission, we must assume that the stars do not have hot coronas, but other processes could drive mass loss. The stars are so large that gravity is weak at their surfaces, and convection in the cool gas can drive shock waves outward and power mass loss. In addition, some giants are so cool that specks of carbon dust condense in their atmospheres, just as soot can condense in a fireplace. The pressure of the star's radiation can push this dust and any atoms that collide with the dust completely out of the star.

Another factor that can cause mass loss is periodic eruptions in the helium-fusion shell. The triple-alpha process that fuses helium into carbon is extremely sensitive to temperature, and that can make the helium-fusion shell so unstable that it may experience eruptions called **thermal pulses.** Every 200,000 years or so, the shell can erupt and suddenly produce energy equivalent to a million times the luminosity of the sun. Almost all of this energy goes into lifting the outer layers of the star and driving gas away from the surface.

Whatever drives this mass loss from giants, it can affect the mass of the star appreciably in a short time. A star expanding as its carbon–oxygen core contracts could lose an entire solar mass in only 10^5 years, which is not a long time in the evolution of a star. Thus, a star that began its existence on the main sequence with a mass of 8 solar masses might reduce its mass to only 3 solar masses in half a million years.

Stellar mass loss confuses our story of stellar evolution. We would like to say that stars more massive than a certain limit will evolve one way and stars less massive will evolve another way. But stars may lose enough mass to alter their own evolution. Thus, we must consider both the initial mass a star has on the main sequence and the mass it retains after mass loss. Because we don't know exactly how effective mass loss is, it is difficult for us to be exact about the mass limits in our discussion of stellar evolution.

The mass loss we have discussed so far has been a gradual process, but a sunlike star with a contracting carbon–oxygen core expands so far that it can lose all of its outer layers in a sudden expulsion. This surface expulsion is helped by the growing pressure of the radiation inside the star and by instabilities in the helium-fusion shell, which can cause periodic bursts of energy production. The expanding layers are visible as small, spherical nebulae.

---

*This mass limit is uncertain, as are many of the masses stated here. The evolution of stars is highly complex, and such parameters are not well known.

a

b

c

d

FIGURE 14-2

Planetary nebulae. (a) The Ring Nebula is a classic planetary nebula, an expanding, hollow shell of gas driven away by the star at the center, which is contracting to become a white dwarf. (b) The Dumbbell Nebula, an older planetary nebula, is asymmetrical and is mixing with the interstellar medium. (c) The Hourglass Nebula was created by a fast stellar wind running into a dense ring of gas ejected earlier. (d) New high-resolution telescopes are revealing that planetary nebulae often have highly complex structures caused by disks of material around the star, jets of gas, and asymmetrical outflows from the central star. (a, NOAO; b, Rudolph Schild; c, Raghvendra Sahai and John Trauger, JPL, WFPC2 Science Team and NASA; d, J. P. Harrington and K. J. Borkowski, University of Maryland and NASA)

Planetary Nebulae

Even a small telescope will reveal small circles of haze scattered among the stars. These are called **planetary nebulae** because they look like the small, greenish disks of planets such as Uranus and Neptune. In fact, they have nothing to do with planets. They are made up of gases expelled by elderly giant stars.

The most popular planetary nebula is called the Ring Nebula (Figure 14-2a), and it is easily visible in small telescopes. Larger telescopes reveal some 1500 planetary nebulae, with diameters ranging from 1000 AU to 1 pc. Many seem to have ring shapes, and for many years astronomers have thought of these nebulae as spherical shells of ejected matter that look like rings because we see them in projection. Newer images, however, show that many planetary nebulae are not true rings but are in fact stretched and distorted. This discovery suggests that the formation of planetary nebulae is a more complex process than simply the ejection of a layer of material from the central star (Figure 14-2c).

Apparently, the aging giant star loses matter for millions of years in a strong stellar wind, which can sometimes create denser rings of material in the equatorial plane of the rotating star or inflate hot cavities in the interstellar medium around the star. Later, when the star expels enough of its surface layers to expose deeper, hotter layers, the intense radiation ionizes the innermost gas and drives it outward to overtake and sweep up the mass

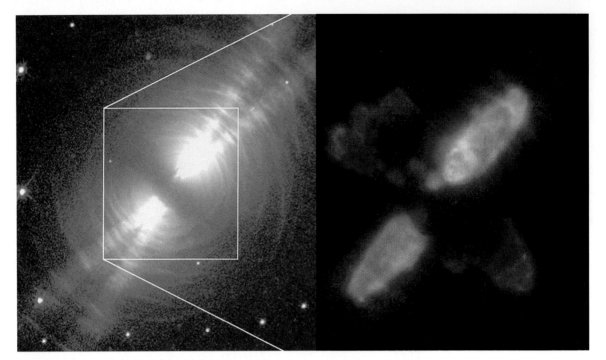

FIGURE 14-3

The Egg Nebula, shown in visible light at left, illustrates the complexity of stellar outflows. Evidently created by an aging star embedded in a thick doughnut of dust and gas, the nebula is illuminated by beams of light escaping from the dense center and illuminating shells of gas previously expelled. At infrared wavelengths (right) high-speed beams of gas and dust are seen emerging from the hidden star and colliding with the outer arcs of gas, while the outer boundary of the dusty doughnut glows with emission from molecular hydrogen. The star and the center of the disk are hidden by thick dust. (Rodger Thompson, Marcia Rieke, Glenn Schneider, Dean Hines, University of Arizona; Raghvendra Sahai, JPL; NICMOS Team; and NASA)

that was lost earlier, much as a snowplow sweeps up the snow in front of it. What shape we see depends on how the gas interacts with any disk of gas lost earlier or with any hot cavity around the star. What we see also depends on the angle at which we view the nebula. In some cases, planetary nebulae appear to have been influenced by the presence of a companion star. Astronomers are just beginning to understand the complex factors that determine the shapes of planetary nebulae (Figure 14-3).

The spectra of planetary nebulae are emission spectra with lines of elements such as H, He, N, O, C, Ne, S, Ar, Cl, Fe, and so on. The nebulae look green because of strong forbidden emission lines of OIII (twice-ionized oxygen) in the green part of the spectrum. At low light levels, our eyes are most sensitive to these wavelengths, but longer-exposure images reveal many colors caused by emission lines from the other atoms (Figure 14-2). Doppler shifts show that the nebulae are expanding at 10–20 km/sec. They mix into the interstellar medium and vanish within about 50,000 years.

Although the ejected gas is only about 10 percent of the star's total mass, it is an important part of a stable star—namely, the insulating blanket that confines the star's internal heat. When the surface puffs away into space, the white-hot core is exposed, and the star emits intense ultraviolet radiation. It is this radiation that ionizes and drives away the gases of the nebula.

If we follow the evolution of such a star as it produces a planetary nebula, we see it move rapidly to the left on the H–R diagram. At this point, it consists of an inert carbon–oxygen core surrounded by helium- and hydrogen-fusing shells and topped by a shallow atmosphere of hydrogen and helium. The surface temperatures of these planetary nebula nuclei range from 25,000–100,000 K or more. But with the insulating surface layers gone, the star loses heat rapidly and continues to contract until it becomes degenerate and enters the white-dwarf region (Figure 14-4).

White Dwarfs

Both low-mass red dwarfs and medium-mass stars eventually become white dwarfs. Our survey of neighboring stars (Chapter 9) showed that most stars have masses less than that of the sun, but these are very long lived stars. Our galaxy is not old enough for any of these red dwarfs to have become white dwarfs. However, white dwarfs are very numerous—our galaxy probably contains billions—so they must be the remains of stars with masses similar to the sun's.

The first white dwarf discovered was the faint companion to Sirius. In that visual binary system, the bright star is Sirius A. The white dwarf, Sirius B, is 10,000 times fainter than Sirius A. The orbital motions of the stars

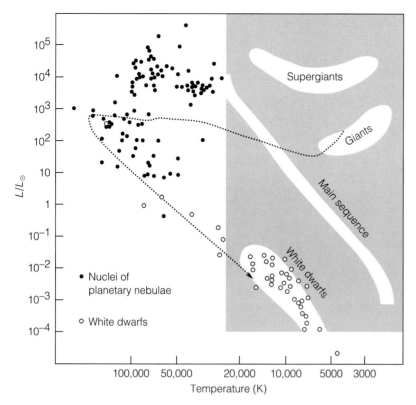

FIGURE 14-4
Our customary H–R diagram must be extended to higher temperatures to show the evolution of a star after it ejects a planetary nebula. Central stars of planetary nebulae (filled circles) are hotter and more luminous than most white dwarfs (open circles). The dotted line shows the evolution of a 0.8-solar-mass star as it collapses from a red giant star toward the white-dwarf region. Presumably, the central stars of planetary nebulae cool to become white dwarfs long after their nebulae have dissipated. (Adapted from diagrams by C. R. O'Dell and S. C. Vila)

(shown in Figure 10-8) tell us that the white dwarf's mass is about 1 solar mass, and its blue-white color tells us that its surface is hot, about 32,500 K. Because its luminosity is low, it must have a small surface area—in fact, it is about 76 percent of the earth's diameter. The mass and size imply that its average density is over 3×10^6 g/cm^3. On the earth, a teaspoonful of Sirius B material would weigh more than 15 tons (Figure 14-5).

A normal star is supported by energy flowing outward from its core, but a white dwarf has no internal energy source, so there is no energy flow to balance gravity. The star contracts until it becomes degenerate (Chapter 13). Thus, a white dwarf is supported not by energy flowing outward but by the refusal of its electrons to pack themselves into a smaller volume.

The interior of a white dwarf is mostly carbon and oxygen ions floating among a whirling storm of degenerate electrons. It is the degenerate electrons that exert the pressure to support the star's weight, but most of its mass is represented by the carbon and oxygen ions. Theory predicts that as the star cools these ions will lock together to form a crystal lattice, so there may be some truth in thinking of aging white dwarfs as great crystals of carbon and oxygen. Near the surface, where the pressure is lower, a layer of ionized gases makes up a hot atmosphere.

The tremendous surface gravity of white dwarfs—100,000 times that of the earth—affects their atmospheres in strange ways. The heavier atoms in the atmosphere tend to sink, leaving the lightest gases at the surface. We see some white dwarfs with atmospheres of almost pure hydrogen, whereas others have atmospheres of nearly pure helium. Still others, for reasons not well

FIGURE 14-5
The degenerate matter from inside a white dwarf is so dense that a lump the size of a beach ball would, transported to the earth, weigh as much as an ocean liner.

understood, have atmospheres that contain traces of heavier atoms. In addition, the powerful surface gravity pulls the white dwarf's atmosphere down into a very shallow layer. If the earth's atmosphere were equally shallow, people on the top floors of skyscrapers would have to wear oxygen masks.

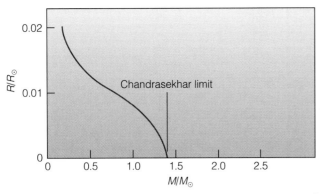

FIGURE 14-6

The more massive a white dwarf, the smaller its radius. Stars more massive than the Chandrasekhar limit of 1.4 solar masses cannot be white dwarfs.

FIGURE 14-7

The IRAS satellite mapped the sky at far infrared wavelengths and found regions of star formation filled with twisting currents of gas and dust. This image includes most of the constellation of Orion. The Great Nebula in Orion (lower right) is only one of the regions where young, hot stars drive away gas and dust and stir the interstellar medium. The large ring (upper right) is composed of gas and dust expanding away from Lambda Orionis (An O8 giant). (NASA/IPAC Courtesy Deborah Levine)

Clearly, a white dwarf is not a normal star. It generates no nuclear energy and is supported by degenerate electron pressure. Instead of calling a white dwarf a "star," we call it a **compact object.** In the next chapter, we will encounter two other compact objects—neutron stars and black holes.

A white dwarf's future is bleak. As it radiates energy into space, its temperature gradually falls, but it cannot shrink any smaller because its degenerate electrons cannot get closer together. This degenerate matter is a very good thermal conductor, so heat flows to the surface and escapes into space, and the white dwarf gets fainter and cooler, moving downward and to the right in the H–R diagram. Because the white dwarf contains a tremendous amount of heat, it needs billions of years to radiate that heat through its small surface area. Eventually, such objects may become cold and dark, so-called **black dwarfs.** Our galaxy is not old enough to contain any of these objects. The coolest white dwarfs in our galaxy are about the temperature of the sun.

Perhaps the most interesting thing about white dwarfs appears in mathematical models. The equations predict that if we added mass to a white dwarf, its radius would *shrink*, because added mass would increase its gravity and squeeze it tighter. If we added enough to raise its total mass to about 1.4 solar masses, its radius would shrink to zero (Figure 14-6). This is called the **Chandrasekhar limit** after Subrahmanyan Chandrasekhar, the astronomer who discovered it.

The Chandrasekhar limit tells us that no white dwarf can exist with a mass greater than 1.4 solar masses. It seems quite likely that stars more massive than this can ultimately collapse into white dwarfs if they lose enough mass when they are giants. We have seen in an earlier section that mass loss is common and very effective (Figure 14-7). A 6-solar-mass star should be able to reduce its mass to 1.4 solar masses before it collapses. Some astronomers have suggested that stars as massive as 9 solar masses may ultimately collapse into white dwarfs with final masses below the Chandrasekhar limit.

CRITICAL INQUIRY

How will the sun die?

We have seen that lower main-sequence stars like the sun die by producing planetary nebulae and then becoming white dwarfs. For the sun, this process will begin in a few billion years, when it exhausts the hydrogen in its core. It will then expand to become a giant star. Although the surface of the sun will become cooler, it will expand until it has such a large surface that it will be about 100 times more luminous than it is now. (Incidentally, this will drive away the earth's atmosphere and oceans.) As a giant star, the sun will have a strong solar wind carrying gas into space and pushing back the interstellar medium. Eventually, the loss of mass will expose deeper layers in the sun. These extremely hot layers will ionize the gas around the sun and propel it outward to scoop up previously expelled gases and produce a planetary nebula. Soon the last remains of the sun will collapse inward and slowly cool to form a small, dense white dwarf. The surface will be very hot, but so small that the sun will be about 100 times fainter than it is now (and any cinder that remains of the earth will fall into a deadly deep freeze).

The story of stellar evolution tells us that our sun will someday create a planetary nebula and become a white dwarf. At that point, what will have become of all the hydrogen the sun once contained? ■

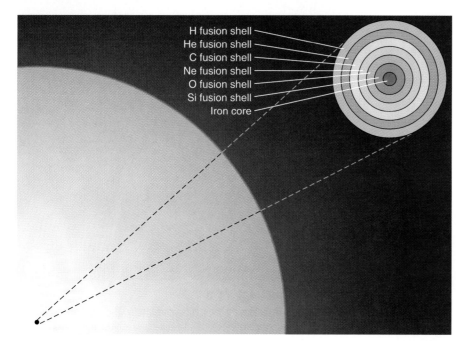

FIGURE 14-8
The schematic interior of a massive star approaching the end of its life. This cool giant star, about the size of the orbit of Jupiter, contains a core about the size of the earth in which concentric layers fuse heavier fuels around the growing iron core. Because the iron cannot produce energy, the iron core must eventually collapse and trigger a supernova explosion.

Labels on figure:
H fusion shell
He fusion shell
C fusion shell
Ne fusion shell
O fusion shell
Si fusion shell
Iron core

Lower main-sequence stars die by gradually producing beautiful planetary nebulae and shrinking to become white dwarfs. But we have seen that no white dwarf with a mass greater than 1.4 solar masses can exist. Then how do massive stars die? We address this question in the next section.

14-2 THE UPPER MAIN SEQUENCE

We have seen that the low-mass red dwarfs and the sunlike, medium-mass stars die relatively quietly as they exhaust their nuclear fuels. The most violent event in the evolution of these stars is the ejection of surface layers to form planetary nebulae. Massive stars, however, end their days in spectacular explosions that destroy the star.

Hydrogen, Helium, and Carbon

The evolution of an upper main-sequence star begins like the evolution of a sunlike star. The star exhausts the hydrogen in its core, and, when the core contracts, a hydrogen-fusing shell ignites, swelling the star into a giant. While it is a giant, it fuses helium in its core and then in a shell, leaving behind a carbon–oxygen core, which contracts and grows hotter.

Mass loss will reduce the mass of the star, but if it still has a mass between 3 and 9 solar masses when its carbon–oxygen core contracts, it develops a potentially lethal problem. The contracting core becomes degenerate, turning off the pressure–temperature thermostat that normally would keep nuclear reactions under control. The carbon–oxygen core in such a star is a bomb,

destined to explode when the temperature reaches 600,000,000 K, the ignition temperature for carbon. The resulting explosion is called the **carbon detonation.**

The carbon detonation may be powerful enough to blow some stars apart, and it could be responsible for some of the violent stellar explosions known as supernovae. As we will see later, supernovae are rare and poorly understood. Other stars may be able to survive carbon detonation because the rising temperature eventually forces the core to expand, and the gas stops being degenerate.

Stars more massive than about 9 solar masses do not face carbon detonation, because their cores are so hot that carbon ignites before the gas becomes degenerate. Thus, the pressure–temperature thermostat is in working order, and carbon fusion turns on gradually. Stars that survive carbon ignition continue their lives for a short while as they fuse heavier elements (see Table 13-3), but they, too, eventually face an explosive end. The development of an iron core spells their absolute and final collapse.

The Iron Core

Some stars do survive carbon ignition, but we cannot be certain of their evolution. Nuclear reactions involving heavier elements are very complex (see Figure 13-11), and stellar models are unable to follow the behavior of stars as they fuse these fuels. Nevertheless, we can predict the general plan of their evolution.

As the star fuses heavy elements, each new fuel ignites first in the core and then in a shell, building layer after layer of heavy elements inside the star (Figure 14-8). Because fusion makes a more massive atomic nucleus by combining small nuclei, each successive stage involves

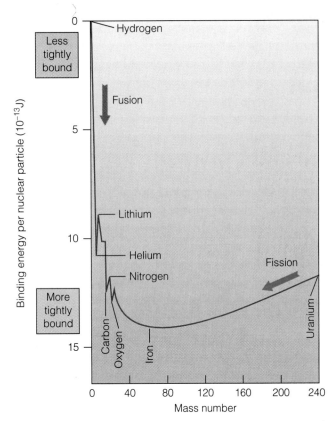

FIGURE 14-9

The binding energy is the energy that holds an atomic nucleus together. Fission reactions (such as those in nuclear power plants on earth) can break heavy nuclei, such as uranium, into lighter fragments and release energy. Fusion reactions (such as those in stars) can combine low-mass nuclei to form more massive nuclei and release energy. Iron has the most tightly bound nucleus, so an iron core inside a star cannot release energy by fusion or fission.

FIGURE 14-10

Supernova explosions do not occur symmetrically. This color graphic, made from a mathematical model of a supernova explosion, shows a cross section of the core of an exploding star only 0.102 second after the in-falling material bounces off of the core. The square is only 150 km on a side, and black arrows show the motion of the gas. The model reveals that the bouncing material becomes violently turbulent (blue) as it rushes outward to meet the rest of the star, which is falling inward (red at upper right). The innermost core of the star is forming a neutron star (red at lower left). To include the surface of the star at this scale, this diagram would need to be about 140 km (90 mi) in diameter. (Courtesy Adam Burrows, John Hayes, and Bruce Fryxell)

fewer particles. But higher temperatures are required for each new fuel. Thus, the reactions are faster and more violent for each new fuel (see Table 13-3).

The accumulation of iron atoms in the star's core forebodes the end of heavy-element fusion and the death of the star. Nuclear-fusion reactions can release energy if the nucleus produced is bound more tightly than the lower-mass nuclei that were fused. Because iron is the most tightly bound nucleus of all (Figure 14-9), no nuclear-fusion reactions can combine iron nuclei and release energy. In fact, any nuclear reactions that do fuse iron into heavier nuclei absorb rather than release energy. Of course, that robs the core of some of the energy it needs to support the weight pressing inward.

When a massive star develops an iron core, nuclear fusion cannot occur in the core to produce energy, and the core contracts and grows hotter. The shells around the core burn outward, fusing lighter elements and leaving behind more iron, which further increases the mass of the core. When the mass of the iron core exceeds 1.3–2 solar masses, the core must collapse.

As the core begins to contract, two processes can make it contract even faster. Heavy nuclei in the core can capture high-energy electrons, thus removing thermal energy from the gas. This robs the gas of some of the pressure it needs to support the crushing weight of the outer layers. Also, in more massive stars, temperatures are so high that many photons have gamma-ray wavelengths and can break more massive nuclei into less massive nuclei. This reversal of nuclear-fusion absorbs energy and allows the core to collapse even faster.

Although a massive star may live for millions of years, its iron core—only about 500 km in diameter—collapses in only a few thousandths of a second. This collapse happens so rapidly that our most powerful computers are not capable of following the details. One thing is clear, however. The collapse of the iron core of a massive star triggers a star-destroying explosion—a **supernova.**

Supernova Explosions

Modern theory predicts that the collapse of a massive star can eject the outer layers of the star to produce a supernova explosion while the core of the star collapses to form a neutron star or a black hole. We will examine neutron stars and black holes in detail in the next chapter; here we concentrate on the process that triggers the supernova explosion.

We can't be sure how a supernova explodes, but recent advances in mathematical-modeling techniques a   b   c

FIGURE 14-11

(a) The Crab Nebula is the remains of the supernova of AD 1054. At visual wavelengths, the glow of the nebula is woven through a network of expanding filaments. (b) From the Doppler shift of the filaments, astronomers have built this 3-D image of the Crab Nebula. Blue filaments are on the near side moving toward us, and yellow and red filaments are on the back of the nebula moving away. (c) By viewing the model from a point in space to the southeast, we can see that the Crab Nebula filaments enclose a hollow volume. (a, Palomar Obs/Caltech; b and c, Gordon MacAlpine)

and increases in computer power and speed are giving us clues. The collapse of the innermost part of the degenerate core allows the rest of the star's interior to fall inward, creating a tremendous "traffic jam" as all the nuclei fall toward the center. It is as if all the residents of Indiana suddenly tried to drive their cars into downtown Indianapolis. There would be a traffic jam not only downtown but also in the suburbs; as more cars arrived, the traffic jam would spread outward. Similarly, as the inner core of the star falls inward, a shock wave (a traffic jam) develops and begins to move outward. Containing about 100 times the energy necessary to destroy the star, such a shock wave was thought to be the cause of supernova explosions.

Recent models, however, show that this shock wave stalls within 0.025–0.040 second. Matter flows inward just as fast as the shock wave spreads outward. If this happens to a star, it presumably will collapse without any visible explosion. However, theory predicts that 99 percent of the energy released in the collapse will appear as neutrinos. In the sun, neutrinos zip outward unimpeded by the gas of the solar layers, but in a collapsing star the density is as high as 10^{12} g/cm^3, as dense as an atomic nucleus. This gas is opaque to neutrinos, and they are partially trapped in the gas. For some years, astronomers thought that these neutrinos could spread energy across the shock wave and, about one-quarter second after the collapse, reaccelerate the shock wave. But the computer models of supernovae still refused to explode. Recent research reveals that temperatures are so high that convection begins within a fraction of a second, and the violent turbulence carries heat across the stalled shock wave (Figure 14-10). In a second or so, the shock wave begins to push outward, and after just a few hours it smashes out

through the surface of the star, producing a supernova explosion.

The supernova we see is the brightening of the star as its distended outer layers are blasted outward. As the months pass, the cloud of gas expands, thins, and begins to fade. But the way it fades in some cases suggests that nuclear reactions in the compressed outer layers have enriched it with short-lived radioactive nuclei such as nickel-56. The gradual decay of these nuclei can keep the gas hot and prevent it from fading rapidly. Thus, the supernova explosion may be violent enough to trigger nuclear fusion in the outer layers of the dying star.

New supercomputers may eventually tell us more about how supernovae explode, but whatever the mechanism, it must generate tremendous energy. A single supernova explosion is equivalent to the explosion of 10^{28} megatons of TNT. (For comparison, this much TNT would amount to about 1 trillion times the mass of the earth, or 3 million times the mass of the sun.)

Observations of Supernovae

In 1054, Chinese astronomers saw a "guest star" appear in the constellation we know as Taurus, the Bull. The star quickly became so bright it was visible in the daytime. After a month's time, it slowly faded, taking almost 2 years to vanish from sight. When modern astronomers turned their telescopes to the location of the guest star, they found a cloud of gas about 1.35 pc in radius, expanding at 1400 km/sec. Projecting the expansion back in time, they concluded that the expansion must have begun about 900 years ago, just when the guest star made its visit. Thus, we think the nebula, now called the Crab Nebula because of its shape (Figure 14-11), marks the site of the 1054 supernova.

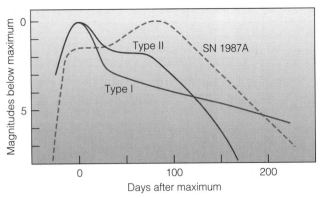

FIGURE 14-12

Type I supernovae decline rapidly at first and then more slowly, but Type II supernovae pause for about 100 days before beginning a steep decline. Supernova 1987A was odd in that it did not rise directly to maximum brightness. These light curves have been adjusted to the same maximum brightness. Generally, Type II supernovae are about 2 magnitudes fainter than Type I.

Supernovae are rare. Only a few have been seen with the naked eye in recorded history. Arab astronomers saw one in 1006, and the Chinese saw one in 1054. European astronomers observed two—one in 1572 (Tycho's supernova) and one in 1604 (Kepler's supernova). Also, the guest stars of 185, 386, 393, and 1181 may have been supernovae. For 383 years following 1604, no naked-eye supernova appeared, and then in February 1987 a star in the southern sky exploded. We will examine the great supernova of 1987 later in this section. Here we will explore supernovae in general

Because supernovae are so rare, most of what we know about them has come from observations of supernovae in other galaxies. These are visible only through telescopes. From the study of such explosions, astronomers have found two types of supernovae. **Type I supernovae** become about 4 billion times more luminous than the sun, decline rapidly at first, and then fade more slowly (Figure 14-12). **Type II supernovae** become only about 0.6 billion times the sun's luminosity and decline in a more irregular way. Spectra of Type II supernovae show hydrogen lines, but spectra of Type I supernovae do not.

Type I supernovae seem to occur when a white dwarf in a binary system gains mass from its companion (a process we will examine in detail later in the chapter) and exceeds the Chandrasekhar limit. This causes the white dwarf to collapse, triggering a star-destroying explosion. Spectra of Type I supernovae lack hydrogen lines because white dwarfs contain little hydrogen. Type II supernovae are believed to occur when the iron core of a massive star collapses. The reason their spectra contain strong hydrogen lines is apparently that the outer layers of massive stars are rich in hydrogen.

Although these are the two basic types of supernovae, astronomers are beginning to recognize variants caused by the nature of individual stars. For example, one subtype, termed "peeled" supernovae, seems to be pro-

a

b

FIGURE 14-13

(a) The Cygnus Loop is a supernova remnant created when a massive star exploded 20,000 years ago (see Figure 11-13). (Palomar Obs/Caltech) (b) This radio map of the 300-year-old remnant Cas-A was made with the VLA radio telescope (see Figure 14-18). (National Radio Astronomy Observatory/Association of Universities, Inc.)

False-Color Images and Reality

Astronomers use computers to add false color to images and reveal things our eyes cannot see. Radio, infrared, ultraviolet, and X-ray radiation are not visible to our eyes, so data collected at these wavelengths can be displayed in images that are given colors visible to our eyes. Of course, the colors used are entirely arbitrary. Radio astronomers might choose to use reds and yellows to color a radio map of a cloud of gas, or they might choose greens and blues. The colors make the radio energy visible to our eyes, but the choice of colors is up to the astronomer analyzing the image.

The variation in color across an image does have meaning, however. Radio astronomers might decide to use reds for the regions of strongest radio signal, yellows for fainter regions, and blues for the faintest. Thus, we could look at the radio map of a gas cloud and immediately pick out the parts of the cloud that were emitting the strongest radio energy.

In addition to using false color in images recorded in wavelengths our eyes cannot see, astronomers often use false color in photographs recorded at visible wavelengths. These images can reveal to our eyes subtle variations in brightness not visible in the original photographs.

False-color images are common in astronomy, but they are also used in other sciences.

Doctors often analyze medical X-ray images, CAT scans, and so on by converting the images to false color. Biologists use false color to analyze microscopic photographs, and geologists use false color to study photographs of the earth recorded by a satellite at various wavelengths.

We might think of false-color images as deceptive, but they actually help us see parts of the universe that would otherwise never register on our limited eyes. ■

---

duced by the collapse of massive stars that have lost their hydrogen-rich outer layers. We will see later how a star could be peeled by a companion in a close binary system.

Although the supernova explosion fades to obscurity in a year or two, an expanding shell of gas marks the explosion site. The gas, originally expelled at 10,000–20,000 km/sec, may carry away one-fifth of the mass of the star. The collision of that expanding gas with the surrounding interstellar medium can sweep up even more gas and excite it to produce a **supernova remnant,** the nebulous remains of a supernova explosion.

Supernova remnants look quite delicate (Window on Science 14-1) and survive only a few tens of thousands of years before they gradually mix with the interstellar medium and vanish. The Crab Nebula, only 900 years old, is a young remnant. The Cygnus Loop (Figure 14-13a) is larger and more diffuse, having originated in a supernova about 20,000 years ago.

The Great Supernova of 1987

Until 1987, astronomers had never looked at a bright supernova through a telescope. Nearly all supernovae are in distant galaxies and thus are very faint. The last supernova visible to the naked eye, Kepler's supernova of 1604, predated the invention of the telescope. Then, in late February 1987, the news raced around the world. Astronomers in Chile had discovered a naked-eye supernova in the Large Magellanic Cloud, a small galaxy very near our Milky Way Galaxy (Figure 14-14). Because the supernova was only 20 degrees from the south celestial pole, it could be studied only from southern latitudes. It was named SN 1987A to denote the first supernova discovered in 1987.

Spectra suggested that the supernova was a Type II, caused by the collapse of the core of a massive star. As the months passed, however, the light curve proved to be

FIGURE 14-14

Supernova 1987A was discovered February 24, 1987. The Tarantula Nebula lies at upper left and the supernova at the right in this photo. Reaching a peak brightness of magnitude 2.9, the supernova was the first visible to the naked eye since Kepler's supernova of 1604, six years before the invention of the astronomical telescope. (National Optical Astronomy Observatories)

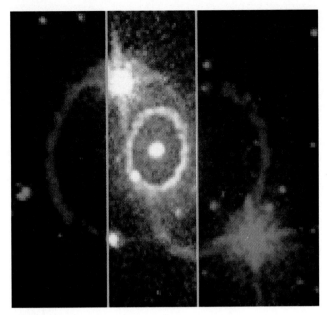

FIGURE 14-15

The Hubble Space Telescope detected a ring of ionized gas surrounding SN 1987A. Ejected about 30,000 years before the star exploded, the gas in the ring is now excited by radiation from the star, as are the larger, fainter ring features. The blast wave will destroy the ring within a century. (Jason Punn, NOAO, SINS Collaboration, and NASA)

FIGURE 14-16

The Hubble Space Telescope Imaging Spectrograph produced a spectrum of the image between the green lines in Figure 14-15. The red rings in the spectrum are images of the bright ring in the light of excited sulfur, and the green ring is produced by oxygen. The relative brightness of the images allows astronomers to find the temperature and density of the gases in the ring. (George Sonneborn, GSFC, and NASA)

odd (Figure 14-12) in that it paused for a few weeks before rising to its final maximum. From photographs of the area made some years before, astronomers were able to determine that the star that exploded, cataloged as Sanduleak −69°202, was not the expected red supergiant but rather a hot, blue supergiant of only 15 solar masses and 50 solar radii, not extreme for a supergiant. Theorists now believe that the star was chemically poor in elements heavier than helium and had consequently contracted and heated up after a phase as a cool, red supergiant, during which it lost mass into space (Figure 14-15). The relatively small size of the supergiant may explain the pause in the light curve. Much of the energy of the explosion went into blowing apart the smaller, denser than usual star and making it expand (Figure 14-16).

The brightening of the supernova after the first few weeks seems to have been caused by the decay of radioactive nickel into cobalt. Theory predicts the production of such nickel atoms in the explosion, and their decay into cobalt would release gamma rays that would heat the expanding shell of gas and make it brighter. About 0.07 solar mass of nickel was produced, about 20,000 times the mass of the earth.

The cobalt atoms are also unstable, but they decay more slowly, so it was not until some time later, after much of the nickel had decayed, that the decay of cobalt into iron began providing energy to keep the expanding gas hot and luminous. Although these processes had been predicted, they were clearly observed in SN 1987A. Gamma rays from the decay of cobalt to iron were detected, and cobalt and iron are clearly visible in the infrared spectra of the supernova.

Two independent observations confirm that SN 1987A probably gave birth to a neutron star. Theory predicts that the collapse of a massive star's core should liberate a tremendous blast of neutrinos that leave the star hours before the shock wave from the interior blows the star apart. Two detectors, one in Ohio and one in Japan, recorded a burst of neutrinos passing through the earth at 2:35:41 AM EST on February 23, 1987, about 18 hours before the supernova was seen. The detectors caught only 19 neutrinos during a 12-second interval, but recall that neutrinos hardly ever react with normal matter. The full flood of neutrinos was immense. Within a few seconds of that time, roughly 20 trillion neutrinos passed harmlessly through your body. The detection of the neutrino blast confirms that the collapsing core gave birth to a neutron star.

The expanding gas shell of the supernova will continue to thin and cool, and eventually astronomers on the earth will be able to peer inside and see what is left of Sanduleak −69°202. Will we see the expected neutron star, or will theories need further revision? For the first time in almost 400 years, astronomers can observe a bright supernova to confirm or disprove their theories.

Radio and X-ray Observations of Supernova Remnants

Supernova remnants are often visible at radio and X-ray wavelengths. In fact, some are invisible at visual wavelengths and can be detected only as radio and X-ray objects. The nature of the radio waves and X rays these nebulae emit can tell us more about the nebulae and the explosions that form them.

Many nebulae emit energy (Figure 14-13b) called **synchrotron radiation**—energy radiated by high-speed electrons spiraling through a magnetic field (Figure 14-17). Low-speed electrons radiate at longer wavelengths, and high-speed electrons radiate at shorter

FIGURE 14-17
Synchrotron radiation is emitted when an electron spirals around a magnetic field line and radiates its energy away as photons. Only a few photons are shown here, for clarity.

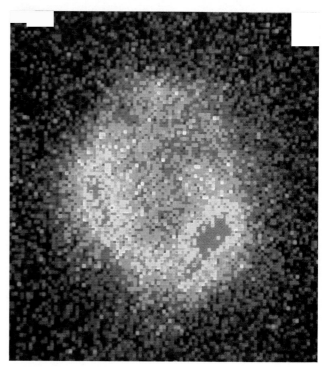

FIGURE 14-18
The supernova remnant Cas-A is visible in this X-ray image from the Exosat orbiting telescope. The X rays are produced by gas that is highly excited as it is compressed by debris expanding away from the explosion. (Compare with Figure 14-13b.) (W. Brinkman and B. Aschenbach of the Max-Planck Institute, J. Davelaar of Exosat Observatory, and the European Space Agency)

wavelengths, so synchrotron radiation is spread over a wide range of wavelengths. In the Crab Nebula, some of the synchrotron radiation has such short wavelengths it falls in the visible part of the spectrum, and we see it as light. Much of its radiation, however, falls in the X-ray region of the spectrum. This tells us that the Crab Nebula contains some very high speed electrons.

As the electrons in a nebula emit synchrotron radiation, they slow down. Thus, the moving electrons that produce synchrotron radiation must be continuously resupplied, or the synchrotron radiation will fade. To emit light, the electrons in the Crab Nebula must be traveling very fast, but over the 900-year history of the nebula, all of those electrons should have slowed down, and the Crab Nebula should have faded long ago. Something in the Crab Nebula must be continuously supplying energy to keep the electrons moving rapidly. We will solve this mystery in the next chapter when we meet the last remains of the star that exploded in 1054.

Orbiting X-ray telescopes have observed a number of supernova remnants. Since human eyes can't see X rays, the X-ray images are reproduced as false-color maps (Figure 14-18). These X rays are synchrotron radiation from gas at very high temperatures, so the X rays coming from supernova remnants tell us that the gas there is highly excited. The X-ray images often show the supernova remnants to be spherical shock waves expanding outward and heating the gas as the shock wave pushes into the interstellar medium. The Crab Nebula is not at all spherical. It is quite a young remnant and is filled with hot gas that is bright at X-ray wavelengths. We will see in the next chapter that this gas is heated by the remains of the star that exploded.

Supernovae are clearly powerful explosions, and we naturally wonder what it would be like to see such an explosion nearby. The sun is a low-mass star and will not die in a supernova explosion, but there are many stars ominously near us in space.

Local Supernovae and Life on the Earth

Although supernovae are rare events, they are very powerful and could affect life on planets orbiting nearby stars. In fact, supernovae explosions long ago may have affected the earth's climate and the evolution of life.

If a supernova occurred within about 50 ly of the earth, the human race would have to abandon the surface and live below ground for at least a few decades. The burst of gamma rays and high-energy particles from the supernova explosion could kill many life forms and cause serious genetic damage in others. The only way we could avoid this radiation would be to move our population into tunnels below the earth's surface. Of course, if a supernova did occur, we would not have time to dig enough tunnels.

Even if we could survive in tunnels long enough for the radioactivity on the earth's surface to subside, we might not like the earth when we emerged. Genetic mutation induced by radioactivity could alter plant and animal life so seriously that we might not be able to support our population. After all, we humans depend almost totally on grass for food. The basic human foods—milk, butter, eggs, wheat, tomatoes, lettuce, and meat (Big Macs, in other words)—are grass and similar vegetation processed into different forms. Seafood is merely

FIGURE 14-19
The most massive star known has a luminosity about 10 million times that of the sun, which suggests that it has a mass of about 100 solar masses. It apparently formed only a few million years ago with a mass of 200 solar masses, but it has expelled roughly half its mass to form the nebula shown in this infrared image. Its powerful solar wind continues to drive mass into space. At a distance of 25,000 ly, we should be safe when it explodes as a supernova within the next few million years. (Don F. Figer, UCLA, and NASA)

processed ocean plankton and plant life, which might also be altered or damaged by a local supernova explosion. Even if surface life survived the radiation, damage to the delicate upper layers of our atmosphere might alter the climate dramatically.

Local supernovae have been suggested as a possible cause of occasional climate changes and extinctions in the earth's past. It is possible that a supernova does occur near the earth every few hundred million years. Such an explosion has been suggested as a speculative explanation for the extinction of the dinosaurs.

However, we seem to be fairly safe for the moment. No star within 50 ly is known to be a massive giant capable of exploding as a supernova. Massive stars are rare and most are far away. The most massive star known is 25,000 ly away (Figure 14-19). Originally some 200 solar masses, the star has expelled half its mass and will probably explode as a supernova within a few million years or so. Massive stars are so rare that Type II supernovae caused by their collapse do not threaten us; but low-mass stars are common. A cooling white dwarf teetering on the edge of the Chandrasekhar limit could collapse as a Type I supernova. Even a nearby white dwarf could be too faint to notice until it exploded.

CRITICAL INQUIRY

Why does a Type II supernova explode?

A Type II supernova occurs when a massive star reaches the end of its usable fuel and develops an iron core. The iron is the final ash produced by nuclear fusion, and it cannot fuse because iron is the most stable element.

When energy generation begins to fall, the star contracts, but since iron can't ignite there is no new energy source to stop the contraction. In seconds, the core of the star falls inward and a shock wave moves outward. Aided by a flood of neutrinos and sudden convection, the shock wave blasts the star apart, and we see it brighten as its surface gases expand into space.

Type II supernovae are easy to recognize because their spectra contain hydrogen lines. Use what you know about Type I supernovae to explain why the spectra of these supernovae do not contain visible hydrogen lines. ■

Supernovae mark the death of massive stars, and white dwarfs mark the death of less massive stars. But so far we have discussed only single stars. At least half of all stars are binary. How do stars that are members of binary systems die? That is the subject of the next section.

14-3 THE EVOLUTION OF BINARY STARS

So far we have discussed the deaths of stars as if they were all single objects that never interact. But more than half of all stars are members of binary star systems. Most such binaries are far apart, and one of the stars can swell into a giant and eventually collapse without affecting the companion star. Some systems, however, are close together. When the more massive star begins to expand, it interacts with its companion star in peculiar ways.

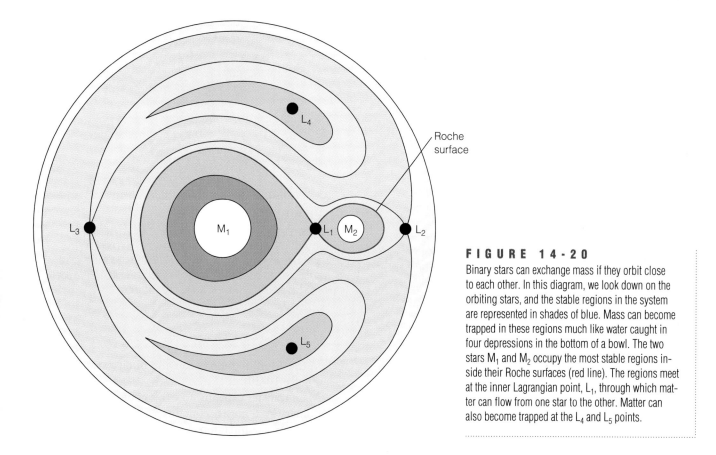

FIGURE 14-20
Binary stars can exchange mass if they orbit close to each other. In this diagram, we look down on the orbiting stars, and the stable regions in the system are represented in shades of blue. Mass can become trapped in these regions much like water caught in four depressions in the bottom of a bowl. The two stars M_1 and M_2 occupy the most stable regions inside their Roche surfaces (red line). The regions meet at the inner Lagrangian point, L_1, through which matter can flow from one star to the other. Matter can also become trapped at the L_4 and L_5 points.

These interacting binary stars are interesting objects themselves, but they are also important because they help us explain observed phenomena such as nova explosions. In the next chapter, we will use them to help us find black holes.

Mass Transfer

Binary stars can sometimes interact by transferring mass from one star to the other. To understand this process, we must understand how the gravity of the two stars controls the matter in the binary system.

Each of the two stars in a binary system is held together by its own gravity, and that gravity, combined with the rotation of the binary system, controls a teardrop-shaped surface around each star called the **Roche surface** (Figure 14-20). Matter inside a star's Roche surface is gravitationally bound to the star. The size of the Roche surface depends on the mass of the star and the separation between the two stars.

Matter in a binary system can flow from one star to another if it can flow through the point where the two Roche surfaces meet. This inner **Lagrangian point**[*] is a balance point between the stars. Other Lagrangian points mark balance points where gas can become trapped. The L_3 and L_2 points are unstable, and gas there tends to leak away; the L_4 and L_5 points are like depressions in the

bottom of a bowl where matter can collect. (We will use these Lagrangian points in later chapters to discuss planetary satellites and the orbits of asteroids.) The L_1 point, the inner Lagrangian point, is the only point through which matter can flow between the stars.

In general, there are only two ways matter can escape from a star and reach the inner Lagrangian point. First, if a star has a strong stellar wind, some of the gas blowing away from the star can pass through the inner Lagrangian point and be captured by the other star. Second, if an evolving star expands so far that it fills its Roche surface, it will be forced to take on the teardrop shape of the Roche surface. We have seen that effect in our study of binary stars (Figure 10-19). If the star expands farther, matter will flow through the inner Lagrangian point (like water in a pond flowing over a dam) and will fall into the other star. Mass transfer driven by a stellar wind tends to be slow, but mass transfer driven by an expanding star can occur rapidly.

Recycled Stellar Evolution

Mass transfer between stars can affect their evolution in surprising ways. In fact, it provides the explanation for a problem that puzzled astronomers for many years.

In some binary systems, the less massive star has become a giant, while the more massive star is still on the main sequence. If more massive stars evolve faster than lower-mass stars, how does the low-mass star in such binaries manage to leave the main sequence first? This is

[*]The Lagrangian points are named after French mathematician Joseph Louis Lagrange, who solved this famous mathematical problem around the time of the French Revolution.

FIGURE 14-21

The evolution of a close binary system. As the more massive star evolves (a), it fills its Roche surface (b) and begins to transfer mass through the inner Lagrangian point to its companion. The companion grows more massive (c), and the star losing mass collapses (in this example) into a white dwarf (d). The companion, now a massive star, evolves into a giant and begins transferring mass back to the white dwarf (e).

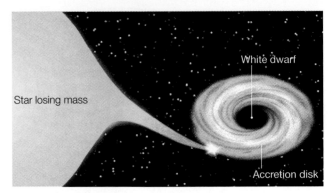

FIGURE 14-22

Matter falling into a compact object forms a whirling accretion disk. Friction and tidal forces can make the disk very hot. Through poorly understood processes within the disk, the matter can slow and eventually fall into the compact object. If the compact object is a white dwarf, a nova explosion can eventually result from this mass transfer.

called the Algol paradox after the binary system Algol (see Figure 10-20).

Mass transfer explains how this could happen. Imagine a binary system that contains a 5-solar-mass star and a 1-solar-mass companion (Figure 14-21). The two stars formed at the same time, so the more massive star will evolve faster and leave the main sequence first. When it expands into a giant, it can fill its Roche surface and transfer matter to the low-mass companion. Thus, the massive star could evolve into a lower-mass star, and the companion could gain mass and become a massive star still on the main sequence. We might find a system such as Algol containing a 5-solar-mass main-sequence star and a 1-solar-mass giant.

The evolution of close binary stars could result in one of the stars having its outer layers "peeled" away. A massive star expanding to become a giant could lose its outer layers to its companion and then collapse to form a lower-mass peculiar star. A few such peculiar stars are known. If the star explodes as a supernova, we would see the peculiar type of supernova called a peeled supernova.

Another exotic result of the evolution of close binary systems is the merging of the stars. We see many binaries in which both stars have expanded to fill their Roche surfaces and spill mass out into space. If the stars are close enough together and expand rapidly enough, theorists believe, the two stars could merge into a single, rapidly rotating giant star. Most giants rotate slowly because they conserved angular momentum as they expanded, but examples of rapidly rotating giants are known. Inside the distended envelope of such a star, the cores of the two stars could continue to orbit each other until friction slows them down and they sink to the center.

Yet another exotic possibility can arise if the massive star in a binary system transfers mass to its companion and then evolves to form a white dwarf. Such a system can later become a site of tremendous explosions when

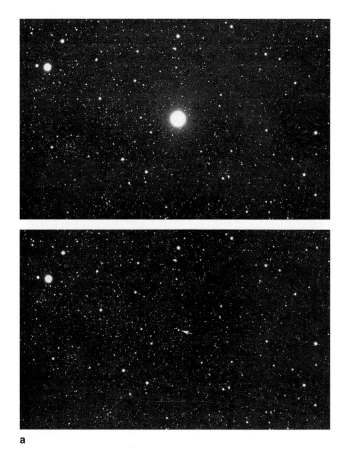

a

FIGURE 14-23

(a) Nova Cygni 1975 near maximum at about second magnitude (top) and later when it had declined to about eleventh magnitude (bottom). (UCO/Lick Observatory) (b) The variable R Aquarii (arrow) suffers nova-like outbursts as a cool giant star spills matter into a white dwarf companion. Eruptions eject jets of gas more than 4×10^{11} km long (toward top and bottom in this Hubble Space Telescope image). (NASA)

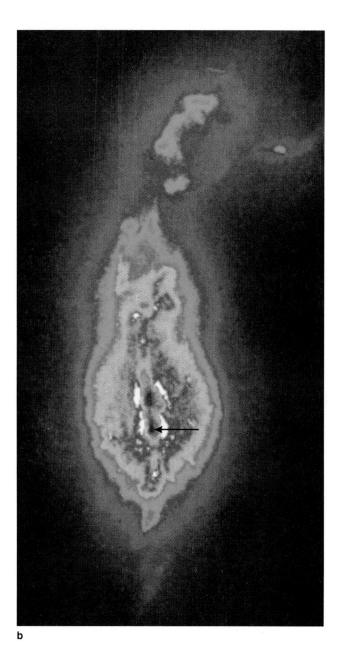

b

the companion begins to transfer mass back to the white dwarf. To see how these explosions occur, we must consider how mass falls into a star.

Accretion Disks

We say that matter flows through the inner Lagrangian point and falls onto the white dwarf, but the matter cannot fall directly into a white dwarf because of the conservation of angular momentum (as described in Chapter 5). Instead, it falls into a whirlpool around the white dwarf. For a common example, consider a bathtub full of water. Gentle currents in the water give it some angular momentum, but its slow circulation is not apparent until we pull the stopper. Then, as the water rushes toward the drain, conservation of angular momentum forces it to form a whirlpool. This same effect forces gas falling into a white dwarf to form a whirling disk of gas called an **accretion disk** (Figure 14-22).

Two important things happen in an accretion disk. First, the gas in the disk grows very hot due to friction and tidal forces. The disk also acts as a brake, ridding the gas of its angular momentum and allowing it to fall into

the white dwarf. The temperature of the gas in the inner parts of an accretion disk can exceed 1,000,000 K, and the gas can emit intense X rays. In addition, the matter falling inward from the accretion disk can cause a violent explosion if it accumulates on the white dwarf.

Novae

Nova is Latin for "new," and in astronomy it refers to the appearance of what seems to be a new star. A nova can appear in the sky, brighten in a few days, and then fade back to obscurity during the next few months (Figure 14-23a). A nova, however, is not a new star but the eruption of an old star, a white dwarf.

Nova explosions appear to be caused by the transfer of matter from a normal star, through an accretion disk, onto the surface of a white dwarf. Because the matter comes from the surface of a normal star, it is rich in unfused fuel, mostly hydrogen, and when it accumulates on the surface of the white dwarf, it forms a layer of unprocessed fuel.

As the layer of fuel grows deeper, it becomes hotter and dense and eventually becomes degenerate, much like the core of a sunlike star approaching the helium flash. In such a gas, the pressure–temperature thermostat does not work, so the layer of unfused hydrogen is a thermonuclear bomb waiting to explode. By the time the white dwarf has accumulated about 100 earth masses of hydrogen, the temperature at the base of the hydrogen layer reaches millions of degrees and the density is 10,000 times the density of water. Suddenly, the hydrogen begins to fuse on the proton–proton chain, and the energy released drives the temperature so high that the CNO cycle begins. With no pressure–temperature thermostat to control the fusion, in seconds the temperature shoots to 100 million degrees. Finally, the temperature is high enough to force the gas to expand, but by then the amount of energy released drives the surface layers of the star into space as a violently expanding shell of hot gas.

At a velocity typically 1500 km/sec, the surface of the white dwarf rushes outward, and the high temperature and rapid increase in surface area make the expanding shell very luminous. A nova can increase in brightness by a factor of 60,000 in a day or two and become 100,000 times more luminous than the sun. Spectra show tremendous Doppler shifts as the shell rushes outward. The shell, containing only about 0.0001 solar mass, eventually thins and cools, and the nova fades.

The explosion hardly disturbs the white dwarf and its companion star. Mass transfer quickly resumes, and a new layer of fuel begins to accumulate. How fast the fuel builds up depends on the rate of mass transfer, although most novae probably require 10,000 to 100,000 years to reach the explosive stage. Thus, novae can explode over and over as long as the companion star can feed mass onto its white dwarf companion.

CRITICAL INQUIRY

How do we explain the Algol paradox?

We see the Algol paradox when we look at binary star systems, such as Algol, that contain a low-mass giant star and a more massive main-sequence star. Since the two stars must have formed together, they are the same age, and the more massive star should have evolved faster and left the main sequence first. But in binary systems such as Algol, the lower-mass star has left the main sequence.

We can explain this paradox if we think about mass transfer. The first star to leave the main sequence will expand to become a giant star, and, if the stars are close enough to each other, the expanding star can fill its Roche surface and transfer mass through the inner Lagrangian point back to its lower-mass companion. The companion can grow more massive, and the giant star can grow less massive. Thus, the low-mass giant star we see in the system may have originally been a more massive star that evolved away from the main sequence first.

Mass transfer can explain the Algol paradox, and it can also explain the violent eruptions called novae. How does mass transfer explain why novae can explode over and over in the same star system? ■

The study of the deaths of stars has led us to discover astonishing objects of unbelievable density, temperature, and violence—all consequences of the victory of gravity over matter. But we have not considered the strangest circumstance of all. What happens when degenerate matter can't support the weight of the dying star, that is, when the mass of the compact object exceeds the Chandrasekhar limit? The answer is in the title of the next chapter. ■

■ *Summary*

When a star's central hydrogen-fusing reactions cease, its core contracts and heats up, igniting a hydrogen-fusing shell and swelling the star into a cool giant. The contraction of the star's core ignites helium first in the core and later in a shell. If the star is massive enough, it can eventually fuse carbon and other elements.

How a star evolves depends on its mass. Stars less massive than about 0.4 solar mass are completely mixed and have very little hydrogen left when they die. They cannot ignite a hydrogen shell or a helium core, so they become white dwarfs—degenerate stars about the size of the earth containing no nuclear reactions. Our galaxy does not seem old enough for any of these stars to have produced white dwarfs.

Medium-mass stars between about 0.4 and 4 solar masses become giants and fuse helium but cannot fuse carbon. They swell very large as their carbon–oxygen cores contract and eventually eject their surface layers to form planetary nebulae. Such nebulae are expanding gas ionized by the high-temperature core of the star left at the center. These planetary nebula nuclei eventually collapse to form white dwarfs.

Because stars lose mass through stellar winds, the mass of a star can decrease as it evolves. Thus, it seems possible for a star with a main-sequence mass of 6 solar masses to lose enough mass to produce a planetary nebula and a white dwarf. No white dwarf can exist if it is more massive than the Chandrasekhar limit of 1.4 solar masses, so the most massive stars cannot collapse into white dwarfs. Presumably, they collapse to form other compact objects—neutron stars or black holes.

The most massive stars can get hot enough to ignite carbon and other nuclear fuels. Eventually, they develop iron cores, which cannot release energy through fusion reactions, and collapse to produce supernova explosions. The reason a collapsing star explodes is not well understood, but it seems to be blasted apart by a shock wave rising from the interior. Type II supernovae appear to be caused by the collapse of a massive star, while Type I supernovae may be caused by the collapse of a white dwarf that has gained enough mass to exceed the Chandrasekhar limit. The ejected envelope of the star produces a supernova remnant.

Evolving stars in close binary systems can transfer matter from one star to the other, producing an accretion disk around the star gaining mass. This can dramatically alter the evolution of the stars. If the star gaining mass is a white dwarf, the added fuel on its surface can explode as a nova.

New Terms

red dwarf

thermal pulse

planetary nebula

compact object

black dwarf

Chandrasekhar limit

carbon detonation

supernova, Type I and Type II

supernova remnant

synchrotron radiation

Roche surface

Lagrangian point

accretion disk

nova

Questions

1. Why can't the lowest-mass stars become giants?

2. Presumably, all the white dwarfs we see in our galaxy were produced by sunlike stars of medium mass. Why couldn't any of the white dwarfs we see have been produced by the deaths of the lowest-mass stars?

3. What leads us to believe that stars can lose mass?

4. What kind of spectrum does the gas in a planetary nebula produce? Where does it get the energy to radiate?

5. The coolest stars we see at the center of planetary nebulae are about 25,000 K. Why don't we see planetary nebulae containing cooler central stars? (HINT: What kind of photons excite the gas in a planetary nebula?)

6. As white dwarfs cool, they move toward the lower right in the H–R diagram (Figures 14-4 and 9-11), maintaining constant radius. Why don't they contract as they cool?

7. All white dwarfs are about the same mass—somewhere between 0.5 and 1 solar mass. Why?

8. Why does a massive star explode when it develops an iron core?

9. What could cause a Type I supernova explosion? a Type II?

10. How do supernova remnants produce synchrotron radiation?

11. Why is a star in a close binary system unable to swell into a full-size giant star? Why can a star become a giant if it is in a binary system with a large separation?

Discussion Questions

1. If we could live for a billion years, we might notice a certain star exploding periodically as a nova and finally destroying itself in a supernova explosion. What could account for this evolution?

2. Given what you know about the lives of the stars, did our ancestors 1000 years ago see the same star patterns we see? What about the first humans 3 million years ago? What about the dinosaurs 65 million years ago?

Problems

1. How long would it take for a star of 0.4 solar mass to produce a white dwarf? (HINT: See Chapter 13.)

2. The Ring Nebula in Lyra is a planetary nebula with an angular diameter of 72 seconds of arc and a distance of 5000 ly. What is its linear diameter? (HINT: Use the small-angle formula.)

3. If the Ring Nebula is expanding at a velocity of 15 km/sec, typical of planetary nebulae, how old is it? (HINT: 1 ly = 9×10^{12} km and 1 y = 3.15×10^7 sec.)

4. Suppose that a planetary nebula is 1 pc in diameter and the Doppler shifts in its spectrum show that it is expanding at 30 km/sec. How old is it? (HINT: See Problem 3.)

5. A planetary nebula photographed 20 years ago and photographed today has increased its radius by 0.6 second of arc. If Doppler shifts in its spectrum show that it is expanding at a velocity of 20 km/sec, how far away is it? (HINT: First figure out how many parsecs it has increased in radius in 20 years. Then use the small-angle formula.)

6. The Crab Nebula is now 1.35 pc in radius and is expanding at 1400 km/sec. About when did the supernova occur?

7. Doppler shifts in the spectra of the Cygnus Loop show that it is now expanding at 70 km/sec. If it is increasing its radius by 0.03 second of arc per year, how far away is it? (HINT: How many kilometers will its radius increase in 1 year? Use the small-angle formula.)

8. The Cygnus Loop is now 2.6° in diameter and lies about 500 pc distant. If it is 50,000 years old, what was its average velocity of expansion? (HINT: Find its radius in parsecs first.)

9. The supernova remnant Cassiopeia A is expanding in radius at a rate of about 0.5 second of arc per year. Doppler shifts show that the velocity of expansion is about 5700 km/sec. How far away is the nebula?

10. Cassiopeia A has a radius of about 2.5 minutes of arc. If it is expanding at 0.5 second of arc per year, when did the supernova explosion occur? (There is no record of a supernova being seen at that time.)

Critical Inquiries for the Web

1. As seen in Figures 14-2 and 14-3, planetary nebulae show an incredible diversity in appearance. Browse the Web for images and information on these dying stars and discuss why we see such a range of shapes of planetary nebulae.

2. Naked-eye supernovae in our galaxy are rare, but astronomers have noted supernovae in other galaxies for years. Search the Web for summaries of observations of recent supernovae. Are similar numbers of Type I and Type II being seen? Compare the number of supernovae seen during the last few years with that of two decades ago. Why are we finding so many more supernovae in recent years than in the past?

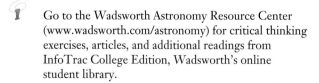 Go to the Wadsworth Astronomy Resource Center (www.wadsworth.com/astronomy) for critical thinking exercises, articles, and additional readings from InfoTrac College Edition, Wadsworth's online student library.

Chapter Fifteen

NEUTRON STARS AND BLACK HOLES

GUIDEPOST

The preceding four chapters have traced the story of stars from their birth as clouds of gas in the interstellar medium to their final collapse. This chapter

finishes that story by discussing the two most extreme of the three end states of stellar evolution. A star must end its existence as a white dwarf, neutron star, or black hole. White dwarfs were discussed in the preceding chapter; this chapter discusses the remaining two end states.

Almost anything is easier to get into than out of.

AGNES ALLEN

To talk about neutron stars and black holes, astronomers must use the language of relativity. Gravity is so strong near neutron stars and black holes that space itself is distorted and light rays bend. Throughout this chapter, remember that only the most sophisticated analysis of general relativity allows astronomers to understand these strange bodies.

This chapter ends the story of individual stars. The next chapter extends the story to the communities that stars live in—the galaxies. ■

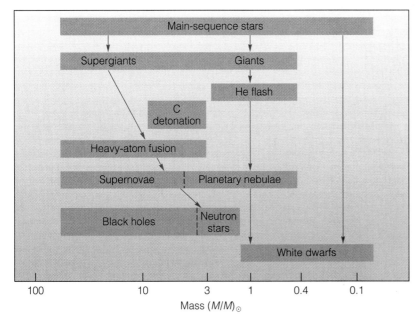

FIGURE 15-1

How a star evolves depends on its mass. The low- and medium-mass stars become white dwarfs. More massive stars can eventually collapse to form neutron stars or black holes, depending on the amount of mass left. Mass loss during the star's evolution can move it to the right in this diagram. The dividing line between stars that form planetary nebulae and stars that form supernovae is not known accurately. (Illustration design by author)

However a star dies, gravity ensures that its last remains must eventually reach one of three final states—white dwarf, neutron star, or black hole. These objects, often called compact objects, are small, high-density monuments to the power of gravity. Every star faces an ultimate collapse into such an object.

We discussed white dwarfs in the preceding chapter. In this chapter, we complete our story of stellar death by discussing the other two forms a dead star can take—neutron stars and black holes. In many ways, we learn more about normal stars by studying the remains of dead stars.

Throughout this chapter, we struggle to compare theoretical predictions with observations. Do neutron stars really exist? Are there really black holes? As we trace the story of this quest, we will see how science depends on evidence to confirm or disprove hypotheses. The search for evidence that these extreme objects really exist is a case study in scientific inquiry. It is also one of the greatest adventures in modern science, and it isn't over yet.

15-1 NEUTRON STARS

The mysterious object called a **neutron star** is the core of a star that has collapsed to a radius of only 10 km and to a density so high that only neutrons can exist. In the early 1960s, astronomers thought of neutron stars as speculative curiosities predicted by theory that might or might not exist. Now all astronomers accept that neutron stars are real. The effort to show that they exist has combined theoretical astrophysics, to predict their properties, with observational astronomy, to locate real objects with the predicted properties. Like partners in a complicated dance, theory and observation have interacted to produce something greater than the sum of the two parts.

Predicting the Properties of Neutron Stars

The existence of neutron stars was predicted over 50 years ago. The neutron was detected in the laboratory in 1932, and the following year Walter Baade and Fritz Zwicky predicted that supernova explosions might leave behind neutron stars.

If a supernova explosion leaves behind a mass between 1.4 and 2–3 solar masses, gravity will be strong enough to force it to collapse into a neutron star (Figure 15-1). The core of the star quickly collapses past the density of a white dwarf. Inside a white dwarf, the weight would be supported by degenerate electrons, but the gravity of the collapsing massive star is strong enough to overcome the pressure of the degenerate electrons. Under the enormous pressure, electrons combine with protons to form neutrons with the emission of a neutrino:

$$e + p \rightarrow n + \nu$$

Thus, in a burst of neutrinos, the core of the star is forced to become a gas of neutrons. Because neutrons can be packed tighter than electrons, the core of the star collapses further, but it eventually halts when the neutrons become degenerate.

The collapsed core of the star reaches stability at a radius of 10 km (Figure 15-2) and a density of about 10^{14} g/cm^3. On the earth, a sugar-cube-size lump of this material would weigh 100 million tons. This is roughly the density of an atomic nucleus, and, to a certain extent, we could think of the neutron star as a gigantic atomic nucleus.

Degenerate neutrons can support the weight of a neutron star up to a mass of 2–3 solar masses. This limit is not known exactly because the theoretical properties of pure neutron matter are not well understood. Nevertheless, theory predicts that a mass greater than 2–3 solar masses would not remain a neutron star but would

FIGURE 15-2

A tennis ball and a road map illustrate the size of a neutron star. Such a star would fit inside the beltway around Washington, D.C., with room to spare yet would contain more than the entire mass of the sun.

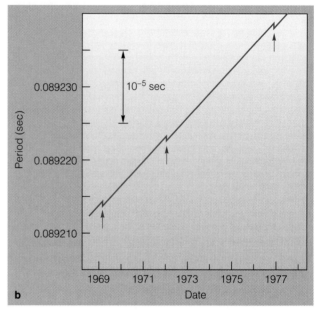

FIGURE 15-3

(a) The first pulsar, CP 1919 , was discovered November 28, 1967, when its regularly spaced pulses (marked by ticks) were noticed in the output of a radio telescope. (b) Careful measurements of the pulsation period of pulsars reveal that most are slowing down. Thus, their periods grow gradually longer. Glitches occur when a pulsar suddenly speeds up by a small amount. Here, three glitches in the pulsation of the Vela pulsar were recorded over an 8-year interval.

collapse into a black hole. Thus, theory predicts that neutron stars cannot be more massive than 2–3 solar masses.

The sudden collapse of a star down to a radius of 10 km can give the star a rapid rotation, a strong magnetic field, and a high temperature. All stars rotate, and as a star collapses it must rotate faster, conserving angular momentum, just as spinning ice skaters rotate faster as they draw in their arms. If the sun collapsed to a radius of 10 km, its rotation would increase from once every 25 days to 1000 times a second. A star collapsing into a neutron star would eject part of its mass and thus lose some angular momentum, but we could expect a newly formed neutron star to rotate 100 times a second.

The collapse to neutron star densities would also produce a powerful magnetic field. The magnetic field in a star is trapped in the ionized gases, and the field would be concentrated and strengthened as the star collapses. We might expect the neutron star to have a magnetic field a billion times stronger than the star had before its collapse. We will measure the strengths of magnetic fields using the unit **Gauss (G).** The average magnetic field at the surface of the earth is about 0.5 G, and the sun's average field is about 1 G. Some main-sequence stars have fields as strong as 1000 G, so a newly formed neutron star might have a magnetic field of 10^{12} G, about 3 million times stronger than the strongest field produced in laboratories on the earth.

We have seen in earlier chapters that the contraction of a star converts gravitational energy into thermal energy and heats the star. The extreme contraction of a star collapsing into a neutron star should heat it to very high temperatures. A surface temperature of 1,000,000 K would be typical.

We might expect such a hot object to be easily visible, but theory predicts otherwise. Neutron stars are small, and their small surface cannot radiate much energy even at such a high temperature. A typical neutron star 10 km in radius with a surface temperature of 1,000,000 K would radiate only about 15 percent as much black body radiation as the sun. Also, most of the energy radiated would be in the X-ray part of the spectrum and thus not visible to telescopes below the earth's atmosphere. Until the mid-1960s, theorists assumed that even if neutron stars did exist they would be very difficult to detect. Progress came not from theory but from observation.

The Discovery of Pulsars

In November 1967, Jocelyn Bell, a graduate student at the University of Cambridge, England, found a peculiar pattern on a paper chart from a radio telescope. Unlike other radio signals from celestial objects, this was a series of pulses (Figure 15-3a) with a highly regular period of 1.33730119 seconds. Bell and Anthony Hewish, the director of the experiment, investigated further and found that the signals could not be local—from an earth satellite, for instance. Day after day, the pulses came from the same place among the stars.

Proof

No scientific theory or hypothesis can be proved correct. We can test a theory over and over by performing experiments or making observations, but we can never prove that the theory is absolutely true. It is always possible that we have misunderstood the theory or the evidence, and the next observation we make might disprove the theory. In that sense, we never learn anything new until we disprove a theory or hypothesis. Only at that point do we discover something we did not know before.

For example, we might propose the theory that the sun is mostly iron. We might test the theory by looking at the iron lines in the solar spectrum, and the strength of the lines would suggest our theory is right. Although our observation has confirmed our theory, we know nothing more than we did before. We might continue to confirm the theory over and over, but we still know nothing new and there is always the danger that we have misunderstood the evidence. Finally, we might study the formation of atomic spectra and discover that iron atoms are very good absorbers of photons, so the strong iron lines in the solar spectrum do not mean that the sun is made mostly of iron. In fact, the hydrogen lines mean that the sun is mostly hydrogen atoms. Now that we have disproved our hypothesis, we have learned something new.

The nature of scientific hypothesis testing can lead to two kinds of mistakes. Sometimes nonscientists will say, "You scientists just want to tear everything down—you don't believe anything." It is the nature of science to test every hypothesis, not because the scientist wants to tear things down, but because the scientist wants to know what is the most dependable description of nature. Others will say, "You scientists are never sure of anything." Again, the scientist knows that no theory can ever be proved correct. That the sun will rise tomorrow is very likely, but in the end it is still a theory.

People will say of an idea they dislike, "That is only a theory," as if a theory were simply a random guess. In fact, a theory can be a well-tested truth in which all scientists have great confidence. Yet we can never prove that any theory is absolutely true. ∎

Another possibility—that the signals were coming from a distant civilization—led the investigators to consider naming it LGM, for Little Green Men. But within a few weeks the team found three more objects in other parts of the sky pulsing with different periods. The objects were clearly natural, and the team dropped the name LGM in favor of **pulsar**—a pulsing radio source.

As more pulsars were found, astronomers argued over their nature. Although the periods were almost as constant as an atomic clock, months of observation showed that many of the pulsars were slowing. Their periods were growing longer by a few billionths of a second per day (Figure 15-3b).

In early 1969, radio astronomers were surprised to discover that a pulsar had changed its period suddenly. The Vela pulsar has a period of about 0.089 second and is gradually slowing. But in early 1969 the pulsar suddenly sped up. It decreased its period by about 10^{-7} second (Figure 15-3b). Known as **glitches,** these sudden changes in period have been seen in a small number of pulsars. Something had to regulate the pulsation precisely, to slow it down gradually, yet be subject to occasional glitches.

Pulsars cannot be normal stars. A normal star is too big to pulse that fast. Nor can a star with a hot spot on its surface spin fast enough to produce the pulses. Some astronomers suggested spinning white dwarfs, but the fastest pulsar then known blinked 30 times a second. Even a white dwarf, the smallest known star, would fly apart if it spun 30 times a second.

The pulses themselves last only about 0.001 second, and that is a clue. If a white dwarf blinked on and then off in that interval, we would not see a 0.001-second pulse. The near side of the white dwarf would be about 6000 km closer to us, and light from the near side would arrive 0.022 second before the light from the bulk of the white dwarf. Thus, its short blink would be smeared out into a longer pulse. This is an important principle in astronomy: an object cannot change its brightness appreciably in an interval shorter than the time light takes to cross its diameter. If pulses from pulsars are no longer than 0.001 second, then the region emitting the energy cannot be larger than 300 km in diameter.

Only a neutron star is small enough to be a pulsar. In fact, a neutron star is so small it can't vibrate slowly enough, but it can spin as fast as 1000 times a second without flying apart. Thus, astronomers began to suspect that pulsars might be spinning neutron stars.

The missing link between pulsars and neutron stars was found in October 1968 when radio astronomers found a pulsar at the heart of the Crab Nebula (Figure 14-11). The Crab Nebula is the remnant of the supernova of 1054, and theory predicts that some exploding stars may leave behind neutron stars. The Crab Nebula pulsar is evidently such an object.

If we reconsider the theoretical properties of neutron stars and combine them with the observed properties of pulsars, we can devise a model of a pulsar.

A Model of a Pulsar

The term *pulsar* is a misnomer. The periodic flashing of pulsars is linked to rotation, not pulsation. The spinning neutron star emits beams of radiation that sweep around the sky. When one of these beams sweeps over us, we detect a pulse, just as sailors see a pulse of light when the beam from a lighthouse sweeps over their ship. In fact, this model is called the **lighthouse theory.**

The lighthouse theory is generally accepted (Window on Science 15-1), but astronomers are still not

FIGURE 15-4

Schematic diagram of a neutron star (yellow-green) with its powerful magnetic field (red). Beams of electromagnetic radiation blast out of the magnetic poles, and the rotation of the neutron star sweeps them around the sky like beams of light from a lighthouse.

confident of the mechanism that produces the beams. The theory suggests that the neutron star spins so fast and its magnetic field is so strong that it acts like a generator and creates an electric field around itself. This field is so intense that it rips charged particles, mostly electrons, out of the surface near the magnetic poles and accelerates them to high velocity. These accelerated electrons emit photons traveling in the same direction as the electrons. Thus, the photons leave the neutron star in narrow beams shining out of the magnetic poles. If the magnetic axis is inclined with respect to the axis of rotation, as is the case with the earth and most of the planets in the solar system that have magnetic fields, the neutron star will sweep the beams around the sky (Figure 15-4).

About 500 pulsars are known, but there are probably many more. Only when a pulsar's beams sweep over the earth do we detect its presence. In most cases, the beams of radio energy never point to the earth, and the pulsar remains almost invisible (Figure 15-5).

Two properties of pulsars support the lighthouse theory. First, many pulsars are slowing down. Their periods are increasing by a few billionths of a second each day. Astronomers can calculate that the energy released by a slowing neutron star, about 10^5 solar luminosities, matches the energy emitted by a pulsar. The agreement between these two numbers is evidence that pulsars are spinning neutron stars. For example, the hazy glow in the Crab Nebula comes from synchrotron radiation, high-speed electrons moving through a magnetic field. But in the 9 centuries since the explosion, the electrons should have slowed down and the nebula should have faded. The energy to keep the nebula glowing must come from the pulsar that lies at the center of the nebula (Figure 15-6). A spinning neutron star could produce enough energy to power the Crab Nebula continuously for many centuries.

The second property of pulsars that supports the lighthouse theory is the glitch—the sudden increase in the pulse rate seen in some pulsars. Two theories have been proposed to explain these changes, and both depend on spinning neutron stars.

One theory suggests that "starquakes" occur on the surface of a slowing neutron star. Models of neutron stars are uncertain because we cannot produce these high temperatures and densities on earth. Theoretical models, however, predict that a neutron star is a spinning body of neutron liquid containing protons and electrons that can move through the liquid with almost no resistance. This would make the liquid an almost perfect electrical conductor—a **superconductor.** This spinning superconductor probably produces the neutron star's magnetic field. Near the center there may be a solid core, and near the surface, where the pressure is lower, atomic nuclei can exist as a rigid crystal layer about 10^{16} times stronger than steel and roughly 1 km thick.

This model could explain glitches because the spinning neutron star would be slightly flattened, and as it slowed, its gravity would squeeze it until the crust broke in a "starquake" allowing the neutron star to become more nearly spherical. When this happened, conserva-

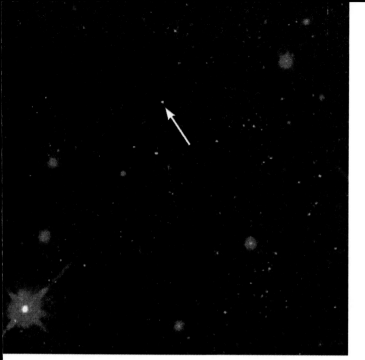

FIGURE 15-5

A solitary neutron star (arrow) glows faintly in this Hubble Space Telescope image. With a temperature of nearly 700,000 K, the neutron star cannot be larger than about 28 km in diameter. It is not known to pulse, perhaps because its beams do not sweep over the earth. (Fred Walter, State University of New York at Stony Brook, NASA)

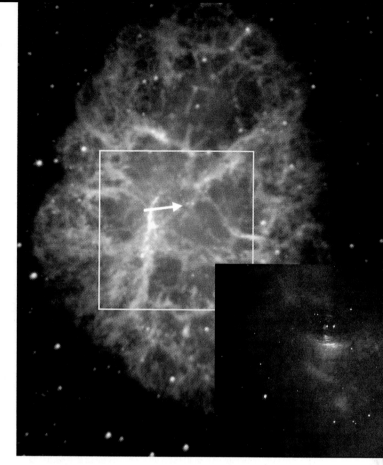

FIGURE 15-6

The Crab Nebula (Figure 14-11) is a hollow shell of filaments enclosing a hazy nebula. A pulsar blinking at the center (arrow) produces energy to power the synchrotron radiation from the nebula. Jets from the pulsar (inset) excite gas near the pulsar, producing changing wisps and waves in the nebula. (Palomar Observatory/Caltech; inset, Jeff Hester and Paul Scowen, Arizona State University, NASA)

tion of angular momentum would cause the spinning neutron star to speed up, and we would see a sudden increase in the pulse rate—a glitch.

A second theory proposes that circulation in the liquid interior, called vortexes, store angular momentum and that the slowing of the neutron star occasionally forces one of these vortexes to transfer its angular momentum to the neutron star as a whole. This would force it to spin a bit faster and produce a glitch.

Notice that both explanations of glitches depend on a pulsar being a spinning neutron star. Although we don't know which explanation is true—and both may be producing glitches—we do have a way for the theoretical object, the neutron star, to produce the observed events, the glitches. This concept gives astronomers further confidence that the lighthouse theory is correct.

Although the physics of neutron stars is not yet well understood, astronomers accept that pulsars are neutron stars. New observations are revealing how the neutron star in the Crab Nebula ejects beams of radiation and gas that illuminate the inner parts of the nebula (Figure 15-6 inset). Such observations are helping us understand how spinning neutron stars are born, how they age, and perhaps, how they die.

The Evolution of Pulsars

Many pulsars are observed to be traveling through space as fast as 450 km/sec—so fast that about half eventually escape from the galaxy. This high velocity must arise in

the power of the supernova explosion itself. Violent convection in the exploding core of the star (Chapter 14) may cause an asymmetry in the explosion that blasts the neutron star away at high velocity. Also, theoretical calculations show that the burst of neutrinos is so powerful that even a 1 percent asymmetry could eject the newborn neutron star at very high velocity.

The violent birth of a neutron star leaves it spinning very fast, perhaps 100 times a second. A newborn neutron star also contains a powerful magnetic field. This spinning magnetic field converts some of the neutron star's energy of rotation into radiation, and the neutron star gradually slows its spin, weakening the magnetic field. The average pulsar is apparently about 2×10^6 years old, and the oldest we can presently detect is about 10^7 years. Presumably, by the time a pulsar gets older than that, it is rotating too slowly to generate detectable beams.

If a pulsar contains a strong magnetic field and spins very fast, it is capable of emitting very strong beams of radiation. In addition, it is capable of emitting shorter-wavelength photons than older, slower pulsars. The Crab Nebula pulsar, the youngest known, emits pulses at

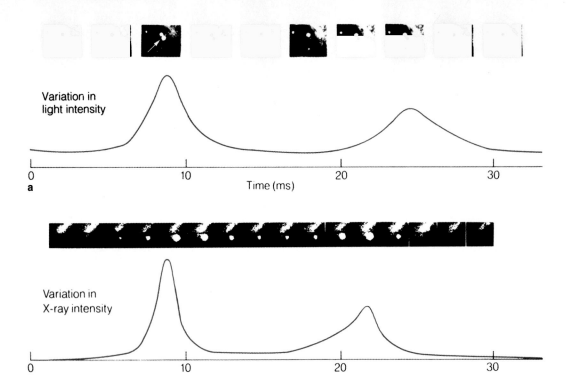

FIGURE 15-7

High-speed images of the Crab Nebula pulsar (arrow) show it pulsing at visible wavelengths (a) and at X-ray wavelengths (b). The period of pulsation is 33 milliseconds (ms), and each cycle includes two pulses as its two beams of unequal intensity sweep over the earth. (a, © AURA, Inc., National Optical Astronomy Observatories, KPNO; b, Courtesy F. R. Harnden, Jr., Harvard-Smithsonian Center for Astrophysics)

radio, infrared, visible, X-ray, and gamma-ray wavelengths (Figure 15-7). In fact, the pulsar has been identified as a star at the center of the nebula long thought to be the remains of a supernova. No one knew it blinked on and off 30 times a second because the blinks blended together when viewed through a telescope or on photographic plates. Not until the star was observed electronically in 1969 were the blinks detected.

At least four pulsars are known that produce visible pulses. All are fast-spinning neutron stars (Table 15-1), and three appear to be young neutron stars located inside supernova remnants. In addition to the Crab Nebula pulsar, the Vela pulsar produces optical pulses with a period of 0.089 second. It has an age of about 11,000 years—young for a pulsar. A pulsar in the Large Magellanic Cloud (LMC) blinks 20 times a second and produces visible pulses. Its age is unknown, but it is located in a supernova remnant. A fourth visible light pulsar, PSR 1937 + 21, pulses very fast but appears to be an old pulsar. (We will discuss PSR 1937 + 21 further below.)

Evidently, we should expect to find the youngest pulsars inside supernova remnants, as in the case of the Crab Nebula pulsar. However, not every supernova remnant contains a pulsar, and not every pulsar is located inside a supernova remnant. Many supernova remnants probably contain pulsars whose beams never sweep over the earth. It will be difficult to detect such pulsars. Also, some pulsars have high proper motions, which suggests that a supernova explosion can occur slightly off center or can disrupt a binary system. Either situation would give a pulsar a high velocity, and it could leave its supernova remnant quickly. Of course, supernova remnants do not survive more than 50,000 years or so before they mix into the interstellar medium. Because the average pulsar is about 2×10^6 years old, its supernova remnant was lost long ago.

As a pulsar grows older, it radiates away its energy and slows down, and its magnetic field should weaken. Thus, we would expect old pulsars to be slow and have weak magnetic fields. But in 1982, astronomers discovered that pulsar PSR 1937 + 21 seems to be old and fast. It pulses about 642 times per second, which is very fast for a pulsar, but it is slowing down only slightly, a sign of an old pulsar with a weak magnetic field. It has been called the millisecond pulsar because its period is roughly a millisecond, but several dozen similar pulsars have been identified, along with traces of at least 75 more in the globular cluster M22. There may be thousands of these millisecond pulsars in our galaxy. The puzzle is how an old pulsar can spin so fast.

TABLE 15-1

Optical Pulsars

Location	Identification	Period (sec)	Age	In Supernova Remnant
Crab	PSR 0531 + 21	0.033	900 years	Yes
Vela	PSR 0833 − 45	0.089	11,000 years	Yes
LMC	PSR 0540 − 69.3	0.050	Unknown	Yes
Vulpecula	PSR 1937 + 21	0.0016	Old	No

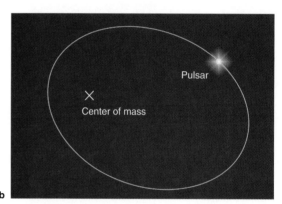

FIGURE 15-8

(a) The period of pulsation of pulsar PSR 1913 + 16 varies over an interval of 7.75 hours. If we treat this interval as the result of the Doppler effect, the resulting radial velocity curve resembles that of a spectroscopic binary. (b) Fitting an orbital solution to the data points reveals that the pulsar revolves in an elliptical orbit around the center of mass of a binary star system. (Adapted from data by Hulse and Taylor)

Two theories have been proposed. First, if an old pulsar with a weak magnetic field lies in a binary system, it can gain mass from its companion star. As that mass swirls around the neutron star and falls onto its surface, it can add angular momentum and spin the neutron star up, like water hitting a mill wheel. A second theory involves a white dwarf in a binary system. Matter transferred to the white dwarf might spin it faster and faster but would not add to its magnetic field. If the white dwarf collapsed into a neutron star, it would look like an old neutron star, but it would be spinning rapidly. The theoretical debate about the origin of millisecond pulsars continues. In any case, it seems that these fast pulsars are actually old objects that have been spun up to high speed, and thus they do not contradict our model of the evolution of pulsars.

The millisecond pulsars seem to be produced in binary systems, and many normal pulsars have also been found orbiting other stars. The study of these binary pulsars has produced some exciting discoveries.

Binary Pulsars

Of the hundreds of pulsars now known, a few are located in binary systems. These pulsars are of special interest because we can learn more about them by studying their orbital motion through the analysis of the arrival times of their pulses. If a pulsar in a binary system is moving toward us, then its pulses will arrive too early, and if it is moving away, the pulses will arrive too late. This is a consequence of the Doppler effect, and the radial velocity of the pulsar in its orbit can be calculated with great accuracy by comparing the arrival time of its pulses with the ticking of an atomic clock.

One of the most exciting discoveries in the study of pulsars was the 1974 discovery of the first binary pulsar by Joseph Taylor and Russell Hulse from the University of Massachusetts. Their detailed analysis of the pulse arrival times revealed that the pulsar was a member of a binary system in which the other object was also a neutron star but not a pulsar. The binary has an orbital period of only 7.75 hours, and the two neutron stars have an average separation roughly equal to the radius of our

sun. This object, PSR 1913 + 16, has become known as the binary pulsar (Figure 15-8), although other pulsars in binary systems have since been found.

As Taylor and Hulse continued to observe the binary pulsar, they were led to an exciting discovery about the nature of gravity—and to a Nobel prize. In 1916, Einstein published his general theory of relativity, which described gravity as a curvature of space-time. Einstein quickly realized that the motion of any mass would produce a disturbance in the curvature of space-time that would radiate outward at the speed of light, a disturbance known as a **gravitational wave.** Theory predicted gravitational waves, but no one had ever detected them. When Taylor and Hulse studied the binary pulsar, they realized that the orbital period is growing shorter and the orbit is shrinking. Evidently, the two stars are losing orbital energy and spiraling closer together. The energy lost

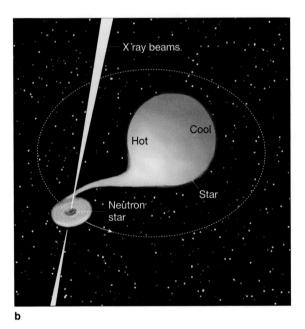

X ray beams

Cool

Hot

Star

Neutron star

a

b

◀ **FIGURE 15-9**

(a) X rays from Hercules X-1 disappear as the pulsar is eclipsed behind its stellar companion every 1.7 days. Insets show the pulsar "off" during an eclipse and "on" between eclipses. (J. Trümper, Max-Planck Institute) (b) A model of Hercules X-1 shows how X rays from the neutron star heat the near side of the companion star to 20,000 K. The rotating system shows us the hot side and then the cool side of the star, varying the visual brightness with the orbital period.

FIGURE 15-10

(a) X-ray bursters emit bursts of X rays that rise to full intensity in a few seconds and then fade in about 20 seconds. (b) The longer the interval since the preceding burst, the brighter the next burst will be.

from the orbiting neutron stars is exactly that predicted by general relativity. The energy lost is apparently carried away by gravitational waves. Although a number of teams around the world are building giant instruments in an attempt to detect gravitational waves, the orbital motion of the binary pulsar shows that such waves actually do exist. Taylor and Hulse won the 1993 Nobel prize in physics for their discoveries.

Binary pulsars can emit strong gravitational waves because the neutron stars contain large amounts of mass in a small volume. This also means that binary pulsars can be sites of tremendous violence because of the strength of gravity at the surface of a neutron star. Matter falling onto a neutron star can release titanic amounts of energy. If you dropped a single marshmallow onto the surface of a neutron star from a distance of 1 AU, it would hit with an impact equivalent to a 3-megaton nuclear warhead. Even a small amount of matter flowing from a companion star to a neutron star can generate high temperatures and release X rays and gamma rays.

Hercules X-1 is such a system, and we can analyze its behavior in some detail to see how astronomers study such systems. X rays from Hercules X-1 arrive in pulses

with a period of 1.2372253 seconds, apparently the period of rotation of a neutron star. But the pulses vanish for 5.8 hours every 1.7 days (Figure 15-9), suggesting that the binary system has an orbital period of 1.7 days and that the pulsar is eclipsed for 5.8 hours when it passes behind its companion star. Optical astronomers know this system as the variable star HZ Hercules, which varies in brightness with a period of 1.7 days. Careful analysis of all the data suggests that the system is a 2-solar-mass star with a temperature of about 7000 K orbiting with a neutron star. Mass flows from the normal star to the neutron star and generates X rays, probably in an accretion disk. The X rays heat one side of the normal star to a temperature of 20,000 K. As the system rotates, we alter-

a

b

c

FIGURE 15-11

(a) The center of globular cluster NGC 6624 is crowded with stars in this blue-light image made with the Hubble Space Telescope. (NASA) (b) The same image in ultraviolet light reveals that one star is intensely hot. It is the X-ray binary 4U1820-30. (NASA) (c) This artist's conception shows that the binary system consists of a neutron star at the center of an accretion disk that is ripping matter away from a companion white dwarf. As the matter accumulates on the neutron star, it erupts in bursts. (Dana Berry/STScI)

nately see the hot side of the star and then the cool side, and its brightness varies. Hercules X-1 is a complex system and is still not well understood.

Binary stars containing neutron stars orbiting normal stars can evolve in peculiar ways. For example, the Black Widow pulsar orbits its companion star with a period of only 9.16 hours—so close that tides squeezing and distorting the companion star may provide its power instead of nuclear fusion. Further, the pulsar is ripping matter from the companion star and consuming it; in just a few hundred million years, the Black Widow will have finished its meal, the companion star will be gone without a crumb remaining, and the pulsar will be alone. This may explain how some millisecond pulsars, which must be in binary systems to reach such high speeds, can be isolated neutron stars without companions. Like the Black Widow, they may have devoured their partners.

Observations first made in the 1970s hint at still more peculiar binary systems. Beginning in the 1970s, X-ray telescopes revealed that some objects emit irregularly spaced bursts of X rays. Typically, bursts that follow a long quiet period are especially large (Figure 15-10), and this suggests that some mechanism is accumulating energy that is released by the bursts. The longer the quiet phase, the more energy accumulates. These **X-ray bursters** were thought to be binaries containing neutron stars. The X-ray burster 4U1820-30 in the globular cluster NGC 6624, for example, appears to be a neutron star pulling mass away from a white dwarf (Figure 15-11). The helium from the white dwarf first flows into an accretion disk and then falls to the surface of the neutron star. The energy released by the in-falling gas heats the disk to roughly 100,000 K and emits a steady glow of X rays. When the helium accumulates on the surface of the neutron star to a depth of about 1 m, it explosively fuses into carbon and produces an X-ray burst. Notice the similarity with the mechanism that produces nova explosions on the surfaces of white dwarfs. Over two dozen X-ray bursters are known, and astronomers suspect that they are all related to binary systems containing neutron stars.

Gamma-ray telescopes above the earth's atmosphere have also detected burst sources called **gamma-ray bursters.** The first were seen in the 1960s by the Vela satellites, which monitored the earth for nuclear

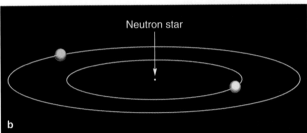

Neutron star

FIGURE 15-12

(a) The period of pulsar PSR 1257 + 12 varies irregularly by a fraction of a billionth of a second (dots). These variations are apparently produced by the motion of two planets orbiting the pulsar and causing it to move about the center of gravity of the system. The blue line in the graph shows the period variations that would result from two planets of the proper masses and orbital periods. (b) As the planets orbit the pulsar, their gravitational pull moves the pulsar about the center of mass by less than 800 km (invisibly small in this figure). Gravitational interactions between the planets change their orbits slightly and cause even smaller changes in the pulsar's period that have been observed, thus confirming that the pulsar is orbited by planets.

explosions in violation of the test-ban treaty. Those data were classified until 1973. More recently, the orbiting Gamma Ray Observatory observed a large number of gamma-ray bursters all over the sky. These bursts of gamma rays last from a few thousandths of a second up to 30 seconds, and their origin is unknown. One hypothesis, however, involves neutron star binaries and is worth discussing here because it illustrates the consequences of gravitational radiation and the potential violence locked in binary neutron stars.

Two neutron stars orbiting each other will emit gravitational radiation and spiral inward as they lose energy. Eventually, they must merge, presumably to form a black hole, and the resulting merger could be the source of the gamma-ray bursts. The final seconds in the life of a neutron star binary create tremendous violence as the two neutron stars orbit each other with a period of milliseconds and tides rip the stars apart. The tremendously hot matter eventually vanishes to form a black

hole, but in that moment of destruction a blast of gamma rays should be emitted. Perhaps this is the origin of the gamma ray bursters.

Incidentally, if such a merger of neutron stars happened at a distance of 1600 ly, the distance to a nearby binary pulsar, the gamma-ray burst would shower the earth with radiation equivalent to a 10,000-megaton nuclear blast. The gamma rays would create enough nitric oxide in the atmosphere to produce intense acid rain. This would also destroy the ozone layer and expose life on the earth to deadly levels of solar ultraviolet radiation. Such nearby gamma-ray bursts may occur as often as once every 100 million years and thus could be one of the causes of the mass extinctions that show up in the fossil record.

Neutron star binaries are rare compared to normal stars, but there may be 30,000 such systems in our galaxy. Counting all the neutron star binaries in other observable galaxies suggests that the Gamma Ray Observatory might detect 1000 bursts a year, roughly what is actually observed. If this hypothesis is true, then each gamma-ray burst marks the merger and final annihilation of a pair of stars that have lived their entire lives in orbit around each other.

What could be more fantastic than X-ray bursters and merging neutron stars? Neutron stars in binaries involve some of the most extreme physics known, and still more surprises are probably hidden in such binary systems. In fact, astronomers now know of at least one pulsar that seems to have planets.

Pulsar Planets

Finding planets orbiting stars other than the sun is very difficult, and only a few are known. Oddly enough, the first such planets found orbit not a normal star but the pulsar PSR 1257 + 12.

By using atomic clocks to time the signals from pulsar PSR 1257 + 12, Aleksander Wolszcan and Dale A. Frail have detected tiny changes in the period of the pulsar (Figure 15-12). Evidently these changes are caused by the Doppler shift, and they show that the pulsar is moving around a center of mass in a complex way. No model with a single stellar companion explains the observations. Rather, the pulsar must be orbited by at least two objects with planetlike masses. The gravitational tugs of the planets are responsible for making the pulsar wobble about the center of mass of the system by no more than about 800 km. Careful analysis reveals that the two planets have masses of 3.4 and 2.8 earth masses and orbit with periods of 66.6 days and 98.2 days. They lie only 0.36 and 0.47 AU from their neutron star sun.

Astronomers greeted this discovery with both enthusiasm and skepticism, but Wolszcan and Frail later announced that they had confirmed the existence of the planets by detecting small changes in their orbits caused by gravitational interaction between the planets. These interactions are predicted by Newton's laws, so the observation of these perturbations removes a great deal of doubt. In fact, the newest observations suggest the presence of a third planet of about 1 earth mass with a period of 1 year and a fourth planet of about one-third the mass of Jupiter with a period of 170 years.

How can a neutron star have a system of planets? Planets orbiting a star would be lost or vaporized when the star exploded as a supernova, and the inner planets orbiting PSR 1257 + 12 are so close that they would have been inside the giant star that exploded. Perhaps these planets are the remains of a stellar companion that was devoured by the neutron star. In fact, the pulsar is very fast (162 pulses per second), so it may have been spun up in a binary system.

We can imagine what these worlds might be like. Formed from the remains of dying stars, they might have chemical compositions richer in heavy elements than the earth. We can imagine visiting these worlds, landing on their surfaces, and hiking across their valleys and mountains. Above us, the neutron star would glitter in the sky, a tiny point of light.

CRITICAL INQUIRY

How can a neutron star be found at X-ray wavelengths?

First, we should remember that a neutron star will be very hot because of the heat released when it contracts to a radius of 10 km. It could easily have a surface temperature of 1,000,000 K, and Wien's law (Chapter 7) tells us that such an object will radiate most intensely at a very short wavelength typical of X rays. However, we know that the total luminosity of a star depends on its surface temperature and its surface area, and a neutron star is so small it can't radiate much energy and would be hard to find even with an X-ray telescope.

There is, however, a second way a neutron star can radiate X rays. If a normal star in a binary system loses mass to a neutron star companion, the inflowing matter will hit with so much energy that it will be heated to very high temperatures. It may form a very hot accretion disk that can radiate intense X rays easily detectable by X-ray telescopes orbiting above the earth's atmosphere.

If you discovered a pulsar, what observations would you make to determine whether it was young or old, single or a member of a binary system, alone or accompanied by planets? ■

Perhaps the strangest planets in the universe are those orbiting pulsars. But however strange a pulsar planet may be, we can imagine going to one. Our next topic, in contrast, seems beyond the reach of even our imaginations.

FIGURE 15-13

In 1572, Tycho Brahe saw a new star in the heavens, and the modern X-ray telescope Exosat reveals an expanding supernova remnant at the site of Tycho's star. But no star or pulsar remains. In some cases, supernova explosions may leave behind objects so massive they collapse into black holes rather than neutron stars. Whether that is true of Tycho's supernova, no one knows. (J. Trümper, Max-Planck Institute)

15-2 BLACK HOLES

If an object with a mass greater than 2–3 solar masses collapses, no known force in nature can stop it (Figure 15-13). The object reaches white dwarf density, but the degenerate electrons cannot support the weight, and the collapse continues. When the object reaches neutron star density, the degenerate neutrons cannot support the weight, and the collapse goes on. The object quickly becomes smaller than an electron. No force remains that can stop gravity from squeezing the object to zero radius and infinite density.

As an object shrinks, its density and the strength of gravity at its surface increase, and when an object shrinks to zero radius, its density and gravity become infinite. Mathematicians call such a point a **singularity**. Physically, it is difficult to think about objects of infinite density and zero radius, but even if such objects exist, they may not be visible to us. Theory predicts that they will be hidden inside a region of space called a **black hole.**

Although black holes are difficult to discuss without general relativity and sophisticated mathematics, we can use common sense and some simple physics to see why they form. Finding the velocity we need to escape from the gravity around a celestial body will help explain how black holes were first predicted theoretically and how they might be detected.

FIGURE 15-14
Escape velocity, the velocity needed to escape from a celestial body, depends on the mass and the radius of the body. The escape velocity at the surface of a small asteroid could be so low that an astronaut could jump off of its surface. But the mass of the sun compressed to the size of a small asteroid would have such a high escape velocity that nothing could escape, not even light.

Escape Velocity

Suppose we threw a baseball straight up. How fast must we throw it if it is not to come down? Of course, gravity would pull back on the ball, slowing it, but if the ball were traveling fast enough to start with, it would never come to a stop and fall back. Such a ball would escape from the earth. As we discovered in Chapter 5, the escape velocity is the initial velocity an object needs to escape from a celestial body (Figure 15-14).

Whether we are discussing a baseball leaving the earth or a photon leaving a collapsing star, the escape velocity depends on two things: the mass of the celestial body and the distance from the center of mass to the escaping object. If the celestial body had a large mass, the gravity at its surface would be strong and we would need a very high velocity to escape. If, for example, we

increased the mass of the earth but didn't change its radius, the escape velocity from its surface would increase because there was more mass inside. But there is another way to increase the earth's escape velocity: we could leave its mass fixed and squeeze it to a smaller radius. Then the surface of the earth would be closer to the center of mass, the gravity at the surface would be stronger, and the escape velocity would be higher. The important factor is the ratio of the mass and the radius. If the mass of the object divided by its radius is big enough, the escape velocity from its surface could be greater than the speed of light.

The Reverend John Mitchell, a British amateur astronomer, was the first person to realize that Newton's laws of gravity and motion contained this implication. In 1783, he pointed out that an object 500 times the radius of the sun but of the same density would have an escape velocity greater than the speed of light. Then "all light emitted from such a body would be made to return towards it." Mitchell had discovered the black hole.

Schwarzschild Black Holes

If the core of a star collapses and contains more than 2–3 solar masses, it will continue to collapse to a singularity and form a black hole. Some theorists believe that a singularity is impossible and that when we better understand the laws of physics we will discover that the collapse halts before diameter zero.

It makes little difference to us, however. If the object becomes small enough, the escape velocity nearby is so high that no light can escape. We can receive no information about the object or about the volume of space near it, and we refer to this volume as a black hole. The boundary of this region is called the **event horizon** (Figure 15-15), because any event that takes place inside the surface is invisible to an outside observer. To see how such a region can exist, we must consider general relativity.

In Chapter 5, we saw how general relativity explains a gravitational field as a curvature of space-time. That is, the presence of mass curves space-time in a way that we experience as gravity. Einstein published the equations that described his theory in 1916, and almost immediately astronomer Karl Schwarzschild found a way to solve the equations to describe the gravitational field around a single, nonrotating, electrically neutral lump of matter. That solution contained the first general relativistic description of a black hole; nonrotating, electrically neutral black holes are now known as Schwarzschild black holes.

Schwarzschild's solution showed that if matter is packed into a small enough volume, space-time is curved back on itself. Objects can still follow paths that lead into the region, but no path leads out, so nothing can escape —not even light. Thus, the inside of the black hole is totally beyond the view of an outside observer. The event horizon is the boundary between the isolated volume of space-time and the rest of the universe, and the radius of

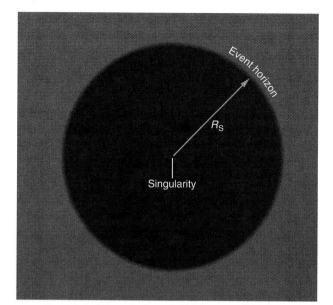

FIGURE 15-15
A black hole forms when an object collapses to a small size (perhaps to a singularity) and the escape velocity in its neighborhood is so great that light cannot escape. The boundary of this region is called the event horizon because any event that occurs inside is invisible to outside observers. The radius of the region is R_s, the Schwarzschild radius.

| **T A B L E** | **15-2** |

The Schwarzschild Radius

Object	Mass ($M_\odot$)	Radius
Star	10	30 km
Star	3	9 km
Star	2	6 km
Sun	1	3 km
Earth	0.000003	0.9 cm

the event horizon is called the **Schwarzschild radius** (**R_s**)—the radius within which an object must shrink to become a black hole.

Although Schwarzschild's work was highly mathematical, his conclusion is quite simple. The Schwarzschild radius depends only on the mass of the object:

$$R_s = \frac{2GM}{c^2}$$

In this simple formula, G is the gravitational constant (6.67×10^{-11} N·m^2/kg^2), M is the mass, and c is the speed of light. A bit of arithmetic shows that a 1-solar-mass black hole will have a Schwarzschild radius of 3 km, a 10-solar-mass black hole will have a Schwarzschild radius of 30 km, and so on (Table 15-2). Even a very massive black hole would not be very large.

Any object could be a black hole if it were smaller than its Schwarzschild radius. For example, if we could squeeze the earth to a radius of about 1 cm, its gravity would be so strong it would become a black hole. Fortunately, the earth will not collapse spontaneously into a black hole because its mass is less than the critical mass of 2–3 solar masses. Only exhausted stellar cores more massive than this can form black holes under the sole influence of their own gravity. In this chapter, we are interested in black holes that might originate from the deaths of massive stars. These would have masses larger than 3 solar masses. In the following chapters, we will meet black holes whose masses might exceed 10^6 solar masses.

Do not think of black holes as giant vacuum cleaners that will pull in everything in the universe. A black hole is just a gravitational field, and at a reasonably large distance its force is quite small. If the sun were replaced by a 1-solar-mass black hole, the orbits of the planets would not change at all.

Black Holes Have No Hair

Theorists who study black holes are fond of saying, "Black holes have no hair." By that they mean that once matter forms a black hole, it loses almost all of its normal properties. A black hole made of a collapsed star will be indistinguishable from a black hole made from peanut butter or fake-fur mittens. Once the matter is inside the event horizon, it retains only three properties—mass, angular momentum, and electrical charge.

The Schwarzschild black hole is represented by a solution to Einstein's equations for the special case where the object has only mass. Schwarzschild black holes do not rotate or have charge. The solutions for rotating or charged black holes (or for rotating, charged black holes) are more difficult and have been found in only the last few decades. Generally, rotating, charged black holes are similar to Schwarzschild black holes.

It seems that astronomers need not worry about charged black holes, because stars, whose collapse presumably forms black holes, cannot have large electrostatic charges. Suppose that we could give the sun a large positive charge. It would begin to repel protons in its corona and attract electrons and would soon return to neutral charge. Thus, we should expect black holes to be electrically neutral.

But everything in the universe seems to rotate, and collapsing stars spin rapidly as they conserve angular momentum. Thus, we should probably expect black holes to have angular momentum. In 1963, New Zealand mathematician Roy P. Kerr found a solution to Einstein's equations that describes a rotating black hole. This is now known as the **Kerr black hole.**

The rotation of a Kerr black hole splits the event horizon into two concentric surfaces, which touch at the poles (Figure 15-16). The region between these surfaces

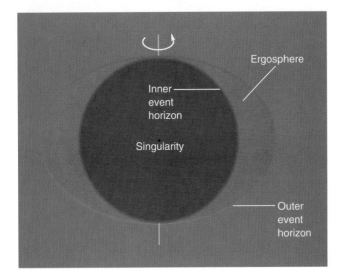

FIGURE 15-16

The Kerr black hole consists of a rotating mass at the center, surrounded by two event horizons. The outer event horizon marks the boundary within which an observer cannot resist being dragged around the black hole with space-time. The inner event horizon marks the boundary from within which an observer cannot escape. The volume between the event horizons is known as the ergosphere.

is known as the **ergosphere.** To understand the nature of the ergosphere, we must imagine that we are approaching a spinning black hole in a spaceship. Far away from the event horizon, all is well, but as we draw closer we begin to feel as if we were spinning around the black hole. We might feel dizzy, although our spaceship is keeping us from rotating compared with the distant stars. What we feel is the space-time near the black hole being dragged around with the spinning mass.

The outer event horizon marks the boundary beyond which we cannot go without being dragged around with the rotating black hole. We might still be able to escape falling into the black hole, but we could not resist being dragged along with space-time within the ergosphere. The inner event horizon marks the boundary from within which we could never escape.

The word *ergosphere* comes from the Greek word *ergo,* meaning "work," because the rotating space-time in the ergosphere can do work on a particle; that is, the particle can gain energy. In particular, the Kerr solution shows that a particle that enters the ergosphere can break into two pieces, one falling into the black hole and the other escaping with more energy than it had when it entered. Thus, energy can be extracted from a rotating black hole, and, as a result, the black hole slows its rotation very slightly.

The Kerr solution is a fascinating bit of theoretical physics, but it has an important application in astronomy. Almost certainly black holes rotate, and matter falling into black holes must pass through the ergosphere. Thus, we may eventually find situations where energy is extracted from rotating black holes.

TABLE 15-3

The Gravitational Red Shift

Object	Red Shift (percent)
Sun	0.0002
White dwarf	0.01
Neutron star	20
Black hole	Infinite

Leaping In

Because we cannot observe a black hole up close, let us examine its peculiar properties by using our imaginations. Let's imagine that we leap, feet first, into a Schwarzschild black hole.

If we were to leap into a black hole of a few solar masses from a distance of an astronomical unit, the gravitational pull would not be very large, and we would fall slowly at first. Of course, the longer we fell and the closer we came to the center, the faster we would travel. Our wristwatches would tell us that we fell for about 65 days before we reached the event horizon. We would not notice crossing the event horizon.

Our friends who stayed behind would see something different. They would see us falling slower as we came closer to the event horizon, because, as explained by general relativity, clocks slow down in curved space-time. This is known as **time dilation.** In fact, our friends would never actually see us cross the event horizon. To them we would fall slower and slower until we seemed hardly to move. Generations later, our descendants could focus their telescopes on us and see us still inching closer to the event horizon. We, however, would have sensed no slowdown and would conclude that we had crossed the event horizon after only about 65 days.

Another relativistic effect would make it difficult to see us with normal telescopes. As light travels out of a gravitational field, it loses energy, and its wavelength grows longer. This is known as the **gravitational red shift.** Although we would notice no effect as we fell toward the black hole, our friends would need to observe at longer and longer wavelengths in order to detect us. Eventually, the gravitational red shift would become so large that the photons would be stretched to nearly infinite wavelengths.

The gravitational red shift, as predicted by general relativity, has been observed in the sun and in white dwarfs. For example, in the spectrum of a typical white dwarf, the observed wavelength is lengthened by about 0.01 percent, and a spectral line with a wavelength of 500 nm would be observed at 500.05 nm. For a neutron star, the gravitational red shift is 20 percent, and, of course, for the event horizon of a black hole, the gravitational red shift is infinite (Table 15-3).

These relativistic effects seem peculiar, but other effects would be quite unpleasant. Imagine again that we are falling feet first toward the event horizon of a black hole. We would feel our feet, which would be closer to the black hole, being pulled in more strongly than our heads. This is a tidal force, and at first it would be minor. But as we fell closer, the tidal force would become very large (Figure 15-17). Another tidal force would compress us as our left side and our right side both fell toward the center of the black hole. For any black hole with a mass like that of a star, the tidal forces would crush us laterally and stretch us longitudinally long before we reached the event horizon. The friction from such severe distortions of our bodies would heat us to millions of degrees, and we would emit X rays and gamma rays.

Some years ago, a popular book suggested that we could travel through the universe by jumping into a black hole in one place and popping out of another somewhere far across space. That makes good science fiction, but tidal forces would make it an unpopular form of transportation even if it worked.

Our imaginary leap into a black hole is not entirely frivolous. We now know how to find a black hole. Look for a strong source of X rays. It may be a black hole into which matter is falling.

The Search for Black Holes

Do black holes really exist? Beginning in the 1970s, astronomers searched for evidence that their theories were correct. They tried to find one or more objects that were obviously black holes. That very difficult search is a good illustration of how the unwritten rules of science help us understand nature (Window on Science 15-2).

A black hole alone is totally invisible, but if gas is rapidly flowing into the black hole it will be heated and can emit X rays before the gas enters the event horizon. A solitary black hole in space would not gain much mass, but a black hole in a close binary system could drain mass from a companion star and form a hot accretion disk that would emit X rays as the gas spirals inward.

Some X-ray binaries, such as Hercules X-1, contain a neutron star and not a black hole. We can tell the difference because neutron stars must contain less than 3 solar masses. Thus, we can search for black holes by looking for X-ray binaries with a compact object that contains more than 3 solar masses.

Cygnus X-1, the strongest X-ray source in Cygnus, was the first black hole candidate, but it isn't conclusive. It consists of a supergiant O star with a strong stellar wind carrying gas into a hot accretion disk around a compact object (Figure 15-18). The compact object is invisible, but Doppler shifts in the spectrum of the O star suggest that the mass of the compact object is between 10 and 15 solar masses, well above the limit for a neutron star. But this evidence isn't conclusive. The O star might not be normal so we can't estimate its mass accurately, and there may be a third star in the system confusing the

FIGURE 15-17

Leaping feet first into a black hole. A person of normal proportions (left) would be distorted by tidal forces (right) long before reaching the event horizon around a typical black hole of stellar mass. Tidal forces would stretch the body lengthwise while compressing it horizontally. Friction from this distortion would heat the matter to high temperatures.

analysis. Cygnus X-1 isn't a conclusive example of a black hole.

X-ray telescopes have found more candidate black holes, and the list has grown to a few dozen. A few of

Fraud and the Publication of Science

Fraud is actually quite rare in science. The nature of science makes fraud difficult, and the way scientists publish their research makes it almost impossible. In fact, we can think of science as a set of unwritten rules of behavior that have evolved to prevent scientists from lying to one another or to themselves, even by accident.

Suppose for a moment that we wanted to commit scientific fraud. We would have to invent data supposedly obtained from experiment or observation. We might invent X-ray data supposedly obtained by observing an X-ray binary star. Or if we were interested in theory, we might invent a fraudulent mathematical calculation of the physics going on in an X-ray binary. We might get away with it for a short time, but one of the most important rules in science is that good results must be reproducible. Other people must be able to repeat our observations, experiments, and calculations. In fact, most scientists routinely repeat

other scientists' work as a way of getting started on a research topic. As soon as someone tries to repeat our fraudulent research, we will be caught. The more important a scientific result is, the sooner other scientists will repeat it, so we don't have much of a chance of getting away with scientific fraud. In this way, science is self-correcting.

Even if we could invent some convincing scientific research, we would probably have difficulty publishing it. When a scientist submits an article to a scientific journal, it is subject to peer review. That is, the editor of the journal sends the article to one or two other experts in the field for comment and suggestions. These reviewers often make helpful suggestions, but they may also point out errors that have to be fixed before the journal can publish the article. In some cases, an article may be so flawed the editor will refuse to publish it at all. If we submitted our fraudulent research on X-ray bina-

ries, the reviewers would almost certainly notice things wrong with it, and it would never get into print.

Scientists know the rules, and they use them. If someone makes a big discovery and is interviewed by the press, scientists will begin asking, "Has this work been published in a peer-reviewed journal yet?" That is, they want to know if other experts have checked the work. Until research is published, it isn't official, and most scientists would treat the results with care.

Fraud isn't impossible in science. There have been cases in some sciences that have been in the national news. Big grant money is a terrible temptation. But in science, "the truth will out." Because of the way scientists reproduce research and because of the way research is published, scientific fraud is quite rare. ■

---

these objects, such as the first two in Table 15-4, contain massive stellar companions, either giants, supergiants, or massive main-sequence stars. Such systems are difficult to analyze because the massive companions dominate the system. For example, LMC X-3 (LMC refers to the Large Magellanic Cloud, a small galaxy near our own) contains a massive B star and a compact object of 4 to 11 solar masses. One reason astronomers think the compact object is so massive is that it distorts the B main-sequence star into an egg shape, and as the system rotates, the light

from the B star varies as we see the side of the egg and then the end. By analyzing the light variation, astronomers can determine the shape of the B star and from that find the mass of the compact object.

Many black hole candidates are binaries in which the normal star is a lower-mass main-sequence star. Such systems don't remain steady X-ray sources but suffer X-ray nova outbursts as matter flows rapidly into the accretion disk. A year or so after the outburst, the flow has stopped, the accretion disk has dimmed, and telescopes can detect the spectrum of the main-sequence star. The spectral type and Doppler motions reveal the mass of the compact object with little uncertainty. The low-mass companion makes these X-ray novae systems easier to analyze than systems with massive companions.

Table 15-4 lists a few examples of these X-ray novae. V616 Mon is an old nova that erupted again in 1975. It contains an ordinary main-sequence K star and a compact object that orbit each other with a period of 7.75 hours. The orbital motion and the distortion of the K star tell us that the compact object must have a mass of 3.3 to 4.2 solar masses. V404 Cygni is a similar system containing a compact object with a mass between 8 and 15 solar masses. One of the best examples is J1655-40 (Nova Scorpii

TABLE 15-4

Six Black-Hole Candidates

Object	Location	Companion Star	Orbital Period	Mass of Compact Object
Cygnus X-1	Cygnus	O Supergiant	5.6 days	10–15 $M_\odot$
LMC X-3	Dorado	B3 main-sequence	1.7 days	4–11 $M_\odot$
V616 Mon	Monocerotis	K main-sequence	7.75 hours	3.3–4.2 $M_\odot$
V404 Cygni	Cygnus	K main-sequence	6.47 days	8–15 $M_\odot$
J1655-40	Scorpius	F-G main-sequence	2.61 days	4–5.2 $M_\odot$
QZ Vul	Vulpecula	K main-sequence	8 hours	5–14 $M_\odot$

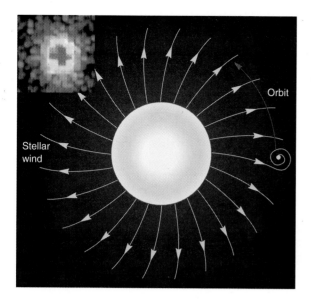

FIGURE 15-18

The X-ray source Cygnus X-1 (inset) is generally believed to be a super-giant O star in a binary system with a black hole. Some of the gas from the O star's stellar winds flows into an accretion disk around the black hole, and the hot accretion disk emits the X rays we detect. (Courtesy J. Trümper, Max-Planck Institute)

FIGURE 15-19

This artist's conception shows a disk of matter swirling around a black hole. As matter falls inward, it is heated to very high temperatures, and the inner parts of the accretion disk eject powerful beams of radiation and gas in opposite directions along the axis of rotation. This resembles the lower-power bipolar flows around protostars and the much more powerful jets ejected from the cores of galaxies. (Dana Berry, Space Telescope Science Institute)

1994). It contains a compact object of 4 to 5.2 solar masses. QZ Vul contains a compact object with a mass of at least 5 solar masses.

These candidates are more conclusive than Cygnus X-1 because their masses are more accurately known. There seems to be no way they could be neutron stars.

Another kind of evidence gives astronomers further confidence that black holes really exist. In a binary system, gas flowing toward a neutron star emits X-rays as it spirals inward through the accretion disk, but it also emits X-rays as it hits the neutron star's surface. Observations show that when the compact object has a mass greater than 3 solar masses, we don't detect the X-rays coming from the impact on the surface. Why? Because the gas has crossed the event horizon and entered a black hole.

Astronomers continue to test the theory of black holes, but the evidence has grown very strong. Black holes really do exist. The problem now is to understand how these objects interact with matter flowing into them through accretion disks to produce X rays, gamma rays, and jets of matter.

Jets of Energy from Compact Objects

Our impression of a black hole might suggest that it is impossible to get any energy out of such an object. In later chapters, however, we will meet galaxies that extract large amounts of energy from objects that may be massive black holes, so we should pause here to see how a compact object can produce energy.

Whether the compact object is a black hole or a neutron star, it has a strong gravitational field. Any material that falls into that field is accelerated inward and picks up speed. The falling matter converts its gravitational energy into energy of motion, and it arrives at the compact object traveling at a high velocity. As we have seen, matter cannot fall directly into a compact object but must usually fall into an accretion disk, and the in-falling material can heat such a disk to very high temperatures. Thus, we should expect a compact object with in-falling matter to possess a hot accretion disk, and if the flow of matter is strong, the disk will be very hot. Thermal and magnetic processes can cause such a system to eject powerful jets of gas and radiation along the axis of rotation of the accretion disk (Figure 15-19). This is very similar to the bipolar flows ejected by protostars, but much more powerful.

One of the best examples of this process is an X-ray binary called SS 433. Its optical spectrum shows sets of spectral lines that are Doppler-shifted by about one-fourth the speed of light, with one set shifted to the red and one set shifted to the blue. The object is both receding and approaching at fantastic speed.

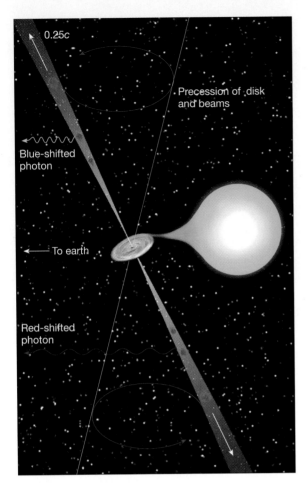

FIGURE 15-20

The generally accepted model of SS 433 includes a compact object with a very hot accretion disk producing beams of radiation with embedded blobs of gas traveling at one-fourth the velocity of light. The precession of the disk swings the beams around a conical path every 164 days.

FIGURE 15-21

The hot supergiant star, η Carinae, one of the most massive stars known, is losing mass asymmetrically. Known since 1677 as an irregular variable star, it erupted in 1843 to become the second brightest star in the sky. It is now hardly visible to the naked eye. This Hubble Space Telescope image reveals two clouds of gas expanding in opposite directions while jets and flows emerge from the central plane between them. Note the faint jet extending from the 5 o'clock position to 11 o'clock. A deeper image (inset) shows outer clouds of gas ejected earlier. Such a massive star may be doomed to explode as a supernova. (Jon Morse, University of Colorado, NASA)

Apparently, SS 433 is a binary system in which a compact object (a neutron star) pulls matter from its companion star and forms an extremely hot accretion disk. The disk is so hot it is blasting radiation and matter away in beams aimed in opposite directions. The precession of the disk sweeps these beams around the sky once every 164 days, and we see light from gas trapped in both beams. One beam produces a red shift, and the other produces a blue shift (Figure 15-20). SS 433 is a prototype that illustrates how the gravitational field around a compact object can produce powerful beams of radiation and matter.

Observations made with the largest and newest telescopes are revealing that stars often lose mass, not as spherical shells but as jets, beams, and oppositely directed flows. For example, the star η Carinae is believed to be as much as 100 times the sun's mass and is losing mass rapidly, but high-resolution images from the Hubble Space Telescope reveal complex, asymmetrical structures including jets (Figure 15-21). The rotation of a star defines an axis, and mass flowing inward or outward can be focused into disks, beams, and jets. Even a black hole has a rotational axis, thus we should not be surprised by jets emerging from compact objects.

CRITICAL INQUIRY

How can a black hole emit X rays?

Once a bit of matter falling into a black hole crosses the event horizon, no light or other electromagnetic radiation it emits can escape from the black hole. The matter becomes lost to our view. But if it emitted radiation before it crossed the event horizon, that radiation could escape, and we could detect it. Furthermore, the powerful gravitational field near a black hole stretches and distorts in-falling matter, and internal friction heats the matter to millions of degrees. Wien's law (Chapter 7) tells us that matter at such a high temperature should emit X rays. Any X rays emitted before the matter crosses the event horizon will escape, and thus we can look for black holes by looking for X-ray sources. Of course, an isolated black hole will probably not have much matter falling in, but black holes in binary systems may have large amounts of matter flowing in from the companion star. Thus, we can search for black holes by looking for X-ray binaries.

The search for black holes has succeeded in finding a few strong candidates, but the problem is being sure the binary system contains a black hole and not a neutron star. What observations would you make of an X-ray binary system to distinguish between a black hole and a neutron star? ∎

Compact objects emitting X rays and producing precessing jets of radiation and gas may not be as unusual as they seem. Many stars collapse to form black holes or neutron stars in binary systems, but these are objects of only a few solar masses. We will see similar phenomena many times more powerful when we explore the galaxies in Chapters 16, 17, and 18. ∎

∎ Summary

If the remains of a star collapse with a mass greater than the Chandrasekhar limit of 1.4 solar masses, the object cannot reach stability as a white dwarf. It must collapse to the neutron star stage with a radius of about 10 km and a density equal to that of an atomic nucleus. Such a neutron star can be supported by the pressure of its degenerate neutrons. But if the mass is greater than 2–3 solar masses, the degenerate neutrons cannot stop the collapse, and the object must become a black hole.

Theory predicts that a neutron star should rotate very fast, be very hot, and have a strong magnetic field. Such objects have been identified as pulsars, sources of pulsed radio energy. Pulsars are evidently spinning neutron stars that emit beams of radiation from their magnetic poles. As they spin, they sweep the beams around the sky, and, if the beams sweep over the earth, we detect pulses. This is known as the lighthouse theory. The spinning neutron star slows as it radiates energy into space, and glitches in its pulsation appear to be caused by starquakes in the crust of the neutron star.

A few pulsars have been found in binary systems. One such binary pulsar is losing orbital energy at the rate predicted by the theory of gravitational waves. In some binary systems, such as Hercules X-1, mass flows into a hot accretion disk around the neutron star and causes the emission of X rays. In other systems, called X-ray bursters, the accumulation of fuel on the neutron star causes periodic outbursts of X rays.

If a collapsing star has a mass greater than 2–3 solar masses, it must contract to a very small size—perhaps to a singularity, an object of zero radius. Near such an object, gravity is so strong not even light can escape, and we term the region a black hole. The surface of this region, called the event horizon, marks the boundary of the black hole. The Schwarzschild radius is the radius of this event horizon, amounting to only a few kilometers for black holes of stellar mass.

If we were to leap into a black hole, we would experience peculiar effects. Our friends who stayed behind would see two

relativistic effects. Our clocks would slow down relative to our friends' clocks because of time dilation in the strong gravitational field. Also, the gravitational red shift would cause the light we emitted to be shifted to longer wavelengths. As we fell into the black hole, we would feel tidal forces that would deform and heat our mass until we grew hot enough to emit X rays. Any X rays emitted before we crossed the event horizon could escape.

To search for black holes, we must look for binary star systems in which mass flows into a compact object and emits X rays. If the mass of the compact object is greater than 2–3 solar masses, the object is presumably a black hole. A few such objects have been located.

SS 433 is an extreme case of an X-ray binary; its accretion disk is hot enough to drive blobs of gas away in its beams of radiation. SS 433 is considered a small prototype of the powerful energy machines found in the centers of some galaxies.

∎ New Terms

neutron star	singularity
Gauss (G)	black hole
pulsar	event horizon
glitch	Schwarzschild radius (R_s)
lighthouse theory	Kerr black hole
superconductor	ergosphere
gravitational wave	time dilation
X-ray burster	gravitational red shift
gamma-ray burster	

∎ Questions

1. Why is there an upper limit to the mass of neutron stars? Why is that upper limit not well known?

2. Explain in detail why we expect neutron stars to be hot, spin fast, and have strong magnetic fields.

3. Why can't we use visual-wavelength telescopes to locate neutron stars?

4. Why does the short length of pulsar pulses eliminate normal stars as possible pulsars?

5. What do we mean when we say "Every pulsar is a neutron star, but not every neutron star is a pulsar"?

6. According to our model of a pulsar, if a neutron star formed with no magnetic field at all, could it be a pulsar? Why or why not?

7. Why did astronomers first assume that the millisecond pulsar was very young?

8. Why do we suspect that only very fast pulsars can emit visible pulses?

9. How does an X-ray burster resemble a nova?

10. If the sun were replaced by a 1-solar-mass black hole, how would the earth's orbit change?

11. What do we mean when we say "Black holes have no hair"?

12. What evidence do we have that black holes exist?

▪ Discussion Questions

1. Has the existence of neutron stars been sufficiently tested to be called a theory, or should it be called a hypothesis? What about the existence of black holes?

2. Why would you expect an accretion disk around a star the size of the sun to be cooler than an accretion disk around a compact object?

3. In this chapter, we imagined what would happen if we jumped into a Schwarzschild black hole. From what you have read, what do you think would happen to you if you jumped into a Kerr black hole?

▪ Problems

1. If a neutron star has a radius of 10 km and rotates 642 times a second, what is the speed of the surface at the neutron star's equator in terms of the speed of light?

2. Suppose that a neutron star has a radius of 10 km and a temperature of 1,000,000 K. How luminous is it? (HINT: See Chapter 9.)

3. A neutron star and a white dwarf have been found orbiting each other with a period of 11 minutes. If their masses are typical, what is their average separation? Compare the separation with the radius of the sun, 7×10^5 km. (HINT: See Chapter 10.)

4. If the accretion disk around a neutron star has a radius of 2×10^5 km, what is the orbital velocity of a particle at its outer edge? (HINT: Use circular velocity, Chapter 5.)

5. What is the escape velocity from the surface of a typical neutron star? How does that compare with the speed of light? (HINT: See Chapter 5.)

6. If the earth's moon were replaced by a typical neutron star, what would the angular diameter of the neutron star be as seen from the earth? (HINT: Use the small-angle formula, Chapter 3.)

7. If the inner accretion disk around a black hole has a temperature of 1,000,000 K, at what wavelength will it radiate the most energy? What part of the spectrum is this in? (HINT: Use Wien's law, Chapter 7.)

8. What is the orbital period of a bit of matter in an accretion disk 2×10^5 km from a 10-solar-mass black hole? (HINT: Use circular velocity, Chapter 5.)

9. If SS 433 consists of a 20-solar-mass star and a neutron star orbiting each other every 13.1 days, what is their average separation? (HINT: See Chapter 10.)

▪ Critical Inquiries for the Web

1. Imagine that you are on a mission to explore one of the pulsar planets noted in this chapter. What would you find there? Look on the Web for information about pulsars and the known pulsar planets and describe what you might encounter on such a mission.

2. What would you experience if you were to pilot a spacecraft near a black hole? Visit related Internet sites to determine what the gravitational effects and general environment would be like. Also, use the Internet to find the limits of human tolerance to strong gravitational forces. (HINT: Look for information about astronaut training and find out how many Gs a human can withstand.) Using these sources, give a brief account of what your voyage would be like.

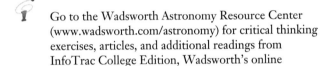 Go to the Wadsworth Astronomy Resource Center (www.wadsworth.com/astronomy) for critical thinking exercises, articles, and additional readings from InfoTrac College Edition, Wadsworth's online student library.

Chapter Sixteen

THE MILKY WAY GALAXY

GUIDEPOST

This chapter plays three parts in our cosmic drama. First, it introduces the concept of a galaxy. Second, it discusses our home, the Milky Way Galaxy, a natural object of our curiosity. Third, it elaborates our story of stars by introducing us to galaxies, the communities in which stars exist.

Science is based on the interaction of theory and evidence, and this chapter will show a number of examples of astronomers using evidence to test theories. If the theories seem incomplete and the evidence contradictory, we should not be disappointed. Rather, we must conclude that the adventure of discovery is not yet over.

We struggle to understand our own galaxy as an example. We will extend the concept of the galaxy in Chapters 17 and 18 on normal and peculiar galaxies. We will then apply our understanding of galaxies in Chapter 19 to the study of the universe as a whole. ■

Jane was watching Mrs. Corry splashing the glue on the sky and Mary Poppins sticking on the stars. . . . "What I want to know," said Jane, "is this: Are the stars gold paper or is the gold paper stars?" There was no reply to her question and she did not expect one. She knew that only someone very much wiser than Michael could give her the right answer.

P. L. TRAVERS

Mary Poppins

a

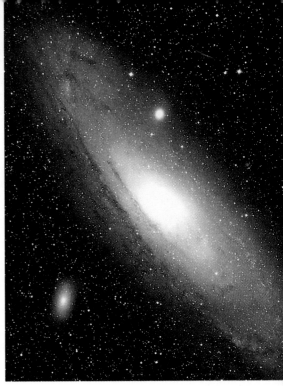

b

FIGURE 16-1

(a) From the inside, we see our galaxy as a faintly luminous path that circles the sky. Invisible from the brightly lit skies of cities, the Milky Way is easily visible from a dark site. This artwork shows the location of a portion of the Milky Way near a few bright winter constellations. (b) If we could leave our galaxy and photograph it from a distance, it would look much like the Great Galaxy in Andromeda, a spiral galaxy a bit over 2 million ly from us. (Bill Scheening and Vanessa Harvey, NSF REU, NOAO, © AURA, Inc.) (See Figure 16-5 for a chart of a region of the summer Milky Way. Also, see the star charts at the end of this book to locate the Milky Way throughout the year.)

Before you plunk down big bucks to go to France and see the Eiffel Tower or to go to Nepal and see the Himalayan Mountains, take a shorter trip and see the biggest attraction of all. Get away from city lights and you can see the Milky Way, the band of light that the Greeks called *galaxies kuklos*, the "milky circle." The Romans changed its name to *via lactea*, the "milky road." Today, we enjoy its beauty and recognize its significance. Seen from the inside, it is the galaxy we live in (Figure 16-1a).

If you leave the light pollution of the city behind and look up at the night sky, almost every object you see is part of our Milky Way Galaxy. Exceptions include the **Magellanic Clouds,** small irregular galaxies located in the southern sky, and the Andromeda Galaxy, just visible to our unaided eyes as a faint patch of light in the constellation Andromeda (Figure 16-1b). Our galaxy probably looks much like the Andromeda galaxy. All the other stars and nebulae we see in the sky are part of our galaxy. Yet only within the last century have astronomers discovered that we live in a galaxy and that there are tens of billions of other galaxies.

P. L. Travers wrote her story about Mary Poppins, the supercalifragilisticexpialidocious nanny, in 1934, and at that time no one could answer the question that Jane asked in the quotation given on the previous page. But by the 1960s, astronomers could assure her that the gold paper was stars, or rather that the atomic elements of which gold paper is made were created inside generation after generation of stars. How the stars made the elements is one of the keys to understand our Milky Way Galaxy.

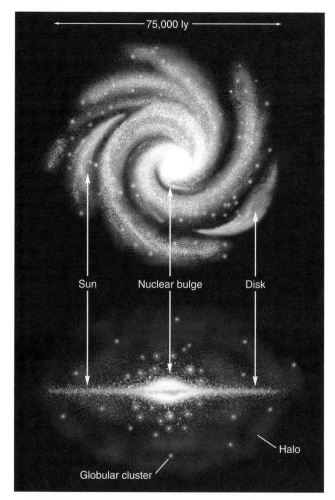

FIGURE 16-2
An artist's conception of our Milky Way Galaxy, seen face-on and edge-on, shows the shape and location of the disk, halo, and nuclear bulge. Note the position of the sun and the distribution of globular clusters in the halo.

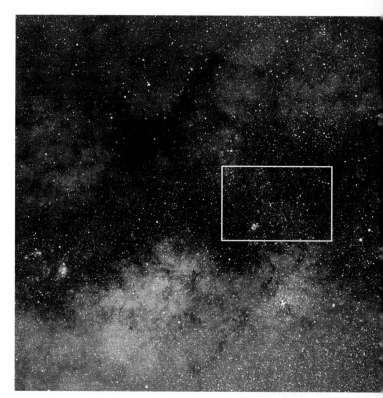

FIGURE 16-3
Vast clouds of gas and dust lie in the disk of the Milky Way Galaxy and block our view when we try to look farther than a few kiloparsecs in the plane of the galaxy. The center of our galaxy is located in the center of the box in this photo, but we receive no visible light from the center because of intervening dust clouds. (Daniel Good)

This chapter is about processes. We are interested in the scientific process by which we learn about our galaxy and in the natural processes that operate within the galaxy. Like themes in a piece of music, these processes intertwine throughout this chapter.

16-1 The Nature of the Milky Way Galaxy

The secret to studying any science lies in the simple question, "How do we know?" Because all scientific knowledge is based on evidence and logical arguments about natural processes, we need to know facts. But we also need to understand how we know. Thus, we set the stage in the next section with a few facts, but our real interest is in how we know what our galaxy is like.

The Structure of Our Galaxy

We live in a disk-shaped galaxy with beautiful spiral arms reaching from its center to its edge. The disk is about 25,000 pc (75,000 ly) in diameter, with the sun located about two thirds of the way from the center to the edge (Figure 16-2).

The disk of the galaxy contains most of the stars and nearly all of the gas and dust. Although most of the stars in the disk are like the sun or fainter, a small number are brilliant O and B stars, and they light up the disk and make it bright. It is difficult to judge just how big the disk is because it is filled with clouds thick with dust that blocks our view. Even when we look toward the center of our galaxy (Figure 16-3), the clouds of gas and dust prevent us from seeing further than a few **kiloparsecs (kpc).** (A kiloparsec, is 1000 parsecs.)

The disk of our galaxy is surrounded by an extended **halo.** This spherical cloud of stars contains many individual stars, about 200 globular clusters, and almost no gas and dust. The **nuclear bulge** at the center of our galaxy has a radius of about 3 kpc. It contains a larger number of stars much like those in the halo.

This quick portrait of our galaxy is just bare facts. It raises fascinating questions. How do we know what our galaxy is like? How did it get this way?

First Studies of the Galaxy

How do astronomers know what our galaxy is like? Humanity has known of the Milky Way, the hazy path of light around the sky, since antiquity, but not until Galileo looked at the sky with his telescope in 1610 did anyone know the Milky Way was made of stars. Little more was known for two centuries after Galileo lived.

By the mid 18th century, astronomers generally understood that the stars were other suns, but they had little understanding of how the stars were distributed in space. One of the first people to study this problem was the English astronomer Sir William Herschel

a

b

c

FIGURE 16-4

(a) William Herschel made many discoveries with the telescopes he built. As commemorated on this stamp, he discovered the plant Uranus, but his most significant work may have been his mapping of the "grindstone" shape of our galaxy. (b) Seen edge-on, his model of the universe appears to be an irregular disk with the sun near the center. (c) By analogy with the stone wheels used to grind flour, the model was known as the "grindstone model."

(1738–1822) (Figure 16-4a). He and his sister Caroline (1750–1848) mapped the stars in three dimensions by assuming that the sun was located inside a great cloud of stars and that they could see to the edges of the cloud in any direction. They believed that by counting the number of stars visible in different directions, they could gauge the relative distance to the edge of the cloud. If their telescope revealed few stars in one direction, they assumed that the edge was not far away, and if they saw many stars in another direction, they assumed that the edge was very distant. Calling their method "star gauges," they counted stars in 683 directions in the sky and outlined a model of the star cloud. Their data showed that the cloud was a great disk with the sun near the center, and using the technology of their day as an analogy, they called it the "grindstone model" (Figure 16-4b, c).

This model of the star system explains the most obvious feature of the Milky Way; it is a glowing path that circles the sky. The grindstone model reproduces this as a disk of stars with the sun near the center. Unfortunately, there was no way for the Herschels to find the diameter of the disk.

Generations of astronomers studied the size of our star system, culminating with Jacobus C. Kapteyn (1851–1922). In the early 20th century, he analyzed the brightness, number, and motion of stars and concluded that our star system is a disk about 10 kpc in diameter and about 2 kpc thick, with the sun near the center.

Why did Kapteyn's model underestimate the diameter of the galaxy? Why did such models mistakenly place the sun near the center of the star system? Because astronomers assumed that space was empty and that they could see to great distances. Today, we know that the disk of our galaxy is filled with gas and dust that dim stars and block our view. We can see only a small region near us in the disk of our galaxy. How we discovered the true size of our galaxy and came to understand the nature of a galaxy, is one of the great adventures of human knowledge. It began about a century ago when a man studying star clusters and a woman studying variable stars unlocked one of nature's biggest secrets—that our star system is a galaxy.

Discovering the Galaxy

It seems odd to say that astronomers discovered something that is all around us, but until early in this century, no one knew that the star system we live in is a galaxy. That began to change in the years following World War I, when a young astronomer named Harlow Shapley (1885–1972) discovered how big our star system really is. Besides being one of the turning points of modern astronomy, Shapley's study illustrates one of the most common techniques in astronomy. If we want to know how we know things in astronomy, then Shapley's story is well worth tracing in detail.

Shapley began by noticing that open star clusters lie along the Milky Way but that more than half of all glob-

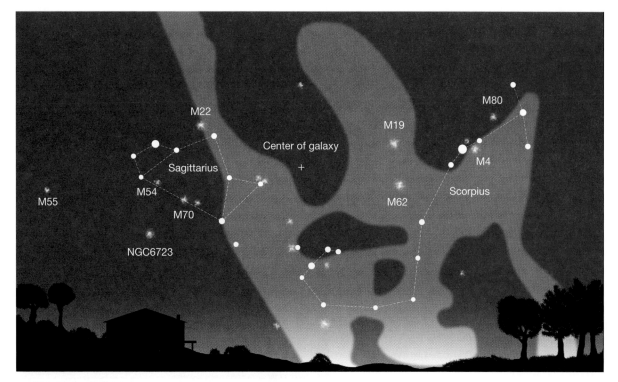

FIGURE 16-5

A few of the brighter globular clusters in the Scorpius-Sagittarius region. The brightest (shown here as hazy spots labeled according to their catalog descriptions) are visible through good binoculars on a dark night. The position of the constellations is shown as they appear above the southern horizon on a summer night from a latitude of 40°N, typical for most of the United States.

ular clusters lie in or near the constellation Sagittarius (Figure 16-5). Shapley suspected that the center of the star system was not near the sun but somewhere in the direction of Sagittarius.

To find the distance to the center of the star system and thus the size of the star system as a whole, Shapley needed to find the distance to the star clusters, but that was difficult to do. The clusters are much too far away to have measurable parallaxes. They do, however, contain variable stars (Chapter 13), and those were the lampposts Shapley needed to find the distances to the clusters.

Shapley knew of the work of Henrietta S. Leavitt (1868–1921), who in 1912 had shown that a Cepheid variable star's period of pulsation was related to its brightness. Leavitt worked on stars whose distances were unknown, so she was unable to find their absolute magnitudes or luminosities, but astronomers realized that the Cepheids could be a powerful tool in astronomy if their true luminosities could be discovered. Ejnar Hertzsprung (co-discoverer of the H–R diagram) made an early attempt to determine the absolute magnitudes of these Cepheids, but Shapley went further and produced the first usefully calibrated period–luminosity diagram (see Figure 13-21). The diagram showed not only the absolute magnitudes of Cepheids, but also the absolute magnitudes of the related variable stars, the RR Lyrae stars, which Shapley had found in the star clusters. The vari-

able stars were now calibrated lampposts that Shapley could use to find the distances to the clusters (Window on Science 16-1).

Finding distance using Cepheid variable stars is so important in astronomy we should pause to illustrate the process. Suppose we studied a star cluster and discovered that it contained a Type I Cepheid with a period of 10 days and an average apparent magnitude of 10.5. How far away is the cluster? From the period–luminosity diagram (Figure 13-21), we see that the absolute magnitude of the star must be about −3. Then the distance modulus is:

$$m - M_v = 10.5 - (-3) = 13.5$$

We could use this distance modulus and a table such as Table 9-1 to estimate the distance to the cluster as somewhere between 1000 and 16,000 pc. We can be more accurate if we solve the magnitude–distance formula (Chapter 9) for distance:

$$d = 10^{\frac{m - M_v + 5}{5}}$$

Now we can substitute the apparent magnitude and absolute magnitude, and we discover that the distance is equal to $10^{3.7}$. A pocket calculator tells us that the distance is about 5000 pc. This is one of the most common calculations in astronomy.

Calibration

Astronomers often say that Shapley "calibrated" the Cepheids for the determination of distance, meaning that he did all the detailed background work so that the Cepheids could be used to find distances. Other astronomers could use the Cepheids without repeating the detailed calibration.

Calibration is actually very common in science. Chemists, for instance, have carefully calibrated the colors of certain compounds against acidity. They can quickly measure the acidity of a solution by dipping into it a slip of paper containing the indicator compound and looking at the color. They don't have to repeat the careful calibration every time they measure acidity.

Engineers in steel mills have calibrated the color of molten steel against its temperature. They can use a hand-held device to measure the color of a ladle of molten steel and then look up the temperature in a table. They don't have to repeat the calibration every time.

Astronomers have made the same kind of color–temperature calibration for stars.

As you read about any science, notice how calibrations are used to simplify common measurements. But notice, too, how important it is to get the calibration right. An error in calibration can throw off every measurement made with that calibration. Some of the biggest errors in science have been errors of calibration. ■

Once Shapley had calibrated the period–luminosity relation, he could use it to find the distance to any cluster in which he could identify variable stars. He could measure the apparent magnitudes of the variable stars from photographic plates, and he could determine the periods of pulsation from a series of photographic plates. He could find the absolute magnitudes of the stars from their periods of pulsation and the period–luminosity diagram. Once he knew both the apparent and the absolute magnitudes, he could calculate the distance to the cluster.

This worked well for the nearer globular clusters, but the variable stars in the more distant clusters are too faint to detect. Shapley estimated the distances to these more distant clusters by calibrating the diameters of the clusters. For the clusters whose distance he knew, he could use their angular diameters and the small-angle formula to calculate linear diameters in parsecs. He found that the nearby clusters are about 25 pc in diameter, which he assumed is the average diameter of all globular clusters. He then used the angular diameters of the more distant clusters to find their distances.

When he plotted the direction and distance to the globular clusters, he found that they comprised a swarm centered thousands of parsecs away in the direction of Sagittarius (Figure 16-6), and he reasoned that this must be the center of the entire star system. The distance to the center of the swarm of globular clusters revealed that our star system is much bigger than anyone had supposed.

Shapley's estimate made our star system 10 times bigger than Kapteyn's model. Earlier astronomers had underestimated the size of the star system because they didn't know that dust dims stars and blocks our view. Herschel was not counting stars to the edge but only as far as he could see through the dust. Shapley looked at globular clusters out in the halo away from the dusty disk. His first estimate was a bit too large. Modern studies fix the diameter of the star system at roughly 25 kpc (often rounded to 75,000 ly). Our sun is about 8.5 kpc from the center with an uncertainty of ±1 kpc.

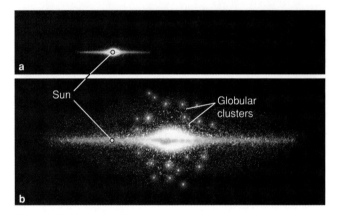

FIGURE 16-6

(a) Kapteyn's model of the star system was small and centered on the sun. We understand now that dust in interstellar clouds spread through the plane of our galaxy, blocks our view, and makes the galaxy seem small and sun-centered. (b) By looking above and below the plane of the galaxy, Shapley was able to see out into the halo and map the distance to the globular clusters. When he found that the center of the swarm of globular clusters was far from the sun, he assumed that center was also the center of the galaxy. Thus, he could find the distance to the center of the galaxy even though gas and dust clouds blocked his view.

The size of our star system was startling, but a bigger surprise awaited. In 1924, Edwin Hubble, namesake of the Hubble Space Telescope, photographed variable stars in the Andromeda Nebula (Figure 16-1b) and then measured its distance. He found that it was far beyond the edge of our star system. It was another star system like our own. Many more such star systems were found scattered through space, and these objects became known as galaxies. Our Milky Way star system was revealed to be a galaxy in a universe filled with galaxies.

Shapley's discovery of the size of our star system and its subsequent recognition as only one of many galaxies was a turning point in astronomy. Older models placed

F I G U R E   1 6 - 7

Just as the inset map projects the entire surface of the earth onto a flat oval, this infrared image from the IRAS satellite projects the entire sky onto a flat map. The plane of our galaxy runs horizontally from left to right with the center of the galaxy at the center of the oval. The red glow in this false color image is produced by dust in the disk of the galaxy warmed by starlight. The thickness of the distribution of dust is evident in this image. (Warmer dust in our solar system is located along the ecliptic and produces the faint blue s-shaped curve.) (NASA)

the sun at the center of a small, isolated star system. Within a decade, astronomers realized that we do not live at the center and that our star system is just a galaxy. Like Copernicus, Shapley helped to change the way we think about our place in the cosmos—not a special place, but rather a typical place. He said, "[It] is a rather nice idea because it means that man is not such a big chicken. He is incidental—my favorite term is 'peripheral.' "*

An Analysis of the Galaxy

We want to know how we know, and we are now ready to explore our modern knowledge of the galaxy. We can survey the galaxy's components, understand how we have learned about them, and begin to understand what they tell us about the nature of our home galaxy.

We can divide our galaxy into two main components whose characteristics will help us understand how our galaxy formed. The **disk component** of the galaxy contains most of the stars, gas, and dust that orbit within the disk. In contrast, the **spherical component** includes the nuclear bulge and the halo.

In addition to stars, the disk of the galaxy contains vast amounts of gas and dust—the interstellar medium. We saw in Chapter 11 how we can learn about this material. It contains mostly hydrogen gas and helium gas and is irregularly distributed as dense clouds separated by less dense regions. The densest of these clouds are the giant molecular clouds within which stars form (Chapter 12), and thus nearly all star formation in our galaxy takes place in the disk. The most massive of these newborn stars are the most luminous, the O and B stars. They light up the disk and make it glow brightly. Some of these stars are hot enough to ionize the nearby gas to produce HII regions, and thus the disk contains all of the emission nebulae in our galaxy.

Although we can survey the location and distance of the stars (Chapter 9), we cannot cite a single number for the thickness of the disk because it lacks sharp boundaries. Stars become less crowded farther from the central plane of the galaxy. Also, the thickness of the disk depends on the kind of object we study. Stars like the sun with ages of a few billion years lie within about 500 pc above and below the central plane. But the youngest stars, including the O and B stars, and the gas and dust from which these young stars are forming are confined to a disk only about 100 pc thick (Figure 16-7). With a diameter of 25 kpc, the disk is thinner than a thin pizza crust.

The most dramatic features of the disk are the spiral arms—long spiral patterns of bright stars, HII regions, star clusters, and clouds of gas and dust. The sun is located on the inside edge of one of these spiral arms. We will see later in this chapter how we can observe the spiral arms.

*Harlow Shapley, *Through Rugged Ways to the Stars* (New York: Scribner, 1969), p. 60.

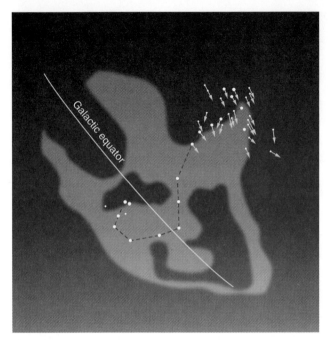

FIGURE 16-8

Many of the stars in the constellation Scorpius are members of an O and B association. They have formed recently from a single gas cloud and are moving together southwest along the Milky Way (shaded). Arrows show observed proper motions.

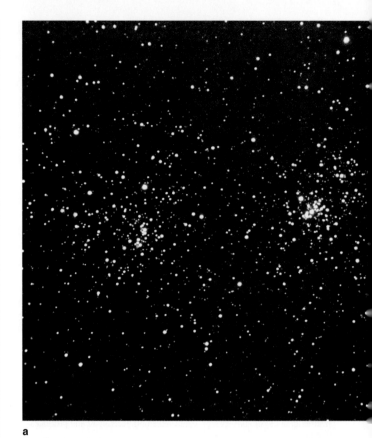

a

b

FIGURE 16-9

Two kinds of star clusters. (a) Open clusters contain a few hundred to a few thousand stars and are found in the disk of our galaxy. This pair of clusters, known as the double cluster in Perseus, is visible in binoculars. (Celestron) (b) Globular clusters are approximately the same size as open clusters, but they contain 50,000 to a million stars and are thus much more stable. Globular clusters are found in the halo of our galaxy. This cluster, known as 47 Tucanae, is one of the largest. (NOAO)

We see two kinds of star groupings in the disk of our galaxy. **Associations** are groups of 10 to a few hundred stars so widely scattered in space that their mutual gravity cannot hold the association together. From the turn-off points in their H-R diagrams (Chapter 13) we can tell they are very young groups of stars. We see them moving together through space (Figure 16-8) because they formed from a single gas cloud and haven't had time to wander apart. Two kinds of association—O and B associations and T Tauri associations—are located along the spiral arms.

The second kind of cluster in the disk is the open cluster (Figure 16-9a), a group of 100 to a few thousand stars in a region about 25 pc in diameter. Because they have more stars in less space than associations, open clusters are more firmly bound by gravity. Although they lose stars occasionally, they can survive for a long time, and the turn-off points in their H-R diagrams give ages from a few million to a few billion years.

In contrast to the disk, the halo of our galaxy contains almost no gas and dust and only a thin scattering of stars and globular clusters. Because the halo contains no dense gas clouds, it cannot make new stars. Halo stars are old, cool, lower main-sequence stars and red giants. It is difficult to judge the extent of the halo, but it could extend far beyond the edge of the visible disk and may be slightly flattened like a thick hamburger bun.

The halo contains roughly 200 globular clusters, each of which contains 50,000 to a million stars in a sphere about 25 pc in diameter (Figure 16-9b). Because they contain so many stars in such a small region, the clusters are very stable and have survived for billions of years. From the turn-off points in their H-R diagrams, we know they are more than 10 billion years old.

The nuclear bulge at the center of the galaxy is the most crowded part of the spherical component. Visible above and below the plane of the galaxy and through gaps in the obscuring dust, the nuclear bulge contains stars that are old and cool like the stars in the halo.

Our analysis of the components of our galaxy leave us wondering about one critical question. How much matter does our galaxy contain? To answer that question, we must watch our galaxy rotate.

The Mass of the Galaxy

To find the mass of an object, astronomers must observe it in orbital motion as, for example, in a binary star system. We don't live long enough to see our galaxy rotate significantly, but astronomers can observe the radial velocities, proper motions, and distances of stars and then calculate their orbits. The results can tell us the mass of the galaxy and give us hints about its origin.

Stars in the disk of the galaxy follow nearly circular orbits in the plane of the disk (Figure 16-10a). The sun, for example, moves at about 220 km/sec in the direction of Cygnus following a nearly circular orbit 8.5 kpc in radius. It must take about 240 million years per orbit.

Halo stars and globular clusters, however, follow highly elongated orbits tipped steeply to the plane of the disk (Figure 16-10b). Although we show these orbits as elliptical, they are more like rosettes because of the influence of the thick bulge. Where halo stars pass through the disk, they are cutting across the regular traffic of disk stars and seem to have very high velocities. Thus, they are called **high-velocity stars.** The dramatic difference between the motion of halo stars and disk stars will be important evidence when we discuss the formation of the galaxy in the next section.

The motion of the sun can tell us the mass of the galaxy. When we studied binary stars in Chapter 10, we saw that the total mass (in solar masses) of a binary star system equals the cube of the separation of the stars a (in AU) divided by the square of the period P (in years):

$$M = \frac{a^3}{P^2}$$

The radius of the sun's orbit is about 8500 pc, and each parsec contains 206,265 AU. Multiplying, we find that the radius of the sun's orbit is 1.75×10^9 AU. The orbital period is 240 million years, so the mass is:

$$M = \frac{(1.75 \times 10^9)^3}{(240 \times 10^6)^2} = 0.93 \times 10^{11} M_\odot$$

This is only a rough estimate because it does not include any mass outside the orbit of the sun. Adding an estimate for the mass outside the sun's orbit gives us a total mass for our galaxy of at least 2×10^{11} solar masses.

The rotation of our galaxy is actually much more interesting than the description above. Stars at different distances from the center revolve around the center of the galaxy with different periods, so three stars near each other will draw apart as time passes (Figure 16-11). This motion is called differential rotation. (Recall that we defined differential rotation in Chapter 8.) To fully

a

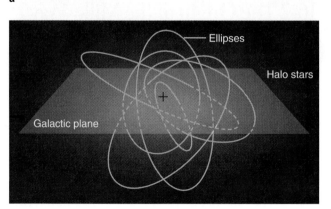

b

FIGURE 16-10

(a) Stars in the galactic disk have nearly circular orbits that lie in the plane of the galaxy. (b) Stars in the halo have randomly oriented, highly elongated orbits.

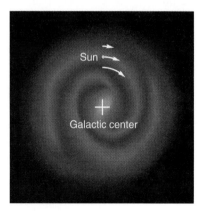

FIGURE 16-11

The differential rotation of the galaxy's disk means that not all stars orbit the galaxy with the same period. Three stars lined up near the sun will not remain aligned as they follow their orbits. The inner star will pull ahead, and the outer star will fall behind.

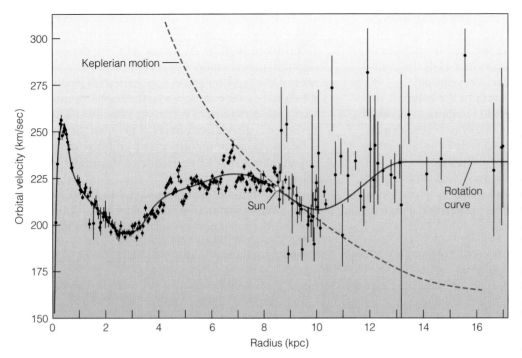

FIGURE 16-12
Radio observations of orbital velocities are plotted here with vertical error bars showing the uncertainty in each measurement. Measurements outside the sun's orbit are more difficult and thus more uncertain. The fitted rotation curve (solid line) does not decline outside the sun's orbit as we would expect for Keplerian motion (dashed line). Rather, the curve is flat at great distances, and this flatness suggests that the galaxy contains significant mass outside the orbit of the sun. (Adapted from a diagram by Françoise Combes)

describe the rotation of our galaxy, we must graph orbital velocity versus radius—a **rotation curve** (Figure 16-12).

If all of the mass of the galaxy were concentrated near its center, then we would expect to see orbital velocities fall as we moved away from the center. This is what we see in our solar system, where nearly all of the mass is concentrated in the sun, and it is called **Keplerian motion** (a reference to Kepler's laws). The best observations of the rotation curve of the Milky Way show orbital velocities rising in the outer disk rather than falling. These higher orbital velocities indicate that the larger orbits enclose more mass and seem to tell us that our galaxy is more massive than it appears to be.

Evidently the galactic halo extends much farther into space and contains more mass than was thought. This **galactic corona** may extend 8 or 9 times farther than the visible disk and could contain 10^{12} solar masses. Clearly our galaxy has secrets yet to be explored.

CRITICAL INQUIRY

If clouds of gas and dust block our view, how do we know how big our galaxy is?

The gas and dust in our galaxy block our view only in the plane of the galaxy. When we look away from the plane of the galaxy, we look out of the gas and dust and can see to great distances. The globular clusters are scattered through the halo with a strong concentration in the direction of Sagittarius. When we look above or below the plane of the galaxy, we can see those clusters, and careful observations reveal variable stars in the clusters. By using Shapley's calibration of the period–luminosity diagram, we can find the distance to those clusters. If we assume that the distribution of the clus-

ters is controlled by the gravitation of the galaxy as a whole, then we can find the distance to the center of the galaxy by finding the center of the distribution of globular clusters.

In the critical analysis of ideas, it is important to ask "How do we know?", and that is especially true in science. For example, what measurements and assumptions tell us the total mass of our galaxy? ∎

The differences between the disk and spherical components extend even to their chemical composition. As we will see in the next section, those differences hold clues to the origin of our galaxy.

16-2 The Origin of the Milky Way

Just as dinosaurs left behind fossilized footprints, our galaxy has left behind a fossil footprint of its youth. The stars of the spherical component are old and must have formed long ago when the galaxy was very young.

Stellar Populations

Near the end of World War II, astronomers realized that there are two families of stars in the galaxy. They form and evolve in similar ways, but they differ in the abundance of atoms that are heavier than helium—atoms that astronomers call **metals**. (Note that this is not the way the word "metal" is used by non-astronomers.) **Population I stars** are metal-rich, containing 2–3 percent metals, whereas **population II stars** are metal-poor, containing only about 0.1 percent metals. The difference is dramatically evident in spectra (Figure 16-13).

Processes

One of the most common organizing themes in science is the concept of a process—a sequence of events that leads to some result or condition. Scientists try to discover the natural processes that shape the universe. Telling the stories of these processes explains why the universe is the way it is. At the same time, recognizing these processes and learning to tell their stories simplify much of science.

For example, astronomers try to assemble the sequence of events that led to the formation of the chemical elements. We know that the first stars formed from nearly pure hydrogen and helium and that they made certain heavier elements in their cores. Supernova explosions made other elements and mixed the elements back into the interstellar medium. New stars formed from this gas, and generation after generation of stars added to the heavier elements that we are made of. Telling that story explains much of the science of the stars.

As you study any science, be alert for processes as organizing themes. Science, at first glance, seems to be nothing but facts, but one of the ways to organize the facts is to fit them into the story of a process. If you study biology and learn the process by which plant cells know whether to become roots, stems, or leaves, you will have organized a great deal of factual information into a story that is not only easy to remember, but also makes sense.

When you see a process in science, ask yourself a few basic questions. What conditions prevailed at the beginning of the process? What sequence of steps occurred? Can some steps occur simultaneously, or must one step occur before another can occur? What is the final state that this process produces? Most important, what evidence do we have that nature really works this way?

Not everything in science is a process, but identifying a process will help you remember and understand a great many details and principles. ∎

Population I stars belong to the disk component of the galaxy and are sometimes called disk population stars. They have nearly circular orbits in the plane of the galaxy and are relatively young stars that formed within the last few billion years. The sun is a population I star, as are the Type I Cepheid variables discussed in Chapter 13.

Population II stars belong to the spherical component of the galaxy and are sometimes called the halo population stars. These stars have randomly tipped, elliptical orbits and are old stars. The metal-poor globular clusters are part of the halo population, as are the RR Lyrae and Type II Cepheids.

Since the discovery of stellar populations, astronomers have realized that there is a gradation between populations. Extreme population I stars, such as the young, hot stars in the constellation Orion, are found only in the spiral arms. Slightly less metal-rich population I stars, called intermediate population I stars, are located throughout the disk. The sun is such a star. Stars even less metal-rich, such as stars in the nuclear bulge, belong to the intermediate population II. The most metal-poor stars are those in the halo, including those in globular clusters. These are extreme population II stars.

Why do the disk and halo stars have different metal abundances? The answer to this question is the key to our understanding of the history of our galaxy.

The Element-Building Process

We can use our understanding of stellar structure and stellar evolution (Chapters 12–14) to understand the origin of the chemical elements heavier than helium—elements that astronomers call metals. When the universe began, it contained only hydrogen and helium atoms. (We will review the evidence that supports this statement in Chapter 19.) All of the remaining chemical elements have been produced by **nucleosynthesis,** the process that fuses hydrogen and helium inside stars to make the heavier elements (Window on Science 16-2).

When our galaxy formed some time after the beginning of the universe, the gas contained about 90 percent hydrogen atoms, 10 percent helium atoms, and almost

FIGURE   16-13

Spectra of a population II star and a population I star illustrate the dramatic difference in metal abundance. The stars are of similar spectra type, so they have similar temperatures, and their hydrogen absorption lines are very similar. But spectral lines formed by atoms heavier than helium are prominent in the spectrum of the population I star (bottom) and are quite weak in the spectrum of the population II star (top). (UCO/Lick Observatory photo)

a

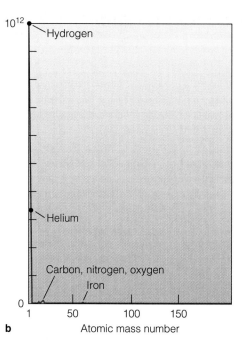

b

F I G U R E   1 6 - 1 4
The abundance of the elements in the universe. (a) Plotted on an exponential scale (see Window on Science 9-1), we see that elements heavier than iron are about a million times less common than iron and that all elements heavier than helium (the metals) are quite rare. (b) The same data plotted on a linear scale provide a more realistic impression of how rare the metals are. Carbon, nitrogen, and oxygen make small peaks near atomic mass 15, and iron is just visible in the graph.

no heavier atoms. Thus, the first stars to form were metal-poor. Succeeding generations of stars manufactured heavier atoms and the metal abundance gradually increased.

The most massive stars fuse helium into carbon, nitrogen, oxygen, and heavier elements up to iron (Chapter 12), and when those stars die in supernova explosions, traces of those elements are spread back into the interstellar medium. Furthermore, the supernova explosion itself can fuse nuclei into atoms heavier than iron. These are rare atoms, such as gold, silver, uranium, and platinum, but they, too, get spread back into space during supernova explosions (Figure 16-14).

When we look at population II stars such as those in the halo, we are looking at the survivors of the early generations of stars in our galaxy. Those stars formed from gas that was metal-poor, and thus their spectra show only weak metal lines. The survivors of these early generations are low-mass, long-lived stars that don't make heavy elements. Furthermore, any elements they might have made are locked in their cores. Thus, the spectrum we see tells us the composition of the gas from which they formed and that the gas was metal-poor. Population I stars, such as the sun, formed more recently, after the interstellar medium had been enriched in metals, and their spectra show stronger metal lines. Stars forming now show the strongest metal lines.

In this way, we can use metal abundance as an index to the ages of the components in our galaxy. The halo is old, and the disk is younger. Such ages can help us understand the formation of our galaxy.

The Age of the Milky Way

Because we know how to find the age of star clusters, we can estimate the age of our galaxy. Nevertheless, uncertainties make our easy answer hard to interpret.

The oldest open clusters are about 7 billion years old. These ages come from the turn-off point in the H–R diagram (see Chapter 13), but finding the age of an old cluster is difficult because clusters change so slowly. Also, the exact location of the turn-off point depends on chemical composition, which differs among clusters. Finally, open clusters are not strongly bound by their gravity, so even older open clusters may have dissipated as their stars wandered away. Thus, open clusters tell us the galactic disk is at least 7 billion years.

The H–R diagrams of globular clusters tell us their age, and conventional data suggest that they are from 13 to 17 billion years old. New data from the Hipparcos satellite, however, provides more accurate distances and luminosities, suggesting that globular clusters may be as young as 11 billion years. Part of the uncertainty reflects the difficulty in measuring the ages of old star clusters, but part of the variation in age seems real. Some globular clusters are older than others. From this information, it seems that the halo is at least 11 billion years old.

Both populations and clusters tell us that the disk is younger than the halo. We can combine these ages with the process of nucleosynthesis to tell the story of our galaxy.

The History of the Milky Way Galaxy

In the 1950s, astronomers began to develop an hypothesis to explain the formation of our galaxy. Recent observations, however, are forcing a reevaluation of that traditional hypothesis.

The traditional hypothesis says that the galaxy formed from a single large cloud of gas 10–15 billion years ago. As gravity pulled the gas together, the cloud began to fragment into smaller clouds, and because the gas was turbulent, the smaller clouds had random velocities. Stars and star clusters that formed from these fragments went into randomly shaped and randomly tipped orbits. Of course, these first stars were metal-poor because no stars had existed earlier to enrich the gas with metals. Thus, the contraction of the large gas cloud produced the spherical component of the galaxy.

The second stage of this hypothesis accounts for the disk component. The contracting gas cloud was roughly spherical at first, but the turbulent motions canceled out, as do eddies in recently stirred coffee, leaving the cloud with uniform rotation. A rotating, low-density cloud of gas cannot remain spherical. A star is spherical because its high internal pressure balances its gravity, but in a low-density cloud, the pressure cannot support the weight. Like a blob of pizza dough spun in the air, the cloud must flatten into a disk (Figure 16-15).

This contraction into a disk took billions of years, with the metal abundance gradually increasing as generations of stars were born from the gradually flattening gas cloud. The stars and globular clusters of the halo were left behind by the cloud when it was spherical, and subsequent generations of stars formed in flatter distributions. The gas distribution in the galaxy now is so flat that the youngest stars are confined to a disk only about 100 parsecs thick. These stars are metal-rich and have nearly circular orbits.

This traditional hypothesis accounts for many of the Milky Way's properties. Advances in technology, however, have improved astronomical observation, and beginning in the 1980s, contradictions arose. For example, many halo stars have metal abundances similar to those of globular clusters and may have escaped from such clusters, but some stars are even more metal-poor. One star contains 30,000 times less metals than the sun. These stars may be older than the globular clusters. Also, some of the oldest stars in our galaxy are in the central bulge, not in the halo. Furthermore, not all globular clusters have the same age, and the younger clusters seem to be in the outer halo. The traditional hypothesis says that the halo formed first and that the clusters within it should have a uniform age.

Ages are a key problem for the traditional hypothesis. The youngest globular clusters are about 11 billion years old, but the disk seems much younger. The oldest open star clusters in the disk are only about 7 billion years old. Also, white dwarfs tend to accumulate over the eons, and there aren't as many white dwarfs in the disk as there should be if the disk were 11 billion years old. A successful hypothesis should account for these ages.

Another problem is that the oldest stars are metal-poor but not metal-free. There must have been at least a few massive stars to create these metals before the formation of the oldest stars we see in the halo.

Can we modify the traditional hypothesis to explain these observations? Perhaps the galaxy began with the contraction of a gas cloud to form the central bulge and the later accumulation of the halo from gas clouds that had been slightly enriched in metals by an early generation of massive stars. This would explain the age of the central bulge and the metals in the oldest stars.

FIGURE 16-15

According to the traditional hypothesis, the galaxy began as a spherical cloud of gas (shaded) in which stars and star clusters (dots) formed. As the rotating gas cloud collapsed into a disk, the halo stars were left behind as a fossil of the early galaxy.

The disk could have formed later as the gas that was already in the galaxy flattened and as more gas fell into the galaxy and settled into the disk. Perhaps entire galaxies were captured by the growing Milky Way galaxy. (We will see in the next chapter that such mergers do occur.) If the galaxy absorbed a few small but partially evolved galaxies, then some of the globular clusters we see in the halo may be hitchhikers. This would explain the range of globular cluster ages.

In a way, the story of the formation of our galaxy is unsatisfying because of the uncertainties, but it is surprising how much of the story we can tell. In the next chapter, our study of galaxies in general will give us further insight into the origin of our own galaxy.

CRITICAL INQUIRY

Why do metal-poor stars have the most elongated orbits?

Of course, the metal abundance of a star cannot affect its orbit, so our analysis must not confuse cause and effect with the relationship between these two factors. Both chemical composition and orbital shape depend on a third factor—age. The oldest stars are metal-poor because they formed before there had been many supernova explosions to create and scatter metals into the interstellar medium. Those stars formed long ago when the galaxy was young and motions were not organized into a disk, and they tended to take up randomly shaped orbits, many of which are quite elongated. Thus, today, we see the most metal-poor stars following the most elongated orbits.

Nevertheless, even the oldest stars we can find in our galaxy contain some metals. They are metal-poor, not metal-free. How does that constrain our theories? ■

The origin of our galaxy is an astronomical work in progress, but that is not the only mystery hidden in the Milky Way. The next two sections discuss two special problems that astronomers are trying to understand—the nature of spiral arms and the secret of the galactic nucleus.

16-3 Spiral Arms

Astronomers commonly refer to our galaxy as a spiral galaxy, but great clouds of dust block our view, so how do we know that our galaxy contains spiral arms? Furthermore, what are spiral arms? What natural process creates these graceful streamers of stars? In this section, we will map the spiral arms and try to understand the process that creates them.

Tracing Spiral Arms

Many of the other galaxies we see around us contain spiral arms illuminated by hot, blue stars (Figure 16-16a). Thus, one way to find evidence of spiral arms in our own galaxy is to locate these stars. Fortunately, this is not difficult because O and B stars are often in associations, and, being very bright, they are easy to detect across great distances. Unfortunately, at these great distances their parallax is too small to measure, so their distances must be found by other means, usually by spectroscopic parallax (Chapter 9).

O and B associations near the sun are not located randomly (Figure 16-16b). They form three bands, indicating that there are three segments of spiral arms near the sun. If we could penetrate the gas and dust, we could locate other O and B associations and trace the spiral arms farther, but, like travelers in a fog, we can see only the region near us.

Objects used to map spiral arms are called **spiral tracers.** O and B associations are good spiral tracers because they are bright and easy to see at great distances. Other tracers include young open clusters, clouds of hydrogen ionized by hot stars (emission nebulae), and certain kinds of young variable stars.

Notice that all spiral tracers are young objects. O stars, for example, live only a few million years. If their orbital velocity is about 250 km/sec, they cannot have moved more than about 500 pc since they formed. This is less than the width of a spiral arm. Because they don't live long enough to move away from the spiral arms, they must have formed there.

The youth of spiral tracers gives us an important clue to the nature of the arms. Somehow the arms are associated with star formation. Before we can follow this clue, however, we must extend our map of spiral arms to show the entire galaxy.

Radio Maps of Spiral Arms

The dust that blocks our view at visual wavelengths is transparent at radio wavelengths because radio waves are much longer than the diameter of the dust particles. When we point a radio telescope at a section of the Milky Way, we receive 21-cm radio signals (Chapter 11) coming from cold clouds of neutral hydrogen in a number of spiral arms at various distances across the galaxy. In most areas of the disk, the signals can be unscrambled using the Doppler shifts of the 21-cm radiation, and thus we can map the distribution of neutral hydrogen gas throughout the disk of the galaxy (Figure 16-17). Only when we look toward the center are we unable to unscramble the radio signals from different spiral arms. Thus, our radio map has a wedge-shaped gap around and beyond the center of the galaxy.

Radio astronomers can also use the strong radio emission from carbon monoxide (CO) to map the location of giant molecular clouds along the Milky Way.

a

b

FIGURE 16-16

(a) We see spiral arms in many other galaxies. Their arms are brightly illuminated by regions of star formation containing hot, blue O and B stars. (Bill Schoening, NOAO, © AURA, Inc.) (b) We see spiral arms in our own galaxy when we plot the location of O and B associations and the ionized nebulae that often surround such associations. These groups of massive stars have very short lives, so we find them only in regions of recent star formation. Dust clouds block our view, and we can see only a few kiloparsecs in the plane of the Milky Way Galaxy. In the region that we can see near the sun, however, O and B associations trace out segments of three spiral arms.

FIGURE 16-17

A map of our galaxy based on 21-cm observations. The sun is located near the top at center, and the center of the galaxy is located in the center of the map. The conical region around and beyond the center of the galaxy is the region where 21-cm radiation from gas clouds at different distances cannot be unscrambled. Such maps show clear evidence that spiral arms exist throughout our galaxy, but the pattern is neither simple nor clear. (Adapted from radio map by Gart Westerhout)

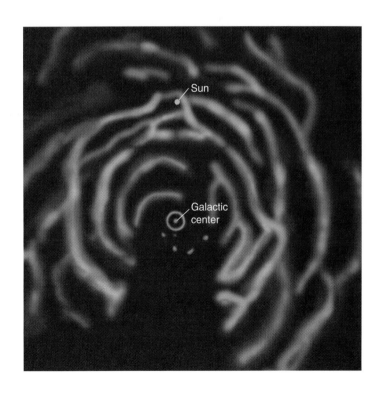

Plotting the distances to the clouds reveals large structures that resemble spiral arms (Figure 16-18).

Radio maps reveal a number of things. First, the spiral pattern we see near the sun continues throughout the disk. Second, the spiral arms are rather irregular and are interrupted by bends, spurs, and gaps. The stars we see in Orion, for example, appear to be a detached segment of a spiral arm. There are significant sources of error in the radio-mapping method, but many of the irregularities along the arms seem real, and photographs of nearby spiral galaxies show similar features. In fact, a few astronomers argue that our galaxy has no overall spiral pattern, only a distribution of broken segments.

a

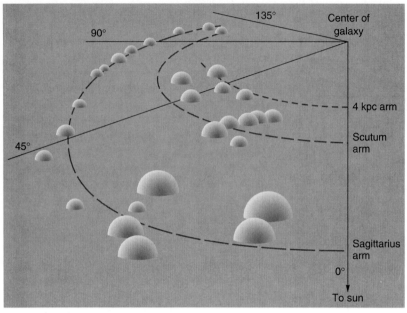

b

FIGURE 16-18

Mapping spiral arms using giant molecular clouds. (a) At the wavelength emitted by the CO molecule we find many overlapping giant molecular clouds along the Milky Way. By finding the distance to each cloud from its radial velocity and the rotation of the galaxy, radio astronomers can use the clouds to map spiral arms. (b) This diagram shows the location of giant molecular clouds in the plane of our galaxy as seen from a location 2 kpc directly above the sun. Now we can see that the molecular clouds are located along spiral arms. Angles in this diagram are galactic longitudes measured clockwise from the sun. (Adapted from a diagram by T. M. Dame, B. G. Elmegreen, R. S. Cohen, and P. Thaddeus)

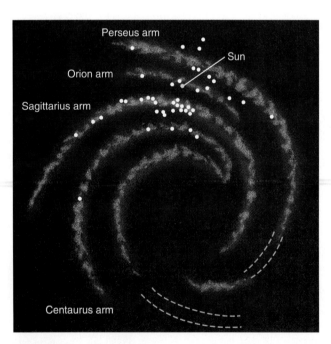

FIGURE 16-19

Combined optical and radio data reveal the overall spiral pattern of the Milky Way.

Other astronomers argue that the spiral pattern is four-armed rather than the "grand-design" two-armed pattern usually assumed.

If we combine optical and radio data, we can construct a general map of our galaxy (Figure 16-19). From this it seems that we live in a spiral galaxy and that the sun is now on the inner edge of a spiral arm segment known as the Orion arm because the bright stars of Orion are part of it. But it is not possible to connect all the arm segments in the map into a single, grand design of two arms. A "best guess" of what our galaxy looks like is shown in Figure 16-20.

The most important feature of the spiral arms is easily overlooked—spiral arms are regions of higher gas density richly populated by bright, young stars. The spiral arms are somehow associated with star formation.

The Density Wave Theory

Just what are spiral arms? We can be sure they are not physically connected structures like bands of magnetic field holding the gas in place. If they were, the strong differential rotation of the galaxy would destroy them within a billion years. They would get wound up and torn apart like paper streamers caught on the wheel of a speeding car. Yet spiral arms are common in galaxies and must last billions of years.

Most astronomers believe that spiral arms are dynamically stable—they retain the same appearance even though the gas, dust, and stars in them are constantly changing. To see how this works, think of the traffic jam behind a slow-moving truck. Seen from an airplane, the traffic jam would be stable, moving slowly down the highway. But any given car would approach from behind, slow down, work its way to the front of the jam, pass the truck, and resume speed. The individual cars in the jam are constantly changing, but the traffic jam itself is dynamically stable.

In the **density wave theory,** the spiral arms are dynamically stable regions of compression that move slowly around the galaxy, just as the truck moves slowly down the highway. Gas clouds, moving at orbital velocity around the galaxy, overtake the slow-moving arms from behind and slam into the gas already in the arms.

a

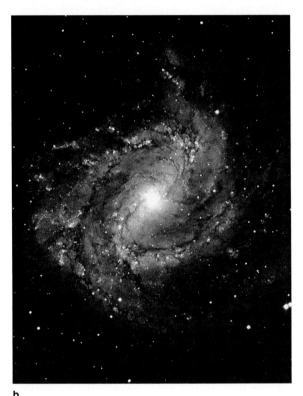

b

FIGURE 16-20

Our galaxy in perspective. (a) An observer a few million light-years away might have this view of our galaxy. The cross at upper right marks the position of the sun. (Painting by M. A. Seeds based on a study by G. De Vaucouleurs and W. D. Pence) (b) Like our own Milky Way, the galaxy M83 appears to have a strong two-armed spiral pattern with branches and spurs. (Anglo-Australian Telescope Board)

FIGURE 16-21

According to the density wave theory, gas clouds overtake the spiral arm from behind and smash into the density wave. The compression triggers the formation of stars. The massive stars (open circles) are so short-lived that they die before they can leave the spiral arm. The less massive stars (dots) emerge from the front of the arm with the remains of the gas cloud.

The sudden compression of the gas can trigger the collapse of the gas clouds and the formation of new stars (see Chapter 12). The newly formed stars and the remaining gas eventually move on through the arm and emerge from the front of the slow-moving arm to resume their travels around the galaxy (Figure 16-21). Thus, whenever we look we see a spiral arm, although the stars, gas, and dust are constantly changing.

Science depends on evidence, so we should ask what evidence we have that spiral arms are caused by spiral density waves. For one thing, the stars in spiral arms fit the theory well. We find stars of all masses forming in spiral arms, but the O and B stars are the brightest. They live such short lifetimes that they die before they can move out of the spiral arm. Thus, the presence of these stars along with gas and dust assures us that spiral arms are sites of star formation. Lower-mass stars, such as the sun, also form in spiral arms. But because they live so long, they can leave the arms and circle the galaxy many times. The sun probably formed as part of a cluster in a spiral arm roughly 5 billion years ago, left that cluster, and has circled the galaxy about 20 times, passing through spiral arms many times.*

Giant molecular clouds and thick dust in the arms are evidence that material is compressed in the arms and spawns active star formation. These clouds can convert up to 30 percent of their mass into stars. The Orion complex and the Ophiuchus dark cloud are just two of these regions filled with infrared protostars. Such a cloud can form stars for 10–100 million years before it is disrupted by heat and light from the most massive new stars. The cloud needs about this long to pass through a spiral arm.

The evidence seems to fit the spiral density wave theory well, but the theory has two problems. First, how is the complicated spiral disturbance started and sustained? Spiral density waves should slowly die out in a billion years or so. Something must regenerate the spiral wave. One possibility is that the galaxy is naturally unstable to certain disturbances, just as a guitar string is unstable to certain vibrations. Any sudden disturbance—the rumble of a passing truck, for example—can set the string vibrating at its natural frequencies. Similarly, minor fluctuations in the galaxy's disk, such as supernova explosions, might regenerate the waves.

Another suggestion is that gravitational disturbances can generate a spiral density wave. Some evidence suggests that the center of our galaxy is not a sphere but a bar. Other galaxies with central bars are known. (See Chapter 17.) The rotation of such a bar could disturb the disk of the galaxy and stimulate the formation of a spiral density wave. A close encounter or collision with a nearby galaxy could also produce a spiral pattern.

The second problem for the density wave theory involves the spurs and branches in the arms of our own and other galaxies. Computer models of density waves produce regular, two-armed spiral patterns. Some galaxies, called grand-design galaxies, do indeed have symmetric two-armed patterns, but others do not. Other galaxies have a great many short spiral segments, giving them a fluffy appearance. These galaxies have been termed **flocculent,** meaning "woolly." Our galaxy is probably an intermediate galaxy. How can we explain these variations in the spiral pattern? Perhaps the answer lies in a process that sustains star formation once it begins.

---

*Some scientists have suggested that the periodic passage of the sun through the denser gas in spiral arms could affect the amount of sunlight reaching the earth and thus cause ice ages. This is interesting but highly speculative.

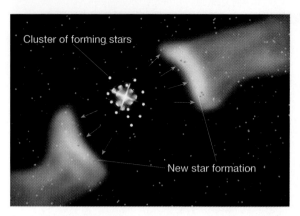

FIGURE 16-22

Self-sustaining star formation may occur when a cluster of forming stars develops a massive member. The energetic birth of a massive star or the sudden expansion of a supernova remnant can compress nearby gas clouds and trigger new star formation. (See Figure 12-4.)

Star Formation in Spiral Arms

Star formation is a critical process in the creation of the spiral patterns that we see. It makes spiral arms visible and may shape the spiral pattern itself.

The spiral density wave creates spiral arms by the gravitational attraction of the stars and gas flowing through the arms. Even if there were no star formation at all, rotating disk galaxies could form spiral arms. But without star formation to make young, hot, luminous stars, the spiral arms would be difficult to see. It is the star formation that lights up the spiral arms and makes them so prominent. Thus, star formation helps determine what we see when we look at a spiral galaxy.

But star formation can also control the shape of spiral patterns if the birth of stars in a cloud of gas can renew itself and continue making more new stars. Consider a newly formed star cluster with a single massive star. The intense radiation from that hot star can compress nearby parts of the gas cloud and trigger further star formation (Figure 16-22). Also, massive stars evolve so quickly that their lifetimes are only an instant in the history of a galaxy. When they explode as supernovae, the expanding gasses can compress neighboring clouds of gas and trigger more star formation. Examples of such a **self-sustaining star formation** have been found. The Orion complex consisting of the Great Nebula in Orion and the star formation buried deep in the dark interstellar clouds behind the nebula is such a region.

Self-sustaining star formation can produce growing clumps of new stars, and the differential rotation of the galaxy can drag the inner edge ahead and let the outer edge lag behind to produce a cloud of star formation with a spiral appearance (Figure 16-23). Mathematical models of galaxies filled with such segments of spiral arms do have a spiral appearance, but they lack the bold, two-armed spiral that astronomers refer to as a grand-design spiral pattern (Figure 16-24). Rather, astronomers suspect that self-sustaining star formation can produce the

FIGURE 16-23

Because of differential rotation, the inner edge of a star-forming cloud of gas will pull ahead of the outer edge (top). If self-sustaining star formation drives the cloud to continue making new stars, it will be drawn out into a spiral segment. Many such segments can create a spiral pattern in a galaxy.

a

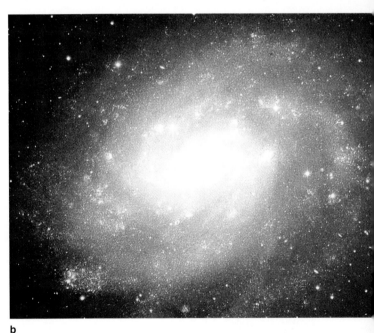

b

FIGURE 16-24

(a) Some galaxies are dominated by two spiral arms, but even in these galaxies, minor spurs and branches are common. The spiral density wave can generate the two-armed, grand-design pattern, but self-sustained star formation may be responsible for the irregularities. (b) Many spiral galaxies do not appear to have two dominant spiral arms. Spurs and branches suggest that star formation is proceeding rapidly in such galaxies. (Anglo-Australian Telescope Board)

FIGURE 16-25

This near-infrared image of the center of the Whirlpool Galaxy has been computer-enhanced and given false colors to reveal the spiral pattern. Hidden by dust at visible wavelengths, the innermost spiral arms are visible at longer wavelengths. The image shows the arms extending much closer to the center than current theory predicts. (See Figure 17-19.) (Courtesy Dennis F. Zaritsky and Hans-Walter Rix)

branches and spurs so prominent in flocculent galaxies, but only the spiral density wave can generate the beautiful two-armed spiral patterns.

Our discussion of star formation in spiral structure illustrates the importance of natural processes. The spiral density wave creates graceful spiral arms, but it is the star formation in the arms that makes them stand out so prominently. Self-sustaining star formation can act in some galaxies to modify the spiral arms and produce branches and spurs. In some galaxies, it can make the spiral pattern flocculent. By searching out and understanding the details of such natural processes, we can begin to understand the overall structure and evolution of the universe around us.

CRITICAL INQUIRY

Why can't we use solar-type stars as spiral tracers?

Sometimes the timing of events is the critical factor in any analysis. In this case, we must think about the evolution of stars and their orbital periods around the galaxy. Stars like the sun live about 10 billion years, but the sun's orbital period around the galaxy is 240 million years. The sun almost certainly formed when a gas cloud passed through a spiral arm, but since then the sun has circled the galaxy many times and has passed through spiral arms often. Thus, the sun's present location has

nothing to do with the location of any spiral arms. An O star, however, lives only a few million years. It is born in a spiral arm and lives out its entire lifetime before it can leave the spiral arm. We find short-lived stars such as O stars only in spiral arms, but G stars are found all around the galaxy. Thus, we can't use solar-type stars as spiral tracers because they live too long.

The spiral arms of our galaxy surely must make it beautiful in photographs taken from a distance, but we are trapped inside it. How do we know that the spiral arms we can trace out near us actually extend across the disk of our galaxy? ■

Spiral arms seem tremendously resilient. Infrared images reveal arms winding inward toward the nucleus where differential rotation is strong (Figure 16-25). These inner spiral arms direct our attention to another mysterious region of our galaxy—its very center.

16-4 The Nucleus

Astronomers are good at imagining unimaginable things, but even some astronomers have found it difficult to accept the evidence about the center of our galaxy. Yet science is founded on evidence, so we must follow where it leads, and it often leads to fantastic conclusions.

Observational Evidence

The center of our galaxy is hidden behind dust clouds that dim its light by 30 magnitudes. Only 1 out of every trillion (10^{12}) photons completes its journey to the earth. Visual-wavelength photos tell us nothing about the nucleus, but infrared and radio observations see through the dust and reveal a region of crowded stars orbiting at high velocity around the center (Figure 16-26).

When radio astronomers map Sagittarius, they find a collection of radio sources with one—**Sagittarius A*** (abbreviated Sgr A* and usually pronounced "sadge A-star")—lying at the expected location of the galaxy's center (Figure 16-27). This single object is smaller than a few astronomical units in diameter but emits five times more radio energy than all of the energy radiated by the sun. Sgr A* is generally thought to be the geometrical center of our galaxy.

Infrared observations show that the stars are tremendously crowded in the central region of the galaxy. Infrared radiation at wavelengths shorter than 2 microns (2000 nm) comes almost entirely from cool stars. Observations at these wavelengths show that the stars are only about 1000 AU apart. Near the sun, stars are about 330,000 AU (1.6 pc) apart. Infrared photons with wavelengths longer than 4 microns (4000 nm) come almost entirely from interstellar dust warmed by stars, and the intensity of this radiation once again confirms that there are large numbers of stars in the central regions.

a

Center
of
Galaxy

b

FIGURE 16-26

(a) The summer Milky Way stretches from horizon to horizon, from Cassiopeia at left to Sagittarius at right. In this visual-wavelength photograph, dark clouds of dust hide the center of the galaxy at extreme right (arrow). (Steward Observatory) (b) An infrared photo penetrates the dust clouds and reveals the crowded stars near the galactic center. This image was made by combining images made at the wavelengths 1350 nm, 1680 nm, and 2260 nm and coloring them blue, green, and red, respectively. Only at the longest of the three wavelengths can we see the star clouds near the center, and thus they look red in this composite image. (Courtesy Bradford W. Greeley, Rhodes College)

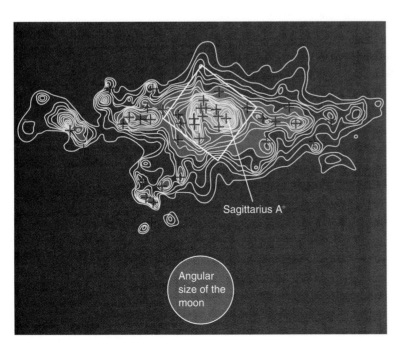

Sagittarius A*

Angular
size of the
moon

FIGURE 16-27

A radio map of the Sagittarius region of the Milky Way (see box in Figure 16-3) reveals an intense radio source, Sagittarius A*, at the expected location of the center. Crosses mark far-infrared sources associated with star formation. Box marks boundaries of Figure 16-28a. Angular size of the moon shown for comparison. (Adapted from observations by W. L. Altenhoff, D. Downes, T. Pauls, and J. Schraml; and by S. F. Odenwald and G. G. Fazio)

Radio observations show that the region around Sgr A* is disturbed. A ring of gas clouds dense enough to contain molecules surrounds the center, extending from a distance of about 25 pc inward to about 6 pc. The inner edge of this ring contains gas ionized by bright ultraviolet radiation from the stars crowded in the central area. Inside this ring of gas lies a cavity in which the gas is 10 to 100 times less dense. Because differences in gas density should smooth out within about 100,000 years, astronomers suspect that a central eruption drove most of the gas outward, forming the cavity, within the last 100,000 years.

The motions of the stars in the inner regions imply that a large amount of mass lies at the very center of the galaxy. For example, infrared emission from ionized neon at the specific wavelength of 12.8 microns (12,800 nm) allows astronomers to use the Doppler shift to measure velocities. Gases near the center are orbiting very rapidly. Only 0.1 pc (20,000 AU) from the center, the orbital velocity is 700 km/sec. These gas clouds must have orbital periods of about 850 years, and Kepler's third law tells us that the orbits must enclose at least one or two million solar masses. It does not appear that such a large amount of mass could exist as stars and escape detection. But what could it be? If the mass is a single object, it must be very small.

What could be so massive and so small? Since the mid-1970s, astronomers have been asking themselves that question. The most interesting possibility is a gigantic black hole.

The Massive Black Hole Hypothesis

To explain the nature of the center of our galaxy, astronomers must explain how a large mass can occupy a small space, and a massive black hole fills the bill quite well. A black hole of 3 million solar masses, for example, would be only about 13 times larger than the sun—quite small in terms of the galaxy. Also, a massive black hole could produce energy if matter dribbled into it through an accretion disk, and it could produce occasional eruptions if an entire star fell in. Before we can decide, we must examine the hypothesis more carefully, compare it with observations, and consider alternatives.

A massive black hole could supply almost unlimited energy. There is no known limit to the mass of a black hole, and the more massive it is, the more matter it could draw inward. Also, other energy sources are limited. A supernova, for example, converts very little of its mass into energy before blowing itself apart, and there is a limit to the number of luminous, massive stars capable of producing supernova that can occupy a small region. For steady energy production from a very small region, a massive black hole works very well.

Such a massive black hole would not be the remains of a single collapsed star but would have developed gradually as gas clouds and stars lost orbital velocity and drifted toward the center of the galaxy. We will see in later chapters that other galaxies may contain massive black holes.

Certain characteristics of Sgr A* are compatible with its being a massive black hole within an accretion disk. Studies of Sgr A* show that it does not seem to be moving at all. While other stars in the central regions rush about their orbits, Sgr A* remains fixed, suggesting that it must have a mass of at least a few hundred solar masses. It can't be a typical compact object of a few solar masses. Radio astronomers using the highest resolution find that Sgr A* is less than 2 AU in diameter and elongated, as we might expect of a massive black hole surrounded by an inclined accretion disk.

Consider how small Sgr A* is in comparison with the entire galaxy. If the Milky Way galaxy were shrunk to the size of Texas, Sgr A* would be only 0.02 inches in diameter—about the size of a mosquito's eye. Yet Sgr A* seems to be a powerful energy source. For example, X-ray observations show clouds of highly ionized gas extending 100 pc on either side of the center, and this suggests that the core erupted with the power of several thousand supernovae only a few centuries ago. The combination of small size and great power suggests accretion into a massive black hole.

Some observations suggest the existence of high-energy jets coming from the core. Radio maps reveal a jet extending 30 pc toward the galaxy's south pole. Such jets are characteristic of accretion disks around black holes, as in the case of SS433, and the strong magnetic field found near the core (Figure 16-28a) suggests that the center of the galaxy contains a rapidly spinning object. Also, the orbiting Gamma Ray Observatory has found gamma rays coming from a region extending several thousand light years from the central region out into the halo. The gamma rays have the energy produced when an electron collides with its antiparticle, the positron. It isn't known what is making these positrons, but they may be coming from high-energy processes in the core, such as a hot accretion disk.

At the highest resolution, the VLA radio telescope can detect a spiral swirl of matter in the central 3 pc of the galaxy, the hole in the disk of gas and dust that whirls around the center. This spiral is not related to the spiral arms but appears to be matter spiraling inward toward the central object (Figure 16-28b, 16-28c). This inflow appears to add a few tenths of a solar mass per year to the central mass, and accretion at that rate onto a massive black hole could easily provide the observed energy.

A supermassive black hole is an exciting idea, but we must notice the difference between adequacy and necessity. A supermassive black hole is adequate to explain the observations, but it is not necessary. A few astronomers have thought of other explanations. For example, gas flowing into the core could form a compact cluster of very massive, highly energetic stars. Thus, the observational evidence we have does not require a black hole at the center of our galaxy. However, astronomers are finding more and more evidence that other galaxies contain supermassive black holes in their nuclei (Chapter 18), so most astronomers take the black hole hypothesis very seriously.

a

b

c

FIGURE 16-28

The center of the galaxy. (a) This VLA map shows filaments of excited gas 50 pc long arcing out of the center as if constrained by a magnetic field. Compare with the boxed region in Figure 16-27. (National Radio Astronomy Observatory) (b) A detailed map of the center (shown roughly by the box in part a) reveals the excited inner edge of a disk of gas and dust with gas flowing into a central object. (N. Killeen and Kwok-Yung Lo) (c) The central spiral appears to lie in the open center of the larger disk of neutral gas. It may show matter flowing into a central black hole.

CRITICAL INQUIRY

Why do we believe that the center of our galaxy contains a large mass?

In science, a belief should be based on evidence, so we must site observations to analyze this problem. The key observation is that orbital velocities are very high close to the center of the galaxy. According to Kepler's third law, those high velocities mean that the small orbits enclose high mass. Of course, we can't observe individual stars orbiting very close to the center, but we can observe the infrared emission lines of neon, and the Doppler shifts tell us that the gas is orbiting at very high velocities.

Observational evidence is the key to science, but notice that we need basic physics to interpret the observations. For example, how do we interpret the strong infrared radiation at wavelengths longer than 4 microns (4000 nm) to imply that vast numbers of stars are crowded near the center of the galaxy? ■

Despite all the observations and theories, we cannot yet be sure what lies at the center of the Milky Way Galaxy. One way to extend our research is to compare our galaxy with others. We begin that strategy in the next chapter. ■

▪ Summary

Our galaxy is a typical spiral galaxy with a disk about 25,000 pc in diameter. The sun is located about two-thirds of the way from the center to the edge. The disk contains nearly all of the gas and dust in the galaxy and most of the star formation. It is illuminated by the brilliant O and B stars. The nuclear bulge at the center of the galaxy is about 3000 pc in radius and contains stars similar to the large spherical halo of stars and globular clusters.

Great clouds of dust and gas in the disk block our view, so until the 20th century, astronomers assumed that we were located at the center of a grindstone-shaped star cloud. Harlow Shapley was able to calibrate variable stars to find the distance to globular clusters, and he concluded that our galaxy was much bigger and that we were not located in the center. Soon after, other astronomers concluded that ours was only one of many galaxies in the universe.

The disk component of our galaxy contains metal-rich population I stars moving in nearly circular orbits in the plane of the disk. The spherical component, made up of the nuclear bulge and the halo, contains metal-poor population II stars moving in randomly tipped and highly elongated orbits. Observations of the rotation curve of our galaxy reveal that orbital velocities do not fall in the outer disk, suggesting that the galaxy contains large amounts of unseen mass commonly thought to be located in an extended halo.

The distribution of populations through the galaxy suggests a top-down hypothesis for the formation of our galaxy from a single cloud of gas and dust that gradually contracted into a disk shape. Evidence suggests that the story is more complicated, however, and many astronomers now suspect that the nuclear bulge and halo formed first and the disk formed later as more gas fell into the galaxy. The galaxy may have absorbed smaller galaxies as well.

The very youngest stars lie along spiral arms within the disk. The most massive live such short lives that they don't have time to move from their place of birth in the spiral arms. Maps of the spiral tracers and cold hydrogen clouds reveal the spiral pattern of our galaxy. The spiral arms are also outlined by giant molecular cloud complexes. These molecular clouds are sites of star formation.

The spiral density wave theory suggests that spiral arms are regions of compression that move through the disk. When an orbiting gas cloud smashes into the compression wave, the gas cloud forms stars. Self-sustaining star formation is an important process that may modify the arms as the birth of massive stars triggers the formation of more stars by compressing neighboring clouds.

Dusty clouds hide the nucleus of our galaxy at visible wavelengths, but radio and infrared observations can penetrate the clouds. Such observations reveal that the central region is crowded with stars orbiting at high velocity. This tells us that the central few light years must contain millions of solar masses. At the very center lies the radio source Sagittarius A*, an object only a few AU in diameter. Theorists suspect that the object is a massive black hole that is drawing in matter through an accretion disk.

▪ New Terms

Magellanic Clouds	disk component
kiloparsec (kpc)	spherical component
halo	association
nuclear bulge	high-velocity star
rotation curve	nucleosynthesis
Keplerian motion	spiral tracers
galactic corona	density wave theory
metals	flocculent
population I star	self-sustaining star formation
population II star	Sagittarius A*

▪ Questions

1. Why is it difficult to specify the dimensions of the disk and halo?
2. Why didn't astronomers before Shapley realize how large the galaxy is?
3. What evidence do we have that our galaxy has a galactic corona?
4. Explain why some star clusters lose stars more slowly than others.
5. Contrast the motion of the disk stars and that of the halo stars. Why do their orbits differ?
6. Why do high-velocity stars have lower metal abundance than the sun?
7. Why are metals less abundant in older stars than in younger stars?
8. Why are all spiral tracers young?
9. Why couldn't spiral arms be physically connected structures? What would happen to them?
10. Why does self-sustaining star formation produce clouds of stars that look like segments of spiral arms?
11. Describe the kinds of observations we can make to study the galactic nucleus.
12. What evidence do we have that the nucleus of the galaxy contains an energy source that is very small in size?

▪ Discussion Questions

1. How would this chapter be different if interstellar dust did not scatter light?
2. Why doesn't the Milky Way circle the sky along the celestial equator or the ecliptic?

▪ Problems

1. Make a scale sketch of our galaxy in cross section. Include the disk, sun, nucleus, halo, and some globular clusters. Try to draw the globular clusters to scale.
2. Because of dust clouds, we can see only about 5 kpc into the disk of the galaxy. What percentage of the galactic disk can we see? (HINT: Consider the area of the entire disk and the area we can see.)
3. If the fastest passenger aircraft can fly 0.45 km/sec (1000 mph), how long would it take to reach the sun? the galactic center? (HINT: 1 pc = 3×10^{13} km.)
4. If a typical halo star has an orbital velocity of 250 km/sec, how long does it take to pass through the disk of the galaxy? Assume that the disk is 1000 pc thick.
5. If the RR Lyrae stars in a globular cluster have apparent magnitudes of 14, how far away is the cluster? (HINT: See Figure 13-21.)

6. If intersellar dust makes an RR Lyrae variable star look 1 magnitude fainter than it should, by how much will we overestimate its distance? (HINT: Use the magnitude–distance formula or Table 9-1.)

7. If a globular cluster is 10 minutes of arc in diameter and 8.5 kpc away, what is its diameter? (HINT: Use the small-angle formula.)

8. If we assume that a globular cluster 4 minutes of arc in diameter is actually 25 pc in diameter, how far away is it? (HINT: Use the small-angle formula.)

9. If the sun is 5 billion years old, how many times has it orbited the galaxy?

10. If the true distance to the center of the galaxy is found to be 7 kpc and the orbital velocity of the sun is 220 km/sec, what is the minimum mass of the galaxy? (HINT: Use Kepler's third law.)

11. What temperature would interstellar dust have to have to radiate most strongly at 100 μm? (HINTS: 1 μm = 1000 nm. Use Wien's law, Chapter 7.)

12. Infrared radiation from the center of our galaxy with a wavelength of about 2 μm (2×10^{-6} m) comes mainly from cool stars. Use this wavelength as λ_{max} and find the temperature of the stars.

13. If an object at the center of our galaxy has a linear diameter of 10 AU, what will its angular diameter be as seen from the earth? (HINT: Use the small-angle formula, Chapter 3.)

■ Critical Inquiries for the Web

1. Henrietta Leavitt discovered the period–luminosity relation for Cepheids while working on the staff at Harvard College Observatory under Edward Pickering. She was one of several women "computers" on staff there a century ago. Search the Web for information on Leavitt, her colleagues, and their work at Harvard. List three of the women employed by Pickering and note their contributions to astronomy. What was life like for a woman in astronomy at the turn of the century?

2. What if we lived near the center of the galaxy? Search the Web for research and information on the distribution of material near the center of our galaxy. Based on what you find, speculate as to how the sky would appear from a planet associated with a star near the galactic center.

Go to the Wadsworth Astronomy Resource Center (www.wadsworth.com/astronomy) for critical thinking exercises, articles, and additional readings from InfoTrac College Edition, Wadsworth's online student library.

Chapter Seventeen

GALAXIES

GUIDEPOST

We can no more understand stars by understanding a single example, the sun, than we could understand humanity by understanding a single person. Whereas the last chapter examined a single galaxy, this chapter generalizes to examine the entire range of objects called galaxies.

We take two lessons from this chapter for use in future chapters. First, galaxies interact with each other in complicated ways, and that will become critical in the next chapter when we study peculiar galaxies. The second lesson is that 90–99 percent of the matter in the universe is invisible. That astonishing conclusion comes from the study of galaxies, but its greatest impact will appear in Chapter 19, where we will see that it determines the structure and fate of the entire universe. ∎

A hypothesis or theory is clear, decisive, and positive, but it is believed by no one but the man who created it. Experimental findings, on the other hand, are messy, inexact things which are believed by everyone except the man who did that work.

HARLOW SHAPLEY

Through Rugged Ways to the Stars

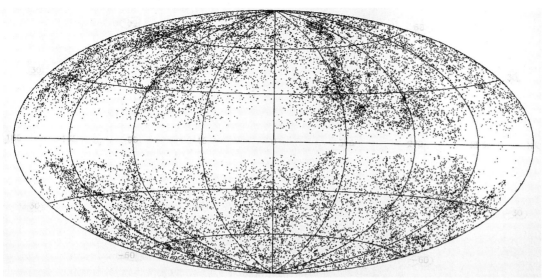

FIGURE 17-1

More than 34,000 galaxies are plotted in this map of the entire sky. The plane of the Milky Way runs along the center. The dust clouds in our galaxy block our view of other galaxies, producing the region of missing galaxies known as the zone of avoidance. Note the clumpy distribution of galaxies. (Nigel A. Sharp)

If an evil wizard turned off all the stars of the Milky Way Galaxy, our sky would be very black. Only the Great Galaxy in Andromeda and the two Magellanic Clouds—small, nearby galaxies—would be visible to the naked eye. Yet the universe is filled with roughly 30 billion galaxies, which cover the entire sky except along the plane of the Milky Way (Figure 17-1). There, in what is known as the **zone of avoidance,** the thick dust clouds of our Milky Way block our view of the distant galaxies.

Previous chapters took us to the stars, out into the spiral arms of our galaxy, and finally into its mysterious core. Yet we have hardly begun our exploration of the universe. If we compare our journey to the farthest galaxies to a flight around the world, our airplane has not yet traveled 1000 feet. In this chapter, we leave behind the familiar Milky Way and penetrate deep into space, out among the vast clouds of galaxies.

We do not, however, leave behind the tools and methods of previous chapters. The tools we developed to determine the properties of stars (Chapter 9) are useful for determining the properties of galaxies. The most important properties are distance, size, luminosity, mass, and motion. As in the case of stars, distance is the key. Once we know the distance to a galaxy, its size and luminosity are easy to find. In some cases, we can even find the masses of galaxies.

Although millions of galaxies are visible on long-exposure photographs of the sky, there are only a few types of galaxies. The shapes of these galaxy types, the kinds of stars they contain, and their distribution in clusters of galaxies are hints to their origin and evolution.

17-1 THE SHAPES OF GALAXIES

The shape of an animal can tell a biologist how a species evolved, and the shapes of galaxies can help modern astronomers understand how galaxies evolve. Our discussion of galaxy shapes begins over a century ago, before people knew there were galaxies.

The Discovery of Galaxies

Galaxies are faint objects, so they were not noticed until telescopes had grown large enough to gather significant amounts of light. In 1845, William Parsons, third Earl of Rosse in Ireland, built a telescope 72 in. in diameter. It was, for some years, the largest telescope in the world. Before the invention of astronomical photography, Parsons had to view the faint nebulae directly at the eyepiece and sketch their shapes. He noticed that some have a spiral shape, and they became known as **spiral nebulae.** Parsons immediately concluded that the spiral nebulae were great spirals of stars. The German philosopher Immanuel Kant had, in 1755, proposed that the universe was filled with great wheels of stars that he called **island universes.** Parsons adopted that term for the spiral nebulae.

Not everyone agreed that the spiral nebulae were island universes. An alternative point of view was that we lived in a great star system. Sir William Herschel had counted stars in different directions and had shown that the star system was shaped approximately like a grindstone (Chapter 16). Beyond the edge of this grindstone universe, space was supposed to be a limitless void. The

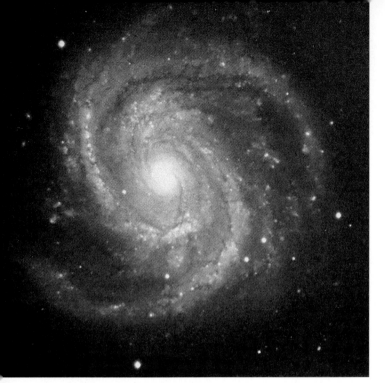

FIGURE 17-2

Modern photographs of spiral galaxies, such as this image of M 100, show bright stars and star clusters clearly. Before World War I, astronomers could see only faint spiral features in some nebulous clouds and could not tell whether these "spiral nebulae" were nearby clouds of gas or distant island universes. (Anglo-Australian Telescope Board)

spiral nebulae, in this view, were vortexes of gas or faint stars within the star system.

The nature of the spiral nebulae could not be resolved in the 19th century because the telescopes were not large enough and the photographic plates were not sensitive enough (Figure 17-2). The spiral nebulae remained foggy swirls on the best photographs, and astronomers continued to disagree on their true nature. In April 1920, two astronomers debated the issue at the National Academy of Science in Washington, D.C. Harlow Shapley of Mt. Wilson Observatory had recently shown that the Milky Way was a much larger star system than had been thought. He argued that the spiral nebulae were nearby objects within the star system. Heber D. Curtis of Lick Observatory argued that the spiral nebu-

lae were island universes. Historians of science mark the debate as a turning point in modern astronomy, but it was inconclusive. Shapley and Curtis cited the right evidence but drew incorrect conclusions. The disagreement was finally resolved with a bigger telescope.

On December 30, 1924, Mt. Wilson astronomer Edwin Hubble (namesake of the Hubble Space Telescope) announced that he had taken photographic plates of a few bright galaxies with the new 100-in. telescope. Not only could he detect individual stars, but he could identify some of them as Cepheid variables with apparent magnitudes of about 18. For the Cepheids, which are supergiants, to look that faint, they had to be very distant, and thus the spiral nebulae were external to the Milky Way star system. They were galaxies.

Over the next few years, Hubble photographed large numbers of galaxies and developed a classification system based on their shapes (Window on Science 17-1). Three broad categories—elliptical, spiral, and irregular—are further subdivided to account for small variations in shape. To organize these classes in an easily remembered system, Hubble arranged them in a **tuning fork diagram** (Figure 17-3). We will consider each main shape category in detail.

Elliptical Galaxies

Elliptical galaxies (Figure 17-4) are very common. They are round or elliptical in shape, have almost no visible gas or dust, lack hot, bright stars, and have no spiral pattern. The stars in elliptical galaxies are more crowded toward the center, and the outer parts of some larger ellipticals are peppered by hundreds of globular clusters (Figure 17-4a).

Elliptical galaxies are identified by the letter *E* followed by a number indicating the apparent shape of the galaxy. If we measure on a photograph the distances *a* and *b*, the largest and smallest diameters through the center of an elliptical galaxy, respectively, then the expression

$$10(a - b)/a$$

rounded to the nearest integer is an index of its shape. For example, if we measure the image of an elliptical

FIGURE 17-3

Modern astronomers use the tuning fork diagram as an organizing framework for galaxy classification. Elliptical (E), spiral (S), barred spiral (SB), and irregular (Irr) galaxies are placed from left to right in approximate order of increasing gas, dust, and star formation.

Classification in Science

Classification is one of the most basic and most powerful of scientific tools. It is often a way to begin studying a topic, and it often leads to dramatic insights. Charles Darwin, for example, sailed around the world with a scientific expedition aboard the ship HMS *Beagle*. Everywhere he went, he studied the living things he saw and tried to classify them. For example, he studied the tortoises, mockingbirds, and finches he saw on the Galápagos Islands. His classifications of these and other animals eventually led him to think about evolution and natural selection. Scientists in many fields depend on carefully designed systems of classification.

Classification is an everyday mode of thought. We all use classifications when we order lunch, buy shoes, catch a bus, and so on. In science, classifications reveal the relationships between different kinds of objects and at the same time save the scientist valuable time by bringing order out of seeming disorder. An economist can classify different kinds of businesses and does not have to analyze each of the millions of businesses as if it were unique. Classifications of minerals, plants, psychological learning styles, modes of transportation, or sandwiches bring order to the world and help us deal with information.

Astronomers use classifications of galaxies, stars, moons, telescopes, and more. Whenever you encounter a scientific discussion, look for the classifications on which it is based. Classifications are the orderly framework on which much of science is built. ■

a

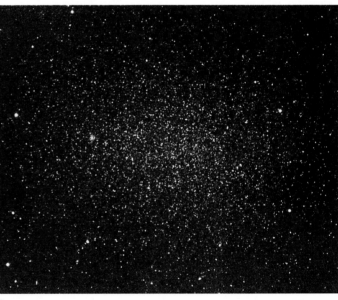

b

FIGURE 17-4

Elliptical galaxies contain little gas and dust and few bright stars. Thus, they are nearly featureless clouds of stars.
(a) M 87 is a giant elliptical galaxy surrounded by a swarm of over 500 globular clusters. Objects to the lower right are smaller galaxies. (National Optical Astronomy Observatories) (b) The Leo I dwarf elliptical is a small, nearby galaxy and is resolved into individual stars. Note the lack of dust, gas, and hot, bright stars. (Anglo-Australian Telescope Board)

galaxy and find that *a* is 6 and *b* is 4, then the expression gives 3.33, which rounds to 3. Thus, the galaxy is an E3. Circular galaxies are E0, and the most elliptical are E7.

Elliptical galaxies generally contain almost no gas and dust and thus contain no newly formed stars. Most of the stars are red giants, which give the galaxies a typically reddish tint, and lower main-sequence stars. It seems that elliptical galaxies have not suffered star formation recently and consist mainly of older stars.

It is difficult to determine the true shape of elliptical galaxies. If they were disks, we would see some face-on, and these would look like E0 galaxies. Some would be inclined and would look like E1 to E7 galaxies. But we should see some edge-on, and none have been so observed. Studies of the observed shapes of the elliptical galaxies combined with spectroscopic studies of the motions of stars in the galaxies tell us that galaxies are ellipsoids, not disks. It is unclear, however, whether the ellipsoids are flattened spheres like thick hamburger buns, stretched spheres like watermelons, or ellipsoids that are longer than they are wide and wider than they

FIGURE 17-5

The shape of elliptical galaxies is not clear from photographs. The galaxies could be slightly flattened spheres (a), slightly stretched spheres (b), or ellipsoids with three different diameters along the three axes (c). All three shapes may occur among the elliptical galaxies.

are thick (Figure 17-5). Observations suggest that all three shapes may be represented among the elliptical galaxies.

The nature of elliptical galaxies is further revealed by the motions of the stars within them. Unlike our galaxy, which rotates about a clearly defined axis, the elliptical galaxies are not dominated by rotational motion about an axis. Rather, the stars each follow their individual orbits around the center like swarming bees.

Another clue to the origin of these galaxies may lie in the dwarf ellipticals (Figure 17-4b). These are small systems containing as few as 10 million stars (about 0.01 percent of the number in our galaxy), and extending only a few hundred parsecs in diameter. In fact, it is not clear if there is a difference between a small dwarf elliptical galaxy and a large globular cluster.

Because these galaxies are small and contain few stars, they are not very luminous and thus are hard to find. Of the approximately two dozen galaxies near the

Milky Way Galaxy, a dozen are dwarf ellipticals. If dwarf ellipticals are that common throughout the universe, they are the most common form of galaxy.

Spiral Galaxies

Spiral galaxies are strikingly beautiful. Their distinguishing characteristic is an obvious disk component that contains gas, dust, and hot, bright stars outlining graceful spiral arms (Figure 17-6).

The gas and dust clouds in spiral galaxies support active star formation, which makes spiral galaxies bright with the light of newly formed stars. Spiral galaxies are so bright they are easy to see at great distances; over two-thirds of all *known* galaxies are spirals, although they probably make up only about 20 percent of all galaxies. Most galaxies (60 percent) are fainter elliptical galaxies, which we do not see as easily as the gaudy spirals.

FIGURE 17-6

(a) NGC 300 is a spiral galaxy containing spiral arms, gas, dust, and many clusters of hot, bright stars. (Anglo-Australian Telescope Board) (b) The dust in this spiral galaxy is dramatically revealed where it is silhouetted against a more distant galaxy. (W. Keel and R. White, NASA)

a

b

FIGURE 17-7

S0 galaxies are clearly disk galaxies, but they contain very little gas or dust and thus are making few new stars. NGC 4526 is a typical S0 galaxy with a prominent nuclear bulge and an extended disk but little dust. (Rudolph E. Schild)

a

b

c

FIGURE 17-8

Barred spiral galaxies: (a) NGC 4639 has an elongated nuclear bulge known as a bar, and its spiral arms spring from the ends of the bar. Notice the many faint galaxies visible in the background. (A. Sandage, A. Saha, G. A. Tamman, L. Labhardt, F. D. Macchetto, N. Panagia, and NASA) (b) M83 is an SBb galaxy with dark dust lanes. (c) NGC 1365 is an SBc galaxy with open spiral arms. (b and c, Anglo-Australian Telescope Board)

Among spiral galaxies we identify three distinct types: S0 galaxies, normal spirals, and barred spirals. Unlike our galaxy, the S0 galaxies show no obvious spiral arms, have very little gas and dust, and contain very few hot, bright stars (Figure 17-7). However, they have an obvious disk component with a large nucleus at the center. They appear to be intermediate between elliptical and other spiral galaxies.

Normal spiral galaxies can be further subclassified into three groups according to the size of their nuclear bulge and the degree to which their arms are wound up (Figure 17-3). Spirals that have little gas and dust, larger nuclear bulges, and tightly wound arms are classified Sa. Sc galaxies have large clouds of gas and dust, small nuclear bulges, and very loosely wound arms. The Sb galaxies are intermediate between Sa and Sc. Because there is more gas and dust in the Sb and Sc galaxies, we find more young, hot, bright stars along their arms. The Andromeda Galaxy (Figure 17-14) and our own Milky Way Galaxy are Sb galaxies.

About 20 percent of spirals have an elongated nuclear bulge with spiral arms springing from the ends of the bar. These **barred spiral galaxies** are classified SBa, SBb, or SBc according to the same criteria listed for normal spirals (Figure 17-8). The elongated shape of the nuclear bulge is not well understood, but some astronomers working with computer models have succeeded in imitating the rotating bar structure. It appears to occur when an instability develops in the stellar distribution within the rotating galaxy. The gravitational field of the bar alters the orbits of the inner stars and generates a

FIGURE 17-9

Irregular galaxies. (a) The Large Magellanic Cloud (LMC) is an irregular galaxy only about 50 kpc away. The large pink emission nebula is the Tarantula Nebula (see Figure 14-14). The bar of the LMC runs from bottom center toward upper right. (Copyright R. J. Dufour, Rice University) (b) Irregular galaxy IC 4182 is typical in that it is small, faint, and contains both red and blue stars but no obvious spiral arms. (George Jacoby and Mike Pierce, NOAO, © AURA, Inc.)

stable, elongated nuclear bulge. Thus, except for the peculiar rotation in the nucleus, barred spirals are similar to normal spiral galaxies. In fact, some theories suggest that barred spirals represent a temporary stage that a spiral galaxy can pass through.

Spiral galaxies contain a mixture of stars—young and old stars, hot and cool stars, and bright and faint stars. However, the short-lived, massive stars produce most of the light, so we see spiral galaxies outlined by the brilliant O and B stars sprinkled along the spiral arms.

Irregular Galaxies

Like spirals, **irregular galaxies** contain a mixture of star types. They contain large clouds of gas and dust mixed with both young and old stars. But unlike spirals, irregular galaxies have no obvious spiral arms or nuclei. They tend to be small and faint and are thus difficult to detect. The Magellanic Clouds are the best-studied examples (Figure 17-9).

First clearly described by the navigator on Magellan's voyage around the world in 1521, the Magellanic Clouds appear to be small galaxies near the Milky Way. They are small, 7 kpc and 3 kpc in diameter, and have low masses, about 20 billion solar masses for the Large Cloud and about 10 percent of that for the Small Cloud. About 15 percent of their mass is present as gas and dust, and this is responsible for the numerous emission nebulae and the hot, young stars that dot the clouds.

We know a great deal about the Magellanic Clouds because they are nearby, only 50–60 kpc away. However, at greater distances the small irregular galaxies, along with the dwarf ellipticals, are difficult to see. From a survey of nearby galaxies, we can estimate that elliptical galaxies are most numerous and that irregular galaxies are about as common as spirals.

Although typical irregular galaxies show no spiral pattern, some astronomers believe that the Large Magellanic Cloud does contain a barred spiral pattern. This has been mapped on long-exposure photographs. Are the

FIGURE 17-10
Studies of the Magellanic Clouds show that they are moving in orbits that led them through the disk of our galaxy about 200 million years ago, drawing out a streamer of stars and gas. The Small Cloud is being pulled apart by the encounter, and both clouds will eventually merge with our galaxy. Measurements of the motion of the Large Cloud reveal that our galaxy contains 5–10 times more matter than is visible. That matter lies in an extended halo (red in this schematic diagram). (Linda C. Owens/*Science*)

Magellanic Clouds really irregular galaxies? The situation is confused by the interaction of the Magellanic Clouds with each other and with our own galaxy. Radio maps show that the clouds are enveloped in a common cloud of neutral hydrogen, and a bridge of gas connects them to the Milky Way (Figure 17-10). It seems possible that they recently passed through the disk of our galaxy. They are being distorted by tidal forces and are distorting the disk of our galaxy, bending one edge down and the other edge up like the brim of a hat.

Radio and optical studies of the Small Magellanic Cloud reveal that it is being pulled apart by its recent encounter with the Milky Way Galaxy. It is, in fact, not a single, small cloud but a long strand of stars and gas 5000 pc wide and 30,000 pc long. It points, like a finger, almost directly at us, so we see the nearest part of the galaxy as the brightest region (Figure 17-9b).

CRITICAL INQUIRY

What color are galaxies?

Different kinds of galaxies have different colors depending mostly on how much gas and dust they contain. If a galaxy contains large amounts of gas and dust, it probably contains lots of young stars, and a few of those young stars will be massive, hot, luminous O and B stars. They will give the galaxy a distinct blue tint. In contrast, a galaxy that contains little gas and dust will probably contain few young stars. It will lack O and B stars, and the most luminous stars in such a galaxy will be red giants. They will give the galaxy a red tint. Because the light from a galaxy is a blend of the light from billions of stars, the colors are only tints. Nevertheless, the most luminous stars in a galaxy determine the color. From

this we can conclude that elliptical galaxies tend to be red and the disks of spiral galaxies tend to be blue.

Of course, the halo of a spiral galaxy would be both dimmer and redder than the disk. Why? ■

We end our discussion of galaxy classification with the Magellanic Clouds, and that is convenient because they lead us to two important questions. First, how do we know the distances to galaxies? We will discuss distance measurements in the next section. The second question is dramatically illustrated by the Small Magellanic Cloud: how do galaxies interact and evolve? In the last section of this chapter, we will try to answer that question and tell the life stories of the galaxies.

17-2 MEASURING THE PROPERTIES OF GALAXIES

Perhaps the most basic problem in astronomy is finding the distances to objects. Until we know an object's distance, we can discover very few of its properties, such as its diameter, luminosity, or mass. Since astronomers first identified galaxies as objects outside the Milky Way Galaxy, they have sought to find the distances to individual galaxies. In this section, we will see how the distances to the galaxies unlock their basic properties.

The Distance Scale

The distances to galaxies are so large that it is not convenient to express them in light-years, parsecs, or even kiloparsecs. Instead we will use the unit **megaparsec (Mpc)**, or 1 million pc. One Mpc equals 3.26 million ly, or approximately 2×10^{19} miles.

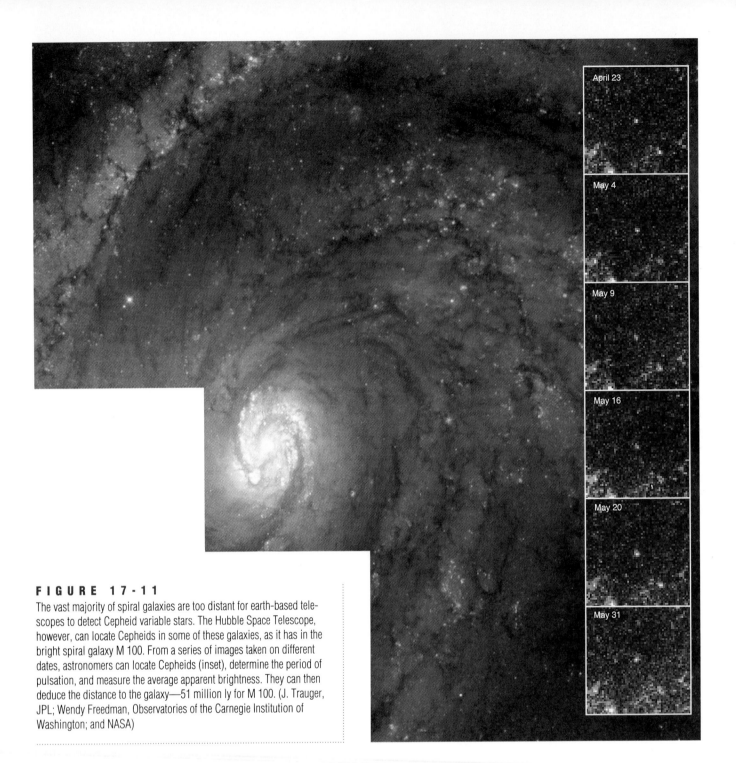

FIGURE 17-11

The vast majority of spiral galaxies are too distant for earth-based telescopes to detect Cepheid variable stars. The Hubble Space Telescope, however, can locate Cepheids in some of these galaxies, as it has in the bright spiral galaxy M 100. From a series of images taken on different dates, astronomers can locate Cepheids (inset), determine the period of pulsation, and measure the average apparent brightness. They can then deduce the distance to the galaxy—51 million ly for M 100. (J. Trauger, JPL; Wendy Freedman, Observatories of the Carnegie Institution of Washington; and NASA)

To find the distance to a galaxy, we must search among its stars for a familiar object whose luminosity is known. Such objects are called **distance indicators** because we can use them to find distance. Distance indicators are also called **standard candles** because they are objects whose luminosity is known.

Because their period is related to their luminosity, Cepheid variable stars are good distance indicators. Chapter 13 explained how a supergiant star evolving through the instability strip in the H–R diagram can pulsate with a period of 1 to 50 days. Because both the period of pulsation and the star's average luminosity

depend on its mass, a period–luminosity relation exists (Figure 13-21). If we know the star's period, we can use the period–luminosity diagram to learn its average absolute magnitude. By comparing its apparent magnitude with its absolute magnitude in the magnitude–distance formula, we can find the star's distance.

If we can locate Cepheids in a galaxy, we can determine the distance to the galaxy, but ground-based telescopes cannot detect Cepheids beyond about 6 MPc. Only a dozen or so galaxies are that close. New CCD cameras and adaptive optics can help a bit, but the only way to see Cepheids at much greater distances is to go above the earth's atmosphere (Figure 17-11). The Hubble Space Telescope can detect Cepheids to about 40 Mpc, over six times farther than ground-based telescopes.

Beyond some limit, Cepheid variable stars are not visible, so astronomers must calibrate other distance indicators. The brightest red supergiants in galaxies can be detected in nearby galaxies whose distances are known from Cepheids. The average absolute magnitude of the brightest red supergiants turns out to be about −9. Thus, the supergiants in a galaxy are calibrated, and we can find the distance to any galaxy in which we can see these stars. Since they are about 15 times brighter than the brightest Cepheids, we can use red supergiants to greater distances.

In this way, astronomers have calibrated many distance indicators. Large globular clusters, for instance, have absolute magnitudes of about −10, and that of the largest HII regions is about −12. Astronomers have also calibrated novae and supernovae at maximum light. Certain kinds of supernovae, for example, reach absolute magnitude −20 at maximum. Supernovae can be seen to great distances, so this distance indicator is very useful, although we might wait 50–100 years to see a supernova explode in any particular galaxy. Even the most luminous planetary nebulae have been calibrated as distance indicators.

To measure the distances to the farthest galaxies, we must calibrate the galaxies themselves as distance indicators. Studies of large numbers of galaxies whose distances are known from more reliable distance indicators tell us the luminosity of the brightest galaxies. Large spiral galaxies, for example, have absolute magnitudes of about −21. If we see a distant group of galaxies, we might assume that the brightest galaxy in the group is as luminous as the most luminous nearby galaxies. Then we can find the distance to the group.

Of course, astronomers would like to use a single distance indicator to measure the distances to all galaxies, but some techniques that work well for nearby galaxies can't be used for more distant galaxies. Thus, astronomers must piece together a **distance scale** based on a number of different methods. This distance scale resembles a pyramid with a broad base of accurately determined distances that are used to calibrate other methods for greater distances. The farther we go up the pyramid, the less accurate the calibration of the distance indicators becomes. We will see in Chapter 19 that this distance scale is critical in the study of the universe as a whole, and much of our understanding of the properties of galaxies depends on the calibrations that make up the distance scale.

The full weight of the distance scale rests on its foundation, the Cepheid distance indicators. Because the Hubble Space Telescope can detect Cepheids out to 40 Mpc, it is able to extend such distance indicators to include the large group of galaxies known as the Virgo cluster (about 17 Mpc distant). This is useful for the study of these galaxies, of course, but it also allows a much more accurate calibration of the entire distance

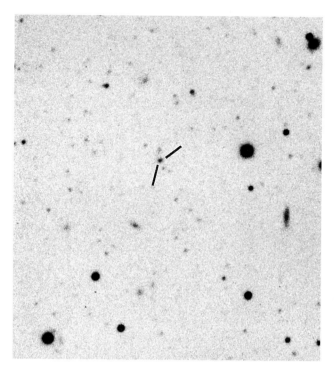

FIGURE 17-12
The visible galaxy associated with radio source 3C 13 is roughly 3000 Mpc (about 10 billion ly) away. At such a great distance, very little detail is visible, and no distance indicators can be found. (Hyron Spinrad)

scale. Thus, the distances to galaxies will become more accurately known at all distances as astronomers use the Hubble Space Telescope to improve the calibration of the distance to the Virgo cluster.

In spite of uncertainties in the distance scale, it is clear that galaxies are far apart and scattered through the universe to tremendous distances. The nearest large galaxy, the Andromeda Galaxy, is 0.66 Mpc (2.2 million ly) distant. For an example of an extremely distant galaxy, we can examine 3C 13, at a distance of 3000 Mpc (about 10 billion ly). Even in the best images, little is visible in a galaxy at such a great distance (Figure 17-12). Radio telescopes detect faint sources of radio energy that are evidently distant galaxies much more luminous in radio energy than in visible light. Thus, radio telescopes can detect galaxies beyond the limit of the largest optical telescopes.

While we are thinking about distances, it is appropriate to note that the vast distances to galaxies produce an effect akin to time travel. When we look at a galaxy billions of light-years away, we see it as it was billions of years ago when its light began the journey toward Earth. Thus, when we look at a distant galaxy, we look back into the past by an amount called the **look-back time,** a time in years equal to the distance to the galaxy in light-years. Although we have expressed most distances in parsecs, it is convenient to express the distances to galaxies in light-years because of this relationship between distance and look-back time.

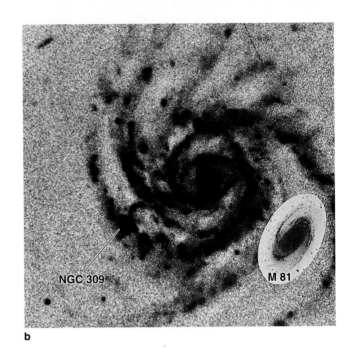

a

b

FIGURE 17-13

Only when we know the distances to galaxies can we compare them. (a) Most of the images in this false-color image are galaxies in the cluster Abell 2029. The central galaxy, a giant elliptical galaxy 100 times brighter than our Milky Way, is one of the largest and most luminous galaxies in the universe. (Juan M. Uson) (b) If the measured distances to NGC 309 and M 81 are correct, then NGC 309 is vastly larger than the giant galaxy M 81. Can NGC 309 be that big? We will not know until we are certain of its distance. (David L. Block and B. Dumoulin; NGC 309 photo by T. Kinman; M 81 image from Palomar Sky Survey)

The look-back time for nearby galaxies is not significant because nearby galaxies are only a few million light-years distant and galaxies change slowly. A few million years is not important in the life of a galaxy. But when we look at more distant galaxies, the look-back time becomes an appreciable part of the age of the universe. We will see evidence in Chapter 19 that the universe is only about 15 billion years old. Thus, we see the most distant visible galaxies as they were when the galaxies and the universe were much younger. This effect will be important in this and the next two chapters.

Diameter and Luminosity

The properties of a galaxy are very hard to judge if we don't know its distance (Figure 17-13). Once we know the distance, finding diameter and luminosity is simple.

The diameter of a galaxy can be found if we know its distance and can measure its angular diameter from an image. All we need is the small-angle formula (Chapter 3). For example, we might see a galaxy that was 100 seconds of arc in diameter and 20 million light-years distant. Then the small-angle formula tells us that the galaxy is about 10,000 ly in diameter.

The luminosity of a galaxy is also easy to find if we know its distance. All we have to do is measure its apparent magnitude and then use the distance to calculate the absolute magnitude.

Such measurements reveal a tremendous range in size and luminosity (Table 17-1). The largest galaxies are 5 times bigger than our Milky Way Galaxy, and the smallest are 100 times smaller. The most luminous are 10 times brighter, and the least luminous are 20,000 times less luminous.

In a sense, the diameter and luminosity of a galaxy are visible to our eyes, and that makes them easy to determine once we know the distance. But the mass of a galaxy is not at all a visible characteristic, and that makes its measurement a challenge.

The Mass of Galaxies

Although the mass of a galaxy is hard to determine, it is an important quantity. It tells us how much matter the galaxy contains, which gives us clues to the origin and evolution of the galaxy. In this section, we examine four ways to find the masses of galaxies. All four methods rely on techniques we have developed in earlier chapters to study the orbital motion of objects such as binary stars (Chapter 10).

One technique is called the **rotation curve method.** We begin by recording the galaxy's spectrum at different points along its diameter and plotting the Doppler-shift velocities in a rotation curve like that in Figure 17-14. This tells us how fast the galaxy is rotating. The sizes of the orbits the stars follow around the galaxy's center are

a

b

FIGURE 17-14

(a) The rotation curve of the Andromeda Galaxy (inset) shows that one side rotates away from us and one side toward us. Because orbital velocity does not fall in the outer parts of the galaxy, we conclude that it contains considerable dark matter in its outer regions. (Lick Observatory) (b) Three examples of rotation curve studies. The galaxies are shown at left, with their spectra at center. Spectral lines (horizontal) are red-shifted on one side of the nucleus and blue-shifted on the other. The rotation curves (only one side shown) reveal no decrease in the outer regions. (Courtesy Vera Rubin)

related to the size of the galaxy, easily found from its angular diameter and its distance. We then ask how massive the galaxy must be to hold stars in orbits of that size at that velocity. This is the same method we used in Chapter 16 to find the mass of our own galaxy from the size of the sun's orbit and its orbital velocity.

The rotation curve method may be the best way to find the masses of galaxies, but it suffers from two shortcomings. First, it can be applied only to the nearer galaxies. More distant galaxies look so small that the astronomer cannot record their spectra at different points along their diameters and thus cannot determine

TABLE 17-1

The Properties of Galaxies*

	Elliptical	Spiral	Irregular
Mass	0.0001–50	0.005–2	0.0005–0.15
Diameter	0.01–5	0.2–1.5	0.05–0.25
Luminosity	0.00005–5	0.005–10	0.00005–0.1

*In units of the mass, diameter, and luminosity of the Milky Way.

a

b

FIGURE 17-15

(a) The Coma cluster of galaxies contains at least 1000 galaxies and is especially rich in E and S0 galaxies. Two giant galaxies lie near its center. If the cluster were visible, it would span eight times the diameter of the full moon in the sky. (L. Thompson, © National Optical Astronomy Observatories) (b) In false colors, this X-ray image of the Coma cluster shows it filled with hot gas. Such hot gas would leak away if it were not held by a strong gravitational field. The cluster must contain up to 90 percent dark matter. (Simon D. M. White, Ulrich G. Briel, and J. Patrick Henry)

their rotation curves. Second, the rotation curve method measures the mass only in the luminous part of the galaxy, the part we can see, and astronomers are finding increasing evidence that galaxies have extended coronas. For example, the rotation curves of some galaxies may not turn downward in Keplerian motion far from the center (Figure 16-12). As we saw in the case of the Milky Way Galaxy, this tells us that the outer parts of the galaxy contain significant amounts of mass. A recent study of the spiral galaxy NGC 3067 shows that 94 percent of the mass lies outside the visible part of the galaxy. The rotation curve method misses this mass, so we must use masses found in this way with some care.

It is not unusual to find pairs of galaxies orbiting each other. In such cases, we analyze them as if they were a pair of binary stars. The only difficulty with this **double galaxy method** is that the galaxies take hundreds of millions of years to orbit once, so over periods of thousands of years we would see no motion. Because we can't observe the shapes of their orbits, we can't determine the orbital inclination, and we can't find their true masses. The solution to this problem is to average the results for many pairs of galaxies. Different orbital inclinations should average out and give good average masses.

The **cluster method** of finding the masses of galaxies depends on the motions of galaxies within a cluster. If we measure the radial velocities of many galaxies in a cluster, we find that some velocities are larger than others because of the orbital motions of the individual galaxies in the cluster. Given the range of velocities and the size of the cluster, we can ask how massive a cluster of

this size must be to hold itself together with this range of velocities. Dividing the total mass of the cluster by the number of galaxies in the cluster yields the average mass of the galaxies. This method contains the built-in assumption that the cluster is gravitationally bound; that is, it is not coming apart. If it is, our result is too large. Only if we assume that most clusters are bound is the method valid.

The fourth technique for measuring a galaxy's mass is the **velocity dispersion method.** It is really a version of the cluster method. Instead of looking at the motions of galaxies in a cluster, we look at the motions of stars in the galaxy and estimate the mass needed to hold the galaxy together. This method, like the rotation curve method, measures only the inner, visible parts of a galaxy and gives generally lower masses.

Measuring the masses of galaxies tells us two things. First, the range of masses is wide—from 10^{-4} as much as that of the Milky Way Galaxy to 50 times more (Table 17-1). And second, there is more to a galaxy than meets the eye.

Dark Matter in Galaxies

Given the size and luminosity of a galaxy, we can make a rough guess as to the amount of matter it contains. We know how much light stars produce, and we know about how much matter there is between the stars, so it is quite possible to estimate the mass of a galaxy. But when astronomers measure the masses of galaxies, they often find that the measured masses are much too large. Mea-

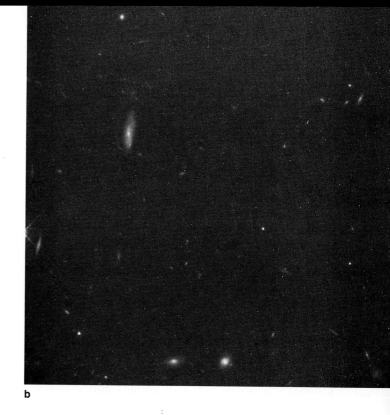

a

b

FIGURE 17-16

If the dark matter in galaxies consists of cool, low-mass stars scattered through extended halos, then the Milky Way Galaxy should contain tremendous numbers of such stars. Any long-exposure photograph with the Hubble Space Telescope should reveal dozens, as simulated here (a). Actual photographs, however, reveal a faint dusting of distant galaxies but almost no cool, low-mass halo stars (b). Such observations seem to rule out normal low-mass objects as the hiding places of the dark matter. (J. Bahcall, Institute for Advanced Study, Princeton, and NASA)

surements show that galaxies contain 10–100 times more mass than we can see. Since this mass is not visible, it is called **dark matter.**

We have seen how the rotation curves of galaxies suggest they contain dark matter, but many other kinds of observations lead to the same conclusion. For example, in many clusters of galaxies, the velocities of the galaxies are so high the clusters would fly apart if they contained only the mass we see. These clusters must contain large amounts of dark matter.

X-ray observations reveal more evidence of dark matter. X-ray images of galaxy clusters show that many of them are filled with very hot, low-density gas that would easily leak away if it were not held by a strong gravitational field. To provide that gravity, the cluster must contain much more matter than what we see. The galaxies near the center of the Coma cluster, for instance, amount to only 11–35 percent of the total mass (Figure 17-15).

Dark matter is not an insignificant issue. Observations of galaxies and clusters of galaxies tell us that 90–99 percent of the matter in the universe is dark matter. The universe we see—the luminous matter—has been compared to the foam on an invisible ocean. Dark matter affects the formation and evolution of galaxies and can affect the evolution of the universe as a whole (Chapter 19).

Dark matter is difficult to detect, and it is even harder to explain. Some astronomers have suggested that dark matter consists of low-luminosity stars and planets scattered through an extended halo around each galaxy. Searches for such objects have not been successful. For example, if the dark matter in our galaxy is hidden in cool, low-mass stars in the halo, then careful searches using the Hubble Space Telescope should reveal many of these stars. The photographs reveal almost none (Figure 17-16), so it seems unlikely that cool, low-mass stars can explain the dark matter in galaxies.

The dark matter can't be hidden in vast numbers of black holes and neutron stars because we don't see the X rays these objects would emit. Remember, we need 10–100 times more dark matter than visible matter, and that many black holes should produce X rays that are easy to detect.

Because observations seem to tell us the dark matter can't be hidden in normal objects, some theorists have suggested that the dark matter is made up of unexpected forms of matter, such as neutrinos with mass, or exotic subatomic particles, such as WIMPs (Chapter 12). So far no laboratory experiments have been able to measure a mass for the neutrino, and no WIMPs have ever been detected. Thus, dark matter remains one of the fundamental unresolved problems of modern astronomy. We will return to this problem in Chapter 19 when we try to

FIGURE 17-17
Edwin Hubble's first diagram of the velocities and distances of galaxies did not probe very deeply into space. It did show, however, that the galaxies are receding from one another. Because of errors in his distances, his first estimate of the Hubble constant was too large.

understand how dark matter affects the nature of the universe, its past, and its future.

For the moment, we must put aside the mystery of dark matter and examine one of the basic laws of astronomy.

The Hubble Law

Because many galaxies look like great whirlpools of stars and because Doppler shifts prove that they are rotating, we might hope to see some motion over the years. Unfortunately, the galaxies are too large and move too slowly to show any visible change even over thousands of years. But the galaxies do have a motion that is detectable and critically important. That motion was discovered almost a century ago.

In 1913, V. M. Slipher at Lowell Observatory reported on the spectra of faint, nebulous objects in the sky. Their spectra seemed to be composed of a mixture of stellar spectra: some had Doppler shifts that suggested rotation, most had red shifts as if they were receding, and the faintest had the largest red shifts. Within two decades, astronomers concluded that the faint objects were galaxies similar to our own Milky Way and that the galaxies are indeed receding from us in a general expansion.

In 1929, Edwin Hubble and Milton Humason published an extensive study of galaxies. Their results on distance and velocity led to a general law of red shifts now known as the **Hubble law.** This law says that a galaxy's velocity of recession equals a constant times its distance:

$$V_r = Hd$$

Thus, the more distant a galaxy is, the faster it recedes from us (Figure 17-17). The constant H, now known as the **Hubble constant,** is very difficult to determine.

One important study of the recession of the galaxies suggests that H equals about 50 km/sec/Mpc.* However, other studies yield values as high as 100 km/sec/Mpc. The uncertainty arises from the difficulty of determining the distances to galaxies. Different groups of astronomers have found distances in different ways and have arrived at different values of H. In Chapter 19, we will see that this uncertainty has important consequences for our understanding of the history of the universe, but here it is sufficient to recognize that the Hubble constant is poorly known and to adopt a provisional value of about 70 km/sec/Mpc.

The Hubble law is important because it is commonly interpreted to show that the universe is expanding. In Chapter 19, we will discuss the implications of this expansion; here we use the Hubble law as a practical way to estimate the distance to a galaxy. A galaxy's distance in megaparsecs equals its radial velocity divided by the Hubble constant. For example, the Virgo cluster of galaxies has a radial velocity of 1180 km/sec. If we assume the Hubble constant is 70 km/sec/Mpc and divide, we get a distance of 1180/70 or 16.9 Mpc.

The Hubble law makes it relatively easy to find galactic distances because large telescopes can photograph the spectrum of a distant galaxy and determine its red shift even though distance indicators such as variable stars are totally invisible. We cannot, however, abandon distance indicators and use the Hubble law exclusively. Because the Hubble constant is poorly known, any distances determined using it are similarly uncertain. The only way to find the Hubble constant more precisely is to improve the methods for finding the distances to galaxies. That is, astronomers must continue to improve the calibration of the distance indicators in the distance scale, and that will tell us more accurately the value of the Hubble constant.

CRITICAL INQUIRY

Why must we know the distance to a galaxy in order to find its mass?

To find the mass of a galaxy, we observe the motions of stars in the outer parts of the galaxy and compare those motions to the size of the orbits the stars follow around the center of the galaxy. That is, we use Kepler's third law. Measuring the velocity of the stars is fairly easy. We need a big telescope with a spectrograph, and we record the spectrum of the galaxy in its outer parts. The Doppler shift tells us the velocity of the stars. We might even create a graph of the rotation curve to show how the stars' velocities depend on their distance from the center of the galaxy. The hard part is finding the size of the orbits. We can measure the distance from the center of the galaxy out to the stars at the outer edge in angular

---

*H has the units of a velocity divided by a distance. These are usually written as km/sec/Mpc, meaning km/sec per Mpc.

FIGURE 17-18

The colliding galaxies NGC 4038 and NGC 4039 are known as The Antennae because their long curving tails resemble the antennae of an insect. Earth-based photos (left) show little detail, but a Hubble Space Telescope image (right) reveals the collision of two galaxies producing thick clouds of dust and raging star formation creating roughly a thousand massive star clusters such as the one at the top (arrow). Such collisions between galaxies are common. (Brad Whitmore, STScI, and NASA)

units such as seconds of arc, but we must use the small-angle formula to convert the angular measurement into a linear measurement of the radius of the orbit. The small-angle formula requires that we know the distance to the galaxy.

Most of what we know about galaxies, including their masses, depends on the calibration of the distance indicators in the distance scale. What would happen to the Hubble constant if astronomers discovered that the Cepheid variable stars were slightly more luminous than had been believed? ▪

From our study of the properties of galaxies, we discover a wide range in the diameters, masses, and luminosities of galaxies. Combined with the different classes of galaxies, this raises the question of the evolution of galaxies. How did galaxies get to be the way they are? As we will see in the next section, that is a complicated question.

17-3 THE LIVES OF THE GALAXIES

In Chapter 19 we will learn that the universe appears to have begun with the so-called big bang. That event produced hydrogen and helium but almost no heavier elements. Thus, we must suppose that the galaxies formed from those primeval clouds of hydrogen and helium. How they formed is one of the deepest mysteries of modern astronomy.

A second mystery concerns the evolution of the galaxies. We see many different kinds of galaxies, and we wonder why one galaxy becomes spiral and another becomes elliptical. Although still not fully understood, that problem seems to be linked to interactions, collisions, and mergers between the galaxies.

Colliding Galaxies

Galaxies should collide fairly often. The average separation between galaxies is only about 20 times their diameter. Like two elephants blundering about at random under a circus tent, galaxies should bump into each other once in a while. Stars, on the other hand, almost never collide. In the region of the galaxy near the sun, the average separation between stars is about 10^7 times their diameter. Thus, collision between two stars is about as likely as collision between two gnats flitting about at random in a football stadium.

Large telescopes reveal hundreds of galaxies that appear to be colliding with other galaxies. One of the most famous pairs of colliding galaxies, NGC 4038 and NGC 4039, are called The Antennae because the long tails resemble the antennae of an insect (Figure 17-18). In addition to having tails, some interacting galaxies are connected by bridges of gas, dust, and stars.

When galaxies collide, tails and bridges are produced by the gravitational fields. These are tidal forces (Chapter 3) because the near side of a galaxy feels a stronger gravitational force than the far side. Whether the galaxies pass through each other or just pass near each other, the stars do not collide because they are too small and too far apart. But tides will deform the shapes of such galaxies.

FIGURE 17-19

A computer simulation of a collision between galaxies produces a pattern very similar to M 51, the Whirlpool Galaxy (bottom), and its small companion. (From "Violent Tides Between Galaxies" by A. Toomre and J. Toomre. Copyright © 1973 by Scientific American, Inc. All rights reserved. National Optical Astronomy Observatories.)

FIGURE 17-20

The Cartwheel Galaxy was once a normal spiral galaxy, but a few hundred thousand years ago one or the other of the two smaller galaxies at the right passed through the center of the larger galaxy. Tidal forces generated an expanding wave of star formation and a central core of newborn stars heavily reddened by dust. Such ring galaxies appear to be the fleeting consequences of bulls-eye collisions. Soon after the collision, the violent star formation exhausts itself, and normal spiral arms may re-form. This is suggested in the Cartwheel Galaxy by the faint spiral structure inside the outer ring. (Kirk Borne, STScI and NASA)

A collision between galaxies can last hundreds of millions of years, but we can watch it happen in computer models. The Whirlpool Galaxy, M51, provides a good example of such a model (Figure 17-19). Models show that the galaxies passed near each other but did not actually penetrate. The smaller galaxy passed behind the larger galaxy.

Unusual **ring galaxies** consist of a bright nucleus surrounded by a ring (Figure 17-20). Models show that they are produced by a galaxy passing roughly perpendicularly through the disk of the larger galaxy. Indeed, many ring galaxies have nearby companions.

As colliding galaxies approach each other, they carry tremendous orbital momentum. Models show that tidal forces can convert some of the orbital motion into random motion among the stars, thus robbing the galaxies of orbital momentum. Such galaxies may lose enough orbital momentum that they fall back together and merge.

A small galaxy merging with a larger galaxy will be pulled apart gradually, and its stars will spread through the larger galaxy in a process called **galactic cannibalism.*** Computer simulations of mergers show that galaxies throw off shells of stars as the nuclei of the two galaxies spiral into each other, and such shells have been observed around real galaxies (Figure 17-21).

---

*The small galaxies that fall victim to cannibalism have been called **missionary** galaxies.

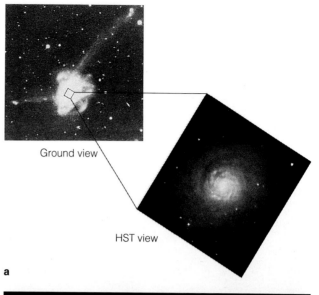

Ground view

HST view

a

b

FIGURE 17-22

Evidence of mergers. (a) The twisted shape of NGC 7252 suggests a collision, and a Hubble Space Telescope image reveals a small, backward-spinning spiral at the center, the product of a merger between oppositely rotating galaxies about a billion years ago. (François Schweizer, Carnegie Institution of Washington, and Brad Whitmore, STScI) (b) Galaxy M64 is shown at left in a radio map where color shows radial velocity—red receding and blue approaching. At the center, the rotation is reversed, evidently the consequence of a merger. (Robert Braun, NRAO)

FIGURE 17-21

Computer simulations suggest that merging galaxies can produce shells of stars as the two galaxies whirl around their common center of mass. Only one galaxy is shown in this sequence. Such shells are seen around galaxies such as NGC 3923 (bottom), shown here as a negative image specially enhanced to reveal low-contrast features. (Model courtesy Francois Schweizer and Alar Toomre; photo courtesy David Malin, Anglo-Australian Observatory)

Evidence of collisions often appears in the motions inside a galaxy. The tails on NGC 7252 (Figure 17-22a) resemble the tails on mice, and when the Hubble Space Telescope imaged the center of the galaxy it found a small, backward-spinning spiral. Evidently, NGC 7252 was created about a billion years ago when two oppositely spinning galaxies collided and merged. The apparently normal galaxy, M64, reveals its secrets in radio maps (Figure 17-22b). The center is rotating backward compared to the outer galaxy—the product of a merger between counter-rotating galaxies.

Evidence of past mergers also appears in the multiple nuclei in some giant elliptical galaxies (Figure 17-23). These galaxies are often located in crowded rich clusters of galaxies, and the extra nuclei appear to be the dense centers of smaller galaxies that have been cannibalized and only partly digested.

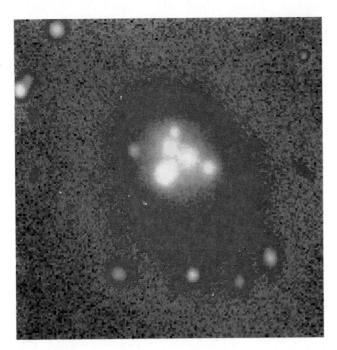

FIGURE 17-23
Giant elliptical galaxies in rich clusters sometimes have multiple nuclei as in this false-color image. These extra nuclei are thought to be the densest parts of smaller galaxies that have been absorbed and partly digested. (Michael J. West)

As new telescopes probe the secrets of the galaxies, we are discovering more and more signs of galaxy collisions and cannibalism. We saw earlier in this chapter that the Magellanic Clouds have passed through the disk of the Milky Way Galaxy and are being distorted and disrupted. Even the big Milky Way Galaxy is disturbed by the encounter, with one edge turned up and the other edge turned down like the brim of a hat. Eventually, both Magellanic Clouds will merge with our galaxy.

Collisions between galaxies are clearly important, not only because they alter the structure of individual galaxies, but also because galaxies are located in groups and clusters where collisions occur often.

Clusters of Galaxies

Single galaxies are rare. Most occur in clusters containing a few to a few thousand galaxies in a volume 1–10 Mpc across (Figure 17-15a). Our Milky Way Galaxy is a member of a cluster containing slightly over three dozen galaxies, and surveys have cataloged over 2700 other clusters within 4 billion ly.

For purposes of our study, we can sort clusters of galaxies into rich clusters and poor clusters. **Rich clusters** contain a thousand or more galaxies, many elliptical, scattered through a volume roughly 3 Mpc (10^7 ly) in diameter. Such a cluster is nearly always condensed; that is, the galaxies are concentrated toward the cluster center. And at the center, such clusters often contain one or more giant elliptical galaxies.

The Virgo cluster is an example of a rich cluster. It contains over 2500 galaxies located about 17 Mpc (55 million ly) away. The Virgo cluster, like most rich clusters, is centrally condensed and contains the giant elliptical galaxy M 87 at its center (Figure 17-4a). Collisions must be common in such a crowded cluster, and some of the material ripped from galaxies by tidal interactions presumably sinks toward the center, increasing the mass of the giant elliptical galaxy at its center.

X-ray observations have found that at least 40 of these rich clusters are filled with a hot gas—an intracluster medium. This gas cannot be left over from galaxy formation, because it is rich in metals. It must have come from supernovae and stellar winds in the galaxies. A galaxy moving through such a medium would feel a tremendous wind stripping the galaxy of its remaining gas and dust. This may explain why these clusters contain more elliptical and S0 galaxies and fewer spirals. A spiral in such a cluster could lose its gas and dust and stop making new stars to illuminate its spiral arms.

Stripping and collisions are much less important in the **poor clusters.** These clusters contain fewer than a thousand (and often only a few) galaxies and are not condensed toward the center. With fewer galaxies, there is less gas in the intracluster medium, so stripping does not rob the galaxies of their gas and dust. Collisions are also less common. This may explain why such clusters usually contain a larger proportion of spiral galaxies, which are rich in dust and gas.

Our Milky Way Galaxy is a member of a poor cluster known as the Local Group (Figure 17-24a). The total number of galaxies in the Local Group is uncertain, but it probably contains a few more than three dozen galaxies scattered irregularly through a volume roughly 1 Mpc in diameter. Of the brighter galaxies, 15 are elliptical, 4 are spiral, and 13 are irregular.

We can't be sure how many galaxies are in the Local Group because the dust in the Milky Way could be hiding some in the zone of avoidance. For example, Maffei I and Maffei II are a large elliptical galaxy and a large spiral galaxy hidden behind the dust in the Milky Way. Radio astronomers in Dwingeloo, the Netherlands, located another spiral galaxy, dubbed Dwingeloo 1, near the Maffei galaxies. Once alerted to its presence, astronomers were able to photograph the galaxy at visual and infrared wavelengths (Figure 17-24c). Dwingeloo 1 is as large as our own galaxy and is only about five times as far away as the Andromeda Galaxy, our nearest large neighbor. Other galaxies are probably hidden behind the dusty Milky Way.

The Local Group illustrates the subclustering found in poor galaxy clusters. The two largest galaxies, the Milky Way and the Andromeda Galaxy, are the centers of two subclusters. The Milky Way is accompanied by the Magellanic Clouds and seven other dwarf galaxies. The Andromeda Galaxy is attended by more dwarf elliptical galaxies (two of which are visible in Figure 16-1) and a small spiral, M 33. The two large galaxies orbit each

a

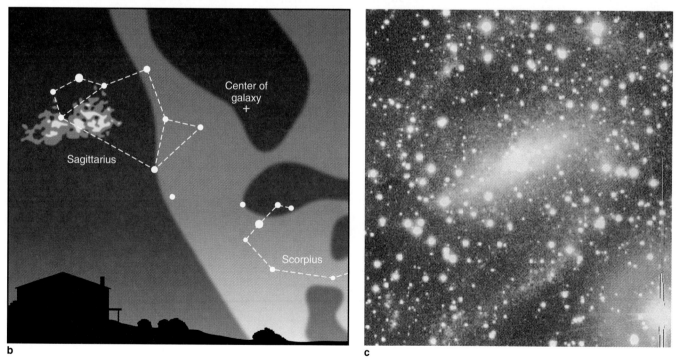

b

c

FIGURE 17-24

(a) Most of the galaxies in the Local Group fall within a sphere of radius of 500 kpc. (b) Some galaxies are hidden behind the clouds in the Milky Way. A dwarf galaxy in Sagittarius just beyond the disk of our galaxy is being ripped apart by our galaxy's gravitational field. If we could see it in the sky, it would be 17 times larger than the full moon. (c) Spiral galaxy Dwingeloo 1, as large as our own galaxy, was not discovered until 1994 because it is hidden behind the Milky Way. Note the faint spiral pattern. Nearly all individual stars here are in our galaxy. (Courtesy Shaun Hughes and Steve Maddox, Royal Greenwich Observatory, and Ofer Lahav and Andy Loan, University of Cambridge)

FIGURE 17-25
The Great Wall, a 150-Mpc-long concentration of galaxies (dashed line), is
revealed in this survey of four slices
of the sky. Our galaxy is located at the
vertex of the slices, which cover only
1/100,000 of the visible universe.
Voids such as the one at left appear
almost empty of galaxies, and other
filamentary structures and voids are
visible in other parts of the map.
(M. J. Geller and J. P. Huchra, Center
for Astrophysics)

other while maintaining gravitational control over the small galaxies.

Just as the clustering of galaxies is repeated on a smaller scale in subclustering, it is repeated on a larger scale in superclustering. Clusters of galaxies seem to be associated with one another in groupings called **super-clusters.** The Local Group is a part of the Local Super-cluster, an approximately disk-shaped swarm of galaxy clusters including the Virgo cluster near its center. It is 25–50 Mpc in diameter.

Until the last decade, astronomers had pictured a universe filled with isolated superclusters, but studies of the three-dimensional distribution of galaxies in space has revealed that the superclusters are linked in a fila-mentary network with great voids between them. (This filamentary structure is visible in the last image in Chap-ter 1.) One study has found the Great Wall (Figure 17-25), a sheet containing thousands of galaxies and extending over 150 million pc. It is the largest known structure in the universe. Clearly, the galaxies, clusters, and superclusters are not scattered uniformly through space; they are linked in a complex structure like the material in a sponge.

The study of this distribution of superclusters is especially difficult because the galaxies are so faint that only the largest telescopes can be helpful in measuring red shifts and estimating distances. In addition, even a small area of the sky contains vast numbers of galaxies, each of which must be studied separately. Surveys have explored a few percent of the sky and have found a num-ber of superclusters and intervening voids at typical dis-tances of 75–150 Mpc. This structure raises a critical question: how did the distribution of galaxies become so clumpy? The answer is related to the origin and evolu-tion of galaxies.

The Origin and Evolution of Galaxies

The test of any scientific understanding is whether or not we can put all the evidence and theory together to tell the history of the objects we study. Can we describe the ori-gin and evolution of the galaxies? Just a few decades ago, it would have been impossible, but new evidence is help-ing astronomers understand the history of the galaxies.

We can eliminate a few older ideas immediately. The tuning fork diagram tempts us to think that galaxies evolve from left to right, from elliptical to spiral to irreg-ular. The evidence, however, shows that elliptical galax-ies contain no gas and dust to make new stars, and they contain lots of old stars. Ellipticals can't be young. Fur-thermore, galaxies can't evolve from right to left in the tuning fork diagram, from irregular to spiral to elliptical; irregular galaxies contain both young and old stars and thus can't be young. The tuning fork diagram is not an evolutionary diagram. It tells us that the galaxies have had different histories of star formation, but it does not tell us how galaxies evolve.

Another old idea held that galaxies that form from rapidly rotating clouds of gas would have lots of angular momentum and would contract slowly to form disk-shaped spiral galaxies. Clouds of gas that rotated less rapidly would contract more rapidly, form stars quickly, use up all the gas and dust, and become elliptical galax-ies. The evidence clearly shows that galaxies didn't form and evolve in isolation. Collisions and mergers dominate the history of the galaxies.

The ellipticals appear to be the product of galaxy mergers, which triggered star formation that used up the gas and dust. In fact, we see many **starburst galaxies** that are very luminous in the infrared because a collision has triggered a burst of star formation that is heating the dust. The warm dust reradiates the energy in the

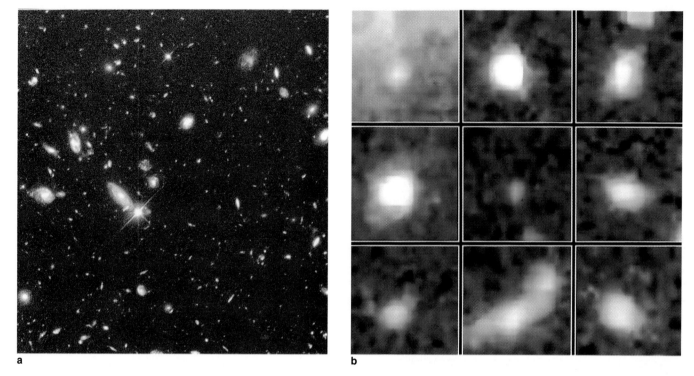

a

b

FIGURE 17-26

(a) An "empty" patch of sky $\frac{1}{30}$ the diameter of the full moon imaged in a 10-day exposure by the Hubble Space Telescope contains roughly 1500 galaxies, some nearby and some so distant we see them soon after the universe began. Presumably the entire sky is filled with galaxies like these. In such images, spirals are more common than they are now, and ellipticals are smaller than elliptical galaxies today. Many of these galaxies are distorted, apparently by interactions. (Robert Williams and the Hubble Deep Field Team, STScI and NASA) (b) Small, blue clouds of gas, dust, and stars appear in the deepest images, and astronomers believe they are the first clouds of matter which later fell together to build the galaxies. The look-back time to these clouds is believed to be about 11 billion years. (Roger Windhorst and Sam Pascarelle, Arizona State University and NASA.)

infrared. Supernovae in such a galaxy may eventually blow away any remaining gas and dust that doesn't get used up making stars. A few collisions and mergers could leave a galaxy with no gas and dust from which to make new stars. Astronomers now suspect that most ellipticals are formed by the merger of at least two or three galaxies.

In contrast, spirals seem never to have suffered major collisions. Their thin disks are delicate and would be destroyed by tidal forces in a collision with a massive galaxy. Also, they retain plenty of gas and dust and continue making stars.

Other processes can alter galaxies. The S0 galaxies may have lost much of their gas and dust in a burst of star formation, but they still managed to remain disk shaped. Also, galaxies moving through the gas trapped in dense clusters of galaxies may have their own gas and dust blown away. For example, X-ray observations show that the Virgo cluster contains thin, hot gas between the galaxies. A galaxy orbiting through that gas would encounter a tremendous wind blowing its gas and dust away. This could explain the dwarf ellipticals, which are too small to be made of merged spirals. In contrast, the irregular galaxies may be small fragments of galaxies ripped apart by collisions.

The evidence for galaxy evolution by merger is quite strong. Observations with the largest telescopes take us to great distances and great look-back times (Figure 17-26a). We see the galaxies as they were long ago, and we discover that there were more spirals then and fewer ellipticals. Also, the ellipticals were smaller than they are now. We can even see that galaxies were closer together long ago; about a third of all distant galaxies are in close pairs, but only 7 percent of nearby galaxies are in pairs. The observational evidence clearly supports the hypothesis that galaxies have evolved by merger.

The Hubble Space Telescope has imaged galaxies so distant that we see them as they were when the universe was less than half its present age. Such images show faint, small, blue clouds of stars, gas, and dust containing roughly 10 times the mass of large globular clusters (Figure 17-26b). Too small to be galaxies, they may be the clouds that eventually fell together to begin building galaxies. In these deep photos, we may be looking back to a time when galaxy formation was beginning.

At the very limit of what we can see, we find the great structures of the universe, the filaments, walls, and voids in the distribution of galaxies. While some astronomers

FIGURE 17-27

A computer simulation of the evolution of a small region of a universe containing dark matter shows small irregularities growing into filaments, walls, and voids. The area visible in each frame remains the same as the universe expands. Galaxy formation presumably occurs at about the third frame. Although this model spans the history of the universe from soon after the big bang to the present, the time steps between frames are not uniform. The first two frames show the young universe, and the last three frames show more recent times. (Courtesy James Gelb and Edmund Bertschinger, MIT)

struggle to map these features and test their reality, others try to understand how they formed. Did these structures form from the first gaseous irregularities in the universe and the gas condense to form galaxies later, or did the galaxies form first and become drawn into the filaments and walls later. Some theoretical models can create these structures (Figure 17-27), but they require special conditions that may not be valid. To understand the origin of these vast structures, we must better understand the general properties of the universe. We will discuss this problem further in Chapter 19.

We can't tell the entire story of the origin and evolution of the galaxies, but modern telescopes give us evidence to test modern theory. That interplay of evidence and theory gives us insight into the story of the galaxies.

CRITICAL INQUIRY

How did elliptical galaxies get that way?

Astronomers don't really understand how galaxies evolve, but there are some hints, and it may be possible to tell a rough story of how an elliptical galaxy became elliptical. It seems that elliptical galaxies have suffered collisions and interactions since they formed, and they have been driven to use up their gas and dust in rapid bursts of star formation. Thus, they now contain little gas and dust and can't make new stars. Collisions with other galaxies or rapid motion through the gas of the intergalactic medium might also strip gas and dust out of a smaller galaxy. The beautiful disk typical of a spiral galaxy is very orderly, with all the stars following similar orbits. When galaxies collide and interact, the stellar orbits get scrambled by tidal forces, and an orderly disk galaxy could be converted into a chaotic swarm of stars typical of elliptical galaxies. Thus, it seems likely that elliptical galaxies have had much more complex histories than spiral galaxies.

What evidence do we have that the story in the preceding paragraph is true? ■

Before we can consider the universe as a whole, we must examine the galaxies from a different perspective. Some galaxies are suffering tremendous eruptions as their cores blast radiation and matter outward. In the next chapter, we will discover that these peculiar galaxies are closely related to collisions and mergers between galaxies. ■

■ *Summary*

Astronomers did not realize that the spiral nebulae visible in larger telescopes were other galaxies until Edwin Hubble detected Cepheid variable stars in a few nearby galaxies. That announcement in 1924 showed that the universe was filled with galaxies much like our own Milky Way.

We can divide galaxies into three classes—elliptical, spiral, and irregular—with subclasses giving the galaxy's shape or the amount of gas and dust present. The galaxy types appear to reflect different histories of star formation. The elliptical galaxies contain little gas and dust and cannot make many new stars. The spiral and irregular galaxies contain large amounts of gas and dust and are still forming stars.

To measure the properties of galaxies, we must first find their distances. For the nearer galaxies, we can judge distances using distance indicators, objects whose luminosities are known. The best of these standard candles are the Cepheid variable stars. Once the distances to some galaxies are known, astronomers can calibrate other distance indicators to piece together a distance scale. Some of the most-used distance indicators are bright red supergiants, globular clusters, HII regions, novae, and supernovae. For the most distant galaxies, astronomers must depend on a calibration of the luminosity of galaxies themselves.

The Hubble law shows that the radial velocity of a galaxy is proportional to its distance. Thus, we can use the Hubble law to estimate distances. The galaxy's radial velocity divided by the Hubble constant is its distance in megaparsecs.

The masses of galaxies can be measured in four ways—the rotation curve method, the double galaxy method, the cluster method, and the velocity dispersion method. The first method is the most accurate, but it is applicable only to nearby galaxies.

Many measurements of the mass of galaxies suggest that galaxies contain many times more mass than what we can see. This dark matter amounts to 90–99 percent of the matter in the universe. While some astronomers suggest that it is made up of neutrinos with mass or exotic atomic particles called WIMPs, no laboratory tests have confirmed that such particles exist.

The rotation and turbulence of the gas clouds from which the galaxies formed may have influenced their final form, but modern studies suggest that collisions between galaxies are also important. At high velocities, colliding galaxies can distort each other through tidal forces, creating tails and bridges. Such collisions can drive a galaxy to consume its gas and dust in rapid star formation. At lower velocities, colliding galaxies can merge to become a single galaxy. Such galaxy cannibalism may be common; our own Milky Way may be consuming the Magellanic Clouds.

Stripping may also affect the evolution of galaxies. When two galaxies collide or when a galaxy moves rapidly through the gas of an intracluster medium, the galaxy can be stripped of its gas and dust. Such galaxies might resemble elliptical and S0 galaxies more than spirals. Thus, we can understand why spirals are less common in rich clusters, where collisions are more common.

Normal elliptical galaxies are apparently formed by the merger of spiral galaxies. The average, bright elliptical galaxy may contain the merged remains of from four to ten spiral galaxies. Dwarf ellipticals are different systems and appear to be small galaxies stripped of gas and dust by their motion through the intracluster gas and by interactions with larger galaxies. According to this theory, the spiral galaxies have not experienced collisions with large galaxies very often since they were formed.

The clusters of galaxies appear to be united in superclusters, which are, in turn, united in a network of filaments surrounding empty voids. How such filaments and voids formed is not well understood.

New Terms

zone of avoidance	rotation curve method
spiral nebula	double galaxy method
island universe	cluster method
tuning fork diagram	velocity dispersion method
elliptical galaxy	dark matter
spiral galaxy	Hubble law
barred spiral galaxy	Hubble constant (H)
irregular galaxy	ring galaxy
megaparsec (Mpc)	galactic cannibalism
distance indicator	rich cluster
standard candle	poor cluster
distance scale	supercluster
look-back time	starburst galaxy

Questions

1. If a civilization lived on a planet in an E0 galaxy, do you think it would have a zone of avoidance? Why or why not?

2. Draw and label a tuning fork diagram. Why can't the evolution of galaxies go from elliptical to spiral? from spiral to elliptical?

3. If all elliptical galaxies had three different diameters, we would never see an elliptical galaxy with a circular outline on a photograph. True or false? Explain your answer. (HINT: Can a football ever cast a circular shadow?)

4. What is the difference between an Sa and an Sb galaxy? between an S0 and an Sa galaxy? between an Sb and an SBb galaxy? between an E7 and an S0 galaxy?

5. Why wouldn't white dwarfs make good distance indicators?

6. Why isn't the look-back time important among nearby galaxies?

7. Explain how the rotation curve method of finding a galaxy's mass is similar to the method used to find the masses of binary stars.

8. Explain how the Hubble law permits us to estimate the distances to galaxies.

9. How can collisions affect the shape of galaxies?

10. What evidence do we have that galactic cannibalism really happens?

11. Describe the future evolution of a galaxy that we now see as a starburst galaxy. What will happen to its interstellar medium?

12. Why does the intracluster medium help determine the nature of the galaxies in a cluster?

Discussion Questions

1. From what you know about star formation and the evolution of galaxies, do you think the Infrared Astronomy Satellite should have found irregular galaxies to be bright or faint in the infrared? Why or why not? What about starburst galaxies? What about elliptical galaxies?

2. Suppose we could observe a cluster of galaxies at a distance of 3 Mpc and we discovered that it contained mostly elliptical galaxies. Would that support or contradict our understanding of the evolution of spiral and elliptical galaxies? Why?

■ *Problems*

1. If a galaxy contains a Type I (classical) Cepheid with a period of 30 days and an apparent magnitude of 20, what is the distance to the galaxy? (HINT: See Figure 13-21.)

2. If you find a galaxy that contains globular clusters with apparent magnitudes no brighter than 18th magnitude, how far away is the cluster? (HINT: What is the absolute magnitude of a globular cluster?)

3. If a galaxy contains a supernova that at brightest has an apparent magnitude of 17, how far away is the galaxy? (HINT: Assume that the absolute magnitude of the supernova at brightest is −19.)

4. A galaxy has been found that is 10 kpc in diameter and whose outer stars orbit the center with a period of 200 million years. What is the mass of the galaxy? On what assumptions does this result depend? (HINT: Use Kepler's third law.)

5. Use the data in Figure 17-14a to calculate a minimum mass for the Andromeda Galaxy. (HINT: See Problem 4.)

6. Among the globular clusters orbiting a distant galaxy, the fastest is traveling 420 km/sec and is located 11 kpc from the center of the galaxy. Assuming the globular cluster is gravitationally bound to the galaxy, what is the mass of the galaxy? (HINT: The galaxy had a slightly faster globular cluster, but it escaped some time ago. What is the escape velocity?)

7. If a galaxy has a radial velocity of 4000 km/sec, how far away is it? (HINT: Use the Hubble law.)

■ *Critical Inquiries for the Web*

1. How far out into the universe can we see Cepheid variables? Research sources on the Internet to find other galaxies whose distances have been found through observation of Cepheids. List the galaxies in which Cepheids have been identified and the distances determined from these data.

2. How does the Milky Way stack up against the other galaxies in the Local Group? Look for information about the other galaxies in our cluster and rank the top six members in order of total mass.

3. In the early 1900s the nature of the "spiral nebulae" was not well understood. In 1920 a "great debate" was held between Harlow Shapley, who held that these objects were relatively nearby swirling clouds of gas, and Heber Curtis, who saw them as distant "island universes." Use the Internet to find information about the debate, outline the lines of evidence used by the two participants to present their views, and explain who was right and who was wrong.

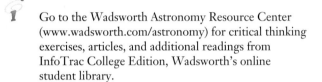

Go to the Wadsworth Astronomy Resource Center (www.wadsworth.com/astronomy) for critical thinking exercises, articles, and additional readings from InfoTrac College Edition, Wadsworth's online student library.

Chapter Eighteen

PECULIAR GALAXIES

GUIDEPOST

It would be wrong to argue that this chapter is unimportant. Certainly it discusses galaxies that are unusual and therefore not representative of galaxies in general. But this chapter draws together ideas from many previous chapters to show how nature uses the same basic rules on widely different scales. Matter flowing into a protostar, into a white dwarf, into a neutron star, or into the heart of a galaxy must obey the same laws of physics, so we see the same geometry and the same phenomena. The only difference is the level of violence.

A second reason why this is an important chapter is that the most distant objects we can see in the universe are the most luminous galaxies, and many of those are erupting in outbursts and are thus peculiar. By studying these galaxies, our attention is drawn out in space to the edges of the visible universe and back in time to the earliest stages of galaxy formation. In other words, we are led to think of the origin and evolution of the universe, the subject of the next chapter. ■

Virtually all scientists have the bad habit of displaying feats of virtuosity in problems in which they can make some progress and leave until the end the really difficult central problems.

M. S. LONGAIR

High Energy Astrophysics

FIGURE 18-1

Spiral galaxy NGC 1566 in the southern sky looks like a normal spiral galaxy, but short-exposure photographs reveal that its nucleus is small and very luminous. The spectrum of the nucleus suggests that highly excited gas is moving at very high velocities. (Anglo-Australian Telescope Board)

Astronomers get used to odd phone calls like this one, which came an hour before dawn.*

Voice: Hello . . . Dr. Seeds? Sorry to wake you up. You don't know me, but you spoke to my church group last week, and I just got up to go to the bathroom, and there is a real bright thing out over my barn and it's flashing colors and it's really bright, and . . .

Seeds: Which direction's your barn?

Voice: Out back . . . uh . . . east mostly.

Seeds: Venus . . . it's Venus.

Voice: Are you sure? It's awful bright, and it looks like it's right above my barn.

Seeds: Venus . . . been rising before the sun for the last few weeks . . . twinkles when it's low on the horizon . . . not above your barn . . . 'bout 100 million miles away.

Voice: You sure? It looks like it's over my barn.

If we don't know how far away an astronomical object is, it is almost impossible to figure out what it is from its appearance. Modern astronomers faced this problem in the early 1960s when they discovered a number of small, starlike objects with totally unrecognizable spectra. Only after a number of years did they realize that the objects were not nearby stars but the most distant galaxies in the visible universe.

Over the last two decades, astronomers have been able to connect these very distant, very luminous galaxies with other peculiar galaxies that are less luminous but closer to us. They all appear to be suffering from tremendous eruptions occurring in their cores—eruptions that

*This is a true story. See Figure 2-18b.

are blasting jets of excited gas outward at high velocity, creating vast clouds of gas emitting radio energy, and illuminating the cores of the galaxies with intense electromagnetic radiation at all wavelengths.

In the quotation that opens this chapter, Longair speaks of "the really difficult central problems" of science. As far as galaxies are concerned, the really difficult central problem lies at the very centers of the galaxies. What is the energy machine? In this chapter, we link our theories of the origin of the Milky Way, the origin and evolution of spiral and elliptical galaxies, the interactions of merging galaxies, and the accretion of matter into black holes and construct a tentative theory to explain the peculiar galaxies.

18-1 ACTIVE GALAXIES

All galaxies, including our Milky Way Galaxy, emit radio energy from neutral hydrogen, molecules, pulsars, and so on, but some galaxies, called **radio galaxies,** emit as much as 10 million times more radio energy from a small region at their centers. Some of these galaxies also emit powerfully at infrared, ultraviolet, and X-ray wavelengths, so the term **active galaxies** is also used. Yet all of this activity seems to originate in a very small region in the nucleus of the galaxy. One of the great adventures of modern astronomy has been the quest to understand these **active galactic nuclei (AGN).**

In our search for the nature of the AGN, we begin by studying the nuclei of certain galaxies that look quite normal in photographs. Then we shift our attention to galaxies that emit powerful radio signals. The clues we discover will lead us to a theory to explain the AGN.

Statistical Evidence

Notice that some scientific evidence is statistical. For example, we might argue that Seyfert galaxies are three times more likely to have a nearby companion than a normal galaxy is. This is statistical evidence because we can't be certain that any single Seyfert galaxy will prove to have a companion. Yet the probability is higher than if it were a normal galaxy, and that leads us to suspect that interactions between galaxies are involved. Such statistical evidence can tell us something in general about the cause of Seyfert eruptions.

Statistical evidence is common in science, but it is inadmissible in most courts of law. The American legal system is based on the principle of reasonable doubt, so most judges would not allow statistical evidence. For example, plaintiffs have had great trouble suing tobacco companies for causing their cases of lung cancer

even though there is a clear statistical link between smoking and lung cancer. The statistics tell us something in general about smoking but cannot be used to prove that any specific case of lung cancer was caused by smoking. There are other causes of lung cancer, so there is always a reasonable doubt as to the cause of any specific case.

We can use statistical evidence in science because we do not demand that the statistics demonstrate anything conclusively about any single example. To continue our example of Seyfert galaxies, we don't demand that the statistics predict that any specific galaxy has a companion. Rather, we use the statistical evidence to gain a general insight into the cause of active galactic nuclei. That is, we are trying to understand galaxies as a whole, not to convict any single galaxy.

Of course, if we surveyed only a few galaxies, our statistics might not be very good, and a critic might be justified in making the common complaint, "Oh, that's only statistics." For example, if we surveyed only four galaxies and found that three had companions, our statistics would not be very reliable. But if we surveyed 1000 galaxies, our statistics could be very good indeed, and our conclusions could be highly significant.

Thus, scientists can use statistical evidence if it passes two tests. It cannot be used to draw conclusions about specific cases, and it must be based on large enough samples so the statistics are significant. With these restrictions, statistical evidence can be a powerful scientific tool. ■

Seyfert Galaxies

In 1943, Mount Wilson astronomer Carl K. Seyfert published his study of spiral galaxies with peculiar nuclei, galaxies now known as **Seyfert galaxies.** Although these galaxies look normal in photographs (Figure 18-1), very short exposures revealed that they have unresolved nuclei. By unresolved we mean that the nucleus looks like a star with no measurable diameter. These small, brilliant nuclei have strange spectra that suggest violent motion.

The light from a normal galaxy comes from stars, and its spectrum is the combined spectra of many millions of stars. Thus, we see the principle absorption lines in stellar spectra, mainly hydrogen Balmer lines and lines of calcium II. But the spectrum of the Seyfert galaxy nuclei contains broad emission lines of highly ionized atoms. Emission lines suggest a hot, low-density gas, and ionized atoms suggest that the gas is very excited. The width of the spectral lines suggests very high velocities such that the Doppler effect would smear the emission lines and make them broad. The velocities at the center of Seyfert galaxies are roughly 10,000 km/sec, about 30 times greater than velocities at the center of normal galaxies.

Hundreds of Seyfert galaxies are now known, about 2 percent of all spiral galaxies, and they are now divided into two categories. Type 1 Seyfert galaxies are very luminous at X-ray and ultraviolet wavelengths and have the typical broad emission lines. Type 2 Seyfert galaxies have only narrow emission lines and are brilliant at

infrared wavelengths but not at X-ray and ultraviolet wavelengths.

In the 1960s, astronomers discovered that the brilliant nuclei of Seyfert galaxies fluctuate rapidly. A Seyfert nucleus can change its brightness by 50 percent in less than a month. As we saw in Chapter 15, an astronomical body cannot change its brightness in a time shorter than the time it takes light to cross its diameter. If the Seyfert nucleus can change in a few weeks, then it cannot be larger in diameter than a few light-weeks. In spite of their small size, the cores of Seyfert galaxies produce tremendous amounts of energy. The brightest emit a trillion (10^{12}) times more energy than the entire Milky Way. Something in the centers of these galaxies not much bigger than our solar system produces a galaxy's worth of energy.

The shapes of Seyferts can give us clues to their energy source. They are three times more common in interacting pairs of galaxies than in isolated galaxies (Figure 18-2a). Also, about 25 percent have peculiar shapes suggesting tidal interactions with other galaxies. This statistical evidence (Window on Science 18-1) hints that Seyfert galaxies may have been triggered into activity by companions. Some Seyferts are expelling matter in oppositely directed jets (Figure 18-2b), typical of matter flowing into a strong gravitational field (Chapter 15).

All of this evidence leads modern astronomers to suspect that the cores of Seyfert galaxies contain supermassive black holes—black holes with masses as high as billions of solar masses. Encounters with other galaxies could throw matter into the black hole and release

a

b

FIGURE 18-2

(a) Seyfert galaxy NGC 7674 is distorted with tails to upper right and upper left in this false-color image. Note the companion galaxies. (John W. Mackenty, Institute for Astronomy, University of Hawaii) (b) Seyfert galaxy NGC 4151 has a brilliant nucleus in visible light images (top). The Space Telescope Imaging Spectrograph reveals Doppler shifts showing that part of the expelled gas has a blue shift (to the right) and is approaching us and part has a red shift (to the left) and is receding. (John Hutchings, Dominion Astrophysical Observatory; Bruce Woodgate, GSFC/NASA; Mary Beth Kaiser, Johns Hopkins University; and the STIS Team)

tremendous energy from a very small region. We will expand this theory later, but first we must search for more evidence of supermassive black holes by turning our attention to galaxies that emit powerful radio signals.

Doubled-Lobed Radio Sources

Beginning in the 1950s, radio astronomers found that some radio sources in the sky consisted of pairs of radio-bright lobes. When optical telescopes studied the locations of these **double-lobed radio sources,** they found galaxies located between the lobes (Figure 18-3). Apparently, the galaxies produced the radio lobes, although in some cases the central galaxies were not emitting detectable radio energy.

Radio lobes are generally much larger than the galaxy they accompany. Many are as large as 60 kpc (200,000 ly) in diameter, twice the size of the Milky Way Galaxy. From tip to tip, the radio lobes span hundreds of kiloparsecs. Records for the largest, smallest, brightest, and so on are broken almost daily in astronomy, but one

FIGURE 18-3

This is 3C 388, a typical double-lobed radio source. The visible galaxy lies at the center and appears to supply energy to the radio lobes on either side of it. Such radio lobes can span millions of parsecs. (National Radio Astronomy Observatory, operated by Associated Universities, Inc., under contract with the National Science Foundation; observers J. O. Burns and W. A. Christiansen)

FIGURE 18-4

Cygnus A is a powerful double-lobed radio source. This radio map produced by the VLA radio telescope shows hot spots as brightenings on the outer edges of the lobes. Note the long, thin jet of matter flowing from the galaxy to the right-hand lobe. The jet is at least 50 kpc long—about twice the diameter of our galaxy. A fainter counterjet is also visible leading to the left lobe. In a visible-light photograph, the deformed central galaxy (inset) is roughly the size of the region between the radio lobes. (National Radio Astronomy Observatory; Palomar Observatory/Caltech)

of the largest known radio galaxies is 3C 236 (the 236th source in the *Third Cambridge Catalogue of Radio Sources*). Its radio lobes span 5.8 Mpc (19 × 10^6 ly).

Radio lobes have two properties that hint at their origin: they radiate synchrotron radiation, and they often have hot spots (regions of intense radio emission) at the edge of the lobe farthest from the central galaxy.

Synchrotron radiation, remember, is produced when high-speed electrons move through a magnetic field (Chapter 14). Although the field in a radio lobe is at least 1000 times weaker than the earth's, it fills a tremendous volume and represents vast stored energy. In addition, the high-speed electrons must be traveling near the speed of light and so must be highly energetic. Such electrons are usually referred to as relativistic electrons. Thus, the total energy stored in a radio lobe is very large, about 10^{53} J—approximately the energy we would get if we converted the mass of a million suns entirely into energy. Clearly, the process that creates radio lobes must involve tremendous power.

The second hint as to the nature of the radio lobes is that many contain **hot spots,** regions of high intensity, on the edge of the lobe farthest from the galaxy (Figure 18-4). This suggests that the lobes are expanding away from the galaxy and that their leading edges are colliding with the intergalactic gas, which would compress the magnetic field and the hot gas, producing stronger radio

emission from that region. In fact, the hot spots can even give us a clue to the source of energy that excites the lobes.

The presence of synchrotron radiation and hot spots and the finely detailed radio images produced by radio telescopes such as the VLA have convinced astronomers that the radio lobes are produced by gas ejected from the central galaxy. This central galaxy is usually a giant elliptical and is often located in a crowded cluster. Many of these galaxies are deformed or otherwise peculiar in their visible appearance, and their spectra contain emission lines of highly excited low-pressure gas. All these properties suggest that the galaxies have suffered some eruption.

Cygnus A (Figure 18-4), the brightest radio source in Cygnus and the second brightest in the sky, was the first double-lobed source discovered (in 1953). It radiates about 10^7 times more radio energy than the Milky Way Galaxy from two lobes containing hot spots. The central galaxy looks peculiar either because it is distorted or because it is partially obscured by clouds of dust, but at its distance of 225 Mpc (730 million ly) we cannot see it well. Recent studies have revealed a thin jet of high-energy matter flowing from the core of the galaxy all the way into the northwest lobe. Such jets are very common in these galaxies.

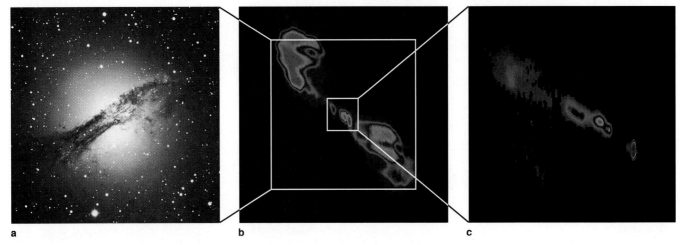

FIGURE 18-5

(a) The radio source Centaurus A consists of a large pair of radio lobes at the center of which optical telescopes reveal the peculiar galaxy NGC 5128. (National Optical Astronomy Observatories) (b) Within this galaxy is a smaller set of radio lobes, shown here in a false-color radio map. (c) Roughly ten times smaller still, a jet of high-energy gas extends out of the core of the galaxy (red). The small jet is evidently responsible for the much larger radio lobes. (National Radio Astronomy Observatory, operated by Associated Universities, Inc., under contract with the National Science Foundation)

FIGURE 18-6

Elliptical galaxy M84 is peculiar in that it contains dust rings and a highly active core ejecting beams of high-energy particles and radio energy. Astronomers using the Hubble Space Telescope placed a narrow slit over the nucleus (inset) and spread the light coming through the slit into a spectrum. A single emission line is shown here, blue shifted above and red shifted below—evidence of a rapidly rotating disk with one side approaching and one side receding. The inner parts of the disk are spinning at very high speeds. This information implies that the core of the galaxy contains at least 300 million solar masses, presumably in a black hole. Note that colors have been artificially applied to this spectral line to show the blue and red shifts. The total width of the line is only about 2 nm. (Gary Bower, Richard Green, NOAO; the STIS Instrument Definition Team; and NASA)

Centaurus A is a double-lobed radio source centered on the giant elliptical galaxy NGC 5128 (Figure 18-5). Radio images show radio lobes spanning 10° in the sky. If we had radio eyes, the lobes would appear 20 times the diameter of the full moon. The galaxy is odd in that it is an elliptical galaxy but contains a dusty band of star formation. Also, a faint jet 40 kpc long has been detected pointing toward the northern radio lobe.

Very long exposures show that this galaxy is surrounded by concentric shells of stars typical of galaxies that have been produced by mergers (Figure 17-22). Doppler shifts show that the bright, spherical component of the galaxy and the dust ring rotate around axes that are perpendicular to each other as if they were separate galaxies. All of this evidence suggests that the galaxy is the product of a merger between an elliptical and a spiral galaxy within the last billion years. We will see later how a merger might trigger outbursts.

High-resolution radio maps show that the galaxy contains an inner pair of lobes and a high-speed jet. Evidently, an eruption at the center has periodically ejected gas and inflated lobes.

The similar elliptical galaxy, M84 (Figure 18-6), gives us further evidence that the radio emission of such galaxies is created by active cores. The core of the galaxy emits brilliant light and ejects beams of high-energy particles that radiate radio energy. Spectra obtained by the Hubble Space Telescope show that the center of the galaxy is surrounded by a disk of gas rotating very rapidly, and the rotational velocity increases closer to the core, reaching 400 km/sec only 26 ly from the center. This data indicates a very high mass at the center, at least 300 million solar masses, and astronomers suspect that the galaxy may contain a supermassive black hole.

FIGURE 18-7

(a) In the double-exhaust model, the hot spots are produced by beams, or jets, of high-speed particles pushing into the intergalactic gas. (b) In this false-color radio map of galaxy 3C 219, an active core (red) ejects a narrow jet pointing toward one of the two radio lobes containing hot spots (yellow). (National Radio Astronomy Observatory—Associated Universities, Inc.; observers A. H. Bridle and R. A. Perley)

These active galaxies have been described as *exploding* galaxies, but that may not be quite the right word. An explosion is a sudden event that is over quickly. Perhaps *erupting* is a better term. Such galaxies evidently release tremendous energy by occasional outbursts in their cores.

Jets from the AGN

Now that we have examined a few examples of double-lobed radio galaxies, can we construct a model that explains their characteristics? We will not yet try to guess the identity of the energy source at the center—we will save that until later in this section. Let's begin by trying to find a way for such a powerhouse to create radio lobes.

Our working hypothesis is that the lobes are created and sustained by jets of hot gas coming from the active galactic nucleus. The **double-exhaust model** supposes that two jets exist, which bore through the interstellar medium forming tunnels leading out of the galaxy. The jets push into the intergalactic medium and inflate cavities of hot gas on either side of the galaxy. We see these cavities as radio lobes, and the points where the jets push back the intergalactic medium we see as hot spots (Figure 18-7a).

This explains how the lobes hold themselves together. Their magnetic and gravitational fields are much too weak, but they are evidently confined by their impact with the intergalactic medium. Their gas cools, of course, and mixes with the intergalactic medium, but as long as the jet is on, it will keep pumping new hot gas into the cavity.

Thanks to the newest high-resolution radio telescopes and orbiting X-ray telescopes, we can see this happening (Figure 18-7b). The jets are very narrow and very hot and are bright at radio and X-ray wavelengths. The gas in some jets (in NGC 5128, for instance) travels about 5000 km/sec, while the gas velocity in other jets (such as Cygnus A) may be a sizable fraction of the speed of light. Such jets have been found in 90 percent of all low-luminosity radio galaxies and may exist in all of them. We will see later that such jets occur in other, higher-luminosity objects.

In many cases, we see two radio lobes but only one jet, and this can be explained by a consequence of relativity. If atoms move at very high velocity and emit radiation, the photons are emitted in the direction of travel. Thus, a jet will look much brighter if it is pointing more or less toward us and much fainter if it is pointing away. We tend to see the jet pointed toward us, and the jet pointed away is usually so faint we can't detect it (Figure 18-4).

The double-exhaust model describes radio lobes as cavities in the intergalactic medium inflated by jets. We can be sure that an intergalactic medium really exists because of galaxies called **head–tail radio galaxies.** When they were first discovered, at fairly low resolution, they looked like speedboats racing across a lake and leaving a wake behind (Figure 18-8a). Higher-resolution observations revealed that they were active galaxies ejecting jets in opposite directions as they moved relative to the intergalactic medium. The ejected material is swept away in a long tail, like the plumes from twin smokestacks carried away by the breeze. This interpretation is supported by the observation that such galaxies occur in clusters of galaxies where the intergalactic medium is relatively dense.

A few head–tail galaxies have complex structures such as blobs or spirals among the two streams of gas (Figure 18-8b). The blobs hint that the jets do not remain "on" all the time, and the spirals suggest that the axis along which the jets are expelled is precessing and spewing relativistic particles out of the galaxy in swirling streams.

FIGURE 18-8

Head–tail radio galaxies. (a) NGC 1265 ejects jets in opposite directions as it moves at about 2000 km/sec through the intergalactic medium. The relative motion blows the jets back to form a tail. (National Radio Astronomy Observatory—Associated Universities, Inc.; C. P. O'Dea and F. N. Owen) (b) The head–tail radio galaxy 3C 129 ejects jets with a spiral pattern. Apparently, the source of the jets is precessing. (Laurence Rudnick and Jack Burns, University of New Mexico)

a

a

b

FIGURE 18-9

(a) M 87, a giant elliptical galaxy, contains a small, bright core. Short-exposure photographs also reveal a jet 1800 pc long extending from the core partway to the edge of the visible galaxy. (Anglo-Australian Telescope Board) (b) This false-color radio map made with the VLA shows that the jet is extremely narrow and contains knots of high intensity. (National Radio Astronomy Observatory, operated by Associated Universities, Inc., under contract with the National Science Foundation)

Astronomers have constructed a convincing scientific argument (Window on Science 18-2) that active galactic nuclei and radio lobes are powered by something in the core of active galaxies. To extend our model further, we must propose a source for the energy.

The Central Energy Source

Clearly, galaxies can be powerful sources of energy, but so far we have ignored the "really difficult central problem." What object at the center of a galaxy could produce such energy? Telltail clues lie at the center of a giant elliptical galaxy.

The giant elliptical galaxy M 87, located at the center of the Virgo cluster, has a peculiar bright core visible at optical, radio, and X-ray wavelengths (Figure 18-9). The impressive feature of M 87, however, appears only on short-exposure photographs, which reveal a tremen-

dous jet of matter 1800 pc (6000 ly) long squirting out of the nucleus at two-thirds the speed of light. The jet emits synchrotron radiation, contains knots, and has been detected at X-ray wavelengths by the Einstein Observatory.

In the late 1970s, astronomers were able to detect a starlike point of light at the center of M 87. This bright core is produced by tremendously crowded stars, and Doppler shifts show that the stars move with high velocities. The astronomers concluded that the core must contain a very high mass if it can hold such rapidly moving stars in such a small region. This evidence is not conclusive, but many astronomers assumed that M 87 had a supermassive black hole at its core. In 1994, the Hubble Space Telescope observed the center of M 87 and found even stronger evidence. The telescope was able to detect a spiral disk of hot gas at the very core of the galaxy (Figure 18-10), and Doppler shifts show that the disk is rotating at 450 km/sec at a distance of only 60 ly from the

Scientific Arguments

An argument can be a shouting match, but another definition of the word is "a discourse intended to persuade." Scientists construct arguments as part of the business of science not because they want to persuade others they are right, but because they want to test their ideas. For example, we can construct a scientific argument to show that the radio lobes that flank some active galaxies have their origin in jets coming from the cores of the galaxies. If we do our best and our argument is not persuasive, we begin to suspect we are on the wrong track. But if our argument seems convincing, we gain confidence that we are beginning to understand radio lobes.

A scientific argument is a logical presentation of evidence and theory with interpretations and explanations that help us understand some aspect of nature. Geologists might construct an argument to explain volcanism on mid-ocean islands, and the argument could include almost anything—maps, for example, or mineral sam-

ples compared with seismic evidence and mathematical models of subsurface magma motions. Such an argument might involve the physics of radioactive decay or observations of the shapes of bird beaks on the islands. The scientists would be free to include any evidence or theory that helps persuade, but they must observe one fundamental rule of scientific argument: they must be totally honest. The purpose of a scientific argument is to test our understanding, not to win votes or sell soap. Dishonesty in a scientific argument is self-deluding, and scientists consider such dishonesty the worst possible behavior.

On the other hand, scientists are human, and they often behave no better than other humans. They sometimes defend their own theories and attack opposing theories more vigorously than the evidence warrants. They may unintentionally overlook contradictory evidence or unconsciously misinterpret data. Of course, if a scientist did these things intentionally, we

would be shocked because he or she would be violating the most serious rules of honesty in science. In the struggle to understand how nature works, however, it is not surprising to see that human passions rise and that scientific debate becomes heated. That is why science is so exciting—it is a struggle. Nonetheless, no matter how scientists behave in the excitement of a controversy, they all aspire to that ideal of the scientific argument—the cool, logical presentation of evidence that leads to a new understanding of nature.

Much of this book or any other scientific book consists of scientific arguments—logical presentations of evidence. As you read about any science, look for the arguments and practice organizing and presenting them to test your own understanding. The "Critical Inquiry" that ends each chapter section in this book is intended to illustrate how arguments are constructed. Use the same plan when you develop your own arguments. ∎

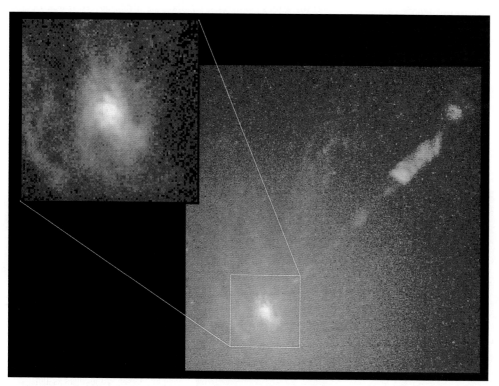

FIGURE 18-10

The Hubble Space Telescope imaged the center of the giant elliptical galaxy M 87 and revealed that the jet originates in a small disk of gas surrounding the starlike core. Spectra of the disk show that it is rotating at high velocity. From the size and rotation of the disk, astronomers conclude that the core of the galaxy must contain 3 billion solar masses. A supermassive black hole seems the only explanation. (NASA/STScI)

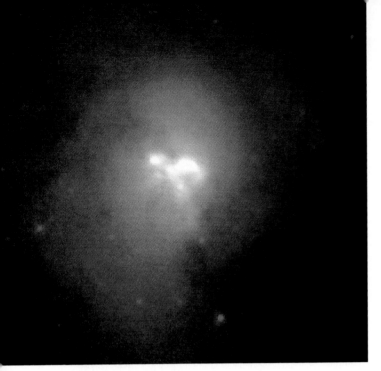

FIGURE 18-11

Ultraluminous galaxy Arp 220 is undergoing a "starburst" producing prodigious amounts of energy. Thick dust in the galaxy absorbs the radiation and reradiates it as infrared radiation. The Hubble Space Telescope observing in the near infrared reveals two clouds, each containing roughly a billion stars, in the galaxy's core. These stars may be the nuclei of two spiral galaxies that collided to form the starburst galaxy. The star cloud at right is partly obscured at the bottom by a disk of gas and dust about 300 ly in diameter. Such a disk around a central mass may be the early stages in the formation of an active galactic nucleus. (Rodger Thompson, Marcia Rieke, Glenn Schneider, University of Arizona; Nick Scoville, California Institute of Technology; and NASA)

A team of radio astronomers probed the core of M 87 by linking radio telescopes in the United States, Italy, Sweden, and Germany to form a very long baseline interferometer nearly the diameter of the earth (Chapter 6). The result was a resolution of 0.00015 second of arc. Their study of the core of M 87 reveals that the jet springs from a point no more than a few light-weeks from the center of the galaxy. This places an even smaller limit on the size of the massive core and increases astronomer's confidence that a massive black hole is the central energy source.

One important clue to the cause of galactic activity dates back to 1929, when astronomers discovered a variable star in the constellation Lacerta and gave it a variable star designation, BL Lac (Figure 18-11). In 1968, radio astronomers discovered that it was a powerful radio source. The visible spectrum of BL Lac is featureless, but by blocking the brilliant light from the center, astronomers were able to record the spectrum of the faint fuzz that surrounds the bright core. The spectrum of that fuzz is typical of a giant elliptical galaxy. Obviously, this star-like object is not a star at all but the powerful core of an active galaxy. A number of these objects have been found, and they are now known as **BL Lac objects** or **blazars**. The most luminous are 10,000 times more luminous than our Milky Way Galaxy and fluctuate rapidly in brightness. X-ray fluctuations can occur in a few hours. From this evidence, astronomers conclude that these objects must contain extremely small but extremely powerful energy sources, probably supermassive black holes.

The cores of active galaxies appear to contain supermassive black holes surrounded by disks of gas and dust (Figure 18-11) which feed the black holes and direct jets of excited particles. Great distance and obscuring dust make such disks difficult to study, but a few have been detected clearly. For example, NGC 4261 is 45 million ly away and has radio lobes spanning 88,000 ly (Figure 18-12a). The Hubble Space Telescope found a disk with a central opening 300 ly in diameter and a small bright

center. Closer to the center, the velocity is even higher. From the size and velocity of rotation, we can conclude that the core of the galaxy must contain as much as 3×10^9 solar masses. Only a black hole could pack that much mass into so small a region.

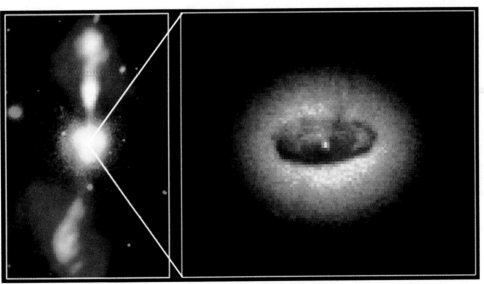

a   b

FIGURE 18-12

(a) Galaxy NGC 4261 looks like a fuzzy blob in an earth-based photograph at visible wavelengths (white), but radio maps reveal radio lobes (orange). (b) A Hubble Space Telescope image of the core of the galaxy reveals a bright central spot surrounded by a disk. The axis of the disk is parallel to the line connecting the radio lobes; astronomers assume the disk feeds matter into a central black hole that ejects jets feeding the radio lobes. (L. Ferrarese, Johns Hopkins University, and NASA) object at its center. The axis of the disk aligns with the axis of the radio lobes. Presumably, we are seeing a disk of matter swirling into a central black hole and ejecting jets along the axis of rotation. The rotation of the disk implies a central body with a mass of 1.2 billion solar masses. That object is not much larger than our solar system.

Black Holes in AGN

The evidence is very strong that supermassive black holes lie in the cores of active galaxies, but many questions remain. What produced these monster black holes? How can a black hole eject jets of hot gas? Why are some active galaxies so different from others?

In previous chapters, we discussed black holes that are the remains of dead stars, but the supermassive black holes can't be produced this way. They are much too massive. Rather, the supermassive black holes seem to have accumulated at the center of galaxies. As the galaxy formed, much of the mass wound up orbiting the center, but some of the mass may not have had enough angular momentum to remain in orbit. This mass would have sunk to the center to form a black hole. Also, collisions between gas clouds and interactions between stars in a galaxy can rob some matter of its orbital motion, and that mass would naturally sink toward the center of the galaxy. Over 10–15 billion years, quite a massive black hole could accumulate in the core of a galaxy.

If our black hole hypothesis is correct, how could it account for energy in the form of jets? Astronomers now compare active galactic nuclei with the peculiar binary star system SS 433 (Chapter 15). That system consists of a neutron star surrounded by a disk of in-falling gas, which ejects powerful jets in both directions along the axis of rotation. To create a radio galaxy with jets, we would have to build an SS 433 source about 100 times more powerful. That would take a supermassive black hole with a mass of a million to a billion solar masses with matter flowing inward.

A study of the Seyfert galaxy NGC 4151 seems to have detected mass flowing into the core. The observations were made by analyzing the shapes of spectral lines as distorted by Doppler shifts. The results show mass flowing into a region about 2 light-days in diameter at a rate consistent with a central mass of 60 million solar masses. If the central mass is a black hole, the inflowing mass could easily release enough energy to power a Seyfert galaxy. The small size of the central region would allow Seyferts to change their brightness rapidly, as they are observed to do.

What could cause an inflow of matter? That process may be related to tidal interactions between galaxies. Theoretical studies show that an encounter between galaxies can not only eject mass from the galaxies to form tails and bridges but can also throw mass into the centers of the galaxies at a rate capable of supporting an AGN (Figure 18-13). This may explain why active galaxies tend

FIGURE 18-13

This computer simulation shows a disk galaxy during an encounter with a smaller galaxy, represented by a blue dot. The disk galaxy is not only distorted by the encounter but is also made unstable, and significant amounts of matter flow into the core. This study suggests that encounters can force enough matter into a galaxy's nucleus to power an active-core energy source. (G. Byrd, M. Valtonen, B. Sundelius, and L. Valtaoja)

a

b

FIGURE 18-14

Interactions between galaxies may trigger eruptions by throwing mass into the supermassive black holes in the galactic cores. (a) In this radio image of 3C 75, two galaxies whip around each other, and both cores are stimulated to eject jets. The visible galaxies here are many times larger than their active cores. (NRAO–Associated Universities, Inc.; F. N. Owen, C. P. O'Dea, and M. Inoue) (b) This Hubble Space Telescope image of the center of elliptical galaxy NGC 6251 shows a dark disk surrounding the brilliant core. An ultraviolet image (blue) shows that at least part of the inner disk is excited to high temperature. The warp in the disk is caused by the matter flowing into it. (Philippe Crane, European Southern Observatory; and NASA)

to have companions and are often distorted as if by encounters (Figure 18-14). Perhaps every galaxy carries at its heart a black hole waiting to be fed. If so, activity and jets in galactic nuclei could be common.

It isn't clear exactly how accretion into a disk around a black hole can eject jets of gas, but we can understand some of what goes on. As matter falls inward, it forms a thick, opaque doughnut or torus of gas and dust a few light-years in diameter (Figure 18-15). As matter slows in the torus, it can flow into the outer accretion disk. This matter is relatively cool, but it is heated by friction in the disk and flows rapidly inward. Any rotating object that contracts must conserve angular momentum and spin faster, and thus the inner parts of the disk must spin very rapidly. Theoretical calculations tell us that the inner disk becomes very hot and thick. The heat, the motion, and the embedded magnetic fields can drive away some of the material above and below the disk to form jets, while the remaining mass moves slowly enough to fall into the black hole.

All of the evidence now points to a single phenomenon to explain the various kinds of active galaxies we see. This **unified model** proposes that active galaxies contain supermassive black holes with gas flowing inward

through disks and ejecting powerful jets in opposite directions (Figure 18-15). The central region, within less than 1 pc of the black hole, is filled with clouds of gas, perhaps stars bloated by the intense radiation from the inner disk. Moving at high orbital velocities, these clouds produce broad emission lines in the spectrum, and the region is known as the **broad line region.** Farther from the black hole, perhaps 10 pc away, gas clouds move more slowly, and those excited by the radiation from the core produce narrow emission lines. Thus, this region is called the **narrow line region.** With a giant torus of opaque gas and dust perhaps 3 pc from the core, this model tells us that what we see depends on the angle at which we observe.

If the disk is tipped so that the jet is pointed right at the earth, we see a blindingly bright blazar with its featureless spectrum produced by synchrotron radiation. We are looking through the nearest radio lobe at these objects, and they look like a brilliant point of light that fluctuates rapidly as the gas in the broad line region races around the black hole. If we happen to see such a disk edge-on, the thick torus of gas and dust blocks our view of the broad line region and we get emission from the gas clouds of the narrow line region. Radio galaxies seen in

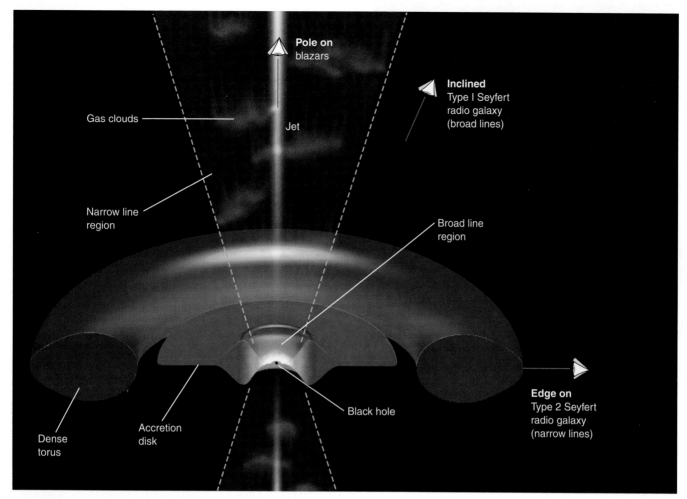

FIGURE 18-15
Matter flowing into the black hole at the heart of an active galaxy flows first through an opaque torus of gas and dust and then into the accretion disk. The inner parts of the accretion disk are very hot, and some of the gas is ejected in oppositely directed jets while the rest of the gas flows into the black hole. The inner parts of the accretion disk produce broad emission lines, while gas clouds above and below the disk produce narrow emission lines. According to the unified model, the angle at which we view this structure determines the kind of radio galaxy we see.

this orientation have double lobes and narrow emission lines. The Type 2 Seyfert galaxies with their narrow emission lines and strong infrared radiation are also seen in this orientation. If the disk is tipped slightly but not pole-on, we can see the broad line region; then we observe radio galaxies with broad emission lines and Type 1 Seyfert galaxies with their high velocities and intense ultraviolet and X-ray radiation.

CRITICAL INQUIRY

Why are there two kinds of Seyfert galaxies?

Astronomers have constructed an argument that what we see depends on the angle at which an active galaxy is tipped to our line of sight. This unified model supposes that a Seyfert galaxy has a massive black hole at its center, and, perhaps because of encounters with another

galaxy, matter begins to flow into the black hole. This matter must create a thick accretion disk, and if the disk happens to be inclined to our line of sight we will be able to see into the central regions. There the gas is very hot and is orbiting the black hole at high velocity, so the spectrum we see will include emission lines broadened by the Doppler effect. We may even detect intense ultraviolet and X-ray radiation from the inner region. Such a galaxy would be a Type 1 Seyfert galaxy.

But if the disk happened to be edge-on to our line of sight, we would not be able to see the hot central regions. We would see narrow emission lines from the slower-moving gas farther from the black hole, and we would detect infrared radiation from the dust. This would be a Type 2 Seyfert galaxy.

The paragraphs above are a quick summary of a scientific argument. Construct an argument to show that galaxy encounters are necessary to trigger active galactic nuclei. ■

FIGURE 18-16
Quasars look starlike in photographs and are not obviously galaxies—thus the name "quasi-stellar objects." In this CCD image of quasar 3C 275.1 (bright image near center) made with the 4-m telescope atop Kitt Peak in Arizona, a small amount of nebulosity is visible around the quasar. (NOAO image by P. Hintzen and W. Romanishin)

The unified model makes sense if at least some galaxies contain supermassive black holes. In fact, there is evidence that both our Milky Way Galaxy and the nearby Andromeda Galaxy contain such objects. To further our understanding of these monsters in the hearts of galaxies, we must look at the most extreme examples—the quasars.

18-2 QUASARS

For decades, the most mysterious objects in the sky were the **quasars** (also called quasi-stellar objects, or QSOs), brilliant sources of energy that seem to lie far away. Some of the mystery has been resolved, and astronomers now think of quasars as the powerful cores of distant active galaxies. Their significance, however, goes far beyond this interpretation. First, they are so distant that their look-back times are tremendous, and they lead us back to a time when the universe was very young. Second, the discovery of quasars in the 1960s revolutionized astronomy. The story of how astronomers untangled the quasar puzzle is one of the great adventures in 20th-century astronomy.

Discovery

The existence of quasars became apparent gradually over a period of a few years. Not until the mid-1960s did astronomers realize that the quasars were a dramatically puzzling phenomenon.

Radio astronomers first detected quasars as sources of radio energy. A large radio telescope can detect myriads of very distant galaxies, many of which are not visible at optical wavelengths. Radio interferometers (Chapter 6) could measure the angular diameters of these objects. Many were double-lobed radio sources. But some radio sources had angular diameters so small they could not be measured. We now know that only 3 percent of the quasars are radio sources, but it was the radio astronomers who first called attention to them because of their small size.

When optical astronomers photographed the locations of these pointlike radio sources, they did not find the broad fuzzy images of galaxies but rather small, starlike points of light (Figure 18-16). The first to be located optically was 3C 273, and 3C 48 was identified soon after. Because the objects looked starlike at optical wavelengths, they were called quasi-stellar objects, or quasars.

Recall from the introduction to this chapter that we cannot judge the true nature of an astronomical body when we do not know its distance. Many astronomers assumed that the quasars were nearby, in our own backyard. The key to the nature of the quasars was hidden in their spectra.

The spectra of quasars were a mysterious combination of a continuous spectrum plus a few unidentifiable emission lines. In the last months of 1963, Maarten Schmidt at Hale Observatories tried red-shifting the hydrogen Balmer lines to see if they could be made to agree with the lines in the spectrum of 3C 273. At a red shift of 15.8 percent, three lines clicked into place (Figure 18-17). Other quasar spectra quickly yielded to this approach, revealing even larger red shifts. The red shift of 3C 48, for instance, was 0.37.

To an astronomer, the red shift, Z, is the change in wavelength $\Delta\lambda$ divided by the unshifted wavelength λ_0:

$$\text{red shift} = Z = \frac{\Delta\lambda}{\lambda_0}$$

Quasars have large red shifts. If these shifts arise from the expansion of the universe, the Hubble law predicts large distances. Thus, the large red shift of a quasar can be interpreted to mean great distance.

The implications of the large red shifts were startling. At the time, the largest observed galaxy red shifts were about 1, but quasars were soon found with much larger red shifts and, presumably, much greater distances. Normal galaxies at such large distances would be invisible, but the quasars were clearly visible on photographs. They had to be 10 to 1000 times more luminous than a large galaxy.

a

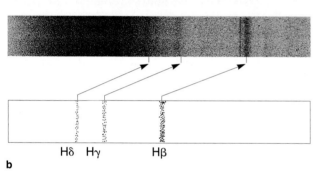

b

FIGURE 18-17

(a) Quasar 3C 273 is the brightest known quasar and was one of the first studied. (Palomar Obs/Caltech) (b) Its spectrum (reproduced here as a negative image) contains three hydrogen Balmer lines red-shifted by 15.8 percent. The drawing shows the unshifted positions of the lines. (Maarten Schmidt)

As astronomers tried to explain the superluminosity of the quasars, another observation complicated the problem. Teams of astronomers discovered that the light from quasars was fluctuating rapidly and erratically. In some cases, a quasar could change its brightness by a large factor in a month, a week, or perhaps even a day. If the quasars could fluctuate rapidly, they had to be small, not more than a light-week or a light-day in diameter. How could quasars generate 10 to 1000 times more energy than a galaxy in a volume only a few light-weeks in diameter?

Since the discovery of quasars, astronomers have found roughly 8000, of which 97 percent are radio-quiet. These objects have large red shifts that translate into staggering distances. Most are much farther away than

visible galaxies, and the most distant are 10 to 15 billion light-years distant.

If quasars are as far away as their large red shifts indicate, their look-back times are very large, and we see them as they were when the universe was significantly younger. In the next chapter, we will see that the universe is believed to be 10–20 billion years old. Thus, we see the most distant quasars as they were when the universe was only a few billion years old.

About 10 percent of quasars have absorption lines as well as emission lines in their spectra. These absorption lines have a smaller red shift than the corresponding emission lines, and, in some cases, multiple sets of absorption lines are observed, each with its individual red shift. Some of the lines are hydrogen lines, but some are lines of metals, heavier elements such as carbon, silicon, and magnesium. Apparently, these absorption lines are formed when the light from the quasar passes through objects on its way to the earth. The hydrogen lines must be formed as the light from the quasar passes through low-density clouds of hydrogen between the galaxies. The metal lines must be formed when the light passes through galaxies on its way to the earth. But the galaxies themselves are so distant that they are not visible, further evidence that quasars are very distant objects.

Quasar Distances

The very large red shifts of the quasars lead to such astonishing conclusions that some astronomers questioned the Hubble law. Before we go further in our discussion, we should review the nature of quasar red shifts.

A great many quasars have red shifts larger than 1, meaning that for any given line in their spectrum the change in wavelength divided by the wavelength is greater than 1. If we use this result in the classical Doppler formula (Chapter 7), we get a radial velocity greater than the speed of light. This is forbidden by the theory of relativity, so we must begin by understanding how a red shift can exceed 1.

The classical Doppler formula given in Chapter 7 is only an approximation. It works quite well as long as the velocity of a star or galaxy is much smaller than the speed of light. Because the velocities of planets, binary stars, and star clusters in our galaxy are hardly ever more than a few hundred kilometers per second, the classical Doppler formula is accurate enough for most astronomical problems.

When the velocity of an object is an appreciable fraction of the speed of light, however, we must use an equation derived by the theory of relativity, the **relativistic Doppler formula**. It relates the radial velocity V_r to the speed of light c and the red shift Z:

$$\frac{V_r}{c} = \frac{(Z+1)^2 - 1}{(Z+1)^2 + 1}$$

The relativistic Doppler formula cannot produce a radial velocity greater than the speed of light no matter

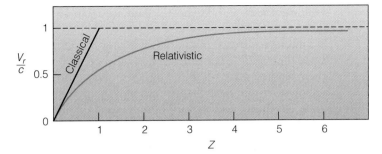

FIGURE 18-18

At high velocities, the relativistic red shift must be used in place of the classical approximation. Note that no matter how large Z gets the speed can never equal the speed of light.

how large Z becomes. For example, suppose a quasar had a red shift of 2. The classical Doppler formula would give the unreasonable result that the radial velocity was twice the speed of light, but when we put 2 into the relativistic Doppler formula we discover that $Z + 1$ is 3 and therefore $(Z + 1)^2$ is 9. Then the radial velocity divided by the speed of light is:

$$\frac{V_r}{c} = \frac{(9) - 1}{(9) + 1} = \frac{8}{10} = 0.8$$

Thus, the quasar has a radial velocity of 80 percent the speed of light.

Figure 18-18 shows how the radial velocity depends on Z. For small red shifts, the velocity is the same in both the classical and the relativistic case, but as Z gets larger the difference between the classical approximation and the true velocity increases. No matter how large Z becomes, the velocity can never quite equal the speed of light.

We will see in the next chapter how the red shifts of the galaxies arise from the expansion of the universe itself, and that adds a further uncertainty to calculating the velocities of recession from the red shift. Nevertheless, it does not change our conclusion here. Quasar red shifts greater than 1 are not unreasonable.

The very large red shifts of the quasars lead to very large radial velocities, and, if we assume that these are due to the expansion of the universe, we can use the Hubble law to find distances. We merely divide the radial velocity by the Hubble constant. Although there continues to be some controversy over the value of the Hubble constant, the uncertainty is only a factor of 2 or so. Thus, we need not be concerned that an uncertainty in the Hubble constant is dramatically misleading us about the distance to the quasars. If their red shifts obey the Hubble law, then they must be very distant indeed.

The vast distances to the quasars lead us to conclude that they must be superluminous, and their rapid fluctuations tell us that they must be small. Faced with the difficulty of explaining how such small objects could produce so much energy, some astronomers proposed what became known as the **local hypothesis.** That is, they suggested that the quasars were local, not distant, and thus did not have to be superluminous. However, if the quasars were nearby, then the expansion of the universe could not give them large red shifts, and if they were whizzing through nearby space as fast as their Doppler

shifts suggested, we should see half approaching and half receding. That is, we should see blue shifts as well as red shifts. In fact, no blue-shifted quasar has ever been observed. Thus, the local hypothesis traded the superluminosity puzzle for a red-shift puzzle.

We refer to red shifts produced by the expansion of the universe as cosmological red shifts, drawing on the word *cosmology*, which refers to the study of the universe as a whole. If the local hypothesis was to succeed, it had to involve noncosmological red shifts. Although the laws of nature did not account for such phenomena, a few astronomers argued that undiscovered laws of nature might be responsible, and they attempted to find examples of noncosmological red shifts—that is, a red shift that was clearly not produced by the expansion of the universe. A single conclusive example would demonstrate that the local hypothesis was possible.

One of the most studied examples was first discussed in 1971. The quasar Markarian 205 is located very near the galaxy NGC 4319, and some astronomers claimed to detect a bridge of matter connecting the two (Figure 18-19). If they are linked, then they are at the same distance. However, the galaxy has a red shift of 0.006, and the quasar has a red shift of 0.07. If they are at the same distance, the red shift of the quasar cannot arise from the expansion of the universe.

Almost all astronomers believe that the quasar is much farther away than the galaxy and only appears to be linked. Some studies of the system have suggested that the apparent link is a photographic effect, and a study in 1985 revealed evidence that the link between the galaxy and the quasar is actually a distorted galaxy with the same distance as the quasar. Although it is still controversial, Markarian 205 is not a conclusive example of a noncosmological red shift.

The local hypothesis has generally been abandoned, not because of any astonishing discovery but because of the lack of any conclusive evidence that quasars are nearby and because of a slow accumulation of evidence that they are very distant and are associated with active galactic nuclei.

The Quasar–Galaxy Connection

A number of lines of evidence show that quasars are objects in very distant galaxies. Unlike the local hypothesis, this hypothesis requires that they be superluminous. We will examine the energy source later.

a

b

FIGURE 18-19

(a) A false-color image of the galaxy NGC 4319 and Markarian object 205 (arrow) reveals what some believe is a link between the galaxy and the quasar. If the link is real, the objects would be at the same distance, and the large red shift of the quasar would be an example of a noncosmological red shift. Most astronomers believe that the link (shown enlarged in b) is not real. (Peter Wehinger)

As electronic imaging such as CCDs (Chapter 6) developed in the 1970s and 1980s, astronomers found that quasars with the smaller red shifts (less than 1), and presumably smaller distances, are often in clusters of galaxies (Figure 18-20). These galaxies are very faint, but they appear to be typical of those we would expect to find in a crowded cluster. The red shifts of these galaxies are the same as that of the quasar they accompany, implying that the cluster and the quasar both lie at the same great distance.

These same observation techniques have allowed the discovery of quasar fuzz. The best images of the low-red-shift quasars often show that they are surrounded by fuzz (Figure 18-20). The spectrum of this fuzz is typical of a galaxy with the same red shift as the quasar. The quasar 3C 273, for example, appears to lie in a giant elliptical galaxy.

Faint galaxies accompanying quasars and quasar fuzz are visible only for low-red-shift quasars. Higher-red-shift quasars are so distant that any galaxies near them would be invisible to our best telescopes and instruments.

FIGURE 18-20

Quasar 0351 + 026 (top) is located near a faint galaxy (bottom) that has the same red shift and thus, presumably, the same distance from the earth as the quasar. The extended red region around the quasar (fuzz) is typical of lower-red-shift quasars and typically reveals the spectrum of a normal galaxy. Thus, this image shows two interacting galaxies, one of which contains a quasar. (National Optical Astronomy Observatories)

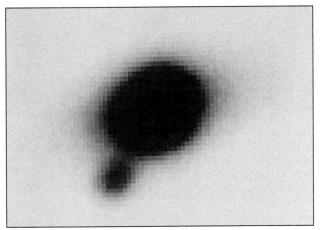

FIGURE 18-21

These two negative photographs show the quasar 1059 + 730. The lower photograph contains an extra image: a supernova occurring in the galaxy containing the quasar. The apparent magnitude of the object and the distance to the quasar based on its red shift give an absolute magnitude for the supernova of −17.6, a value consistent with the supernova explanation. (Bruce Campbell)

a

b   c

FIGURE 18-22

The gravitational lens effect. (a) A distant quasar can appear to us as multiple images if its light is deflected and focused by the mass of an intervening galaxy. The galaxy, being much fainter than the quasar, is not easily visible from the earth. (b) The two images of quasar 0957 + 561 appear as black blobs in this computer-enhanced false-color image. When the upper image is subtracted from the lower image (c), we are able to detect the faint image of the giant elliptical galaxy (arrow) whose gravity produces the gravitational lens effect. (Alan Stockton, Institute for Astronomy, University of Hawaii)

Astronomers studying the quasar QSO 1059 + 730 discovered an image very near the quasar that did not appear on other plates. After eliminating other possibilities, they concluded that the object was a supernova explosion in the galaxy that hosts the quasar (Figure 18-21). This seems to confirm the view that otherwise normal galaxies can be the site of quasars.

A discovery made in 1979 further supports the belief that quasars are very distant and not local. The object 0957 + 561 lies just a few degrees west of the bowl of the Big Dipper and consists of two quasars separated by only 6 seconds of arc. The optical spectra of these objects proved to be nearly identical, with the same red shift (1.40) and the same relative strengths of lines. Quasar spectra are as different as fingerprints, so when two quasars so close together proved to have the same spectra and the same red shift, it was evident that they were separate images of the same quasar.

The two images are formed by an intervening galaxy. The gravitational field of the galaxy deflects the light of the quasar and focuses it into two images (Figure 18-22a). The galaxy itself is much too far away to be easily visible against the glare of the quasar, but computer manipulation of a CCD image reveals an image of the galaxy, probably a giant elliptical (Figure 18-22b).

The discovery of the double nature of 0957 + 561 is significant for two reasons. First, the effect, known as the **gravitational lens effect,** was predicted in 1936 by Einstein but had never been seen before. Thus, the discovery further confirms general relativity. Second, the gravitational lens effect is important because it shows that the quasar cannot be local. It must be much farther away than the galaxy, and the galaxy is a giant elliptical galaxy that is just barely bright enough for us to detect. The galaxy must be fairly distant itself. Thus, the quasar cannot be local.

Quite a number of gravitational lenses have been found, and in each case light from a distant quasar is deflected by the gravitational field of a nearer galaxy and multiple images of the quasar are formed. These discoveries, along with other evidence that quasars are associated with distant galaxies, have been a serious blow to the local hypothesis. At least some quasars are very distant. Even if some quasars are local, we are still faced with explaining the ones that are very distant and therefore superluminous.

Superluminal Expansion

A few astronomers have objected that quasars cannot be located in very distant galaxies because, if the quasars are really that far away, a few examples seem to be expanding at speeds greater than the velocity of light, and that is forbidden by the theory of relativity.

Radio astronomers discovered this **superluminal expansion** by using very long baseline interferometry (VLBI)—the interconnection of radio telescopes in different parts of the world. Such a network of antennae can resolve details about 0.002 second of arc in diameter. At the distance of quasar 3C 273, for example, this corresponds to a few parsecs.

For a few quasars these radio maps show small blobs near the quasars, and maps made over a few years show these blobs moving away from the quasar. Quasar 3C 273 has a blob that is moving about 0.0008 second of arc farther from the quasar each year (Figure 18-23). If we believe that the quasar's red shift of 0.16 arises from the expansion of the universe, the quasar is about 960 Mpc distant, and the small-angle formula tells us that the blob is moving away from the quasar at a speed of 3.8 pc per year. This is the same as 12 ly per year, which is impossible. The theory of relativity clearly shows that nothing can travel faster than the velocity of light.

This superluminal expansion has been found in a number of quasars, and it is also seen in blazars. For quasars the apparent velocities of expansion are usually greater than five times the velocity of light, and for blazars the velocity is usually less than five times the velocity of light. As we have seen earlier in this chapter, we suspect that blazars are active galaxy cores in which one of the jets is pointed almost directly at us. This is a clue to how to explain superluminal expansion.

One way to solve the problem is to assume that the quasar is local. If it is only 10 Mpc away, then the blob is moving at a velocity less than 15 percent the velocity of light. But, of course, if quasars are local, we can't explain their large red shifts.

The most commonly discussed hypothesis to explain this superluminal expansion without making the quasar a local object is the **relativistic jet model.** It supposes that the quasar is ejecting a jet of matter at nearly the velocity of light and that the jet is pointed nearly toward the earth. Then a blob of matter in the jet could appear to move away from the quasar at velocities greater than the velocity of light.

FIGURE 18-23

Superluminal expansion. Very high resolution radio maps of quasar 3C 273 show a blob separating from the quasar at a rate of 0.0008 second of arc per year. If the quasar is at the distance indicated by its red shift, the blob appears to be moving at about 12 times the velocity of light. (Timothy Pearson and *Nature*)

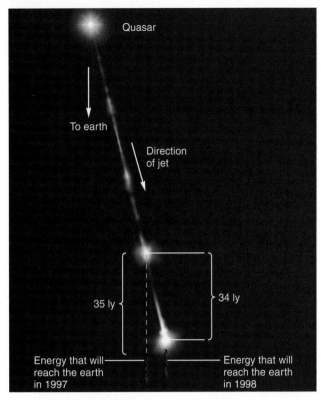

FIGURE 18-24

The relativistic jet model explains superluminal expansion by supposing that the quasar emits a jet that points nearly at the earth. If a blob of gas in the jet, traveling at nearly the velocity of light, emits radio energy toward the earth at two different times separated by 35 years, the first bundle of energy will have only a 1 ly lead. As the energy arrives at the earth, the blob will appear to move in 1 year a distance that actually took 35 years.

Labels on figure: Quasar; To earth; Direction of jet; 35 ly; 34 ly; Energy that will reach the earth in 1997; Energy that will reach the earth in 1998

To see how this can happen, consider Figure 18-24. A blob of matter in the jet emits radiation that will reach the earth in 1997. In the next 35 years, that radiation travels 35 ly toward the earth, but the blob, traveling at 98 percent the velocity of light, travels 34 ly closer to the earth. Radiation that it emits from this second position is only 1 ly behind the radiation it emitted 35 years before. The newly emitted radiation will arrive on the earth in 1998, and we will see the blob move in only 1 year a distance that actually took 35 years. This is a simplified explanation that omits certain relativistic effects, but it clearly shows that a blob of gas in a relativistic jet can appear to travel faster than light if the jet points nearly toward the earth.

In fact, this interesting bit of physics has been observed closer to home. The jet in M 87 has been observed to have an apparent velocity along a portion of its length of 2.5 times the speed of light. Evidently, the jet twists and turns as it leaves the nucleus of the galaxy, and one of those turns points the jet almost directly at the earth. Although the gas is moving at slightly over 90 percent the speed of light, along that segment pointing toward us the gas appears to move 2.5 times faster than light. Even closer to home, within our own galaxy, a binary star con-

taining a neutron star or a black hole has been observed ejecting material at speeds of about 1.25 times the speed of light. The gas is actually traveling at only 92 percent the speed of light, but the material was ejected within 19° of our line of sight, and that produces the apparent superluminal motion.

A Model Quasar

From the evidence we have gathered, we can construct a model of a quasar that relates it to the unified model of active galaxies. In many ways, quasars resemble active galaxies. Some quasars, for example, are ejecting jets of material (Figure 18-17), and a few are flanked by radio lobes (Figure 18-25). Just as for active galaxies, there is evidence that quasars occur in galaxies that are interacting with neighboring galaxies (Figure 18-20). Thus, whatever model we build for a quasar, it should be a more violent version of the energy machine in the cores of active galaxies.

The most widely accepted model of a quasar was suggested in 1973. Like the unified model of active galaxies, it proposes that a quasar is produced by a supermassive black hole at the center of a galaxy. Matter flows into the black hole at a rate of 0.01–10 solar masses a year and forms a large, thick disk around the black hole. Friction in the disk heats the inner regions of the disk to very high temperatures. This emits high-energy radiation and ejects jets of material along the axis of rotation.

The unified model suggests that what we see depends on the orientation of the disk (Figure 18-15). If we happen to see the disk edge-on, we cannot see the violence of the inner disk. We see mostly the jets emerging from the disk, any radio lobes that are inflated, and narrow line emission from gas excited by the intense radiation from the inner disk. Such a quasar would probably look like one of the radio-emitting quasars. If we happen to see the disk tipped at some angle, we can see into the central core, and we see both the broad spectral lines produced by the high-velocity gas there and the intensely bright radiation produced in the core. If we happen to see a disk that is tipped nearly pole-on to us, then the jet points right at us, and we see intensely bright radiation that fluctuates rapidly and may be subject to superluminal expansion. Quasar 3C 273 fluctuates rapidly and exhibits superluminal expansion much like a blazar. In both cases, the unified model predicts that we are looking directly down the monster's throat.

The rapid fluctuations of the quasars set limits on the sizes of the regions emitting energy. Much of a quasar's light appears in a continuous spectrum that can fluctuate in a few weeks. That region must be no larger than a few light-weeks in diameter, about 30 times the size of our solar system. The continuous radiation, both light and radio, appears to be synchrotron radiation produced by high-speed electrons from the inner disk spiraling through a weak magnetic field.

The X-ray brightness of a quasar, on the other hand, can fluctuate in a time as short as 200 seconds, so we must

FIGURE 18-25

Some quasars exhibit jets and radio lobes. Quasar 4C 32.69 consists of a bright energy source, at the center of this map, with a jet (yellow) extending toward the radio lobe (red) at the right. Another radio lobe (red) lies to the left. (Robert Potash)

assume that these energetic photons come either from the innermost parts of the disk or from the jets, a region no larger than the orbit of Venus.

The emission lines in a quasar spectrum do not appear to fluctuate rapidly at all, so our model can create them in a larger region of gas clouds ionized by the intense synchrotron radiation streaming out of the inner regions.

Finally, the absorption lines we see in quasar spectra are produced as light from the quasar passes through galaxies and gas clouds on its way to the earth. Although there do not seem to be enough galaxies to produce all of the metal lines we see, we must remember the evidence that galaxies are surrounded by extensive coronae. If galaxies are that big, there may be enough to produce the observed absorption lines.

If quasars are powered by black holes, we can draw interesting inferences from their association with distant galaxies. Quasars seem to occur in regions where galaxies are crowded, and the galaxies near them are often distorted. This might mean that the quasar was triggered into existence when interacting galaxies fed matter into a massive black hole in one of the galaxies (Figure 18-26). But we must also recall the enormous look-back times to quasars. If they are associated with galaxies, we see those galaxies as they were long ago, and thus those galaxies are young. We do not see quasars near us, at small look-back times. Rather, most quasars are distant and have red shifts of 2 or 3, suggesting that they formed when the universe and galaxies were younger. At that age in the universe, quasars were at least 1000 times more common than they are now, suggesting that the quasar phenomenon is related to the formation and early evolution of galaxies.

Another question is, Where are all the dead quasars? If quasars formed naturally when galaxies formed or when young galaxies in crowded clusters interacted, why don't we see them now? Some astronomers believe that we do see them. If a quasar is a galaxy core into which mass is flowing, then when the gas and dust are exhausted the core stops looking like a quasar, and we would see a normal galaxy. Many of the so-called normal galaxies near us may hide massive black holes that were quasars long ago. Furthermore, some of the active galaxies we see now may have been much more active in the past when their cores gobbled mass at a great rate and created a quasar. In fact, observations of Cygnus A (Figure 18-4)

FIGURE 18-26

These three quasars appear to illustrate three stages in the triggering of a quasar by the tidal interaction between galaxies. Quasar IRAS 00275-2859 (top) is located in a galaxy interacting with a relatively distant companion (not visible in this image) and thus represents an early stage. In the galaxy containing quasar PG 1613 (middle), the galaxies are closer together. The nucleus of the companion is visible at the 2 o'clock position, and the outline of the interacting galaxies is distorted. Markarian 231 (bottom) is located in a severely distorted galaxy containing radial streaks and a tail to the lower right. Such features are common when galaxies merge, suggesting that Markarian 231 represents a late stage in the merger of two galaxies. (National Optical Astronomy Observatories) show that the core of that very active galaxy is emitting a spectrum very much like that of a quasar. We cannot see the core directly because of thick dust clouds surrounding the nucleus, but astronomers were able to see the light from the quasarlike core scattered toward the earth by dust particles high above the plane of the galaxy. Thus, the relatively nearby active galaxy Cygnus A may conceal in its heart the powerful remains of a quasar.

CRITICAL INQUIRY

Why are most quasars so far away?

Two factors determine that most of the quasars we see lie at great distances. First, quasars are rare, and few galaxies contain quasars. In order to sample a large number of galaxies in our search, we must extend our search to great distances. Most of the galaxies lie far away from us because the amount of space we search increases rapidly with distance. Just as most seats in a baseball stadium are far from home plate, most galaxies are far from our Milky Way Galaxy. Thus, most of the quasars we see lie at great distances.

But a second factor is much more important. The farther we look into space, the farther back in time we look. It seems that there was a time in the distant past when quasars were at least 1000 times more common, and thus we see most of those quasars at large look-back times, meaning at large distances. Thus, most quasars lie at great distances.

If our model of quasars is true and quasars are the cores of highly active galaxies, we might suspect that they are triggered into eruption by encounters with other galaxies. Cite observational evidence to construct an argument that quasars must be stimulated into eruption. ■

Our study of peculiar galaxies had led us far out in space and back in time to quasars. The light now arriving from the most distant quasars left them when the universe was only about one-fourth its present age. The quasars stand at the very threshold of the study of the universe itself. ■

▪ Summary

Some galaxies have peculiar properties. Seyfert galaxies are spirals with small, highly luminous cores. Double-lobed radio galaxies emit radio energy from the lobes on either side of the galaxies. Some have jets extending from the core of the galaxy into the radio lobes. Some giant elliptical galaxies have small, energetic cores, some of which have jets of matter rushing outward.

The proposed explanation for double-lobed radio galaxies is called the double-exhaust model. This model assumes that the galaxy ejects jets of relativistic particles in two beams that push into the intergalactic medium and blow up the radio lobes like balloons. The impact of the beams on the intergalactic medium produces the hot spots detected on the leading edges of many radio lobes. We can see instances where jets are blown back by the intergalactic medium to produce head–tail galaxies.

The source of energy in the active galaxies is believed to be supermassive black holes in their centers. M 87, a giant elliptical galaxy with a small, active core and a high-velocity jet, contains very high velocity motions close to its center. This implies that its small core contains billions of solar masses, probably in the form of a supermassive black hole that has accumulated there since the formation of the galaxy.

Active galactic nuclei seem to occur in galaxies involved in collisions or mergers. Seyfert galaxies, for example, are three times more common in interacting galaxies than in isolated galaxies. This suggests that an encounter between galaxies can throw matter into the central regions of the galaxies, where it can feed a black hole and release energy. Galaxies not recently involved in collisions will not have matter flowing into their central black hole and will not have active galactic nuclei.

The unified model of active galaxies supposes that what we see depends on the orientation of the black hole and its accretion disk. If we can see into the core, we see broad spectral lines and rapid fluctuations. If we see the disk edge-on, we see only narrow spectral lines. If the jet points directly at us, we see a blazar.

The quasars appear to be related objects. Their spectra show emission lines with very large red shifts. Their large red shifts imply that they are very distant, and the fact that they are visible at all implies that they are superluminous. Because they fluctuate rapidly, we conclude that their cores must be very small. A typical quasar can be 100 times more luminous than the entire Milky Way Galaxy but only a few times larger than our solar system.

Because quasars lie at great distances, we see them as they were long ago. The look-back time to the most distant quasar is about 15 billion years. They may be young interacting galaxies distorting each other and funneling mass into black holes, or they may be young galaxies in the early stages of formation. In any case, heavy mass flow into a central black hole could heat the gas and produce the observed synchrotron radiation and emission lines.

A growing mass of evidence suggests that the quasars are the active cores of distant galaxies. The discovery of the gravitational lens effect shows that at least a few quasars lie at cosmological distances. Also, the spectrum of the fuzz that surrounds some quasar images looks like the spectrum of a normal galaxy with the same red shift as the quasar.

Superluminal expansion appears as blobs of material that appear to be rushing away from some quasars at speeds as high as 12 times the speed of light. Of course, it is impossible for the matter to exceed the speed of light, but we can understand it as matter traveling at slightly less than the speed of light in a jet directed almost exactly toward the earth. Then we would see the illusion of superluminal expansion. Thus, superluminal expansion does not contradict our understanding of quasars.

▪ New Terms

radio galaxy	unified model
active galaxy	broad line region
active galactic nucleus (AGN)	narrow line region
Seyfert galaxy	quasar
double-lobed radio source	relativistic Doppler formula
hot spot	local hypothesis
double-exhaust model	gravitational lens effect
head–tail radio galaxy	superluminal expansion
BL Lac object	relativistic jet model
blazar	

Questions

1. What is the difference between a Type 1 Seyfert galaxy and a Type 2 Seyfert galaxy? How does the unified model explain these types?

2. What evidence do we have that radio lobes are inflated by jets from active galactic nuclei (AGN)?

3. Why is it significant that most head–tail galaxies occur in clusters of galaxies?

4. What evidence do we have that AGN contain black holes?

5. What properties of SS 433 resemble the energy source in an AGN?

6. If you located a galaxy that had not interacted with another galaxy recently, would you expect it to be active or inactive? Why?

7. Seyfert galaxies are three times more common in close pairs of galaxies than in isolated galaxies. What does that suggest?

8. Why do we conclude that quasars must be small? How do quasars resemble the AGN in Seyfert galaxies?

9. How does our model quasar account for the different components in a quasar's spectrum?

10. Why have most astronomers abandoned the local hypothesis?

Discussion Questions

1. Why do quasars, active galaxies, SS 433, and protostars have similar geometry?

2. Considering all the galaxies we can see, only about 0.1 percent contain a quasar. Does that mean that all galaxies contained quasars for a short time long ago, or does it mean that only certain galaxies could contain quasars?

3. Why are there few quasars near us?

Problems

1. The total energy stored in a radio lobe is about 10^{53} J. How many solar masses would have to be converted into energy to produce this energy? (HINTS: Use $E = m_0 c^2$. One solar mass equals 2×10^{30} kg.)

2. If the jet in NGC 5128 is traveling 5000 km/sec and is 40 kpc long, how long will it take for gas to flow from the core of the galaxy out to the end of the jet?

3. Cygnus A is 225 Mpc away, and its jet is about 50 seconds of arc long. What is the length of the jet in parsecs? (HINT: Use the small-angle formula.)

4. Use the small-angle formula to find the linear diameter of a radio source with an angular diameter of 0.0015 second of arc and a distance of 3.25 Mpc.

5. If the active core of a galaxy contains a black hole of 10^6 solar masses, what will the orbital period be for matter orbiting the black hole at a distance of 0.33 AU? (HINT: See circular velocity, Chapter 5.)

6. If a quasar is 1000 times more luminous than an entire galaxy, what is the absolute magnitude of such a quasar? (HINT: The absolute magnitude of a bright galaxy is about -21.)

7. If a quasar in Problem 6 were located at the center of our galaxy, what would its apparent magnitude be? (HINT: Use the magnitude–distance formula.)

8. What is the radial velocity of 3C 48 if its red shift is 0.37?

9. The hydrogen Balmer line Hβ has a wavelength of 486.1 nm. It is shifted to 563.9 nm in the spectrum of 3C 273. What is the red shift? (HINT: What is $\Delta\lambda$?)

10. The quasar images in Figure 18-22b are separated (center to center) by 6 seconds of arc. What is the angular diameter of the giant elliptical galaxy? Assume, conservatively, that this galaxy has a linear diameter of about 30 kpc, and use the small-angle formula to estimate its distance.

Critical Inquiries for the Web

1. What object currently holds the distinction as the farthest known galaxy? Search the Web for information on this distant object and find out its redshift and distance. What is the look-back time for this object?

2. The text discusses the superluminal expansion of material from 3C 273, but how common is this faster-than-light effect? Search the Internet for information on superluminal motion related to quasars and blazars and list several examples—with the expansion speeds implied by their motion. What is the highest superluminal expansion rate you can find?

 Go to the Wadsworth Astronomy Resource Center (www.wadsworth.com/astronomy) for critical thinking exercises, articles, and additional readings from InfoTrac College Edition, Wadsworth's online student library.

Chapter Nineteen

COSMOLOGY

GUIDEPOST

This chapter marks a watershed in our study of astronomy. From Chapter 1, our discussion has focused on learning to understand the universe. On our outward journey, we have discussed the appearance of the night sky, the birth and death of stars, and the interactions of the galaxies. Now we reach the limit of our journey in space and time—the origin and evolution of the universe as a whole.

The Universe, as has been observed before, is an unsettlingly big place, a fact which for the sake of a quiet life most people tend to ignore.

DOUGLAS ADAMS

The Restaurant at the End of the Universe

The ideas in this chapter are the biggest and the most difficult in all of science. Our imaginations can hardly grasp ideas such as the edge of the universe and the first instant of time. Perhaps it is fitting that the biggest questions are the most challenging.

But this chapter is not an end to our story. Once we complete it, we will have a grasp of the nature of the universe, and we will be ready to focus on our place in that universe—the subject of the rest of this book. ∎

What is the biggest number? A billion? How about a billion billion? That is only 10^{18}. How about 10^{100}? That very large number is called a googol, a number that is at least a billion billion times larger than the total number of atoms in all of the galaxies we can observe. Can you name a number bigger than a googol? Try a billion googols, or, better yet, a googol of googols. You can go even bigger than that—ten raised to a googol. (That's a googolplex.) No matter how large a number you name, we can immediately name a bigger number. That is what infinity means—big without limit.

If the universe is infinite in size, then you can name any distance you want, and the universe is bigger than that. Try a googol to the googol light-years. If the universe is infinite, there is more universe beyond that distance.

Is the universe infinite or finite? In this chapter, we try to answer that question and others, not by playing games with big numbers, but through **cosmology,** the study of the universe as a whole.

If the universe is not infinite, we face the edge–center problem. Suppose that the universe is not infinite, and you journey out to the edge of the universe. What do you see: a wall of cardboard; a great empty space; nothing, not even space? What happens if you try to stick your head out beyond the edge? These are almost nonsense questions, but they illustrate the problem we have when we try to think about an edge to the universe. Modern cosmologists believe that the universe cannot have an edge. In this chapter, we will see how the universe might be finite but have no edge.

If the universe has no edge, it can have no center. We find the centers of things—galaxies, globular clusters, oceans, pizzas—by referring to their edges. With no edge, there can be no center.

Finally, we must answer a third question: How old is the universe? If it is not eternal (infinite in age), it must have had a beginning, and we must think about how the universe began. Are a beginning and an ending like edges in time?

These are arguably the most challenging questions in human knowledge, and we start our search for answers by trying to understand the structure of the universe. That leads us, as is always the case in astronomy, to compare observations with theories.

19-1 THE STRUCTURE OF THE UNIVERSE

We begin our study of cosmology by considering the most basic property of the universe—its geometry. By understanding the geometry of space-time, we will see how we might discover whether the universe is infinite or finite.

Why Does It Get Dark at Night?

We have all noticed that the night sky is dark. However, reasonable assumptions about the geometry of the universe can lead us to the conclusion that the night sky should glow as brightly as a star's surface. This conflict between observation and theory is called **Olbers' paradox** after Heinrich Olbers, a Viennese physician and astronomer, who discussed the paradox in 1826.

However, Olbers' paradox is not Olbers', and it isn't a paradox. The problem of the dark night sky was first discussed by Thomas Digges in 1576 and was further analyzed by astronomers such as Johannes Kepler in 1610 and Edmund Halley in 1721. Olbers gets the credit through an accident of scholarship on the part of modern cosmologists who did not know of previous discussions. What's more, Olbers' paradox is not a paradox. We will be able to understand why the night sky is dark by revising our assumptions about the nature of the universe.

To begin, let's state the so-called paradox. Suppose we assume that the universe is static, infinite, and eternal. Suppose we also assume that it is uniformly filled with stars. (The aggregation of stars into galaxies makes no difference to our argument.) If we look in any direction, our line of sight must eventually reach the surface of a star (Figure 19-1). Consequently, every point on the surface of the sky should be as bright as the surface of a star, and it should not get dark at night.

Of course, the most distant stars would be much fainter than the nearer stars, but there would be more distant stars than nearer stars. The intensity of the light from a star decreases according to the inverse square law, so distant stars would not contribute much light. However, the farther we look in space, the larger the volume we survey. Thus, the number of stars we see at any given distance increases as the square of the distance. The two effects cancel out, and the stars at any given distance contribute as much total light as the stars at any other distance. Then given our assumptions, every spot on the sky must be occupied by the surface of a star, and it should not get dark at night.

Imagine the entire sky glowing with the brightness of the surface of the sun. The glare would be overpowering. In fact, the radiation would heat the earth and all other celestial objects to the average temperature of the surface of the stars, at least 1000 K. Thus, we can pose Olbers' paradox in another way: why is the universe so cold?

Kepler and many other astronomers of his time saw no paradox in the dark night sky, because they believed in a finite universe. Once they looked beyond the sphere of stars, there was nothing to see except, perhaps, the dark floor of heaven. But soon Western astronomers accepted that the universe was infinite, and thus they had to explain why the sky is dark between the stars.

Olbers assumed that the sky was dark because clouds of matter in space absorbed the radiation from distant stars. But this interstellar medium would gradually heat

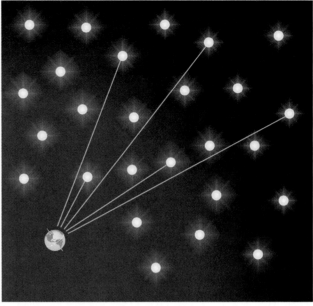

a

b

FIGURE 19-1

(a) Every direction we look in a forest eventually reaches a tree trunk, and we cannot see out of the forest. (Janet Seeds)
(b) If the universe is infinite and uniformly filled with stars, then any line from the earth should eventually reach the surface of a star. This predicts that the night sky should glow as brightly as the surface of the average star, a puzzle commonly referred to as Olbers' paradox.

up to the average surface temperature of the stars, and the gas and dust clouds should be glowing as brightly as the stars.

Today, cosmologists believe they understand why the sky is dark. Olbers' paradox makes the incorrect prediction that the sky should be bright because it is based on two incorrect assumptions. The universe is neither static nor infinitely old.

In Chapter 17, we saw that the galaxies are receding from us. The distant stars in these galaxies are receding from the earth at high velocity, and their light is Doppler-shifted to long wavelengths. We can't see the light from these stars because their light is red-shifted, and the energy of the photons is reduced to levels we cannot detect. Expressed in another way, the universe is cold because the photons from very distant stars arrive with such low energy they cannot heat up objects. Although this explains part of the problem, the red shifts of the distant galaxies are not enough to make the sky as dark as it appears.

The second part of the explanation was first stated by Edgar Allan Poe in 1848. He proposed that the night sky was dark because the universe was not infinitely old but had been created at some time in the past. The more distant stars are so far away that light from them has not reached us yet. That is, if we look far enough, the look-back time is greater than the age of the universe, and we look back to a time before stars began to shine. Thus, the night sky is dark because the universe is not infinitely old.

This is a powerful idea because it clearly illustrates the difference between the universe and the observable universe. The universe is everything that exists, and it could be infinite. But the **observable universe** is the part that we can see. We will learn later that the universe is 15–20 billion years old. In that case, the observable universe has a radius of 15–20 billion ly. Do not confuse the observable universe, which is finite, with the universe as a whole, which could be infinite.

The assumptions that we made when we described Olbers' paradox were at least partially in error. This illustrates the importance of assumptions in cosmology and serves as a warning that our commonsense expectations are not dependable. All of astronomy is reasonably unreasonable; that is, reasonable assumptions often lead to unreasonable results. That is especially true in cosmology, so we must examine our assumptions with special care.

Basic Assumptions

Although we could make many assumptions, three are basic—homogeneity, isotropy, and universality.

Homogeneity is the assumption that matter is uniformly spread throughout space. Obviously this is not true on the small scale, because we can see matter concentrated in planets, stars, and galaxies. Homogeneity refers to the large-scale distribution. If the universe is homogeneous, we should be able to ignore individual galaxies and think of matter as an evenly spread gas in which each particle is a cluster of galaxies. Recent observations suggest that the universe is homogeneous only on the largest scales.

Isotropy is the assumption that the universe looks the same in every direction, that it is isotropic. On the small scale this is not true, but if we ignore local variations such as galaxies and clusters of galaxies, the universe should look the same in any direction. For example, we should see roughly the same number of galaxies in every direction. Again, observations suggest that the universe is isotropic on the largest scales.

The most easily overlooked assumption is **universality,** which holds that the physical laws we know on the earth apply everywhere in the universe. Although this may seem obvious at first, some astronomers challenge universality by pointing out that when we look out in space we look back in time. If the laws of physics change with time, we may see peculiar effects when we look at distant galaxies. For now we will assume that the physical laws observed on the earth apply everywhere in the universe.

The assumptions of homogeneity and isotropy lead to an assumption so fundamental it is called the **cosmological principle.** According to this principle, any observer in any galaxy sees the same general features of the universe. For example, all observers should see the same kinds of galaxies. As in previous assumptions, we ignore local variations and consider only the overall appearance of the universe, so the fact that some observers live in galaxies in clusters and some live in isolated galaxies is only a minor irregularity.

Evolutionary changes are not included in the cosmological principle. If the universe is expanding and the galaxies are evolving, observers living at different times may see galaxies at different stages. The cosmological principle says that once observers correct for evolutionary changes, they should see the same general features.

The cosmological principle is actually an extension of the Copernican principle. Copernicus found that the earth is not in a special place; it is just one of a number of planets orbiting the sun. The cosmological principle says that there are no special places in the universe. Local irregularities aside, one place is just like another. Our location in the universe is typical of all other locations.

If we accept the cosmological principle, we may not imagine that the universe has an edge or a center. Such locations would not be the same as all other locations.

Once we establish our basic assumptions, we face a choice between two methods of attack. We can observe the universe and try to deduce its properties from its appearance, or we can build a theoretical universe based on our assumptions and the laws of nature and then compare our theory with reality. Both methods are valid.

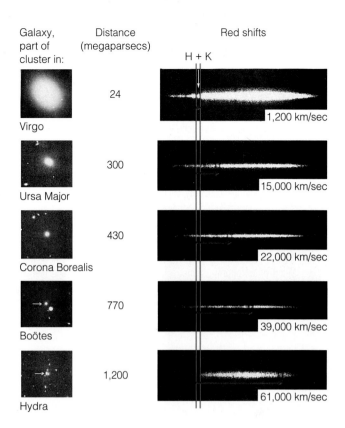

FIGURE 19-2

Galaxies at different distances are shown at left, with their spectra shown at the right. The two vertical blue lines mark the unshifted location of the H and K lines of once-ionized calcium. The expansion of the universe causes a red shift that moves the spectral lines to longer wavelengths, as shown by the red arrows. More distant galaxies have larger red shifts, expressed as radial velocities in kilometers per second. (Palomar Obs/Caltech)

The Expansion of the Universe

All of cosmology is based on a single observation. The spectra of galaxies contain red shifts that are proportional to their distances (Figure 19-2). These red shifts are commonly referred to as Doppler shifts due to the recession of the galaxies, which is why we say the universe is expanding.

Edwin P. Hubble discovered the velocity–distance relationship in 1929 using the spectra of only 46 galaxies (Figure 17-17). Since then the Hubble law, as it has come to be known, has been confirmed for hundreds of galaxies out to great distances. This law (Chapter 17) is clear evidence that the universe is expanding uniformly and has no center.

To see how the Hubble law implies uniform, centerless expansion, we can think about an analogy (Window on Science 19-1). Imagine that we make a loaf of raisin bread (Figure 19-3). As the dough rises, the expansion pushes the raisins away from one another. Two raisins that were originally separated by only 1 cm move apart

Reasoning by Analogy

The economy is overheating and it may seize up," an economist might say. Economists like to talk in analogies because economics is often abstract, and one of the best ways to think about abstract problems is to find a more approachable analogy. Rather than discuss the details of the national economy, we might make conclusions about how the economy works by thinking about how a gasoline engine works. Of course, if our analogy is not a good one, it can mislead us.

Much of astronomy is abstract, and cosmology is the most abstract subject in astron-omy. Furthermore, cosmology is highly mathe-matical, and unless we are prepared to learn some of the most difficult mathematics known, we must use analogies, such as an ant on an orange, to discuss the curvature of space-time and the nature of the universe.

Reasoning by analogy is a powerful tech-nique. An analogy can reveal unexpected in-sights and lead us to further discoveries. But carrying an analogy too far can be misleading. We might compare the human brain to a com-puter, and that would help us understand how data flow in and are processed and how new data flow out. But our analogy is flawed. Data in computers are stored in specific locations, but memories are stored in the brain in a distrib-uted form. No single brain cell holds a specific memory. If we carry the analogy too far, it can mislead us. Whenever we reason by analogy, we should be alert for potential problems.

As you study any science, be alert for analogies. They are tremendously helpful, but we must always take care not to carry them too far. ∎

a

b

FIGURE 19-3
The uniform expansion of the uni-verse can be represented by raisins in a loaf of raisin bread (a). As the dough rises (b), raisins originally near each other move apart more slowly, and raisins originally farther apart move away from each other more rapidly. A colony of bacteria living on any raisin will find that the velocities of recession of the other raisins are proportional to their distances.

rather slowly, but two raisins that were originally sepa-rated by 4 cm of dough are pushed apart more rapidly. Thus, the uniform expansion of the dough causes the raisins to move away from each other at velocities pro-portional to their distances. According to the Hubble law, the larger the distance between two galaxies, the faster they recede from each other. This is exactly the result we expect from uniform expansion.

The raisin bread also illustrates that the expansion has no identifiable center. A colony of bacteria living on one of the raisins would see themselves surrounded by a universe of receding raisins. (Here we must assume that they cannot see to the edge of the loaf; that is, they must not be able to see to the edge of their universe.) It does not matter on which raisin the bacteria live. The bac-terial astronomers would measure the distances and velocities of recession and derive a bacterial Hubble law showing that the velocities of recession are proportional to the distances. As long as they cannot see the edge of the loaf, they cannot identify any raisin as the center of the expansion.

Similarly, we see galaxies receding from us, but we cannot identify any galaxy or point in space as the center of the expansion. Any observer in any galaxy should see the same expansion that we see.

Although astronomers and cosmologists commonly refer to these red shifts as Doppler shifts and often speak of the recession of the galaxies, relativity provides a more elegant explanation. Einstein's general theory of relativ-ity explains the expansion of the universe as an expansion of space-time itself. A photon traveling through this space-time is stretched as space-time expands, and the photon arrives with a longer wavelength than it had when it left (Figure 19-4). Photons from distant galaxies travel for a longer time and are stretched farther than photons from nearby galaxies. Thus, the expansion of the universe is an expansion of the geometry and not just a simple recession of the galaxies. In fact, this relativistic explana-tion will make no difference to our discussion, but it reminds us that nature is often more elegant than our nonmathematical language.

FIGURE 19-4

General relativity explains that the red shift of light from distant galaxies is produced by the stretching of space-time (shown here as an expanding grid) between the time the photon is emitted (a) and the time it arrives at the earth (e). (Notice that the wavelength in frame e is longer that that in frame a.) Photons from distant galaxies spend more time traveling through space and are thus stretched more than photons from nearby galaxies. Thus, distant galaxies have a larger red shift.

FIGURE 19-5

An ant confined to the two-dimensional surface of an orange could explore the entire surface without coming to an edge. Were it to leave dirty footprints, it might realize that its two-dimensional universe was finite but unbounded.

The Geometry of Space-Time

How can the universe expand if it does not have any extra space to expand into? How can it be finite if it doesn't have an edge? The properties we have ascribed to the universe seem to violate common sense, so we must proceed with caution. We must think carefully, because our everyday intuition can be misleading. For example, everything we touch and see every day has an edge, so thinking about an edgeless universe presents a challenge. One powerful tool is reasoning by analogy, but we must be careful not to carry analogies too far. Also, we must use words carefully so they do not mislead us (Window on Science 19-2). With such cautions in mind, we can use general relativity to understand the geometry of space-time.

In Chapter 5, we saw that the presence of mass can curve space-time, and we sense that curvature as a gravitational field. This explains the earth's gravity and also that of a black hole, but it also predicts that the mass of the universe can produce a general curvature that determines the motion of the universe. Thus, modern cosmological theories are based on general relativity and curved space-time.

To discuss curvature, we can use a two-dimensional analogy for our three-dimensional universe. Suppose an ant lived on an orange and was a perfectly two-dimensional creature (Figure 19-5). That is, the ant could move in two dimensions over the surface of the orange but could not move in the third dimension, perpendicular to the surface of the orange. Furthermore, suppose that light could not travel perpendicular to the surface of the orange. Then the ant would never know there was a third dimension, and it could wander over the surface of the orange thinking it was moving on a horizontal surface that had no edge. If the ant left dirty footprints, however, it might eventually say, "I have visited every square centimeter in my two-dimensional universe, so it must be finite, but I can't find any edge or any center. My universe is finite but unbounded." Of course, if the ant lived on a sheet of paper, it would discover the edges and

Words Lead Thoughts

There are certain words we should never say, not even as a joke. We should never call our friend "fool," for example, because we might begin to think of our friend as a fool. Words lead thoughts. It works in advertising and politics, and it can work in science, too. Using words carelessly can lead us to think carelessly.

In science, there are certain ways to say things, not because scientists are sticklers for good grammar, but because if we say things wrong we begin to think things wrong. For example, a biologist would never let us say a

beehive knows it must store food. "No, don't say it that way," the biologist would object. "The hive is just a collection of individual creatures and it can't 'know' anything. Even the individual bees don't really 'know' things. The instinctive behavior of individual bees causes food to accumulate in the hive. That's the way to say it."

All scientists are careful of language because careless words can mislead us. We would never refer to the center of the universe, for example, but we must also be careful not to

say things like "galaxies flying away from the big bang." Those words imply a center to the expansion of the universe, and we know the expansion must be centerless. Rather, we should say "galaxies flying away from each other."

Whatever science you study, notice the customary ways of using words. It is not just a matter of convention. It is a matter of careful thought. ∎

FIGURE 19-6
In two-dimensional space, curvature distorts the area of a circle; in three-dimensional space, it distorts the volume of a sphere. We could measure distortion and detect curvature by counting galaxies.

conclude that it lived in a finite, bounded universe. Our two-dimensional analogy shows that curvature can allow a universe to be finite but have no edge. If we expand our analogy to three dimensions, we see that our universe could be finite in volume if it is curved such that it is unbounded.

To continue our discussion of curved space-time, we can extend our analogy of the ant. There are three possible ways that a universe could be curved. It could be flat (zero curvature), it could be spherical (positive curvature), or it could be saddle-shaped (negative curvature). If an ant lived on a curved two-dimensional universe, it might be unaware of any curvature, but it could detect the curvature of its universe if it could draw circles and measure their areas (Figure 19-6). On the flat surface, a circle would always have an area of exactly πr^2, but on

the positively curved surface its area would be less than πr^2, and on the negatively curved surface it would be more. Drawing small circles would not suffice, because the area of small circles would not differ noticeably from πr^2, but if the ant could draw big enough circles, it could actually measure the curvature of its two-dimensional universe.

We are three-dimensional creatures, but our universe might still be curved, and we could measure that curvature by measuring the volume of spheres. If space-time is flat, then no matter how big a sphere we measure, its volume will always be $\frac{4}{3}\pi r^3$. But in positively curved space-time, spheres would have less volume than this; in negatively curved space-time, they would have more. The spheres must be many megaparsecs in radius for the difference to be detectable.

We could measure the volume of such spheres by counting the number of galaxies within a certain distance r from the earth. If galaxies are homogeneously scattered through space, the number within distance r should be proportional to the volume of the sphere. If, as we count to greater and greater distances, we find the number of galaxies increasing proportional to $\frac{4}{3}\pi r^3$, space-time is flat. However, if we find an excess of distant galaxies, space-time is negatively curved, and if we find a deficiency, space-time is positively curved.

By thinking of the two-dimensional analogy shown in Figure 19-6, astronomers usually refer to a universe with positive curvature as a **closed universe.** A zero-curvature universe is said to be a **flat universe,** and a universe with negative curvature is an **open universe.** Notice that flat and open universes are infinite, but closed universes are finite.

A closed universe has a finite volume but no edge. Like the ant in Figure 19-5, we might explore our universe and visit every location but never find an edge. Of course, an open universe, being infinite, also has no edge. Thus, we can answer our question about the edge of the universe. Whether it is open or closed, the universe can have no boundary.

We must not try to visualize the expansion of the universe as an outer edge moving into previously unoccupied space. Open, flat, or closed, the universe has no edge, so it does not need additional room to expand. The universe contains all of the volume that exists, and the expansion is a change in the nature of space-time that causes that volume to increase.

CRITICAL INQUIRY

Why do we say the universe can't have an edge or center?

The quick answer to this question is that an edge or a center would violate the cosmological principle, which says that every place in the universe is similar in its general properties to every other place. A place at the edge of the universe or at its center would be different, so there can be no edge and no center.

Of course, this answer is only an appeal to the arbitrary authority of the cosmological principle. What is the real reason we conclude that there can be no edge and no center? Imagine that the universe had an edge and you went there. What would you see? Of course, you might imagine an edge to the distribution of matter with empty space beyond, but that would not be a real edge. Imagine an edge beyond which there is no space. What would happen if you stuck your head beyond the edge and looked around? Could you do that if there was no space beyond the edge? These questions seem senseless because thinking about an edge to the universe leads us to paradoxes we cannot resolve. From this we conclude that an edge is impossible. Of course, if there is no edge, then we can't find a center, because we define the centers of things by reference to their edges.

We believe the universe has no edge and no center for good, logical reasons, not because of some arbitrary rule. But why do we believe the cosmological principle is correct? What evidence or assumptions support it? ■

According to all the evidence and consistent with general relativity, we live in an expanding universe. If the universe were a videotape, we could run it backward and see the universe contracting. Gradually, the galaxies would approach one another until they began to merge. The total volume of the universe would decrease until all the matter and energy that exist would be trapped in a very high temperature, high-density state—a state called the big bang.

19-2 THE BIG BANG

The **big bang theory** proposes that our universe began in a moment of tremendous violence and continues to expand, cool, and evolve from that state. The very name,

FIGURE 19-7
The population of Cosmos, Minnesota, is apparently in flux; someone has painted out the numbers on the city limits sign. This sign reminds us of the big bang theory. Although the theory seems at first sight difficult to believe and is sometimes the object of jokes and unwarranted criticism, it is currently the only widely accepted theory for the origin of the universe. The evidence that there was a big bang is overwhelming. (Richard Fluck)

big bang, was coined by cosmologist Fred Hoyle as a sarcastic comment on the believability of the theory, and it is often the butt of jokes and cartoons (Figure 19-7). At first glance, it is difficult to imagine how the big bang could have occurred, but it is the dominant theory for the origin of the universe. Our challenge is to stretch our imaginations to encompass the beginning of everything.

The Beginning

The big bang occurred long ago, and our instinct is to think of it as a historical event, like the Gettysburg Address. There is no way we can travel back and see Lincoln deliver the Gettysburg Address, but the big bang is different. In this section, we will discover that the big bang isn't over and that we can indeed see it happening.

Because light travels at a finite speed, we see distant galaxies not as they are now, but as they were when the light left them to begin its journey to the earth. The Andromeda Galaxy is about 2 million light-years away, so the look-back time to the Andromeda Galaxy is 2 million years. That isn't a long time in the history of a galaxy, but the look-back time to more distant galaxies is larger. New telescopes are so large and so sensitive they can detect galaxies that are extremely distant, and the look-back time to those galaxies is a large fraction of the age of the universe (Figure 19-8). Thus, the look-back time allows us to see what the universe was like when it was young.

Suppose that we look even farther, past the most distant objects, backward in time to the fiery clouds of matter that filled the universe during the big bang. From these great distances, we receive light that was emitted by the hot gas soon after the universe began.

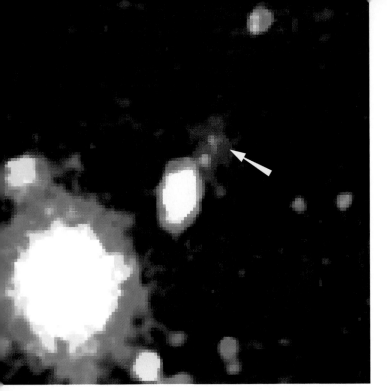

FIGURE 19-8

One of the most distant galaxies ever photographed (arrow) has a red shift of 4.25. The look-back time to this galaxy is so large we see it as it was when the universe was only 7–10 percent of its present age. First mapped by radio astronomers, the galaxy was imaged using the 10-m Keck Telescope atop Mauna Kea in Hawaii. The two brightest objects in the field of view are less distant galaxies (H. Spinrad, A. Dey, and J. Graham, University of California, Keck Observatory)

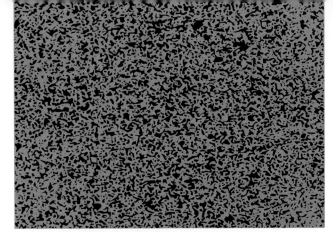

a   A region of the universe during the big bang

b   A region of the universe now

c   The present universe as it appears from our galaxy

FIGURE 19-9

Three views of a small region of the universe centered on our galaxy. (a) During the big bang explosion, the region is filled with hot gas and radiation. (b) Later, the gas forms galaxies, but we can't see the universe this way because the look-back time distorts what we see. (c) Near us we see galaxies, but farther away we see young galaxies (dots), and at a great distance we see radiation (arrows) coming from the hot clouds of the big bang explosion.

Although our imaginations try to visualize the big bang as a localized event, we must keep firmly in mind that the big bang did not occur at a single place but filled the entire volume of the universe. The matter of which we are made was part of that big bang, so we are inside the remains of that event, and the universe continues to expand around us. We cannot point to any particular place and say, "The big bang occurred over there." The big bang occurred everywhere, and, no matter what direction we look, at great distances we look back to the age when the universe was filled with hot gas (Figure 19-9).

The radiation that comes to us from this great distance has a tremendous red shift. The most distant visible objects are quasars, which have red shifts up to almost 5. In contrast, the radiation from the hot gas of the big bang has a red shift of about 1000. Thus, the light emitted by the big-bang gases arrives at the earth as infrared radiation and short radio waves. We can't see it with our eyes, but we should be able to detect it with infrared and radio telescopes. Thus, unlike the Gettysburg Address, the big bang isn't over, and we should be able to see it happening if we can detect the radiation it emitted.

Primordial Background Radiation

In the mid-1960s, two Bell Laboratories physicists, Arno Penzias and Robert Wilson, were using a horn antenna to measure the radio brightness of the sky (Figure 19-10a). Their measurements showed a peculiar noise in the system. After rebuilding the receiver and still detecting the noise, they decided the problem lay with a pair of pigeons living inside the antenna. They trapped the pigeons, relocated them, and began cleaning out the droppings. Perhaps they would have enjoyed the job more if they had known they would win the 1978 Nobel prize for physics for the discovery they were about to make.

When the antenna was cleaned, they measured the brightness of the sky at radio wavelengths and again found the low-level radio noise. The pigeons were innocent, but what was causing the signal?

The explanation of the radio noise goes back to 1948, when George Gamow predicted that the early stages of the big bang would have been very hot and would have emitted large amounts of black body radiation. A year later, physicists Ralph Alpher and Robert Herman pointed out that the very large red shift would lengthen the wavelengths and make the gas clouds of the big bang seem very cold. In the late 1940s, there was no way to detect this radiation, but in the mid-1960s Robert Dicke at Princeton developed his own theories and concluded that the radiation was just strong enough to detect. Dicke and his team began building a receiver to search for it.

When Penzias and Wilson heard of Dicke's work in a chance phone conversation, they recognized the noise they had detected as radiation coming from the big bang, the **primordial background radiation.**

Since its discovery, the primordial background radiation has been measured at many wavelengths (Figure 19-10b). The most critical measurements were those near the peak of the curve, because they defined the shape of the curve and the temperature of the equivalent black body. Unfortunately, the peak of the curve falls at a wavelength of about 1 mm, and observations at such wavelengths must be made from high-flying balloons or rockets. In 1989, the Cosmic Background Explorer (COBE) satellite was launched, and by January 1990 the results were in. Observing from orbit, COBE showed that the primordial background radiation follows a black body curve with a temperature of 2.735 K.

It may seem strange that the hot gas of the big bang seems to have a temperature of only 2.7 K, but recall the tremendous red shift. When we look back to the big bang, we see light that has a red shift of about 1000—that is, the wavelengths of the photons are about 1000 times longer than when they were emitted. The gas clouds that emitted the photons had a temperature of about 3000 K, and they emitted black body radiation with a λ_{max} of about 1000 nm (Chapter 7). Although this is in the near infrared, the gas would also have emitted enough visible light to glow orange-red. But the red shift has made the

a

$T = 2.735 \pm 0.06$ K

Intensity

0.05   0.1   0.5   1

Wavelength (cm)

b

FIGURE 19-10

(a) Robert Wilson (left) and Arno Penzias pose before the horn antenna with which they discovered the primordial background radiation. (AT&T Archives) (b) The primordial background radiation from the big bang is observed in the infrared and radio part of the spectrum. Until recently, the critical measurements near the peak of the curve had to be made from balloons and rockets, but the COBE satellite observed from orbit. These COBE data show that the radiation fits a black body curve with a temperature of 2.735 K.

wavelengths about 1000 times longer, so λ_{max} is about 1 million nm (equivalent to 1 mm). Thus, the hot gas of the big bang seems to be 1000 times cooler, about 3 K, or, more precisely, 2.735 K.

Although the radiation is almost perfectly isotropic, observations show a small departure from complete isotropy. The primordial background radiation seems to have slightly shorter wavelengths—it seems hotter—in the direction of the constellation Leo. In the opposite direction, the radiation seems slightly cooler. This difference is evidently due to the motion of our galaxy through space. The Milky Way Galaxy is moving about 540 km/sec in the direction of Leo, causing a slight blue

FIGURE 19-11

If we could see galaxies in the sky, we would quickly notice that they are not evenly distributed. The great Virgo cluster (center) contains over a thousand galaxies and, being only about 17 Mpc away, looks large in our sky. Above and slightly to the left in this diagram is the more distant Coma cluster and to the right is the Leo cluster. These clusters of galaxies are linked to form even larger systems called superclusters. Here, the local supercluster includes the Virgo Cluster and the South-East Extension at the lower left linked to other clusters out of the diagram to upper right. The more distant Coma supercluster includes clusters extending from northern Bootes across into Leo.

shift in the background radiation that makes it appear slightly hotter. Radiation from the opposite side of the universe is slightly red-shifted and looks cooler.

The isotropy of the background radiation implies that the early universe was very homogeneous. Today, however, we see galaxies grouped together in clusters and superclusters (Figure 19-11), and recent observations have suggested that there are great voids between the superclusters, voids as large as 50 Mpc in diameter (Figure 19-12). If the early universe was as homogeneous as the background radiation suggests, how did it get to be so clumpy today? This problem is related to the formation of the galaxies and clusters of galaxies. We will return to the background radiation and the origin of galaxies later in this chapter.

The greatest importance of the primordial background radiation lies in its interpretation as radiation from the big bang. If that interpretation is correct, the radiation is evidence that a big bang really did occur. That evidence is so strong that an alternative theory of the universe has been abandoned.

The Steady State Theory

Through the 1950s and most of the 1960s, astronomers had an alternative to the big bang theory. The **steady state theory** proposed that the universe did not evolve,

that it always had the same general properties. Stars and galaxies might age and die, but new stars and new galaxies would be born to take their place and thus preserve the general properties of the universe.

One of the most distinctive features of the steady state theory was caused by the expansion of the universe. As the universe expands, its average density should decrease. The steady state theory, however, held that the universe did not change, so it proposed that matter was created continuously to maintain the density of the universe. This newly created matter, presumably in the form of hydrogen atoms, collected in great clouds between the receding galaxies and eventually gave birth to new galaxies to take the place of aging galaxies no longer able to make new stars.

The steady state theory was a highly controversial though exciting idea in astronomy, and the competition with the big bang theory was fierce. The discovery of the primordial background radiation, however, was a fatal blow. By the late 1960s, the background radiation had been so widely accepted as evidence of a big bang that the steady state theory was abandoned. Thus, the steady state theory is now only of historical interest, illustrating the fate of a theory that is contradicted by a single dramatic observation.

The big bang theory is now the accepted theory of the universe among nearly all cosmologists. We might

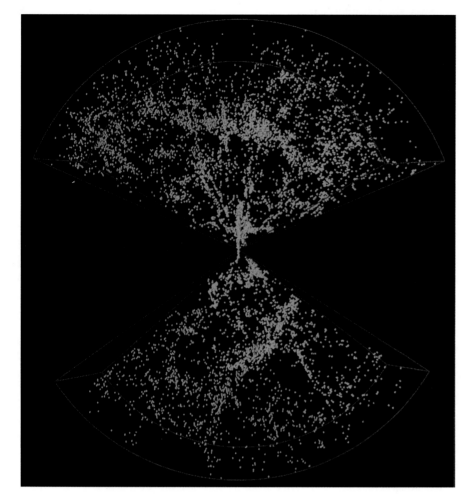

FIGURE 19-12
This diagram of the distribution of galaxies within 100 Mpc was created by measuring red shifts and converting them into distances. Random motions in clusters of galaxies cause such clusters to appear as fingers pointing toward us at the center. The Virgo cluster, about 17 Mpc away, is the finger just above center (see Figure 19-11). These two slices look north and south out of our galaxy and reveal great walls and voids up to 50 Mpc in diameter in the distribution of galaxies. The Great Wall stretches horizontally across the northern slice (see Figure 17-25), and the Southern wall crosses the southern slice diagonally. Each wall contains thousands of galaxies. (Margaret J. Geller, Luis A. N. da Costa, John P. Huchra, and Emilio E. Falco, Smithsonian Astrophysical Observatory © 1994)

compare it to the geologists' theory that the earth is round. Certainly, geologists struggle to understand details of the earth's structure, but they are not likely to revise their conclusion that the earth is round. Similarly, cosmologists still struggle to understand the history of the universe, but they no longer deny that there was a big bang. The primordial background radiation means we can see it happening.

A History of the Big Bang

Modern cosmologists have been able to reconstruct the history of the early universe to reveal how energy and matter interacted as the universe began. As we review that history, remember that the big bang did not occur in a specific place. The big bang filled the entire volume of the universe from the first moment.

We cannot begin our history at time zero, because we do not understand the physics of matter and energy under such extreme conditions, but we can come close. If we could visit the universe when it was very young, only 10 millionths of a second old, for instance, we would find it filled with high-energy photons having a temperature well over 1 trillion (10^{12}) K and a density greater than 5×10^{13} g/cm^3, nearly the density of an atomic nucleus. When we say the photons had a given temperature, we mean that the photons were the same as black body radi-

ation emitted by an object of that temperature. Thus, the photons in the early universe were gamma rays of very short wavelength and therefore very high energy. When we say that the radiation had a certain density, we refer to Einstein's equation $E = m_0 c^2$. We can express a given amount of energy in the form of radiation as if it were matter of a given density.

If two photons have enough energy, they can collide and convert their energy into a pair of particles—a particle of normal matter and a particle of antimatter. In the early universe, photons had sufficient energy to produce proton–antiproton pairs and neutron–antineutron pairs. When a particle collides with its antiparticle, however, the particles are annihilated, and the mass is converted into a pair of gamma-ray photons. Thus, the early universe was a soup of energy continuously switching from photons to particles and back again.

While all of this went on, the universe was expanding, and the wavelengths of the photons were lengthened by the expansion. This lowered the energy of the gamma rays, and the universe cooled. By the time the universe was 0.0001 second old, its temperature had fallen to 10^{12} K. By this time, the average energy of the gamma rays had fallen below the energy equivalent to the mass of a proton or a neutron, so the gamma rays could no longer produce such heavy particles. The particles combined with

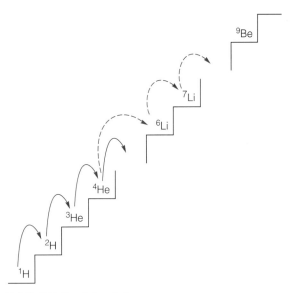

FIGURE 19-13

Cosmic element building. During the first few minutes of the big bang, temperatures and densities were high, and nuclear reactions built heavier elements. Because there are no stable nuclei with atomic weights of 5 or 8, the process built very few atoms heavier than helium.

their antiparticles and quickly converted most of the mass into photons.

It would seem from this that all the protons and neutrons should have been annihilated with their antiparticles, but for quantum-mechanical reasons a small excess of normal particles existed. For every billion protons annihilated by antiprotons, one survived with no antiparticles to destroy it. Thus, we live in a world of normal matter, and antimatter is very rare.

Although the gamma rays did not have enough energy to produce protons and neutrons, they could produce electron–positron pairs, which are about 1800 times less massive than protons and neutrons. This continued until the universe was about 4 seconds old, at which time the expansion had cooled the gamma rays to the point where they could no longer create electron–positron pairs. Thus, the protons, neutrons, and electrons of which our universe is made were produced during the first 4 seconds of its history.

This soup of hot gas and radiation continued to cool and eventually began to form atomic nuclei. High-energy gamma rays can break up a nucleus, so the formation of such nuclei could not occur until the universe had cooled somewhat. By the time the universe was about 2 minutes old, protons and neutrons could link to form deuterium, the nucleus of a heavy hydrogen atom, and, by the end of the next minute, further reactions began converting deuterium into helium. But no heavier atoms could be built because no stable nuclei exist with atomic weights of 5 or 8 (in units of the hydrogen atom). Cosmic element building during the big bang had to proceed rapidly, step by step, like someone hopping up a flight of stairs (Figure 19-13). The lack of stable nuclei at atomic weights of

5 and 8 meant there were missing steps in the stairway, and the step-by-step reactions could not jump over these gaps.

By the time the universe was 30 minutes old, it had cooled sufficiently that nuclear reactions had stopped. About 25 percent of the mass was helium nuclei, and the rest was in the form of protons—hydrogen nuclei. This is the cosmic abundance we see today in the oldest stars. (The heavier elements, remember, were built by nucleosynthesis inside many generations of massive stars.) The cosmic abundance of helium was fixed during the first minutes of the universe.

For the first million years, the universe was dominated by radiation (Figure 19-14). The gamma rays interacted continuously with the matter, and they cooled together as the universe expanded. The gas was ionized because it was too hot for the nuclei to capture electrons to form neutral atoms, and the free electrons made the gas very opaque. A photon could not travel very far before it collided with an electron and was deflected. Thus, radiation and matter were locked together.

When the universe reached an age of about 1 million years, it was cool enough for nuclei and electrons to form neutral atoms. The free electrons were captured by atomic nuclei, the gas became transparent, and the radiation was free to travel through the universe. The temperature at the time of this **recombination** was about 3000 K. We see these photons arriving now as primordial background radiation.

After recombination, the universe was no longer dominated by radiation. Instead, matter was free to move under the influence of gravity, so we say the universe after recombination is dominated by matter. How the matter cooled and collected into clouds and eventually gave birth to galaxies in clusters and superclusters is not well understood.

The End of the Universe: A Question of Density

Will the universe ever end? Will it go on expanding forever with stars burning out, galaxies exhausting their star-forming gas and becoming cold, and dark systems expanding forever through an endless, dark universe? How else could the universe end?

Earlier we decided that the universe could be open, flat, or closed. The geometry of the universe determines how it will end, so we must consider these three possible universes.

The general curvature of the entire universe is determined by its density. If the average density of the universe is equal to the **critical density** of 4×10^{-30} g/cm^3, space-time will be flat. If the average density of the universe is less than the critical density, the universe is negatively curved and open. If the average density is more than the critical density, the universe is positively curved and closed.

We decided earlier that an open universe and a flat universe are both infinite. Both of these universes will

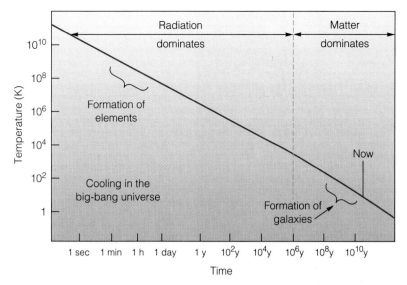

FIGURE 19-14

During the first few minutes of the big bang, some hydrogen became deuterium and helium and a few heavier atoms. Later, when radiation was no longer dominant, galaxies formed, and nuclear reactions inside stars made the rest of the chemical elements.

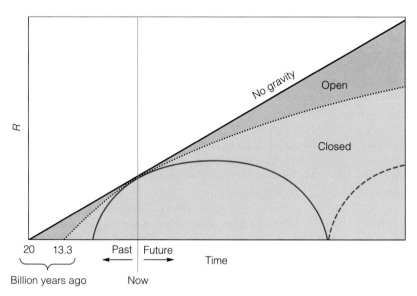

FIGURE 19-15

The expansion of the universe as a function of time. Models of the universe can be represented as curves in a graph of R, a measure of the size of the universe, and time. Open-universe models expand without end, and the corresponding curves fall in the region shaded orange. Closed models expand and then contract back to a high-density state (solid curve) from which they might expand again. Curves representing closed models fall in the region shaded blue. The dotted line represents a flat universe, the dividing line between open and closed models. Note that the estimated age of the universe depends on the rate at which the expansion is slowing down. (This figure assumes $H = 50$ km/sec/Mpc.)

expand forever. The gravitational field of the material in the universe, present as a curvature of space-time, will cause the expansion to slow; but if the universe is open, it will never come to a stop (Figure 19-15). If the universe is flat, it will slow to a stop after an infinite time. Thus, if the universe is open or flat, it will expand forever, and the galaxies will eventually become black, cold, solitary islands in a universe of darkness.

If the universe is closed, however, its fate will be quite different. In a closed universe, the gravitational field, present as curved space-time, is sufficient to slow the expansion to a stop and make the universe contract. Eventually, the contraction will compress all matter and energy back into the high-energy, high-density state from which the universe began. This end to the universe, a big bang in reverse, has been called the big crunch. Nothing in the universe could avoid being destroyed.

Some theorists have suggested that the big crunch will spring back to produce a new big bang and a new expanding universe. This theory, called the **oscillating**

universe theory, predicts that the universe undergoes alternate stages of expansion and contraction. Recent theoretical work, however, suggests that successive bounces of an oscillating universe would be smaller and smaller until the oscillation ran down. In addition, we have no theoretical reason to explain how a big crunch could be converted into a new big bang.

The big question of course is, What is the density of the universe? And that very difficult question brings us back to the problem of the dark matter.

The Density of the Universe

Whether the universe is open, closed, or flat depends on its density, but it is quite difficult to measure the density of the universe. We could count galaxies in a given volume, multiply by the average mass of a galaxy, and divide by the volume, but we are not sure of the average mass of a galaxy, and many galaxies are too small to see even if

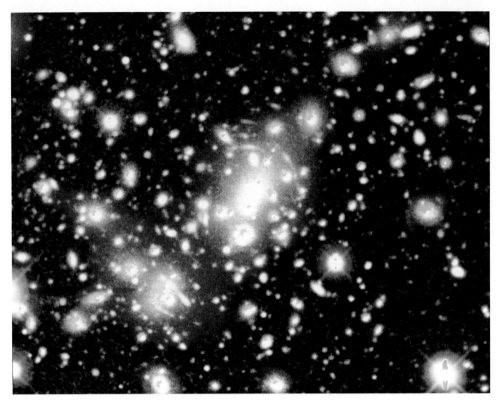

FIGURE 19-16
Galaxy cluster Abell 2218 (orange images) distorts the images of background galaxies to produce short blue arcs. This gravitational lens effect outlines regions of high mass concentration in the galaxy cluster and tells astronomers that over 90 percent of the matter in the cluster is invisible—it is dark matter. (Courtesy Tony Tyson, Bell Labs)

they are nearby. We would also have to include the mass equivalent to the energy in the universe, because mass and energy are related. The best attempts yield a density of about 5×10^{-31} g/cm³, about 10 percent of the mass needed to close the universe.

However difficult this measurement may be, it is almost pointless because it does not include dark matter, and, as we saw in Chapter 17, many observations suggest that much of the mass in the universe has not been detected. Galaxies have invisible, massive coronae, and clusters of galaxies contain more mass than the sum of the visible galaxies. This is not a small correction. Apparently, 90–99 percent of the mass of the universe is dark matter. This has a big effect on our calculations. To correct for the dark matter, we would not add a small percentage to the observed density. We would have to multiply the mass of the detectable matter by a *factor* between 10 and 100.

If we have any doubt as to how insubstantial the visible universe is, we have only to look at gravitational lensing of distant galaxies. In Chapter 18, we saw how light from distant quasars could be bent by the gravitational field of an intervening galaxy to create false images of the quasar. Using the largest telescopes in the world, astronomers are finding that light from the most distant galaxies can be bent by the mass of a nearby cluster of galaxies. The distant galaxies are extremely faint and blue, perhaps because of rapid star formation. As the light from these faint galaxies passes through a nearby cluster of galaxies, the images of the distant galaxies can be distorted into short arcs like blue rainbows (Figure 19-16).

From such arcs, astronomers can calculate the mass of the cluster of galaxies. For instance, the arcs in Figure 19-16 show that over 90 percent of the mass in the cluster is dark matter.

Looking at the universe of visible galaxies is like looking at a ham sandwich and seeing only the mayonnaise. Most of the universe is invisible, and most astronomers now believe that invisible matter is not the normal matter of which you and the stars are made. If dark matter were hot gas or cold gas, we would see X rays or infrared radiation. Black holes or neutron stars would betray their presence with X rays, and large numbers of faint, cool stars would emit large amounts of infrared radiation. We see none of these things, so dark matter is evidently not normal matter made of protons and neutrons. These particles belong to a family of atomic particles called **baryons,** so normal matter is often called **baryonic matter.** Many observations show that dark matter is not baryonic.

The density of a certain isotope of hydrogen can tell us about the density of baryonic matter. During the first few minutes of the big bang, nuclear reactions converted some protons into helium and a very small amount of other, heavier elements. How much of these elements was created depends critically on the density of the material, so astronomers have tried to measure their abundance. Deuterium, for example, is an isotope of hydrogen in which the nucleus contains a proton and a neutron. This element is so easily converted into helium that none can be produced in stars. Stars destroy what deuterium they have by converting it into helium. Nevertheless, ultraviolet observations of the interstellar medium made with the Hubble Space Telescope reveal that about 15 out of every million hydrogen atoms in space are actually deuterium. That deuterium, plus the deuterium that has been destroyed in stars, must have been made during the first minutes of the big bang, and that places limits on the present density of baryonic matter in the universe.

If the universe now contained enough baryonic matter to be closed, then during the big bang it would have been so dense it would have destroyed most of the deuterium and would have produced isotopes such as lithium-7. That we have deuterium in the interstellar medium and little lithium-7 seems to indicate that the universe cannot contain enough baryonic matter to be closed. But many observations show that dark matter is abundant, so most astronomers now believe that dark matter cannot be composed of baryonic matter. That is, dark matter must not be made of protons and neutrons.

Theorists have proposed a wide range of exotic subatomic particles that could make up the dark matter, including axions, photinos, and WIMPs. But none of these theoretical particles has ever been detected in the laboratory. Another theory holds that neutrinos are not massless. Because there are over 10^8 neutrinos in the universe for every normal particle, even a tiny mass for the neutrino could add up to significant dark matter. As yet, however, no one has been able to conclusively measure any mass for the neutrino.

The neutrinos detected coming from the explosion of supernova 1987A give us a further clue. If neutrinos are perfectly massless, they will all travel at the speed of light, but if they have some mass, however small, they must travel more slowly. Also, lower-energy neutrinos must travel more slowly than high-energy neutrinos. The neutrino burst from supernova 1987A was spread over about 10 seconds, but there was little tendency for the first arriving neutrinos to have higher energies. This statistical evidence suggests that the mass of the neutrino cannot exceed 0.00003 of an electron and could be zero.

But recall that the problem of the solar neutrinos could be solved if neutrinos can oscillate between three different states (Chapter 9). An exciting consequence of that theory is that if neutrinos can oscillate between states they must have a nonzero mass. Thus, a conclusive solution to the solar neutrino problem might resolve the problem of the dark matter as well. Recent laboratory experiments may have detected this oscillation, and they suggest a neutrino mass of about 0.00001 of an electron. If this is confirmed, neutrinos might make up about 20 percent of the mass in the universe.

A contrasting theory holds that each galaxy is orbited by swarms of objects dubbed **MACHOs** for Massive Compact Halo Objects. In contrast to the exotic WIMPs, MACHOs are simply very low mass stars or planets too faint to see. One way to try to detect them would be to watch stars in a distant galaxy and hope that a MACHO in the halo of our own galaxy passes in front of one of the stars. Theory predicts that, rather than dimming the star, the MACHO would act as a gravitational lens and bend

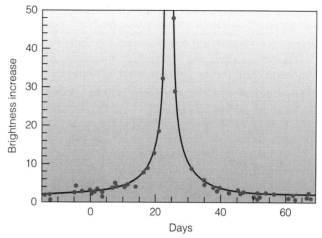

FIGURE 19-17

The brightness of a distant star was observed to increase and then decrease exactly as predicted by general relativity for a star gravitationally lensed by a MACHO, a Massive Compact Halo Object such as a low-mass star or planet. Such an object passing between the earth and the star could focus the light of the star and make it temporarily brighten. In this example, the star brightened by a factor over 40. The growing number of MACHO detections suggests that they make up part of the dark matter.

and focus the light from the star, and we would see the star grow brighter for a period of a few tens of days. To have any chance of seeing such a rare event, we would have to monitor the brightness of vast numbers of stars. Two teams have done exactly that, and their early results indicate that they are detecting sudden brightenings in distant stars (Figure 19-17). The observations show that this form of gravitational lensing does occur. But they are not seeing enough of these events to provide more than 20 percent of the dark mass believed to lie in our galaxy's halo. Thus, MACHOs can't solve all of the dark matter problem.

Of course, there is good evidence that the dark matter can't be made up entirely of MACHOs. The deuterium in space seems to show that the dark matter can't be made up entirely of baryonic matter, and MACHOs are normal baryonic matter. Whatever dark matter is, astronomers haven't found it yet.

CRITICAL INQUIRY

How do we know there was a big bang?

The primordial background radiation consists of photons emitted by the hot gas of the big bang, so when we detect those photons we are "seeing" the big bang. Of course, all scientific evidence must be interpreted, so we must understand how the big bang could produce isotropic radiation before we can accept the background radiation as evidence. First, we must recognize that the big bang event filled all of the universe with hot, dense gas. The big bang didn't happen in a single place; it happened everywhere. At recombination, the expansion of the universe reached the point where the matter became transparent and the radiation was free to travel through space. Today we see that radiation from the age of recombination arriving from all over the sky. It is isotropic because we are part of the big bang event, and as we look out into space to great distance, we look back in time and see the hot gas in any direction we look. We can't see the radiation as light because of the large red shift that has lengthened the wavelengths by a factor of 1000 or so, but we can detect the radiation as infrared and radio photons.

With this interpretation, the primordial background radiation is powerful evidence that there was a big bang. That tells us how the universe began. How can the density of the universe tell us how it will end? ∎

Whether the universe is open or closed, the big bang theory has been phenomenally successful in explaining the origin and evolution of the universe. At present, no other theory is a serious contender. Yet there are important questions about the big bang theory to be resolved.

19-3 REFINING THE BIG BANG: THEORY AND OBSERVATION

Many problems remain to puzzle cosmologists, but the three we examine here will illustrate some important trends in future research. We can phrase these problems as questions: Why was there a big bang? How old is the universe? How does the universe evolve?

The Quantum Universe

A theory developed in 1980 combining general relativity and quantum mechanics may be able to tell us how the big bang began and why the universe is in the particular state that we observe. Some theorists claim that the new theory will explain why there was a big bang in the first place. To introduce the new theory, we can consider two unsolved problems in the current big bang theory.

One of the problems is called the **flatness problem.** The universe seems to be balanced near the boundary between an open and a closed universe. That is, it seems nearly flat. Given the vast range of possibilities, from zero to infinite, it seems peculiar that the density of the universe is within a factor of 10 of the critical density that would make it flat. If dark matter is as common as it seems, the density may be even closer than a factor of 10.

Even a small departure from critical density when the universe was young would be magnified by subsequent expansion. To be so near critical density now, the density of the universe during its first moments must have been within 1 part in 10^{49} of the critical density. So the flatness problem is, Why is the universe so nearly flat?

Another problem with the big bang theory is the isotropy of the primordial background radiation. When we correct for the motion of our galaxy, we see the same background radiation in all directions to at least 1 part in 1000. Yet when we look at background radiation coming from two points in the sky separated by more than a degree, we look at two parts of the big bang that were not causally connected when the radiation was emitted. That is, when recombination occurred and the gas of the big bang became transparent to the radiation, the universe was not old enough for any signal to have traveled from one of these regions to the other. Thus, the two spots we look at did not have time to exchange heat and even out their temperatures. Then how did every part of the entire big bang universe get to be so nearly the same temperature by the time of recombination? This is called the **horizon problem** because the two spots are said to lie beyond their respective light-travel horizons.

The key to these two problems and to others involving subatomic physics may lie with the newer theory. It has been called the **inflationary universe** because it predicts a sudden expansion when the universe was very young, an expansion even more extreme than that predicted by the big bang theory.

To understand the inflationary universe, we must recall that physicists know of only four forces—gravity, the electromagnetic force, the strong force, and the weak force. We are familiar with gravity, and the electromagnetic force is responsible for making magnets stick to refrigerator doors and cat hair stick to wool sweaters charged with static electricity. The strong force holds atomic nuclei together, and the weak force is involved in certain kinds of radioactive decay.

For many years, theorists have tried to unify these forces; that is, they have tried to describe the forces with a single mathematical law (Figure 19-18). A century ago, James Clerk Maxwell showed that the electric force and the magnetic force were really the same effect, and we now count them as a single electromagnetic force. In the 1960s, theorists succeeded in unifying the electromagnetic force and the weak force in what they called the electroweak force, effective only for processes at very high energy. At lower energies, the electromagnetic force and the weak force behave differently. Now theorists have found ways of unifying the electroweak force and the strong force at even higher energies. These new theories are called **grand unified theories,** or **GUTs.**

Studies of GUTs suggest that the universe expanded and cooled until about 10^{-35} second after the big bang, when it became so cool that the forces of nature began to separate. This released tremendous energy, which suddenly inflated the universe by a factor between 10^{20} and 10^{30}. At that time the part of the universe that we can see

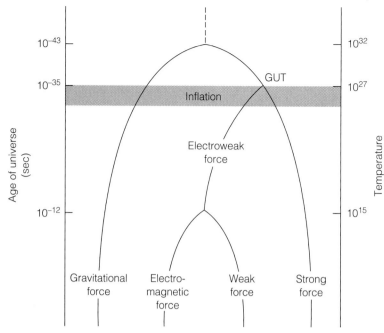

FIGURE 19-18

When the universe was very young and hot (top), the four forces of nature were indistinguishable. As the universe began to expand and cool, the forces separated and triggered a sudden inflation in the size of the universe.

now, the entire observable universe, was no larger than the volume of an atom, but it suddenly inflated to the volume of a cherry pit and then continued its slower expansion to its present extent.

That sudden inflation can solve the flatness problem and the horizon problem. The sudden inflation of the universe would have forced whatever curvature it had toward zero, just as inflating a balloon makes a small spot on its surface flatter. Thus, we now see the universe nearly flat because of that sudden inflation long ago. In addition, because the part of the universe we can see was once no larger in volume than an atom, it had plenty of time to equalize its temperature before the inflation occurred. Now we see the same temperature for the background radiation in all directions.

The inflationary universe is based, in part, on quantum mechanics, and a slightly different aspect of quantum mechanics may explain why there was a big bang at all. Theorists believe that a universe totally empty of matter could be unstable and decay spontaneously by creating pairs of particles until it was filled with the hot, dense state we call the big bang. This theoretical discovery has led some cosmologists to believe that the universe could have been created by a chance fluctuation in space-time. In the words of physicist Frank Wilczyk, "The reason there is something instead of nothing, is that 'nothing' is unstable."

The inflationary universe and other theoretical phenomena predicted by quantum mechanics have become an important and exciting part of modern cosmology. But another area of excitement and controversy is centered on an aspect of traditional observational cosmol-

ogy that began with Hubble's discovery of the expanding universe.

The Age of the Universe

In recent decades, astronomers have been battling over a single number, the Hubble constant. Some say it is large and some say it is small, but all agree it is important. The value of the Hubble constant tells us the approximate age of the universe. The 1990s may see the resolution of this controversy, which began with Edwin Hubble himself.

In 1929, Edwin Hubble announced that the universe was expanding and reported the rate of expansion, the number now known as the Hubble constant (H). Within a few years, astronomers had proposed the idea that the universe had a violent beginning and had used H to extrapolate backward to find the age of the universe. But the value of H that Hubble reported, 530 km/sec/Mpc, led them to conclude that the universe was only half as old as the earth. Something was wrong.

In the decades that followed, astronomers figured out that Hubble's distances to galaxies were systematically too small and thus his value of H was too large (Figure 17-17). Modern measurements of H yield an age of the universe that is at least twice the age of the earth, but controversy still rages.

The calculation of the age of the universe from its expansion is really very simple. We make such calculations whenever we take a drive and divide distance by speed. We say, "It is 100 miles to the city and I can average 50 miles an hour, so I can get there in 2 hours." We know the distance to a galaxy and its radial velocity, so we can divide distance by velocity to find out how long the expansion of the universe has taken to separate the galaxies to their present distance. That is the age of the universe.

The Hubble constant makes this calculation easy. We need to divide distance by velocity, and the Hubble constant is just velocity divided by distance. The reciprocal of the Hubble constant ($\frac{1}{H}$) must be proportional to the age of the universe. The only problem is the units. Astronomers always express H as a velocity in kilometers per second divided by a distance in megaparsecs. To make the division give us an age, we must convert megaparsecs to kilometers, and then the distances will cancel out and we will have an age in seconds. To get years, we must divide by the number of seconds in a year. If we make these simple changes in units, the age of the universe in years is approximately 10^{12} divided by H in its normal units, km/sec/Mpc.

$$T \approx \frac{1}{H} \times 10^{12} \text{ years}$$

This is known as the **Hubble time.** For example, if H is 70 km/sec/Mpc, then the age of the universe can be no older than $10^{12}/70$ which equals 14 billion years.

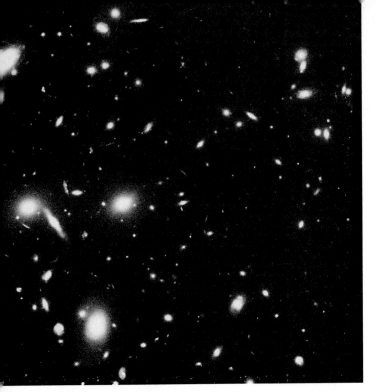

FIGURE 19-19

Long time exposures with the Hubble Space Telescope reveal that the sky is filled with distant clusters of galaxies. The look-back time to this cluster is two-thirds the age of the universe, so we see these galaxies as they were only a few billion years after the big bang. Such clusters are crowded with galaxies of all types, but there were once many more spiral galaxies in these clusters than we see today. Furthermore, the spiral galaxies at large look-back times are often distorted by collisions, interactions, and mergers. All of this evidence suggests that the structure we see in the universe developed quickly after the big bang and has continued to evolve to the present day. (Alan Dressler, Carnegie Institution, and NASA)

When we calculate the Hubble time, we are assuming the galaxies have receded at a constant velocity since the big bang, but we know that gravity slows the expansion of the universe. Then the galaxies must have traveled faster in the past, and the Hubble time is just an upper limit. The universe is no older than one Hubble time, but it could be younger. To know how much younger, we must know the density of the universe. If the density exactly equals the critical density (if the universe is flat), then the true age equals two-thirds of a Hubble time $\left(\frac{2}{3}\frac{1}{H}\right)$. If the density of the universe is less than the critical density, the universe is older; and if the density is higher, the universe is younger.

Some astronomers measured H and obtained numbers near 50 km/sec/Mpc, which yields a Hubble time of 20 billion years, and two-thirds of that would be 13.3 billion years. Other astronomers used other methods to measure H and obtained numbers near 100 km/sec/Mpc, which suggests the Hubble time is 10 billion years and

two-thirds of that is only 6.6 billion years. Although these results mark the extremes, most observations suggest a Hubble constant and an age near the middle of these ranges. The Hubble constant is uncertain because it is hard to measure the distances to galaxies accurately, so perhaps it isn't surprising that different estimates do not agree. It is reassuring, however, that the most recent results seem to be converging.

Another reason H is uncertain is the motions of the galaxies themselves. Our Milky Way Galaxy and the Local Group of galaxies in which we live seem to be falling toward the massive Virgo and Hydra–Centaurus galaxy clusters, which themselves appear to be falling toward a more distant concentration of galaxies called the Great Attractor. These streaming motions can confuse attempts to measure H, so astronomers must try to correct their observations for streaming motions.

Controversy rages because astronomers have strong evidence that the universe is older than the Hubble constant suggests. The oldest globular star clusters in our galaxy are 11 to 14 billion years old. One study suggested that the universe must be at least 17 billion years old to allow time for nucleosynthesis in massive stars to build the present abundance of heavy elements. It is obviously impossible for the universe to be younger than its oldest stars, so astronomers are struggling to measure the Hubble constant and the density of the universe. If H is larger than 70 or 80 and if the universe is flat, as inflationary cosmology predicts, then there is something dramatically wrong with our estimates of ages.

The Hubble Space Telescope and the new generation of very large telescopes are giving astronomers more accurate ways to measure the distance to galaxies and, thus, more accurate values for the Hubble constant. One study used the Hubble Space Telescope to locate Cepheid variable stars in galaxies within the Virgo cluster and produced a value for H of 80 km/sec/Mpc. Cepheid variables are accurate distance indicators (Chapter 16), but uncertainties in the location of the galaxies within the giant Virgo cluster introduce some uncertainty. Another study used supernova in distant galaxies to get distances and produced an H of 64 km/sec/Mpc. Astronomers studying brightness fluctuations in the gravitationally lensed quasar QSO 0957+561 (Figure 18-22) found H to be 63 km/sec/Mpc. Still another team used the Hubble Space Telescope to find globular clusters in a distant giant elliptical galaxy, and they obtained an H of 65 km/sec/Mpc.

Recent results suggest an H of about 65 km/sec/Mpc, and a Hubble time of a bit over 15 billion years. If the universe contains enough dark matter to close the universe and make it exactly flat, then the age of the universe is $\frac{2}{3}$ of 15 billion years, which equals 10 billion years. These numbers still conflict with the measured ages of the oldest star clusters, so astronomers continue to measure H more accurately, to measure the density of the universe, and to understand how stars are born, evolve, and die. Only when we solve all of these problems, can we be sure how old our universe is.

The Origin of Structure

Modern cosmologists are working to understand how the structure in the universe could have formed from the gas of the big bang. By *structure* we mean not only individual galaxies, but also the clusters and superclusters, plus the vast clouds of galaxies such as the Great Wall (Figure 19-12). The farther we look into space, the more clusters of galaxies we see (Figure 19-19); the wider the area we survey, the larger the structures we discover.

This structure is a problem for cosmologists because the primordial background radiation is extremely uniform and thus the gas of the big bang must have been extremely uniform at the time of recombination. If the universe was so uniform when it was young, how did it get to be so lumpy now?

The problem of the origin of structure has developed slowly since the 1960s, when the background radiation was first discovered. Then the most distant galaxies visible had red shifts no greater than 1. These galaxies had look-back times no greater than 4 billion years, so it was not difficult to imagine how the uniform gas of the big bang could eventually form galaxies. That is, there seemed to be plenty of time between recombination and the origin of galaxies.

Through the 1970s and 1980s, however, that age of mystery narrowed. Astronomers recognized quasars as the active nuclei of galaxies much farther away than 4 billion ly. Galaxies at such great distances are almost beyond detection, but the brilliant quasars assure us that such galaxies exist (Figure 19-20). These galaxies must have formed early in the history of the universe. At present, the most distant quasars have look-back times that are 93 percent of the age of the universe. Thus, the age of mystery when the galaxies formed must be less than 7 percent of the age of the universe.

The discovery of the Great Wall and other immense structures makes the problem more puzzling. If it is difficult to explain how galaxies formed quickly from the uniform gas of the big bang, it is even more difficult to explain vast walls.

How did all of this structure originate? The universe had to be very uniform during recombination and then clump quickly to form the structure we see. Cosmologists now work on a two-pronged solution: dark matter provides the gravitation, and galaxy seeds provide nuclei around which structure grows.

We have seen clear evidence that dark matter must exist, and cosmologists are now trying to build models that show how the gravitation of dark matter could cause the matter of the early universe to clump quickly into large structures. One kind of dark matter is called hot dark matter. If it exists, it would consist of particles that travel at nearly the speed of light, such as neutrinos with nonzero masses. Cold dark matter, on the other hand,

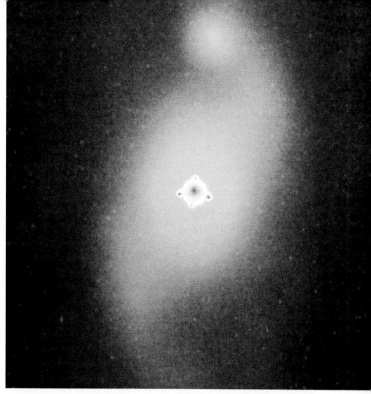

FIGURE 19-20

We can contrast a galaxy and a quasar in this image of the Einstein Cross, a gravitational lens. The galaxy is yellow in this computer-enhanced image, and the core of the galaxy is almost exactly aligned with a distant quasar. Four gravitational images of the quasar are visible bracketing the core of the galaxy. At a distance of 8 billion ly, the quasar is 20 times farther away than the galaxy. The existence of quasars at such great distances assures us that galaxies formed early in the history of the universe. (NASA)

would consist of slow-moving particles such as WIMPs, axions, and similar as yet undetected particles. Cold dark matter, because its particles travel more slowly, seems better able to clump together quickly.

Galaxy seeds around which this clumping occurs may be defects in the fabric of space-time. As the forces of nature separate from one another in what cosmologists call a phase transition, it is possible that defects occur between regions with different properties. We might compare such defects with the flaws we see in ice cubes. When the water freezes, beginning at the outer edges, different parts of the ice cube have different crystal orientations, and as the water freezes inward toward the center, these different regions eventually meet and form defects that we see inside the ice cube as lines and sheets. Defects in space-time would possess strong gravitational fields and could act as galaxy seeds around which dark matter clumps to create galaxies, clusters, filaments, walls—all of the structure we see.

Most cosmologists believe that these phase transitions occurred very early (as in the inflationary universe theory), so any resulting structure should be visible as irregularities in the background radiation. Until 1992, the best observations could detect no irregularities in the

FIGURE 19-21

This all-sky map based on data from the COBE satellite shows the distribution of the primordial background radiation. Red areas are slightly warmer and blue areas slightly cooler than the average. These variations suggest that the gas of the big bang was not perfectly uniform. Some regions were slightly denser than other regions. The gravitational influence of dark matter could have forced denser regions to grow into the clouds of galaxies we see today. Thus, the structure we see in the form of clusters, filaments, and voids could have had its beginning in the irregularities in the hot gas of the big bang. (NASA/Goddard Space Flight Center)

background radiation, and that made astronomers wonder where the structure in the universe could have come from. But in the spring of 1992, astronomers announced that the data from the COBE satellite reveal irregularities in the background radiation of about 0.0006 percent (Figure 19-21). This discovery is of tremendous importance because it tells us that the seeds of cosmic structure were present as irregularities in the big bang at the time of recombination. The great clouds of galaxies we see today may have had their origin when a moment of inflation created defects in space-time that grew because of the gravitational influence of dark matter. Furthermore, the deepest photographs from the Hubble Space Telescope show that galaxies formed quickly after the big bang but then evolved through mergers to produce the elliptical and spiral galaxies we see today (Figure 19-22).

The three questions we have reviewed here—the inflationary universe, the age of the universe, and the origin of structure—in no way invalidate the big bang theory. They are refinements that extend our understanding of how our universe began and how it has evolved from a first moment we call the big bang.

CRITICAL INQUIRY

How does dark matter affect the growth of structure?

According to the inflationary theory, an instant after the big bang the forces of nature separated and caused the universe to inflate suddenly by a tremendous factor. That same inflation may have created defects in space-time that would have a gravitational field, and matter may have collected around such defects to begin the formation of structure, meaning galaxies, clusters, sheets, filaments, and voids. However, the visible matter in the universe does not appear to have enough gravity to have pulled itself together quickly, so the dark matter may have been important in causing the first contractions. Hot dark matter is made up of such rapidly moving particles that it would not help form condensations quickly, so some theorists believe cold dark matter is the answer. Once galaxies and galaxy clusters formed, the presence of dark matter stabilized them as the galaxies continued to collide, merge, and evolve. One of the unknown fac-

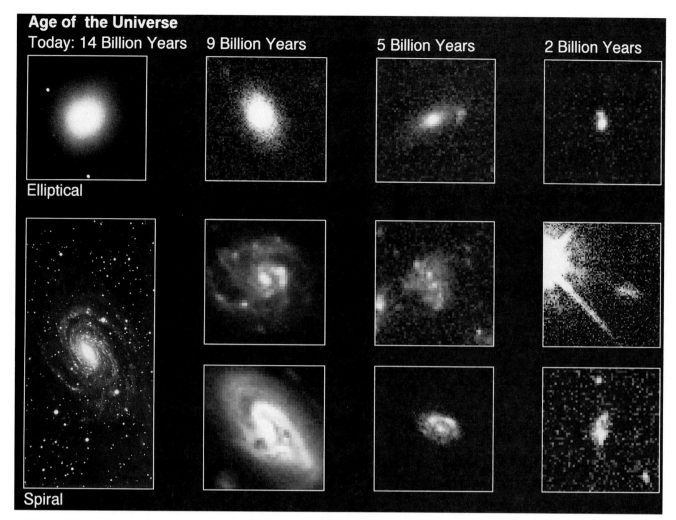

Age of the Universe
Today: 14 Billion Years | 9 Billion Years | 5 Billion Years | 2 Billion Years

Elliptical

Spiral

FIGURE 19-22

Because modern telescopes such as the Hubble Space Telescope can photograph very distant galaxies, we can look back in time and see that the galaxies formed quickly after the big bang and have evolved since then. Images at the left represent an elliptical and a spiral galaxy as they are today. Images to the right show typical galaxies when the universe was younger than it is today. Images at the far right show galaxies only 2 billion years after the big bang. Elliptical galaxies appear to have formed quickly and to have taken on a modern appearance, but early spiral galaxies are often distorted, further evidence that galaxies have evolved as the universe has aged. (A. Dressler, Carnegie Institutions of Washington; M. Dickinson, STScI; D. Macchetto, ESA/STScI; M. Giavalisco, STScI; and NASA)

tors in this story is the age of the universe. The best estimates of the Hubble constant seem to make the universe too young for the matter to have contracted to form the structure we see.

Cosmology is definitely a work in progress, but astronomers have made dramatic progress in understanding the history of the universe, and they have done so by comparing observation and theory. As an example, how would you compare observation with the inflationary theory to explain the horizon problem? ∎

Although we have traced the origin of the universe, the origin of the elements, and the birth and death of stars, we have left out one important class of objects—

planets. How do the earth and other planets fit into this grand scheme of origins? That is the subject of the next unit. ∎

∎ *Summary*

The fact that the night sky is dark shows that the universe is not infinitely old. If it were infinite in extent and age, every spot on the sky would glow as brightly as the surface of a star. This problem, known incorrectly as Olbers' paradox, illustrates how important assumptions are in cosmology.

The basic assumptions of cosmology are homogeneity, isotropy, and universality. Homogeneity says the matter is spread uniformly through the universe. Isotropy says the universe looks the same in any direction. Both deal only with general features. Universality assumes that the laws of physics known on the earth apply everywhere. In addition, the cosmological principle asserts that the universe looks the same from any location.

The Hubble law implies that the universe is expanding. Tracing this expansion backward in time, we come to an initial high-density state commonly called the big bang, the explosion that started the expansion. From the Hubble constant, we can conclude that the expansion began 10–20 billion years ago.

The universe can be infinite or finite, but it cannot have an edge or a center. Such regions would violate the cosmological principle. Instead, we assume that the universe occupies curved space-time and thus could be finite but unbounded. Depending on the average density of the universe, it could be open, flat, or closed. Measurements of the density of the universe are uncertain because of the presence of dark matter, which we cannot detect easily.

During the first few minutes of the big bang, about 25 percent of the matter became helium, and the rest remained hydrogen. Very few heavy atoms were made. As the matter expanded, instabilities caused the formation of clusters of galaxies, which are still receding from one another.

Whether the universe expands forever or slows to a stop and falls back depends on the amount of matter in the universe. If the average density is greater than the critical density of 4×10^{-30} g/cm^3, it will provide enough gravity to slow the expansion and force the universe to collapse. The collapse will smash all matter back to a high density from which a new big bang may emerge. Such a universe is termed closed because the curvature of space is positive and it has a finite volume. If the average density is less than 4×10^{-30} g/cm^3, gravity will be unable to stop the expansion, and it will continue forever. Such a universe is termed open because the space curvature is negative.

The inflationary universe is a theory that combines quantum mechanics, general relativity, and cosmology. It predicts that the universe underwent a sudden inflation when it was only 10^{-35} second old. This seems to explain why the universe is so flat and why the primordial background radiation is so isotropic.

We can estimate an upper limit to the age of the universe from the Hubble constant, but we need to know the average density of the universe to find a true age. Also, the value of the Hubble constant is not precisely known. It is believed to be 50–100 km/sec/Mpc. If it is more than about 70 km/sec/Mpc, the age of the universe will be less than the ages determined for some star clusters.

Throughout the universe we see structure, ranging from galaxies up to vast clouds of galaxy clusters such as the Great Wall. This structure must have formed from the clouds of gas created during the big bang. Yet the primordial background radiation is very uniform, and this tells us that the gas of the big bang was also uniform. How did this uniform gas coalesce to form the vast structures we see around us? Data from the COBE satellite show that the primordial background radiation is not perfectly uniform. Its tiny irregularities suggest that the gas of the big bang was not perfectly uniform and could have grown around galaxy seeds, centers of condensation that may be related to defects in space-time, to form the structure we see today.

■ New Terms

cosmology

Olbers' paradox

observable universe

homogeneity

isotropy

universality

cosmological principle

closed universe

flat universe

open universe

big bang theory

primordial background
   radiation

steady state theory

recombination

critical density

oscillating universe theory

baryon, baryonic matter

MACHOs

flatness problem

horizon problem

inflationary universe

grand unified theories
   (GUTs)

Hubble time

galaxy seeds

■ Questions

1. Would the night sky be dark if the universe was only 1 billion years old and was contracting instead of expanding? Explain your answer.

2. How can we be located at the center of the observable universe if we accept the Copernican principle?

3. Why can't an open universe have a center? Why can't a closed universe have a center?

4. What evidence do we have that the universe is expanding? that it began with a big bang?

5. Why couldn't atomic nuclei exist when the universe was younger than 3 minutes?

6. Why is it difficult to determine the present density of the universe?

7. How does the inflationary universe theory resolve the flatness problem? the horizon problem?

8. If the Hubble constant is really 100 km/sec/Mpc, much of what we understand about the evolution of stars and star clusters must be wrong. Explain why.

9. Why do we conclude that the universe must have been very uniform during its first million years?

10. What is the difference between hot dark matter and cold dark matter? What difference does it make to cosmology?

■ Discussion Questions

1. Do you think Copernicus would have accepted the cosmological principle? Why or why not?

2. What observations would you recommend that the Hubble Space Telescope make to help us choose among an open, flat, or closed universe?

■ Problems

1. Use the data in Figure 19-2 to plot a velocity–distance diagram, find H, and determine the approximate age of the universe.

2. If a galaxy is 8 Mpc away from us and recedes at 456 km/sec, how old is the universe, assuming that gravity is not slowing the expansion? How old is the universe if it is flat?

3. If the temperature of the big bang had been 1,000,000 K at the time of recombination, what wavelength of maximum would the primordial background radiation have as seen from the earth?

4. If the average distance between galaxies is 2 Mpc and the average mass of a galaxy is 10^{11} solar masses, what is the average density of the universe? (HINTS: The volume of a sphere is $\frac{4}{3}\pi r^3$. The mass of the sun is 2×10^{33} g.)

5. If the value of the Hubble constant were found to be 60 km/sec/Mpc, how old would the universe be if it were not slowed by gravity? if it were flat?

6. Hubble's first estimate of the Hubble constant was 530 km/sec/Mpc. If his distances were too small by a factor of 7, what answer should he have obtained?

7. What is the maximum age of the universe predicted by Hubble's first estimate of the Hubble constant?

8. High-resolution radio observations show that a galaxy contains a supernova remnant that is expanding 0.00004 second of arc per year, and spectra show that the remnant is expanding at 10,000 km/sec. How far away is the galaxy?

■ *Critical Inquiries for the Web*

1. Will the universe go on expanding forever? Search the Web for information on recent investigations that shed light on the question of the density of matter in the universe. What preductions do these studies make about the fate of the universe? What kinds of observtions were necessary to make these predictions?

2. Our ability to view galaxies at great distances allows us to see the universe as it was billions of years ago. Look for images of galaxies at the frontiers of the observable universe and compare the types, shapes, relative sizes, and distributions of early galaxies to galaxies that are closer to us in space and time. Were galaxies in the early universe similar to today's galaxies?

Go to the Wadsworth Astronomy Resource Center (www.wadsworth.com/astronomy) for critical thinking exercises, articles, and additional readings from InfoTrac College Edition, Wadsworth's online student library.

Chapter Twenty

THE ORIGIN OF THE SOLAR SYSTEM

What place is this?

Where are we now?

CARL SANDBURG

Grass

GUIDEPOST

The preceding 19 chapters have described the origin, structure, and evolution of the physical universe, but they have neglected one important class of objects—planets. In this chapter, we can look back on what we have learned and find our place in the universe. We live on a planet. What does that mean? Where do we fit?

Each time we have studied a new object, we have asked how it formed and how it evolved to its present state. We have done that with stars and galaxies and the universe, so it is appropriate to begin our discussion of the solar system by considering its origin.

Another reason for discussing the origin of the solar system here is to give ourselves a framework into which we can fit the planets as we discuss them in the chapters that follow. Without a theoretical framework, science is nothing but a jumble of facts. With a good framework in hand, we will be ready to make sense of the solar system through the next six chapters. ■

Microscopic creatures live in the roots of your eyelashes. Don't worry. Everyone has them, and they are harmless.* They hatch, fight for survival, mate, lay eggs, and die in the tiny spaces around the roots of our eyelashes without doing us any harm. Some live in renowned places—the eyelashes of the Queen of England, of a glamorous movie star, of a famous athlete—but the tiny beasts are not self-aware; they never stop to say, "Where are we?" Humans are much more intelligent; we can wonder where we are in the universe and how we came to be here.

In this chapter, we begin exploring our solar system: its sun, nine planets, and some scattered gas, dust, and debris. Our planetary system occupies no more than 10^{-30} of the volume of the observable universe. If the words in this book represent the volume of the universe, the solar system is less than a trillionth of a trillionth of one letter. However, the significance of the solar system—for us at least—far exceeds its relative size.

There are many reasons for studying the solar system. Not the least of these reasons is that understanding how planets form and evolve teaches us more about our own planet. Our planet now has so many people that we are altering our environment in dramatic ways, and it is possible that we could make Earth uninhabitable. Understanding the origins of the barren wastes of Mars and the sulfuric acid fogs of Venus may teach us how to preserve our more comfortable climate.

Astronomers need no justification for studying the solar system, for it is the only collection of nonluminous bodies in the universe that we can study in detail. There are planets orbiting other stars, but we have been able to detect only a few, and we cannot observe any of them in detail. Yet there are probably more planets in the universe than stars, so we should study this class of objects with care. Thus, the objects in our solar system are representatives of an important component of the universe.

An intriguing reason for studying the planets of our solar system is the search for life beyond the earth. As if enchanted, some matter on the earth's surface lives and is aware of its existence. To search for other living beings, we must examine the surfaces of planets—stars are too hot for the evolution of life forms, and space is too cold. As far as we know, life seems to require the moderate conditions found only on the surfaces of some planets. To focus our search, we must understand how planets form and evolve.

Above all, we study the solar system because it is our home in the universe. Because we are an intelligent species, we have the right and the responsibility to wonder what we are and where we are. Our kind have inhabited this solar system for at least a million years, but only within our lifetime have we begun to understand what a solar system is. Like sleeping passengers on a train, we waken, look out at the passing scenery, and mutter to ourselves: "What place is this? Where are we now?"

* *Demodex folliculorum* has been found in 97 percent of individuals with healthy skin.

20-1 THE GREAT CHAIN OF ORIGINS

We are linked through a great chain of beginnings that leads backward through time to the first instant of the big bang. The gradual discovery of the links in that chain is one of the most exciting adventures of the human intellect. In earlier chapters, we discussed some of that story: the origin of the universe, the galaxies, the stars, and the elements. Here we will explore further to consider the formation of planets.

Early Hypotheses

The earliest theories for the origin of the earth are myths and folktales that go back beyond the beginning of recorded history. In addition, almost all religions contain an account of the origin of the world. It was not until about the time of Galileo that philosophers began searching for rational explanations for natural phenomena. While people like Copernicus, Kepler, and Galileo tried to find logical explanations for the motions of the earth and other planets, other philosophers began thinking about the origin of the planets.

The first rational theory for the origin of the earth was proposed by the French philosopher and mathematician René Descartes (1596–1650). Since he lived and wrote before the time of Newton, Descartes did not have the concept of gravitation as the dominant force in the universe. Rather, he believed that force was communicated by contact between bodies and that the entire universe was filled with vortices of whirling particles. In 1644, he proposed that the sun and planets formed when a large vortex contracted and condensed. Thus, his hypothesis explained the general properties of the solar system known at the time.

A century later, in 1745, the French naturalist Georges-Louis de Buffon (1707–1788) proposed the alternative hypothesis that the planets were formed when a passing star collided with or passed close to the sun and pulled matter out of the sun and the star. The matter condensed to form the planets, and they fell into orbit around the sun (Figure 20-1a). This **passing star hypothesis** was popular off and on for two centuries, but it contains serious flaws. First, stars are very small compared to the distances between them, and thus they collide very infrequently. In the entire history of our galaxy, stars have probably collided only once. More important, the gas pulled from the sun and the star would be much too hot to condense to make planets. Even if planets formed, they would not go into stable orbits.

The hypotheses of Descartes and Buffon fall into two broad categories. Descartes' hypothesis is **evolutionary** in that it calls upon common, gradual events to produce the sun and planets. If it is correct, stars with planets are very common. Buffon's hypothesis, on the

a

b

FIGURE 20-1

(a) The passing star hypothesis was catastrophic. It proposed that the sun was hit by or had a very close encounter with a passing star and that matter torn from the sun and the star formed planets orbiting the sun. The theory is no longer accepted. (b) Laplace's nebular hypothesis was evolutionary. It suggested that a contracting disk of matter conserved angular momentum, spun faster, and shed rings of matter that formed planets. The hypothesis could not explain the sun's low angular momentum and was never fully accepted, even though the modern theory shares some of its features.

other hand, is **catastrophic.** It calls upon unlikely, sudden events to produce the solar system, and thus it implies that solar systems are very rare. While our imaginations may be tempted by colliding stars, modern hypotheses for the origins of the planets are evolutionary, with, as we will see, a few astonishing catastrophes thrown in (Window on Science 20-1).

Our modern theory of the origin of the solar system had its true beginning with Pierre-Simon de Laplace (1749–1827), the brilliant French astronomer and mathematician. In 1796, he combined Descartes' vortex with Newton's gravity to produce a model of a rotating cloud of matter contracting under its own gravitation and flattening into a disk—the **nebular hypothesis.** As the disk

Evolution and Catastrophe

Many theories in science can be classified as either evolutionary, in that they involve gradual processes, or catastrophic, in that they depend on specific, unlikely events. Scientists have generally preferred evolutionary theories. Nevertheless, catastrophic events do occur.

Something in us prefers catastrophic theories, perhaps because we like to see spectacular violence from a safe distance, which may explain the success of movies that include lots of car crashes and explosions. Also, cataclysmic theories resonate with Old Testament accounts of catastrophic events and special acts of creation. Thus, we have an understandable interest in catastrophic theories.

Nevertheless, most scientific theories are evolutionary. Such theories do not depend on unlikely events or special acts of biblical creation, and thus they are open to scientific investigation. For example, geologists much prefer theories of mountain building that are evolutionary, with the mountains being pushed up slowly as centuries pass. All the evidence of erosion and the folding of rock layers show that the process is gradual. Because most such natural processes are evolutionary, scientists sometimes find it difficult to accept any theory that depends on catastrophic events.

We will see in this and later chapters that catastrophes do occur. The earth, for example, is bombarded by debris from space, and some of those impacts are very large. As you study astronomy or any other natural science, notice that most theories are evolutionary but that we must allow for the possibility of unpredictable catastrophic events. ■

---

grew smaller, it had to conserve angular momentum and spin faster and faster. Laplace reasoned that, when it was spinning as fast as it could, the disk would shed its outer edge to leave behind a ring of matter. Then the disk could contract further, speed up again, and leave another ring. Thus, the contracting disk would leave behind a series of rings, each to become a planet circling the newborn sun at the center of the disk (Figure 20-1b).

According to the nebular hypothesis, the sun should be spinning very rapidly, or, to put it another way, the sun should have most of the angular momentum of the solar system. (Recall from Chapter 5 that angular momentum is the tendency of a rotating object to continue rotating.) As astronomers studied the planets and the sun, however, they found that the sun rotated slowly and that the planets moving in their orbits had most of the angular momentum in the solar system. In fact, the rotation of the sun contains only about 0.3 percent of the angular momentum of the solar system. Because the nebular hypothesis could not explain this **angular momentum problem,** it was never fully successful, and astronomers toyed with various versions of the passing star hypothesis for over a century.

In contrast to the astronomy of earlier centuries, 20th-century astronomy has applied modern physics to the stars and galaxies and so has illuminated our origins. We can now trace that great chain of origins that began with the big bang and led to the matter of which we are made. We will review the story of the origin of matter, and then we will add to it the story of how that matter formed our world.

A Review of the Origin of Matter

The mass of which your body is made came into existence within minutes of the big bang explosion. The protons, neutrons, and electrons were created by the energy present as gamma rays; by the time the universe was only 3 minutes old, these subatomic particles had combined to form atomic nuclei.

About 25 percent of the mass became helium, an inert chemical element, which is not present in your body. Most of the matter in the big bang, however, remained in the form of single protons, the nuclei of hydrogen atoms. Today roughly 75 percent of the mass in the universe is still hydrogen, including the hydrogen atoms in the water molecules in your body. These hydrogen nuclei have survived unchanged since the big bang.

Within a billion years or so after the big bang, matter began to collect into galaxies and form stars. Nuclear fusion reactions inside the stars arranged the protons and neutrons into new partnerships and created heavier elements. The carbon, nitrogen, and oxygen so important throughout your body, the calcium in your bones, and all of the most common heavy elements up to the iron in your blood have been cooked up by nuclear fusion inside many generations of stars.

The elements heavier than iron were created by rapid nuclear fusion, which occurred during supernova explosions. These elements are much less common than elements lighter than iron, but some of them, such as the iodine in your thyroid gland, play important roles in your body chemistry. The supernova explosions also spread these newly created elements back into the interstellar medium, where they could be incorporated in the formation of new stars. Indeed, the expanding supernova remnants sometimes trigger the formation of new stars.

The sun is one of the stars that formed from these clouds of gas enriched in heavy elements. How a tiny fraction of that matter came together to form the earth, the oceans, and your body is a story we will explore in the rest of this book. As we explore, we must remember the great chain of origins that produced the chemical elements. As geologist Preston Cloud remarked, "Stars have died that we might live."

FIGURE 20-2

(a) Because the solar nebula was rotating, (b) it contracted into a disk, and (c) the planets formed with orbits lying in nearly the same plane.

The Solar Nebula Hypothesis

The stars gave birth to the heavy atoms of which our world is made, and the modern theory of the origin of the planets is based on star formation. The **solar nebula hypothesis** proposes that the planets were formed from the disk of gas and dust that surrounded the sun as it formed (Figure 20-2).

As stars form in contracting clouds, they remain surrounded by cocoons of dust and gas, and the rotation of the cloud causes that dust and gas to form a spinning disk around the protostar. When the center of the star grows hot enough to ignite nuclear reactions, its surface quickly heats up, becomes more luminous, and blows away the gas and dust cocoon.

Modern theories suppose that planets form in the rotating disks of gas and dust around young stars. Infrared observations of T Tauri stars, for instance, show that some are surrounded by gas clouds rich in dust, and spectra show that these stars are blowing away their nebulae

at speeds up to 200 km/sec. Astronomers find many very young stars ejecting jets in opposite directions, evidence that the outflow was focused by a disk around the star. Our own planetary system probably formed in such a disk-shaped cloud around the sun. When the sun became luminous enough, the remaining gas and dust were blown away into space, leaving the planets orbiting the sun. Notice that the solar nebula hypothesis also suffers from the angular momentum problem. If we are to adopt the theory, we must understand why the sun now rotates so slowly, a puzzle we will consider later in this chapter.

If the solar nebula hypothesis is correct, then our earth and the other planets of the solar system formed billions of years ago as the sun condensed from the interstellar medium. If that is true, then planets form as a by-product of star formation, and most stars should have planets. Before we go further, we should ask if there is any way to check this prediction against the evidence. Can we see planets orbiting other stars?

Extra-Solar Planets

Until recently, the only planets astronomers could see were those that circle the sun. Nevertheless, astronomers had good theoretical reasons for believing that planets are common. Now new observations can detect planets orbiting other stars—extra solar planets.

A planet orbiting another star is difficult to detect directly. Planets are small and shine by reflected light, so even a planet as big as Jupiter orbiting the nearest star (4 ly away) would be only about 22nd magnitude. Also, it would lie only 4 arc seconds from its star. Thus, seeing this faint object so close to its star would be like looking into a searchlight and detecting a bug crawling on the bulb. Current telescopes can't image extra-solar planets directly.

But astronomers are confident that such planets are common because theory predicts that planets are born from disks of gas and dust around young stars; such disks have been found.

These disks have been detected at visual and radio wavelengths, but they are most commonly studied in the infrared. The amount of infrared radiation emitted by a cool object, such as a planet, depends on its surface area. A few planets orbiting a distant star would have relatively small surface area and emit little infrared radiation. But any dust left over from the formation of the planets can have an enormous surface area. A single chunk of rock 10 km in diameter, for example, ground into dust would have twice the surface area of the sun. Thus, the planets may be invisible, but the dust left over from the formation of the planets emits large amounts of infrared radiation. Stars with more infrared radiation than expected must be surrounded by dust disks. Such dust disks with little or no gas have been found around stars Beta Pictoris, Alpha Lyrae (Vega), Alpha Piscis Austrini, and Epsilon Eridani, and are suspected around over 100 more.

FIGURE 20-3

Planets are believed to form in disks of gas and dust around young stars. This false-color image of Beta Pictoris reveals an edge-on dust disk centered on the star (hidden behind the black strip at center). The dust disk is much larger than our solar system, as shown by the orbit of Pluto drawn to scale. The inner part of the disk (below) shows a number of asymmetries which could be produced by the gravitational effects of planets orbiting Beta Pictoris. (Al Schultz, CSC/STScI; Sally Heap; GSFC/NASA; and NASA)

The Beta Pictoris disk, for example, is about ten times the diameter of our solar system and contains roughly one-millionth of a solar mass of dust with a temperature of about 100 K. With a small screen in the telescope to block the light of the star, the nearly edge-on disk can be seen extending beyond the blocking screen (Figure 20-3a). Astronomers suspect that planet formation may have taken place in the inner part of the disk, because that region seems to contain much less dust. Also, a warp in the inner part of the disk suggests that at least one planet orbits there. Systems like Beta Pictoris may represent the last cold debris left behind when planets form. Our own solar system is believed to contain such debris in its outer regions.

Radio astronomers have found other disks that seem to be hotter and denser than the Beta Pictoris disk (Figure 20-3b). These radio disks are believed to be much younger systems in which planet formation may be taking place as we watch. Thus, observations reveal disks in different stages of planet formation.

If we could travel back in time and observe the sun and planets as they formed, we would probably see something that looks much like the Herbig–Haro object HH30 (Figure 20-4). A Hubble Space Telescope image reveals that it is a newly forming star surrounded by a disk that is nearly edge-on to our view. We see the top and bottom of the disk illuminated by the star, but the thick central part of the disk hides the star from direct view. The image also shows a jet of gas about 1000 AU long flowing away from the star. Such jets are typical of disk structures.

Observations made with the Hubble Space Telescope indicate that these planet-forming disks may be very common. When the telescope looked into the Orion

FIGURE 20-4

Herbig–Haro object HH30 is a newly forming star surrounded by a disk of gas and dust. This is much like the solar nebula that gave birth to the sun and planets. In HH30, we cannot see the star because it is hidden inside its edge-on disk, but we see the top and bottom of the disk illuminated by the star. We also see jets of gas flowing away from the star perpendicular to the disk. The left jet in this image is about 1000 AU long. (C. Burrows, STScI & ESA, WFPC 2 Investigation Definition Team, NASA)

Nebula, it found that 50 percent of the stars studied had disks of gas and dust (Figure 20-5). A young star is visible at the center of each disk, and astronomers estimate that the disks contain at least a few times the earth's mass in a region a few times larger in diameter than our solar system. The disks are visible because they are brightly illuminated by the hot stars in the Orion Nebula, and some of the disks are seen silhouetted against the bright nebula. The Orion star-forming region is only a few million years old, so it does not seem likely that planets could have formed in these disks yet. Nevertheless, the presence of so many disks is evidence confirming our theory that planetary systems are common in the universe.

Theory predicts that planets should be common, and observations of disks around protostars confirm the theory. Not until the 1990s, however, were extra-solar planets found. They can't be seen, but they can be detected by the motion of their star around the center of gravity of the system. Just as someone walking an excited dog gets tugged about, the star moves slightly as the planet tugs on it gravitationally.

FIGURE 20-5

This Hubble Space Telescope image of the central part of the Orion Nebula reveals that many young stars there are surrounded by disks of gas and dust (insets). Most of the disks glow brightly because they are illuminated by the hottest stars in the nebula, but a few, such as the dark disk seen nearly edge-on (lower inset), are silhouetted against the bright nebula. Note that the central star in the silhouetted nebula is visible by light scattered by dust above and below the nebula. This particular disk is the largest detected, with a diameter about 17 times that of our solar system. (C. R. O'Dell, Rice University, NASA; lower inset, Mark McCaughrean, Max-Planck Institute for Astronomy, C. R. O'Dell, and NASA)

Skeptical Storytellers

"Scientists are just a bunch of skeptics who don't believe in anything." That is a common complaint about scientists, but it misinterprets the fundamental characteristic of the scientist. Yes, scientists treat new discoveries with skepticism, but scientists do hold strong beliefs about how nature works, and their skepticism is the tool they use to test their beliefs. Ultimately, scientists try to tell stories about how nature works, and if those stories are to be correct, then every idea must be tested over and over.

When the discovery of a planet orbiting the star 51 Pegasi was first announced, astronomers were skeptical—not because they didn't expect other stars to have planets and not because they thought the observations were wrong, but because that is the way science works. Every observation is tested, and every discovery must be confirmed. Only an idea that survives many tests begins to be accepted as a scientific truth, what a scientist would call a law of nature.

This makes scientists seem like irritable skeptics to nonscientists, but among scientists it is not bad manners to ask, "Really, how do you know that?" or "Why do you think that?" or "What evidence do you have?"

The goal of science is to tell stories about how nature works, stories that are sometimes called theories or hypotheses. A story by John Steinbeck can be a brilliant work of art invented to help us understand some aspect of human nature, and another story, such as a TV script, can be little more than entertaining chewing gum for the mind. But a scientist tries to tell a story that is different in one critical way—it is the truth. To create such stories about nature, scientists cannot be artistic and create facts to fit their story, and they cannot invent a plot just to be entertaining. Every single link in the scientist's story must be based on evidence and clear, logical steps. Of course, evidence can be misunderstood, and logical slipups happen all too often. To test every aspect of the story, the scientist must be continuously skeptical. That is the only way to discover those scientific stories that help us understand how nature works.

Skepticism is not a refusal to hold beliefs. Scientists often believe sincerely in their theories and hypotheses once they have been tested over and over. Rather, skepticism is the tool scientists use to find those natural principles worthy of belief. ■

The first of these planets were revealed by Doppler shifts in the pulses from the spinning neutron star PSR 1257+12, but this seems a very special case. Astronomers are trying to understand how the remains of a dead star could have planets in such small orbits. These are extra-solar planets, but they don't seem typical.

In 1995, Doppler shifts in the spectrum of the G5 V star 51 Pegasi revealed a planet half the mass of Jupiter orbiting with a period of only 4.2 days. It is only 0.05 AU from its sun. Astronomers greeted the announcement with cheers and skepticism (Window on Science 20-2). Not until the observations were confirmed by a second observatory was the discovery widely accepted (Figure 20-6). An alternative hypothesis, that there is no planet and that the star is pulsating, was studied with care but abandoned. The planet seems very real.

Roughly 10 such planets have been identified orbiting other stars. For example, a planet over 6.8 Jupiter masses orbits the star 70 Virginis in an elliptical orbit and should have a moderate temperature that might permit liquid water. The star 47 Ursae Majoris has a planet over 2.4 Jupiter masses, and 16 Cygni B has a planet over 1.7 Jupiter masses. More such extra-solar planets will probably turn up in the future.

Most of these planets are massive and have short period orbits. We should expect this because low-mass planets and planets with long periods are difficult to detect; we only notice the massive planets with short periods. Some of these massive planets are so close to their suns that they have been dubbed hot Jupiters. At first, astronomers wondered how such massive planets

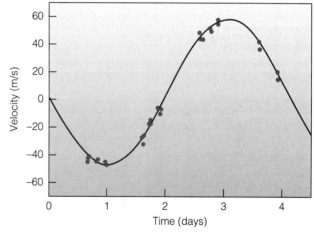

FIGURE 20-6

The radial velocity of the sunlike star 51 Pegasi shows Doppler motions with a period of 4.2 days and an amplitude of only 50 meters per second. This wobble in the star is apparently produced by a planet of about half the mass of Jupiter orbiting only 0.05 AU from the star. This is the first planet found to orbit a normal star other than our sun. (Adapted from data by Geoff Marcy and Paul Butler)

could have formed so close to their stars, but it now seems that they might have formed farther out and then spiraled inward as they swept up mass from a disk of planetesimals during planet formation. Interactions between massive planets could also send a massive planet into a small orbit and might also explain the elliptical

Reconstructing the Past

Scientists often face problems in which they must reconstruct the past. Some of these reconstructions are obvious, such as an archaeologist excavating the ruins of a burial tomb, but others are less obvious. In each case, success requires the interplay of hypotheses and evidence to re-create a past that no longer exists.

The reconstruction of the past is obvious when we use the chemical abundance of stars to reconstruct the story of the formation of our galaxy, but a biologist studying a centipede is also reconstructing the past. How did this creature come to have a segmented body with so many legs? How did it develop the metabolism that allows it to move quickly and hunt prey? Although the problem might at first seem to be one of mere anatomy, the scientist must reconstruct an environment that no longer exists.

Geologists are fond of saying "the present is the key to the past," and that is true for any scientific problem that requires us to reconstruct the past. Because the past is over, we cannot observe it directly, so we must use the evidence of the present to test hypotheses about the nature of the past.

The grand interplay of evidence and theory is the distinguishing characteristic of science, and nowhere do we see it more clearly than when scientists reconstruct the past. ■

---

orbits we see. Thus, the theory of planet formation seems able to account for these hot Jupiters.

Extra-solar planets have been found, and the theory that planets form as a by-product of star formation is confirmed. No doubt we will learn more about our own solar system as we discover more extra-solar planets.

CRITICAL INQUIRY

Why does the solar nebula hypothesis imply that planets are common?

If the solar nebula hypothesis is correct, the planets of our solar system formed from the disk of gas and dust that surrounded the sun as it condensed from the interstellar medium. If that is true, then it should be a common process. Most stars should form with disks of gas and dust around them and so might have planets.

The older, close encounter hypothesis suggested that the planets formed when the sun and another star interacted. Such close encounters are very rare and in the history of our galaxy may have occurred only once. If our solar system formed in this way, then planetary systems would be very rare indeed. Ours might be the only planetary system in our galaxy.

This suggests a way we could test our theories. If we look at a large number of stars and find lots of planets, the solar nebula theory is confirmed, and the close encounter hypothesis is contradicted. What observational evidence do we have on this issue? ■

So far, only a few other stars are known to have a planet. We can expect more such worlds to be discovered, and that raises the question of just how planets form. To answer that question, we begin by surveying our own solar system to discover the distinguishing characteristics of planetary systems.

20-2 A SURVEY OF THE SOLAR SYSTEM

To test our theories, we must search the present solar system for evidence of its past (Window on Science 20-3). In this section, we survey the solar system and compile a list of its most significant characteristics, potential clues to how it formed.

We begin with the most general view of the solar system. It is, in fact, almost entirely empty space (Figure 20-7). Imagine that we reduce the solar system until Earth is the size of a grain of table salt, about 0.3 mm (0.01 in.) in diameter. The moon is a speck of pepper about 1 cm (0.4 in.) away, and the sun is the size of a small plum 4 m (13 ft) from Earth. Mercury, Venus, and Mars are grains of salt. Jupiter is an apple seed 20 m (66 ft) from the sun, and Saturn is a smaller seed over 36 m (120 ft) away. Uranus and Neptune are slightly larger than average salt grains, and Pluto, the farthest planet (on average), is a speck of pepper over 150 m (500 ft) from the central plum. We would need a powerful microscope to detect the asteroids and the nuclei of comets. These tiny specks of matter scattered around the sun are the remains of the solar nebula.

Revolution and Rotation

The planets revolve around the sun in orbits that lie close to a common plane. The orbit of Mercury, the closest planet to the sun, is tipped 7° to Earth's orbit, and Pluto's orbit is tipped 17.2°. The rest of the planets' orbital planes are inclined by no more than 3.4°. Thus, the solar system is basically disk-shaped.

The rotation of the sun and planets on their axes also seems related to this disk shape. The sun rotates with its equator inclined only 7.25° to Earth's orbit, and most of the other planets' equators are tipped less than 30°. The rotations of Venus, Uranus, and Pluto are peculiar, however. Venus rotates backward compared with the other

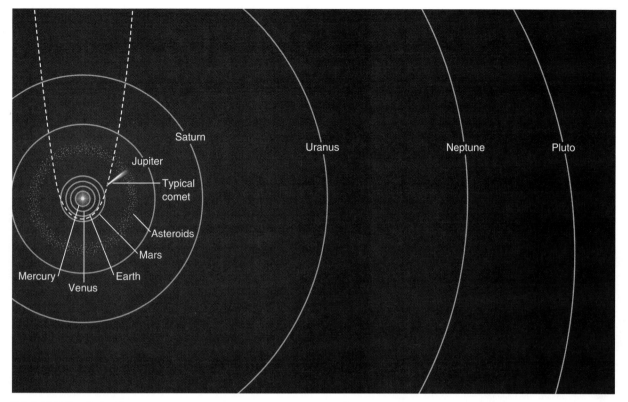

FIGURE 20-7

The solar system. The radii of the orbits are drawn to scale here. Note the eccentricity of Pluto's orbit. At this scale, only the sun would be visible. The planets are too small to be seen.

planets, and both Uranus and Pluto rotate on their sides (with their equators almost perpendicular to their orbits). We will discuss these planets in detail in Chapters 23 and 25, but later in this chapter we will try to understand how they could have acquired their peculiar rotations.

Apparently, the preferred direction of motion in the solar system—counterclockwise as seen from the north—is also related to its disk shape. All the planets revolve counterclockwise around the sun, and with the exception of Venus, Uranus, and Pluto they rotate counterclockwise on their axes.* Thus, with only a few exceptions, most of which are understood, revolution and rotation in the solar system follow a disk theme.

Two Kinds of Planets

Perhaps the most important clue we have to the origin of the solar system is the division of the planets into two categories: terrestrial, or Earth-like, and Jovian, or Jupiter-like (Figure 20-8). The **terrestrial planets** are small, dense, rocky worlds with less atmosphere than the Jovian planets. The terrestrial planets—Mercury, Venus,

*Recall from Chapter 2 that astronomers distinguish between the words *revolve* and *rotate*. A planet revolves around the sun but rotates on its axis.

FIGURE 20-8

The relative sizes of the planets compared with the solar disk.

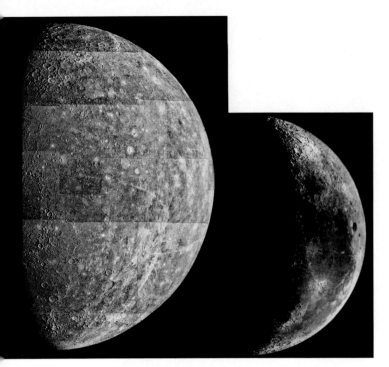

FIGURE 20-9

Mercury, like Earth's moon, shown here to the same scale, is heavily cratered and lacks an atmosphere. The youngest craters are bright with rays of material ejected by the impacts that formed them. Although it is not readily apparent from this photograph, some regions are old, heavily cratered terrain, and other regions are younger and have fewer craters. (Mercury, NASA; moon, UCO/Lick Observatory photograph)

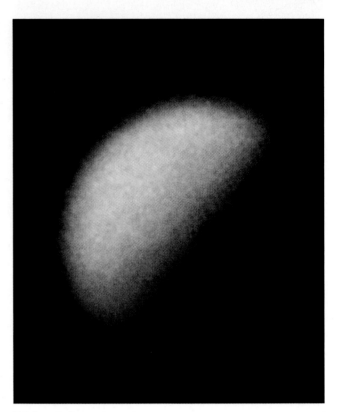

FIGURE 20-10

Venus is covered by a thick layer of clouds that never allows us to see its surface. In this photo, taken through an earth-based telescope, the planet is visible at quarter phase. Radar mapping from Earth and from robotic spacecraft orbiting Venus have been able to penetrate the clouds and reveal the planet's rocky terrain. (NASA)

Earth, and Mars—lie in the inner solar system. By contrast, the **Jovian planets** are large, gaseous, low-density worlds. The Jovian planets—Jupiter, Saturn, Uranus, and Neptune—lie in the outer solar system beyond the asteroids. Note that Pluto does not fit either category very well. It is small like the terrestrial planets but lies far from the sun and has a low density like the Jovian planets.

All the terrestrial planets are scarred by craters. Mercury's surface looks much like that of Earth's moon, with thousands of overlapping craters (Figure 20-9). Because Mercury is very hot and only 40 percent larger than our moon, it has held no atmosphere. Venus is nearly as large as Earth and has a thick atmosphere. In fact, its surface is perpetually hidden below a dense layer of clouds (Figure 20-10), so it is only by mapping its surface with radar that astronomers have found craters on Venus. Earth, too, has some craters, although erosion rapidly wears away such features. Mars is about half Earth's diameter and has a much thinner atmosphere, which permits us to see its surface easily. Space probe photos show that some of the markings visible from Earth are, in fact, craters (Figure 20-11). In addition, planetary probes have found craters on the natural satellites of Mars, Jupiter, Saturn, Uranus, and Neptune. Photographs taken by the Hubble Space Telescope and by spacecraft flying past asteroids have revealed that they, too, are heavily cratered. Although no

spacecraft has ever visited Pluto and its moon, we can guess that it is probably cratered as well.

These craters are the scars of meteorite impacts, and every surface in the solar system that is capable of retaining craters appears to be marked by them. Any theory we propose must explain this extensive bombardment of the planets.

The terrestrial planets have densities ranging from about 3–5 g/cm³. Typically, they contain cores of iron and nickel surrounded by a mantle of dense rock. In contrast, the Jovian planets are rich in hydrogen and helium, so their average density is low—less than 1.75 g/cm³. Saturn's very low density, 0.7 g/cm³, is less than the density of water, so the planet would float if we could find a bathtub big enough. Nevertheless, all the Jovian planets are massive. Jupiter is over 318 times as massive as Earth, and Saturn is about 95 Earth masses. Uranus and Neptune are smaller—15 and 17 Earth masses, respectively.

Although photographs of the Jovian planets reveal swirling cloud patterns (Figure 20-12), their atmospheres are not very deep compared to their radii. In fact, if these so-called gas giants were shrunk to a few centimeters in diameter, their gaseous hydrogen and helium atmospheres would be no deeper than the fuzz on a badly worn tennis ball.

FIGURE 20-11

Earth, Venus, and Mars to scale. Earth (left) is mostly a water planet with large masses of exposed rocky surface and an atmosphere of changing cloud patterns. The surface of Venus is perpetually hidden below thick clouds, so we must depend on radar images to reveal its surface (right), a dry desert marked by impact craters, volcanism, lava flows, and faults. Mars (center) is a smaller world with impact craters and old volcanoes on its rocky surface. (NASA)

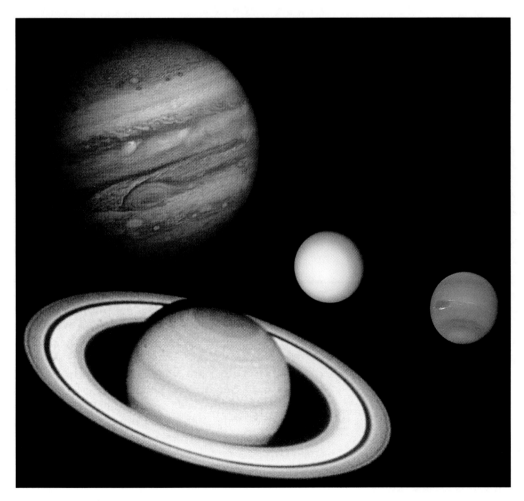

FIGURE 20-12

Of the Jovian planets (reproduced here to scale), Jupiter and Saturn are the largest. Jupiter is marked by dramatic cloud belts and the Great Red Spot, while Saturn has a bright ring system. Uranus looks featureless, but faint cloud features are visible in computer-enhanced photos. The dark blue color of Neptune and the lighter blue of Uranus are caused by atmospheric methane, which absorbs red photons. (Adapted from NASA photographs)

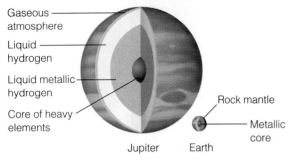

FIGURE 20-13

Cross sections of Earth and Jupiter illustrate the differences between the interiors of terrestrial and Jovian planets. Earth contains a metallic core with a rock mantle. Jupiter, 11.18 times larger in diameter, is composed mostly of liquid hydrogen. A core composed of heavy elements such as iron, nickel, and silicon lies at the center.

a

b

FIGURE 20-14

Jupiter and Saturn through a small telescope. (a) Jupiter has at least 16 satellites, but only the 4 Galilean moons are visible in small telescopes. In this photo, the planet was overexposed to show 3 of the moons. (b) The icy rings of Saturn are so reflective that they are visible through even a small telescope. Saturn also has a large satellite system. (Grundy Observatory)

Mathematical models predict that the interiors of Jupiter and Saturn are mostly hydrogen in two different states (Figure 20-13). Beneath the cloud belts of these planets are layers where pressure forces the hydrogen into a liquid state. Even deeper, the pressure is so high that the hydrogen atoms can no longer hold their electrons. With the electrons free to move about, the material becomes an excellent electrical conductor and is known as **liquid metallic hydrogen.** The centers of Jupiter and Saturn are believed to be hot cores of heavy elements slightly larger than Earth. These are often called "rocky cores," although the heat and pressure would not allow rock to exist as it does on Earth.

Uranus and Neptune, being smaller, have less pressure inside and consequently cannot force hydrogen into the liquid metallic state. Rather, model calculations suggest they contain heavy-element cores surrounded by deep oceans of liquid water with dissolved minerals. Above that is a very dense, deep atmosphere made mostly of hydrogen gas. Although Uranus and Neptune have low densities, hydrogen-rich compositions, and atmospheric features similar to those of Jupiter and Saturn, the two smaller Jovian worlds are clearly different kinds of planets.

Another characteristic of the Jovian planets is their large satellite systems. Jupiter has at least 16 known moons; 4 of them, the **Galilean satellites** (named after Galileo, who discovered them in 1610), are visible through small telescopes or even binoculars (Figure 20-14a). Saturn, Uranus, and Neptune also have systems of moons.

All of the Jovian planets have ring systems. Saturn has bright rings made of ice particles (Figure 20-14b), and Jupiter has a thin ring of dark, rocky dust. Uranus and Neptune also have thin rings composed of myriads of small particles.

We will examine the satellites and rings of the Jovian planets in later chapters. Here it is sufficient to note that the Jovian planets have large numbers of satellites and that all four have ring systems. In contrast, the terrestrial planets have few satellites and no rings at all. These

observations will help us understand how planets were formed in the solar nebula. For now, however, we turn our attention to the smaller members of our solar system.

Space Debris

The sun and planets are not the only remains of the solar nebula. The solar system is littered with three kinds of space debris: asteroids, comets, and meteoroids. Although these objects represent a tiny fraction of the mass of the system, they are a rich source of information about the origin of the planets.

The **asteroids*** are small rocky worlds, most of which orbit the sun in a belt between the orbits of Mars and Jupiter. Roughly 20,000 asteroids have been identified, of which about 1000 follow orbits that bring them into the inner solar system, where they can occasionally collide with a planet. Earth has been struck many times

*Also called minor planets.

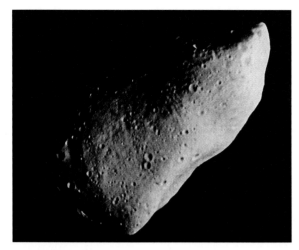

FIGURE 20-15
The asteroid Gaspra as photographed by the Galileo spacecraft. The asteroid is 20 by 12 by 11 km and rotates once every 7.04 hours. Its irregular shape and numerous craters testify that it has suffered many impacts. The color here is exaggerated to reveal subtle color variation. Gaspra would look gray to the human eye. (NASA/Galileo Imaging Team)

in its history. Some asteroids are located in Jupiter's orbit, some have been found beyond the orbit of Saturn, and a growing number are being discovered among and even beyond the outermost planets.

About 200 asteroids are more than 100 km (60 mi) in diameter, and over 2000 are more than 10 km (6 mi) in diameter. There are probably 500,000 that are larger than 1 km (0.6 mi) and billions that are smaller. Because even the largest are only a few hundred kilometers in diameter, earth-based telescopes can detect no details on their surfaces.

We do, however, have clear evidence that asteroids are irregularly shaped, cratered worlds. The Galileo spacecraft, on its way to Jupiter, flew past two asteroids, Gaspra and Ida, and photographed their surfaces. Both asteroids are irregular lumps of rock pocked by impact craters (Figure 20-15). Furthermore, the Hubble Space Telescope has been able to image some of the larger asteroids and confirm that their surfaces are irregular. We will discuss these observations in detail in Chapter 26, but in our quick survey of the solar system we can note that all evidence suggests the asteroids have suffered many impacts from collisions with other asteroids.

Older theories proposed that the asteroids are the remains of a planet that broke up, but modern astronomers recognize the asteroids as the debris left over by a planet that failed to form at a distance of 2.8 AU from the sun. When we discuss the formation of planets later in this chapter, we must be prepared to explain why material in the solar nebula failed to form a planet at a distance of 2.8 AU.

In contrast to asteroids, the brightest **comets** are impressively beautiful objects (Figure 20-16a). Most comets are faint, however, and are difficult to locate even at their brightest. A comet may take months to sweep

a

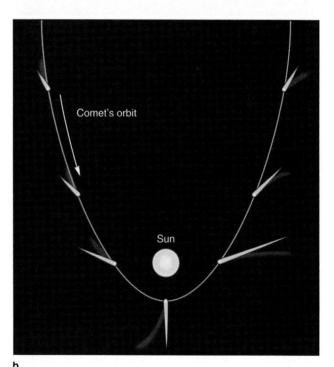

Comet's orbit

Sun

b

FIGURE 20-16
(a) A comet may remain visible in the evening or morning sky for weeks as it moves through the inner solar system. Comet West was in the sky during March 1976. (b) A comet in a long, elliptical orbit becomes visible when the sun's heat vaporizes its ices and pushes the gas and dust away in a tail. (Celestron International)

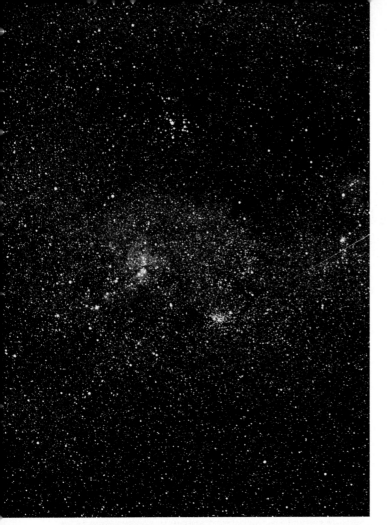

FIGURE 20-17
A meteor is the streak of glowing gases produced by a bit of material falling into the earth's atmosphere. Friction with the air vaporizes the material about 80 km (50 mi) above the earth's surface. (Daniel Good)

the formation of the Jovian planets, and we will discuss comets in more detail in Chapter 26.

Unlike the stately comets, **meteors** flash across the sky in momentary streaks of light (Figure 20-17). They are commonly called "shooting stars." Of course, they are not stars but small bits of rock and metal falling into the earth's atmosphere and bursting into incandescent vapor about 80 km (50 mi) above the ground because of friction with the air. This vapor condenses to form dust, which settles slowly to the earth, adding about 40,000 tons per year to the earth's mass.

Technically, the word *meteor* refers to the streak of light in the sky. In space, before its fiery plunge, the object is called a **meteoroid,** and any part of it that survives its fiery passage to the earth's surface is called a **meteorite.** Most meteoroids are specks of dust, grains of sand, or tiny pebbles. Almost all the meteors we see in the sky are produced by meteoroids that weigh less than 1 g. Only rarely is one massive enough and strong enough to survive its plunge and reach the earth's surface.

Thousands of meteorites have been found, and we will discuss their particular forms in Chapter 26. We mention meteorites here for one specific clue they can give us concerning the solar nebula: meteorites can tell us the age of the solar system.

The Age of the Solar System

If the solar nebula theory is correct, the planets should be about the same age as the sun. The most accurate way to find the age of a celestial body is to bring a sample into the laboratory and determine its age by analyzing the radioactive elements it contains.

When a rock solidifies, it incorporates known percentages of the chemical elements. A few of these elements are radioactive and can decay into another element, called the daughter element. For example, the isotope of uranium ^{238}U can decay into an isotope of lead, ^{206}Pb. The **half-life** of a radioactive element is the time it takes for half of the atoms to decay. The half-life of ^{238}U is 4.5 billion years. Thus, the abundance of a radioactive element gradually decreases as it decays (Figure 20-18), and the abundance of the daughter element gradually increases. If we know the abundance of the elements in the original rock, we can measure the present abundance and find the age of the rock. For example, if we studied a rock and found that only 50 percent of the ^{238}U remained and the rest had become ^{206}Pb, we would conclude that one half-life must have passed and the rock was 4.5 billion years old.

Uranium isn't the only radioactive element used in radioactive dating. Potassium (^{40}K) decays with a half-life of 1.3 billion years to form calcium (^{40}Ca) and argon (^{40}Ar). Rubidium (^{87}R) decays to strontium (^{87}Sr) with a half-life of 47 billion years. Any of these elements can be used as a radioactive clock to find the age of mineral samples.

through the solar system, during which time it appears as a glowing head with an extended tail of gas and dust.

According to the **dirty snowball model** of comets, the tail, which can be over an astronomical unit in length for a bright comet, is produced by the nucleus, a ball of dirty ices (mainly water and carbon dioxide) only a few dozen kilometers in diameter. When the nucleus enters the inner solar system, the sun's radiation begins to vaporize the ices, releasing gas and dust. The pressure of sunlight and the solar wind pushes the gas and dust away, forming a long tail. The motion of the nucleus along its orbit, the pressure of sunlight, and the outward flow of the solar wind can create comet tails that are long and straight or gently curved, but in either case the tails of comets always point generally away from the sun (Figure 20-16b).

The nuclei of comets are icy bodies left over from the origin of the planets. Thus, we must conclude that at least some parts of the solar nebula were rich in ices. We will see later in this chapter how important ices were in

Of course, to find a radioactive age, we need a sample in the laboratory, and the only celestial bodies from which we have samples are Earth, the moon, and meteorites.

The oldest earth rocks so far discovered and dated are about 3.9 billion years old. That does not mean that Earth formed 3.9 billion years ago. The surface of Earth is active, and the crust is continually destroyed and re-formed from material welling up from beneath the crust (see Chapter 21). Thus, the age of these oldest rocks tells us only that Earth is *at least* 3.9 billion years old.

One of the most exciting goals of the Apollo lunar landings was bringing lunar rocks back to Earth's laboratories, where they could be dated. Because the moon's surface is not being recycled like Earth's, some parts of it might have survived unaltered since early in the history of the solar system. Dating the rocks showed the oldest to be 4.48 billion years old. Thus, the solar system must be *at least* 4.48 billion years old.

Another important source for determining the age of the solar system is meteorites. Radioactive dating of meteorites yields a range of ages, with the oldest about 4.6 billion years old. This figure is widely accepted as the age of the solar system.

One last celestial body deserves mention: the sun. Astronomers estimate the age of the sun to be about 5 billion years, but this is not a radioactive date because they cannot obtain a sample of solar material. Instead, they estimate the sun's age from the radioactive ages of Earth, the moon, and meteorites. Computer models of the sun give only approximate ages, but they generally agree with the age of Earth.

Apparently, all the bodies of the solar system formed at about the same time some 4.6 billion years ago. This completes our survey of the solar system. We have found five significant characteristics (Table 20-1). We must now try to understand how the formation of the planets in the solar nebula explains these significant characteristics.

In what ways is the solar system a disk?

First, the general shape of the solar system is that of a disk. The planets follow orbits that lie in nearly the same plane. The orbit of Mercury is inclined 7° to the plane of Earth's orbit, and the orbit of Pluto is inclined a bit more than 17°. Thus, the planets follow orbits confined to a thin disk with the sun at its center.

Second, the motions of the sun and planets also follow this disk theme. The sun and most of the planets rotate in the same direction, counterclockwise as seen from the north, with their equators near the plane of the solar system. Also, all of the planets revolve around the sun in that same direction. Thus, our solar system seems to prefer motion in the same direction, which further reflects a disk theme.

One of the basic characteristics of our solar system is its disk shape, but another dramatic characteristic is

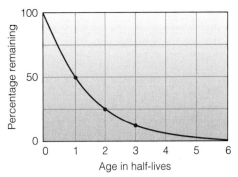

FIGURE 20-18

The number of nuclei remaining in a radioactive sample decreases such that 50 percent are left after one half-life, 25 percent after two half-lives, 12.5 percent after three half-lives, and so on.

TABLE 20-1

Characteristic Properties of the Solar System

1. Disk shape of the solar system
   Orbits in nearly the same plane
   Common direction of rotation and revolution

2. Two planetary types
   Terrestrial—inner planets; high density
   Jovian—outer planets; low density

3. Planetary ring systems for Jupiter, Saturn, Uranus, and Neptune

4. Space debris—asteroids, comets, and meteors
   Composition
   Orbits

5. Common ages of about 4.6 billion years for Earth, the moon, mete-orites, and the sun

the division of the planets into two groups. What are the distinguishing differences between the terrestrial and Jovian planets?   ■

We have completed our survey of the solar system and gathered the preliminary evidence we can use to evaluate the solar nebula hypothesis. Now it is time to ask how the planets formed from the solar nebula and took on the characteristics we see today.

20-3 THE STORY OF PLANET BUILDING

The challenge for modern planetary astronomers is to compare the characteristics of the solar system with the solar nebula theory and tell the story of how the planets formed.

TABLE 20-2

Observed and Uncompressed Densities

Planet	Observed Density (g/cm³)	Uncompressed Density (g/cm³)
Mercury	5.44	5.4
Venus	5.24	4.2
Earth	5.52	4.2
Mars	3.93	3.3
(Moon)	3.36	3.35

TABLE 20-3

The Condensation Sequence

Temperature (K)	Condensate	Planet (Estimated Temperature of Formation; K)
1500	Metal oxides	Mercury (1400)
1300	Metallic iron and nickel	
1200	Silicates	
1000	Feldspars	Venus (900)
680	Troilite (FeS)	Earth (600) Mars (450)
175	H_2O ice	Jovian (175)
150	Ammonia–water ice	
120	Methane–water ice	
65	Argon–neon ice	Pluto (65)

The Chemical Composition of the Solar Nebula

Everything we know about the nature of the solar system and the nature of star formation suggests that the solar nebula was a fragment of an interstellar gas cloud. Such a cloud would have been mostly hydrogen with some helium and tiny traces of the heavier elements.

This is precisely what we see in the composition of the sun (see Table 7-3). Analysis of the solar spectrum shows that the sun is mostly hydrogen, with 25 percent of its mass being helium and only about 2 percent being heavier elements. Of course, nuclear reactions have fused some hydrogen into helium, but this happens in the sun's core and has not affected its surface composition. Thus, the composition revealed in its spectrum is essentially the composition of the gases from which it formed.

This must have been the composition of the solar nebula, and we can see that composition reflected in the chemical compositions of the planets. One factor, however, can dramatically alter the composition of a planet. We must recognize that planets grow by two processes. Planets begin growing by the sticking together of solid bits of matter. Only when a planet has grown to a mass of about 15 earth masses does it have enough gravitation to begin capturing gas directly from the solar nebula in a process called **gravitational collapse.** The Jovian planets began forming by the aggregation of bits of rock and ice. Once they accumulated enough mass, they began growing by gravitational collapse. That is, they could capture large amounts of gas, mostly hydrogen and helium, directly from the solar nebula. Jupiter and Saturn grew rapidly and captured the most hydrogen and helium, while Uranus and Neptune grew more slowly and never became massive enough to capture gas as rapidly as Jupiter and Saturn. Thus, the Jovian worlds grew to be low-density, hydrogen-rich planets, with Jupiter and Saturn the largest and least dense.

The terrestrial planets, in contrast, contain very little hydrogen and helium. These small planets have such low masses they were unable to keep the hydrogen and helium from leaking into space. Indeed, because of their small masses, they were probably unable to capture very

much of these lightweight gases from the solar nebula. The terrestrial planets are dense worlds because they are composed of the heavier elements from the solar nebula.

Although we can see how the chemical composition of the solar nebula is reflected in the present composition of the sun and planets, a very important question remains: how did gas and dust in the solar nebula come together to form the solid matter of the planets? We must answer that question in two stages. First, we must understand how gas and dust formed billions of small solid particles, and then we must explain how these particles built the planets.

The Condensation of Solids

The key to understanding the process that converted the nebular gas into solid matter is the variation in density among solar system objects. We have already noted that the four inner planets are high-density, terrestrial bodies, whereas the outermost planets are low-density, giant planets (except for Pluto, which is low in density but not a giant). This division is due to the different ways gases condensed into solids in the inner and outer regions of the solar nebula.

Even among the four terrestrial planets, we find a pattern of subtle differences in density. Merely listing the observed densities of the terrestrial planets does not reveal the pattern, because Earth and Venus, being more massive, have stronger gravity and have squeezed their interiors to higher densities. We must look at the **uncompressed densities**—the densities the planets would have if their gravity did not compress them. These densities (Table 20-2) show that the closer a planet is to the sun, the higher its uncompressed density.

This density variation probably originated when the solar system first formed solid grains. The kind of matter that condensed in a particular region would depend on the temperature of the gas there. In the inner regions, the temperature may have been 1500 K or so. The only materials that could form grains at this temperature are compounds with high melting points, such as metal oxides and pure metals, which are very dense. Farther out in the nebula it was cooler, and silicates (rocky material) could condense. These are less dense than metal oxides and metals. In the cold outer regions, ices of water, methane, and ammonia could condense. These are low-density materials.

The sequence in which the different materials condense from the gas as we move away from the sun is called the **condensation sequence** (Table 20-3). It suggests that the planets, forming at different distances from the sun, accumulated from different kinds of materials. Thus, the inner planets formed from high-density metal oxides and metals, and the outer planets formed from low-density ices.

We must also remember that the solar nebula did not remain the same temperature throughout the formation of the planets but may have grown progressively cooler. Thus, a particular region of the nebula may have begun by producing solid particles of metals and metal oxides but, after cooling, began producing particles of silicates. Allowing for the cooling of the nebula makes our theory much more complex, but it also makes the processes by which the planets formed much more understandable.

The Formation of Planetesimals

In the development of a planet, three groups of processes operate. First, grains of solid matter grow larger, eventually reaching diameters ranging from a few centimeters to kilometers. These objects, called **planetesimals,** are believed to be the bodies that the second group of processes collects into planets. Finally, a third set of processes clears away the solar nebula. The study of planet building is the study of these three groups of processes.

According to the solar nebula theory, planetary development in the solar nebula began with the growth of dust grains. These specks of matter, whatever their composition, grew from microscopic size by two processes: condensation and accretion.

A particle grows by **condensation** when it adds matter one atom at a time from a surrounding gas. Thus, snowflakes grow by condensation in the earth's atmosphere. In the solar nebula, dust grains were continuously bombarded by atoms of gas, and some of these stuck to the grains. A microscopic grain capturing a gas atom increases its mass by a much larger fraction than a gigantic boulder capturing a single atom. Thus, condensation can increase the mass of a small grain rapidly, but as the grain grows larger, condensation becomes less effective.

The second process is **accretion,** the sticking together of solid particles. In building a snowman, we roll

FIGURE 20-19
Gravitational instabilities in the rotating disk of planetesimals may have forced them to collect in clumps, accelerating their growth.

a ball of snow across the snowy ground so that it grows by accretion. In the solar nebula, the dust grains were, on the average, no more than a few centimeters apart, so they collided frequently. Their mutual gravitation was too small to hold them to each other, but other effects may have helped. Static electricity generated by their passage through the gas could have held them together, as could compounds of carbon that might have formed a sticky surface on the grains. Ice grains might have stuck together better than some other types. Of course, some collisions might have broken up clumps of grains; on the whole, however, accretion must have increased grain size. If it had not, the planets would not have formed.

There is no clear distinction between a very large grain and a very small planetesimal, but we can consider an object a planetesimal when its diameter becomes a kilometer or so. Objects this size and larger were subject to new processes that tended to concentrate them. One important effect may have been that the growing planetesimals collapsed into the plane of the solar nebula. Dust grains could not fall into the plane because the turbulent motions of the gas kept them stirred up, but the larger objects had more mass, and the gas motions could not have prevented them from settling into the plane of the spinning nebula. This would have concentrated the solid particles into a thin plane about 0.01 AU thick and would have made further planetary growth more rapid.

This collapse of the planetesimals into the plane is analogous to the flattening of a forming galaxy. However, an entirely new process may have become important once the plane of planetesimals formed. Computer models show that the rotating disk of particles should have been gravitationally unstable and would have broken up into small clouds (Figure 20-19). This would further concentrate the planetesimals and help them coalesce into objects up to 100 km (60 mi) in diameter. Thus, the theory predicts that the nebula became filled

FIGURE 20-20

At left, a planet grows slowly by the accretion of planetesimals of similar composition. After the protoplanet has formed, the accumulation of heat from radioactive decay in its interior melts it and allows it to differentiate. An improved theory, right, proposes that the solar nebula cooled as the planets formed. The first particles to accumulate were metals, but later particles were silicates. Also, it seems likely that the planets formed fast enough to prevent the loss of the heat of formation, which melted the protoplanets and allowed them to differentiate as they formed.

with trillions of solid particles ranging in size from pebbles to tiny planets. As the largest began to exceed 100 km in diameter, new processes began to alter them, and a new stage in planet building began, the growth of protoplanets.

The Growth of Protoplanets

The coalescing of planetesimals eventually formed **protoplanets,** massive objects destined to become planets. As these larger bodies grew, new processes began making them grow faster and altered their physical structure.

If planetesimals collided at orbital velocities, it is unlikely that they would have stuck together. The average orbital velocity in the solar system is about 30 km/sec (67,000 mph). Head-on collisions at this velocity would have pulverized the material. However, the planetesimals were moving in the same direction in the nebular plane and thus avoided head-on collisions. Instead, they merely rubbed shoulders at low relative velocities. Such collisions would have been more likely to fuse them than to shatter them.

In addition, some adhesive effects probably helped. Sticky coatings and electrostatic charges on the surfaces of the smaller planetesimals probably aided formation of larger bodies. Collisions would have fragmented some of the surface rock, but if the planetesimals were large enough, their gravity would have held on to some fragments, forming a layer of soil composed entirely of crushed rock. Such a layer on the larger planetesimals may have been effective in trapping smaller bodies.

The largest planetesimals would grow the fastest because they had the strongest gravitational field. Not only could they hold on to a cushioning layer to trap fragments, but their stronger gravity could attract additional material. These planetesimals probably grew quickly to protoplanetary dimensions, sweeping up more and more material. When massive enough, they trapped some of the original nebular gas to form primitive atmospheres. At some point, they crossed the boundary between planetesimals and protoplanets.

To trace the growth of a protoplanet, we can think of the formation of Earth. In its simplest form, our theory of protoplanet growth supposes that all the planetesimals had about the same chemical composition. The planetesimals accumulated gradually to form a planet-size ball of material that was of homogeneous composition throughout (Figure 20-20). Once the planet formed, heat began to accumulate in its interior from the decay of short-lived radioactive elements, and this heat eventually melted the planet and allowed it to differentiate. **Differentiation** is the separation of material according to density. When the planet melted, the heavy metals such as iron and nickel settled to the core, while the lighter silicates floated to the surface to form a low-density crust.

In this simple form, the theory predicts that Earth's present atmosphere was not its first. The first atmosphere consisted of gases trapped from the solar nebula—

mostly hydrogen and helium. They were later driven off by the heat, aided perhaps by outbursts from the infant sun, and new gases that baked from the rocks formed a secondary atmosphere. This creation of a planetary atmosphere from the planet's interior is called **outgassing.**

We can improve this simple theory of planet formation in two ways (Figure 20-20). First, it seems likely that the solar nebula cooled during the formation of the planets, so they did not accumulate from planetesimals of common composition. As planet building began, the first particles to condense in the inner solar system were metals and metal oxides, so the protoplanets may have begun by accreting metallic cores. Later, as the nebula cooled, more silicates could form, and the protoplanets added silicate mantles. A second improvement in the theory proposes that the planets grew so rapidly that the heat released by the in-falling particles, the **heat of formation,** did not have time to escape. This heat rapidly accumulated and melted the protoplanets as they formed. If this is true, then the planets would have differentiated as they formed.

Our improved theory suggests that our present atmosphere was not produced entirely by outgassing. If Earth formed in a molten state, then there was never a time when it had a primitive atmosphere of hydrogen and helium accumulated from the solar nebula. Rather, the gases of the atmosphere were released by the molten rock as the protoplanet grew. Those gases would not have included much water, however, so some astronomers now think that Earth's water and much of its present atmosphere accumulated late in the formation of the planets as Earth swept up volatile-rich planetesimals forming in the cooling solar nebula. Such icy planetesimals may have formed in the outer parts of the solar nebula and been scattered by encounters with the Jovian planets in a bombardment of comets.

The details of the formation of the planets are still being unraveled, but it seems clear that the solar nebula hypothesis can explain the formation of planets around the forming sun. But can it explain the distinguishing characteristics of the solar system that we compiled earlier in this chapter?

Explaining the Characteristics of the Solar System

Table 20-1 contains the list of distinguishing characteristics of the solar system. Any theory of the origin of the solar system should explain these characteristics.

The disk shape of the solar system is inherited from the solar nebula. The sun and planets revolve and rotate in the same direction because they formed from the same rotating gas cloud. The orbits of the planets lie in the same plane because the rotating solar nebula collapsed into a disk, and the planets formed in that disk.

The solar nebula hypothesis is evolutionary in that it calls on continuing processes to gradually build the planets. To explain the rotation of Venus, Uranus, and Pluto, however, we must introduce catastrophic events. Apparently, these planets acquired their peculiar rotations when they were struck by very large planetesimals late in their formation. Such impacts may be unusual, but some planets probably experienced them, and an off-center impact by a planetesimal as large as Mars might have made Venus rotate backward. A similar event might have forced Uranus to rotate on its side. In Chapter 25, we will discuss Pluto, and we will see that such a catastrophic event may have been involved in its formation as well.

The second item in Table 20-1, the division of the planets into terrestrial and Jovian worlds, can be understood through the condensation sequence. The terrestrial planets formed in the inner part of the solar nebula, where the temperature was high and most compounds remained gaseous. Only compounds such as the metals and silicates could condense to form solid particles. Planets must begin forming by the sticking together of solid particles because a small blob of matter does not have enough gravitation to capture gas. Thus, the planets that began growing in the inner solar system had to form mainly from metals and silicates. They became the small, dense terrestrial planets.

In contrast, the Jovian planets formed in the outer solar nebula, where the lower temperature allowed the gas to form large amounts of ices, perhaps three times more ices than silicates. Thus, the Jovian planets grew rapidly and became massive. Some theoretical models indicate that Jupiter formed from this abundance of material in less than a thousand years, although most theories predict millions of years for the formation of the Jovian planets.

The heat of formation (the energy released by infalling matter) was tremendous for these massive planets. Jupiter must have grown hot enough to glow with a luminosity of about 1 percent that of the present sun. However, because it never got hot enough to generate nuclear energy as a star would, it cooled. Jupiter is still hot inside. In fact, both Jupiter and Saturn radiate more heat than they absorb from the sun, so they are evidently still cooling.

Although planets must begin growing by the accretion of solid particles, once they become a few times more massive than Earth they have enough gravity to capture gas directly from the solar nebula. Thus, the Jovian planets grew by attracting vast amounts of nebular gas and grew rich in light gases such as hydrogen and helium. The terrestrial planets could not do this because they never became massive enough and because the gas in the inner nebula was hotter and more difficult to capture.

A glance at the solar system suggests that we should expect to find a planet between Mars and Jupiter at the present location of the asteroid belt. Evidently, we have asteroids there and not a planet because Jupiter grew into such a massive planet it was able to gravitationally disturb the motion of nearby planetesimals. Thus, the bodies that should have formed a planet just inward in the solar system from Jupiter were broken up, thrown into the sun, or ejected from the solar system. The asteroids we see today are the last remains of those objects.

The large satellite systems of the Jovian worlds may contain two kinds of moons. Some moons may have formed in orbit around the forming planet in a miniature of the solar nebula. But some of the smaller moons may be captured planetesimals and asteroids. The large masses of the Jovian planets would have made it easier for them to capture satellites.

In Table 20-1, we noted that all four Jovian worlds have ring systems, and we can understand this by considering the large mass of these worlds and their remote location in the solar system. A large mass makes it easier for a planet to hold onto orbiting ring particles, and being farther from the sun the ring particles are not as easily swept away by the pressure of sunlight and the solar wind. It is hardly surprising, then, that the terrestrial planets, low-mass worlds located near the sun, have no planetary rings.

The last entry in Table 20-1 is the common ages of solar system bodies, and the solar nebula hypothesis has no difficulty explaining that characteristic. If the hypothesis is correct, then the planets formed at the same time as the sun and thus should have roughly the same age. Although we have mineral samples for radioactive dating only from Earth, the moon, and meteorites, so far the ages seem to agree.

The solar nebula hypothesis can account for the distinguishing characteristics of the solar system, but there is yet another test we should apply to the hypothesis. What about the problem that troubled Laplace and his nebular hypothesis? What about the angular momentum problem? If the sun and planets formed from a contracting nebula, the sun should have been left spinning very rapidly. That is, it should have most of the angular momentum in the solar system, and instead it has very little. The solar nebula hypothesis allows us to explain the angular momentum problem by thinking about the relationship between the origin of planetary systems and star formation. What we know about star formation tells us that the sun must have had a strong stellar wind when it was young. Also, a rapidly spinning star is likely to generate a strong magnetic field. Like a great paddle wheel, the sun's magnetic field must have stuck out into space, and gas flowing away from the sun would have been dragged around with the rotating magnetic field. This would have slowed the sun and left it rotating slowly. Astronomers now believe that the angular momentum problem is no longer an objection to the solar nebula hypothesis.

Our general understanding of the origin of the solar system also explains the origin of asteroids, meteors, and comets. They appear to be the last of the debris left behind by the solar nebula. These objects are such important sources of information about the history of our solar system that we will discuss them in detail in Chapter 26. But for now we ask: what happened to the solar nebula?

Clearing the Nebula

The planets appear to have grown in the solar nebula at the same time that the sun was forming, about 4.6 billion years ago. Indeed, much of the planetary growth may have taken place in darkness because the sun may not have become luminous yet. But once the sun became hot and luminous, it began to clear the nebula of gas and dust and brought planet building to a halt.

Four effects helped clear the nebula. The most important was **radiation pressure.** When the sun became a luminous object, light streaming from its surface pushed against the particles of the solar nebula. Large bits of matter like planetesimals and planets were not affected, but low-mass specks of dust and individual gas atoms were pushed outward and eventually driven from the system. This is not a sudden process, and it may not have occurred at the same time everywhere in the nebula. Before sunlight could begin clearing the outer nebula, it first had to push its way through the inner nebula.

The second effect that helped clear the nebula was the solar wind, the flow of ionized hydrogen and other atoms away from the sun's upper atmosphere. This flow is a steady breeze that rushes past Earth at about 400 km/sec (250 mi/sec). When the sun was young, it may have had an even stronger solar wind, and irregular fluctuations in its luminosity, like those observed in T Tauri stars, may have produced surges in the wind that helped push dust and gas out of the nebula.

The third effect for clearing the nebula was the sweeping up of space debris by the planets. All of the old, solid surfaces in the solar system are heavily cratered by meteorite impacts (Figure 20-21). Earth's moon, Mercury, Venus, Mars, and most of the moons in the solar system are covered with craters. A few of these craters have been formed recently by the steady rain of meteorites that falls on all the planets in the solar system, but most of the craters we see appear to have been formed roughly 4 billion years ago in what is called the **heavy bombardment,** as the last of the debris in the solar nebula was swept up by the planets. We will find more evidence of this bombardment in later chapters. Thus, many of the last remaining solid objects in the solar nebula were swept up by the planets soon after they formed.

The fourth effect was the ejection of material from the solar system by close encounters with planets. If a small object such as a planetesimal passes close to a planet, it can gain energy from the planet's gravitational field and be thrown out of the solar system. Ejection is most probable in encounters with massive planets, so the Jovian planets were probably very efficient at ejecting the icy planetesimals that formed in their region of the nebula.

Together these four effects cleared away the solar nebula. This may have taken considerable time, but eventually the solar system was relatively clear, the planets could no longer gain mass, and planet building ended.

a

b

FIGURE 20-21
Every old, solid surface in the solar system is scarred by craters. (a) Earth's moon is scarred by craters ranging from basins hundreds of kilometers in diameter down to microscopic pits. (b) The surface of Mercury as photographed by a passing spacecraft shows vast numbers of overlapping craters. Throughout the solar system, the heavy bombardment cratered all the solid worlds as the last of the debris in the solar nebula was swept up. (NASA)

CRITICAL INQUIRY

Why are there two kinds of planets in our solar system?

Planets begin forming from solid bits of matter, not from gas. Consequently, the kind of planet that forms at a given distance from the sun depends on the kind of compounds that can condense out of the gas to form solid particles. In the inner parts of the solar nebula, the temperature was so high that most of the gas could not condense to form solids. Only metals and silicates could form solid grains, and thus the innermost planets grew from this dense material. Much of the mass of the solar nebula consisted of hydrogen, helium, water vapor, and other gases, and they were present in the inner solar nebula but couldn't form solid grains. The small terrestrial planets couldn't grow from these gases, so the terrestrial planets are small and dense.

In the outer solar nebula, the composition of the gas was the same, but it was cold enough for water vapor to condense to form ice grains, and because hydrogen and oxygen are so abundant there was lots of ice available. Thus, the outer planets grew from solid bits of metal and silicate combined with large amounts of ice. The outer planets grew so rapidly that they became big enough to capture gas directly, and thus they became the hydrogen- and helium-rich Jovian worlds.

The condensation sequence combined with the solar nebula hypothesis gives us a way to understand the formation of the planets. Can we extend our insight to other stars? What would a planetary system be like if it formed around a star that was slightly hotter than the sun? ■

The great chain of origins leads from the first instant of the big bang through the birth of galaxies, the formation of stars, the origin of our solar system, and finally to us. One of the most elegant and complex links in that chain is planet building. As we explore the solar system in detail in the following chapters, we must stay alert for further clues to the birth of the planets. ■

■ *Summary*

René Descartes proposed that the solar system formed from a contracting vortex of matter, and Buffon later suggested that a passing star pulled matter out of the sun to form the planets. While Buffon's hypothesis was catastrophic, Laplace's nebular hypothesis was evolutionary. The nebular hypothesis required a contracting nebula to leave behind rings that formed each planet, but it could not explain the low angular momentum of the sun. The modern solar nebula hypothesis suggests that the sun was slowed by its magnetic field and mass loss. Disks of gas and dust have been found around young stars, so astronomers suspect that planetary systems are common. Two or three peculiar, earth-size planets are known to be orbiting a pulsar, and a planet half the mass of Jupiter has been found orbiting the star 51 Pegasi.

The solar nebula hypothesis proposes that the solar system began as a contracting cloud of gas and dust that flattened into a rotating disk. The center of this cloud eventually became the sun, and the planets formed in the disk of the nebula. The properties of this nebula and the processes that built the planets explain many of the characteristic properties of the solar system.

One of the most striking properties of the solar system is its disk shape. The orbits of the planets lie in nearly the same plane, and they all revolve around the sun in the same direction, counterclockwise as seen from the north. This is also true of many of the satellites of the planets. With only three exceptions, the planets rotate counterclockwise around axes roughly perpendicular to the plane of the solar system. This disk shape and the motion of the planets appear to have originated in the solar nebula.

Another striking feature of the solar system is the division of the planets into two families. The terrestrial planets, which are small and dense, lie in the inner part of the system. The Jovian planets are large, low-density worlds in the outer part of the system. In general, the closer a planet lies to the sun, the higher its uncompressed density.

The solar system is now filled with smaller bodies such as asteroids, comets, and meteors. The asteroids are small, rocky worlds, most of which orbit the sun between Jupiter and Mars. They appear to be material left over from the formation of the solar system.

Another important characteristic of the solar system bodies is their similar ages. Radioactive dating tells us that Earth, the moon, and meteorites are no older than about 4.6 billion years. Thus, it seems our solar system took shape about 4.6 billion years ago.

According to the condensation sequence, the inner part of the nebula was so hot that only high-density minerals could form solid grains. The outer regions, being cooler, condensed to form icy material of lower density. The planets grew from these solid materials, with the denser planets forming in the inner part of the nebula and the lower-density Jovian planets forming farther from the sun.

Planet building began as dust grains growing by condensation and accretion into planetesimals ranging from a kilometer to hundreds of kilometers in diameter. These planetesimals settled into a thin plane around the sun and accumulated into larger bodies, the largest of which grew the fastest and eventually became protoplanets.

Once a planet had formed from a large number of planetesimals, heat from radioactive decay could have melted it and allowed it to differentiate into a dense metallic core and a lower-density silicate crust. In fact, it is possible that the solar nebula cooled as the protoplanets grew so that the first planetesimals were metallic and later additions were silicate. It is also likely that the planets grew rapidly enough that the heat of formation released by the in-falling material melted the planets and allowed them to differentiate as they formed.

The Jovian planets probably grew rapidly from icy materials and became massive enough to attract and hold vast amounts of nebular gas. Heat of formation raised their temperatures very high when they were young, and Jupiter and Saturn still radiate more heat than they absorb from the sun.

Once the sun became a luminous object, it cleared the nebula as its light and solar wind pushed material out of the system. The planets helped by absorbing some planetesimals and ejecting others from the system. Once the solar system was clear of debris, planet building ended.

■ New Terms

passing star hypothesis	half-life
evolutionary hypothesis	gravitational collapse
catastrophic hypothesis	uncompressed density
nebular hypothesis	condensation sequence
angular momentum problem	planetesimal
solar nebula hypothesis	condensation
terrestrial planet	accretion
Jovian planet	protoplanet
liquid metallic hydrogen	differentiation
Galilean satellites	outgassing
asteroid	heat of formation
comet	radiation pressure
dirty snowball model	heavy bombardment
meteor, meteoroid, meteorite	

■ Questions

1. What were the objections to the passing star hypothesis? to the nebular hypothesis?

2. What produced the helium now present in the sun's atmosphere? in Jupiter's atmosphere? in the sun's core?

3. What observational evidence do we have that stars commonly form with disks of gas and dust surrounding them?

4. If a thorough search among all the nearest stars found no planetary systems, the solar nebula theory would be put in doubt. Why?

5. According to the solar nebula theory, why is the sun's equator nearly in the same plane as that of Earth's orbit?

6. If you visited another planetary system circling a G star like our sun, would you expect the outer planets to be Jovian, or might this be peculiar to our solar system?

7. Why is almost every solid surface in the solar system scarred by craters?

8. If you visited the planetary system described in Question 6 and determined its age by radioactive dating, would you expect to get 4.6 billion years or some other number? Why?

9. What is the difference between condensation and accretion?

10. Why are the terrestrial planets so much less massive than the Jovian planets?

■ Discussion Questions

1. In your opinion, if planets formed with one of the first stars to form in our galaxy, how would those planets differ from the planets in our solar system? (HINT: To what population would the star belong?)

2. If the solar nebula hypothesis is correct, then there are probably more planets in the universe than stars. Do you agree? Why or why not?

■ Problems

1. If you observed the solar system from the nearest star (1.3 pc), what would the maximum angular separation be between Jupiter and the sun? (HINT: Use the small-angle formula.)

2. The brightest planet in our sky is Venus, which is sometimes as bright as apparent magnitude −4 when it is at a distance of about 1 AU. How many times fainter would it look from a distance of 1 parsec (206,265 AU)? What would its apparent magnitude be? (HINTS: Remember the inverse square law, Chapter 5; see Chapter 2.)

3. What is the smallest-diameter crater you can identify in Figure 20-9? (HINT: See the table on the inside back covers, Properties of the Planets, to find the diameter of Mercury in kilometers.)

4. A sample of a meteorite has been analyzed, and the result shows that out of every 1000 nuclei of ^{40}K originally in the meteorite, only 100 have not decayed. How old is the meteorite? (HINT: See Figure 20-18.)

5. In Table 20-3, which object's observed density differs least from its uncompressed density? Why?

6. What composition might we expect for a planet that formed in a region of the solar nebula where the temperature was about 100 K?

7. Suppose that Earth grew to its present size in 1 million years through the accretion of particles averaging 100 g each. On the average, how many particles did Earth capture per second? (HINT: See Appendix A, Table A-6 to find Earth's mass.)

8. If you stood on Earth during its formation as described in Problem 7 and watched a region covering 100 m², how many impacts would you expect to see in an hour? (HINTS: Assume that Earth had its present radius. The surface area of a sphere is $4\pi R^2$.)

9. The velocity of the solar wind is roughly 400 km/sec. How long does it take to travel from the sun to Pluto?

■ Critical Inquiries for the Web

1. The process of protoplanetary accretion is still not well understood. Search the Web for current research in this field. From the results of your search, outline the basic steps in the formation of a protoplanet through accretion. What specific factors are important in these models of planet building? Do these models produce planetary systems similar to the ones we know to exist?

2. How is radioactive dating carried out on meteorites and rocks from surfaces of various bodies in the solar system? Look for Web sites on the details of radioactive dating and summarize the methods used to uncover the abundances of radioactive elements in a particular sample. (HINT: Try looking for information on how a particular meteorite—for example, the Martian meteorite ALH 84001—was studied, what age range was determined, and what radioactive elements were used to arrive at the age.)

Go to the Wadsworth Astronomy Resource Center (www.wadsworth.com/astronomy) for critical thinking exercises, articles, and additional readings from InfoTrac College Edition, Wadsworth's online student library.

Chapter Twenty-One

PLANET EARTH

GUIDEPOST

Astronomy has been described as the science of everything above the clouds. Planetary astronomers, however, must also think about what lies below the clouds, because the earth is the basis for comparison with all other earthlike planets. We know the earth well, and we can apply what we know about the earth to other worlds.

There is another reason for studying the earth in an astronomy course. Astronomy is really about us. Astronomy is exciting and fascinating because it helps us understand what we are and where we are in the universe. Thus, we cannot omit the earth from our discussion—it is where we are.

The next two chapters will discuss the earthlike planets, but that will not end our thoughts on the earth. The moons of the giant outer worlds will seem earthlike in strange ways, and our discussion of the smaller bodies of our solar system will alert us to the dangers the earth faces. Throughout the rest of this book, we will remain painfully aware of the fragile beauty of our planet. ■

Nature evolves. The world was

different yesterday.

PRESTON CLOUD

Cosmos, Earth and Man

Stars die, but planets don't. That should give us a feeling of security, but we know that the survival of our planet is limited by the life expectancy of our star. Furthermore, although planets don't die, they do evolve, and our planet has changed dramatically since it formed. Mountains rise and weather away, the climate warms and then cools, and the continents themselves drift and grind against each other. To fully understand how our own planet works, we need to know how planets in general evolve. Why is Venus hot and Mars cold? Why does Mercury have many craters and Earth only a few?

To answer such questions, we must use **comparative planetology,** the study of planets through comparison and contrast, and Earth is our basis of comparison. One reason why we begin with Earth is that it is the planet we know best. We can study it carefully and then draw comparisons to other worlds. But another reason is that our home is a planet of extremes (Data File Two). Its interior is molten and generates a magnetic field. Its crust is active, with moving sections that push against each other and trigger earthquakes, volcanoes, and mountain building. Even Earth's atmosphere is extreme. Processes have altered Earth's air from its original composition to a highly unusual, oxygen-rich sea of gas. Once we understand Earth's complex properties, the remaining planets in our system should be easier to comprehend.

In our study of Earth, we will find a four-stage history of planetary development. The moon and all the terrestrial planets have passed through these stages, although differences in the way the planets were altered by these stages have produced dramatically different worlds. The moon, for example, is much like Earth, but its evolution has been dramatically altered by its smaller size. Thus, as we explore the solar system, we will discover not entirely new processes but rather familiar effects working in slightly different ways.

21-1 THE EARLY HISTORY OF EARTH

Like all the terrestrial planets, Earth formed from the inner solar nebula about 4.6 billion years ago. As it took form, it began to change, passing through four developmental stages (Figure 21-1). Processes that occurred during these four stages determined the present condition of Earth's interior, crust, and atmosphere.

Four Stages of Planetary Development

The first stage of planetary evolution is *differentiation*, the separation of material according to density. The earth now has a dense core and a lower-density crust, and that structure must have originated very early.

Differentiation would have occurred easily if the earth was molten when it was young. Two sources of heat

EARTH

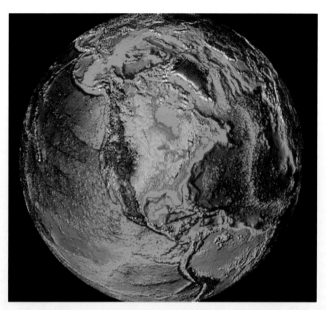

A shaded relief map of Earth with the oceans removed reveals the geological features on the continents and under the oceans. (National Geophysical Data Center)

Average distance from the sun	1.00 AU (1.495979 × 10⁸ km)
Eccentricity of orbit	0.0167
Maximum distance from the sun	1.0167 AU (1.5210 × 10⁸ km)
Minimum distance from the sun	0.9833 AU (1.4710 × 10⁸ km)
Inclination of orbit to ecliptic	0°
Average orbital velocity	29.79 km/sec
Orbital period	1.00 y (365.26 days)
Period of rotation (with respect to the sun)	24ʰ00ᵐ00ˢ
Inclination of equator to orbit	23°27′
Equatorial diameter	12,756 km
Mass	5.976 × 10²⁴ kg
Average density	5.497 g/cm³
Surface gravity	1.0 earth gravities
Escape velocity	11.2 km/sec
Surface temperature	−50° to 50°C (−60° to 120°F)
Average albedo*	0.39
Oblateness	0.0034

*Albedo is the fraction of light striking an object that the object reflects. The albedo of a perfect reflector is 1.

FIGURE 21-1

The four stages of planetary development. First: Differentiation into core and crust as dense material sank to the center and low-density material floated to the surface. Second: Cratering of the newly formed crust by the heavy bombardment of meteorites in the young, debris-filled solar nebula. Third: Flooding of lowlands by lava from the interior and later by water from the cooling atmosphere. Fourth: Slow surface evolution due to geological activity and erosion.

could have heated the earth. First, heat of formation would be created by in-falling material. A meteorite hitting the earth at high velocity converts most of its energy of motion into heat, and the in-falling of a large number of meteorites could release tremendous heat. If the earth formed rapidly, this heat would have accumulated much

more rapidly than it could leak away, and the earth may have been molten when it formed. A second source of heat requires more time to develop. The decay of radioactive elements trapped in the earth releases heat gradually, but as soon as the earth formed, that heat would have begun to accumulate and could have helped melt the earth to facilitate differentiation. Most of the earth's radioactive elements are now concentrated in the crust, where they continue to warm and soften the rock layers.

The earth formed by material falling together, but meteorites could have left no trace until a crust solidified. Once the earth had a hard surface, the meteorites could form craters. This second stage in planetary evolution, *cratering*, was violent. The early heavy bombardment was intense because the solar nebula was filled with rocky and icy debris, and the young earth was battered by meteorites that pulverized the newly forming crust. The largest meteorites blasted out crater basins hundreds of kilometers in diameter. As the solar nebula cleared, the amount of debris decreased, and the level of cratering fell to its present low level. Although meteorites still occasionally strike the earth and dig craters, cratering is no longer the dominant influence on the earth's geology. As we compare other worlds with the earth, we will discover traces of this intense period of cratering on every old surface in the solar system.

The third stage, *flooding*, no doubt began while cratering was still intense. The fracturing of the crust and the heating caused by radioactive decay allowed molten rock to well up through fissures and flood the deeper basins. We will discuss such flooded basins on other worlds, such as the moon, but all traces of this early lava flooding have been destroyed by later geological activity in the earth's crust. On the earth, flooding continued as the atmosphere cooled and water fell as rain, filling the deepest basins to produce the first oceans. Notice that on the earth flooding involves both lava and water, a circumstance that we will not find on most worlds.

The fourth stage, *slow surface evolution*, has continued for the last 3.5 billion years or more. The earth's surface is constantly changing as sections of crust slide over each other, push up mountains, and shift continents. Also, moving air and water erode the surface and wear away geological features. Almost all traces of the first billion years of the earth's geology have been destroyed by the active crust and erosion.

Earth as a Planet

We are studying the earth in this chapter in order to use it as a basis for comparison with other worlds. Of course, four of the worlds in our solar system, the Jovian planets, are dramatically different from the earth, so we will not try to compare them with the earth. But the four terrestrial planets are rocky worlds much like the earth, and many of the moons in the solar system have geology that we can compare with the earth and its four stages of development.

a   b

FIGURE 21-2

The earth is a water world with 75 percent of its surface covered by deep oceans of liquid water. A typical photograph of the surface of the earth would show nothing but ocean (a). The water evaporates to form clouds and falls as rain and snow, which produce powerful erosion. Compression and faulting in the earth's crust can push up high mountain ranges (b), but the presence of water can erode the mountains away completely in a small fraction of the earth's total age. (a, Wm. K. Hartmann; b, Janet Seeds)

All terrestrial planets pass through these four stages, so in that respect the earth is a good basic reference planet for comparative planetology. Some planets have emphasized one stage over another, and some planets have failed to progress fully through the four stages. Nevertheless, the earth is a good standard of comparison. Every major process on any rocky world in our solar system is represented in some form on the earth. Clearly, the earth is a highly complex world; for that reason, it makes a good standard of comparison.

On the other hand, the earth is peculiar in two ways. First, it has large amounts of liquid water on its surface (Figure 21-2a). Fully 75 percent of its surface is covered by this liquid, and no other planet in our solar system is known to have such extensive liquid water. Not only does water fill the oceans, but it evaporates into the atmosphere, forms clouds, and then falls as rain. Water falling on the continents flows downhill to form rivers that flow back to the sea, and, in so doing, the water produces extensive erosion (Figure 21-2b). Entire mountain ranges can literally dissolve and wash away in only a few tens of millions of years, less than 1 percent of the earth's total age. We will not see such intense erosion on most worlds. Liquid water is, in fact, a rare material on most planets.

Our planet is special in a second way. Some of the matter on the surface of this world is alive, and a small part of that living matter is aware. We do not know how the presence of living matter has affected the evolution of the earth, but this process seems to be totally missing from other worlds in our solar system. Furthermore, the thinking part of the life on the earth, humankind, is actively altering our planet. It does not seem likely that we have made major changes in the earth yet, but our study of the earth as a standard of comparison with other worlds will also give us new insight into our own planet and how we may be altering it.

CRITICAL INQUIRY

Why do we think the earth went through a stage of cratering?

Recall from the previous chapter that the planets formed by the accretion of planetesimals from the solar nebula. The protoearth may have been molten as it formed, but as soon as it grew cool enough to form a solid crust, the remaining planetesimal impacts would have formed craters. Thus, we can reason from the solar nebula hypothesis that the earth should have been cratered. But we can't use a theory as evidence to support some other theory. To find real observational evidence, we need only look at the moon. The moon has craters, and so does every other old surface in our solar system. There must have been a time, when the solar system was young, when there were large numbers of objects striking all the planets and moons and blasting out craters. If it happened to other worlds in our solar system, it must have happened to the earth, too.

Of course, we don't see those craters on the earth today. Why not? ■

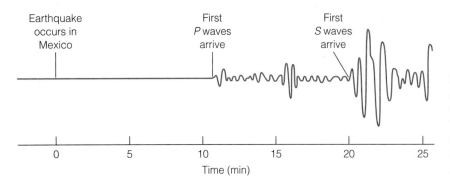

FIGURE 21-3

A seismograph in northern Canada made this record of seismic waves from an earthquake in Mexico. The first vibrations, *P* waves, arrived 11 minutes after the quake, but the slower *S* waves took 20 minutes to make the journey.

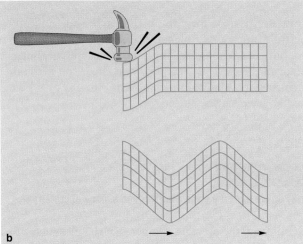

FIGURE 21-4

(a) *P*, or pressure, waves, like sound waves in air, travel as a region of compression. (b) *S*, or shear, waves, like vibrations in a bowl of jelly, travel as displacements perpendicular to the direction of travel. *S* waves tend to travel more slowly than *P* waves and cannot travel through liquids.

As we explore our solar system in the chapters that follow, the earth will be our standard of comparison. Our goal in this chapter is to study the earth's interior, crust, and atmosphere in order to establish a basis for comparison with other planets. Only by understanding the earth in detail can we understand planets in general.

21-2 THE SOLID EARTH

Although we might think of the earth as made of solid rock, it is in fact neither entirely solid nor entirely rock. The thin crust seems solid, but it floats and shifts on a semi-liquid layer of molten rock just below the crust. Below that lies a deep, rocky mantle surrounding a core of liquid metal. Much of what we see on the surface of the earth is determined by its interior.

The Earth's Interior

The theory of the origin of planets from the solar nebula predicts that the earth should have melted and differentiated into a dense metallic core with a low-density silicate crust. But did it differentiate? Clearly, the earth's surface is made of silicates, but what of the interior?

High temperature and tremendous pressure in the earth's interior make any direct exploration impossible. Even the deepest oil wells extend only a few kilometers down and don't reach through the crust. It is quite impossible to drill far enough to sample the earth's core. Yet earth scientists have studied the interior and found clear proof that the earth did differentiate.

This analysis of the earth's interior is possible because earthquakes produce vibrations called **seismic waves,** which travel through the interior and eventually register on sensitive detectors, known as **seismographs,** all over the world (Figure 21-3). Two kinds of seismic waves are important to us here. The **P,** or **pressure, waves** are much like sound waves in that they travel as a region of compression. As a *P* wave passes, particles of matter vibrate back and forth parallel to the direction of wave travel (Figure 21-4a). In contrast, the **S,** or **shear, waves** move as displacements of particles perpendicular to the waves' direction of travel (Figure 21-4b). Thus, *S* waves distort the material but do not compress it. Normal sound waves are pressure waves, whereas the vibrations in a bowl of jelly are shear waves. Because *P* waves are compression waves, they can move through a liquid, but *S* waves cannot. A glass of water can't shimmy like jelly because a liquid does not have the rigidity required to transmit *S* waves.

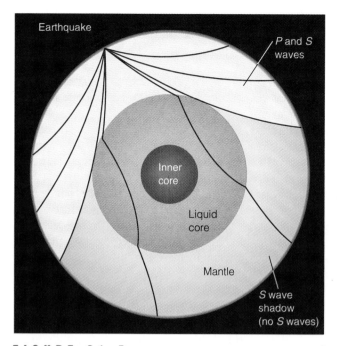

FIGURE 21-5

P and *S* waves give us clues to the earth's interior. That no direct *S* waves reach the far side shows that the earth's core is liquid. The size of this *S*-wave shadow tells us the size of the core.

The *P* and *S* waves caused by an earthquake do not travel in straight lines or at constant speed within the earth. The waves may reflect off boundaries between layers of different density, or they may be refracted as they pass through such a boundary. In addition, the gradual increase in density toward the earth's center means the speed of sound is greater toward the center. Sound waves approaching such a region tend to curve away (refract), so seismic waves do not travel in straight lines through the earth's interior but are refracted (Figure 21-5). Earth scientists can use the arrival times of reflected and refracted seismic waves from distant earthquakes to construct a picture of the earth's interior.

Such studies show that the interior consists of three parts: a central core, a thick mantle, and a thin crust. *S* waves give us an important clue to the nature of the core, because no such waves are observed to travel through the core. When an earthquake occurs, no direct *S* waves register on seismographs on the opposite side of the earth, as if the core were casting a shadow (Figure 21-5). This shows that the core is mostly liquid, and the size of the shadow fixes the size of the core at about 55 percent of the earth's radius. Theoretical calculations predict that the core is hot (about 6000 K), dense (about 14 g/cm^3), and composed of iron and nickel.

Although the core is hot, it is not entirely molten. Nearer the center the material is under higher pressure, which in turn raises the melting point so high that the material cannot melt at the existing temperature (Figure 21-6). Thus, there is an inner core of solid iron and nickel. Estimates suggest the inner core's radius is about 22 percent that of the earth.

The **mantle** is the layer of dense rock and metal oxides that lies between the molten core and the crust. The paths of seismic waves in the mantle show that it is not molten, but it is not precisely solid either. Mantle material behaves like a **plastic,** a material with the properties of a solid but capable of flowing under pressure. The asphalt used in paving roads is a common example of a plastic. It shatters if struck with a sledgehammer, but it bends under the weight of a heavy truck. Just below the earth's crust, where the pressure is less than at greater depths, the mantle is most plastic.

The earth's rocky crust is quite thin, about 60 km (40 mi) under the continents and only 10 km (6 mi) under the oceans. Owing to its lower density (2.5–3.5 g/cm^3), this material floats on the denser mantle (3.5–5.8 g/cm^3). Unlike the mantle, the crust is brittle and breaks much more easily than the plastic mantle. We will see that this is a characteristic property of many planetary crusts.

Perhaps no region is more immediate yet more inaccessible than the core of the earth. Only a few thousand kilometers from where we sit, the earth's interior is beyond our immediate reach. Yet we can send our

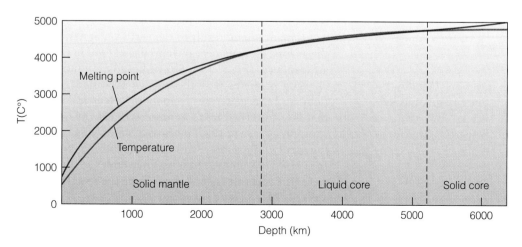

FIGURE 21-6

Theoretical models combined with observations of the velocity of seismic waves reveal the temperature inside the earth (blue line). The melting point of the material (red line) is determined by its composition and by the pressure. In the mantle and in the inner core, the melting point is higher than the existing temperature, and the material is not molten.

Seeing the Unseen

One of the most fascinating aspects of science is its power to show us the unseen. That is, it gives us an opportunity to learn about regions we can never visit. We saw this in earlier chapters when we discussed the inside of the sun and stars, the surface of neutron stars, the event horizon around black holes, the cores of active galaxies, and more. In this chapter, we have seen the core of the earth.

Not only does science take us to places we can never visit, but it takes us to scales we can never explore. For example, physicists can explore the inside of an atom, and biologists can study the structure of a virus. Not only can we stretch the scale of space, but we can also stretch the scale of time. Geologists can watch mountain ranges rising slowly from the plains and eroding away to grit. Astronomers can watch the 200-million-year-long whirl of two galaxies colliding.

In this respect, astronomy, geology, and psychology are particularly similar. All three sciences study realms that cannot be explored in person. The insides of stars, the ancient history of the earth, and the workings of the human mind lie beyond our immediate experience.

Only by careful observation and careful thought can scientists learn to understand these aspects of the natural world.

Science can reveal the unseen because it can combine human imagination with the predictive power of the laws of nature and the precision of mathematics. This calls on scientists to be as creative as artists and as precise as surgeons to produce reliable theories and models, but the reward is one of the great thrills of science—exploring beyond the limits of normal human experience. ■

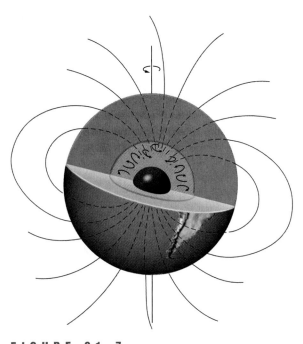

FIGURE 21-7
The dynamo effect couples convection in the liquid core with the earth's rotation to produce electric currents that are believed to be responsible for the earth's magnetic field.

imagination where our bodies can never go (Window on Science 21-1). The earth's seismic activity gives us evidence of the earth's innermost secrets. But there is another source of evidence about the earth's interior—its magnetic field.

The Magnetic Field

Apparently, the earth's magnetic field is a direct result of its rapid rotation and its molten metallic core. The internal heat forces the liquid core to churn with convection while the rotation of the earth turns the core about an axis. This core is a highly conductive iron–nickel alloy, a better electrical conductor than copper, the material commonly used for electrical wiring. The rotation of this convecting, conducting liquid generates the earth's magnetic field in a process called the dynamo effect (Figure 21-7). This is believed to be the same process that generates the solar magnetic field in the convective layers of the sun, and we will meet it again when we explore other planets.

The earth's magnetic field protects it from the solar wind. Blowing outward from the sun at about 400 km/sec, the solar wind consists of ionized gases carrying a small part of the sun's magnetic field. When the solar wind encounters the earth's magnetic field, it is deflected like water flowing around a boulder in a stream. The surface where the solar wind is first deflected is called the **bow shock,** and the cavity dominated by the earth's magnetic field is called the **magnetosphere** (Figure 21-8). High-energy particles from the solar wind leak into the magnetosphere and become trapped within the earth's magnetic field to produce the **Van Allen belts** of radiation. We will see in later chapters that all planets that have magnetic fields have bow shocks, magnetospheres, and radiation belts.

One dramatic result of the earth's magnetic field is the aurora, glowing rays and curtains of light in the upper atmosphere (Figure 21-8b). The solar wind carries charged particles past the earth's extended magnetic field, and this generates tremendous electrical currents that flow into the earth's atmosphere near the north and south magnetic poles. The currents ionize gas atoms in the earth's atmosphere, and when the ionized atoms capture electrons and recombine, they emit light as if they were part of a vast "neon" sign. Thus, the spectrum of an aurora is an emission spectrum.

Although we are confident that the earth's magnetic field is generated within its molten core, many mysteries

a **b**

FIGURE 21-8

(a) The earth's magnetic field dominates space around the earth by deflecting the solar wind around a region called the magnetosphere and by trapping high-energy particles to form radiation belts. (b) Electrical currents in the earth's magnetic field flow downward into the earth's atmosphere and stimulate atmospheric atoms to emit photons, which we see as auroral displays. The general red glow in this photograph and the "curtain" effect are both examples of auroras. (b, Courtesy David Miller)

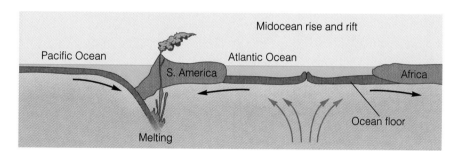

FIGURE 21-9

The solid crust of the earth is about 60 km thick under the continents but only 10 km thick under the oceans. It floats on the plastic mantle, and convection currents in the mantle spread the seafloor open and move the continents.

remain. For example, rocks retain traces of the magnetic field in which they solidify, and some contain fields that point backward. That is, they imply that the earth's magnetic field was reversed at the time they solidified. Careful analysis of such rocks indicates that the earth's field has reversed itself every million years or so, with the north magnetic pole becoming the south magnetic pole and vice versa. These reversals are poorly understood, but they may be related to changes in the core convection.

Convection in the earth's core is important because it generates the magnetic field. As we will see in the next section, convection in the mantle constantly remakes the earth's surface.

The Earth's Crust

The earth's surface is active. It is continually destroyed and renewed as large sections of crust move about. Geologists refer to this phenomenon as **plate tectonics.** *Plate* refers to the sections of moving crust, and *tectonics* comes from the Greek word for "builder." Plate tectonics is the builder of the earth's surface, because interactions be-

tween plates destroy old terrain, push up mountains, and create new crust.

The plates consist of an upper layer of very-low-density rock plus the uppermost layers of the mantle. These layers make up the **lithosphere,** literally the rock-sphere at the earth's surface. At the base of the lithosphere, at a depth of about 100 km, the melting point is low enough that the rock becomes highly plastic. That is, it deforms and flows easily. Thus, the sections of lithosphere—the tectonic plates—move easily on this plastic layer or mantle.

The energy that moves the plates comes from the hot interior. Convection currents of hot mantle material rise from deep layers and spread out under the crust, while cooler regions sink. This circulation carries energy out of the earth's interior and drives the motions of the plates at speeds of a few centimeters per year (Figure 21-9), a motion that radio astronomers can now measure as they compute the distances between antennae in different parts of the world. Radio telescopes in North America, for instance, are moving away from radio telescopes in Europe at about 3 cm per year, all because of convection currents in the earth's mantle.

FIGURE 21-10

If we could drain the Atlantic Ocean, we would see the midocean rift snaking along the crest of the midocean rise from Iceland into the South Atlantic. The ocean floor spreads apart along the rift, pushing the continents apart. As magma rises into the rift and solidifies, it records the alternation of the earth's magnetic field (inset). These alternating bands of normal (dark) and reversed (white) magnetism in the ocean floor are clear evidence of seafloor spreading. (Alcoa)

The key to understanding plate tectonics lies hidden in the ocean floor. The crust there is thin and composed primarily of dense **basalt**, rock characteristic of solidified lava. These seafloors are not at all flat. Running down the center of the ocean beds are undersea mountain ranges called **midocean rises,** and splitting these rises are chasms called **midocean rifts** (Figure 21-10). The rocks of the midocean rises are young, but that only hints at the explanation. The real evidence appeared when scientists measured the residual magnetism of the seafloor and found that some sections retained fields that were the reverse of the earth's present field. Evidently, these regions solidified from molten rock during the earth's periodic magnetic reversals. That these regions show alternating, symmetric bands of magnetism running parallel to the midocean rifts indicates that the earth is

creating new crust along the rifts while the seafloors spread outward.

The Atlantic Ocean serves as a good example (Figure 21-10). Molten material wells up along the midocean rift and solidifies to form young basalt. As the crustal plates spread apart at about 2–4 cm/y, the midocean rift opens, and new material rises to form more crust. Thus, midocean rises are composed of young rock. Parallel magnetic bands form because the earth's field reverses every million years or so, and the solidifying rock records these changes in alternating strips (Figure 21-10 inset) just as the moving tape in a tape recorder makes a record of the changing magnetic field in the recording head.

If the earth is adding crust in one region, it must be destroying it in another. Crust vanishes in trenches along the coasts of some continents where the spreading sea-

FIGURE 21-11

A color-coded relief map of the earth's continents and ocean floors reveals deep trenches in the ocean floor where plates descend (for example, near Japan and Chile) and midocean rises where the ocean floor is spreading apart (for example, the mid-Atlantic). The Hawaiian–Emperor chain of extinct undersea volcanoes extends northwest from Hawaii. (Compare with Figure 21-10.) (National Geophysical Data Center)

floor slides downward (Figures 21-9 and 21-11). The collision of moving plates often crumples the crust and pushes up mountains. In addition, the descending seafloor melts along with its accumulated sediment and releases low-density molten rock that rises to the surface and produces volcanism. The Andes Mountains along the west coast of South America and the associated volcanism are due to the descent of the Pacific Ocean floor beneath the continent. The same process produces the volcanoes (including Mount St. Helens) and earthquakes of Washington state, western Canada, Alaska, and Japan.

One of the telltale marks of plate tectonics is **folded mountain chains,** long, linear ranges of mountains that look much like folds in a rug that has been pushed up against a wall. These mountain ranges are caused by the collision of plates and the compression of the earth's crust. The Allegheny Mountains are typical folded

mountain ranges. We will search for such folded mountain ranges on other planets.

When a seafloor plate collides with a continental plate, it does not always slip below the continent. The floor of the Atlantic Ocean is locked to North and South America and is pushing the continents westward. Tracing this continental drift backward in time, earth scientists find that North and South America were in contact with Europe and Africa only 200 million years ago (Figure 21-12).

This motion continues today. The plates drift at the urging of the currents in the mantle. They split apart to form new seas, come together to crumple the crust into mountains, and slip against each other to generate earthquakes. The first sign of a continent's splitting is a long, straight, deep depression called a **rift valley.** Africa has recently split from Arabia, opening a rift valley now filled

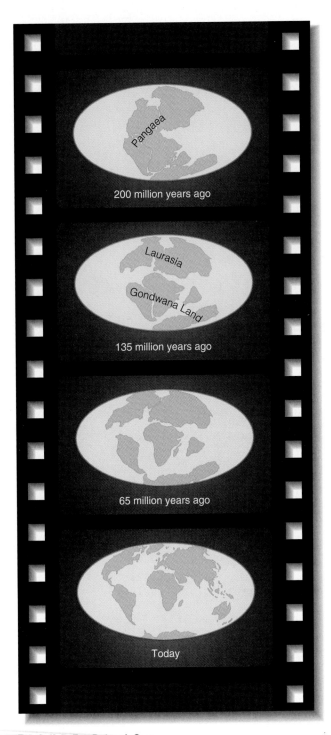

FIGURE 21-12

Continental drift has broken the original land mass Pangaea first into Laurasia and Gondwanaland and eventually into the continents we know. (Adapted from R. S. Dietz and J. C. Holden. *Scientific American* 223, April 1970)

addition, sections of crust in direct contact may slip past each other, as in the case of the Pacific plate, which is carrying part of southern California northward along the San Andreas Fault and generating frequent earthquakes.

Plotting the location of earthquakes and volcanism on a world map reveals the edges of the moving plates (Figure 21-13). The Pacific plate, bounded by a ring of earthquake zones, descends under the Eurasian plate along the North Pacific and Japanese islands. This active zone of circum-Pacific earthquakes and volcanoes is often called the "ring of fire." Earthquakes and volcanoes also occur along the midocean rises where new crust forms and the seafloor spreads.

The Astronomer's Geology

As astronomers, we want to use the earth as a basis for comparison with other worlds, and thus we must take an astronomer's view of geology. As we study geological features on other worlds, we will discover that the earth is astonishing for its geological activity, both through plate tectonics and erosion.

The surface of the earth seems to change very slowly, but that is only because we live such short lives. Fifteen minutes at a bus stop seem an eternity, and a night sleeping in an airport can be an eon. But in truth, a human lifetime is less than an eye blink in geological terms.

The Grand Canyon (Figure 21-14a) seems a vast and imposing geological feature, and we are not surprised to hear that the Colorado River took 10 million years to dig the canyon. The bottom of the canyon is over a mile (1.6 km) below the rim, and it takes several hours to hike to the bottom. The Grand Canyon is a geological paradise, but the astronomer sees it as a flicker in the history of the earth. When we say that the Grand Canyon is 10 million years old, it sounds imposing, but remember that the earth is 4.6 billion years old. If we express 10 million in units of billions, we discover that the river began carving the canyon only 0.01 billion years ago, only a tiny fraction of the history of the earth (Figure 21-14b). Although the human race has dammed the river and temporarily reduced its erosive power, the dam cannot survive more than a few centuries, and once humanity is done playing with the river, it will resume its business in the canyon. The entire canyon will probably be gone in a few times 0.01 billion years.

The astronomer takes a long view of geology. Entire mountain ranges can be pushed up and eroded away to sand in a few times 0.01 billion years. The continents drift and jostle against each other, and we study the breakup of Pangaea 200 million years ago. That is only 0.2 billion years ago. The oldest horizontal sedimentary layers at the bottom of the Grand Canyon were laid down 570 million years ago, and they lie on top of the remains of tilted and eroded layers that were deposited over a hundred million years before and then almost entirely eroded away. All of that happened within the last 1 billion years.

by the Red Sea. Other rift valleys extend southward from the Red Sea across eastern Africa. In contrast, the collision of Africa with the east coast of North America roughly 250 million years ago folded the crust into the Appalachian Mountains, just as the collision of India with southern Asia is now building the Himalaya Mountains. In

FIGURE 21-13

Earthquakes, plotted as red dots, outline the edges of the earth's crustal plates (yellow). Volcanism follows the same pattern, and much of the Pacific is ringed by volcanic peaks that extend from Chile through Central America, Alaska, and south through Japan—the "ring of fire." The earthquakes and volcanism in Hawaii are produced where a hot spot in the mantle penetrates upward through the middle of a plate. (National Geophysical Data Center)

a

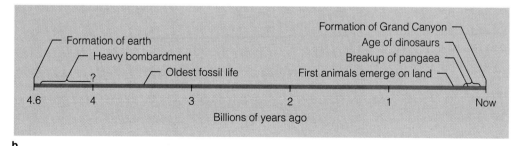

b

FIGURE 21-14

(a) The Grand Canyon illustrates the impermanence of geological features on the earth. The Colorado River began cutting the canyon only 0.01 billion years ago, and the oldest horizontal sedimentary layers at the bottom of the canyon were laid down in the bottom of an ocean 0.57 billion years ago. (M. A. Seeds) (b) On a time line of the geological history of the earth, the cutting of the Grand Canyon can be represented as a narrow line. Many of the geological features we know on the earth and the rise of complex living things have been recent events.

As we visit the other terrestrial planets, we will discover dry river beds 2 billion years old, lava flows 3 billion years old, and craters 4 billion years old. Almost nowhere on the earth can we find geological features with such great ages. The plate tectonics and erosion have wiped away almost all traces of the early crust.

As astronomers, we see the geology of the earth dominated by two dramatic forces. Just below the thin crust of solid rock lies a churning molten interior that rips the crust to fragments and pushes the pieces about like bits of algae on a pond. The second force modifying the crust is water. It falls as rain and snow and tears down the mountains, erodes the river valleys, and washes any raised ground into the sea. Like angry children, tectonics and erosion build mountains and continents and then rip them to nothing.

CRITICAL INQUIRY

What evidence do we have that the earth has a molten metallic core?

Since we can never visit the earth's deep interior, we must depend on observations compared with theory to tell us about the earth's core. Seismic waves from distant earthquakes pass through the earth, and by timing their arrival geologists can build a model of the earth's interior. Because *S* waves cannot travel through the core, scientists conclude that the core must be molten. This is consistent with theoretical predictions based on radioactive decay heating the interior. The earth's magnetic field gives us further evidence of a metallic core. The theory for the generation of magnetic fields, the dynamo effect, requires a liquid, convective, conducting core. If the core were not a liquid, it should not be able to generate a magnetic field. Thus, we have two different kinds of evidence to tell us our planet has a liquid core.

If a theory is to be successful, it must have supporting evidence. What evidence do we have to support the theory of plate tectonics? ■

For the astronomer, accustomed to the longest view of cosmic time, the geology of the earth seems not so much the history of slow-moving rock as it does the churning surface of a pot of boiling soup. And just as steam rises from boiling soup, vapors rise from the crust of the earth. Our atmosphere not only acts to erode the rocky surface, but some of it came from the crust as a by-product of geological activity.

21-3 THE ATMOSPHERE

We cannot complete our four-stage history of the earth without mentioning the atmosphere. Not only is it necessary for life, but it is also intimately related to the crust. It affects the surface by eroding geological features through wind and water, and in turn the chemistry of the earth's surface affects the composition of the atmosphere.

Origin of the Atmosphere

Until recent years, earth scientists believed that the earth's atmosphere had developed slowly by outgassing, the release of gases from a planet's interior. According to this picture, the earth formed by the slow accumulation of planetesimals into a large, cold ball of matter whose composition was uniform from surface to center (Figure 20-20). The early earth might have attracted gases such as hydrogen, helium, and even methane and ammonia from the solar nebula to form a **primeval atmosphere.** The slow decay of uranium, thorium, and potassium heated the earth's interior, melted it, caused it to differentiate into core, mantle, and crust, and drove out gases that diluted the primeval atmosphere. These products of outgassing became a **secondary atmosphere** rich in CO_2 and H_2O.

A more recent understanding of how the solar system formed suggests that the earth formed rapidly and that the heat of formation released by the in-falling planetesimals melted the earth as it formed. Thus, any volatiles in the rock, such as CO_2 and H_2O, would have been driven out immediately, and the atmosphere would have grown rich in these gases from the beginning.

Furthermore, modern astronomers now suspect that the cooling solar nebula condensed volatile-rich materials late in the development of the planets. Thus, the earth may have later been bombarded by lower-density, volatile-rich planetesimals that built the atmosphere and oceans. It is fascinating to speculate that these icy planetesimals may have formed in the outer solar system and been scattered into the inner solar system by the growing Jovian planets. Such icy planetesimals moving through the inner solar system are now known as comets, so we might wonder if the earth's oceans and atmosphere were brought to the earth by a bombardment of comets. If this speculation is true, the night sky must have been illuminated with comets falling through the inner solar system.

The earth may have existed for half a billion years before the condensation of the volatile-rich planetesimals, but once the earth had an atmosphere, the mix of gases began to change. The young atmosphere must have been rich in water vapor, carbon dioxide, and other volatiles. As the atmosphere cooled, the water condensed to form the first oceans, and as the oceans grew bigger, carbon dioxide began to dissolve in the water. Carbon

TABLE 21-1

The Earth's Atmosphere

Gas	Percent by Weight
N_2	75.5
O_2	23.1
Ar	1.29
CO_2	0.05
Ne	0.0013
He	0.00007
CH_4	0.0001
Kr	0.0003
H_2O (vapor)	1.7–0.06

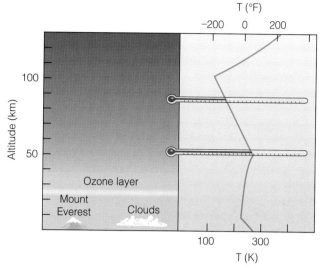

FIGURE 21-15

Where we live, in the lower few kilometers of the earth's atmosphere, the temperature is moderate thanks in part to the small amount of CO_2 in the atmosphere. But not far above the clouds, the temperature of the atmosphere is quite low. The ozone layer lies about 25 km above the earth's surface.

dioxide is soluble in water—which is why carbonated beverages are so easy to manufacture—but the first oceans could not have absorbed all of the atmospheric carbon dioxide if the gas had not reacted with dissolved substances in the seawater to form silicon dioxide, limestone, and other mineral sediments. Thus, chemical reactions in the oceans transferred the carbon dioxide from the atmosphere to the seafloor.

Any methane or ammonia in the atmosphere would have been broken up by ultraviolet radiation from the sun. The earth's lower atmosphere is now protected from ultraviolet radiation by an **ozone layer** about 25 km above the surface. However, the atmosphere of the young earth did not contain free oxygen, so an ozone layer could not form, and the sun's ultraviolet radiation penetrated deep into the atmosphere. The energetic ultraviolet photons broke up water, methane, and ammonia molecules, and the hydrogen escaped to space. Carbon and oxygen formed carbon dioxide and dissolved in the oceans, leaving nitrogen to accumulate in the atmosphere (Table 21-1).

The origin of the atmospheric oxygen is linked to the origin of life, the subject of Chapter 27, but it is sufficient here to note that life must have originated within a billion years of the earth's formation. This life did not significantly alter the atmosphere, however, until nature invented photosynthesis about 3.3 billion years ago. Photosynthesis absorbs carbon dioxide from the air or water and utilizes it for plant growth, releasing oxygen. Because oxygen is a very reactive element, it combines easily with other compounds, and thus the oxygen abundance grew slowly at first. Apparently, the development of large, shallow seas along the continental margins half a billion years ago allowed ocean plants to manufacture oxygen faster than chemical reactions could consume it. Atmospheric oxygen then increased rapidly, and it is still increasing at the rate of about 1 percent every 36 million years.

The climate on the earth is critically sensitive to many different factors. For example, the **albedo** of a world is the fraction of the sunlight that hits it that gets reflected away. An albedo of 1 would be a perfectly white world, and an albedo of 0 would be perfectly black. The earth's overall albedo is 0.39, meaning it reflects back into space 39% of the sunlight that hits it. Much of this reflection is caused by clouds, and the formation of clouds depends critically on the presence of water vapor in the upper atmosphere, the temperature of the upper atmosphere, and the patterns of atmospheric circulation. Even a small change in any of these factors could change the earth's albedo and thus its climate. We are just beginning to understand all of the factors that have combined to assure our comfortable lives on the surface of our world.

Human Effects on the Earth's Atmosphere

The earth is comfortable for human life because of our atmosphere. The temperature is moderate at sea level, but if we climb to the top of a high mountain, we find the temperature to be very low (Figure 21-15). Most clouds form at such altitudes. Higher still we find the air much colder and so thin it cannot protect us from the intense ultraviolet radiation in sunlight. We can walk the surface of the earth in safety because our atmosphere protects us, but we are altering our atmosphere in at least two serious ways: we are adding carbon dioxide (CO_2), and we are destroying ozone.

FIGURE 21-16

The greenhouse effect. (a) Short-wavelength light can enter a greenhouse and heat its contents, but the longer-wavelength infrared (heat) cannot get out. (b) The same process can heat a planet's surface if its atmosphere contains CO_2. (c) The concentration of CO_2 has been rising since it was first measured. The waves are seasonal variations.

The concentration of CO_2 in our atmosphere is important to us because CO_2 can trap heat in a process called the **greenhouse effect** (Figure 21-16a). When sunlight shines through the glass roof of a greenhouse, it heats the benches and plants inside. The warmed interior radiates infrared radiation, but the infrared cannot get out through the glass. Heat is trapped within the greenhouse, and the temperature climbs until the glass itself grows warm enough to radiate heat away as fast as the sunlight enters. In the earth's atmosphere, the CO_2 admits sunlight, and the surface grows warm. However, the CO_2 is not transparent to infrared radiation, so some of the heat is trapped, and that helps keep our planet warm (Figure 21-16b).

The problem is that we are increasing the amount of CO_2 in our atmosphere and are thus making the earth warmer still. For over 4 billion years, the earth has buried carbon in the form of coal, oil, and natural gas. Our civilization is rapidly digging up that carbon, burning it to get energy, and releasing CO_2 into the atmosphere. At the same time, we are cutting down the great forests on our planet that would absorb large amounts of CO_2. The resulting increase in CO_2 concentration has been monitored since 1957 (Figure 21-16c). The global climate has warmed by 0.3° to 0.6°C in the last century, and models predict an increase of as much as 5°C by the year 2050.

World leaders are now seriously concerned about the greenhouse effect, because even a small rise in the average temperature of the earth could cause dramatic changes in our ability to feed our population. Also, a rise of a few degrees would melt part of the polar ice caps and raise sea levels by as much as 50 m. This would flood many low coastal regions such as Florida, the Atlantic coast of North America, and most of the Mississippi River valley. A rise of only 8 m would flood the White House.

To avoid a planetwide disaster, governments will need to limit the amount of CO_2 released to the atmosphere, and that will mean using energy sources that are more efficient and cleaner than the burning of fossil fuels. We must also conserve our forests, because they cleanse the air of excess CO_2.

In addition to the CO_2 problem, we are destroying ozone in the earth's atmosphere. Ozone is an unstable molecule formed of three oxygen atoms (O_3). Ozone is produced in auto emissions, and it is a pollutant in city air. But ozone is produced naturally in our atmosphere at an altitude of about 25 km, as we have seen, and there it is beneficial. This ozone layer absorbs ultraviolet photons and prevents them from reaching the ground.

Certain chemicals called chlorofluorocarbons (CFCs), used in industrial processes and in refrigeration and air conditioning, can destroy ozone. As these CFCs escape into the atmosphere, they become mixed into the ozone layer and convert the ozone into normal oxygen molecules. Oxygen does not block ultraviolet radiation, so depleting the ozone layer will cause an increase in ultraviolet radiation at the earth's surface. In small doses, ultraviolet radiation can produce a suntan, but in larger doses it can cause skin cancers.

The ozone layer over the Antarctic may be especially sensitive to CFCs. Since the late 1970s, the ozone concentration has been falling over the Antarctic, and a hole in the ozone layer now develops over the continent each October—spring for Antarctica. Satellite and ground-based measurements show that the same thing is happening at higher northern latitudes (Figure 21-17) and that the amount of ultraviolet radiation reaching the ground

FIGURE 21-17

Measurements of the ozone over Toronto, Canada, show a seasonal variation, but there is also a gradual decrease in the average. This decrease in ozone has been correlated with an increase in the amount of ultraviolet radiation reaching the ground. (Reprinted with permission from *Science*, vol. 262, 12 Nov. 1993, p. 1032. Copyright American Association for the Advancement of Science, J. B. Kerr and C. T. McElroy.)

is increasing. This is an early warning that human civilization is modifying the earth's atmosphere in a potentially dangerous way.

The CO_2 problem and the ozone problem in the earth's atmosphere are paralleled on Venus and Mars. When we study Venus in Chapter 23, we will discover a runaway greenhouse effect that has made the planet into a thermal Hades. On Mars we will discover an atmosphere without an ozone layer. A few minutes of sunbathing on Mars would kill you. Once again, we will learn more about our own planet by studying other planets.

CRITICAL INQUIRY

Why is our atmosphere poor in carbon dioxide and rich in oxygen?

We might expect our atmosphere to be very rich in carbon dioxide, CO_2. Outgassing from the earth's crust releases mostly CO_2 and water vapor. Luckily for us, CO_2 is highly soluble in water, and the earth's surface temperature is just right for it to be covered with liquid water. The CO_2 dissolves in the oceans and combines with minerals in seawater to form deposits of silicon dioxide, limestone, and other mineral deposits. Thus, the CO_2 is removed from our atmosphere and buried in

the earth's crust. Oxygen, in contrast, is highly reactive and forms oxides so easily it could not remain in our atmosphere very long unless it was continually replenished. We depend on life in the form of green plants to release oxygen into our atmosphere faster than chemical reactions can remove it. Were it not for liquid-water oceans, we would be choking in a thick CO_2 atmosphere with no free oxygen.

Why would an excess of CO_2 and a deficiency of free oxygen be harmful to all life on the earth in ways that go beyond mere respiration? ■

The earth is a rich and beautiful world, and when we visit Venus and Mars, both of which have CO_2-rich atmospheres, we will appreciate the earth even more. But before we rush to compare the earth with planets that have atmospheres, we should examine two worlds that have no atmosphere at all—the moon and Mercury. As usual, the comparison of worlds will be revealing. ■

■ *Summary*

Comparative planetology is the study of planets by comparison and contrast. We use the earth as the basis for our study of planets because we know it best and because it is a planet of extremes. It has an active crust and a water-rich atmosphere.

Like all terrestrial planets, the earth has passed through a four-stage history: (1) differentiation; (2) cratering; (3) flooding by lava, water, or both; and (4) slow surface evolution.

We know that the earth differentiated because seismic waves traveling through its interior reveal a core of nickel and iron, a thick mantle of dense rock, and a thin crust of low-density rock. Except for the very center, the core is molten, and circulation within the conducting liquid combined with the rotation of the earth generates the planet's magnetic field through the dynamo effect.

Circulation in the plastic rock of the upper mantle moves plates of crustal rock and makes the earth's surface active. The crust splits open in midocean rifts where upwelling lava makes new rock, and old crust plunges into deep trenches along the coasts of some continents, where it is remelted. This plate tectonics causes volcanism and earthquakes and constantly recycles the earth's crust. Thus, we see few old rocks and few impact craters.

The earth's atmosphere has developed from gases and ices that were trapped in the in-falling planetesimals that built the earth. Other gases have been outgassed by volcanism, and the oxygen has been added by plant life.

Lowland regions must have been flooded by molten lava when the earth was young, but on earth, water can form a liquid, so the lowlands are now filled with oceans.

The human race is altering the earth's atmosphere in two dramatic ways: we are increasing the greenhouse effect, and we are reducing the ozone layer. The greenhouse effect is produced by carbon dioxide in the atmosphere, which traps heat and makes the earth warmer. As we burn fossil fuels and cut down forests, we add carbon dioxide to the atmosphere and reduce the plant life that cleans the air. Thus, we appear to be making the earth a warmer planet, and that could eventually alter our climate, melt the polar ice caps, flood low regions, and reduce our ability to grow food.

Certain waste gases called chlorofluorocarbons are thinning the ozone layer 25 km above the earth's surface. Ozone absorbs ultraviolet radiation from the sun, and as we thin the ozone layer, we admit more ultraviolet radiation. Such high-energy photons can cause skin cancers.

■ New Terms

comparative planetology
seismic waves
seismograph
pressure (*P*) waves
shear (*S*) waves
mantle
plastic
bow shock
magnetosphere
Van Allen belts
plate tectonics

lithosphere
basalt
midocean rise
midocean rift
folded mountain chain
rift valley
primeval atmosphere
secondary atmosphere
ozone layer
albedo
greenhouse effect

■ Questions

1. How do we know that the earth differentiated?

2. What evidence do we have that the earth's metallic core is molten?

3. How are earthquakes near midocean rifts different from those in southern California?

4. If the earth's core were not an electrical conductor, how would the magnetic field differ?

5. What evidence do we have that the Atlantic Ocean was once a rift valley?

6. How is our concept of the earth's first atmosphere related to the speed with which the earth formed from the solar nebula?

7. What has produced the oxygen in our atmosphere?

8. Why would we expect an increase in CO_2 in the earth's atmosphere to cause a rise in sea levels?

9. Why would a decrease in the density of the ozone layer cause public health problems?

■ Discussion Questions

1. If we visited a planet in another solar system and discovered oxygen in its atmosphere, what might we guess about its surface?

2. If liquid water is rare on the surface of planets, then most terrestrial planets must have CO_2-rich atmospheres. Why?

■ Problems

1. In Figure 21-3, the earthquake occurred 7440 km from the seismograph. How fast did the *P* waves travel in km/sec? in mph? How fast did the *S* waves travel?

2. What percentage of the earth's volume is taken up by its metallic core?

3. If the Atlantic seafloor is spreading at 3 cm/yr and is now 6400 km wide, how long ago were the continents in contact?

■ Critical Inquiries for the Web

1. The process of plate tectonics has modified Earth's surface significantly over the years. Use Web sites that trace the motion of the plates in the past to determine what the surface of Earth looked like 600 million years ago, 400 million years ago, and 200 million years ago. How did North America look at these times?

2. Earth's magnetic poles are not coincident with its axis of rotation, nor are their positions constant in time. Search the Web for information on the locations and variations in positions of the poles. What are the current locations of the northern and southern poles? Why do their positions change?

 Go to the Wadsworth Astronomy Resource Center (www.wadsworth.com/astronomy) for critical thinking exercises, articles, and additional readings from InfoTrac College Edition, Wadsworth's online student library.

Chapter Twenty-Two

THE MOON AND MERCURY: AIRLESS WORLDS

GUIDEPOST

The two preceding chapters have been preparation for the exploration of the planets. In this chapter, we begin that detailed study with two goals in mind. First, we search for evidence to test the solar nebula hypothesis for the formation of the solar system. Second, we search for an understanding of how planets evolve once they have formed. The moon is a good place to begin because people have been there. This is an oddity in astronomy in that astronomers are accustomed to studying objects at a distance. In fact, many of the experts on the moon are not astronomers but geologists, and much of what we will study about the moon is an application of earthly geology.

While no one has visited Mercury, we will recognize it as familiar territory. It is much like the moon, so our experience with lunar science will help us understand Mercury as well as the other worlds we will visit in the chapters that follow. ∎

That's one small step for man . . .

one giant leap for mankind.

NEIL ARMSTRONG
On the Moon

Beautiful, beautiful. Magnificent desolation.

EDWIN E. ALDRIN, JR.
On the Moon

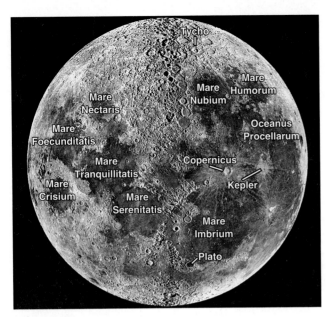

FIGURE 22-1
The moon is tidally locked to Earth and rotates to keep the same familiar side facing us as it follows its orbit. We never see the back side from Earth. The lunar maria are dark and smooth and have picturesque names such as Mare Imbrium, the Sea of Rains, and Mare Tranquillitatis, the Sea of Tranquillity. In fact, there is no water on the moon, and the maria are ancient lava plains. The largest craters are named after famous scientists and philosophers such as Kepler, Copernicus, Tycho, and Plato. This photo is inverted (north is at the bottom), as the moon would appear through an astronomical telescope. (UCO/Lick Observatory)

When Neil Armstrong stepped out onto the moon on July 20, 1969, he spoke words for the history books. Edwin Aldrin, the second man to walk on the moon, spoke more directly about what he saw. The lunar landscape is indeed "magnificent desolation." Had Armstrong and Aldrin landed on Mercury, they would have had much the same view: an airless, rocky surface blasted to dust by over 4 billion years of impacts.

In this chapter, we begin our study of other worlds by applying comparative planetology to the study of the moon and Mercury. As we compare them with each other and with Earth, we will discover three important themes of planetary astronomy. First, we will see the importance of impact cratering. The surfaces of the moon and Mercury have been scarred by the impact of uncountable meteorites in a process that must have affected all the planets. Second, we will observe the importance of internal heat in worlds smaller than Earth. Third, we will discover the possibility of giant impacts so large they could shatter planets. By concentrating on these three principles, we will be able to organize the flood of details astronomers have learned about the moon and Mercury.

Our ultimate goal in the study of any world is to tell its story. How did the planet form, how has it evolved, and why is it the way it is? Each time we study a new planet, we will conclude our discussion by trying to combine what we have learned to tell the planet's story. For the moon and Mercury, we will use our three themes—cratering, internal heat, and giant impacts—to try to understand how small worlds can become such magnificently desolate places.

A purist might argue that the moon and Mercury can't be similar because one is a satellite and one is a planet. But both are worlds. They have had complex geological histories that we can compare with each other and with Earth's history. The moon and Mercury illustrate the fate of small, airless worlds.

22-1 THE MOON

Only 12 people have stood on the moon, but planetary astronomers know it well. The photographs, measurements, and samples brought back to Earth paint a picture of an airless, ancient, battered crust.

The View from Earth

A few billion years ago, the moon must have rotated faster, but Earth is four times larger in diameter than the moon (Data File Three), and its tidal forces on the moon are strong. Earth's gravity raised tidal bulges on the moon, and friction in the bulges slowed the moon until it now rotates once each orbit to keep the same side facing Earth in what is called **tidal coupling.** This means that we who live on Earth always see the same side of the moon; we never see the back. The moon's familiar face has shone down on Earth since long before there were humans (Figure 22-1).

Even from Earth, we can tell that the moon is an airless world. How do we know? First, our understanding of gravity tells us that a world as small as the moon must have a low escape velocity (Data File Three), and gas atoms near its surface escape easily into space. Furthermore, we can see dramatic shadows near the **terminator,** the dividing line between daylight and darkness. There is no air on the moon to scatter light and soften shadows. We also see stars disappear behind the **limb** of the moon —the edge of its disk—without dimming. Clearly, the moon is an airless (and soundless) world.

The most dramatic features on the moon, regions that are visible even to our unaided eyes, are **mare,** Latin for "sea." The dark maria (plural of *mare*) that make up the face of the man in the moon were so named because early observers thought the great smooth expanses were seas. Even a small telescope will show that the maria are marked by ridges, faults, smudges, and craters and therefore can't be water. The maria are actually great lowland plains covering 17 percent of the moon's surface.

Not many decades ago, astronomers argued about the nature of the maria. Some said that they were great solidified lava flows, but others argued that they were vast

a

b

FIGURE 22-2

(a) The edge of Mare Imbrium is marked by high mountains and the sinuous rille called Hadley Rille (arrow). Such rilles were caused by lava flowing through a channel. Other rilles, straight faults, and many small craters are visible in the surface of the mare. (b) Apollo 15 landed near the 1200-foot-deep rille (circle in part a) and used a battery-powered rover to visit the rille and nearby mountains and craters. (NASA)

DATA FILE THREE

THE MOON

Earth   Moon

The moon's diameter is roughly 27 percent that of Earth.

Average distance from Earth	384,400 km (center to center)
Eccentricity of orbit	0.055
Maximum distance from Earth	405,500 km
Minimum distance from Earth	363,300 km
Inclination of orbit to ecliptic	5°9′
Average orbital velocity	1.022 km/sec
Orbital period (sidereal)	27.321661 days
Orbital period (synodic)	29.5305882 days
Inclination of equator to orbit	6°41′
Equatorial diameter	3476 km
Mass	7.35×10^{22} kg (0.0123 $M_\oplus$)
Average density	3.36 g/cm^3 (3.35 g/cm^3 uncompressed)
Surface gravity	0.167 earth gravities
Escape velocity	2.38 km/sec (0.21 $V_\oplus$)
Surface temperature	−170° to 130°C (−274° to 266°F)
Average albedo	0.07

oceans of dust blasted from the lunar mountains by billions of years of meteorites and drawn to the lowlands by the lunar gravity. Science fiction writers wrote fascinating stories about the dust oceans. For example, *A Fall of Moon Dust* by Arthur C. Clarke tells of tourists in a lunar sightseeing boat trapped when their boat sinks in the dust to the bottom of a mare. Although it is a wonderful rescue story, it can never happen. We know now that the maria are great lava flows with only a few centimeters of loose dust on their surfaces.

These lava flows suggest volcanism, but only small traces of past volcanic activity are visible from Earth. No major volcanic peaks are visible, and no active volcanism has been detected. The lava flows that created the maria happened long ago and were much too fluid to build peaks. We can see a few small domes pushed up by lava below the surface, and we can see long, winding channels called **sinuous rilles** (Figure 22-2), often found near the edges of the maria. These channels were evidently cut by flowing lava. In some cases, such a channel may have had

a

b

FIGURE 22-3

(a) The lunar highlands are the oldest parts of the moon and are heavily battered by craters. (b) The lunar lowlands are filled by solidified lava flows containing few craters and long wrinkle ridges. The lava flows occurred after the end of the heavy bombardment and thus escaped most cratering impacts. (Lick Observatory)

a roof of solid rock, forming a lava tube. When the lava drained away, meteorite impacts collapsed the roof to form a sinuous rille. Thus, the view from Earth gives us only hints of volcanic activity associated with the maria.

Besides the maria, the most striking feature of the moon is the millions of impact craters. The lighter parts of the moon's face are heavily cratered highlands. Although they look mountainous at first glance, there are no folded mountain ranges on the moon. Its crust has never wrinkled into folded mountains as Earth's crust has. The mountainous regions of the highlands are actually millions of meteorite craters, one on top of the other (Figure 22-3a). We will discuss the process that forms craters in the next section.

The crater density in different regions of the moon gives us a way of estimating the ages of different features, important information if we are to tell the story of the moon. The highlands must be old to have accumulated so many craters, but the maria contain only a few craters, so they can't be that old (Figure 22-3b). Most of the craters we see on the moon were produced during the first half billion years of the moon's existence by the heavy bombardment at the end of planet formation (Chapter 20). As the debris left over from the solar nebula was gradually swept up, the rate of cratering fell rapidly

(Figure 22-4), and craters from the early heavy bombardment were probably totally destroyed by later cratering. The craters we see in the highlands were made by the last of the heavy bombardment about 3.8 billion years ago. The lava flows that fill the maria, which have very few craters, must have formed after the end of the heavy bombardment.

A few minutes at a telescope will reveal clear cases of craters that partially cover other craters. Evidently, older craters were partially destroyed by newer craters. We can also find craters that have been partially buried by maria. Evidently, these craters are older than the maria. These are called **relative ages** because we can tell which feature formed first but we can't give either feature an age in years. By comparing the number of craters on a given surface with the rate of cratering as given in Figure 22-4, planetary astronomers can estimate **absolute ages** in years. Of course, an even more accurate way to find the ages of lunar features is to go to the moon and bring back rocks that can be dated radioactively. We will discuss these radioactive ages later.

The view from Earth tells us that the moon has an ancient crust that was heavily cratered when the solar system was young. The maria are slightly younger than the highlands, but all parts of the moon are old and scarred

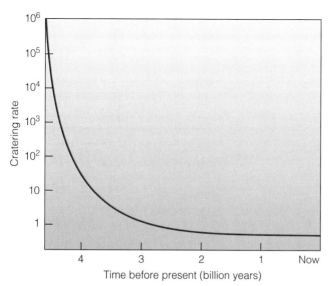

FIGURE 22-4

The cratering rate was high during the first half billion years after the formation of the solar system, but as the debris was swept up, the cratering rate fell. The maria formed after the end of the heavy bombardment, and the cratering rate today is very low. Notice that the exponential scale on the vertical axis makes the curve seem less extreme than it really is. By 4 billion years ago the cratering rate had already fallen to 10,000 times less than when the planets formed, and by 3 billion years ago it had fallen to 1 million times less.

by craters. Our goal is to tell the story of the moon (Window on Science 22-1), and to do that we must understand how craters are formed.

Impact Craters

When a meteorite strikes a planet, it delivers tremendous energy and can excavate an impact crater many times its own diameter. Meteorites strike the surface of a planet with velocities that range from 15 to 60 km/sec, and even a small meteorite can carry tremendous energy. A meteorite with the mass of a house would hit with energy equivalent to a large nuclear bomb.

When a meteorite strikes, it penetrates the surface, deforms the rock layers, and converts some of its energy into enough heat to vaporize itself and nearby rock (Figure 22-5). The deformed rock rebounds, and shock waves move upward and outward, driven in part by the heat of the impact. These rebounding shock waves blast away the surface to create a crater. The resulting scar is a circular crater over ten times the diameter of the impacting meteorite and about 10 percent as deep as it is wide. Rock blasted out of the crater falls to form a rim that rises slightly above the surrounding surface.

When we look at the moon through even a small telescope, we see many large craters that contain a central mountain and terraces along the inner walls of the crater. The central mountains are produced by the rebound of the fractured rock under the center of the

FIGURE 22-5

The formation of a crater. An approaching meteorite (a) strikes the lunar surface (b) and penetrates deeply. The meteorite vaporizes, and shock waves fracture the rock and blast the surface away (c) to form a circular crater. (d) Ejected material forms a raised crater rim. The rebound of the rock can form a raised central peak, and slumping along the crater wall can form terraces. The rock below the crater is severely fractured.

Lost in the Details of Science

Like any technical subject, science includes a mass of details, facts, figures, measurements, and observations. It is easy to be overwhelmed by the flood of details, but one of the most important characteristics of science comes to our rescue. The goal of science is not to discover more details, but to explain the details with a unifying hypothesis or theory. A good theory is like a basket that makes it easier for us to carry a large assortment of details.

This is true of all the sciences. When a psychologist begins studying the way the human eye and brain respond to moving points of light, the data are a sea of detailed measurements and observations. Once the psychologist forms a hypothesis about the way the eye and brain interact, the details fall into place as parts of a logical story. If we understand the hypothesis, the details all fit together and make sense, and thus we can remember the details without blindly memorizing tables of facts and figures. The goal of science is understanding, not memorization.

Scientists are in the story-telling business. The stories are often called hypotheses or theories, but they are really just stories to explain how nature works. The difference between scientific stories and works of fiction lies in the use of facts. Scientific stories are constructed to fit all known facts and are then tested over and over against new facts obtained by observation and experiment.

When we try to tell the story of each planet in our solar system, we pull together all the hypotheses and theories and try to make them into a logical history of how the planet got to be the way it is. Of course, our stories will be incomplete because astronomers don't understand all the factors in planetary astronomy. Nevertheless, our story of each planet will draw together the known facts and details and attempt to make them a logical whole.

Memorizing a list of facts can give us a false feeling of security, just as when we memorize the names of things without understanding them (Window on Science 2-2). Rather than memorizing facts, we should search for the unifying hypothesis that pulls the details together into a single story. Our goal in studying science is understanding nature, not just remembering facts. ■

---

crater. The terraces are produced by slumping as material slides back down into the crater from the inner walls. In even larger craters, we see an inner ring of mountains, and astronomers refer to these features as **multiringed basins.** We will find multiringed basins on the moon and other worlds.

Debris blasted out of the crater is called **ejecta;** it falls back to the surface to create an ejecta blanket surrounding the crater. Some ejecta falls back into the crater, but some can be blasted far across the surface. On the moon, we can see the ejecta blankets around young craters because the rough surface scatters sunlight and looks bright. Ejecta shot out along specific directions can fall back to the surface and form bright **rays.** The crater Copernicus is the center of a vast ray system (Figure 22-1), and rays from the crater Tycho extend halfway around the moon. Older craters must have had rays, but exposure to sunlight and a continuous blasting from micrometeorites smaller than grains of sand gradually darken and smooth the lunar surface and obliterate the rays. Thus, the presence of bright, rayed craters shows that at least a few craters have been formed recently. Tycho, for example, may be only 100 million years old.

Impacts on the moon are tremendously powerful. Chunks of ejecta can travel large distances and fall back to the surface to form more craters called **secondary craters.** Some secondary craters are even big enough to form their own secondary craters. Some lunar impacts have been violent enough to blast rocks completely off the moon, and a few such rocks have been found on Earth. They must have been ejected from the moon and fallen to Earth as meteorites (Figure 22-6).

FIGURE 22-6
The meteorite ALHA81005 was found in Antarctica in January 1982 and later identified as a fragment of the lunar surface. (NASA, Ursula B. Marvin)

In the introduction to this chapter, we identified three themes of planetary astronomy that we would meet in studying the moon and Mercury. The first was impact cratering, and the blasted face of the moon tells us of a time when meteorites rained down on the lunar surface. We will see evidence of this early bombardment on other worlds.

The view from Earth is stark and fascinating, but we can learn only so much by looking. The only way to really understand the moon is to go there.

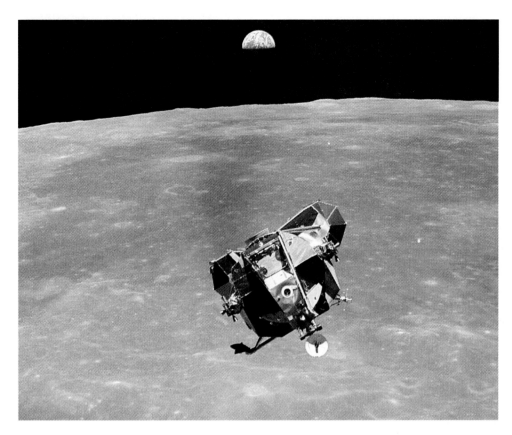

FIGURE 22-7
The Apollo astronauts took two spaceships to the moon. While one astronaut stayed in orbit in the command
module, the other two rode the lunar
lander down to the surface. (NASA)

The Apollo Missions

On May 25, 1961, President John Kennedy addressed Congress and committed the United States to landing a human being on the moon by 1970. The reasons for that decision are complicated and related more to economics and the stimulation of technology than to science, but the Apollo program became a fantastic adventure in science, a flight to the moon that changed how we think about Earth.

Flying to the moon is not particularly difficult; with powerful enough rockets and enough food, water, and air, it is a simple trip. Landing on the moon is more difficult but not impossible. The moon's gravity is only one-sixth that of Earth, and there is no atmosphere to disturb the trajectory of the spaceship. Getting to the moon is simple and landing is possible; the difficulty is doing both on one trip. The spaceship must carry food, water, and air for a number of days in space plus fuel and rockets for midcourse corrections and for a return to Earth. All of this adds up to a ship that is too massive to make a safe landing on the lunar surface. The solution was to take two spaceships to the moon, one to ride in and one to land in (Figure 22-7).

The command module was the long-term home and command center for the trip. Three astronauts had to live in it for a week, and it had to carry all the life-support equipment, navigation instruments, computers, power packs, and so on for a week's jaunt in space. The lunar landing module (LM for short) was tacked to the front of the command module like a bicycle strapped to the front of the family camper. It carried only enough fuel and supplies for the short trip to the lunar surface, and it was built to minimize weight and maximize maneuverability.

The weaker gravity of the moon made the design of the LM simpler. Landing on Earth requires reclining couches for the astronauts, but the trip to the lunar surface involved smaller accelerations. In an early version of the LM, the astronauts sat on what looked like bicycle seats, but these were later scrapped to save weight. The astronauts had no seats at all in the LM, and once they began their descent and acquired weight, they stood at the controls held by straps, riding the LM like two daredevils riding a rocket surfboard.

Lifting off from the lunar surface, the LM saved weight by leaving the larger descent rocket and support stage behind. Only the astronauts, their instruments, and the cargo of rocks returned to the command module orbiting above. Again, the astronauts in the LM blasted up from the lunar surface standing at the controls. The rocket engine that lifted them back into lunar orbit was not much bigger than a dishwasher.

The most complicated parts of the trip were the rendezvous and docking between the tiny remains of the LM and the command module, aided by radar systems and computers. The astronauts rejoined their colleague in the command module, transferred their moon rocks, and jettisoned the LM. Only the command module returned to Earth.

The first manned lunar landing was made on July 20, 1969. While Michael Collins waited in orbit, Neil Armstrong and Edwin Aldrin, Jr., took the LM down to the

TABLE 22-1

Apollo Lunar Landings

Apollo Mission*	Astronauts: Commander LM Pilot CM Pilot	Date	Mission Goals	Sample Weight (kg)	Typical Samples	Ages (10^9 y)
11	Armstrong Aldrin Collins	July 1969	First manned landing; Mare Tranquillitatis	21.7	Mare basalts	3.48–3.72
12	Conrad Bean Gordon	November 1969	Visit Surveyor 3; sample Oceanus Procellarum (mare)	34.4	Mare basalts	3.15–3.37
14	Shepard Mitchell Roosa	February 1971	Fra Mauro, Imbrium ejecta sheet	42.9	Breccia	3.85–3.96
15	Scott Irwin Worden	July 1971	Edge of Mare Imbrium and Apennine Mountains, Hadley Rille	76.8	Mare basalts; highland anorthosite	3.28–3.44 4.09
16	Young Duke Mattingley	April 1972	Sample highland crust; Cayley formation (ejecta); Descartes	94.7	Highland basalt; breccia	3.84 3.92
17	Cernan Schmitt Evans	December 1972	Sample highland crust; dark halo craters; Taurus–Littrow	110.5	Mare basalt; highland breccia; fractured dunite	3.77 3.86 4.48

*The Apollo 13 mission suffered an explosion on the way to the moon and did not land.

a

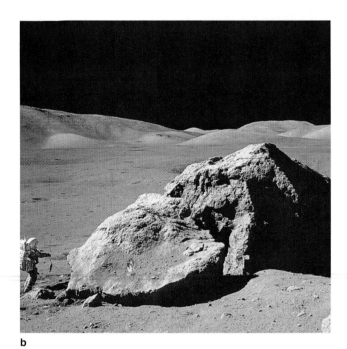

b

FIGURE 22-8

(a) Edwin E. Aldrin, Jr., one of the three Apollo 11 astronauts, stands on the dusty surface of Mare Tranquillitatis in the lunar lowlands, site of the first lunar landing. The flat lava plain is almost featureless, and the horizon is straight. (b) In the lunar highlands at Taurus–Littrow, the surface is mountainous and irregular. Note the horizon. Harrison Schmitt, a scientist–astronaut on Apollo 17, is shown passing huge boulders. (NASA) surface. Although much of the descent was controlled by computers, the astronauts had to override a number of computer alarms and take control of the LM to avoid a boulder-strewn crater bigger than a football field. Climbing down the ladder, Armstrong and then Aldrin stepped onto the lunar surface.

Between July 1969 and December 1972, 12 people reached the lunar surface and collected 380 kg (840 lb) or rocks and soil (Table 22-1). The flights were carefully planned to visit different regions and thus develop a comprehensive history of the lunar surface.

The first flights went into relatively safe landing sites (Figure 22-8a)—Mare Tranquillitatis for Apollo 11 and Oceanus Procellarum for Apollo 12. Apollo 13 was aimed at a more complicated site, but an explosion in an oxygen tank on the way to the moon ended all chances of a landing and nearly cost the astronauts their lives. They succeeded in using the life support in the LM to survive.

The last four Apollo missions, 14 through 17, sampled geologically important places on the moon. Apollo 14 visited the Fra Mauro region, which is covered by ejecta from the impact that dug the basin now filled by Mare Imbrium. Apollo 15 visited the edge of Mare Imbrium at the foot of the Apennine Mountains and examined Hadley Rille (Figure 22-2b). Apollos 16 and 17 visited highland regions to sample the older parts of the lunar crust (Figures 22-8b and 22-9). Almost all of the lunar samples from these six landings are now held at the Planetary Materials Laboratory at the Johnson Space Center in Houston. They are a national treasure containing clues to the beginnings of our solar system.

Moon Rocks

Of the 382 kg (842 lb) of rock samples that the Apollo astronauts carried back to Earth, every one is igneous. That is, they formed by the solidification of molten rock. No sedimentary rocks were found, which is consistent with the moon never having had liquid water on its sur-

FIGURE 22-9

The landing sites of the six Apollo missions that reached the moon. For the names of lunar features, see Figure 22-1.

face. In addition, the rocks were extremely dry. Almost all earth rocks contain 1–2 percent water either as free water trapped in the rock or as water molecules chemically bonded with certain minerals. But moon rocks contain no water at all.

Rocks from the lunar maria are dark-colored, dense basalts much like the solidified lava produced by the Hawaiian volcanoes. These rocks are rich in heavy elements such as iron, manganese, and titanium, which give them their dark color. Some of the basalts are **vesicular,** meaning that they contain holes caused by bubbles of gas in the molten rock (Figure 22-10a). Like bubbles in a carbonated beverage, these bubbles do not form while the

a

b

c

FIGURE 22-10

Moon rocks. (a) A vesicular basalt contains numerous bubbles frozen into the rock when it solidified from a lava flow on the surface. (b) A breccia is formed from two kinds of rocks cemented together. (c) Rocks exposed on the surface of the moon for long periods become pitted with small micrometeorite craters. (NASA)

Follow the Energy

One of the best ways to think about a scientific problem is to follow the energy. According to the principle of cause and effect, every effect must have a cause, and every cause must involve energy (Window on Science 5-3). Energy moves from regions of high concentration to regions of low concentration and in doing so produces changes. For example, we may burn coal to make steam in a power plant, and the steam passes through a turbine and then escapes into the air. In flowing from the burning coal to the atmosphere, the heat spins the turbine and makes electricity.

It is surprising how commonly scientists use energy as a key to understanding nature. A biologist might ask where certain birds get the energy to fly thousands of miles, and an economist might ask where the economic energy to support the creation of new investments comes from. Energy is everywhere, and when it moves, whether it is birds, money, or molten magma, it causes change. Energy is the cause in cause and effect.

In earlier chapters, we used the flow of energy from the inside of a star to its surface to understand how stars, including the sun, work. We saw that the outward flow of energy supports the star against its own weight, drives convection currents that produce magnetic fields, and causes surface activity such as spots, prominences, and flares. We were able to understand stars because we could follow the flow of energy outward from their interiors.

We can think of a planet by following the energy. The heat in the interior of a planet may be left over from the formation of the planet, or it may be heat generated by radioactive decay, but it must flow outward toward the cooler surface, where it is radiated into space. In flowing outward, the heat can cause convection currents in the mantle, magnetic fields, plate motions, quakes, faults, volcanism, mountain building, and more.

When you think about any world, be it a small asteroid or a giant planet, think of it as a source of heat that flows outward through the planet's surface into space. If we can follow that energy flow, we can understand a great deal about the world. A planetary astronomer once said, "The most interesting thing about any planet is how its heat gets out." ∎

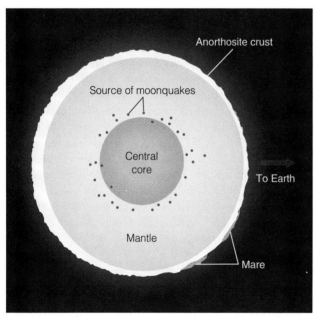

FIGURE 22-11

A cross section of the moon shows an anorthosite crust thinner on the side facing Earth and a deep mantle of brittle rock. Moonquakes occur deep in the mantle and suggest that the moon has a central core of rock that may be slightly less brittle than the mantle. The low density of the moon and its lack of a magnetic field rule out any iron core more than a few hundred kilometers in diameter.

magma is under pressure. Only when the rock flows out onto the surface, where the pressure is low, do bubbles appear. Thus, the presence of vesicular basalts assures us that these rocks formed in lava flows that reached the surface and did not solidify underground. The ages of the mare basalts can be found from the radioactive atoms they contain (Chapter 20), and these ages range from about 3.1–3.8 billion years. Thus, the lava flows that formed these rocks must have happened some time after the formation of the moon.

Rocks from the highlands are light-colored, lower-density **anorthosite.** This rock is rich in calcium, aluminum, and oxygen and is the kind of rock we would expect to form first and float to the top of cooling molten magma. It seems that the anorthosite is the true crustal rock of the moon, while the basalts are lava flows that filled the lowlands. The ages of the anorthosite range from 4.0–4.5 billion years, significantly older than the mare basalts.

A large fraction of the lunar rocks are **breccias,** rocks that are made up of fragments of earlier rocks cemented together by heat and pressure (Figure 22-10b). Evidently, meteorite impacts have broken up the rocks and fused them together time after time.

Both the highlands and the lowlands of the moon are covered by a layer of powdered rock fragments called the **regolith.** It is about 10 m deep on the maria but over 100 m deep in certain places in the highlands. About 1 percent of the regolith is meteoric fragments; the rest is the smashed remains of moon rocks that have been ground down to dust by the constant blasting of meteorites. The smallest meteorites, the **micrometeorites,** do the most damage by their constant sand blasting of the lunar sur-

FIGURE 22-12
The entire surface of the moon has been blasted by meteorite impacts and is fractured to a depth of about 2 km. The uppermost layers of the surface are rock dust produced by meteorite impacts. The large rock visible here is ejecta from a meteorite impact that must have occurred far beyond the horizon. (NASA)

face (Figure 22-10c). Thus, we see that impact cratering dominates the lunar surface and is even responsible for the lunar regolith.

The moon rocks are old, dry, igneous, and badly shattered by impacts. We can use these facts, combined with what we know of lunar features, to reconstruct the moon's past.

The History of the Moon

The ultimate goal in the study of any world is to describe its history, to explain how it evolved from its formation to its present state (Window on Science 22-1). We have seen that the surface of the moon has been severely battered by impact craters, but the maria tell of a different theme in lunar history—the moon's internal heat. Like Earth, the moon formed with heat trapped in its interior, and radioactive decay released more heat. Unlike Earth, the moon is small, and that has sealed its fate. Because it is small, it could not retain its internal heat, it aged quickly, and it is now a dead world (Window on Science 22-2).

There is no evidence that the interior of the moon now contains significant heat. The moon has no magnetic field of its own, so it cannot contain a core of molten iron. In fact, the moon's average density is 3.36 g/cm³, so low we can conclude that the moon cannot contain more than a tiny iron core a few hundred kilometers in diameter. The astronauts left seismographs on the lunar surface, which have detected moonquakes about as large as big firecrackers deep in the interior near the boundary between the brittle rock of the mantle and a more plastic rock in a central core (Figure 22-11).

The moon is now a cold, dead, geologically inactive world, but it began as a sea of molten rock. The lack of

rocks older than about 4.3 billion years and the chemical composition of the anorthosite have convinced planetary astronomers that the surface of the moon was covered by a sea of magma when it was very young. The energy to melt the rock probably came from the heat of formation (Chapter 20) released by in-falling meteorites.

In the first of the four stages of planetary evolution, the moon cooled and differentiated as the denser minerals sank into the interior and the lower-density minerals floated to the top, like slag in an open-hearth furnace. Given the present chemical composition of the moon, we would expect this slag to cool and form an anorthosite crust within a few hundred million years of the moon's formation.

The second stage of planetary evolution is cratering, and the moon was heavily battered by meteorites. The early heavy bombardment must have blasted the newly formed crust over and over. Most rocks older than 4 billion years are smashed to fragments, and the entire crust of the moon is fractured to a depth of a few kilometers. The largest meteorites excavated vast basins. The mare Oceanus Procellarum may overlie one of the oldest and largest basins on the moon, but later impacts and lava flows have partially obliterated its circular outline. Mare Imbrium was excavated by the impact of a meteorite the size of Rhode Island (roughly 65 km across). The resulting basin was 1300 km across, and the impact smashed the crust into mountain ranges up to 5 km high. Ejecta from this impact covers 16 percent of the lunar surface (Figure 22-12). Apollo 14 landed at Fra Mauro to sample it.

The Imbrium Basin is hard to study because it is flooded with later lava flows, but a similar basin, Mare Orientale, lies at the western limb of the moon and has not been flooded. Mare Orientale is a great multiringed basin consisting of three circular mountain ranges that were smashed into the lunar crust by the impact. The Imbrium Basin has similar rings, although they are mostly buried under successive lava flows (Figure 22-13).

Our story of the moon might suggest that it was a violent place during this cratering phase, but large impacts were in fact rare and, except for a continuous rain of smaller meteorites, the moon was, for the most part, a peaceful place even during the heavy bombardment. Centuries might pass between major impacts. Of course, a large impact might have threatened to bury us under ejecta or jolted us by seismic shocks. We could have felt the Imbrium or Orientale impacts anywhere on the moon, but had we been standing on the side of the moon opposite one of these impacts, we would have been at the focus of seismic waves traveling around the moon from different directions. When the waves met under our feet, the surface would have jerked up and down by as much as 10 m. The places on the moon opposite the Imbrium Basin and the Orientale Basin are both strangely disturbed landscapes called **jumbled terrain.** We will find similar effects of large impacts on other worlds.

By the end of the heavy bombardment, the moon's crust was heavily cratered and fractured, and the moon had cooled significantly. Below the crust, however, the rock was molten, and the decay of radioactive elements added heat. The fractures below the largest basins allowed some of that magma to reach the surface (Figure 22-14), and the resulting lava flows filled the basins with layer after layer of dense magma that solidified to form the maria. Some maria are round, such as Mare Crisium, following the outline of the basin, but others are irregular in shape, showing that the lava overflowed the basin boundaries.

This flooding, the third stage of planetary evolution, did not occur all at once. Studies show that maria consist of successive lava flows one on top of another (Figure 22-15). In many maria, the weight of the lava pressed the crust downward and created wrinkle ridges (see Figure 22-3b). In other cases, the lava melted channels or tunnels through the rock, features we see today as sinuous rilles (see Figure 22-2).

Near Side

a

FIGURE 22-13

(a) A topographic map of the moon's near side shows altitude as color. Blue regions lie about 3 km below the average radius of the moon and red regions about 3 km above. The lunar maria are deep, generally round basins later flooded by lava. In this map from the Clementine spacecraft, north is at the top, and Mare Imbrium is the large round basin at the upper left. (NASA/JPL) (b) Major impacts produce multi-ringed basins and break the crust to great depth. Magma can follow these fractures upward to flood the basins. (Adapted from William Hartmann, *Astronomy: The Cosmic Journey,* Belmont, Calif.; Wadsworth Publishing Company, 1978)

300 m
ejecta

3000 m
lava

|— 200 km —|

b

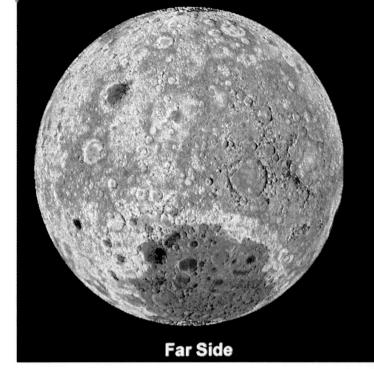

Far Side

FIGURE 22-15

Lava flows occurred repeatedly on the lunar near side but flooded only a few of the small basins on the lunar far side. Nevertheless, the far side of the moon is marked by some very deep basins including the South Pole–Aitken basin (purple). Although it is very deep, it has an old, cratered floor and appears to have suffered little flooding. (NASA/JPL)

FIGURE 22-14

The evolution of the lunar surface. Mare Imbrium began when a meteorite as large as Rhode Island excavated a great crater basin. Lava welled up to flood the basin, starting about 3.8 billion years ago and lasting until no less than 2 billion years ago. Since then only scattered meteorites have modified Mare Imbrium. (Don Davis)

Almost all of this flooding occurred on the side of the moon facing Earth. The back side of the moon is heavily cratered, but only a few of the largest craters contain any lava flows. Even the multiringed basin Mare Orientale, near the west limb, is only slightly flooded. The difference is due to the thickness of the crust. On the side toward Earth, the lunar crust is only about 60 km thick, but on the far side it is more than 100 km thick (Figure 22-11). This asymmetry may have resulted from the immense impact that formed the Oceanus Procellarum Basin and weakened the near-side crust.

The fourth stage of planetary evolution, slow surface evolution, is hardly evident on the moon. It is a small world, and small worlds cool rapidly because they have a large ratio of surface area to volume, just as a small cupcake cools more rapidly than a large cake. The moon has thus lost much of its internal heat. We saw in the case of Earth that geological activity is driven by energy flowing upward from within. But the moon has lost much of its internal energy, and the crust has grown thick and has never divided into moving plates. There are no rift valleys or folded mountains on the moon. The moon is too small to retain an atmosphere, so it has never had liquid water on its surface to cause erosion. The only slow surface evolution on the moon is the constant blast of micrometeorites that grinds the surface rocks down to dust. The footprints of the Apollo astronauts will last for millions of years, but they will eventually be erased by the constant flux of meteorites.

a

b

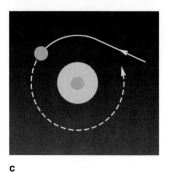

c

FIGURE 22-16
Three traditional theories for the moon's origin. (a) Fission theories suppose that Earth and the moon were once one body and broke apart. (b) Co-accretion theories suppose that the moon formed from material near Earth. (c) Capture theories suggest that the moon formed elsewhere and was captured by Earth. None of these theories explains all the facts.

On the other hand, the overall terrain on the moon is almost fixed. A billion years from now, plate tectonics will have totally changed the map of Earth, erosion will have worn away the Rocky Mountains, the descent stage of the Apollo 17 lunar lander will be nothing but a peculiar chemical contamination in the regolith at Taurus–Littrow, but the lunar mountains will look much the same.

Our story of the moon's history is as complete as comparative planetology can make it, but it omits one initial detail. Where did Earth get such a large satellite?

The Origin of the Moon

The Apollo missions to the moon told us a great deal, but they failed to answer the most basic question: how did the moon form? The three traditional theories do not work, but a newer hypothesis may solve the problem.

The first of the three traditional theories, the **fission hypothesis,** supposes that the moon formed by the fission of Earth. If the young Earth spun fast enough, tides raised by the sun might make it break into two parts (Figure 22-16a). If this separation occurred after Earth differentiated, the moon would have formed from crust material. Thus, the theory explains the moon's low density.

But the fission theory has problems. No one knows why the young Earth should have spun so fast, nearly ten times faster than today, nor where all that angular momentum went after the fission. In addition, the moon's orbit is not in the plane of Earth's equator, as it would be if it had formed by fission.

The second traditional theory is the **co-accretion** (or double-planet) **hypothesis.** It supposes that Earth and the moon condensed as a double planet from the same cloud of material (Figure 22-16b). However, if they formed from the same material, they should have the same chemical composition and density, which they don't. The moon is very poor in iron and related elements such as titanium, but it also contains almost exactly the same ratios of oxygen isotopes as does Earth's mantle. Co-accretion cannot explain these compositional differences.

The third theory is the **capture hypothesis.** It supposes that the moon formed somewhere else and was later captured by Earth (Figure 22-16c). We might imagine the moon forming inside the orbit of Mercury, where the heat would prevent the condensation of solid metallic grains and only high-melting-point metal oxides could have solidified. A later encounter with Mercury could "kick" the moon out to Earth.

The capture theory is not popular, because it requires highly unlikely events involving interactions with Mercury and Earth to move the moon from place to place. Scientists are always suspicious of explanations that require a chain of unlikely coincidences. Also, upon encountering Earth, the moon would have been moving so rapidly that Earth's gravity would have been unable to capture it without ripping the moon to fragments through tidal forces.

Until recently, astronomers were thus left with no acceptable theory to explain the origin of the moon, and they occasionally joked that the moon could not exist. But during the 1980s, planetary astronomers developed a new theory that combines the best aspects of the fission hypothesis and the capture hypothesis.

The **large-impact hypothesis** supposes that the moon formed from debris ejected into a disk around Earth by the impact of a large body (Figure 22-17). The impacting body may have been as large as Mars. Instead of saying that Earth was hit by a large body, we may be more nearly correct in saying that Earth and the moon resulted from the collision and merger of two very large planetesimals. The resulting large body became Earth, and the ejected debris formed the moon.

This hypothesis would explain a number of things. The collision had to occur at a steep angle to eject enough matter to make the moon. That is, the objects did not collide head-on. Such a glancing collision would have spun the resulting material rapidly, which would explain the high angular momentum in the earth–moon system. If the colliding planetesimals had already differentiated, the ejected material would be mostly iron-poor mantle and crust. Calculations show that the iron core of the impacting body would have fallen into Earth within 4 hours. This would explain why the moon is so poor in iron and why the abundances of other elements are so similar to rocks from Earth's mantle. Also, the material that eventually became the moon would have remained in a disk long enough for volatile elements, which the moon lacks, to be lost to space.

Thus, the moon may be the remains of a giant impact. Until recently, astronomers have been reluctant to consider such catastrophic events, but a number of lines of evidence suggest that planets have been affected by giant impacts. Consequently, the third theme we identified in the introduction to this chapter, giant impacts, has the potential to help us understand other worlds.

CRITICAL I N Q U I R Y

If the moon was intensely cratered by the heavy bombardment and then formed great lava plains, why didn't the same thing happen on Earth?

In fact, the same thing did happen on Earth. Although the moon has more craters than Earth, the moon and Earth are the same age, and both were battered by meteorites during the heavy bombardment. Some of those impacts on Earth must have been large and dug giant multiringed basins. Lava flows must have welled up through Earth's crust and flooded the lowlands to form great lava plains much like the lunar maria.

Earth, however, is different from the moon in one important respect. It is larger. Because it is a larger world, Earth has more internal heat, and that heat escapes more slowly than the moon's heat did. The moon is now geologically dead, but Earth is very active, with heat flowing outward from the interior to drive plate tectonics. The moving plates long ago erased all evidence of the cratering and lava flows dating from Earth's youth. Earth's surface was almost certainly cratered and flooded by lava flows, as was the moon's, but our planet has continued to evolve, and the moon has not.

Comparative planetology is a powerful tool in that it allows us to see similar processes occurring under different circumstances. For example, suppose you visited a world in our solar system that was even smaller than the moon and so had lost its heat even faster. What kind of surface features would you find, and what kinds of rocks would you pick up? More important, what features and rocks would be missing? We will find such small worlds among the asteroids. ∎

The moon's evolution has been restricted by its size. It is too small to have an atmosphere, and it cooled so rapidly that it never developed plate motion. Another world in our solar system has suffered from its diminutive size. Mercury, like the moon, is a small, rocky world, cratered by impacts and flooded by ancient lava flows driven by its internal heat.

F I G U R E   2 2 - 1 7
A Mars-sized planetesimal strikes the young, molten Earth, blasting iron-poor mantle material into orbit. In this simulation, the iron core of the planetesimal falls back into Earth within 4 hours, and the moon begins to form within about 10 hours.

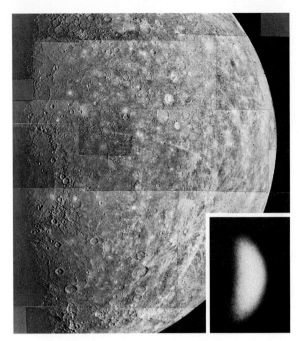

FIGURE 22-18
Photomosaic of Mercury made by the Mariner 10 spacecraft. Caloris Basin is in shadow at left. Almost no surface detail is visible from Earth (inset). (NASA, inset courtesy Lowell Observatory)

22-2 MERCURY

Earth's moon and Mercury are similar in a number of ways. Their rotation has been altered by tides, their surfaces are heavily cratered, their lowlands are flooded in places by ancient lava flows, and both are small and airless and have ancient, inactive surfaces. Yet the differences between them are impressive and will help us understand the nature of these airless worlds.

Mercury is the innermost planet in the solar system (Data File Four), and thus its orbit keeps it near the sun in the sky. It is sometimes visible near the horizon in the evening sky after sunset or in the dawn sky just before sunrise. An earth-based telescope shows Mercury passing through phases like the moon, but little detail is visible (Figure 22-18). Most of what we know about the surface of Mercury is based on photographs and measurements taken by the Mariner spacecraft when it looped through the inner solar system in 1974 and 1975. Until then, Mercury was a mystery world.

Rotation and Revolution

During the 1880s, the Italian astronomer Giovanni Schiaparelli sketched the faint features he thought he saw on the disk of Mercury and concluded that the planet was tidally coupled to the sun just as the moon is tidally coupled to Earth. That is, he concluded that Mercury rotated on its axis to keep the same side facing the sun through-

out its orbit. This was actually a very good guess because, as we will see, tidal coupling between rotation and revolution is common in the solar system. But the rotation of Mercury is more complex than Schiaparelli thought.

In 1962, radio astronomers detected radio emissions from the planet and concluded that the dark side was not as cold as it should have been if the planet kept one side perpetually in darkness. In 1965, radio astronomers made radar contact with Mercury. They used the 305-m Arecibo dish (Figure 6-19) to transmit a pulse of radio energy at Mercury and then waited for the reflected signal to return. Doppler shifts in the reflected radio pulse showed that the planet was rotating with a period of only about 59 days, much shorter than the orbital period of 87.969 days.

Apparently, Mercury is tidally coupled to the sun such that it rotates not once per orbit but 1.5 times per orbit (Figure 22-19). Thus, its period of rotation is two-thirds its orbital period, or 58.65 days. This means that a mountain on Mercury directly below the sun at one place in its orbit will point away from the sun one orbit later and toward the sun after the next orbit.

If you flew to Mercury and landed your spaceship in the middle of the day side, the sun would be high overhead, and it would be noon. Your watch would show that almost 44 earth days passed before the sun set in the west, and a total of 88 earth days would have passed before the sun reached the midnight position. In those 88 earth days, Mercury would have completed one orbit around the sun (Figure 22-19). It would require another entire orbit of Mercury for the sun to return to the noon position overhead. So a full day on Mercury is two Mercury years long!

The complex tidal coupling between the rotation and revolution of Mercury is an important illustration of the power of tides. Just as the tides in the earth–moon system have slowed the moon's rotation and locked it to Earth, so have the sun–Mercury tides slowed the rotation of Mercury and coupled its rotation to its revolution. We can refer to such a relationship as a **resonance.** We will see many such resonances as we explore the solar system.

Like its rotation, Mercury's orbital motion is also peculiar. In Chapter 5, we discussed the orbital motion of Mercury and saw that its elliptical orbit precesses faster than can be explained by Newton's laws but at just the rate predicted by Einstein's theory of general relativity. Thus, the orbital motion of Mercury is taken as a confirmation of the curvature of space-time as predicted by general relativity.

The Surface of Mercury

Because the days are so long on Mercury and because it is close to the sun and too small to keep an atmosphere, the temperatures on Mercury are extreme. The highest daytime temperatures can exceed 600 K (330°C), although 500 K is a more usual high temperature. In the

FIGURE 22-19

Mercury's rotation is tidally coupled to the sun, but unlike the moon, which rotates to keep the same face toward Earth, Mercury rotates 1.5 times during each orbit. A mountain facing the sun (a) will face away from the sun after one orbit (e) and will face the sun again after the next orbit.

MERCURY

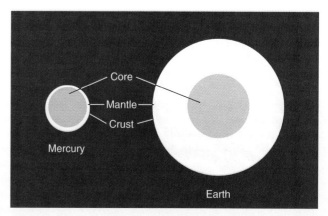

In proportion to its size, Mercury has a very large metallic core.

Average distance from the sun	0.387 AU (5.79×10^7 km)
Eccentricity of orbit	0.2056
Maximum distance from the sun	0.467 AU (6.97×10^7 km)
Minimum distance from the sun	0.306 AU (4.59×10^7 km)
Inclination of orbit to ecliptic	7°00′16″
Average orbital velocity	47.9 km/sec
Orbital period	87.969 days (0.24085 y)
Period of rotation	58.646 days (direct)
Inclination of equator to orbit	0°
Equatorial diameter	4878 km (0.382 $D_\oplus$)
Mass	3.31×10^{23} kg (0.0558 $M_\oplus$)
Average density	5.44 g/cm³ (5.4 g/cm³ uncompressed)
Surface gravity	0.38 earth gravities
Escape velocity	4.3 km/sec (0.38 $V_\oplus$)
Surface temperature	−173° to 330°C (−279° to 620°F)
Average albedo	0.1
Oblateness	0

a

b

c

FIGURE 22-20

(a) Only part of the giant Caloris Basin on Mercury is visible in this
Mariner 10 photomosaic. The center of the image is in darkness just
beyond the lower left corner of the photo, and the outer ring is marked by
a dashed line. This giant crater basin was produced by a large impact
and subsequent lava flows. (b) Seismic waves from the impact were
focused on the far side of Mercury and produced the jumbled terrain
shown in part c. (NASA)

dark nights, with no atmosphere to retain heat, the surface can cool to 100 K ($-173°C$). Thus, the surface of
Mercury cycles from very high temperatures to very low
temperatures each Mercury day.

In appearance, Mercury looks much like Earth's
moon. It is heavily battered, with craters of all sizes,
including some large basins. Some craters are obviously
old and degraded; others seem quite young and have
bright rays of ejecta. However, a quick glance at photos
of Mercury shows no large, dark maria like the moon's
flooded basins.

The largest basin on Mercury is called Caloris Basin
after the Latin word for "heat," recognition of its location at one of the two "hot poles," which face the sun
at alternate perihelions. At the times of the Mariner
encounters, the Caloris Basin was half in shadow (Figure
22-20a). Although half cannot be seen, the low angle of
illumination is ideal for the study of the lighted half
because it produces dramatic shadows.

The Caloris Basin is a gigantic impact basin 1300 km
in diameter with concentric mountain rings up to 3 km
high. The impact threw ejecta 600–800 km across the
planet, and the focusing of seismic waves on the far side
produced peculiar terrain (called "weird" terrain) that
looks much like the jumbled surface of the moon (Figure
22-20b and c). Like Mare Orientale on the moon, the
Caloris Basin is partially filled with lava flows. Some of
this lava may be material melted by the energy of the
impact, but some may be lava from below the crust that
leaked up through cracks. The weight of this lava and the
response of the crust have produced deep cracks in the
central lava plains. The geophysics of such large, multiringed crater basins is not well understood at present, but
Caloris Basin seems to be the same kind of structure as
Mare Orientale and the Imbrium Basin on the moon.

When planetary scientists began looking at the
Mariner photographs in detail, they discovered something not seen on the moon. Mercury is marked by great
curved cliffs called **lobate scarps** (Figure 22-21). These
seem to have formed when the planet cooled and shrank
by a few kilometers, wrinkling its crust as a drying apple
wrinkles its skin. Some of these scarps are as high as 3 km
and reach hundreds of kilometers across the surface.
Other faults in Mercury's crust are straight and may have
been produced by tidal stresses generated when the sun
slowed Mercury's rotation.

The Plains of Mercury

The most striking difference between Mercury and the
moon is that Mercury lacks the great dark lava plains so
obvious on the moon. Under careful examination, the
Mariner 10 photographs show that Mercury has plains,
two different kinds, in fact, but they are different from
the moon's. Understanding the differences is the key to
understanding the history of Mercury.

Much of Mercury's surface is old, cratered terrain
(Figure 22-22), but other areas called **intercrater plains**

FIGURE 22-21
A lobate scarp (arrow) crosses craters, indicating that Mercury cooled and shrank, wrinkling its crust, after many of its craters had formed. (NASA)

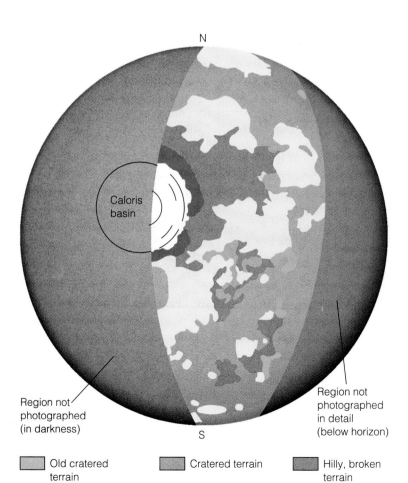

N

Caloris
basin

Region not
photographed
(in darkness)

S

Region not
photographed
in detail
(below horizon)

	Old cratered terrain		Cratered terrain		Hilly, broken terrain
	Intercrater plains		Caloris mountains		Smooth plains

FIGURE 22-22
A geological map of part of Mercury shows the Caloris Basin at left and the hilly, broken terrain associated with that impact. The intercrater plains are interrupted by smooth plains and more heavily cratered terrain.

are less heavily cratered. These plains are marked by meteorite craters less than 15 km in diameter and secondary craters produced by chunks of ejecta from larger impacts. Unlike the heavily cratered regions, the intercrater plains are not totally saturated with craters. This suggests that the intercrater plains were produced by lava flows that buried older terrain.

In contrast, smaller regions called **smooth plains** appear to be even younger than the intercrater plains. They have fewer craters and appear to be ancient lava flows. Much of the region around the Caloris Basin is composed of these smooth plains (Figure 22-22), and they appear to have formed soon after the Caloris impact.

Given the available evidence, most planetary astronomers believe that the plains of Mercury are solidified lava flows, in which case they are much like the maria on the moon. Unlike the maria, Mercury's lava plains are not significantly darker than the rest of the crust. Except for a few bright crater rays, Mercury's surface is a uniform gray with an albedo of only about 0.1. Thus, Mercury's lava plains are not as dramatically obvious on photographs as the much darker maria on our own moon.

The Interior of Mercury

One of the most striking differences between Mercury and the moon is the difference between their interiors. We have seen that the moon is a low-density world that contains no more than a small core of metals. Yet evidence suggests that Mercury has a large metallic core.

Mercury is over 60 percent denser than the moon. Yet Mercury's surface appears to be normal rock, so we must conclude that its interior contains a large core of dense metals, probably iron. In proportion to its size, Mercury must have a larger metallic core than Earth (see diagram in Data File Four).

If Mercury has a large metallic core that remains molten, then the dynamo effect should generate a magnetic field (Chapter 21). The Mariner 10 spacecraft, however, found a magnetic field only about 0.5 percent as strong as Earth's, and this weak field makes it difficult to understand the nature of the interior. Since Mercury is a relatively small world, about halfway between the size of Earth and the size of the moon, it should have lost most of its internal heat long ago. But if its metallic core is not molten, we would not expect to detect a magnetic field as strong as 0.5 percent of Earth's field. The verdict is still out on the nature of Mercury's large metallic core and weak magnetic field.

Similarly, it is difficult to explain the proportionally large size of the metallic core inside Mercury. In Chapter 20, we saw that planets that formed near the sun should incorporate more metals and thus be denser, but some models suggest that this cannot account for all of Mercury's large metallic core. One hypothesis that has been proposed involves a giant impact when Mercury was young, an impact much like the planet-shattering impact proposed to explain the origin of the moon. If the forming planet had differentiated and was then struck by a large planetesimal, the impact could have shattered the crust and mantle and blasted much of the lower-density material into space. The denser core could have survived, re-formed, and attracted some of the lower-density debris to form a thin mantle and so leave Mercury with a deficiency of low-density crust and mantle rock. Thus, both Mercury and the moon may be products of giant impacts.

A History of Mercury

Having reviewed the evidence, we must now try to tell the story of Mercury. Presumably, it formed in the innermost part of the solar nebula, and, as we have seen, a giant impact may have robbed it of some of its lower-density rock and left it a small, dense world with a large metallic core.

Like the moon, Mercury suffered heavy cratering by debris in the young solar system. We don't know the absolute ages of features on Mercury, but we can assume that cratering, the second stage of planetary evolution, occurred over the same period as did the cratering on the moon—during the heavy bombardment that lasted from formation until about 4 billion years ago. This intense cratering declined rapidly as the planets swept up the last of the debris left over from planet building.

The cratered surface of Mercury is not exactly like that of the moon. Because of Mercury's stronger gravity, the ejecta from a crater on Mercury would be thrown only about 65 percent as far as on the moon, and thus the ejecta from an impact does not blanket as much of the surface. Also, the intercrater plains appear to have formed when lava flows occurred during the early bombardment, burying the older surface, and then accumulated more craters. Sometime near the end of cratering, a planetesimal over 100 km in diameter smashed into the planet and blasted out the great multiringed Caloris Basin. Only parts of that basin have been flooded by lava flows. Finally, as cratering rapidly declined, the cooling core contracted, and the crust broke to form the lobate scarps.

The third stage, flooding, began with lava flows filling some lowlands. The Caloris impact may have been so big it fractured the crust and triggered still more outpourings of molten lava. These later lava flows formed the smooth plains. Nevertheless, the age of lava flooding ended quickly, perhaps because the shrinkage of the planet squeezed off the lava channels to the surface. Mercury lacks a true atmosphere, so we would not expect flooding by water, but radar images show a bright spot at the planet's north pole (Figure 22-23) that may be caused by ice trapped in perpetually shaded crater floors where the temperature never exceeds 60 K (−350°F).

The fourth stage in the story of Mercury, slow surface evolution, is now limited to micrometeorites, which grind the surface to dust; rare larger meteorites, which leave bright rayed craters; and the slow but intense cycle of heat and cold, which weakens the rock at the surface.

The planet's crust is now thick, and although its core may be partially molten, the heat flowing outward is unable to drive plate tectonics that would actively erase craters and build folded mountain chains.

CRITICAL INQUIRY

Why don't Earth and the moon have lobate scarps?

At first glance, we might suppose that any world with a large metallic interior should have lobate scarps. When the metallic core cools and contracts, the world should shrink, and the contraction should wrinkle and fracture the brittle crust to form lobate scarps.

Of course, Earth has a fairly large metallic core, but it has not cooled very much and the crust is not brittle. In fact, the heat flowing out of Earth's interior keeps the crust thin, flexible, and active. If any lobate scarps ever formed on Earth, they would have been quickly destroyed by plate tectonics.

The moon, unlike Mercury, does not have a large metal component in its interior. It may be that the rocky interior did contract slightly when the moon lost its internal heat, but that smaller contraction may not have produced major lobate scarps. Also, any surface features that formed early in the moon's history would have been destroyed by the heavy bombardment.

We can, in a general way, understand lobate scarps, but not all materials contract when they cool and solidify. Water, for instance, expands when it freezes. What kinds of features would you expect to find on small, cold, icy worlds that once had interiors of liquid water? Such icy worlds exist as the moons of the outer planets, and we will explore them in Chapter 24. ∎

Mercury, like the moon, is now an inactive world. Impact cratering, internal heat, and possible giant impacts have left their marks on these worlds, but they are now nearly dead. To find planets where the heat flowing outward from the interior still drives geological activity, we must look at larger worlds such as Venus and Mars, the subjects of the next chapter. ∎

∎ *Summary*

The moon is, at first glance, dramatically different from Earth. We see two distinct kinds of terrain on the moon. The maria, named after seas, are lowland plains with few craters. They are great lava flows and are younger than the highlands. The highlands are lighter in color and heavily cratered.

Samples returned by the Apollo missions show that the maria are dark basalts, whereas the highlands are light anorthosite. Many samples are breccias, which show how severely the moon's crust has been fractured.

The ages of the mare samples range from about 3.1–3.8 billion years, but the highland samples are about 4.0–4.3 billion years old. Few samples are older than 4.3 billion years.

FIGURE 22-23
A radar map of Mercury reveals a bright region at its north pole (top). Observers suspect that ice deposits may survive just under the surface or in regions that never receive direct sunlight. (Martin Slade, JPL)

This suggests that the moon's surface was molten until about 4.4 billion years ago.

As the anorthosite crust solidified, cratering battered it and excavated great basins. Later, molten rock rose through fractures in the crust and flooded the basins to produce the maria. By the time the maria began to form, cratering had declined, so the maria are marked by few craters.

The lunar surface now is evolving very slowly. Because the moon is small, it has cooled rapidly, its crust has grown thick, and it has never divided into moving plates. Now only meteorites alter the surface.

The large-impact hypothesis suggests that the earth–moon system formed in the collision of two large planetesimals. The differentiated bodies formed a single large object that became the higher-density, iron-rich Earth. Ejected mantle material, poor in iron, formed a disk and eventually condensed to form the lower-density, iron-poor moon.

Mercury is a small world, only about 40 percent larger than the moon, and, like the moon, it has lost any permanent atmosphere it might have had, and its interior has cooled rapidly. Debris from the formation of the solar system cratered Mercury's surface as it did the moon's, and lava flows buried parts of that terrain under the intercrater plains. The Caloris Basin, formed near the end of cratering, is the largest crater basin. Soon after the Caloris impact, lava flows created the smooth plains. These lava plains have about the same color as the intercrater plains, so the contrast between plains and highlands is not as obvious on Mercury as it is on the moon.

The interior of Mercury must contain a large metallic core to account for the planet's high density. A slight shrinkage in the diameter of the planet at the time of the cooling of the core may have led to wrinkling of the crust. This would account for the lobate scarps that mark all of the photographed parts of Mercury.

New Terms

tidal coupling
terminator
limb
mare
sinuous rille
relative age
absolute age
multiringed basin
ejecta
ray
secondary crater
vesicular basalt
anorthosite

breccia
regolith
micrometeorite
jumbled terrain
fission hypothesis
co-accretion hypothesis
capture hypothesis
large-impact hypothesis
resonance
lobate scarp
intercrater plain
smooth plain

Questions

1. How do we find the relative ages of the moon's maria and highlands?

2. How do we know that Copernicus is a young crater?

3. Why did the first Apollo missions land on the maria? Why were the other areas of more scientific interest?

4. How does the range of ages of lunar samples suggest that the moon formed with a molten surface?

5. Why are so many lunar samples breccias?

6. What do the vesicular basalts tell us about the evolution of the lunar surface?

7. What evidence would we expect to find on the moon if it had been subjected to plate tectonics? Do we find such evidence?

8. Cite objections to the fission, co-accretion, and capture hypotheses.

9. How does the large-impact hypothesis explain the moon's lack of iron?

10. How does the tidal coupling between Mercury and the sun differ from that between the moon and Earth?

11. What is the difference between the intercrater plains and the smooth plains in terms of time of formation?

12. What evidence do we have that Mercury has a partially molten, metallic core?

Discussion Questions

1. Old science-fiction paintings and drawings of colonies on the moon often showed very steep, jagged mountains. Why did the artists assume that the mountains would be more rugged than mountains on Earth? Why are lunar mountains actually less rugged than mountains on Earth?

2. From your knowledge of comparative planetology, propose a description of the view that astronauts would have if they landed on the surface of Mercury.

Problems

1. Calculate the escape velocity of the moon from its mass and diameter. (HINT: See Chapter 5.)

2. Why do small planets cool faster than large planets? Compare surface area to volume.

3. The smallest detail visible through earth-based telescopes is about 1 second of arc in diameter. What size is this on the moon? (HINT: See Chapter 3.)

4. Midocean rifts and the trenches where Earth's seafloor slips downward are 1 km or less wide. Could earth-based telescopes resolve such features on the moon? Why are we sure that such features are not present on the moon?

5. The Apollo command module orbited the moon about 200 km above the surface. What was its orbital period? (HINT: See Chapter 5.)

6. From a distance of 200 km above the surface of the moon, what is the angular diameter of an astronaut in a space suit? Could we have seen the astronauts from the command module? (HINT: See Chapter 3.)

7. If we transmitted radio signals to Mercury when it was closest to Earth and waited to hear the radar echo, how long would we wait?

8. Suppose we sent a spacecraft to land on Mercury and it transmitted radio signals to us at a wavelength of 10 cm. If we saw Mercury at its greatest angular distance west of the sun, to what wavelength would we have to tune our radio telescope to detect the signals? (HINTS: See Data File Four and the Doppler shift in Chapter 7.)

Critical Inquiries for the Web

1. What did the Apollo astronauts do while they were on the moon? Search for Web sites that describe the experiments performed during the astronauts' extra-vehicular activities (EVAs). What types of experiments/activities were performed? What did we learn from them?

2. Recent observations have indicated the possibility of the existence of ice on the Moon and on Mercury. Where does the ice exist on these bodies, why do astronomers believe that it is there, and how do they propose that it got there?

3. What would it be like to walk on the lunar surface? Apollo astronauts visited six different locations on the moon exploring the variations in lunar terrain. Describe the horizons and general relief of the landing locations of the different missions by exploring Web sites that provide lunar surface photography from the missions. What differences do you see between images from landings in highlands and in maria?

Go to the Wadsworth Astronomy Resource Center (www.wadsworth.com/astronomy) for critical thinking exercises, articles, and additional readings from InfoTrac College Edition, Wadsworth's online student library.

Chapter Twenty-Three

VENUS AND MARS

GUIDEPOST

The previous chapter grouped Earth's moon and Mercury together because they are similar worlds. This chapter groups Venus and Mars together because we might expect them to be similar. They are earthlike in their size and location in the solar system, so it is astonishing to see how different they actually are. Much of this chapter is aimed at understanding how Venus and Mars evolved to their present states.

Neither Venus nor Mars can tell us much about the formation of the planets. Both planets have evolved since they formed. Nevertheless, we find further hints that the solar system was a dangerous place, with major impacts smashing the surfaces of the planets, a process we first suspected when we studied the moon and Mercury.

This chapter concludes our exploration of the earthlike worlds. In the next two chapters, we will visit planets that give new meaning to the word "unearthly." ■

There wasn't a breath in that land

of death . . .

ROBERT SERVICE

The Cremation of Sam McGee

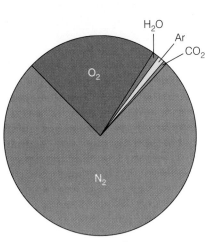

Earth's atmosphere   Venus's atmosphere

FIGURE 23-1
The chemical composition of the atmosphere of Venus is dramatically different from that of Earth. The fraction of water in the atmosphere of Venus is much too small to show here.

Without a space suit, you could live only about 30 seconds on the surface of Mars. The temperature might not be too bad. On a hot summer day at noon, the temperature might reach 20°C (68°F). But the air is 96 percent carbon dioxide, is deadly dry, and contains almost no oxygen. Also, the air pressure at the surface of Mars is only 0.01 that at the surface of Earth. Stepping out onto the surface of Mars without a space suit would be about the same as opening the door of an airplane flying at 30,000 m (100,000 ft) above the earth. (Commercial jets do not generally fly above 12,000 m, or 38,000 ft.)

No space suit would save you on Venus. The air pressure is 90 times Earth's, and the air is almost entirely carbon dioxide, with traces of various acids. Worse yet, the surface is hot enough to melt lead.

Why are Venus and Mars such rotten places for picnics? In many ways, they resemble Earth. They are made of silicates and metals, have atmospheres, and lie in the inner solar system. We might expect them to be similar. How did they get to be so different from each other and from Earth? Comparative planetology will give us some clues.

When we used comparative planetology to study the moon and Mercury in the previous chapter, we saw the importance of impact cratering and the outward flow of heat. As we study Venus and Mars, we will see the same effects, but they will be given new significance because of the larger size of Venus and Mars. In this chapter, we will discover the important principle that the size of a world determines how fast it loses its internal heat. Venus, roughly the size of Earth, has lost internal heat slowly, but Mars, intermediate in size between Earth and the moon, is nearly as cold and dead as the moon. Furthermore, these larger worlds have enough gravity to hold onto an atmosphere, and that alters the evolution of surface features.

Venus and Mars are more than just two more planets to add to our gallery of weird worlds. They give us an entirely new perspective on comparative planetology and thus a new perspective on Earth.

23-1 VENUS

An astronomer once became annoyed when someone referred to Venus as a planet gone wrong. "No," she argued, "Venus is probably a fairly normal planet. It is the earth that is peculiar. The universe probably contains more planets like Venus than like the earth." To understand how unusual, how special Earth is, we have only to compare it with Venus.

In many ways, Venus is a twin of Earth, and we might expect the two planets to be quite similar (Data File Five). Venus is nearly 95 percent the diameter of Earth and has a similar average density. It formed in the same general part of the solar nebula, so we might expect it to have a composition similar to Earth's. Planets of Earth's size cool slowly, so we might expect Venus to have a molten metallic interior and an active crust.

Unfortunately, we cannot see the surface of Venus because of a perpetual layer of thick, white clouds that completely envelops the planet. From the time of Galileo to the early 1960s, astronomers could only speculate about Earth's twin, and science fiction writers supposed that Venus was a steamy swamp planet inhabited by strange creatures.

During the last few decades, astronomers have peered below the clouds using radar, 18 spacecraft have flown past or orbited Venus, and 17 probes have landed on its surface. The resulting picture of Venus is dramatically different from the murky swamps of fiction. In fact, the surface of Venus is drier than any desert on Earth and two times hotter than a hot kitchen oven.

If Venus is not a planet gone wrong, it is certainly a planet gone down a different evolutionary path. How did Earth's twin become so different?

The Rotation of Venus

One way Venus differs from Earth is that it rotates backward. Except for Uranus and Pluto, whose rotations are also peculiar, all other planets in the solar system rotate from west to east (counterclockwise as seen from the north). In 1962, radar observations showed that Venus rotates from east to west (retrograde) with a period of 243.01 days.

This rotation may be linked to Earth. As Venus revolves around the sun and rotates on its axis, it turns nearly the same face toward Earth each time the two planets come near each other. This might arise if Venus were tidally locked to Earth, but a perfect resonance would require a period of rotation of 243.16 days. Venus rotates slightly too slowly to be tidally coupled with Earth.

An alternate explanation is that the rotation of Venus was altered during the final accretion of the planet. A very large planetesimal striking the planet off center could have altered its rotation. Such large impacts are becoming more acceptable among astronomers. In the previous chapter, we saw that a large impact may have given birth to the moon and that a similarly large impact may have ejected part of Mercury's crust. In the next chapter, we will discover evidence that some of the satellites in the outer solar system have suffered from major impacts.

The Atmosphere of Venus

Although Venus is Earth's twin in size, the composition, temperature, and density of its atmosphere make it the most inhospitable of planets. About 96 percent of its atmosphere is carbon dioxide, and 3.5 percent is nitrogen. The remaining 0.5 percent is water, sulfuric acid (H_2SO_4), hydrochloric acid (HCl), and hydrofluoric acid (HF) (Figure 23-1). In fact, the thick clouds that hide the surface are believed to be composed of sulfuric acid droplets and microscopic sulfur crystals.

Soviet and American spacecraft have dropped probes into the atmosphere of Venus, and those probes have radioed data back to Earth as they fell toward the surface. These studies show that the cloud layers are much higher and much more stable than those on Earth. The highest layer of clouds, the layer we see from Earth, extends from 68–58 km above the surface (Figure 23-2). For comparison, the highest clouds on Earth do not extend higher than about 16 km. Another cloud layer lies from 56–52 km, and a third, the densest, from 52–48 km. Thin haze extends down to about 33 km, below which the air is surprisingly clear.

These cloud layers are highly stable because the atmospheric circulation on Venus is much more regular than that on Earth. The heated atmosphere at the **subsolar point,** the point on the planet where the sun is directly overhead, rises and spreads out in the upper atmosphere. Convection circulates this gas toward the dark side of the planet and the poles, where it cools and sinks. This circulation produces 300-km/h jet streams in the upper atmosphere, which carry the atmosphere from east to west (the same direction the planet rotates) so rapidly that the entire atmosphere seems to rotate with a

VENUS

Radar-based globes of Venus and Earth show their similar size, but although this globe of Venus uses blue to color the low, rolling plains, Venus is a deadly dry desert world hidden below a perpetually cloudy atmosphere. (NASA)

Average distance from the sun	0.7233 AU (1.082×10^8 km)
Eccentricity of orbit	0.0068
Maximum distance from the sun	0.7282 AU (1.089×10^8 km)
Minimum distance from the sun	0.7184 AU (1.075×10^8 km)
Inclination of orbit to ecliptic	2°23′40″
Average orbital velocity	35.03 km/sec
Orbital period	0.61515 y (224.68 days)
Period of rotation	243.01 days (retrograde)
Inclination of equator to orbit	177°
Equatorial diameter	12,104 km (0.95 $D_\oplus$)
Mass	4.870×10^{24} kg (0.815 $M_\oplus$)
Average density	5.24 g/cm³ (4.2 g/cm³ uncompressed)
Surface gravity	0.903 earth gravities
Escape velocity	10.3 km/sec (0.92 $V_\oplus$)
Surface temperature	745 K (472°C, or 882°F)
Albedo (cloud tops)	0.76
Oblateness	0

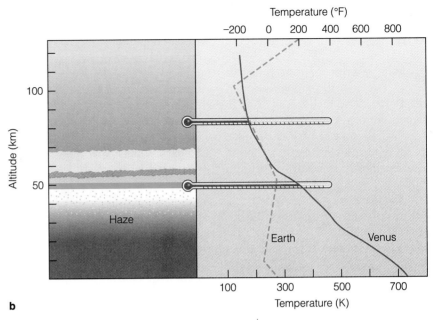

a **b**

FIGURE 23-2

(a) The clouds of Venus are composed not of water droplets but of sulfuric acid droplets. This ultraviolet image made with the Hubble Space Telescope shows typical weather patterns in the clouds, but no part of the surface is visible. (L. Esposito, University of Colorado, Boulder, and NASA) (b) The four principal cloud layers in the atmosphere of Venus are shown at left. Thermometers inserted into the atmosphere at different levels would register the temperatures indicated by the red line in the graph at the right. The blue dashed line shows the temperatures in Earth's atmosphere for comparison. Note the difference in surface temperatures.

period of only 4 days. As evidence of these circulations, we can see a distinctive Y pattern in ultraviolet images of the cloud pattern (Figure 23-2a).

The high-speed winds in the upper atmosphere were studied in June 1985 when the Vega 1 and Vega 2 spacecraft, on their way to rendezvous with Comet Halley, flew past Venus and dropped instrumented balloons into the upper atmosphere. Floating at an altitude of 55 km, the 3.5-m-diameter balloons found winds as high as 240 km/h. In the 46 hours the balloons survived, they were carried one-third of the way around Venus by the high-altitude winds.

The details of this atmospheric circulation are not well understood, but it seems that the slow rotation of the planet is an important factor. On Earth, large-scale circulation patterns are broken into smaller cyclonic disturbances by Earth's rapid rotation. Because Venus rotates more slowly, its atmospheric circulation is not broken up into small cyclonic storms but instead is organized as a planet-wide wind pattern.

Although the upper atmosphere is cool, the lower atmosphere is quite hot (Figure 23-2b). Instrumented probes that have reached the surface report that the temperature is 745 K (880°F) and the atmospheric pressure is 90 times that of Earth. Earth's atmosphere is 1000 times less dense than water, but on Venus the air is only 10 times less dense than water. If we could survive the unpleasant composition, intense heat, and high pressure, we could strap wings to our arms and fly.

The present atmosphere of Venus is extremely dry, but there is evidence that it once had significant amounts of water. As a Pioneer Venus spacecraft descended through the atmosphere of Venus in 1978, it detected about 150 times more deuterium per hydrogen atom than on Earth. This abundance of deuterium, the heavy isotope of hydrogen, could develop if ultraviolet radiation in sunlight broke water molecules into hydrogen and oxygen. The hydrogen would escape into space, but the heavier deuterium would escape more slowly than normal hydrogen. Thus, the excess of deuterium is evidence that Venus once had more water. Calculations suggest that it may have once had enough water to make a planet-wide ocean up to 25 meters deep. (For comparison, the water on Earth would make a uniform planet-wide ocean 3000 meters deep.) Without an ozone layer to protect the atmosphere from ultraviolet radiation, the water was broken up and much of the hydrogen was lost to space. The oxygen presumably formed oxides in the soil. Thus, Venus is now a deadly dry world with only enough water vapor to make an ocean 0.3 meter deep.

Even if Venus once had lots of water, it was probably too warm for that water to fall from the atmosphere as rain and form oceans, and that lack of oceans is the biggest difference between Earth and Venus. Earth's oceans have made it a living paradise. The lack of oceans on Venus, through the greenhouse effect, has made it a hell.

FIGURE 23-3
This image of Venus is made up of a mosaic of radar maps made by the Magellan spacecraft. The maps reveal many impact craters as well as large volcanoes such as Beta Regio (upper right center), which is made up of two volcanic peaks, Rhea Mons and Theia Mons. Lava flows and faults in the crust appear bright in the radar images. (NASA)

The Venusian Greenhouse

We saw in Chapter 21 how the greenhouse effect warms Earth. Carbon dioxide (CO_2) is transparent to light but opaque to infrared (heat) radiation. Thus, energy can enter the atmosphere as light and warm the surface, but the surface cannot radiate an equal amount of energy back to space because of the CO_2. Like Earth, Venus has a greenhouse effect, but on Venus the effect is fearsome. Whereas Earth's atmosphere contains only about 0.03 percent CO_2, the atmosphere of Venus contains 96 percent CO_2, and the surface of Venus is nearly twice as hot as a hot kitchen oven.

When Venus was young, it may have been cooler than it is now, but because it formed 30 percent closer to the sun than did Earth, it was warmer than Earth, and that unleased processes that made it even hotter. As we saw in the previous section, even if Venus once had water, it probably never had large liquid-water oceans. CO_2 is highly soluble in water, but without large oceans to dissolve CO_2 and remove it from the atmosphere, Venus was trapped with a CO_2-rich atmosphere, and the greenhouse effect made the planet warmer. As the planet warmed, any small oceans that did exist evaporated, and Venus lost the ability to cleanse its atmosphere of CO_2. Volcanoes on Earth vent gases that are mostly CO_2 and water vapor, and presumably the volcanoes on Venus did the same. The water vapor was broken up by ultraviolet radiation, and the CO_2 accumulated in the atmosphere. The high temperature baked even more CO_2 out of the surface, and the atmosphere became even less transparent to infrared radiation. Consequently, the temperature rose further. This runaway greenhouse effect has made the surface so hot that even sulfur, chlorine, and fluorine have baked out of the rock and formed sulfuric, hydrochloric, and hydrofluoric acid vapors.

Earth avoided this runaway greenhouse effect because it was farther from the sun and cooler. Thus, it could form and preserve liquid-water oceans to absorb the carbon dioxide, which left a nitrogen atmosphere that was relatively transparent in some parts of the infrared. If all of the carbon dioxide in Earth's sediments was put back into the atmosphere, our air would be as dense as that of Venus, and we would suffer from a severe greenhouse effect. Recall from Chapter 21 how our use of fossil fuels and our destruction of forests is increasing the CO_2 concentration in our atmosphere and warming the planet. Venus warns us of what a greenhouse effect can do.

The Surface of Venus

The clouds of Venus are opaque to visible light, but they are transparent to radio waves, so astronomers can map the surface using radar. The first radar maps, made from Earth, had very low resolution, but the Pioneer Venus spacecraft orbited Venus in 1978 and made maps with a resolution of 25 km. Later the Soviet Venera 15 and 16 spacecraft mapped the north polar region at a resolution of about 2 km. In 1990, the Magellan spacecraft orbited Venus and created a radar map showing features as small as 100 m (Figure 23-3).

Such radar maps can tell us a number of things about the surface. If we transmit a radio signal down through

FIGURE 23-4

Although it is nearly 1000 years old, this lava flow near Flagstaff, Arizona, is still such a rough jumble of rock that it is dangerous to venture onto its surface. Rough surfaces are very good reflectors of radio waves and look bright in radar maps. Lava flows on Venus show up as bright regions in the radar maps because they are rough. (M. A. Seeds)

FIGURE 23-5

This Magellan radar map of Venus covers all but the polar regions, with a resolution up to 1 km. This is an altitude map with colors indicating height. The highest mountaintops are red, and the lowest basins are blue. Although large parts of this map are blue, recall that Venus has no oceans and is, in fact, a desert planet. The highlands consist mainly of Ishtar Terra in the north, Aphrodite Terra at right in this map, and Lada Terra in the south. Beta Regio is a pair of volcanic peaks (see Figure 23-3), and Maxwell Montes is the highest peak on Venus. (NASA)

the clouds and measure the time until we hear the echo coming back up through the clouds, we can measure the altitude of the surface. Part of the Magellan data is a detailed altitude map of the surface of Venus. We can also measure the amount of the radio signal that is reflected from each spot on the surface. Rough surfaces such as lava flows are highly reflective (Figure 23-4) and stand out brightly in the radar maps. Other surfaces, because they are smoother or because they have different mineral compositions, are less reflective and appear

a

b

FIGURE 23-6

(a) This radar map of Lakshmi Planum and Maxwell Montes is based on radar data from both Soviet and American spacecraft, with ground-based radar data added (color). Violet and purple regions are the smoothest, and orange the roughest. Lakshmi Planum appears to be a volcanic plain produced by lava flooding from vents such as Collette and Sacajawea. The folded ranges of mountains around Lakshmi Planum suggest limited crustal movement. (USGS) (b) Maxwell Montes, highest mountain range in the solar system, might look like this computer-generated landscape derived from Magellan radar data if we could remove the effects of the orange light filtering through the heavy clouds. (Copyright © 1993, David P. Anderson, Southern Methodist University)

darker. Thus, part of the Magellan data consists of detailed maps of surface features related to roughness and composition. Many of these Magellan maps are reproduced in a bright orange color to mimic the orange light that filters down through the thick clouds. Photographs made by robotic landers show that the surface rock is dark gray.

The radar maps made by the Pioneer and Magellan spacecraft reveal that the surface is approximately 60 percent **rolling plains,** 24 percent highlands, and about 16 percent very high volcanic peaks (Figure 23-5). It is tempting to think of the highlands as continents and the rolling plains as seafloors, but that is unwarranted. Although many map makers insist on color coding the lowlands blue, we must remember that Venus has never had major oceans of water as has the earth. The geology of the highlands and rolling plains may be entirely

different from that of the continents and seafloors on Earth.

One of the highland regions, Ishtar Terra,* is larger than the United States and contains some of the most dramatic terrain in the solar system (Figure 23-6). At its eastern edge, the mountain called Maxwell Montes thrusts up 12 km. (Mount Everest is only 8.8 km high.) The neighboring Lakshmi Planum, a great plateau rising 4 km above the rolling plains, appears to have formed from lava flows from vents such as Collette and Sacajawea. North and west of Lakshmi Planum are ranges of wrinkled mountains that suggest some limited compression in the crust.

Volcanism is critically important on Venus. Beta Regio (Figure 23-3) is a great volcanic uplift containing

---

*Ishtar was the Babylonian goddess of love. Except for Maxwell Montes, Alpha Regio, and Beta Regio, most of the place names on Venus are female.

Graphical Data Manipulation

Because so much modern scientific data is stored inside computers, it is common to see data manipulated to exaggerate certain characteristics and suppress others. In some fields, this would be dishonest, but the goals of science make such manipulation valuable. By purposely distorting data, we may reveal things we would not have seen otherwise.

Computers make it easy to manipulate the colors in an image and produce false-color images. Astronomers commonly change and exaggerate the colors in images and maps to reveal details invisible to our eyes. Radio maps of galaxies and X-ray maps of supernova remnants are only two examples. As long as we remember what real galaxies and real nebulae look like, we can learn a great deal from false-color images.

Geologists, too, manipulate data. Geological maps have been color-coded for over a century, and now computers allow faster and more precise use of false color in maps and images. But geological data can also be manipulated to reveal other features. A relief map of the United States would look as smooth as glass if the map maker did not exaggerate the height of mountain ranges, and having such data in a computer makes it easy to "stretch" the heights to make them more visible. Astronomers often do the same with radar maps of Venus and surface maps of Mars.

Such data distortion would be inadmissible in a court of law. In fact, photographic evidence is generally inadmissible in court because it is so easy to fake a photograph and show the defendant in almost any incriminating situation. Faking a photograph in journalism is at least disreputable, and in some cases it is libelous. In fact, modern computer graphics make it possible to put Humphrey Bogart into a modern TV commercial with no trace of fakery, so photographic manipulation is unlikely ever to become acceptable in court or in journalism.

In science, however, our quest is not for a first impression but for detailed analysis. Data can be manipulated and distorted any way that is useful as long as the form of the manipulation is specified. Then viewers are free to form their own opinions as to the meaning of the data. Of course, this assumes that the viewer is alert and takes into consideration the nature of any data manipulation. Thus, when we look at a relief map of the United States or a radar image of a mountain on Venus, it is our responsibility to take into account any data manipulation. if we don't, we might think that the Rocky Mountains are 100 miles high and that mountains on Venus are 20 times steeper than mountains on Earth. ■

FIGURE 23-7

(a) Volcano Sapas Mons, lying along a major fracture zone, is topped by two lava-filled calderas and flanked by rough lava flows. The orange color of this radar map mimics the orange light that filters through the thick atmosphere. (NASA) (b) Seen by light typical of Earth's surface, Sapas Mons might look more like this computer-generated landscape. Volcano Maat Mons rises in the background. The vertical scale has been exaggerated by a factor of 15 to reveal the shape of the volcanoes and lava flows. (Copyright © 1992, David P. Anderson, Southern Methodist University)

two peaks on a major fault. Sapas Mons is another volcanic peak located along a fracture zone (Figure 23-7). It seems likely that strains in the crust created the faults, opening a path for magma to rise from below and thereby triggering the volcanism. Such faults are widespread on Venus.

Notice that many images created from the Magellan data are reproduced with their vertical scales stretched. This is easy to do with computer-based data (Window on Science 23-1), and it makes the surface features easier to study; however, it makes the mountains look much steeper than they really are (Figure 23-8). Volcanic peaks on Venus are **shield volcanoes,** which form from highly fluid lava and have extremely shallow slopes. Shield volcanoes, such as Mauna Loa in Hawaii, are common on Earth.

a

b

FIGURE 23-8

(a) A computer-generated image based on Magellan radar data shows volcanoes Gula Mons (left) and Sif Mons (right) with their associated lava flows. The vertical scale has been stretched to make surface details stand out. The true profile of the volcanoes resembles that of normal shield volcanoes (inset). (NASA) (b) In this computer model, details of the true shape of the mountain are dramatically revealed by a vertical magnification of 10 times (top) but are invisible when the model is reproduced with no vertical stretching (bottom). (Model by M. A. Seeds)

a

b

The volcanic surface of Venus contains roughly 900 impact craters (Figure 23-9), significantly fewer than the moon but many more than Earth. This shows that the surface of Venus is not as rapidly renewed as is Earth's surface, where erosion and plate tectonics recycle the crust quickly. Nevertheless, volcanism on Venus does eventually bury craters under ash and lava. Counts of craters show that the average age of the crust of Venus is 300–500 million years.

The radar images show that craters on Venus are different from those on the moon. The thick atmosphere tends to destroy the smaller meteorites before they can hit the surface. Consequently, there are no craters smaller than about 3 km in diameter. There is also a deficiency of craters smaller than about 25 km in diameter,

FIGURE 23-9

Impact craters are common in the Magellan radar images. The largest crater here is 50 km (30 mi) in diameter. Note the central peaks and ejecta sheets typical of impact craters. The numerous craters on Venus tell us that the surface is not recycled as rapidly as is Earth's active crust. At lower right, however, volcanic peaks remind us that Venus is a volcanic world. (NASA)

a

b

FIGURE 23-10
Evidence of subsurface magma on Venus. (a) Coronae are circular bulges bordered by circular and radial faults apparently caused by rising currents of magma below the crust. Light gray areas are lava flows, and volcanic cones are visible at lower left. These two coronae, Bahet (left) and Onatah (right), may be related. The vertical dark bar is caused by missing data. (b) Pancake domes of lava up to 65 km in diameter formed when viscous lava leaked through cracks on the surface. These features are roughly 1 km thick. (NASA)

which is interpreted to mean that a medium-size meteorite tends to break into fragments in the thick atmosphere and thus produces a cluster of very small craters or no craters at all instead of a single large crater. The ejecta around craters on Venus does not extend as far as it does on the moon, a consequence not only of the stronger gravity but also of the thicker atmosphere. The high temperature means that rock melted by impact does not cool rapidly, so some ejecta remains soft enough to ooze downhill, producing flow patterns around impact craters.

Much of Venus is marked by circular features that are clearly not impact craters. These features, called **coronae**, are circular bulges bordered by fractures, volcanoes, and lava flows (Figure 23-10a). Up to 2100 km in diameter, coronae appear to be caused by rising convection currents of molten magma below the crust. Some experts believe that the largest coronae may be sites where new crust is created and spreads or where old crust sinks and is destroyed.

a

b

FIGURE 23-11
The color of the surface of Venus appears orange in the original Venera 13 photograph (a), but analysis shows that the color is produced by the thick atmosphere. The corrected image (b) reveals slabs of gray rock and dark soil typical of iron oxides at high temperatures. (C. M. Pieters through the Brown/Vernadsky Institute to Institute Agreement and the U.S.S.R. Academy of Sciences)

Many examples of volcanism are visible in the Magellan radar maps. Small volcanoes are common, and lava flows may extend 100 km across the surface. In places, less fluid lava has built great pancakes of rock on the surface (Figure 23-10b). The crust is broken by numerous faults and by sunken regions produced when subsurface magma drained away and allowed the surface to collapse.

Although some ridges, such as those around Lakshmi Planum, seem to be the result of horizontal motion in the crust, plate tectonics is not important on Venus. We see no spreading centers like Earth's midocean rifts, and we see no subduction trenches where crust descends. More important, Venus lacks long chains of folded mountains such as the Rockies and the Andes. The surface of Venus may be too hot and flexible to form the rigid plates necessary for plate tectonics. Rather, Venus appears to be dominated by rising currents of molten rock below its crust. Some of its volcanic regions are probably still active.

No one has ever stood on the surface of Venus, but a few robot spacecraft have landed there. The heat and atmospheric pressure disabled the landers and their instruments in about an hour, but in that hour they transmitted data and pictures back to Earth. Only four of the spacecraft carried cameras, but those revealed a rocky desert world (Figure 23-11).

In addition to photographing the surface, seven Soviet landers have carried instruments to analyze the soil. Although the results are difficult to match with known rock types on Earth, it is clear that all of the probes landed on basalt surfaces. Most of the basalts seem similar to those of Earth's seafloors.

Because basalts are produced by solidified magma, these results suggest that the surface of Venus has been dominated by volcanism and that most of the surface is covered by basaltic lava flows of various ages. The history of Venus has evidently been volcanic.

A History of Venus

The four-stage history of Venus is not well understood. We are not sure how the planet formed and differentiated, how it was cratered and flooded, or how it evolved a surface.

Venus formed only slightly closer to the sun than did Earth, so we might expect it to be a similar planet differentiated into a silicate mantle and a molten iron core. The density and size of Venus tell us that it must have a dense interior much like Earth's, but if the core is molten, then we might expect the dynamo effect to generate a magnetic field. No spacecraft has ever detected such a field around Venus. The magnetic field must be at

FIGURE 23-12
By analogy with Earth, the interior of Venus should contain a molten core (estimated here), but no spacecraft has detected a planetary magnetic field. Thus, Venus is unprotected from the solar wind, which strikes the planet's upper atmosphere and is deflected into an ion tail.

least 25,000 times weaker than Earth's. Some theorists believe the planet rotates too slowly for the dynamo effect to produce a magnetic field; others wonder if the core of the planet is solid. If it is solid, then we must wonder how Venus got rid of its internal heat so much faster than Earth did.

Because the planet lacks a magnetic field, the solar wind is deflected by the uppermost layers of the planet's atmosphere, thus forming a bow shock where the solar wind is slowed and deflected (Figure 23-12). The magnetic field carried by the solar wind drapes over Venus like seaweed over a fish hook, forming a long tail within which ions flow away from the planet. We will see in Chapter 26 that comets, which also lack magnetic fields, interact with the solar wind in the same way.

The interaction of Venus with the solar wind is clear evidence that the planet has little or no magnetic field. Thus, we know little about the differentiation of the planet into core and mantle. The size of the core shown in Figure 23-12 is estimated by analogy with Earth's.

We know a bit more about the cratering history of Venus. The Magellan radar maps reveal no small craters, apparently because the dense atmosphere breaks up and vaporizes smaller meteorites. Also, the larger craters are about 15 percent more common than craters on the lunar maria. This means that the surface of Venus is, on the average, a bit older than Earth's active surface but not nearly as old as the surface of the moon. Planetary astronomers are now trying to understand how Venus managed to erase all traces of its older crust, and some experts now discuss an age of violent volcanism that may have resurfaced Venus roughly half a billion years ago.

The third stage of planetary evolution is flooding by lava and water, and volcanism has certainly flooded the surface of Venus with lava. Even the chemical composition of the surface rocks resembles the basalts of earthly lava flows. But flooding by water may never have occurred on Venus. As we saw earlier, Venus may have been too warm for liquid water to exist on its surface, and the ultraviolet radiation in sunlight could easily break up any water and allow the hydrogen to escape to space. Thus, Venus suffered from lava floods but not from water floods.

Because Venus never had liquid water on its surface, it could not cleanse its atmosphere of CO_2, and that gas accumulated to form a thick blanket that traps solar energy in a dramatic example of the greenhouse effect. Had Earth been just 5 percent closer to the sun, it might never have formed oceans to dissolve CO_2 from the air and thus might have suffered a Venusian greenhouse death.

The last stage of planetary evolution is slow surface evolution. On Earth, that means erosion and plate tectonics, but, with no water, erosion is not important on Venus. Since we see no long, folded mountains, no rift valleys outlining plates, and no long features like Earth's midocean ridges and rift valleys, it also seems very unlikely that plate tectonics is important on Venus.

Fully 70 percent of the heat from Earth's interior flows outward through the midocean rifts. But Venus lacks crustal rifts, and its numerous volcanoes cannot carry much of the heat out of the interior. Rather, Venus seems to get rid of its interior heat through large currents of hot magma that rise beneath the crust. We see coronae above such currents. Planetary scientists now speculate that cooler magma may sink back into the interior and cause minor horizontal motions in the crust that could account for the folded mountains near Ishtar Terra.

True plate tectonics is not important on Venus. For one thing, the crust is very dry and is consequently about 12 percent less dense than Earth's crust. This low-density crust is more buoyant than Earth's crust and thus resists being pushed into the interior. Also, the crust is so hot it is halfway to the melting point of rock. Such hot rock is not very stiff, so it cannot form the rigid plates typical of plate tectonics on Earth.

Some planetary astronomers speculate that we might be seeing an entirely different tectonic style on Venus. Rather than experiencing gradual plate motion, as does Earth, Venus may suffer periodic overturning as large fractions of its hot, pliable crust sink rapidly into the mantle and new crust forms from lava flows and volcanism. The resurfacing that seems to have occurred half a billion years ago may be simply the most recent of these overturnings. Such a tectonic style could allow Venus to lose large amounts of its internal heat and might explain how the planet could have a solid core. If this is true, then the dramatic differences between Venus and Earth can be blamed on the thick atmosphere that keeps the crust of Venus hot and flexible.

It will take decades to fully analyze the Magellan radar maps, but our study of Venus already illustrates two important principles of comparative planetology. First, large planets cool slowly and remain geologically active much longer than small planets. The moon and Mercury are dead, but Earth and Venus are active. Second, even a small difference in the temperature of a world can have a dramatic effect if it affects the stability of liquid water. Earth has water oceans and Venus does not, and that has made them dramatically different worlds.

CRITICAL INQUIRY

What evidence do we have that plate tectonics does not occur on Venus?

When we talked about Earth in Chapter 21, we identified certain geological features that are characteristic of plate tectonics: rift valleys, subduction regions, and folded mountain chains. We don't see these features on Venus even in the highest-resolution radar maps made by the Magellan spacecraft. There are no rift valleys or subduction zones outlining plates, and there are no major chains of folded mountains. Rather, we see evidence of volcanism and lava flows, and the small regions of folded mountains suggest only limited horizontal motion in the crust. Rather than being dominated by the motion of rigid crustal plates, Venus may have a crust made more flexible by the heat, a crust dominated by rising plumes of molten rock that strain the crust to produce coronae and that break through to form volcanoes and lava flows.

At first glance, comparative planetology suggested that Earth and Venus should be sister worlds, but it seems they can be no more than distant cousins. We can blame the thick atmosphere of Venus for altering its geology, but that raises another question: Why doesn't Earth have a similar atmosphere? ■

Venus is certainly an unearthly world, but the familiar principles of comparative planetology help us understand it. As we turn our attention to Mars, we will see the same principles in action, but we will discover that Mars is a world that lacks what Venus has in abundance—internal heat and air.

23-2 MARS

Mercury and the moon are small. Venus and Earth are, for terrestrial planets, large. But Mars occupies an intermediate position (Figure 23-13). It is twice the diameter of the moon but only 53 percent Earth's diameter. Its small size has allowed it to cool faster than Earth, and much of its atmosphere has leaked away. Its present carbon-dioxide atmosphere is only 1 percent as dense as Earth's.

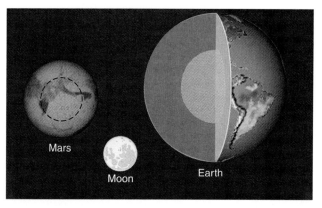

FIGURE 23-13

Mars is a medium-size world smaller than Earth but larger than the moon. Earth's core is hot and molten, while the moon is cold and may not have a molten core at all. The nature of the interior of Mars is unknown. Unlike the moon, Mars has an atmosphere, but it is much thinner than Earth's atmosphere.

In some ways, Mars is much like Earth. A day on Mars is nearly the same length as an Earth day—24 hours and 40 minutes—and its year lasts 1.88 Earth years. Also, just as Earth's axis is tipped 23.5°, that of Mars is tipped 24°. As the northern and southern hemispheres turn alternately toward the sun, seasonal changes are visible even through a small telescope. As spring comes to the southern hemisphere, the white polar cap shrinks and the grayish surface markings grow darker and, according to some observers, greener. At one time, these seasonal changes led some to believe that plant life flourished on Mars when spring thawed the polar cap. We will see later that this is not so.

In other ways, Mars is very different from Earth. Its surface has never broken into moving plates like those that create the topography on Earth. Instead, Mars is a one-plate planet, with its crust frozen into a solid layer but bearing signs of geological activity that include some of the largest volcanoes in the solar system.

The Canals of Mars

Even before the space age brought us photographs of the Martian surface, the planet Mars was a mysterious landscape in the public mind. In the century following Galileo's first astronomical use of the telescope, astronomers discovered dark markings on Mars as well as bright polar caps (Data File Six). Timing the motions of the markings, they concluded that a Martian day was about 24 hours long, and the similarity with Earth's day further supported the belief that Mars was another Earth.

Apparent proof that Mars was peopled by intelligent beings appeared late in the summer of 1877, when Italian astronomer Giovanni Virginio Schiaparelli, using a telescope only 8.75 in. in diameter, thought he glimpsed fine, straight lines visible when Earth's atmosphere was

a

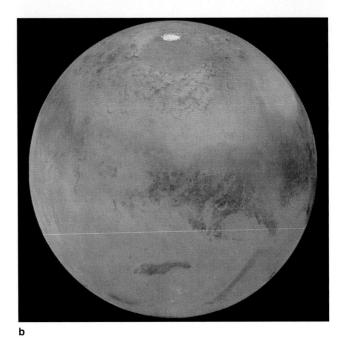

b

FIGURE 23-14

(a) Early in the 20th century, Percival Lowell mapped canals over the face of Mars and concluded that intelligent life resided there. (b) Modern images recorded by spacecraft reveal a globe of Mars with no canals. Instead, the planet is marked by craters and, in some places, volcanoes. Both of these images are reproduced with south at the top, as they appear in telescopes. Lowell's globe is inclined more nearly vertically and is rotated slightly to the right compared with the modern globe. (a, Lowell Observatory; b, U.S. Geological Survey)

unusually still. When he reported his discovery, he used the Italian word *canali*, which means "channel." But the word was translated into English as "canal," an artificially dug channel, and the "canals of Mars" were born. Many astronomers could not see the canals, whereas others drew maps showing hundreds (Figure 23-14).

In the decades that followed Schiaparelli's discovery, many people assumed that the canals were real water courses built by an intelligent race on Mars to carry water from the polar caps to the lower latitudes. Much of this excitement was generated by Percival Lowell, a wealthy Bostonian who founded Lowell Observatory in 1894, principally for the study of Mars. He not only mapped hundreds of canals but also publicized his results. Although some astronomers claimed the canals were merely illusions, by 1907 the general public was so sure that life existed on Mars that the *Wall Street Journal* could suggest that the most extraordinary event of the previous year had been "the proof by astronomical observations . . . that conscious, intelligent human life exists upon the planet Mars." Further sightings of bright clouds and flashes of light on Mars strengthened this belief, and some urged that gigantic geometrical diagrams be traced in the Sahara Desert to signal to the Martians that Earth, too, is inhabited. All seemed to agree that the Martians were older and wiser than humans.

This fascination with men from Mars was not a passing fancy. Beginning in 1912, Edgar Rice Burroughs wrote a series of 11 novels about the adventures of the earthman John Carter lost on Mars. Burroughs made the geography of Mars, named by Schiaparelli after Mediterranean lands both real and mythical, into household words. He also gave his Martians green skin.

By Halloween night of 1938, people were so familiar with life on Mars that they were ready to believe that Earth had been invaded. When a radio announcer repeatedly interrupted dance music to report the landing of a spaceship in New Jersey, the emergence of monstrous beings, and their destruction of whole cities, thousands of otherwise sensible people rushed to their cars and fled in panic, not knowing that Orson Welles and other actors were dramatizing H. G. Wells's book *The War of the Worlds*.

The fascination with Mars, its canals, and its little green men lasted right up to July 15, 1965, when Mariner 4, the first spacecraft to fly past Mars, radioed back photos of the surface and proved that there are no canals and no Martians. Apparently, the canals are optical illusions produced by the human brain's astounding ability to assemble a field of disconnected marks into a coherent image. If your brain could not do this, the photos on these pages would be nothing but swarms of dots, and the images on the screen of a television set would never make sense. The brain of an astronomer looking for something at the edge of visibility is capable of connecting random markings on Mars into the straight lines of canals.

Even today, Mars holds a fascination for the general public. The grocery store tabloids regularly run stories about a giant face carved on Mars by an ancient race.

Although planetary scientists recognize it as nothing more than chance shadows in a photograph and dismiss the issue as a silly hoax, the stories persist. A hundred years of speculation have given us high expectations for Mars. If there really was intelligent life on Mars and its representatives came to Earth, they would probably be a big disappointment to the readers of the tabloids.

The Atmosphere of Mars

If we visited Mars, our first concern, even before we opened the door of our spaceship, would be the atmosphere. Is it breathable? Even for the astronomer observing safely from Earth, the atmosphere of Mars is a major concern. The gases that cloak Mars are critical to our understanding of the history of the red planet.

The air on Mars is 95 percent carbon dioxide, with a few percent each of nitrogen and argon. It contains almost no water vapor or oxygen, and its density at the surface of the planet—only 1 percent that of Earth's atmosphere—does not provide enough pressure to prevent liquid water from boiling into vapor. If you stepped outdoors on Mars without a space suit, your own body heat would make your blood boil.

Although the air is thin, it is dense enough to have weather patterns, and the Martian wind often picks up dust from the surface in great dust storms. This red dust, suspended in the atmosphere, produces a pink sky as seen from the surface of Mars (Figure 23-15). The reddish soil is evidently caused by iron oxides (rusts), and that tells us that the oxygen we humans would prefer in the atmosphere is actually locked in chemical compounds with iron in the soil.

To understand Mars, we must ask why its atmosphere is so thin and dry and why the surface is rich in oxides. The answers to these questions lie in the origin and evolution of the Martian atmosphere.

Presumably, the gases in the Martian atmosphere were mostly outgassed from its interior. Volcanism on terrestrial planets typically releases carbon dioxide and water vapor plus other gases. Because Mars formed farther from the sun, we might expect that it incorporated more volatiles when it formed. But Mars is smaller than Earth, so it has had less internal heat to drive geological activity, and we might suspect that it has not outgassed as much as Earth. In any case, the outgassing occurred early in the planet's history, and Mars, being small, cooled rapidly and now releases little gas.

How much atmosphere a planet has depends on how rapidly it releases internal gas and how rapidly it loses gas from its atmosphere, and the rate at which a planet loses gas depends on its mass and temperature. The more massive the planet, the higher its escape velocity (Chapter 5) and the more difficult it is for gas atoms to leak into space. Mars has a mass less than 11 percent that of Earth, and its escape velocity is only 5 km/sec, less than half Earth's. Thus, gas atoms can escape from it much more easily than they can escape from Earth.

MARS

This Hubble Space Telescope view of Mars shows the northern polar cap at the top, impact craters, and dark surface features. The spot near the left edge is the volcano Ascraeus Mons, 25 km (16 mi) high, poking up through morning clouds of water-ice crystals in the thin CO_2 atmosphere. (Philip James, University of Toledo; Steven Lee, University of Colorado, Boulder; and NASA)

Average distance from the sun	1.5237 AU (2.279×10^8 km)
Eccentricity of orbit	0.0934
Maximum distance from the sun	1.6660 AU (2.492×10^8 km)
Minimum distance from the sun	1.3814 AU (2.066×10^8 km)
Inclination of orbit to ecliptic	1°51′09″
Average orbital velocity	24.13 km/sec
Orbital period	1.8808 y (686.95 days)
Period of rotation	$24^h37^m22.6^s$
Inclination of equator to orbit	23°59′
Equatorial diameter	6796 km (0.53 $D_\oplus$)
Mass	0.6424×10^{24} kg (0.1075 $M_\oplus$)
Average density	3.94 g/cm³ (3.3 g/cm³ uncompressed)
Surface gravity	0.379 earth gravities
Escape velocity	5.0 km/sec (0.45 $V_\oplus$)
Surface temperature	−140° to 20°C (−220° to 68°F)
Average albedo	0.16
Oblateness	0.009

FIGURE 23-15

The sky of Mars is not blue but red, due to reddish dust suspended in the atmosphere. The red surface is caused by iron oxides in the soil and thus tells us that oxygen on Mars lies not in the atmosphere but in the soil. This photograph of the Martian surface was made by one of the two robotic Viking landers that reached Mars in 1976. (NASA)

The temperature of a planet's atmosphere is also important. If the gas is hot, its molecules have a higher average velocity and are more likely to exceed escape velocity. Thus, a planet that is near the sun and very hot is less likely to retain an atmosphere than a more distant, cooler planet. The velocity of a gas molecule, however, also depends on the mass of the molecule. On average, a low-mass molecule travels faster than a massive molecule. Thus, a planet loses its lowest-mass gases more easily because those molecules travel fastest.

We can see this principle of comparative planetology if we plot a diagram such as that in Figure 23-16. The points show the escape velocity versus temperature of the larger objects in our solar system. The temperature we use is the temperature of the gas that is in a position to escape. For the moon, which has essentially no atmosphere, this is the temperature of the sunlit surface. For Mars, we use the temperature at the top of the atmosphere. The lines in Figure 23-16 show the typical velocity of the fastest-traveling examples of various molecules. At any given temperature, some water molecules, for example, travel faster than others, and it is the highest-velocity molecules that escape from a planet. The diagram shows that the Jovian planets have escape velocities so high that very few molecules can escape. Earth and Venus can't hold hydrogen, and Mars can hold only the more massive molecules. Earth's moon is too small to keep any of the common gases. We will refer to this diagram again as we study the atmospheres of other worlds.

Over the 4.5 billion years since Mars formed, it has lost some of the lower-mass gases. Water molecules are massive enough for Mars to keep, but ultraviolet radiation can break up the water molecules. On Earth, the ozone layer protects water vapor from ultraviolet radia-

tion, but Mars never had an oxygen-rich atmosphere so it never had an ozone layer. Ultraviolet photons from the sun can penetrate deep into the atmosphere and break up molecules such as water. The hydrogen escapes, and the oxygen, a very reactive element, forms oxides in the soil—the oxides that make Mars the red planet. Thus, molecules too massive to leak into space can be lost if they break into lower-mass fragments.

The argon in the Martian atmosphere is evidence of a denser blanket of air in the past. Argon atoms are massive, almost as massive as a CO_2 molecule, and would not be lost easily. In addition, argon is inert and would not form compounds in the soil. The 1.6 percent argon in the atmosphere of Mars is evidently left over from an ancient atmosphere that was 10–100 times denser than the present Martian air.

Finally, we must consider the interaction of the solar wind with the atmosphere of Mars. This is not an important process for Earth because Earth has a magnetic field that deflects the solar wind. Because Mars has no magnetic field, the solar wind interacts directly with the Martian atmosphere, and detailed calculations show that significant amounts of CO_2 could have been carried away by the solar wind over the history of the planet. This process would have been most efficient long ago, however, when the sun was more active and the solar wind was stronger. We must also wonder if Mars had a magnetic field when it was younger and still retained significant internal heat. If it had a molten, conducting core, it may have had a magnetic field to protect it from the solar wind.

Although we remain uncertain as to how much of an atmosphere Mars has had in its past and how much it has lost, it is a good example for our study of comparative planetology. When we look at Mars, we see what

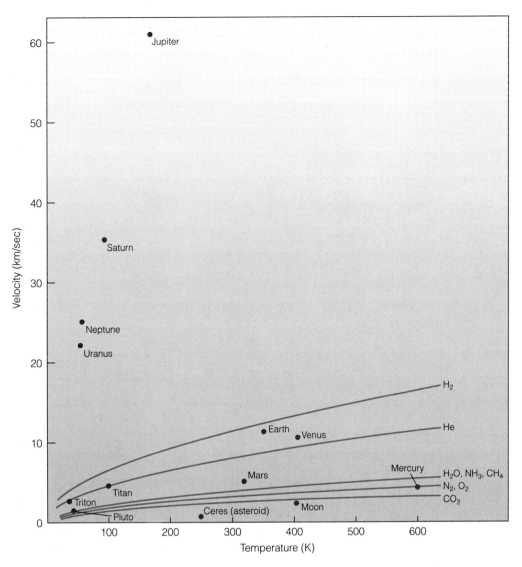

Velocity (km/sec) vs Temperature (K)

FIGURE 23-16
The loss of planetary gases. Dots represent the escape velocity and temperature of various solar-system bodies. The lines represent the typical highest velocities of molecules of various masses. The Jovian planets have high escape velocities and can hold on to even the lowest-mass molecules. Mars can hold only the more massive molecules, and the moon has such a low escape velocity that even the most massive molecules can escape.

can happen to the atmosphere of a medium-size world. Like its atmosphere, the geology of Mars is typical of intermediate-size worlds.

The Geology of Mars

The surface of Mars has been photographed by a number of spacecraft, ranging from Mariner 4, which flew by in 1965, to the Viking spacecraft, which orbited Mars in 1976. These photos show that Mars is a cratered world like the moon but that it also has gigantic volcanoes, deep canyons, dry channels, and vast dust storms (Figure 23-17).

FIGURE 23-17
This mosaic of photographs taken by the orbiting Viking spacecraft shows the reddish surface of Mars marked by circular impact craters and larger, darker volcanoes (left). The canyon system Valles Marineris stretches across the planet for a distance greater than the distance from New York to Los Angeles. (NASA/USGS, courtesy Alfred S. McEwen)

Volcanic plains   Basin ejecta   Polar deposits   Cratered plains   Volcano

FIGURE 23-18

A geological map of Mars. Most of the southern hemisphere is old cratered terrain, whereas most of the northern hemisphere is younger volcanic plains. Note the volcanic areas, Tharsis and Elysium, and the large basins, Hellas and Argyre.

The seasonal dust storms are intrinsically interesting because they explain the apparent seasonal changes in the dark markings. Apparently, yearly dust storms cover outcrops of dark rock, and as spring begins the winds gradually strip away the dust, and the markings grow darker. Because the entire planet has a red color, the dark markings look greenish in contrast.

Although the canals and seasonal changes caused by vegetation proved illusory,* Mariner 9 and Vikings 1 and 2, which arrived in orbit in June 1976, have shown that Mars is geologically fascinating. The southern hemisphere is old and heavily cratered, giving it a lunar appearance (Figure 23-18). The northern hemisphere, however, has few craters, and those few are much sharper and less eroded and thus younger. The region has been smoothed by repeated lava flows that have buried the original surface, and two areas of volcanic cones have been found. In addition, the northern hemisphere is the site of deformed crustal sections in the form of uplifted blocks and collapsed depressions that suggest geological activity (Window on Science 23-2).

By counting the number of craters in a given area, planetary astronomers can estimate the age of the region. These studies show that the old cratered southern hemisphere is 2–3 billion years old. The lava flows of the northern hemisphere, however, have fewer craters and are about 1 billion years old. Some regions may have been flooded by lava within the last few hundred million years. This confirms our understanding of Mars. It is larger than the moon, so its volcanism has continued longer than did the moon's. But Mars is smaller than Earth, so some of its ancient cratered terrain has survived undamaged by volcanism.

Martian volcanoes are shield volcanoes, which have shallow slopes showing that the lava flowed easily. Nevertheless, the largest volcano, Olympus Mons, is a vast structure (Figure 23-19). Its base is 600 km (370 mi) in diameter, and it towers 25 km (16 mi) above the surface. In contrast, the largest volcano on Earth, Mauna Loa in Hawaii, rises only 10 km (6 mi) above its base on the Pacific Ocean floor, and its base is only 225 km (140 mi) in diameter. The weight of Mauna Loa is too great to be supported by the seafloor. It has pushed the seafloor down, just as a bowling ball resting on a bed depresses the surface. This depression is visible as an undersea moat ringing Mauna Loa, but no depression is visible ringing Olympus Mons, even though it is 2.5 times higher than Mauna Loa. Evidently the crust of Mars is thicker than Earth's.

Other evidence shows that the Martian crust is thinner than the moon's. The various Mariner spacecraft

*We will return to the search for life on Mars (and elsewhere) in Chapter 27.

The Present Is the Key to the Past

Geologists are fond of saying "The present is the key to the past." By that they mean that we can learn about the history of the earth by looking at the present condition of the earth's surface. The position and composition of various rock layers in the Grand Canyon, for example, tell us that the western United States was once at the floor of an ocean. This principle of geology was astonishing when it was first formulated in the late 1700s, and it continues to be relevant today as we try to understand other worlds, such as Venus and Mars.

In the late 18th century, naturalists recognized that the present gave them clues to the history of the earth. This was astonishing because most people assumed that the earth had no history. That is, they assumed either that the earth had been created in its present state as described in the Old Testament or that the earth was eternal. In either case, people commonly assumed that the hills and mountains they saw around them had always existed more or less as they were. The 18th-century naturalists began to see evidence that the hills and mountains were not eternal but were the result of past processes and were slowly changing. That gave birth to the idea that the earth had a history.

As the naturalists of the 18th century made the first attempts to thoughtfully and logically explain the nature of the earth by looking at the evidence, they were inventing modern geology as a way of understanding the earth. What Copernicus, Kepler, and Newton did for the heavens in the 1600s, the first geologists did for the earth beginning in the late 1700s. Of course, the invention of geology as the study of the earth led directly to our modern attempts to understand the geology of other worlds.

Geologists and astronomers share a common goal: they are attempting to reconstruct the past (Window on Science 20-3). Whether we study Earth, Venus, or Mars, we are looking at the present evidence and trying to reconstruct the past history of the planet by drawing on observations and logic to test each step in the story. How did Venus get to be covered with lava, and how did Mars lose its atmosphere? The final goal of planetary astronomy is to draw together all of the available evidence (the present) to tell the story (the past) of how the planet got to be the way it is. Those first geologists of the late 1700s would be fascinated by the stories planetary astronomers tell today. ∎

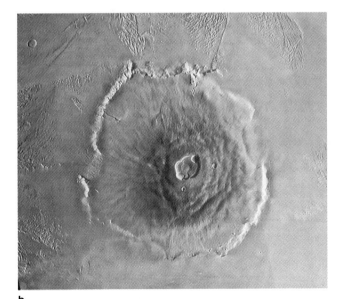

b

FIGURE 23-19

(a) Olympus Mons, the tremendous shield volcano on Mars, is much larger than Mauna Loa, the largest volcano on Earth. Mauna Loa has sunk into the crust, producing a moat around its base. (b) Olympus Mons has no moat, suggesting that the Martian crust is stronger than Earth's. (NASA)

discovered a vast valley at least 4000 km (2500 mi) long and, at its widest, 600 km (400 mi) wide. At its deepest point, it is over 6 km (4 mi) deep, making it over four times deeper than the Grand Canyon. On Earth it would stretch farther than from New York to Los Angeles. The valley, now named Valles Marineris (Figure 23-17), is apparently a block of crust that has dropped downward along parallel faults. We see no such features on the moon, so we can conclude that the crust of Mars is thinner than the moon's crust. This is what we would expect for a planet that is larger than the moon but smaller than Earth.

Valles Marineris begins in an area of volcanism called the Tharsis bulge and may be related to the stresses created in the crust by this volcanic activity. Olympus Mons is only one of a number of volcanoes in the Tharsis region. The area bulges about 10 km (6 mi) higher than the average surface of Mars and has few craters, showing that it is geologically young (Figure 23-20). Some of the volcanoes in this region, such as Olympus Mons, appear to be less than a billion years old, but others may be very old volcanoes whose flanks have been buried by lava from younger volcanoes. The Tharsis bulge itself may be almost as old as the surface of the planet, but recent volcanism has wiped out surface features from more than about 1 billion years ago.

A similar uplifted volcanic plain, the Elysium region, lies almost halfway around the planet. It is more heavily cratered and appears to be slightly older than the Tharsis region. The origin of these uplifted regions is not well understood, but they are probably connected to rising currents of magma in the mantle. On Venus, we see these

Mars

Mars to scale

FIGURE 23-20
Earth, Venus, and Mars are shown in these maps, with the highest regions shaded beige and the lowest blue. Note the large volcanic highland, the Tharsis bulge, at left center on the Mars map and the Elysium volcanic uplift at extreme right. Two large impact basins on Mars, Argyre and Hellas, are visible at lower left and lower right. The Mars map is reproduced to scale at upper right. (NASA)

Venus

Earth

FIGURE 23-21
(a) On Earth, a rising current of magma breaks through the crust to form a chain of volcanoes because the moving plate carries older volcanic peaks away. The Hawaiian–Emperor island chain formed in this way, and most of the older peaks have now been worn down below sea level. (b) On Mars, volcanoes such as Olympus Mons have grown to great size, suggesting that the crust is not moving. Successive eruptions occur at the same place and add layer after layer to the bulk of the volcano.

Plate motion

a

Stationary plate

b

a   b   c

FIGURE 23-22

Evidence of water on Mars. (a) Dry river channels flow downhill through a cratered Martian plain. (b) Signs of erosion due to a sudden flood mark this cratered region on Mars. (c) Dry stream beds meander into a lowland desert. The largest crater in this image is surrounded by a "splash" of ejected material, suggesting that the crust was rich in water at the time of the impact. (NASA)

regions as coronae, but the thick crust on Mars may prevent such scars from forming.

Volcanoes form where the rising magma breaks through the crust, but volcanoes on Earth do not grow as large as they do on Mars, evidently because Earth's crust is in motion. On Earth, a rising current of magma, like that below the Hawaiian islands, breaks through the crust and forms a volcano. The motion of the crustal plate, however, moves the accumulated lava aside, and the next eruption produces a new volcano (Figure 23-21). Thus, the Hawaiian–Emperor island chain extends 3800 km (2300 mi) northwest across the Pacific seafloor because the crust is moving (see Figure 21-11). On Mars, the magma seems to have erupted over and over in the same place and built massive volcanoes such as Olympus Mons, the largest volcano in our solar system. The vast size of Olympus Mons is evidence that the crust of Mars is not separated into moving plates.

We have no evidence that Martian volcanoes are still active, although that is a possibility. Craters in the youngest lava flows on Olympus Mons show that the volcano may have been active as recently as a few hundred million years ago. The Elysium volcanoes appear to be inactive.

Water on Mars

While geologists were fascinated to see signs of an active surface on Mars, they were especially excited to see two kinds of erosional features that suggested water once flowed over the surface (Figure 23-22). The evidence is hard to evaluate, but it suggests that the climate on Mars may be much more interesting than it seems.

One type of erosional feature on Mars resembles the traces left behind by violent floods that scour the surface in a matter of hours or days. Some of these eroded areas are so large they must have been formed by floods 10,000 times greater than the flow of the Mississippi. One way to account for these features is to suppose that water was frozen in the soil as permafrost. Volcanic heating, a large meteorite impact, or internal pressure might have melted the ice to produce sudden floods. Photos of Mars show jumbled valleys typical of the withdrawal of subsurface water and craters surrounded by "splashes" of ejecta, suggesting that the soil was water-laden at the time of impact (Figure 23-23).

The second type of feature is meandering drainage valleys winding through nearly flat terrain. These features suggest rain or snow, but that seems impossible given current conditions on Mars. The present atmospheric pressure is not high enough to keep water from boiling away long before it could form a meandering drainage pattern. Any water that did not evaporate would freeze solid. Sudden floods might scour the landscape faster than the water could vaporize, but the thin atmosphere makes even that difficult.

By counting craters in the flow features, astronomers can estimate that they formed between 1 and 3 billion years ago, with the flood features being youngest. Evidently, the atmosphere was denser and warmer then, and

a

b

FIGURE 23-23

(a) A computer-enhanced photomosaic of the north polar cap of Mars shows the complexity of the frozen layers. (NASA/USGS) (b) A summer photo taken near the north pole shows frozen water left behind when the dry ice vaporized. The varied extent of the layered terrain (arrows) is apparently caused by periodic changes in the Martian climate. (NASA) (c) In places, different sets of layers are superimposed. Layers are 10–15 m thick. (Adapted from a diagram by J. A. Cutts, K. R. Blasius, G. A Briggs, M. H. Carr, R. Greeley, and H. Masursky)

c

liquid water could survive on the Martian surface. At least some of the flood features were formed by water frozen in the crust, but it is also possible that bodies of liquid water existed on the surface.

Some astronomers have found features on Mars that they interpret to be shorelines and related features produced by oceans. By counting craters, they conclude that the oceans might have existed 2.5–3.5 billion years ago. To explain why the water remained liquid, they propose that volcanic eruptions released enough gases to produce a strong greenhouse effect. Many experts think this hypothesis is highly speculative.

A different team of astronomers believes they have sighted scars made by glaciers in the old southern hemisphere of Mars. To make a glacier, a planet must have a body of liquid water so that evaporation can supply snow to the glacier. Thus, glaciers on Mars imply an ocean, but the glacial scars seem to be only 1–2 billion years old.

Most experts do not accept the idea of ancient oceans or glaciers. The conventional interpretation of the flow features is that they were cut by water 1–3 billion years ago, when the atmosphere was thicker. In fact, Mars may go through relatively short periods when the atmosphere is denser and warmer. On average, over the past 3 billion years, Mars may have looked much as it does today.

Evidence of climatic variation is visible near the polar caps (Figure 23-24). The polar caps consist of frozen carbon dioxide (dry ice) with frozen water beneath. When spring begins in a hemisphere, the polar cap there begins to shrink as the carbon dioxide turns directly into gas. The water, however, remains behind in a smaller, permanent polar cap. The region around this permanent cap is marked by layered terrain, evidently deposited by wind-borne dust that accumulates on the ice cap. As the cap shrinks, the material accumulates in layers. Each layer is a few dozen meters thick and may take about 10,000 years to accumulate. The variation in the layering suggests that some periodic change in climate, perhaps due to orbital changes, may affect the frequency and intensity of planet-wide dust storms and thus alter the rate at which material is deposited. Recall that Earth's climate may go through cycles caused by variations in its orbital motion. (See Chapter 2.)

One astonishing bit of evidence of water on Mars is the analysis of rock samples from the planet. Of course, no astronaut has ever visited Mars and brought back a rock, but over the history of the solar system occasional impacts by asteroids have blasted giant craters on Mars and ejected bits of rock into space. A few of those bits of rock have fallen to Earth as meteorites, and nearly a

a

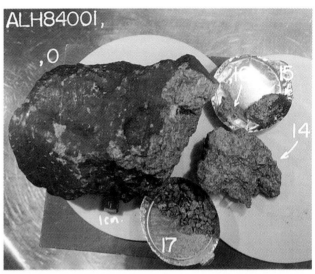

b

FIGURE 23-24

(a) The impact basin Argyre (filling the lower left quarter of this photo) is 740 km in diameter and has a rim composed of uplifted crustal blocks. The large crater at the left is Galle, with a diameter of 210 km. Such large impacts may occasionally eject fragments of the crust into space, and a few of these fragments have fallen to Earth as meteorites. (b) Meteorite ALH84001, found in Antarctica, has been identified an an ancient chunk of the Martian crust blasted into space by a meteorite impact on Mars. Minerals found in the meteorite were deposited in water and, thus, suggest that the Martian crust was once richer in water than it is now. (This meteorite also contains possible signs of ancient bacterial life. See Chapter 27.) (NASA/JSC)

dozen have been found and identified (Figure 23-25). These rocks are basalts, as we might expect from a planet so heavily covered by lava flows, and chemical analysis shows that the magma from which the rocks solidified must have contained about 1.4 percent water. If all of the lava flows on Mars contained that much water and it was all outgassed, it could create a planet-wide ocean 200 m deep. That isn't quite enough to explain all the flood features on Mars, but it does assure us that the planet once had more water. Now, all of the water on Mars would make a planet-wide layer of frost only 0.003 mm deep. However much water Mars once had, it is now a desert world.

The controversy over water on Mars continues. We can only estimate how much water there was, how much has been lost to space, and how much remains frozen in the soil. To better understand the history of Mars, we need to explore its geology directly.

Exploring Mars

On July 4, 1997, the spacecraft Pathfinder entered the Martian atmosphere, popped its parachute, dropped its heat shield, inflated cushioning air bags, fired retro rock-

ets only 100 meters above the surface, and then cut itself free to fall the last few meters and bounce across the rocky surface. Pathfinder landed, almost to the day, 21 years after the Viking spacecraft landed on Mars, and a new period of Mars exploration began.

Soon after the air bags deflated, Pathfinder opened to expose its solar panels to sunlight and to free Sojourner, a 6-wheeled robotic vehicle about the size of a microwave oven. Sojourner rolled down a ramp and, directed by controllers on earth, began visiting nearby rocks to test their composition (Figure 23-25). The lander section was renamed the Carl Sagan Memorial Station.

Pathfinder landed at Ares Vallis in the mouth of an ancient flood plain. At least once in the past, a torrent of water 100 meters deep rushed over Ares Vallis at 100 km/hour. Scientists hoped this would have transported a wide variety of rocks into the area for Sojourner to examine, and photos taken by the Sagan Station and by Sojourner show that the region contains everything from large boulders all leaning in the same direction down to small fragments. Layering seems to be visible in the soil and on the sides of distant hills, and the soil contains a whitish deposit that may be salt from the flooding.

FIGURE 23-25

After Pathfinder bounced to a stop on the surface of Mars, it opened and the rover, Sojourner, rolled down the ramp at lower left to explore the surface. It is shown here pressing its spectrometer against the large bolder, Yogi (red arrow), to analyze its composition. Blue arrows point to rocks Cradle (above) and Barnacle Bill (below). The inset shows white deposits in the dusty soil (white arrow) exposed by Sojourner's wheels. Note the irregular horizon. (NASA)

Photographs taken by the lander show an irregular horizon marked by two prominent hills. This, too, adds to scientists' expectation that a variety of rocks litter the area. Rock debris from the hills might naturally have reached the landing site. Also, some of the irregularities on the horizon are crater rims, so some rock from the craters may have been thrown into the area.

Crawling less than half a meter per minute, Sojourner could visit rocks up to 500 m distant and analyze their composition by placing a probe against the rock. The probe carried a radioactive source that excites atoms in the rock to emit X rays. Scientists can identify the composition of the rock by the energies of the X rays.

The first rock Sojourner visited, dubbed Barnacle Bill, proved to be rich in silica. Such a rock on earth is produced in volcanic activity when plates descend below other plates. In fact, Barnacle Bill seems to resemble andesite, a silica-rich rock found in the Andes of South America. On earth, andesite is formed as ocean plates slide under the South American plate producing volcan-

ism. Thus, the very first rock Sojourner visited raises new questions about the geology of Mars.

While Sagan Station and Sojourner were still exploring the surface, a second spacecraft arrived. Mars Global Surveyor took up a highly elliptical orbit around Mars because such an orbit minimizes rocket fuel, but it quickly began reducing its orbit to circular by dipping into the atmosphere of Mars once each orbit. This aerobraking process is difficult but economical. Roughly four months are needed to circularize the orbit.

Mars Global Surveyor will photograph the surface of Mars repeatedly over a full Mars year of 687 days watching for meteorological and geological changes as the Martian seasons pass and, thus, expanding our understanding of the geology of Mars. Its cameras are able to detect objects as small as 1.4 meters, about the size of an automobile. Such detailed mapping will help planetary astronomers search for signs of past flooding, volcanism, and other geological activity.

A new age of Mars exploration is just beginning. Pathfinder and Mars Global Surveyor are small, inexpensive probes. Pathfinder, for example, cost each taxpayer only $0.50 per year over 4 years. Seven more spacecraft will reach Mars through the year 2005. Eventually, humans will walk across the surface, studying the rocks to try to understand the history of this peculiar world.

A History of Mars

Our goal in studying any planet is to understand how the planet got to be the way it is. Can we tell the story of Mars? To do so, we must combine what we know of the origin of the solar system, the evolution of planetary atmospheres, and the geology of the terrestrial planets.

The four-stage history of Mars is a case of arrested development. The planet began by differentiating into a crust and core. Mars Global Surveyor was able to detect a magnetic field 800 times weaker than earth's. Mars is large enough to be molten at its center, and it spins fast enough for the dynamo effect to generate a strong magnetic field. The lack of a strong field suggests that Mars did not form a core of conducting iron compounds.

During the second stage of development, impacts battered the newly formed crust and cratered it, as we see in the old southern hemisphere. The largest impacts blasted out great basins like Hellas and Argyre, and they were probably flooded by lava. One team of astronomers has suggested that a single major impact caused the volcanism so common in the northern hemisphere. Others argue that the formation of the core also created the Tharsis bulge and its associated volcanism. In any case, we know that cratering was succeeded by flooding by lava.

Whether the third stage, flooding, included flooding by water is controversial. Some astronomers argue that Mars has suffered only periodic episodes of surface water as its climate has oscillated, and other astronomers claim to find traces of ancient lakes, oceans, or glaciers. The stream channels and flood features, combined with the analysis of meteorites from Mars, give us strong evidence that Mars was once much wetter than it is now. Much of its water has apparently been lost to space, and some has been trapped in the polar caps and in permafrost.

The crust of Mars has been stressed and has broken into faults to form Valles Marineris. The small planet cooled rapidly, however, and the crust never divided into moving plates. Valles Marineris might be the beginning of a rift valley, but it does not outline a plate. Nor do we see the folded mountain chains that mark colliding plates. Even the vast size of Olympus Mons is evidence that the crust is not moving.

Lacking plate tectonics, the fourth stage in the history of Mars has been one of slow decline. As the small world cooled, its crust grew thicker and volcanism declined.

In addition, the climate seems to have changed. Recently calculated models suggest that Mars once may

have rotated at a much steeper angle to its orbit, as much as 45°. This could have made the climate warmer and kept more of the CO_2 from freezing out at the poles. The rise of the Tharsis bulge could have tipped the axis to its present 25° and cooled the climate. That might account for the warmer, wetter climate implied by the flow features. Some calculations suggest that Mars goes through cycles as its rotational inclination fluctuates, and this may cause short-term variations in climate much like the ice ages on Earth. That might explain the layered terrain near the polar caps.

Our history of Mars is far from complete, but it does illustrate how the intermediate size of Mars has influenced both its atmosphere and its geology. Neither small nor large, Mars is a medium world.

CRITICAL INQUIRY

Why doesn't Mars have coronae like those on Venus?

The coronae on Venus are believed to have been caused by rising currents of molten magma in the mantle pushing upward under the crust and then withdrawing to leave the circular scars called coronae. Earth, Venus, and Mars have had significant amounts of internal heat, and there is plenty of evidence that they have had rising convection currents of magma under their crusts. Of course, we wouldn't expect to see coronae on Earth because its surface is rapidly modified by erosion and plate tectonics. Furthermore, the mantle convection on Earth seems to produce plate tectonics rather than coronae. Mars, however, is a smaller world and must have cooled faster. We see no evidence of plate tectonics, and we do see giant volcanoes that suggest rising plumes of magma erupting up through the crust at the same point over and over. Perhaps we see no coronae on Mars because the crust of Mars rapidly grew too thick to deform easily over a rising plume. On the other hand, perhaps we could think of the entire Tharsis bulge as a single giant corona.

Planetary astronomers haven't explored enough planets yet to see all the fascinating combinations nature has in store for us. Perhaps after we have visited a few hundred terrestrial planets, we will recognize the geology of Mars as typical of medium-size worlds. Of course, Mars is not medium in terms of its location. Of the terrestrial planets, Mars is the farthest from the sun. How has that affected the evolution of its atmosphere? ∎

Our history of Mars must include a phenomenon that was not part of our histories of Mercury and Venus. Unlike those worlds, Mars has moons. These moons, and the moons we will study in the outer solar system, contain important clues to the origin of the planets.

a

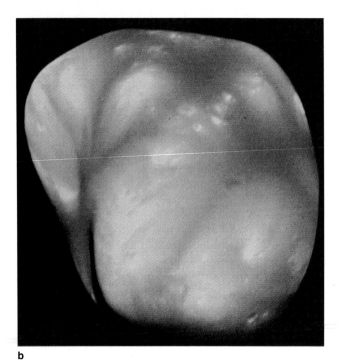

b

FIGURE 23-26

The moons of Mars are small, rocky worlds too small to pull themselves into spherical shape. They may be captured asteroids, or they may have formed with Mars. (a) Phobos, only 28 km (17 mi) in its largest dimension, is irregular in shape, heavily cratered, and marked by grooves that might be related to the impact that formed Stickney, the largest crater. (Damon Simonelli and Joseph Veverka, Cornell University/NASA) (b) Deimos, smaller than Phobos, looks more uniform because of dusty soil covering the smaller features. (NASA)

23-3 THE MOONS OF MARS

If we could camp overnight on Mars, we might notice its two small moons, Phobos and Deimos. Phobos, shaped like a flattened loaf of bread 20 km × 23 km × 28 km (Figure 23-26a), would appear less than half as large as Earth's full moon. Deimos, only 12 km in diameter and three times farther from Mars, would look only 1/15 the diameter of Earth's moon (Figure 23-26b).

Both moons are tidally locked to Mars, keeping the same side facing the planet as they orbit. Also, both moons revolve around Mars in the same direction that Mars rotates, but Phobos follows such a small orbit that it revolves faster than Mars rotates. Thus, Phobos rises in the west and sets in the east.

Origin and Evolution

Deimos and Phobos are typical of the small, rocky moons in our solar system. They are dark gray, with albedos of only about 0.06, and they have low densities, about 2 g/cm³.

Many of the properties of these moons hint that they are captured asteroids. In the outer parts of the asteroid belt, almost all asteroids are dark, low-density objects. Massive Jupiter, orbiting just outside the asteroid belt, can scatter such bodies throughout the solar system, so we should not be surprised if Mars, the next planet inside the asteroid belt, has captured a few of these as satellites.

However, capturing a passing asteroid into a closed orbit is not so easy that it happens often. The asteroid approaches the planet along a hyperbolic orbit and, if it is unimpeded, swings around the planet and disappears back into space. To convert the hyperbolic orbit into a closed orbit, the planet must slow the asteroid as it passes. Tidal forces might do this, but they would be rather weak. Interactions with other moons or grazing collisions with a thick atmosphere might also slow the asteroid.

Both satellites have been photographed by the Mariner and Viking spacecraft, and those photos show that the satellites are heavily cratered. Such cratering could have occurred while the moons were either still in the asteroid belt or in orbit around Mars. In any case, the heavy battering has broken the satellites into irregular chunks of rock, and they cannot pull themselves into smooth spheres because their gravity is too weak to overcome the structural strength of the rock. We will discover that low-mass moons are typically irregular in shape, whereas more massive moons are more spherical.

Photos of Phobos reveal a unique set of narrow, parallel grooves (Figure 23-26a). Averaging 150 m wide and 25 m deep, the grooves run from Stickney, the largest crater, to an oddly featureless region on the opposite side of the satellite. One theory suggests that the grooves are deep fractures produced by the impact that formed the crater. The featureless region opposite Stickney may be similar to the jumbled terrain on Earth's moon and the weird terrain on Mercury. All were produced by the focusing of seismic waves from a major impact on the far side of the body. High-resolution photographs show that the grooves are lines of pits, suggesting that the pulverized rock material on the surface has drained into the fractures or that gas, liberated by the heat of impact, escaped through the fractures and blew away the dusty soil.

Deimos not only has no grooves, it also looks smoother because of a thicker layer of dust on its surface (Figure 23-26b). This material partially fills craters and covers minor surface irregularities. It seems certain that Deimos experienced collisions in its past, so fractures may be hidden below the debris.

The debris on the surfaces of the moons raises an interesting question. How can the weak gravity of small bodies hold any fragments from meteorite impacts? Escape velocity on Phobos is only 12 m/sec. An athletic astronaut could almost jump into space. Certainly, most fragments from impacts should escape, but some could fall back and accumulate on the surface.

Deimos, smaller than Phobos, has a smaller escape velocity. But it has more debris on its surface because it is farther from Mars. Phobos is close enough to Mars that most ejecta from impacts on Phobos will be drawn into Mars. Deimos, being farther from Mars, is able to keep a larger fraction of its ejecta. Phobos is so close to Mars that tides are making its orbit shrink, and it will fall into Mars or be ripped apart by tidal forces within about 100 million years.

Deimos and Phobos illustrate three principles. First, some satellites are probably captured asteroids. Second, small satellites tend to be irregular in shape and heavily cratered. And third, tidal forces can affect small moons and gradually change their orbits. We find even stronger tidal effects in Jupiter's satellite system (Chapter 24).

CRITICAL INQUIRY

Why would we be surprised to find volcanism on Phobos or Deimos?

In discussing the moon, Mercury, Venus, Earth, and Mars, we have seen illustrations of the principle that the larger a world is the more slowly it loses its internal heat. It is the flow of that heat from the interior through the surface into space that drives geological activity such as volcanism and plate motion. A small world, like the moon, cools quickly and remains geologically active for a shorter time than a larger world like Earth. Phobos and Deimos are not just small, they are tiny. However they formed, any interior heat would have leaked away very quickly; with no energy flowing outward, there can be no volcanism.

Some futurists suggest that our first human missions to Mars will be not to land on the surface but to build a colony on Phobos or Deimos. These plans speculate that there may be water deep inside the moons that colonists could use. What would happen to water released in the sunlight on the surface of such small worlds? ■

Venus is nearly the same size as Earth, but it is unlikely that humans will ever colonize its surface. The smaller and colder Mars, however, may someday be home to large colonies. In the next chapter, we turn our attention to Jupiter and Saturn, giant worlds so totally unlike Earth that we will need entirely new principles of comparative planetology to describe them. ■

■ *Summary*

All terrestrial planets pass through a four-stage history: (1) differentiation; (2) cratering; (3) flooding of the crater basins by lava, water, or both; and (4) slow surface evolution. The importance of each of these processes in the evolution of a planet depends on the planet's mass and temperature.

Venus is Earth's twin in size and density, but the planet has evolved along divergent lines because it is slightly closer to the sun. The higher temperature evaporated any early oceans and prevented the absorption of carbon dioxide from the atmosphere. The accumulating carbon dioxide created a greenhouse effect that produced very high surface temperatures.

The crust of Venus is marked by low, rolling plains and higher uplands with some very high volcanoes. Coronae appear to be produced by rising currents of molten rock below the crust. Where the magma breaks through, it builds volcanoes and floods the surface with lava flows. The surface is marked by more impact craters than Earth but fewer than the moon. Venus appears to erase craters by burying them under lava flows and not by plate tectonics or erosion as on Earth.

A number of spacecraft have landed on the surface of Venus, but none has survived for more than an hour or two. The surface temperature is about 745 K (880°F), and the atmosphere contains traces of sulfuric, hydrochloric, and hydrofluoric acids. Four Russian spacecraft have taken photographs on the surface and transmitted them back to Earth. These photos show rocky terrain of broken and weathered rock. Analysis of samples at the landing sites suggests that the rocks are basalts.

The interior of Venus should resemble that of Earth, but no spacecraft has ever detected a magnetic field around Venus. It should have a field if it has a molten metallic core like Earth's. Another mystery associated with Venus is its retrograde rotation. This may have been produced by a major impact during accretion.

Mars is smaller than Earth but larger than Mercury. Although early astronomers saw canals on Mars and thought it to be inhabited, we now know that it is a dry desert world. Its southern hemisphere is a heavily cratered, old region; the northern hemisphere has been resurfaced by lava flows. Much of Mars has been altered by volcanism, and some of its volcanoes are very large. Some may still be active.

Mars may once have had liquid water flowing over its surface. Photographs returned by Mariner 9 and the Viking orbiters show extensive dry streambeds. Most of these seem to have been cut by sudden flooding rather than by gradual runoff. Whatever water Mars still has is now frozen in the soil or in the polar caps.

The Martian atmosphere is now very thin because the low gravity has been unable to retain most gases. But features such as the streambeds and the layered terrain near the polar caps suggest that the Martian climate may have been warmer and wetter in the past and that it may change cyclically, perhaps due to orbital changes.

Deimos and Phobos, the two small moons of Mars, may be captured asteroids. Like most small moons, they are irregular in shape because their gravities are too weak to deform their rock into spheres.

▪ New Terms

subsolar point   rolling plain
shield volcano   corona

▪ Questions

1. Why would we expect Venus and Earth to be similar?

2. What evidence do we have that Venus and Mars once had more water than at present? Where did that water come from? Where did it go?

3. What features will you look for in future radar maps of Venus to search for plate tectonics?

4. Why doesn't Mars have mountain ranges like those on Earth? Why doesn't Earth have large volcanoes like those on Mars?

5. What were the canals on Mars? How do they differ from the dry streambeds on Mars?

6. Propose two different explanations for the nearly pure carbon-dioxide atmospheres of Venus and Mars. How did Earth avoid such a fate?

7. What evidence do we have that the climate on Mars has changed?

8. How can we estimate the ages of the streambeds on Mars?

9. Why are Phobos and Deimos nonspherical? Why is Earth's moon much more spherical?

▪ Discussion Questions

1. From what you know of Earth, Venus, and Mars, do you expect the volcanoes on Venus and Mars to be active or extinct? Why?

2. If humans someday colonize Mars, the biggest problem may be finding water and oxygen. With plenty of solar energy beating down through the thin atmosphere, how might colonizers extract water and oxygen from the Martian environment?

▪ Problems

1. How long would it take radio signals to travel from Earth to Venus and back if Venus were at its farthest point from Earth? Why are such observations impractical?

2. The Pioneer Venus Orbiter circled Venus with a period of 24 hours. What was its average altitude above the surface? (HINT: See Chapter 5.)

3. The smallest feature visible through an earth-based telescope has an angular diameter of about 1 second of arc. If a canal on Mars was just visible when Mars was at its closest to Earth, what would be its linear diameter? (HINT: Use the small-angle formula.)

4. What is the maximum angular diameter of Phobos as seen from Earth? What surface features should we be able to see using earth-based telescopes? (HINT: Use the small-angle formula.)

5. Tides are making the orbit of Earth's moon larger, but tides are making the orbit of Phobos smaller. Explain this by redrawing Figure 3-9 to show Phobos orbiting Mars faster than Mars rotates. Do the tidal bulges lead or lag the satellite?

6. Assume that Deimos is 12 km in diameter and has a density of 2 g/cm^3. What are its mass and escape velocity? (HINTS: The volume of a sphere is $\frac{4}{3}\pi r^3$. See Chapter 5.)

▪ Critical Inquiries for the Web

1. Have you ever wondered how the various features on the planets get their names? Surface features on Venus are (mostly) named for female figures from history and mythology. Who decides how planetary features are named? Look for information on planetary nomenclature and summarize the way different types of features on Venus are assigned names.

2. "Martians" have fascinated humans for the last century or more. Many online sources chronicle the representation of life on Mars by Earthlings throughout history and literature. Read about the Martians as represented by a particular literary work or nonfiction account and discuss to what extent they are (or are not) based on realistic views of the nature of Mars—in terms of our current understanding as well as the views of the period in which the work was written.

Go to the Wadsworth Astronomy Resource Center (www.wadsworth.com/astronomy) for critical thinking exercises, articles, and additional readings from InfoTrac College Edition, Wadsworth's online student library.

Chapter Twenty-Four

JUPITER AND SATURN

GUIDEPOST

As we begin this chapter, we leave behind the psychological security of planetary surfaces. We can imagine standing on the moon, on Venus, or on Mars, but Jupiter and Saturn have no surfaces. Thus, we face a new challenge—to use comparative planetology to study worlds so unearthly we canot imagine being there.

One reason we find the moon and Mars of interest is that we might go there someday. Humans may become the first Martians. But the outer solar system seems much less useful, and that gives us a chance to think about the cultural value of science.

This chapter begins our journey into the outer solar system. In the next chapter, we will visit worlds out in the twilight at the edge of the sun's family. ■

There is something fascinating about science.

One gets such wholesale returns of conjecture

out of such a trifling investment of fact.

MARK TWAIN

Life on the Mississippi

The sulfuric acid fogs of Venus seem totally alien, but compared with the planets of the outer solar system, Venus is a tropical oasis. Of the five planets beyond the asteroid belt, four have no solid surface, and one, Pluto, is so far from the sun that most of its atmosphere lies frozen on its surface. As we explore these strange worlds, we will discover new principles of comparative planetology. We begin with Jupiter and Saturn because they are the largest of the Jovian planets.

These distant worlds can be studied from Earth, but much of what we know has been radioed back by space probes. Pioneer 10 and Pioneer 11 explored Jupiter and Saturn in the mid-1970s, but their instruments were not very sensitive. The Voyager 1 and Voyager 2 spacecraft were launched in 1977 on a mission to visit all four Jovian worlds, concluding when Voyager 2 flew past Neptune in 1989. In December, 1995, the Galileo spacecraft reached Jupiter. While an instrumented probe dropped into Jupiter's atmosphere, the main Galileo spacecraft went into orbit to begin a long-term study of the planet and its moons. Throughout this discussion of the Jovian worlds, you will find images and data returned to Earth by these robotic explorers.

Jupiter and Saturn are unearthly, but because they obey the same laws of nature, we can understand their peculiar personalities. We begin with the closest and largest Jovian planet, named for the ruler of the Roman gods.

24-1 JUPITER

Jupiter is the most massive of the Jovian planets, containing nearly three-fourths of all the planetary matter in our solar system. This high mass accentuates some processes that are less obvious or nearly absent on the other Jovian worlds. Just as we used Earth as the basis of comparison for our study of the terrestrial planets, we will examine Jupiter in detail so we can use it as a standard in our comparative study of the other Jovian planets.

The most striking feature of Jupiter is its varicolored cloud belts, which make it look like a child's rubber ball (Figure 24-1). We will examine these belts later and see that they may be related to liquid circulations deep inside the planet. However, first we must survey Jupiter in general.

Surveying Jupiter

Jupiter is interesting because it is big, massive, mostly liquid hydrogen, and very hot inside. The preceding facts are common knowledge among astronomers, but we should demand an explanation of how they know these facts. Often the most interesting thing about a fact isn't the fact itself, but how we know it.

We know that Jupiter is a big planet because it looks big. At their closest points to the earth, Jupiter is about 8 times farther away than Mars, but even a small telescope will reveal that the disk of Jupiter is more than two times

FIGURE 24-1

The cloud belts in Jupiter's atmosphere are filled with turbulent circulation, driven in part by energy flowing outward from the planet's interior. The Great Red Spot is a circulating storm much like a hurricane on Earth, but the Great Red Spot alone is larger than Earth. Jupiter is the largest planet in our solar system, over 11 times larger in diameter than Earth, but its low density tells us that it is mostly liquid hydrogen. (NASA)

bigger than the disk of Mars. If we use the small-angle formula, we can compute the diameter of Jupiter—a bit over 11 times the diameter of the earth (Data File Seven). That's a big planet!

We know that Jupiter is massive because its moons race around it at high speed. Io is the innermost of the four Galilean moons, and its orbit is just a bit larger than the orbit of our moon around Earth. Io streaks around its orbit in less than two days, whereas Earth's moon takes a month. Jupiter has to be a very massive world to hold on to such a rapidly moving moon. In fact, we can put the radius of Io's orbit and its orbital period into Newton's version of Kepler's third law (Chapter 5) and calculate the mass of Jupiter—almost 318 times the mass of the earth.

The size and mass of Jupiter are easy to find, but we might wonder how astronomers know that it is made of hydrogen. That fact is just a little bit harder to find, but the first step is to divide the mass by the volume and find Jupiter's average density, about 1.34 g/cm^3. Of course, it is more dense at the center and less dense near the surface, but this average density tells us that Jupiter can't be made totally of rock. Rock has a density of 2.5 to 4 g/cm^3, so Jupiter must contain less dense material such as hydrogen.

Spectra recorded from Earth and from spacecraft visiting Jupiter tell us that the composition of Jupiter is

TABLE 24-1

Composition of Jupiter and Saturn (by mass)

Molecule	Jupiter (%)	Saturn (%)
H_2	78	88
He	19	11
H_2O	0.0001	—
CH_4	0.2?	0.6
NH_3	0.5?	0.2

JUPITER

A theoretical model of the interior of Jupiter. Note the oblateness of the planet.

Labels: Liquid hydrogen, Rock/Ice core, Liquid metallic hydrogen, Earth, Jupiter

Average distance from the sun	5.2028 AU (7.783×10^8 km)
Eccentricity of orbit	0.0484
Maximum distance from the sun	5.455 AU (8.160×10^8 km)
Minimum distance from the sun	4.951 AU (7.406×10^8 km)
Inclination of orbit to ecliptic	1°18′29″
Average orbital velocity	13.06 km/sec
Orbital period	11.867 y (4334.3 days)
Period of rotation	$9^h50^m30^s$
Inclination of equator to orbit	3°5′
Equatorial diameter	142,900 km (11.20 $D_\oplus$)
Mass	1.899×10^{27} kg (317.83 $M_\oplus$)
Average density	1.34 g/cm³
Gravity at base of clouds	2.54 earth gravities
Escape velocity	61 km/sec (5.4 $V_\oplus$)
Temperature at cloud tops	−120°C (−180°F)
Albedo	0.51
Oblateness	0.0637

much like that of the sun—it is mostly hydrogen and helium. This fact was confirmed when a probe from the Galileo spacecraft fell into the atmosphere in 1995 and radioed its results back to Earth. Jupiter is mostly hydrogen and helium with traces of heavier atoms forming molecules such as methane (CH_4), ammonia (NH_3), and water (Table 24-1).

Astronomers build mathematical models of the interior of Jupiter by using equations that describe basic physics and the properties of matter. The models tell us that a planet as big and massive as Jupiter, having a low density and composed mostly of hydrogen and helium, must contain only a small core of heavy elements at its center, a so-called "rocky core." The models also tell us that the pressure below the atmosphere is so great that hydrogen cannot exist as a gas; most of Jupiter is liquid hydrogen with some helium surrounding a dense core a bit larger than Earth.

Even the shape of the planet is revealing. By watching the cloud belts, we can tell that Jupiter rotates rapidly, in about 10 hours. Simple measurements on a photograph show that it is slightly flattened; it is 6.37 percent larger in diameter, through its equator, than it is through its poles. Astronomers refer to this by saying the **oblateness** of Jupiter is 0.0637. The amount of flattening depends on the speed of rotation and on the rigidity of the material. A planet that is mostly liquid will be more oblate than a planet that is mostly rocky. Thus, astronomers can use the observed oblateness of Jupiter to check their models of the size of its heavy core.

It is easy to think of the heavy element core as a ball of rock, but astronomers are quick to point out that it is very hot, about 30,000 K. That is 5 times hotter than the surface of the sun, but the pressure is so great that it keeps the material confined to the core of the planet. How do we know it is so hot? Because infrared observations made by the Voyager spacecraft show that Jupiter emits twice as much energy as it receives from the sun. It must be very hot inside to have so much heat flowing outward into space.

The models of Jupiter's interior tell us that ordinary liquid hydrogen cannot support the pressure deep inside

the planet. It is compressed into a state in which the electrons are free to move easily from atom to atom, and although the hydrogen can flow like a liquid, it can conduct electricity like a metal. This **liquid metallic hydrogen** does not occur naturally on Earth, but it has been created under tremendous pressure in laboratories.

Basic observations and the known laws of physics can tell us a great deal about Jupiter. And its vast magnetic field can tell us even more.

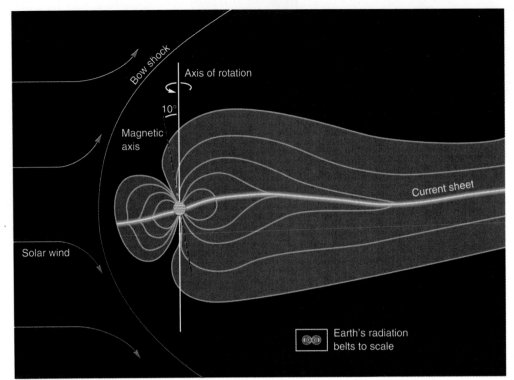

FIGURE 24-2

Jupiter's magnetic field traps particles from the solar wind to form powerful radiation belts. The rapid rotation of the planet forces the slightly inclined magnetic field to wobble as it rotates and confines most of the high-energy particles to a thin plane called the current sheet. Earth's magnetosphere and radiation belts are shown to scale (inset).

Jupiter's Magnetic Field

Astronomers detected a magnetic field around Jupiter as early as the 1950s, which is surprising considering that magnetic fields are invisible to the eye. In fact, Jupiter's giant magnetic field was first detected by radio.

In 1955, radio astronomers detected bursts of synchrotron radio waves coming from Jupiter. They called it **decameter radiation** because its wavelength was tens of meters. Synchrotron radiation is electromagnetic radiation from fast electrons spiraling in a magnetic field, so it was obvious that Jupiter had a magnetic field.

Because the radio signals fluctuated with a period of 9 hours 55.5 minutes, astronomers concluded that this period was the true period of rotation of Jupiter. This result is what we would expect from the dynamo theory for the generation of magnetic fields, which requires an electrically conducting fluid, convection driven by heat flowing outward, and rapid rotation. The liquid metallic hydrogen in Jupiter is a very good electrical conductor, and we have seen clear evidence that heat flows outward from inside Jupiter. That, coupled with Jupiter's rapid rotation, should generate a powerful dynamo effect. Thus, the fluctuation of the radio signals tells us the true rotation period of the planet's liquid interior.

By the early 60s, radio astronomers had discovered short wavelength **decimeter radiation**, with wavelengths a few tenths of a meter long. Although this type of radiation is also synchrotron radiation, radio telescopes revealed that it comes from a region at least 3 times larger than the disk of Jupiter. Detailed studies of the decimeter radiation showed that Jupiter's magnetic field was at least 10 times stronger than Earth's, and, like Earth's field, was inclined about 10° to the axis of rotation.

Radio studies were helpful, but nothing is as good as a visit. In 1973 and 1974, two Pioneer spacecraft flew past Jupiter, followed in 1979 by two Voyager spacecraft. The Galileo spacecraft entered orbit around Jupiter in late 1995. These spacecraft found Jupiter surrounded by a magnetosphere over 100 times larger than Earth's. If we could see Jupiter's magnetosphere, it would appear to be more than 3 times larger in diameter than the full moon.

The magnetic field deflects the solar wind and traps high-energy particles in radiation belts much more intense than Earth's. The radiation is more intense because Jupiter's magnetic field is stronger and can trap and hold higher-energy particles, mostly electrons and protons. The spacecraft passing through the radiation belts received radiation doses equivalent to a billion chest X rays—at least 100 times a lethal dose for a human. Some of the electronics on the Pioneer spacecraft were damaged by the radiation.

Jupiter's rapid rotation has flattened its magnetosphere, and the charged particles are concentrated in a thin layer called the **current sheet** (Figure 24-2). Because the magnetic field is inclined to the axis of rotation, the magnetosphere wobbles up and down as Jupiter rotates.

Some of the heavier ions in the radiation belts appear to come from the inner Galilean moon, Io. As we will see later in this chapter, Io has active volcanoes that spew sulfur. Because Io orbits with a period of 1.8 days, and Jupiter's magnetic field rotates in only 10 hours, the wobbling magnetic field rushes past Io at high speed, sweeping up stray particles, accelerating them to high energy, and spreading them around Io's orbit in a doughnut of ionized gas called the **Io plasma torus.**

Jupiter's powerful magnetic field interacts with Io to produce a powerful electric current (about a million amperes) that flows through a curving tube of ionized gas

FIGURE 24-3

(a) At the left a visible-light image of Jupiter and its moon Io shows the location of its invisible tube of electric current called the Io flux tube. Roughly a million amperes of electrical current flow through the tube. The UV image at the right shows Jupiter's aurora circling its magnetic poles and the tiny spots of aurora where the flux tube intersects Jupiter's atmosphere. (b) These UV images show the aurora around Jupiter's magnetic poles on two different dates. Note the inclination of the magnetic axis. (J. T. Clarke, G. E. Ballester, J. Trauger, and R. Evans; NASA)

from Jupiter out to Io and back to Jupiter (Figure 24-3a). This **Io flux tube** explains a mystery. Back in 1955, when radio astronomers discovered the decametric radio signals, they noticed that the signals were stronger when Io was in certain parts of its orbit. Now we can understand that the electrical current flowing through the Io flux tube is at least partially responsible for generating the decametric radio signals.

We have seen that Earth's magnetosphere produces aurora, and the same is true on Jupiter. Charged particles in the radiation belts leak downward at Jupiter's magnetic poles to produce aurora 1000 times more powerful than those on Earth (Figure 24-3b). Spots of bright aurora lie at the two points where the Io flux tube enters Jupiter's atmosphere. Fluctuations in the aurora tell us that the solar wind buffets Jupiter's magnetosphere, but some of the fluctuations seem to be caused by changes in the magnetic dynamo deep inside the planet. Thus, studies of the aurora on Jupiter may help astronomers learn more about its liquid depths.

Much of the power and fascination of Jupiter is invisible to our eyes, but the swirling cloud belts are beautiful in their complexity.

Jupiter's Atmosphere

If you parachuted into Jupiter with a small rubber boat, expecting to go sailing on the liquid hydrogen ocean, you would be disappointed. Mathematical models of Jupiter tell us that there is no surface. Deep below the clouds, the temperature and pressure exceed the **critical point** for hydrogen, the temperature and pressure above which liquid hydrogen and gaseous hydrogen have the same density and are indistinguishable. As you parachuted down through Jupiter's atmosphere, the temperature, pressure, and density would rise, and gradually the gas would become a fluid. You would never splash down and your rubber boat would be useless. Below the clouds of Jupiter lies the largest ocean in the solar system—and it has no surface and no waves.

When we look at Jupiter, we don't see a surface. We see those parts of the atmosphere that contain clouds. These cloud layers lie deep inside a nearly transparent atmosphere of hydrogen and helium. We can detect this atmosphere by noticing that Jupiter has limb darkening just as the sun does (Chapter 8). When we look near the limb of Jupiter (the edge of its disk), the clouds are much dimmer (Figure 24-1) because it is nearly sunset or sunrise along the limb. If we were on Jupiter at this location, we would see the sun just above the horizon and it would be dimmed by the atmosphere. In addition, light reflected from clouds must travel out at a steep angle through the atmosphere to reach Earth, dimming the light further. Jupiter is brighter near the center of the disk because the sunlight shines nearly straight down and the cloud layers look brighter.

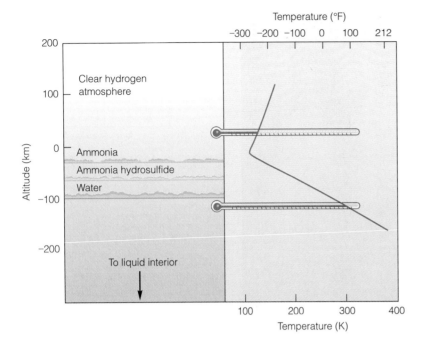

FIGURE 24-4

The three main cloud layers in Jupiter's atmosphere are shown at the left. Thermometers inserted into the Jovian atmosphere would register temperatures as in the graph at the right. We can estimate the temperature of the cloud layers by extending the layers to the right until they touch the line on the graph. Then, we can read the temperature in Kelvins from the scale at the bottom of the graph. Compare this figure with Figure 21-15 for the Earth and Figure 24-18 for Saturn.

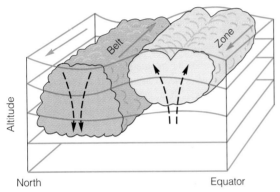

FIGURE 24-5

(a) The bright zones on Jupiter appear to be high-pressure regions of rising gas, which form clouds higher in the atmosphere where they receive more sunlight. Belts are low-pressure regions of sinking gas, and their clouds are lower and less brightly illuminated. (b) The bright zones and dark belts are bounded by high-speed winds and thus resemble the circulation of Earth's atmosphere around high- and low-pressure regions.

Jupiter's atmosphere is not deep. The Galileo probe fell through a transparent hydrogen atmosphere and then into the clouds; it was in total darkness by the time it reached 80 km below the highest clouds. It was destroyed by heat and pressure at a depth of 130 km, and models tell us that the gas of the atmosphere merges into the liquid interior about 1000 km below the clouds, which is not as deep as it seems when we recall that Jupiter's radius is 72,000 km. The entire atmosphere is only about 1 percent of the planet's radius. Earth's atmosphere is about 2.5 percent of Earth's radius. Furthermore, the actual clouds we see on Jupiter are confined to a layer less than 100 km deep.

Spectroscopic observations combined with models of the atmosphere suggest that Jupiter's clouds lie in three layers (Figure 24-4). The highest layer is composed of ammonia crystals at a temperature of about 150 K ($-180°F$). Through holes in these clouds, we can glimpse a deeper layer of ammonia hydrosulfide (NH_4SH) crystals with a temperature of about 200 K, and deeper still there may be a layer of water crystals at a temperature of about 280 K. (Make special note of the temperatures of these cloud layers. We will find similar clouds at similar temperatures when we visit Saturn.)

The clouds of Jupiter are organized into dark **belts** and bright **zones.** Observations made by the Pioneer and Voyager spacecraft show that the zones are cooler and higher than the belts. The zones appear to be high-pressure regions of rising gas that cools as it rises and forms clouds higher in the atmosphere where they receive more sunlight and look brighter (Figure 24-5a). The belts are low-pressure regions with sinking gas and lower clouds that are not as brightly lit.

On Earth, high- and low-pressure regions are bounded by high-speed winds, and the same is true on Jupiter. Winds blowing hundreds of kilometers per hour separate the belts and zones (Figure 24-5b). The clouds

FIGURE 24-6

The Great Red Spot, in its longest dimension twice the diameter of Earth, is a giant storm of rising gas carrying heat upward from deep in Jupiter's atmosphere and circulating counterclockwise with a period of about 6 days. The colors in this Galileo IR image are related to altitude, with red and white regions being highest and blue and dark regions being lowest. This image not only shows rising gas in the Great Red Spot, but it shows thunderheads produced by smaller currents of rising gas. These thunderheads reach over three times higher than thunderheads on Earth. Note the bright white thunderhead at upper left. (NASA)

a   b

FIGURE 24-7

(a) A computer model of the Jovian atmosphere that depends only on the heat of the sun can generate belt–zone circulation (Gareth Williams.) (b) An alternative model links the belts and zones in Jupiter's atmosphere with cylindrical circulations deep in the liquid interior. The Galileo probe traced atmospheric winds deep into the atmosphere and that supports the circulation theory.

on Jupiter are twisted and swirled by the winds, but the belts and zones are highly stable. Although their colors and brightness change sometimes, they have not changed their position since humans have been mapping them.

The colors of the clouds cannot be caused by ammonia, ammonia hydrosulfide, or water, which form colorless, white crystals. Rather, astronomers think that the colors are caused by traces of compounds containing sulfur or phosphorus modified by sunlight.

Mixed within the belt–zone circulation are light and dark spots, a few larger than Earth. Some appear and disappear in days; others last for years. Dark spots appear to be openings through which we see deeper, darker clouds, and white spots appear to be higher cloud circulations. The largest is the Great Red Spot, which has been in one of the southern zones for more than 330 years (Figure 24-6). In its longest dimension, it is twice the diameter of Earth. Pioneer, Voyager, and Galileo observations show that the Great Red Spot is cooler and higher than its surroundings. It is evidently formed by rising gas carrying heat upward from deep below the clouds.

On Earth, rising currents of hot gas can produce thunderheads that tower 15 kilometers and spawn powerful winds and tremendous lightning. The same is true on Jupiter. Voyager photos of the night side of Jupiter revealed lightning bolts among the clouds, and Galileo images show thunderheads up to 50 km high (Figure 24-6).

The stability of the belt–zone circulation on Jupiter has intrigued astronomers. Some models suggest that the circulation is driven by solar heat entering the atmosphere from outside (Figure 24-7a), but other models suggest that the atmosphere is dominated by circulations in the deep interior of the planet. In the laboratory, scientists have imitated Jupiter by placing a fluid in a rotating plastic sphere 30 cm in diameter. Under conditions like those in Jupiter, the fluid divides into concentric cylinders, and where those cylinders touch the surface of the sphere, belt–zone circulation appears (Figure 24-7b). Models of both solar heating and internal circulation have been successful in creating belts and zones.

The controversy may have been settled by the

a

b

FIGURE 24-8

In the summer of 1995, the Galileo spacecraft released a probe that entered Jupiter's atmosphere on December 7, 1995. (a) This artist's conception shows the probe as its heat shield falls away and it begins its hour-long plunge through Jupiter's clouds. The probe transmitted measurements back to Earth before it was crushed by the increasing pressure. (Hughes Aircraft Company) (b) The site of the probe entry is shown by the arrow in this Jupiter image. (Reta Beebe, New Mexico State University, and NASA)

Galileo probe (Figure 24-8). As it fell deeply into Jupiter's atmosphere, it found that the high-speed winds that bound the belts and zones are not limited to the upper cloud layers exposed to bright sunlight. Winds of 650 km/sec extend at least as deep as 130 km below the cloud tops. These deep roots to the winds suggest that the atmospheric circulation is dominated by the circulation of the liquid interior.

Understanding Jupiter not only helps us better understand Earth, but it also suggests that some planets are fundamentally different from our world (Window on Science 24-1). Jupiter differs from Earth in another peculiar way. It has a ring.

Jupiter's Ring

Astronomers have known for centuries that Saturn has rings, but it was not until 1979, when the Voyager 1 spacecraft sent back photos, that Jupiter's ring was discovered. Less than 100 times as bright as Saturn's rings, the ghostly ring around Jupiter is a puzzle. What is it made of? Why is it there? A few simple observations help us understand.

Saturn's rings are made of bright ice chunks, but the particles in Jupiter's ring are very dark and reddish. We can guess that the ring is rocky rather than icy.

FIGURE 24-9

This false-color image of Jupiter's ring was recorded while the Galileo spacecraft passed through Jupiter's shadow. The ring is bright when illuminated from behind, a property called forward scattering, which tells us the ring particles are microscopic. A mist of small particles is visible in a halo extending above and below the rings. Jupiter's upper atmosphere is visible at the right, and the ring appears to end where it enters Jupiter's shadow. (NASA)

We can also conclude that the ring particles are mostly microscopic. Photos of the ring show that it is very bright when illuminated from behind (Figure 24-9) —it is scattering light forward. Efficient **forward scattering** occurs when particles have diameters roughly the same as the wavelength of light, a few millionths of a

Science, Technology, and Engineering

Sending spacecraft to another world is expensive, and it may seem pointless when that world seems totally hostile to human life. What practical value is there in sending a space probe to Jupiter? To resolve that question, we need to consider the distinction between science, technology, and engineering.

Science is nothing more than the logical study of nature, and the goal of science is understanding. Although much scientific knowledge proves to have tremendous practical value, the only goal of science is a better understanding of how nature works. Technology, in contrast, is the practical application of scientific knowledge to solve a specific problem. People trying to find a faster way to paint automobiles might use all the tools and techniques of science, but if their goal has some practical outcome, we should more properly call it technology rather than science. Engineering is the most practical form of technology. An engineer is likely to use well-understood technology to find a practical solution to a problem.

Of course, there are situations in which science and technology blur together. For example, humanity has a practical and tragic need to solve the AIDS problem; we need a cure and a preventative. Unfortunately, we don't understand the AIDS virus itself or viruses in general well enough to design a simple solution, and thus much of the work involves going back to basic science and trying to better understand how viruses interact with the human body. Is this technology or science? It is hard to decide.

We might describe science that has no known practical value as basic science or basic research. Our exploration of worlds such as Jupiter would be called basic science, and it is easy to argue that basic science is not worth the effort and expense since it has no known practical use. Of course, the problem is that we have no way of knowing what knowledge will be of use until we acquire that knowledge. In the middle of the 19th century, Queen Victoria is supposed to have asked physicist Michael Faraday what good his experiments with electricity and magnetism were. He answered, "Madam, what good is a baby?" Of course, Faraday's experiments were the beginning of the electronic age. Many of the practical uses of scientific knowledge that fill our world—transistors, vaccines, plastics—began as basic research. Basic scientific research provides the raw materials that technology and engineering use to solve problems.

Basic scientific research has yet one more important use that is so valuable it seems an insult to refer to it as merely practical. Science is the study of nature, and as we learn more about how nature works, we learn more about what our existence in this universe means for us. The seemingly impractical knowledge we gain from space probes to other worlds tells us about our own planet and our own role in the scheme of nature. Science tells us where we are and what we are, and that knowledge is beyond value. ■

---

meter. Large particles do not scatter light forward, so a ring filled with basketball-size particles would look dark when illuminated from behind. Forward scattering tells us that the ring is made of particles about the size of those in cigarette smoke.

The location of these particles near the planet is understandable. They orbit inside the **Roche limit,** the distance from a planet within which a moon cannot hold itself together by its own gravity. If a moon stays far from a planet, its gravity is much greater than the tidal forces caused by the planet, but if the moon comes inside the Roche limit, the tidal forces will pull it apart. The Space Shuttle can orbit inside the Roche limit because it is held together by bolts and welds, but a moon held together by its own gravity cannot survive inside the Roche limit. If a planet and its moon have similar densities, the Roche limit is 2.44 planetary radii, and Jupiter's rings (and those of Saturn, Uranus, and Neptune) lie inside this limit.

Now we can understand the dust near Jupiter. If a dust speck gets knocked off of a larger rock inside the Roche limit, the rock's gravity cannot hold the dust speck. And the billions of dust specks in the ring can't pull themselves together to make a new moon because of tidal forces inside the Roche limit.

We can be sure that the ring particles are not old. The pressure of sunlight and the powerful magnetic field alter the orbits of the particles, and they gradually spiral into the planet. Images show a gossamer ring extending down toward the cloud tops and a halo above and below the ring formed of dust knocked out of the thin ring. Also, the intense radiation will grind the dust specks down to nothing in a century or so.

Thus, we can conclude that the rings are continuously resupplied with fresh dust. Small moons that orbit near the outer edge of the rings may lose dust particles as they are hit by micrometeorite impacts. The moons Adrastea and Metis are each only 40 km in diameter, but they could provide enough ring particles. Furthermore, the moons' gravitation may keep the ring from spreading beyond its sharp outer edge. We will learn more about this kind of interaction between rings and moons when we study Saturn's rings.

Our exploration of Jupiter reveals that it is much more than just a big planet. It is the gravitational and magnetic center of an entire community of objects. Occasionally the community suffers an intruder.

Comet Impact on Jupiter

Comets are very common in the solar system, and Jupiter, because of its strong gravity, probably gets hit by comets more often than most planets. But no one had ever seen it happen until 1994, when fragments from a disrupted comet smashed into Jupiter.

The comet, named Shoemaker–Levy 9 after its dis-

a

b

c

FIGURE 24-10

(a) Comet Shoemaker–Levy 9, broken into fragments by a previous encounter with Jupiter, falls toward the giant planet. Distances have been compressed in this combination of photos. Note the moon Io and its shadow on Jupiter. (NASA) (b) Fragments of the comet struck Jupiter's southern hemisphere over a period of 6 days and created tremendous fireballs (such as impact R at extreme left), leaving hot spots visible at infrared wavelengths. (University of Hawaii) (c) At visible wavelengths, the impact sites show as dark smudges. The largest scar at the extreme right is the size of Earth. (NASA)

coverers, Eugene and Carolyn Shoemaker and David H. Levy, was captured into orbit around Jupiter sometime in the past. In 1992, it passed inside Jupiter's Roche limit and was pulled into at least 21 pieces that looped out away from Jupiter in long elliptical orbits. At this point, the objects were discovered on a photograph. Drawn out into a long strand of small comets, the pieces fell back

and slammed into Jupiter over a period of 6 days in July 1994 (Figure 24-10).

The fragments were no bigger than a kilometer or so in diameter and contained a fluffy mixture of rock and ices. They hit Jupiter at speeds of about 60 km/sec and released energy equivalent to a few million megatons of TNT. (At the height of the cold war, all the nuclear weapons on Earth totaled about 80,000 megatons. The Hiroshima bomb was only 0.15 megaton.)

The impacts occurred just over Jupiter's horizon as seen from Earth, but they produced fireballs almost 3000 km high, and the rapid rotation of Jupiter brought the impact points within sight of Earth only 15 minutes later. Infrared telescopes easily spotted the glowing scars where the comets hit (Figure 24-10b), and, as they cooled, the impact sites became dusty, dark spots at visible wavelengths. Visible through even small telescopes, some of these dark smudges were larger than Earth (Figure 24-10c).

The comet impact was an astonishing spectacle, but what can we learn from it? We can use the event in two different ways. First, astronomers used the impacts as probes of Jupiter's atmosphere. By making assumptions about the nature of Jupiter's atmosphere and by using the most powerful computers, astronomers created models of a high-velocity projectile penetrating into Jupiter's upper atmosphere. By comparing the observed impacts with the models, astronomers were able to fine tune the models to better represent Jupiter's atmosphere. This method of comparing models with reality is a critical part of science; therefore, the impacts on Jupiter resulted in a better understanding of the nature of Jupiter's atmosphere.

Second, the spectacle helps us in a more general way. It reminds us that planets are hit by large objects such as asteroids and the heads of comets. Jupiter probably is hit every century or so. In 1690, the Italian astronomer Cassini observed a dark spot that appeared on Jupiter and changed in a pattern over a period of days. We now recognize that pattern as the scar of a comet impact. We have seen evidence for large impacts on Earth, the moon, Venus, Mercury, and Mars, and we will see further evidence as we continue our exploration of the solar system.

Comets hit planets all the time. The impact in 1994 was just the first such event to occur since the rise of modern astronomy. We will see in Chapter 26 that a comet or asteroid impact on Earth may have changed the climate and killed the dinosaurs. A solar system is a dangerous place to put an inhabited planet.

The History of Jupiter

Our goal in studying any planet is to be able to tell its story—to describe how it got to be the way it is. While we understand part of the story of Jupiter, we still have much to learn.

If the solar nebula theory for the origin of the solar system is correct, then Jupiter formed from the colder gases of the outer solar nebula, where ices were able to condense. Thus, Jupiter grew rapidly and eventually became massive enough to trap hydrogen and helium gas directly from the solar nebula. The denser material, such as rock and metal, sank to the center to form a core, and we now see Jupiter as a low-density world of liquid hydrogen surrounding a hot core of heavier elements.

In the deeper interior of Jupiter, the hydrogen takes the form of liquid metallic hydrogen, which is a very good electrical conductor. The planet's rapid rotation, coupled with the outward flow of heat from its hot interior, drives a dynamo effect that produces a powerful magnetic field. That vast magnetic field traps high-energy particles from the solar wind to form intense radiation belts.

The rapid rotation and large size of Jupiter cause the weather patterns to take the form of belt–zone circulation. Heat flowing upward from the interior causes rising currents in the bright zones, and cooler gas sinks in the dark belts. As on Earth, winds blow at the margins of these regions, and large spots appear to be cyclonic disturbances.

Although the age of planet building is long past, Jupiter continues to be hit by meteorites and occasional comets, as do all the planets. Any debris left over from the formation of Jupiter would have been blown away long ago by the solar wind, so the dust we find trapped in Jupiter's thin ring must be young. It probably comes from meteorites hitting the innermost moons.

Our study of Jupiter has been challenging because Jupiter lacks any surface—it is difficult to imagine being there. Most of the surface features and processes we found on the terrestrial planets are missing on Jupiter, but, as the prototype of the Jovian worlds, it earns its place as the ruler of the solar system.

CRITICAL INQUIRY

How do we know Jupiter is hot inside?

Critical analysis hinges on testing ideas, and sometimes it is helpful to test even the most basic ideas. Doing so can give us confidence in what we know, and it can sometimes help us discover new ideas. We know that something is hot if we touch it and it burns our fingers, but we can't touch Jupiter. We also know something is hot if it is glowing bright red—it is red hot. But Jupiter is not glowing red hot. We can tell that something is hot if we can feel heat when we hold our hand near it. That is, we can detect infrared radiation with our skin. In the case of Jupiter, we need greater sensitivity than the back of our hand, but infrared telescopes tell us that Jupiter is a source of infrared. It really is glowing in the infrared. Sunlight would warm it a little bit, but it is emitting twice as much infrared as it should. That means it must be hot inside. From models of the interior, astronomers can tell us that the center must be about 30,000 K in order to heat the surface of the planet and make it glow in the infrared.

Astronomical understanding is usually based on simple observations. For example, how do we know Jupiter has a low density? ■

The study of Jupiter is more meteorology than geology, but when we examine its moons, we find all the excitement of impact cratering and active volcanism.

FIGURE 24-11
The Galilean moons of Jupiter, clockwise from upper left: Io, Europa, Callisto, and Ganymede. Callisto and Ganymede are 50 percent water, and Europa contains an icy crust over a water mantle. Io is dry and sulphurous. Earth's moon is shown at center for scale. (NASA and Lick Observatory)

24-2 JUPITER'S FAMILY OF MOONS

We can divide Jupiter's satellites into two groups. The smaller moons resemble the moons of Mars and may be captured asteroids. The four larger moons, the moons seen by Galileo, are clearly not captured asteroids but complicated worlds in their own right (Figure 24-11).

Our study of the moons of Jupiter will illustrate three important principles in comparative planetology. First, a body's composition depends on the temperature of the material from which it formed. This is illustrated by the prevalence of ice as a building material in the outer solar system where sunlight is weak. The second principle is that cratering can tell us the age of a surface. Finally, we will see that internal heat has a powerful influence over the geology of these larger moons.

Callisto: The Ancient Face

The outermost of Jupiter's four large moons, Callisto is half again as large in diameter as Earth's moon (Figure 24-12). Like all of Jupiter's larger satellites, Callisto is tidally locked to its planet, keeping the same side forever facing Jupiter. From its gravitational influence on pass-ing spacecraft, astronomers can calculate Callisto's mass, and dividing by its volume tells us that its density is 1.79 g/cm^3. Ice has a density of about 1 and rock 2.5 to 4 g/cm^3, so Callisto must be a mixture of rock and ice.

Photographs from the Voyager and Galileo space-craft show that the surface of Callisto is dark, dirty ice heavily pocked with craters. Old, icy surfaces in the solar system become dark because of dust added by meteorites and because meteorite impacts vaporize water leaving the rocky dust to accumulate in a dirty crust.

The Galileo spacecraft has found that Callisto has no magnetic field and is not layered. It has never had an internal heat source and has never differentiated into core and mantle. It isn't clear how such a large moon could form without heating enough to differentiate. Certainly, it has not been heated since it formed, and it remains an ancient moon, far from its massive planet.

Ganymede: A Hidden Past

The next Galilean moon inward is Ganymede, larger than Earth's moon, larger than Mercury, and over three-quarters the diameter of Mars. It's density is 1.9 g/cm^3, and its influence on the Galileo spacecraft reveals that it has a rocky core, an icy mantle, and a crust of ice 500 km thick. It may even have a small iron core. It is large

FIGURE 24-12

The heavily cratered surface of Jupiter's satellite Callisto is shown in this composite of images made by the Voyager 2 spacecraft. Nearly all solid surfaces in the solar system are blasted with impact scars. The largest feature here is a vast bulls-eye of an impact feature named Valhalla. It is roughly 3000 km in diameter. (NASA)

enough for radioactive decay to have melted its interior when it formed, allowing iron to drain to its center.

The surface hints at an active past. Although a third of the surface is old, dark, and cratered, the rest is marked by lighter parallel grooves. Because this **grooved terrain** (Figure 24-13) contains fewer craters, it seems young. Highly detailed Galileo images show systems of parallel grooves lying on top of other systems of grooves, suggesting that the icy crust broke repeatedly and allowed water to flood out and freeze.

The Galileo spacecraft found that Ganymede has a magnetic field about 10 percent as strong as Earth's, indicating that Ganymede is still warm inside. An active dynamo effect in its molten core or in a water-rich mantle could generate the magnetic field. But Ganymede should have cooled since its formation and radioactive decay is not enough to keep it molten. Thus, it must have some other heat source. If Ganymede were in an elliptical orbit, Jupiter's gravity would flex it, and friction would heat it by **tidal heating** (Figure 24-14a). Yet the big moon's orbit is nearly circular and tidal heating is negligible.

Astronomers now speculate that within the last billion years or so gravitational interactions among Jupiter's moons may have temporarily forced Ganymede into a more elliptical orbit. The resulting tidal heating melted

a   b

FIGURE 24-13

The surface of Ganymede is divided into old, dark cratered terrain and bands of bright grooves. (See Figure 24-11.) In this Voyager image, the dark terrain is marked by a few bright craters where more recent impacts have broken through the dirty crust into clean ice below. (b) A Galileo image shows how a system of icy grooves covered an older region at the bottom while an even more recent system of grooves at the top covered the previous grooves. The grooves suggest that water flooded from below through cracks, but the age of the grooved terrain is uncertain. (NASA)

FIGURE 24-14

Two effects on planetary satellites. (a) Tidal heating occurs when changing tides cause friction within a moon. (b) The focusing of meteoroids exposes satellites in small orbits to more impacts than satellites in larger orbits receive.

a

its interior, drove geological activity to produce the grooved terrain, and left behind enough heat to generate the observed magnetic field. This theory hinges on the age of the grooved terrain. It seems a few billion years old, but ages are difficult to estimate from crater counts because Jupiter's gravity focuses meteorites inward (Figure 24-14b). If the grooves represent a more recent episode of geological activity, then leftover heat could explain the magnetic field. Ganymede may be hiding an active past behind its grooved surface.

Europa: A Hidden Ocean

The outer two Galilean moons are large and icy, but Europa is smaller (Figure 24-11) and has a density of 3 g/cm³. Thus, it must contain only a small percentage of water. Spectra reveal traces of water on its surface but that water seems to lie in an icy crust less than 100 km thick. As it raced past Europa, the orbital motion of the Galileo spacecraft revealed that the little moon is mostly rock and may have a small metallic core. Its bland face hides a deep secret, however. Data from Galileo suggest that a 200 km deep ocean of water lies below the crust.

Europa is obviously youthful. Its icy surface contains almost no large craters and is bright with an albedo of 0.69. (It reflects 69 percent of the light that hits it.) As we have seen, old, icy surfaces are very dark and cratered, so we can guess that Europa's surface is recently renewed.

The long lines crossing the surface suggest a renewal process (Figure 24-15a). They look like cracks in the icy crust where water from below has welled up and frozen. Some areas look like great blocks of ice that have been shifted in a nearly frozen sea (Figure 24-15b). Other, seemingly older, regions contain numerous small craters

b

FIGURE 24-15

(a) Long lines on the surface of Europa appear to be cracks in its icy crust where water from below has refrozen. Note the bright surface and lack of craters—signs of youth. (b) This Galileo image shows ice blocks that have moved like a disturbed puzzle resembling icebergs floating in pack ice on Earth. (NASA)

that suggest ages of 10 million years or so. Astronomers have speculated about ice volcanoes and lava flows of liquid water, but the details remain unknown.

The youthful surface on Europa suggests that it should contain a source of heat. It is too small to have retained heat from its formation, and radioactive decay could not keep such a small world active. It contains only a very weak magnetic field, so a molten core is ruled out. Tidal heating is more effective on Europa than on Ganymede (because it is closer to Jupiter) and may provide enough energy to drive its activity.

FIGURE 24-16

(a) Jupiter's innermost Galilean moon, Io, has no craters and is richly colored as shown in this image from the Galileo spacecraft. Sulfur-rich volcanoes, such as Pele at the center of the red oval debris shield, bury craters as fast as they can form. Note the changes in Pele between Voyager 1, Voyager 2 only four months later, and the Galileo image 17 years later. (b) A computer-enhanced image shows a volcano erupting at the edge of the disk. (c) Io and its orbit are surrounded by a cloud of glowing sodium vapor. (a and b, NASA; c, JPL, Bruce A. Goldberg)

a

b

c

Io: Bursting Energy

Geological activity is driven by heat flowing out of a planet's interior, and nothing could illustrate this principle better than Io, the innermost of Jupiter's Galilean moons. Photographs from the Voyager and Galileo spacecraft show a colorful surface with no impact craters at all. No subtle deductions are needed to explain the missing craters. Over half a dozen active volcanoes are visible on the moon's surface blasting sulfurous gas and ash out over the surface to bury any newly formed craters (Figure 24-16 a, b). Unlike the dead or dormant outer Galilean moons, Io is bursting with energy.

The colors of Io have been compared to those of a badly made pizza, but we can recognize the yellows, oranges, browns, blacks, and reds as typical of sulfur compounds. Spectra confirm that the surface of Io is rich in sulfur and sulfur compounds. Spectra also tell us that Io has a tenuous atmosphere of gaseous sulfur and oxygen, but those gases can't be permanent.

The erupting volcanoes pour out about 1 ton of gases per second, but, because of Io's low escape velocity, the gases leak into space. Also, any gas atoms that become ionized are swept away by Jupiter's rapidly rotating magnetic field to produce a cloud of sulfur and sodium ions in a doughnut enclosing its orbit (Figure 24-16c). The neighborhood of Io's orbit is dirty (and probably smelly) with sulfur. In fact, the reddish color of the small inner moon Amalthea may be caused by sulfur from Io.

We can use basic observations to deduce the nature of Io's interior. From the density, 3.55 g/cm³, we can conclude that it is mostly rock. Spectra reveal no trace of water at all, so there is no ice. The oblateness of Io, caused by its rotation and by the slight distortion due to Jupiter's gravity, gives astronomers clues to its interior. Models suggest that it has a modest metallic core and a deep rocky mantle, topped by a layer of low-density, sulfur-rich compounds that have been cooked out of the interior. The interior is hot, as evidenced by the volcanism, but the Galileo spacecraft detected nothing more than a weak magnetic field that might be produced by the interaction of the moon with Jupiter's much stronger magnetosphere. Thus, it is not clear that a molten core with a dynamo effect is in operation, but heat is certainly flowing outward.

The volcanism on Io arises as molten sulfur and perhaps molten rock burst through the crust to produce volcanoes and lava flows (Figure 24-17a). Infrared astronomers have known for decades that Io undergoes occasional **infrared outbursts**—its infrared brightness suddenly increases. These outbursts can now be traced to the sudden eruption of fresh volcanoes.

In fact, although individual volcanoes come and go, volcanism is continuous on Io. In 1979, the Voyager spacecraft detected 8 active volcanoes on Io, but when Galileo arrived in 1995, some of those volcanoes had shut down and new ones had appeared. Even the giant volcano Pele changes its debris pattern over periods as short as a month (Figure 24-16a).

Earth's volcanoes eject lava and ash because of the pressure of the water that is dissolved in the lava. As rising lava reaches Earth's surface, it is ejected, and the sudden decrease in pressure is like popping a champagne

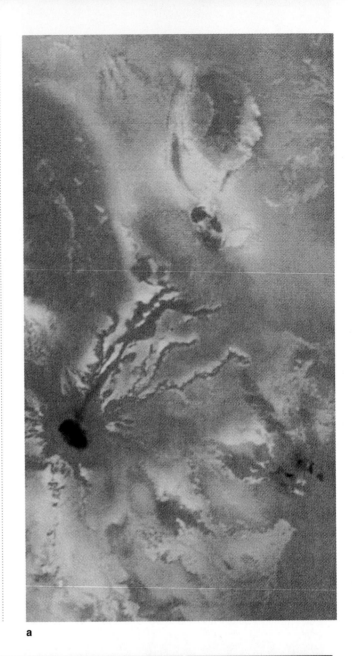

a

FIGURE 24-17

(a) A volcano on Io has not only spewed out ash deposits but has also produced molten magma lava flows. (b) This Galileo spacecraft photo combines images recorded at three different wavelengths to produce a false-color map of Io. Red areas are very young, and green areas are rich in sulfur. Blue areas are sulfur dioxide frosts. The image at the lower right removes intensity variations in the image to make the color differences easier to see. (NASA)

b

cork. The water flashes into vapor and blasts material out of the volcano, a process that was responsible for the Mount St. Helens explosion in 1987. But Io is dry. Its volcanoes appear to be powered by sulfur dioxide dissolved in the magma. When the pressure is released, the sulfur dioxide boils out of solution and blasts sulfurous ash high above the surface in arching plumes up to 300 km high. Ash falling back to the surface produces debris layers around the volcanoes, such as that around Pele, and deposits of sulfur dioxide frost are common (Figure 24-17b).

What powers Io? It is bursting with energy, but it is only a few percent bigger than Earth's moon, which is cold and dead. Io is too small to have retained heat from its formation or to remain hot from radioactive decay. In fact, the energy blasting out of its volcanoes is about three times more energy than it could make by radioactive decay in its interior.

Io is heated by the same kind of tidal heating that is suspected in Ganymede's past and in Europa. Because Io is so close to Jupiter, tides are powerful, and gravitational interactions with Europa and Ganymede force its orbit to be elliptical. As its distance from Jupiter varies, tides squeeze it, and its surface rises and falls by about 100 m. The resulting friction is enough to at least partially melt the interior and drive volcanism. In fact, the energy flowing outward is continually recycling Io's crust. Deep layers melt, are spewed out through the volcanoes, fall back to cover the surface, and are later covered themselves until they are buried so deeply that they are again melted.

Io is the most active world in our solar system because it lives so close to Jupiter. The other Galilean moons are more distant and thus have had less dramatic histories of tidal heating and geological activity.

The History of the Galilean Moons

The goal of astronomy is to tell the story of the universe. Now we must test our understanding of the Galilean moons by trying to tell their story. To do that we not only draw on what we have learned about the moons, but we also draw on what we have learned about Jupiter and about the origin of the solar system (Chapter 20).

The minor moons of Jupiter are probably captured asteroids, but the Galilean moons seem to be primordial. That is, they formed with Jupiter. Also, they seem to be a set in that they share common features and their densities are related to their distance from Jupiter (Table 24-2). This information suggests that the four moons formed in a disk-shaped nebula around Jupiter—a mini-solar nebula. We know that Jupiter was very hot when it formed, so, just as the terrestrial and Jovian planets formed in the solar nebula, the inner moons, Io and Europa, formed from rocky material, and the outer moons incorporated more ices.

But we must also consider the trend in geological activity, a trend that suggests tidal heating. Io is so severely heated that any water it might have had has been driven off. In contrast, the outermost moon, Callisto, has

TABLE 24-2

The Galilean Satellites*

Name	Radius (km)	Density (g/cm³)
Io	1816	3.55
Europa	1563	3.04
Ganymede	2638	1.93
Callisto	2410	1.81

*For comparison, the radius of Earth's moon is 1738 km, and its density is 3.36 g/cm³.

never suffered from tidal heating and is rich in ices. Ganymede and Europa have less ice and more geological activity in their past, so we must suspect that gravitational interactions among the inner three moons have produced periods of tidal heating and geological upheaval.

CRITICAL INQUIRY

What produces Io's internal heat?

Part of critical thinking is recognizing and using basic principles. In this case, we understand that small worlds lose their internal heat quickly and become geologically inactive. Io is a bit smaller than Earth's moon, which is cold and dead, but Io is bursting with energy flowing outward.

Clearly, Io must have a powerful source of heat inside, and that heat source is tides. Io's orbit is slightly elliptical, so it is sometimes closer to Jupiter and sometimes farther away. That means that Jupiter's powerful gravity sometimes squeezes Io more than at other times, and the flexing of the little moon's interior produces heat through friction. Such tides would rapidly force Io's orbit to become circular—and then tidal heating would end and the planet would become inactive—except that the gravitation of the other moons keeps Io's orbit elliptical. Thus, it is the influence of its companions that keeps Io in such an active state.

Io has almost no impact craters, but Callisto has many. What does the distribution of craters on the Galilean satellites tell us about their history? ■

We have now applied the principles of comparative planetology not only to planets but also to moons. With the insight we have gained from Jupiter and its satellite system, we are ready to visit the most beautiful of the Jovian worlds.

24-3 SATURN

Saturn has played second fiddle to its own rings since Galileo first saw it in 1610. He didn't recognize the rings for what they are, but we know them now as one of the great sights in the solar system. Nevertheless, Saturn itself, only slightly smaller than Jupiter (Data File 8), is a fascinating planet with a few mysteries of its own.

Our discussion of Saturn can make use of the principles we have learned from Jupiter, but we have less information about Saturn than we did about Jupiter. Pioneer 2 and both Voyager spacecraft flew past Saturn, but no orbiter circles Saturn as Galileo circles Jupiter. Thus, our discussion of Saturn is based on less evidence and lower quality data than we have for Jupiter.

Planet Saturn

The basic characteristics of Saturn tell us about its composition and interior. Only about a third of the mass of Jupiter and 16 percent smaller in radius, Saturn has an average density of 0.69 g/cm³. It is less dense than water—it would float! Although Saturn rotates about as fast as Jupiter, it is almost twice as oblate, so we can conclude that must not contain a large core of heavy elements. Spectra show that its atmosphere is rich in hydrogen and helium (Table 24-1), and models predict that it is mostly liquid hydrogen.

Infrared observations show that Saturn is radiating almost twice as much energy as it receives from the sun, telling us that heat is flowing out of its interior. It must be hot inside. In fact, it is hotter than we would expect. It should have lost more heat since it formed. Astronomers suspect that helium in the liquid hydrogen interior is condensing into droplets and falling inward. The falling droplets, releasing energy as they pick up speed, heat the planet. This heating is similar to the heating produced when a star contracts.

We can learn more about Saturn's interior from its magnetic field. The Voyager spacecraft found that Saturn's magnetic field is about 20 times weaker than Jupiter's. It also has correspondingly weak radiation belts. Models predict that the pressure inside Saturn, being weaker than that inside Jupiter, produces a smaller mass

a

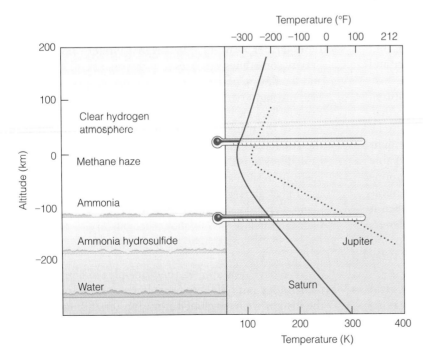

FIGURE 24-18

(a) At ultraviolet wavelengths the Hubble Space Telescope can photograph rings of aurora surrounding Saturn's north and south magnetic poles. (Compare with Figure 24-3b, c.) Caused by charged particles trapped in the magnetic field that leak down into the atmosphere, the aurora extends 2000 km above the clouds. Note the lack of any inclination between the magnetic axis and the rotation axis. (J. T. Trauger, JPL and NASA) (b) Because Saturn is farther from the sun, its atmosphere is colder than Jupiter's, and the clouds form deeper in the hazy atmosphere. Notice that the three cloud layers on Saturn form at about the same temperatures as do the three cloud layers on Jupiter.

of liquid metallic hydrogen. Heat flowing outward must cause convection in this conducting core, and the rapid rotation drives a dynamo effect to produce the magnetic field. Unlike most magnetic fields, Saturn's field is not inclined to its axis of rotation, something we notice in ultraviolet images that show a ring of aurora around Saturn's north pole (Figure 24-18a). This alignment between the magnetic axis and the axis of rotation is peculiar and isn't understood.

Saturn's atmosphere is rich in hydrogen and contains belt–zone circulation, which appears to arise in the same way as that on Jupiter. Zones are higher clouds formed by rising gas, and belts are lower clouds formed by sinking gas.

But the clouds are not very distinct on Saturn. Measurements from Voyager spacecraft tell us that Saturn's atmosphere is much colder than Jupiter's—something we would expect because Saturn is twice as far from the sun and receives only one-fourth as much solar energy per square meter. The clouds on Saturn form at about the same temperature as the clouds on Jupiter, but those temperatures are deeper in Saturn's cold atmosphere. Compare the cloud layers in Figure 24-18b with those in Figure 24-4. The atmospheres of Jupiter and Saturn are very similar once we remember that Saturn is colder.

One dramatic difference between Jupiter and Saturn concerns the winds. On Jupiter, winds bound each of the belts and zones, but on Saturn the pattern is not the same. Saturn has fewer such winds, but they are much stronger. The eastward wind at the equator of Saturn, for example, blows at 500 m/sec (1100 mph), roughly five

SATURN

Average distance from the sun	9.5388 AU (14.27×10^8 km)
Eccentricity of orbit	0.0560
Maximum distance from the sun	10.07 AU (15.07×10^8 km)
Minimum distance from the sun	9.005 AU (13.47×10^8 km)
Inclination of orbit to ecliptic	2°29'17"
Average orbital velocity	9.64 km/sec
Orbital period	29.461 y (10,760 days)
Period of rotation	$10^h39^m25^s$
Inclination of equator to orbit	26°24'
Equatorial diameter	120,660 km (9.42 $D_\oplus$)
Mass	5.69×10^{26} kg (95.147 $M_\oplus$)
Average density	0.69 g/cm³
Gravity at base of clouds	1.16 earth gravities
Escape velocity	35.6 km/sec (3.2 $V_\oplus$)
Temperature at cloud tops	−180°C (−292°F)
Albedo	0.61
Oblateness	0.102

Saturn, like Jupiter, is a liquid hydrogen planet with belt–zone circulation visible in its atmosphere. The white storm clouds shown here occur only rarely. The rings are made of ice particles. (NASA)

Who Pays for Science?

Science is an expensive enterprise, and that raises the question of payment. Some science has direct technological applications, but some basic science is of no immediate practical value. Who pays the bill?

In Window on Science 24-1, we saw that technology is of immediate practical use; thus, business funds much of this kind of scientific research. Auto manufacturers need inexpensive, durable, quick-drying paint for their cars, and they find it worth the cost to hire chemists to study the way paint dries. Many industries have large research budgets, and some industries, such as pharmaceutical manufacturers, depend exclusively on scientific research to discover and develop new products.

If a field of research has immediate potential to help society, it is likely that government will supply funds. Much of the research on public health is funded by government institutions such as the National Institutes on Health and the National Science Foundation. The practical benefit of finding new ways to prevent disease, for example, is well worth the tax dollars.

Basic science, however, has no immediate practical use. That doesn't mean it is useless, but it does mean that the practical-minded stockholders of a company will not approve major investments in such research. Digging up dinosaur bones, for instance, is very poorly funded because no industry can make a profit from the discovery of a new dinosaur. Astronomy is another field of science that has few direct applications, and thus very little astronomical research is funded by industry.

The value of basic research is twofold. Discoveries that have no known practical use today may be critically important years from now. Thus, society needs to continue basic research to protect its own future. But basic research, such as studying Saturn's rings or digging up dinosaur bones, is also of cultural value because it tells us what we are. Each of us benefits in intangible ways from such research, and thus society needs to fund basic research for the same reason it funds art galleries and national parks—to make our lives richer and more fulfilling.

Because there is no immediate financial return from this kind of research, it falls to government institutions and private foundations to pay the bill. The Keck Foundation has built two giant telescopes with no expectation of financial return, and the National Science Foundation has funded thousands of astronomy research projects for the benefit of society. Debates rage as to how much money is enough and how much is too much, but, ultimately, funding basic scientific research is a public responsibility that society must balance against other needs. There isn't anyone else to pick up the tab. ∎

---

times faster than the eastward wind at Jupiter's equator. The reason for this difference is not clear.

Saturn harbors a number of mysteries including great storms that burst up through the clouds when it is summer in Saturn's northern hemisphere. Also, Saturn does not have a great spot corresponding to Jupiter's Great Red Spot. In fact, Saturn has few spots. We will learn more about these mysteries when the Cassini spacecraft goes into orbit around Saturn in 2004 (Window on Science 24-2). It will give us more and better information, not only about Saturn, but also about its rings and moons.

Saturn's Rings

Looking at the beauty and complexity of Saturn's rings, an astronomer once said, "The rings are made of beautiful physics." We might add that the physics is actually rather simple, but the result is one of the most astonishingly beautiful sights in our solar system.

In 1609, Galileo was the first to see the rings of Saturn, but perhaps because of the poor optics in his telescopes, he did not recognize the rings as a disk. He drew Saturn as three objects—a central body and two smaller ones on either side. In 1659, Christian Huygens realized that the rings were a disk surrounding but not touching the planet.

The rings are labeled with letters of the alphabet (Figure 24-19a), with the A and B rings being easily visible even through a small telescope. The A ring lies outside a gap called Cassini's division (named after its discoverer), and the B ring lies inside that gap. The C ring, or Crepe ring (so named after the thin black cloth traditionally associated with mourning), is a faint ring lying inside the B ring. Fainter rings are also identified by letters. At its outer edge, the A ring has a diameter about 21 times that of Earth.

The discovery that the rings are made of particles is a story of both theory and observation. Astronomers knew that the rings were transparent because they could see stars through the rings. In 1859, James Clerk Maxwell (for whom a large mountain on Venus is named) proved theoretically that the rings could not be solid. Because of Kepler's third law, the outer parts of the rings would have to orbit faster than the inner parts, so solid rings could not remain in orbit without ripping themselves apart. Much later, the American astronomer James Keeler measured Doppler shifts in the rings and proved that they did indeed obey Kepler's third law. Thus, they had to be made of particles. Astronomers knew that the rings had to be very thin because they vanish when we view them edge-on as Earth passes through the ring plane. They could not be more than 2 km thick. With the development of modern infrared spectrographs, astronomers were able to detect the presence of solid water in the rings.

Cassini's division and a few other gaps in the rings that are visible from Earth were understood by an elegant application of Newtonian physics. In 1867, Daniel Kirkwood demonstrated that resonances between the

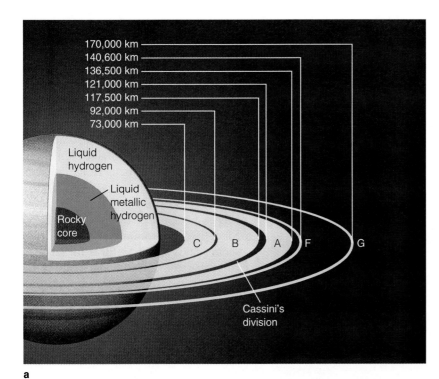

170,000 km
140,600 km
136,500 km
121,000 km
117,500 km
92,000 km
73,000 km

Liquid
hydrogen

Liquid
metallic
hydrogen

Rocky
core

C   B   A   F   G

Cassini's
division

a

F I G U R E   2 4 - 1 9

(a) The A and B rings of Saturn are easily visible from Earth even with a small telescope, but the C ring is more difficult to observe. The F and G ring and a few fainter rings were photographed by the Voyager spacecraft.
(b) Composed of billions of ice particles, the rings of Saturn are quite transparent. Note where the rings cross in front of the planet at the lower left in the Voyager photograph. The rings are generally pale and colorless. They are marked by many gaps and ringlets produced by dynamic effects among the orbiting ring particles and Saturn's moons. (NASA)

b

orbital motion of asteroids and Jupiter could produce gaps in the asteroid belt (Chapter 26). The same theory predicted that a gap should occur in the rings of Saturn where a ring particle would orbit Saturn twice while the moon Mimas orbits once. On alternate orbits, the particle would feel the gravitational attraction of Mimas pulling it outward, and the orbit of the particle would be changed until it was no longer in resonance. Thus, the few gaps visible could be related to resonances with moons.

By 1979, theory and observation accounted for the known properties of Saturn's rings, and as the Voyager 1

a

b

c

FIGURE 24-20

(a) Radial spokes in the bright B ring are visible in these Voyager images. Note the motion of the spokes (arrow) between these two frames. The spokes are apparently produced by Saturn's magnetic field. (b) A Voyager 1 photo of the F ring shows a number of strands with twists and kinks. (c) Another image reveals small shepherd satellites (arrows) orbiting just outside and just inside the F ring. These moons confine the ring and produce its irregularities. (NASA)

spacecraft approached Saturn, astronomers expected to see thin, uniform rings of ice particles with a few well-understood gaps. What the Voyager photos actually revealed were entirely new phenomena in the rings. They were made up of roughly 1000 ringlets no wider than 2 km (Figure 24-19b), some were elliptical, and some contained kinks and knots. New phenomena called for new explanations.

For one thing, we knew that the rings were not ruled by gravitation alone. Since 1896, astronomers have reported seeing faint radial features in Saturn's rings that are hard to understand. If the rings obeyed Kepler's laws, differential rotation should erase radial features. The Voyager images revealed that the **spokes** were real (Figure 24-20a).

Two clues are critical: The spokes are bright in forward scattering, and they sweep around Saturn at the same rate that the magnetic field rotates. Evidently, some dust in the rings becomes electrically charged, and the magnetic field sweeps the dust up out of the rings to form the temporary dust clouds that we see as spokes.

Photographs and measurements reveal the existence of other rings around Saturn. The E ring, detected from Earth, was traced from 3 to 8 Saturn radii and shown to be densest near the orbit of the satellite Enceladus. This moon may be a source of new ring particles as existing particles slam into the moon and blast new particles free. Voyager images also revealed a D ring inside the C ring and two narrow rings labeled F and G.

The F ring posed a difficult problem. Voyager 1 photos showed it was narrow and irregularly twisted (Figure 24-20b). Planetary rings should gradually spread out due to collisions between particles and should not develop irregular twists if the particles move in simple orbits. The mystery was eventually resolved when photos showed two small satellites orbiting, one inside and one outside the F ring (Figure 24-20c).

These **shepherd satellites** interact gravitationally with wandering particles and return them to the ring, thus keeping it confined. The distortions in the F ring appear to be caused by gravitational interactions when the inner shepherd moon overtakes and passes the outer moon. Another satellite at the edge of the A ring ushers straying particles back and thus keeps the edge of the A ring sharp.

The discovery of the shepherd satellites gave astronomers hope of understanding the hundreds of ringlets within the main rings by supposing that small moons are embedded within the rings. Although these moons would be inside the Roche limit of Saturn, they could resist being pulled apart by tides if they were small and made of strong material. Voyager 1 did not record any images of moons within the rings. However, recent studies of irregularities in the rings predicted a moon in a gap called Encke's Gap. The moon, only 20 km in diameter, was found in the summer of 1990 on photographs taken by Voyager 2. Other moons of similar size probably also lurk within the rings.

Some of the grooves seen in the rings have their origin in instabilities that can occur in material rotating differentially in a disk. The spacing of some ringlets shows that spiral density waves (the same phenomena seen in spiral galaxies) can be set up by satellite resonances. These waves can clear gaps and create entire families of ringlets.

Spiral density waves move slowly outward, but other waves, called bending waves, move inward from resonances. These bending waves cause the surface of the rings to be corrugated. This observation helps resolve a question concerning the thickness of the rings. Some photographs by the spacecraft show that the rings are as thin as 10 m, yet observations from Earth yield a thickness of 1–2 km. Apparently, the amplitude of the bending waves makes the rings look much thicker than they really are, just as corrugated cardboard is much thicker than the paper in the corrugations.

The observations by Voyagers 1 and 2 were astounding in many ways, but there was one thing the spacecraft could not detect—individual ring particles. Only a few experiments involving the forward scattering of light, the transmission of radio signals through the rings, and radar observations from Earth can help us understand what the particles are like. Dust particles are very common, especially in the B ring and Cassini's division, making these features bright in forward-scattered light. Particles a few centimeters in diameter are common in some regions, such as the C ring, and almost absent from others, such as Cassini's division. Chunks the size of a person are much rarer, and boulders the size of a house are separated by many meters.

An astronaut in a space suit could probably swim through the densest parts of the ring by pushing against the larger particles and drifting from particle to particle (Figure 24-21). This would not be dangerous because the orbiting particles in any given region travel at nearly the same speed. When particles do collide, they merely nudge each other. Occasionally, more violent collisions

FIGURE 24-21

Swimming through Saturn's rings might be possible. These computer-generated views of particles in the A ring (left) and the C ring (right) show how particle size and number differ through the rings. Particles smaller than 2 cm in diameter are not shown. The actual particles are believed to be irregular ice chunks and not the perfect spheres shown. Astronaut included for scale. (NASA–JPL) may produce a few bits of dust that slowly drift inward through the rings and finally fall into the planet. In the thinner parts of the rings, the particles are so widely spaced that ring swimming would be difficult.

The different distributions of particle sizes through the rings are responsible for differences in the appearance of the A, B, and C rings. But no one presently understands what forces sort the particles by size in the different regions. Nor do we understand the processes that must break up particles and re-form them.

Like a beautiful flower, the rings of Saturn are controlled by many different natural processes. When the Cassini spacecraft reaches orbit around Saturn in 2004, we will learn even more about the rings.

The History of Saturn

The farther we journey from the sun, the more difficult it is to understand the history of the planets. Any fully successful history of Saturn should explain its low density, its peculiar magnetic field, and its beautiful rings. We can't tell that fully successful story yet, but we can understand a few of the principles.

Certainly, we can understand that Saturn formed in the outer solar nebula, where ice particles were stable. It grew rapidly and became massive enough to capture hydrogen and helium gas directly from the nebula. The denser elements sank to the middle to form a denser core, and the hydrogen formed a liquid mantle containing liquid metallic hydrogen. The outward flow of heat from the core is believed to drive convection currents in this mantle that, coupled with the rapid rotation of the planet, produce its magnetic field. Because Saturn is smaller than Jupiter, it has less liquid metallic hydrogen, and its magnetic field is weaker.

We might be tempted at first to suppose that the rings of Saturn are primordial, made of material left over from the formation of the planet. That seems unlikely, however. Saturn, like Jupiter, should have been very hot when it formed, and that heat would have vaporized and driven off ring material. Also, such a hot Saturn would have had a very distended atmosphere, which would have slowed ring particles by friction and caused the particles to fall into the planet.

Planetary rings do not seem to be stable over 4.6 billion years, so we must suppose that the ring material is more recent. One suggestion is that an icy moon may have come within Saturn's Roche limit, and tides pulled the moon apart. At least some of the debris would settle into the ring plane. Another suggestion is that comets sweeping through the outer solar system sometimes collide with Saturn's moons. Because both comets and moons are rich in ice, such a collision might replenish the rings with fresh ice particles.

The history of Saturn is very complex, and although the Voyager spacecraft returned a treasure of data, we may not fully understand the planet's story until Cassini reaches Saturn.

CRITICAL INQUIRY

Why do the belts and zones on Saturn look so bland?

One of the most powerful tools of critical thought is simple comparison and contrast. We can understand Saturn by comparing and contrasting it with Jupiter. In the atmosphere of Jupiter, the dark belts form in regions where gas sinks, and zones form where gas rises. The rising gas cools and condenses to form icy crystals of ammonia, which we see as bright clouds. Clouds of ammonia hydrosulfide and water form deeper, below the ammonia clouds, and are not as visible. Saturn is twice as far from the sun as Jupiter, so sunlight is four times dimmer. (Remember the inverse square law from Chapter 5.) The atmosphere is colder, and rising gas currents reach colder levels and form clouds deeper in Saturn's atmosphere than in Jupiter's atmosphere. Because the clouds are deeper, they are not as brightly illuminated by sunlight and look dimmer. A layer of methane ice–crystal haze high above the ammonia clouds makes the clouds even less distinct.

We can compare Saturn's atmosphere to Jupiter's atmosphere, but we can also learn by comparing the ring systems. How does our understanding of Jupiter's rings help us understand Saturn's? ∎

In any contest, Saturn should win the ribbon for most beautiful planet. Its subtly marked disk and its delicate rings are both massive and graceful. Its extensive system of moons only adds to the mystery of this beautiful and curious world.

24-4 SATURN'S MOONS

Ⓢaturn has nearly two dozen known satellites (Figure 24-22)—far too many for us to examine individually—but these moons share characteristics common to icy worlds. Most of them are small and dead, but one is big enough to have an atmosphere and perhaps even oceans.

Titan

Saturn's largest satellite is a giant ice moon with a thick atmosphere and a mysterious surface. From Earth it is only a dot of light, and the cameras aboard the Voyager spacecraft were unable to penetrate its hazy atmosphere. Nevertheless, a few basic observations tell us a great deal about this strange world.

Titan is an icy world a bit larger than the planet Mercury and almost as large as Jupiter's moon Ganymede. Unlike these worlds, Titan has a thick atmosphere. Its escape velocity is low, but it is so far from the sun it is very cold. Thus, most gas atoms don't move fast enough

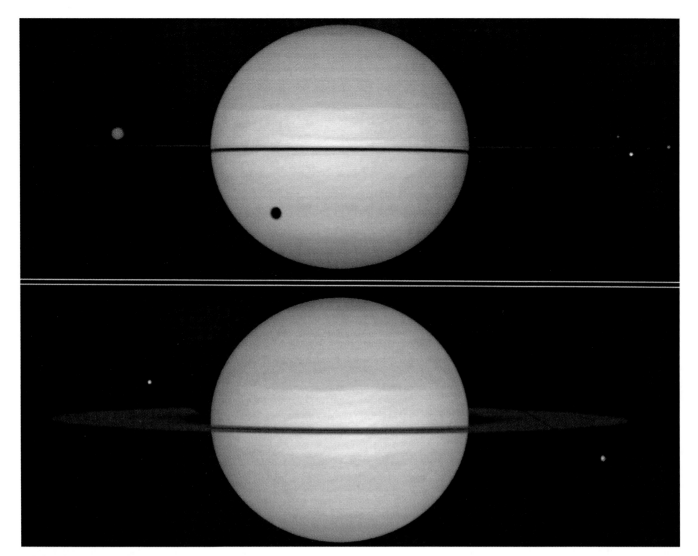

FIGURE 24-22
Hubble Space Telescope images of Saturn show a few of its many moons. The upper image was recorded as the rings were edge-on to Earth. The reddish moon at the upper left is smoggy Titan casting its shadow on Saturn. At the right are the icy moons Mimas, Tethys, Janus, and Enceladus (left to right). The lower image was recorded as the rings were illuminated nearly edge-on to the sun. Tethys is at upper left and Dione at the right. Note that Dione casts a long shadow across the rings. (E. Karkoschka, LPL, and NASA)

to escape. (See Figure 23-16.) Its density is 1.9 g/cm^3, but its uncompressed density (Chapter 20) is only 1.2 g/cm^3. Although it must have a rocky core, it must also contain a large amount of ices.

The surface of Titan is not visible in photographs because its atmosphere is filled with haze (Figure 24-23a). Methane was detected spectroscopically in 1944, as were various hydrocarbons in 1970, but most of Titan's air is nitrogen with only a few percent methane. Sunlight converts methane (CH$_4$) into the gas ethane (C$_2$H$_6$) plus a collection of organic molecules (Table 24-3).* Some of these molecules produce the smoglike haze, and as the smog particles have gradually settled to

---

*Organic molecules are common in living things on Earth but do not have to be derived from living things. One chemist defines an organic molecule as "any molecule with a carbon backbone."

TABLE 24-3

Organic Compounds Detected on Titan by a Voyager Infrared Spectrometer

C$_2$H$_6$	Ethane
C$_2$H$_2$	Acetylene
C$_2$H$_4$	Ethylene
C$_3$H$_4$	Methylacetylene
C$_3$H$_8$	Propane
C$_4$H$_2$	Diacetylene
HCN	Hydrogen cyanide
HC$_3$N	Cyanocetylene
C$_2$N$_2$	Cyanogen

a

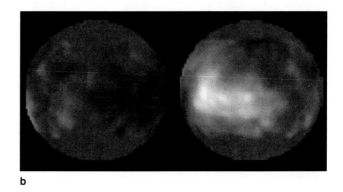

b

FIGURE 24-23

(a) A Voyager photograph of Saturn's moon Titan at visible wavelengths shows nothing but a smoggy atmosphere and a circulation pattern near the north pole. The surface is not visible. (b) Hubble Space Telescope images in the infrared can see through the haze. Here images have been combined to show the side of Titan that faces Saturn (left), and the side that faces forward in its orbit (right). Bright and dark areas assure us that the surface is not a uniform ocean of ethane and methane but also consists of at least some solid surface with differing geology. (NASA)

the surface, they may have deposited a layer of smelly, organic goo 10 m to 1 km thick.

This goo is exciting because such organic molecules may have been the precursors of life on Earth. Thus, similar chemical reactions on Titan might have created life there. Although some scientists argue that the temperature on Titan is too low to create such molecules, future explorers will want to visit Titan to look for life. We will consider this further in Chapter 27.

Before you imagine floundering through this gooey landscape, we must add a bit of chemistry. Voyager measurements show that the pressure at the surface is 50 percent greater than the air pressure on earth, and the temperature is 94 K ($-290°$F). Under these conditions, methane and ethane form liquids, and in the 4.6 billion years since the solar system formed, sunlight could have produced enough ethane to make a layer 1 km deep—an ethane ocean flavored with liquid methane. (On Earth, methane is common natural gas. Also, both methane and ethane are flammable—but only on planets with free oxygen.)

Recent observations tell us that Titan is not entirely covered by such an ocean. Radar signals from Earth have penetrated the atmosphere and found some regions to be more reflective than others. Infrared measurements show that Titan has bright regions, and Hubble Space Telescope images recorded in the infrared show bright

regions, one as large as Australia (Figure 24-23b). One hypothesis about these regions is that methane rain falls in the upper atmosphere, washes the goo off of the highlands and leaves them brighter than the lowlands. The goo doesn't get washed off the lowlands because the methane raindrops, about the size of marbles, fall slowly and evaporate before they reach the lower atmosphere.

Whatever the explanation, the irregular surface and thick atmosphere suggest that Titan has had an interesting history. Because the small amount of methane now in the atmosphere would be totally converted to ethane in only a million years, the methane is being constantly replenished. Ammonia and methane are probably released from the icy mantle, thus explaining the abundance of nitrogen—it could be produced by the breakdown of ammonia (NH_3). The hydrogen would escape to space and the heavier nitrogen would be left in the atmosphere. The irregular surface suggests heat flow from the interior could have generated geological activity to produce highlands and lowlands. Titan is big enough to have had significant internal heat, and it should have differentiated. A later episode of tidal heating could have produced significant geological activity.

Does Titan have vast oceans or small lakes? Is there organic goo on the surface or is it just ice? Is the surface cratered or are there signs of geological activity? We will know soon. The Cassini spacecraft will reach Saturn in

FIGURE 24-24

(a) Heavily cratered Tethys appears to be a dead world with a long fault in its icy crust (arrows). (b) The leading side of Iapetus (facing upper right) is much darker than the trailing side due to dark deposits on its surface. (c) Some portions of Enceladus have fewer craters and must have been resurfaced recently. This suggests that Enceladus has been active since the end of the age of cratering. (NASA)

2004, and it will map Titan by radar and with infrared cameras. Furthermore, the spacecraft will drop a probe into Titan's atmosphere. Named Huygens after Christian Huygens who discovered Titan in 1655, the probe will study the atmosphere and photograph the surface for more than two hours as it descends by parachute. It may survive for an hour or so on the surface and tell us by radio whether it landed on ice, on rock, on goo, or on waves.

Titan is an exciting world to think about because it might harbor life in its organic deposits, but it is also exciting because it is so totally unearthly.

The Smaller Moons

In addition to Titan, the two Voyager spacecraft imaged eight smaller Saturnian moons. They have low densities and are evidently formed of rock and ice.

These smaller moons are heavily cratered and appear geologically dead (Figure 24-24), but a few show signs of past activity. Tethys is marked by a crack 3 km deep and 1500 km long; it stretches three-quarters of the way around the moon. Similarly, the inner moon Mimas is marred by a crack 2 km deep. Such cracks may have formed when the interiors of the icy moons froze and expanded, breaking the crust—the opposite of what happened when the metallic core of Mercury solidified and contracted to form lobate scarps.

Because most of the moons are heavily cratered, we can conclude that they are geologically dead, but Enceladus may be an active moon (Figure 24-24c). Its surface contains areas of fewer craters, and some regions contain parallel grooves similar to those on Ganymede. These may be cracks filled by frozen water from the interior. The light yellow color of Enceladus also suggests that its surface is not old. It seems possible that eruptions of liq-

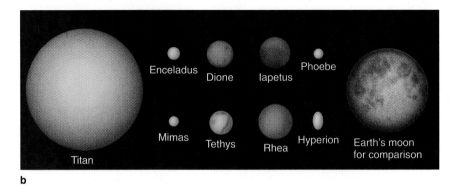

FIGURE 24-25

(a) Hyperion is a small satellite that has not been able to pull itself into a spherical shape. (b) Larger moons have enough gravity to overcome the strength of the rock and ice and force themselves into spheres.

uid water occasionally resurface parts of Enceladus with fresh ice. This would explain why the ice crystals of Saturn's extended E ring are more common near the orbit of Enceladus.

Whenever we see an active moon, we must wonder about the source of its internal energy. Enceladus is too small to have been heated significantly by radioactive decay, so we can suspect that tidal heating is responsible.

Iapetus is an asymmetric moon. Its trailing side is about as bright as dingy snow, but its leading side has an albedo of only 5 percent—about as dark as fresh asphalt on a highway (Figure 24-24b). The origin of this dark material is unknown, but theorists suggest that the little moon has swept up dark, silicon- and carbon-rich material on its leading side. This dust could be produced by meteorites striking the outermost moon, Phoebe.

Saturn's moon Hyperion is odd because it is not spherical. It is shaped like a battered hockey puck with diameters of 250 km, 230 km, and 110 km (Figure 24-25). Mimas and Enceladus are only slightly larger and are quite spherical, so we must suppose that the gravitational field of a larger moon is sufficient to overcome the structural strength of ice and squeeze the moon into spherical shape. A small moon like Hyperion, located far from its parent planet where tidal heating is unimportant, may remain cold, rigid, irregular, and undifferentiated.

The Origin of Saturn's Satellites

Jupiter's four Galilean satellites seem clearly related to one another, and we concluded that they probably formed with Jupiter. No such systematic relationship links Saturn's satellites. We may be forced to conclude that they are captured icy planetesimals left over from the solar nebula or that the impacts of comets have so badly fractured them that they no longer show evidence of their common origin.

Understanding the origin of Saturn's moons is also difficult because the moons interact gravitationally in complicated ways, and thus the orbits they now occupy may differ from their earlier orbits. These interactions were revealed dramatically by the Voyager spacecraft.

We have previously discussed the small satellite that shepherds the outer edge of the A ring and the pair of moons that shepherds the F ring. An even more peculiar pair of moons is known as the coorbital satellites. These two irregularly shaped moonlets have orbits separated by only 100 km. Because one moon is about 200 km in diameter and the other about 100 km, they cannot pass in their orbits. Instead, the innermost moon gradually catches up with the outer moon. As the moons draw closer together, the gravity of the trailing moon slows the leading moon and makes it fall into a lower orbit. Simultaneously, the gravity of the leading moon pulls the trailing moon ahead, and it rises into a higher orbit (Figure 24-26). The higher orbit has a longer period, so the trailing moon begins to fall behind the leading moon, which is now in a smaller, faster orbit. Thus, the moons exchange orbits and draw apart only to meet again and again. It seems very likely that these two moons are fragments of a larger moon destroyed by a major impact.

In addition to waltzing coorbital moons, the Voyager spacecraft discovered small moonlets trapped at the L_4 and L_5 Lagrangian points (Chapter 14) in the orbits of Dione and Tethys. These points of stability lie 60° ahead of and 60° behind the two moons, and small moonlets can become trapped in these regions. (We will see in Chapter 26 that asteroids are trapped in the Lagrangian points of Jupiter's orbit around the sun.) This gravitational curiosity is surely not unique to the Saturn system, and it warns us that interactions between moons can dramatically alter their orbits.

Some of Saturn's larger moons, especially Titan, may have formed with the planet, but the orbital relationships and intense cratering suggest that the moons have interacted and may have collided in the past. Many of the moons may be icy bodies that Saturn has captured since the formation of the solar system. As we continue our exploration of the outer solar system, we must be alert for the presence of such small, icy worlds.

CRITICAL INQUIRY

What features on Enceladus suggest that it has been active?

The smaller moons of Saturn are icy worlds battered by impact craters, and thus we suspect that they are cold and old. Small worlds lose their heat quickly, and with no internal heat, there is no geological activity to erase impact craters. A small, icy world covered with craters is exactly what we would expect in the outer solar system. Enceladus, however, is peculiar. Although it is small and icy, its surface is not very dark, and some areas contain fewer craters than we would expect. In fact, some regions seem almost free of craters. Parallel grooves mark some regions of the little moon and suggest motion in the icy crust. These features should have been destroyed long ago by impact cratering, so we must suppose that the moon has been geologically active since the end of heavy cratering at the conclusion of planet building.

Of course, if Enceladus has been active, then it must have had a source of heat in its interior. Only the outward flow of energy can drive geological activity. Since Enceladus is much too small to have kept itself warm from internal radioactive decay, we might suspect tidal heating, although that is not understood at present.

If we look at Titan, Saturn's largest moon, we see quite a different world. Although Titan is only about 50 percent bigger than Earth's moon, it has a substantial atmosphere. How can such a small world keep an atmosphere? ∎

Jupiter and Saturn reign over the solar system like their namesake ancient gods. Beyond them, out where the sunlight is dim and temperatures deadly cold, orbit three more planets that were unknown to ancient astronomers. In the next chapter, we will add these worlds to our study of comparative planetology. ∎

∎ Summary

The Jovian planets are so massive that they have been able to retain much of their hydrogen and helium. This excess of light elements gives the Jovian planets densities much lower than those of the terrestrial planets. In their interiors, the tremendous pressure forces the hydrogen into a liquid state, and in Jupiter and Saturn the pressure is so high that hydrogen becomes a liquid metal, which is a good conductor of electricity. Electric currents within this material produce magnetic fields that trap particles from the solar wind to form radiation

(Not to scale)

FIGURE 24-26
As the trailing coorbital satellite overtakes the leading satellite, gravitational forces make the moons trade orbits. Because the inner orbit is slightly faster, the innermost moon always gains on the outermost.

belts around the planets. Both Jupiter and Saturn are known to have radiation belts similar to Earth's Van Allen belts, and Jupiter's belts are very intense. Both planets have aurora around their magnetic poles.

Jupiter is encircled by atmospheric dark belts, regions of sinking gas, and bright zones, regions of rising gas. Unknown trace chemicals give the clouds their colors. Although Saturn has the same cloud layers, its atmosphere is colder and the cloud layers form deeper. They are less distinct.

The Great Red Spot on Jupiter is an anti-cyclonic disturbance produced by rising gas. Larger than Earth, it has lasted for at least three centuries. Jupiter's atmosphere contains similar smaller spots that are short lived, but such spots are rare on Saturn.

Jupiter's ring is dark in color but bright in forward scattering, so we know it is made up of dusty, rocky material. The dust falls into Jupiter or is destroyed by radiation in less than 100 years, so it must be constantly resupplied as meteorites chip material off of the innermost moons.

Saturn's rings are bright, icy, and highly complex. Much of the structure in the rings appears to be created by the gravitational influence of Saturn's moons, by waves that move through the rings, and by Saturn's magnetic field which produces spokes. A few small moons have been located shepherding the narrow F ring and the outer edge of the A ring. It is most likely that the icy rings are composed of debris that is created by the breakup of one or more satellites.

Jupiter has a number of small moonlets that are probably captured asteroids. But the four Galilean satellites seem to have formed at about their present locations. The density of these moons decreases with increasing distance from Jupiter, suggesting that they formed in a small nebula around the condensing planet. The heat from Jupiter would have driven off volatiles in the inner nebula and made the inner moons denser than the outer moons. This is a miniature of the formation of the planets from the solar nebula. This process does not seem to have occurred in the Saturn system, perhaps because Saturn is less massive and did not become as hot as Jupiter.

Jupiter's satellites have clearly been affected by tidal heating. Io is violently active with sulfur volcanoes. Europa is less active, but the lack of craters on its surface suggests that its icy crust is still active enough to erase craters. Ganymede and Callisto are both heavily cratered, inactive worlds, but the grooved terrain on Ganymede seems to date from a period when it was more active. How much tidal heating contributed to the activity of the outer satellites is not known.

Titan, Saturn's largest moon, is able to keep a thick atmosphere of nitrogen and methane because it is so cold. In fact, sunlight converts the methane into ethane and other organic materials that form a smog in the atmosphere and hide the moon's surface. Ethane and methane fall to the surface and probably exist as liquid oceans or lakes, while falling smog particles create a deposit of organic goo.

Some of Saturn's moons are clearly inactive. Mimas, Tethys, and Dione are heavily cratered, icy bodies. Nevertheless, the presence of long, deep valleys formed by cracks in the icy crust shows that the worlds once had more internal heat.

Enceladus seems to have had an active history. It is cratered, but some regions are obviously much younger. Lava flows of water may have resurfaced portions of Enceladus. The E ring of Saturn is most dense near the orbit of Enceladus, and the ring particles may originate on the icy moon.

The debris in space craters the surfaces of moons most heavily on the side facing forward in their orbits. The trailing side is less cratered. Iapetus is bright on the trailing side and very dark on the leading side. Whether the black deposits are debris from some outside source or material from the interior is not known.

Many of Saturn's moons are small, irregular bodies that are either captured objects or the fragments of collisions. Hyperion is very irregular in shape.

The orbital dynamics of the smaller moons is complex. The coorbital moons share nearly identical orbits, whereas other moons occupy the Lagrangian points in the orbits of larger moons.

■ New Terms

oblateness	zone
liquid metallic hydrogen	forward scattering
decameter radiation	Roche limit
decimeter radiation	grooved terrain
current sheet	tidal heating
Io plasma torus	infrared outburst
Io flux tube	spoke
critical point	shepherd satellite
belt	

■ Questions

1. Why is Jupiter oblate? Do you expect all Jovian planets to be oblate? Why or why not?

2. How do the interiors of Jupiter and Saturn differ? How does this affect their magnetic fields?

3. Describe two theories to explain the persistence of the Jovian belts and zones.

4. How can we be certain that Jupiter's ring does not date from the formation of the planet?

5. If Jupiter had a satellite the size of our own moon orbiting outside the orbit of Callisto, what would you predict for its density and surface features?

6. Why are there no craters on Io and few on Europa? Why should we expect Io to suffer more impacts per square kilometer than Callisto?

7. Why are the belts and zones on Saturn less distinct than those on Jupiter?

8. If Saturn had no moons, what do you suppose its rings would look like?

9. Suggest two theories for the origin of Saturn's rings.

10. How can Titan keep an atmosphere when it is smaller than airless Ganymede?

11. If you piloted a spacecraft to visit Saturn's moons and wanted to land on a geologically old surface, what features would you look for? What features would you avoid?

12. Why does the leading side of some satellites differ from the trailing side?

13. Why do we expect small satellites to be irregular in shape? Give examples and exceptions.

■ Discussion Questions

1. Some astronomers argue that Jupiter and Saturn are unusual, while other astronomers argue that all solar systems should contain one or two such giant planets. What do you think? Support your argument with evidence.

2. Why don't the terrestrial planets have rings?

■ Problems

1. What is the maximum angular diameter of Jupiter as seen from Earth? What is the minimum? Repeat this calculation for Saturn and Titan. (HINT: Use the small-angle formula.)

2. The highest-speed winds on Jupiter are in the equatorial jet stream, which has a velocity of 150 m/sec. How long does it take for these winds to circle Jupiter?

3. What are the orbital velocity and period of a ring particle at the outer edge of Jupiter's ring? at the outer edge of Saturn's A ring? (HINT: See Chapter 5.)

4. What is the angular diameter of Jupiter as seen from the surface of Callisto? (HINT: Use the small-angle formula.)

5. What is the escape velocity from the surface of Ganymede if its mass is 1.5×10^{26} g and its radius is 2628 km? (HINT: See Chapter 5.)

6. If we photograph the spectrum of Saturn and its rings, we should find light from one edge of the rings red-shifted and light from the other edge blue-shifted. If we observe at a wavelength of 500 nm, what difference in wavelength should we expect between the two edges of the rings? (HINT: See Problem 3 and Chapter 7.)

7. What is the difference in orbital velocity between particles at the outer edge of Saturn's B ring and particles at the inner edge of the B ring? (HINT: See Data File Eight and Chapter 5.)

8. What is the difference in orbital velocity between the two coorbital satellites if the semimajor axes of their orbits are 151,400 km and 151,500 km? The mass of Saturn is 5.7×10^{26} kg. (HINT: See Chapter 5.)

■ Critical Inquiries for the Web

1. Saturn has more known satellites than any planet in our solar system. Locate information on these moons, including discovery dates (make sure you use an updated source that includes discoveries from the past few years). Excluding the satellite discoveries associated with the Voyager data, what observational circumstance do the discoveries of almost all of the moons of Saturn have in common?

2. The collision of Comet Shoemaker-Levy 9 with Jupiter was one of the first events to be publicized directly on the Web. Search for Web sites dedicated to this comet and its spectacular demise. How long were the effects of the impacts visible in the atmosphere of Jupiter? How might they have appeared differently if this event had occurred at Saturn?

3. What is the status of our understanding of the Jovian moon Europa? This moon was studied extensively by the Galileo mission. Visit Galileo-related Web sites to find the results of this exploration and discuss our current ideas regarding the possibility of liquid water beneath Europa's crust. How might future missions to Europa probe such an underground ocean?

 Go to the Wadsworth Astronomy Resource Center (www.wadsworth.com/astronomy) for critical thinking exercises, articles, and additional readings from InfoTrac College Edition, Wadsworth's online student library.

Chapter Twenty-Five

URANUS, NEPTUNE, AND PLUTO

A good many things go around in the dark besides Santa Claus.

HERBERT HOOVER

GUIDEPOST

In the three previous chapters, we have used our tools of comparative planetology to study other worlds, and we continue that theme in this chapter. A second theme running through this chapter is the nature of astronomical discovery. Unlike the other planets in our solar system, Uranus, Neptune, and Pluto were discovered, and the story of their discovery helps us understand how science progresses.

As we probe the outer fringes of our planetary system in this chapter, we see strong evidence of smaller bodies that fall through the solar system and impact planets and satellites. The next chapter will allow us to study these small bodies in detail and will give us new evidence that our solar system formed from a solar nebula. ■

Out in the darkness beyond Saturn, out where sunlight is about 1000 times fainter than on Earth, there are things going around that Aristotle, Galileo, and Newton never imagined. They knew about Mercury, Venus, Mars, Jupiter, and Saturn, but our solar system includes three more planets, planets that were not discovered until after the invention of the telescope.

Uranus and Neptune are Jovian planets and are nearly the same size, about four times Earth's diameter. Thus, they are about half the size of Saturn. Through earth-based telescopes, we see nothing but tiny disks, and until recently what little we knew of these planets was deduced from a few observations compared with Jupiter and Saturn. But in January 1986 the Voyager 2 spacecraft flew past Uranus and returned a wealth of photographs and measurements. In August 1989, Voyager 2 flew past Neptune revealing a blue world with complex weather, rings, and moons. In just a few years, Uranus and Neptune have gone from obscure points of light to known worlds.

Pluto, however, is a more difficult subject. Through an earth-based telescope, it looks like nothing more than a faint point of light, and even through the Hubble Space Telescope it is a tiny disk. Voyager 2 left the solar system without passing near Pluto, so we have no good photos of the little world's surface. Nevertheless, we may be able to guess its nature by combining the limited observations with the principles of comparative planetology.

Uranus, Neptune, and Pluto differ from the other planets because they were not known from antiquity. They had to be discovered. The stories of these discoveries highlight the nature of scientific discovery.

25-1 URANUS

On March 1781, Benjamin Franklin was in France raising money, troops, and arms for the American Revolution. George Washington and his colonial army were only 6 months away from the defeat of Cornwallis at Yorktown and the end of the war. In England, King George III was beginning to show signs of madness. And a German-speaking music teacher in the English resort city of Bath was about to discover the planet Uranus.

The Discovery of Uranus

William Herschel came from a musical family in Hanover, Germany, but immigrated to England as a young man and eventually obtained a prestigious job as the organist at the Octagon Chapel in Bath.

To compose exercises for his students and choral and organ works for the chapel, Herschel studied the mathematical principles of musical harmony from a book by Professor Robert Smith of Cambridge. The mathematics was so interesting that Herschel searched out other books by Smith, including a book on optics. Of course, an 18th-century book on optics written by a professor in England relied heavily on Isaac Newton's optics and described the principles of the reflecting telescope (see

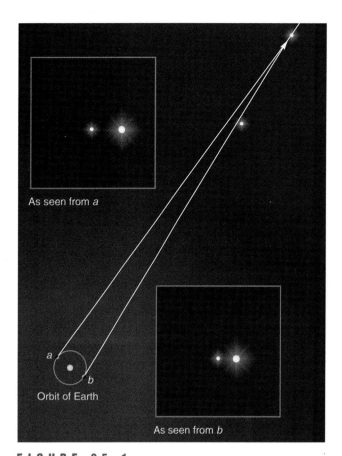

FIGURE 25-1

A double star consisting of a nearby star and a more distant star could be used to detect stellar parallax. The separation between the stars would be different as seen from different places in Earth's orbit. The effect is actually much too small to detect by eye.

Chapter 6). Herschel and his brother Alexander began building telescopes, and William went on to study astronomy in his spare time.

Herschel's telescopes were similar to Newton's in that they had metal mirrors, but they were much larger. Newton's telescope had a mirror about 1 in. in diameter, but Herschel developed ways of making much larger mirrors. He soon had telescopes as long as 20 ft, although one of his favorite telescopes was 7 ft long with a mirror 6.2 in. in diameter. Using it, he began a research project that led to the discovery of Uranus.

Herschel did not set out to search for a planet; he was trying to detect stellar parallax produced by the motion of the earth around the sun. No one had yet detected this effect, although by the 1700s nearly all astronomers accepted the motion of the earth. Galileo had pointed out that parallax might be detected if a nearby star and a very distant star lay so nearly in the same direction that they looked like a very close double star through a telescope. In such a case, the orbital motion of the earth would produce a parallactic shift in the position of the nearby star with respect to the more distant star (Figure 25-1). Herschel began to examine all stars brighter than

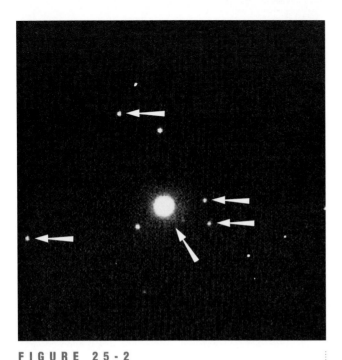

FIGURE 25-2

Through a telescope, Uranus is a featureless disk never more than 3.7 seconds of arc in diameter. This CCD image, made with the 4-m telescope at Cerro Tololo Inter-American Observatory in Chile, shows the disk of Uranus and its five major satellites (arrows). The other images are faint stars. (Cerro Tololo Inter-American Observatory)

eighth magnitude to search for double stars, which might show parallax. That project alone took over 2 years.

On the night of March 13, 1781, Herschel set up his 7-ft telescope and continued his work. He later wrote, "In examining the small stars in the neighbourhood of H Geminorum, I perceived one that appeared visibly larger than the rest." As seen from Earth, Uranus is never larger in angular diameter than 3.7 seconds of arc, so Herschel's detection of the disk illustrates the quality of his telescope and his eye (Figure 25-2). At first he suspected that the object was a comet, but other astronomers quickly showed that it was a planet orbiting the sun beyond Saturn.

The discovery of Uranus made Herschel world famous. Since antiquity, astronomers had known the five planets—Mercury, Venus, Mars, Jupiter, and Saturn—and they supposed that the universe was complete. Herschel's discovery extended the classical universe. The English public accepted Herschel as their astronomer–hero, and, having named the new planet *Georgium Sidus* (George's Star) after King George III, Herschel received a royal pension. The former music teacher was welcomed into court society, where he later met and married a wealthy widow and took his place as one of the great English astronomers in history. With his new financial position he built large telescopes on his estate and, with his sister Caroline, a talented astronomer herself, attempted to map the extent of the universe.

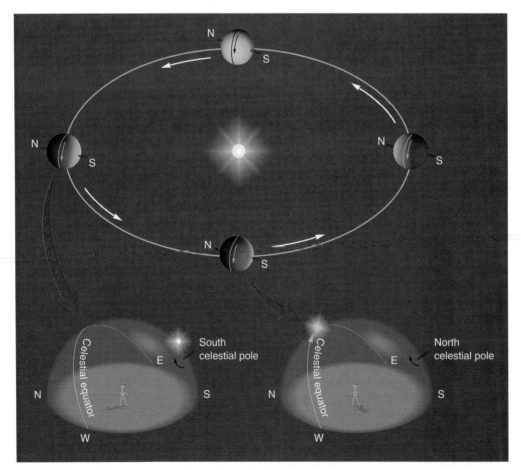

FIGURE 25-3

Uranus rotates on an axis that is tipped 97.9° from the perpendicular to its orbit, so its seasons are peculiar. When one of its poles is pointed nearly at the sun (a solstice), a citizen of Uranus would see the sun near a celestial pole, and it would never rise or set. As it orbits the sun, the planet maintains the direction of its axis in space, and thus the sun moves from pole to pole. At the time of an equinox on Uranus, the sun would be on the celestial equator and would rise and set with each rotation of the planet. Compare with Figure 2-16.

Continental astronomers were less than thrilled that an Englishman had made the great discovery, and even some professional English astronomers called Herschel's discovery a lucky accident. Herschel responded in his own defense by making three points. First, he had built some of the finest-quality telescopes then in existence. Second, he had been conducting a systematic research project and would have found Uranus eventually because he was inspecting all of the brighter stars visible with his telescope. And third, he had great experience seeing fine detail with his telescopes. As a musician, he knew the value of practice and applied it to the business of astronomical observing. In fact, astronomers had seen Uranus more than 17 times before Herschel, but each time they had plotted it on their charts as a star.

This illustrates one of the ways in which scientific discoveries are made. Often, discoveries seem accidental, but on closer examination we find that the scientist has earned the right to the discovery through many years of study and preparation. To quote a common saying, "Luck is what happens to people who work hard."

Continental astronomers, especially the French, insisted that the new planet not be named after an English king. The French stubbornly called the planet Herschel, as did many other non-English astronomers. Some years later, German astronomer J. E. Bode suggested the name Uranus, the oldest of the Greek gods.

Over the half-century following the discovery of Uranus, astronomers noted that Newton's laws did not quite predict the observed position of the planet. Tiny variations in the orbital motion of Uranus eventually led to the discovery of Neptune, a story we will take up later in this chapter.

The Motion of Uranus

Uranus follows a slightly elliptical orbit with an average distance from the sun of 19.18 AU and an orbital period of 84.013 years (Data File Nine). Compare this with the orbital period of Saturn, 29.461 years. Unlike Saturn, Uranus rotates on its side.

Earth rotates approximately upright in its orbit. That is, its axis of rotation is inclined only 23.5° from the perpendicular to its orbit. Uranus rotates on an axis that is inclined 97.9° from the perpendicular to its orbit (Figure 25-3). This peculiar axis of rotation produces strange seasons on the planet and may hint at an important process in the formation of the planets.

In 1986, when Voyager 2 flew past Uranus, the planet was positioned in its orbit such that its south pole faced the inner solar system. Thus, its southern hemisphere was bathed in sunlight, and a creature living on Uranus (an unlikely possibility, as we will discover later) would have seen the sun near the planet's south celestial pole (Figure 25-3). In 21 Earth years, Uranus will have moved one-quarter of the way around its orbit, and the sun will be on its celestial equator (Figure 25-3). Our Uranian creature would see the sun rise and set with each rotation of the planet. In another 21 Earth years, the sun

URANUS

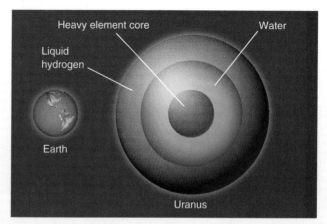

About four times the diameter of Earth, Uranus is believed to contain a liquid interior and dense core.

Average distance from the sun	19.18 AU (28.69×10^8 km)
Eccentricity of orbit	0.0461
Maximum distance from the sun	20.1 AU (30.0×10^8 km)
Minimum distance from the sun	18.3 AU (27.4×10^8 km)
Inclination of orbit to ecliptic	0°46′23″
Average orbital velocity	6.81 km/sec
Orbital period	84.013 y (30,685 days)
Period of rotation	17^h14^m
Inclination of equator to orbit	97°55′
Equatorial diameter	51,118 km (4.01 $D_\oplus$)
Mass	8.69×10^{25} kg (14.54 $M_\oplus$)
Average density	1.29 g/cm^3
Gravity	0.919 earth gravities
Escape velocity	22 km/sec (1.96 $V_\oplus$)
Temperature above cloud tops	−220°C (−364°F)
Albedo	0.35

will have reached a point near the planet's north celestial pole, and the southern hemisphere will be experiencing a lightless winter lasting 21 Earth years. Thus, the ecliptic as seen from Uranus passes very near the planet's north and south celestial poles.

The high inclination of the axis of rotation has never been fully explained. Planetary astronomers generally

a

b

c

d

FIGURE 25-4

(a) The atmosphere of Uranus was nearly featureless in images recorded by Voyager 2 in 1986. Extreme image enhancement reveals (b) atmospheric banding concentric with the axis of rotation and (c) small white clouds that parallel the equator. (d) Near-infrared images made from Earth in the early 1990s suggest a dark spot (arrow) and other features near the planet's pole at the left of the disk. These observations suggest that the atmosphere of Uranus may not always be as featureless as it was in 1986. (a, b, and c, NASA; d, David Wittman)

propose that as Uranus formed from the icy planetesimals of the outer solar nebula, it had grown to nearly full size when it was struck at a sharp angle by a very large planetesimal, perhaps as large as Earth. That impact could have altered the axis of rotation. Such impacts may have been much more common in the outer solar system, where large amounts of ices could condense. We will see more evidence for large-impact events when we consider the satellites of Uranus. But first we must examine the atmosphere and interior of the planet.

The Atmosphere of Uranus

Like Jupiter and Saturn, Uranus has no surface. The gases of its atmosphere—mostly hydrogen, 15 percent helium, and a few percent methane, ammonia, and water vapor—blend gradually with a fluid interior. Also like Jupiter and Saturn, the atmosphere of Uranus is rich in hydrogen and helium.

Seen from Earth, Uranus is a small, featureless disk. As Voyager 2 drew closer to the planet in 1985, astrono-

mers examined the photos to see if Uranus had belt–zone circulation like Jupiter and Saturn. Early photos showed no clouds at all. Uranus is far from the sun, and its atmosphere is so cold that rising currents of gas condense and form clouds at lower levels in its atmosphere. Water clouds form at the lowest levels, with ammonia and then methane clouds at higher and colder levels. But even the highest and coldest methane clouds lie below a deep atmosphere of hydrogen. Small amounts of methane absorb red light, giving the planet its characteristic blue color (Figure 25-4a).

Although no cloud belts were obvious, Voyager photographs did reveal evidence of belt–zone circulation. Methane gas absorbs strongly at a wavelength of 0.619 μm, and photographs made at that wavelength could be computer-enhanced to reveal cloud patterns concentric with the axis of rotation (Figure 25-4b). Other images were enhanced to reveal clouds near the equator, allowing scientists to measure the rotation of the atmosphere (Figure 25-4c). The winds near the equator blow slower than the planet's interior rotates, but at higher latitudes the winds blow faster.

This appears to be a form of belt–zone circulation, which is a bit surprising. Because Uranus is so highly inclined to its orbit, it experiences extreme seasons, and its poles receive, on the average, more solar energy than does the equator. Thus, we might expect its atmospheric circulation to be different from that of Jupiter and Saturn. That its atmospheric circulation is concentric with its axis of rotation suggests that solar energy, although it supplies the energy for the circulation, does not determine its geometry. Rather, the nature of the planet's weather patterns is determined by its rotation.

Apparently, solar heating is not very important to the Uranian seasons. Sunlight is almost 400 times weaker at Uranus than at Earth. Also, atmospheric circulation distributes thermal energy around the planet and prevents seasonal temperature changes of more than a few degrees Kelvin. Infrared observations reveal that the atmosphere above the planet's poles is nearly the same temperature, although the north pole has been in darkness for some years.

Is Uranus always as featureless as it was when Voyager 2 flew past? The Hubble Space Telescope has detected changing clouds on Uranus. Furthermore, in 1993, astronomers at the University of Arizona recorded images of the planet in the near infrared; they detected a dark spot about 35° south of the equator and other faint features at the limit of visibility (Figure 25-4d). If confirmed, these observations suggest that Uranus may suffer outbursts like that seen on Saturn (Figure 24-18a). Another possibility is that Uranus has faint belts and spots that are visible only in the infrared.

If we could descend into the Uranian atmosphere in a spaceship, we would find the outer layers rich in hydrogen and helium. Just above the highest methane clouds, the temperature is −220°C (−364°F), and the pressure is about one-tenth of Earth's atmosphere at sea level. Temperature and pressure would increase as we descended. Below the methane clouds, our spaceship would pass through ammonia clouds and, deeper still, water clouds.

Outside the porthole of our spaceship, the temperature and pressure would continue to grow higher as we descended. Soon the temperature and pressure would pass the critical point where gas and vapor are indistinguishable, and our spaceship would become a submarine descending through a fluid of H_2O with hydrogen, helium, methane, and ammonia mixed in—the fluid interior of the planet.

The Interior of Uranus

To describe the interior of Uranus, astronomers must combine a few limited observations with models describing the behavior of planetary interiors. The result suggests that Uranus is a mostly fluid world.

The average density of Uranus, 1.29 g/cm³, shows that the planet is richer in hydrogen and helium than a terrestrial planet. Nevertheless, it is poorer in these lightweight atoms than Jupiter and Saturn. To produce a world of the size and density of Uranus, models of the interior must include large amounts of denser materials, such as rock and water.

Although Uranus contains hydrogen, the pressure in its interior is not high enough to produce liquid metallic hydrogen as in Jupiter and Saturn. The strong magnetic fields of Jupiter and Saturn are believed to be produced by the dynamo effect acting in the highly conducting liquid metallic hydrogen, so we might expect Uranus to lack a strong magnetic field. Nevertheless, Uranus has a magnetic field about 75 percent as strong as Earth's. Furthermore, the magnetic field is tipped 60° to the axis of rotation and is offset from the center by about 30 percent of a planetary radius (Figure 25-5). Theorists suggest that the dynamo effect is operating nearer the surface of the planet, which would require a large fluid mantle, perhaps made up of water with dissolved ammonia and

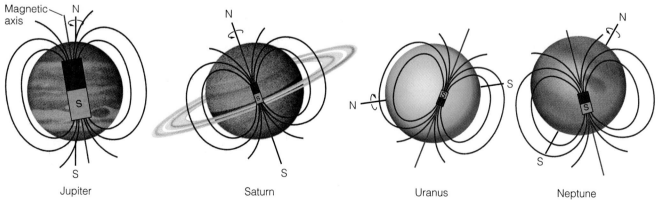

FIGURE 25-5

The magnetic fields of Uranus and Neptune are peculiar. While the magnetic axis of Jupiter is tipped 10° from its axis of rotation and the magnetic axis of Saturn is not tipped at all, the magnetic axes of Uranus and Neptune are tipped at large angles. Furthermore, the magnetic fields of Uranus and Neptune are offset from the center of the planet. This suggests that the dynamo effect operates differently in Uranus and Neptune than in Jupiter and Saturn.

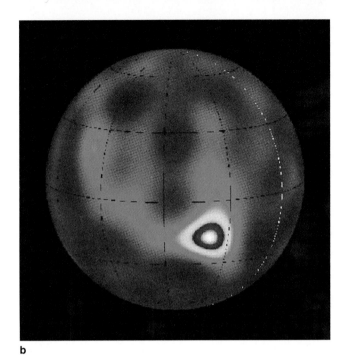

a

b

FIGURE 25-6

Auroras on Uranus were detected in the ultraviolet by the Voyager 2 spacecraft when it flew by in 1986. These maps of opposite sides of Uranus show the location of auroras near the magnetic poles. Recall that the magnetic field is highly inclined and offset from the planet's center, so the magnetic poles do not lie near the poles defined by rotation. In fact, the north magnetic pole (right) falls near the planet's equator. Also recall that Uranus is highly inclined and that in 1986 its south pole, located at the bottom of these diagrams, was pointed nearly at the sun. The white dashed line marks zero longitude. (Courtesy Floyd Herbert, LPL)

methane. Such a fluid would be a very good electrical conductor and might allow the operation of the dynamo effect. Thus, the mere presence of a magnetic field gives us clues to the nature of the planet's interior.

The high inclination of the magnetic field and its offset from the center of Uranus give us a good estimate of the true rotation period of the planet, otherwise difficult to estimate because of the lack of cloud features and high atmospheric winds. The magnetic field deflects the solar wind and traps some charged particles to create weak radiation belts trapped in a magnetosphere. High-speed electrons spiraling along the magnetic field produce synchrotron radio emission just as on Jupiter, and the Voyager 2 spacecraft detected this radiation fluctuating with a period of 17.24 hours—the period of rotation of the magnetic field and, presumably, the planetary interior.

The magnetic field and the high inclination of the planet produce some peculiar effects. As is the case for all planets, the solar wind deforms the magnetosphere and draws it out into a long tail extending away from the planet in the direction opposite the sun. The rapid rotation of Uranus and its high inclination make the magnetosphere and its long extension take on a corkscrew shape, which was partially mapped by Voyager 2. At the time Voyager 2 flew past in 1986, the south pole of Uranus was pointed nearly at the sun, and once each rotation the

solar wind could pour directly down into the south magnetic pole. This interaction produced strong auroras at both magnetic poles, which were detected by the spacecraft in the ultraviolet (Figure 25-6). In the years since, Uranus has moved farther around its orbit, so the geometry of its interaction with the solar wind is now different. Unfortunately, we have no spacecraft bound for Uranus, and thus we have no way to observe this interaction in detail.

The true rotation period of Uranus is an important clue to the nature of its interior when we combine it with the observed oblateness (0.024). Models of the interior of the planet cannot produce such a large oblateness at the observed rotation period if most of the heavy elements are concentrated at the center of the planet. Rather, the models predict that the planet has a small core of heavy elements surrounded by a deep layer of water mixed with heavier material, such as rock, in a slushy stew. The top of this dirty ocean merges with an overlying atmosphere made up mostly of hydrogen and helium.

Even the temperature of Uranus can tell us something about its interior. Jupiter and Saturn are warmer than we would expect given the amount of heat they receive from the sun, and we conclude that heat is leaking out from their hot interiors. Uranus, in contrast, is about the temperature we would expect for a world at its

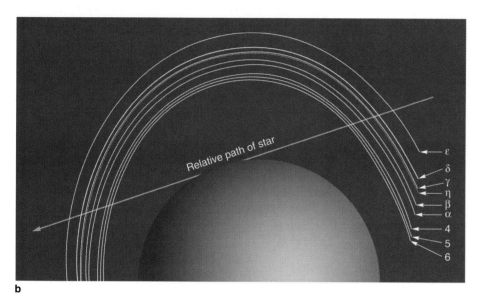

FIGURE 25-7

The rings of Uranus were discovered when the planet crossed in front of a star and the dimming of the starlight revealed the rings (a). The original charts show the intensity of the star's light interrupted by the five principal rings as Uranus approached the star (left) and after it had occulted the star (right). The calculated radius of the rings is given at the bottom of the chart. Note that the ε ring is eccentric. It lies at different distances from the center of Uranus on opposite sides of the planet. (b) As Uranus moved in front of the star, the relative path of the star crossed through the rings. Although only five rings were discovered in the first occultation, a total of nine were detected in subsequent occultations, and a tenth was found by Voyager 2. The η ring and the γ ring are drawn here as a single line. (Adapted from diagrams by James Elliot)

distance from the sun. Apparently, it has lost much of its interior heat. Yet it must have some internal heat to cause convection in the fluid mantle and thus drive the dynamo effect. The decay of natural radioactive elements would generate heat, but some astronomers have suggested that the slow settling of heavier elements through the fluid mantle could release sufficient energy to warm the interior.

The interior of a world like Uranus seems at first glance to lie forever beyond our ken, but a few critical observations, combined with carefully constructed models, give us a glimpse of a fluid world. Unlike the interior, the rings of Uranus are exposed to view. The only problem is detecting the faint sunlight they reflect.

The Rings of Uranus

The rings of Uranus are made up of dark, unreflective particles and were not seen from Earth until 1977, when Uranus crossed in front of a star. During this **occultation,** astronomers saw the star flicker before and again after the planet occulted the star, and the pattern of the flickers implied five rings (Figure 25-7). These rings have since been imaged in detail by the Voyager 2 spacecraft, and a total of ten are now known. The rings lie in the plane of the planet's equator and have widths ranging from 10–100 km (Figure 25-8a).

The ring particles reflect only about 4 percent of the light that falls on them, making them about as dark as lumps of coal. The dark color may arise because the ring particles are composed of a methane-rich mixture of ices trapped in rings within the radiation belts. Radiation could decompose some of the methane to leave behind carbon, which would make the ring particles very unreflective. We will see that this same process may darken the surfaces of the Uranian moons.

The narrowness of the rings tells us that small moons must be present to confine the ring particles. Collisions between ring particles would cause rings to spread outward to become a uniform disk of particles too tenuous to detect. If rings have sharp edges or are very narrow, then shepherd moons must be herding wandering particles back into the rings. Voyager 2 detected two shepherd moons no more than 25 km in diameter inside and outside the ε ring (Figure 25-8b). The eccentricity of the ε ring and its variation in width around the planet are caused by the eccentricity of the two shepherd moons. Other moons are believed to shepherd the other rings, but these moons are too small to detect. Because these moons orbit inside the Roche limit of Uranus, they must

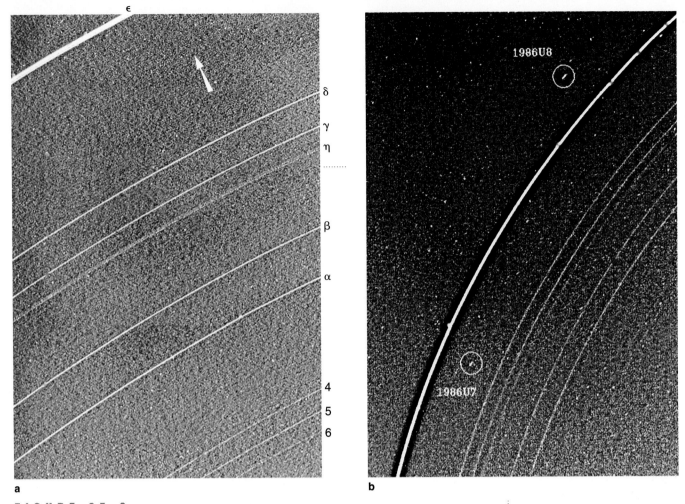

FIGURE 25-8

(a) The narrow rings of Uranus are identified by Greek letter or by number. Voyager images reveal another faint ring (arrow) halfway from the ε ring to the next innermost (δ) ring. Tip this page and view the faint ring along its length. Try to locate the η ring in the occultation data (Figure 25-7a). (b) Two shepherd satellites confine the ε ring into a narrow band. Because the satellites have slightly elliptical orbits, the ring is elliptical and slightly wider where it is farther from the planet. (NASA)

be structurally strong chunks of rock in order to withstand the tidal forces that would otherwise pull them apart.

The rings of Uranus illustrate the importance of satellites in confining planetary rings. The narrow rings are anchored to the small shepherd satellites, and the natural tendency of the rings to spread pushes the satellites outward. Of course, the satellites are much more massive than the tenuous rings and respond slowly. The entire ring system of Uranus, for example, would make an icy satellite only 15 km in radius. Nevertheless, over long periods the small satellites would be pushed outward by the spreading of the rings were it not for the presence of the larger satellites. Many of the small satellites are believed to be locked into resonances with larger satellites, and this anchors the small moons. The spreading pressure of the rings is transferred through resonances to the larger and more massive satellites, and thus the rings can be confined for longer periods. Without satellites, no planet could keep a ring system for long.

It may seem fortuitous that small moons happen to take up orbits at the edges of the rings just where they are needed, but this puts the cosmic cart before the horse. Ring particles tend to become trapped in stable locations within the system of small satellites. Thus, rings exist between closely orbiting satellites where the orbits of the ring particles are most stable.

Clearly, the rings of Uranus are related to its satellite system. Not only are the rings confined by satellites, but the origin of the rings may be linked to the origin of the moons.

The Moons of Uranus

The five largest moons of Uranus, from the outermost inward, are Oberon, Titania, Umbriel, Ariel, and Miranda. Umbriel and Ariel are names from Alexander Pope's *The Rape of the Lock*, and the rest are from Shakespeare's *A Midsummer Night's Dream* and *The Tempest* (an Ariel also appears in *The Tempest*). Spectra show that they contain frozen water, and their surfaces are very dark, so astronomers assumed they were made of dirty ices. Little more was known of the moons before Voyager 2 flew through the system.

Before the Voyager 2 cameras could begin photographing the surfaces of the five known moons, Voyager discovered ten additional satellites orbiting in the planet's equatorial plane inside the orbit of Miranda (Figure 25-9). The first, identified as 1986U1 but now known as Puck (after the character from *A Midsummer Night's Dream*), is about 170 km in diameter and spherical. The remaining nine new satellites, including the ε ring shepherd satellites, are smaller, 40–80 km in diameter. Like Puck, they are as dark as coal. Apparently they, like the ring particles, have been darkened by the radiation in the radiation belts.

The five large satellites are all tidally locked to Uranus, and thus their south poles were pointed toward the sun in 1986; it was these hemispheres that were photographed by Voyager 2. Their densities suggest that they contain relatively large rock cores surrounded by icy mantles (Figure 25-10).

Oberon, the outermost of the large moons, has a dark, cratered surface, but it was once an active moon (Figure 25-11a). A large fault crosses the sunlit hemisphere, and dark material, perhaps dirty water "lava," appears to have flooded the floors of some craters.

Titania is the largest of the five moons and has a heavily cratered surface, but it lacks the largest craters (Figure 25-11b). This suggests that soon after it formed, after most of the large debris had been swept up, Titania underwent an active phase in which internal melting flooded the surface with water and erased the early craters. Since then, more craters have formed, but few are as large as those that were erased. Another clue to Titania's past activity is the network of faults that crosses its surface.

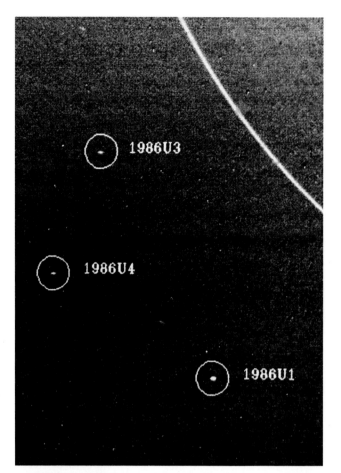

FIGURE 25-9

Three of the ten newly discovered, small moons of Uranus orbit just outside the ε ring. 1986U1, known officially as Puck, is the largest, with a diameter of 170 km, and is roughly spherical. The other new satellites are smaller. The images here are elongated slightly due to motion during the time exposure needed to record the black moons. (NASA)

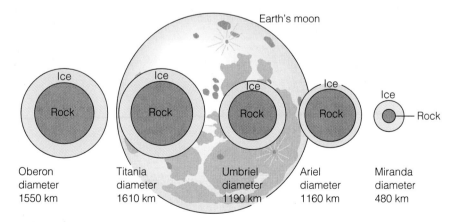

FIGURE 25-10

The densities of the five major satellites of Uranus suggest that they contain relatively large rocky cores with mantles of dirty ice. The size of Earth's moon is shown for comparison.

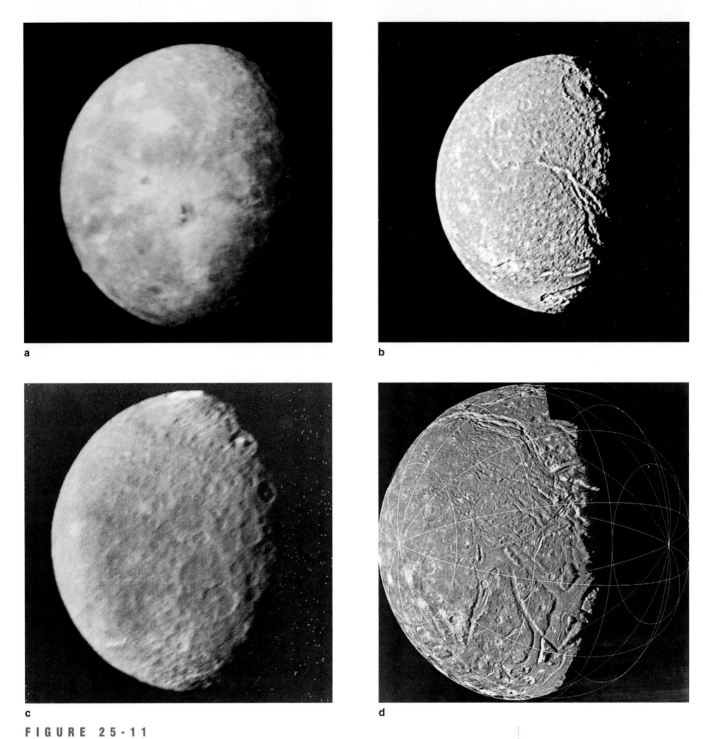

FIGURE 25-11

(a) Oberon appears to have an old, inactive, heavily cratered surface. (b) Titania has a cratered surface crossed by deep valleys up to 100 km across—signs of an active past. (c) Umbriel is a dark, cratered world with no sign of faults or valleys. (d) This mosaic of images of Ariel shows it crossed by faults and broad valleys that appear to have been smoothed of earlier craters. (a, b, and c, NASA; d, U.S. Geological Survey, Flagstaff, Arizona)

a

b

FIGURE 25-12

Miranda, the smallest of the five major moons of Uranus, is only 480 km in diameter, but its surface shows signs of activity. This photomosaic of Voyager 2 images (a) reveals that it is marked by great oval systems of grooves. In the map (b), the smallest features detected are about 1.5 km in diameter. The cliff marked by an arrow near the top of the map is 20 km high. (U.S. Geological Survey, Flagstaff, Arizona; map by P. Bridges)

Umbriel, the third moon inward, is a very dark, cratered world with no sign of faults or surface activity (Figure 25-11c). It is the darkest of the moons, with an albedo of only 0.16 compared with 0.25–0.45 for the other moons. Planetary astronomers suspect that Umbriel's crust is a dark mixture of rock and ice. A bright crater floor in one region suggests that some clean ice may lie at shallow depths in some regions.

Ariel has the brightest surface of the five major satellites and shows clear signs of geological activity. It is crossed by faults over 10 km deep, and some regions appear to have been smoothed by resurfacing (Figure 25-11d). Crater counts show that the smoothed regions are younger than the other regions. Apparently, past melting in the interior has caused geological activity on the moon's surface. Ariel may have been subject to tidal heating caused by an orbital resonance with Miranda and Umbriel.

Miranda, the smallest of the five satellites discovered from Earth, is the most unusual. In fact, its surface is so peculiar that one astronomer referred to it as a "moon designed by a committee." Miranda is marked by oval patterns of grooves, which at first were dubbed *circi maximi* (singular, *circus maximus*) after the ancient Roman chariot-racing tracks (Figure 25-12). Now known as ovoids, these features were originally thought to have been produced when Miranda was shattered by a major impact and the mutual gravity of the fragments pulled them back together to reassemble the little moon. More recent studies of the ovoids show that they are associated with faults, ice-lava flows, and rotated blocks of crust. These features suggest that a major impact was not involved. Rather, large convection currents in Miranda's icy mantle may have forced material upward to create the ovoids.

Certainly, Miranda has had a violent past. Near the equator, a huge cliff rises 20 km. An astronaut who slipped from the top of the cliff would fall for 10 minutes before hitting the surface. The cliff and the ovoids are old, however. Miranda is no longer geologically active, and we can read the history of its past on its disturbed surface.

The peculiar geology of Miranda illustrates the problem astronomers face in understanding the geology of the Jovian satellites. Has it been dominated by interior or exterior factors? If the ovoids were caused by convection in Miranda's icy mantle, then heat rising from its interior must be the dominant factor, and we must suppose that its interior was heated by tidal stress as in Jupiter's moon Io. Miranda is too small to have kept its heat of formation or to have been severely heated by radioactive decay. Tidal heating could have heated Miranda if its orbit was disturbed by a resonance with other satellites. Thus, we can understand how internal factors could have driven slow convection currents rising through Miranda's icy mantle to create its ovoids.

But external factors are also a possibility. Comets sweep through the solar system, and they must occasionally strike the planets and moons. The Jovian planets would retain no permanent scars of such impacts, as Earth's astronomers saw dramatically in the summer of 1994 when a comet struck Jupiter. The moons in the outer solar system, however, could be battered or even disrupted by such impacts. Although astronomers now prefer internal heat to explain Miranda's geology, we must accept that external factors such as major impacts do occur. Thus, Miranda reminds us that modern astronomy recognizes the importance of catastrophic events in the history of the solar system. (See Window on Science 20-1.) We will see more suggestions of catastrophic events when we discuss the moons of Neptune.

A History of Uranus

The challenge of comparative astronomy is to tell the story of a world, and Uranus may present the biggest challenge of all the objects in our solar system. Not only is it so distant that it is difficult to study, but it is also peculiar in many ways.

Presumably, Uranus formed from the solar nebula, as did the other Jovian planets, but it must have grown more slowly, perhaps because the solar nebula was less dense beyond Saturn. It captured some hydrogen and helium from the solar nebula, but it has lost some of its hydrogen and is now a mostly liquid world.

The highly inclined axis of Uranus may have originated late in its formation when it was struck by a planetesimal as large as Earth. That impact may have disturbed the interior of the world and caused it to lose heat rapidly. It now contains little interior heat, although there must be enough heat flowing outward to generate the peculiar offset magnetic field in its slushy mantle.

Impacts may also have been very important in the history of the moons and rings of Uranus. The satellites show signs of very large impacts, and Miranda may even have been disrupted and then re-formed to produce the ovoids. Impacts by icy planetesimals and the nuclei of comets on the larger moons may have produced enough debris to become trapped between the smaller moons and produce the narrow rings. It seems unlikely that the rings could have lasted since the formation of the planet. In any case, when the final history of Uranus is written, it will surely include large impacts.

CRITICAL INQUIRY

How do we know what the interior of Uranus is like?

Obviously, we can't see inside the planet, so we are limited to a few basic observations that we must connect with models in a chain of inference to describe the interior. First, we can measure the size of Uranus in photographs, and we can find its mass by observing the orbital radius and period of its moons. Its mass divided by its volume gives us its density, which implies that, for a Jovian planet, it must contain a higher proportion of dense material such as ice and rock. To produce a planet of this size and density, models predict a core of heavy elements and a mantle of icy slush. Two other observations we can make are the oblateness of the planet and the rotation period, derived from the radio noise produced in the magnetic field. Models cannot make the planet as flattened as is observed if the heavy elements are highly concentrated toward the core, so we must assume that the mantle contains denser material, such as rock, mixed into the water and ice. We can also observe the highly inclined and offset magnetic field, which is believed to be generated in the mantle, and that tells us that the mantle must be a good electrical conductor, such as water with dissolved ammonia.

Going from observable properties to unobservable properties is the heart of astronomical research. For example, what observational properties of the rings of Uranus argue that small moons must orbit among the rings? ■

When Voyager 2 left Uranus, Earth's astronomers were left grappling with new planetary puzzles. Over the following years, some of the puzzles were explained, but in August 1989 Voyager 2 flew past Neptune, and the game began all over again.

25-2 NEPTUNE

Uranus and Neptune are often discussed together. They are about the same size and density and are probably very similar planets in some ways. They do differ, however, in certain respects. Neptune has a peculiar ring and a peculiar satellite system. Even the discovery of Neptune was different from the discovery of Uranus.

The Discovery of Neptune

In 1843, the young English astronomer John Couch Adams (1819–1892) completed his degree in astronomy and immediately began the analysis of one of the great problems of 19th-century astronomy. William Herschel had discovered Uranus in 1781, but earlier astronomers had seen the planet as early as 1690 and had plotted it on charts as a star. When 19th-century astronomers tried to combine this data, they found that no planet obeying Newton's laws of motion and gravity could follow such an orbit.

Some astronomers had suggested that the gravitational attraction of an undiscovered planet was causing the discrepancies. Adams began with the variation in the positions of Uranus, never more than 2 minutes of arc, and by October 1845 had calculated the orbit and position of the undiscovered planet. He sent his prediction to the Astronomer Royal, Sir George Airy, who passed it on to an observer. Neither Airy nor his observer took the prediction very seriously, and they were very dilatory in making the search.

Meanwhile, the French astronomer Urbain Jean Leverrier (1811–1877) made the same calculations and sent his prediction to Johann Galle at the Berlin Observatory. Galle received Leverrier's prediction on the afternoon of September 23, 1846, and, after searching for 30 minutes that evening, found Neptune. It was only 2° from Adams's predicted position.

When the English announced Adams's work, the French suspected that he had plagiarized the calculations, and the controversy was bitter. In fact, both men had performed the calculations correctly. But they could have been beaten to the discovery had Galileo paid less attention to Jupiter and more attention to what he saw in the background.

Recent studies of Galileo's notebooks show that he saw Neptune on December 24, 1612, and again on January 28, 1613, but he plotted it as a star in the background of drawings of Jupiter. It is interesting to speculate about the response of the Inquisition had Galileo proposed that a planet existed beyond Saturn. Unfortunately for history (perhaps fortunately for Galileo), he did not recognize Neptune as a planet, and its discovery had to wait another 234 years.

The discovery of Neptune was a triumph for Newtonian physics: the three laws of motion and the law of gravity had proved sufficient to predict the position and orbit of an unseen planet. Thus, the discovery of Neptune was fundamentally different from the discovery of Uranus—the existence of Neptune was predicted from basic laws. Discovery by prediction is important in science because confirmation of the prediction serves as confirmation of the basic laws.

The Atmosphere and Interior of Neptune

Little was known about Neptune before the Voyager 2 spacecraft swept past it in August 1989. Seen from Earth, Neptune is a tiny, blue-green dot never more than 2.3 seconds of arc in diameter. Astronomers knew it was almost four times the diameter of Earth and about 4 percent smaller in diameter than Uranus (Data File Ten). Spectra revealed that its blue-green color was caused by methane in its hydrogen-rich atmosphere. Methane absorbs red light, making Neptune look blue-green. Neptune's density showed that it was a Jovian planet rich in hydrogen, but almost no detail is visible from Earth, so even its period of rotation was poorly known.

Voyager 2 passed only 4905 km above Neptune's cloud tops, closer than any spacecraft had ever come to one of the Jovian planets. The Voyager photos revealed that Neptune is marked by dramatic belt–zone circulation parallel to the planet's equator. Voyager 2 saw at

NEPTUNE

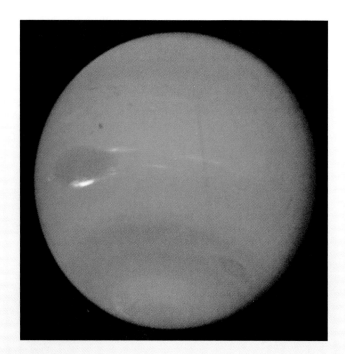

A 1989 Voyager image of Neptune shows the blue color of its methane-rich atmosphere and the Great Dark Spot at the left. (NASA)

Average distance from the sun	30.0611 AU (44.971×10^8 km)
Eccentricity of orbit	0.0100
Maximum distance from the sun	30.4 AU (45.4×10^8 km)
Minimum distance from the sun	29.8 AU (44.52×10^8 km)
Inclination of orbit to ecliptic	1°46′27″
Average orbital velocity	5.43 km/sec
Orbital period	164.793 y (60,189 days)
Period of rotation	16^h3^m
Inclination of equator to orbit	28°48′
Equatorial diameter	49,500 km (3.93 $D_\oplus$)
Mass	1.030×10^{26} kg (17.23 $M_\oplus$)
Average density	1.66 g/cm³
Gravity	1.19 earth gravities
Escape velocity	25 km/sec (2.2 $V_\oplus$)
Temperature at cloud tops	−216°C (−357°F)
Albedo	0.35
Oblateness	0.027

FIGURE 25-13

Two Hubble Space Telescope images of Neptune show weather on opposite sides of the planet in 1996. High-altitude clouds appear white, and the highest clouds appear red/yellow in these images. The Great Dark Spot seen by Voyager in 1989 is not present, and comparison with other Hubble Space Telescope images shows that cloud patterns come and go on Neptune. (Lawrence Sromovsky, University of Wisconsin-Madison; and NASA)

least four cyclonic disturbances. The largest, dubbed the Great Dark Spot, looked similar to the Great Red Spot on Jupiter (Figure 25-13a). The Great Dark Spot was located in the southern hemisphere and rotated counterclockwise, with a period of about 16 days. It appeared to be caused by gas rising from the planet's interior. When the Hubble Space Telescope began imaging Neptune in 1994, the Great Dark Spot was gone, and new cloud features were seen appearing and disappearing in Neptune's atmosphere (Figure 25-13). Thus, the cyclonic disturbances on Neptune are not nearly as long-lived as Jupiter's Great Red Spot.

The Voyager 2 images reveal other cloud features standing out against the deep blue of the methane-rich atmosphere. The white clouds are made up of crystals of frozen methane and range up to 50 km above the deeper layers. At such high altitudes, the clouds catch the sunlight and look bright (Figure 25-14a). Besides the Great Dark Spot, other, smaller spots were visible in the Voyager 2 photographs, and they, too, were presumably related to convection in the atmosphere (Figure 25-14b).

The Hubble Space Telescope has been able to monitor weather on Neptune, although the telescope's resolution is not as good as that of Voyager 2. Hubble photos show changing weather patterns on Neptune and a hint that this activity varies with solar activity. Further studies of clouds and spots on Neptune will help us understand this complex atmosphere and the way it is influenced by solar activity.

As on the other Jovian worlds, high winds circle Neptune parallel to its equator, but the winds blow at very high velocities and tend to blow backward—against the rotation of the planet. Why Neptune should have such high winds is not understood, and it is part of the larger problem of the belt–zone circulation. Now that we have seen at least traces of belts and zones on all four of the Jovian planets (if we may take the faint clouds on Uranus as belt–zone features), we can ask what drives this circulation. Since the belts and zones remain parallel to the planet's equator even when the planet rotates at a high inclination, as in the case of Uranus, it seems reasonable to believe that the atmospheric circulation is dominated by the rotation of the planet and not by solar heating.

The same observations that helped define the interior of Uranus can be applied to Neptune. The interior is believed to contain a rocky core, smaller than that in Jupiter, surrounded by a deep mantle of slushy ice mixed with heavier material such as rock. Neptune's magnetic field is a bit less than half as strong as Earth's and is tipped 47° from the axis of rotation. It is offset 55 percent of the way to the surface (Figure 25-5). As in the case of Uranus, the field is probably generated by the dynamo effect acting in the conducting fluid mantle.

a

b

FIGURE 25-14

(a) Cloud features seen on Neptune are caused by high, white clouds about 50 km above the darker layer of methane clouds. High-speed winds stretch the clouds into streaks parallel to the equator. (b) Spots on Neptune appear to be caused by circulations in the atmosphere. The spot shown at bottom center is D2, for Dark Spot 2. It was photographed by the Voyager 2 spacecraft in 1989. However, observations by the Hubble Space Telescope show that the spots seen by Voyager 2 were not permanent features like the Great Red Spot on Jupiter but rather are more like weather disturbances that come and go. (NASA)

a

b

FIGURE 25-15

(a) The brilliant disk of Neptune is hidden behind the central black bar in this Voyager 2 photo of the rings. Although the rings look uniform here, they actually contain short arcs where the ring particles are denser. (NASA) (b) This computer-enhanced image shows the arcs in Neptune's rings. A mathematical model reveals that the arcs are created by the gravitational influence of Neptune's moon Galatea. Arrows point to fainter arcs whose existence was predicted by the model. (NASA image courtesy of Carolyn C. Porco)

Neptune is a tantalizing world just big enough to be imaged by the Hubble Space Telescope but far enough away to make it difficult to study. The data from Voyager 2 revealed one detail invisible from Earth—rings.

The Rings of Neptune

No rings orbiting Neptune are directly visible from Earth, but every few years as it moves along its orbit Neptune crosses in front of (occulted) distant stars. Sometimes astronomers noticed the light of the star dim for a few seconds before or after the main occultation. This suggested that the planet had rings, but the stars did not always dim as expected. Astronomers proposed that Neptune had ring arcs rather than complete rings. The truth was revealed when Voyager 2 flew past in August 1989.

The nature of the rings and ring particles can be deduced from the Voyager 2 images. All of the rings lie within Neptune's Roche limit. Two narrow rings have widths of only a few dozen kilometers, with a fainter, broader ring in between. Another broad, faint ring with a width of 2500 km lies closer to the planet (Figure 25-15a). The rings are bright in forward-scattered light, so we can conclude that they contain significant amounts of dust. Because the rings are not very reflective, we can guess that the material is rocky rather than icy or that it

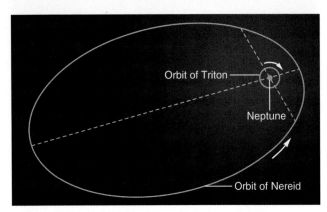

FIGURE 25-16

Nereid has a large, elliptical orbit around Neptune, but Triton follows a small, circular, retrograde orbit. At a distance of 30 AU from Earth, Triton is never more than 16 seconds of arc from the center of Neptune.

is darkened by exposure to radiation in Neptune's magnetosphere. The narrow widths of the two brightest rings suggest they are confined by shepherd satellites, and observations have revealed that the small moons Despina and Galatea shepherd the innermost narrow ring. There are probably other, even smaller moons helping to confine the other rings.

Voyager 2's photographs revealed that Neptune's rings are continuous loops but that the outermost ring includes short arcs containing more particles than average (Figure 25-15b). These are the arcs observed from Earth. Theoretical calculations showed that the gravitational influence of the small moon Galatea could cause the bunching of particles to create arcs in the observed positions. In fact, the calculations predicted even smaller arcs in specific places along the ring, which were then located in photos. Thus, the confirmation of the theory's prediction supports the theory.

Compared to the brilliant rings of Saturn, the dark, tenuous rings of Neptune and Uranus are hardly impressive. Yet their origin may depend on a spectacular process. Since Neptune's rings, like those of Uranus, are too tenuous to have survived since the formation of the planet, we must suppose that some process replenishes them with fresh particles. Comets are quite common, and collisions of the nuclei of comets with small moons may occasionally supply fresh particles to the rings. Certainly, the rings and moons of Neptune are related.

The Moons of Neptune

Neptune has only two moons that are visible from Earth, but Voyager 2 discovered six more small moons. The two largest moons, Triton and Nereid, have been a puzzle for years because of their peculiar orbits.

Triton has a nearly circular orbit, but it travels backward—clockwise as seen from the north. This makes Triton the only large satellite in the solar system with a retrograde (backward) orbit. The other satellite visible from Earth, Nereid, goes in the right direction, but its orbit is highly elliptical and very large (Figure 25-16). Nereid takes 359.4 days to orbit Neptune once. Many astronomers speculated that the orbital motions of Triton and Nereid were disturbed long ago by an encounter with a massive planetesimal.

Voyager 2 found six more moons orbiting Neptune, but they all lie close to the planet among the rings. No new moons were found beyond the orbit of Triton. Some astronomers have suggested that Triton, in its retrograde orbit, would have consumed any moons near it.

The Voyager 2 photographs show that Triton is a highly complex moon. Although it is only 1360 km in radius (78 percent the radius of Earth's moon), it is so cold (34.5 K, or −398°F) that it can hold an atmosphere of nitrogen with some methane. Its air is 10^5 times less dense than Earth's atmosphere. Although a few wisps of haze are visible in the photographs, the atmosphere is transparent, and the surface is easily visible (Figure 25-17).

The surface of Triton is evidently composed of ices. Early analysis suggested that the ices were mostly water ice with methane and ammonia ice mixed in. However, recent studies reveal that the surface ice is dominated by frozen nitrogen with some methane, carbon monoxide, and carbon dioxide. This is consistent with the nitrogen-rich atmosphere. Some regions of the surface look dark and may be older terrain that has grown dark as sunlight has converted methane into dark organic compounds. The south pole of Triton had been turned toward the sun for 30 years when Voyager 2 flew past, and deposits of nitrogen frost in a large polar cap appeared to be vaporizing there and refreezing in the darkness of the north pole (Figure 25-17a). Thus, the cycle of nitrogen on Triton resembles in some ways the cycle of carbon dioxide on Mars.

Many features on Triton suggest that it has had an active past. Few craters mark the surface, suggesting that whatever craters formed in the early history of the moon have been erased by later activity. Also, long linear features appear to be fractures in the icy crust where ices have welled up and frozen. Some roughly round basins appear to have been flooded time after time by liquids from the interior (Figure 25-17b). Most exciting are the dark smudges visible in the bright nitrogen ices near the south pole. These may be caused by nitrogen ice beneath the crust. Warmed slightly by the sun, the nitrogen ice can change from one form of nitrogen ice to another and release heat that could vaporize some of the nitrogen. This nitrogen vapor could vent through the crust to form nitrogen gas geysers. Sunlight could convert some of the methane carried along with the nitrogen into dark organic material, which would fall to the surface and account for the dark smudges on the ice.

Active worlds must have a source of energy, so we may wonder at the activity we see on icy Triton. While radioactive decay in the rocky interior and tidal heating are possible sources of internal energy, another possibility exists for Triton. Its thin atmosphere allows sunlight to fall on the crust, which is composed of ices that are partially transparent to light. As the light penetrates into the ice and is gradually absorbed, it is converted to heat. The ices, however, are not very transparent to infrared radiation, and the heat is trapped in the crust. Thus, Triton may be heated, in part, by an icy form of the greenhouse effect.

The History of Neptune

Can we tell the story of Neptune? In some ways it seems to be a simple world, but its magnetic field is peculiar, and its moons and rings demand careful attention.

Neptune is much like Uranus, and we can assume that it formed from the solar nebula much as did Uranus. It has lost some of its original light elements and is now a mostly liquid world with a dense core. Its hazy atmosphere, marked by changing cloud patterns, hides a mantle of partially frozen water where astronomers suspect the dynamo effect generates its off-center magnetic field.

Neptune's satellite system suggests a peculiar history. Triton, the largest moon, revolves around Neptune in a retrograde orbit, while Nereid's long-period orbit is highly elliptical. These orbital oddities suggest that the satellite system may have been disturbed by an encounter with some other object, such as a large planetesimal. We have seen evidence in other satellite systems of impacts with large objects, so such an encounter is not unreasonable.

Six smaller moons orbit Neptune, but they lie near its ring system. Because the rings are bright in forward scattering, we can conclude that they are rich in dust, and the shepherding of small satellites could confine their width and produce the observed arcs. Such delicate rings could not have survived since the origin of the planets, so we must suppose that the rings are occasionally supplied with fresh dust particles generated by collisions between comets and Neptune's moons. Once again, the evidence we see of major impacts in the solar system's history assures us that such impacts do occur.

CRITICAL INQUIRY

Why is Neptune blue but its clouds white?

When we look *at* something, we really turn our eyes toward it and receive light *from* the object. The light we receive from Neptune is sunlight that is reflected from various layers of Neptune and journeys to our eyes. Since sunlight contains a distribution of photons of all

a

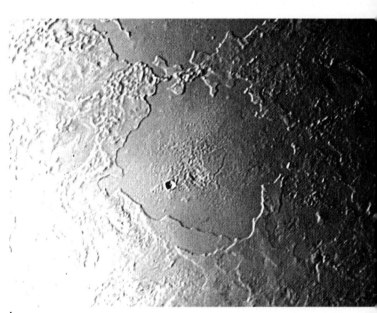

b

FIGURE 25-17

(a) Triton's south pole (bottom) had been in sunlight for 30 years when Voyager 2 flew past in 1989. The frozen nitrogen in the polar cap appears to be vaporizing, perhaps to refreeze in the darkness at the north pole. The dark smudges are produced when liquid nitrogen in the crust vaporizes and drives nitrogen geysers. Methane carried along in the nitrogen is turned to black compounds by sunlight and makes the black smudges. (b) Roughly round basins on Triton may be old impact basins flooded repeatedly by liquids from the interior. Notice the small number of craters on Triton, a clue that it is a partially active world. (NASA)

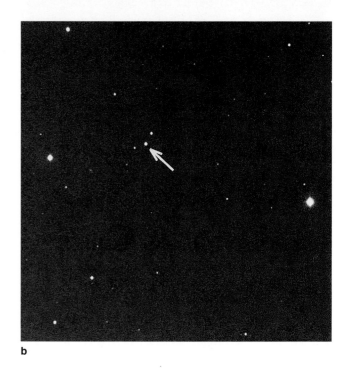

a   b

FIGURE 25-18

Pluto is small and far away, so its image is indistinguishable from that of a star on most photographs. Clyde Tombaugh discovered the planet in 1930 by looking for an object that moved relative to the stars on a pair of photographs taken a few days apart. (Lick Observatory photographs)

visible wavelengths, it looks white to us, but sunlight entering Neptune's atmosphere must pass through hydrogen gas that contains a small amount of methane. While hydrogen is transparent, methane is a good absorber of longer wavelengths, so red photons are more likely to be absorbed than blue photons. Once the light is scattered from deeper layers, it must run this methane gauntlet again to emerge from the atmosphere, and again red photons are more likely to be absorbed. The light that finally emerges from Neptune and eventually reaches our telescopes is poor in longer wavelengths and thus looks blue.

The methane-ice-crystal clouds lie at high altitudes, so sunlight does not have to penetrate very far into Neptune's atmosphere to reflect off the clouds, and thus it loses many fewer of its red photons. The clouds look white.

This discussion shows how a careful, step-by-step analysis of a natural process can help us better understand how nature works. For example, explain where the energy comes from to keep Triton active. ∎

Neptune was the last planet visited by Voyager 2, which is now journeying out of our solar system. It will not pass near any other worlds unless, millions of years from now, it enters some distant solar system. Nevertheless, we have one more planet to analyze.

25-3 PLUTO

Pluto, the farthest outpost of our solar system, is a small, icy world. Discovered through a combination of ingenuity, perseverance, and luck, it is now known to have a satellite.

The Discovery of Pluto

Percival Lowell (1855–1916) was fascinated with the idea that an intelligent race built the canals he thought he could see on Mars (Chapter 23). Lowell founded Lowell Observatory primarily for the study of Mars. Later, to improve the reputation of his observatory, he began to search for a planet beyond Neptune.

Lowell used the same method that Adams and Leverrier had used to predict the position of Neptune. Working from the observed irregularities in the motion of Neptune, Lowell predicted the location of an undiscovered planet of about 7 Earth masses, which, he concluded, would look like a 13th-magnitude stellar image in eastern Taurus. Lowell searched for the planet photographically until his death in 1916.

In the late 1920s, a 22-year-old amateur astronomer, Clyde Tombaugh, began using a homemade 9-in. telescope to sketch Jupiter and Mars from his family's wheat farm in western Kansas. He sent his drawings to Lowell Observatory, and the observatory director hired him

Scientific Discoveries

In 1928, Alexander Fleming noticed that bacteria in a culture dish were avoiding a spot of mold. He went on to discover penicillin. In 1895, Conrad Roentgen noticed a fluorescent screen glowing in his laboratory when he experimented with other equipment. He discovered X rays. In 1896, Henri Becquerel happened to place a mineral containing uranium on a photographic plate safely wrapped in black paper. The plate was fogged, and Becquerel discovered natural radioactivity. Like many discoveries in science, these seem to be accidental, but as we have seen in the case of the discovery of Uranus, Neptune, and Pluto, "accidental" doesn't quite describe what happened.

The most important discoveries in science are those that totally change the way we think about nature, and it is very unlikely that anyone would predict such discoveries. For the most part, scientists work within a paradigm (Window on Science 4-1), a set of models and expectations about nature, and it is very difficult to imagine natural events that lie beyond that paradigm. Ptolemy, for example, could not have imagined a universe filled with galaxies because they lie totally beyond his geocentric paradigm. That means that the most important discoveries in science are almost bound to arise unexpectedly.

An unexpected discovery, however, is not the same as an accidental discovery. Fleming discovered penicillin in his culture dish not because he was the first to see it, but because he had studied bacterial growth for many years, so when he saw what many others must have seen, he recognized it as important. Roentgen recognized the importance of the glowing screen in his lab, and Becquerel didn't simply discard the fogged photographic plate. Long years of experience prepared them to recognize the significance of what they saw.

A study of discoveries made with new telescopes from Galileo to the present revealed that each time astronomers built a new telescope that significantly surpassed existing telescopes, they discovered new and wonderful things, such as the moons of Jupiter, binary stars, spiral galaxies, and quasars. But nearly all the discoveries made with these telescopes were unpredicted.

While it is possible to predict the presence of Neptune from the orbital motion of Uranus, this is not typical of scientific discoveries. Most important scientific discoveries lie outside the current paradigm and so must be made by scientists who are searching for something else or studying some other phenomenon. That means that scientists pursuing true basic research are rarely able to explain the potential value of their work, just as astronomers can't predict what they will discover with a new telescope. That's why science is so exciting. ∎

(without an interview) to resume Lowell's search for the undiscovered planet.

Tombaugh began photographing the sky along the ecliptic, obtaining two 14 × 17-in. glass plates exposed 2–3 days apart. To compare the plates, he mounted them in a blink comparator, which allowed him to examine star images, first on one plate and then on the other, while looking through a microscope eyepiece. Star images do not move, but a planet would reveal itself by its motion (Figure 25-18).

For almost a year, Tombaugh photographed the sky and blinked the plates, which contained as many as 400,000 stars per plate. Then on February 18, 1930, a quarter of the way through a pair of plates, he found a 15th-magnitude image that moved. He later remembered, "Oh, I thought I had better look at my watch, this could be an historic moment. It was within about two minutes of four PM [MST]." The discovery was announced on March 13, the 149th anniversary of the discovery of Uranus and the 75th anniversary of the birth of Percival Lowell. The planet was named Pluto after the god of the underworld and, in a way, after Lowell, because the first two letters in Pluto are the initials of Percival Lowell.

The discovery of Pluto seemed a triumph of discovery by prediction, but Tombaugh sensed something was wrong from the first moment he saw the image. It was moving in the right direction by the right amount, but it was 2.5 magnitudes too faint. Clearly, Pluto was not the 7-Earth-mass planet that Lowell had predicted. The faint image implied that Pluto was a small planet with a mass too low to seriously alter the motion of Neptune.

Later analysis has shown that the variations in the motion of Neptune, which Lowell used to predict the location of Pluto, were random errors of observation and could not have led to a trustworthy prediction. The discovery of the new planet only 6° from Lowell's predicted position was apparently an accident, which proves that if you search long enough and know what to expect, you are likely to find something (Window on Science 25-1).

Pluto and Charon

Pluto has the most elliptical orbit of any planet in the solar system. Its distance from the sun varies from 29.64 AU to 49.24 AU (Data File Eleven). Although it can come closer to the sun than Neptune, the two planets will never collide because Pluto's orbit is inclined 17° to the plane of the solar system. In fact, starting January 21, 1979, and lasting until March 14, 1999, Pluto will be closer to the sun than Neptune (see page 5, Figure 1-7). Even at its closest, however, Pluto is difficult to study. It has an angular diameter of only about 0.1 second of arc, so even the best photographs reveal only a dot whose diameter is not apparent.

a

b

FIGURE 25-19

(a) A high-quality ground-based photo shows Pluto and its moon, Charon, badly blurred by seeing. (NASA) (b) The Hubble Space Telescope image clearly separates the planet and its moon and allows more accurate measurements of the position of the moon. (R. Albrecht, ESA/ESO Space Telescope European Coordinating Facility, NASA)

Pluto's only known moon was discovered photographically in 1978. It is very faint, and even in photographs from the Hubble Space Telescope it is impossible to measure its diameter directly (Figure 25-19). The moon (named Charon after the mythological ferryman who transports souls across the river Styx into the underworld) is tidally locked with Pluto, and the two worlds forever face each other as they orbit with a period of 6.387 days and an average separation of 19,640 km. Just as with a binary star, astronomers can use the period and separation to calculate the masses of the objects. Pluto has a mass of about 6.5×10^{-9} solar masses, about 12 times the mass of Charon. Pluto contains only 0.002 Earth mass.

The best photographs can detect hazy features on Pluto, and we can only assume that it looks similar to the icy world Triton (Figure 25-20). In fact, the more we learn about Pluto, the more closely it seems to be related to Triton.

From February 1987 through September 1991, the orbit of Charon carried it repeatedly across the face of Pluto and then behind the planet (Figure 25-21). Astronomers used these events to find the diameters of the two worlds. Pluto is about 2300 km in diameter (68 percent of the diameter of Earth's moon), and Charon is about 1300 km in diameter, roughly half the size of Pluto. Once astronomers knew the diameters, they could calculate the densities of Pluto and Charon. The density of Pluto is about 2.0 g/cm³, and the density of Charon is about 1.8g/cm³. Pluto must contain about 20 percent ice and 80 percent rock, while Charon must contain significantly less rock than ice.

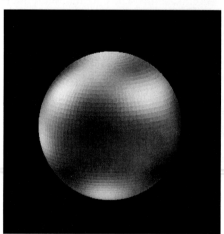

a

b

FIGURE 25-20

The resolving power of the Hubble Space Telescope is just able to reveal bright and dark regions in these two images of Pluto. Some of these regions may be older, darker terrain and younger, brighter impact basins. Many of the brighter regions, however, including the bright northern polar cap (top), are believed to be frosts deposited out of the nitrogen-methane atmosphere. (Alan Stern, Southwest Research Institute, Marc Buie, Lowell Observatory, NASA, and ESA)

PLUTO

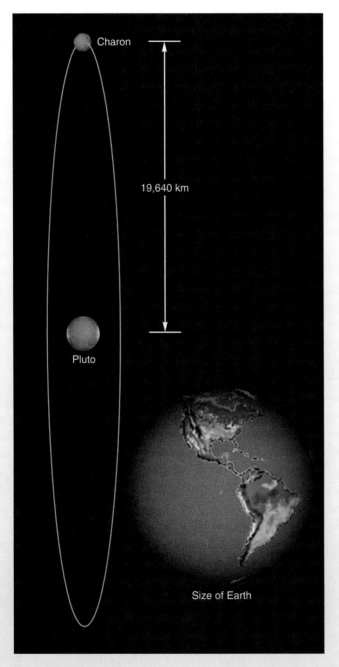

Charon orbits Pluto in a highly inclined, circular orbit only a few times bigger than Earth.

Average distance from the sun	39.44 AU (59.00 × 10⁸ km)
Eccentricity of orbit	0.2484
Maximum distance from the sun	49.24 AU (73.66 × 10⁸ km)
Minimum distance from the sun	29.64 AU (44.34 × 10⁸ km)
Inclination of orbit to ecliptic	17°9′3″
Average orbital velocity	4.73 km/sec
Orbital period	247.7 y (90,465 days)
Period of rotation	6ᵈ9ʰ21ᵐ
Inclination of equator to orbit	122°
Equatorial diameter	2300 km (0.19 $D_\oplus$)
Mass	1.2×10^{22} kg (0.002 $M_\oplus$)
Average density	2.0 g/cm³
Surface gravity	0.06 earth gravities
Escape velocity	1.2 km/sec (0.11 $V_\oplus$)
Surface temperature	−230°C (−382°F)
Albedo	0.5

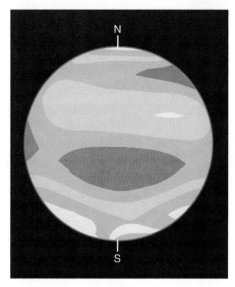

FIGURE 25-22

This map of Pluto was constructed by observing brightness variations during a large number of occultations by its moon, Charon. The bright areas are presumably deposits of brighter ice, and the brightness at the south pole (bottom) is probably a polar cap of methane ice. (Adapted from a map by Richard P. Binzel)

FIGURE 25-21

This computer simulation shows Pluto's moon, Charon, traveling from north to south across the disk of Pluto. Although the disks of Pluto and Charon were hardly resolved as seen from Earth, changes in the total brightness allow the calculation of the diameters and colors of the bodies and the general distribution of light and dark areas on Pluto. Such observations reveal that Pluto has bright polar caps. (Marc Buie)

The diameter of Pluto can also be found by observing occultations. As Pluto crosses in front of a star, the total time the star remains invisible times the velocity of Pluto along its orbit gives us the diameter of the planet. Some corrections are necessary because the center of the planet may not cross directly over the star. These occultations do not give precise diameters for a surprising reason: the star does not wink out suddenly as Pluto covers it but rather fades away over a period of seconds. Evidently, Pluto has an atmosphere!

Although it is small and has a low escape velocity, Pluto is very cold, about 37 K ($-393°$F) at the surface. At this temperature, the molecules move so slowly that Pluto can keep a thin atmosphere about as dense as Earth's atmosphere 80 km above the surface (see Figure 23-16). Charon, on the other hand, is too small to retain much gas. The occultation data show that the atmosphere of Pluto is thin and mostly transparent.

Spectra reveal that Pluto's atmosphere is mostly nitrogen with traces of methane gas. Infrared spectra reveal that the surface is covered by 97 percent nitrogen ice and 1–2 percent carbon monoxide ice. Thus, the atmosphere and surface share an abundance of nitrogen. When Charon repeatedly crossed the disk of Pluto, astronomers recorded the changes in brightness to create a map of the planet that reveals a bright polar cap at the south pole and a fainter cap at the north pole (Figure 25-22). These bright regions seem to be methane ice. Because Pluto reflects less blue than red, astronomers say it is reddish, although it would look very dark to our eyes. The red tint is probably caused by sunlight converting methane ice into red-colored compounds. Charon is darker than Pluto and is gray, not red. Spectra of Charon reveal that it seems to contain dirty water ice on its surface. Both Pluto and Charon must contain large amounts of frozen water, but the nitrogen ice on Pluto may have covered much of the water ice.

Pluto goes through dramatic seasons, and the methane on Pluto goes through a seasonal cycle. Pluto rotates on its side, like Uranus, and its distance from the sun varies dramatically. When it is warmest, as it was in 1989, solar heat vaporizes some methane and enriches the atmosphere. Then, as Pluto recedes from the sun, the methane freezes out to form fresh ice deposits in bright polar caps.

The Origin of Pluto

Although we are learning more about its surface and atmosphere, the origin of Pluto is still a mystery. The differences in density show that Pluto and Charon did not form together, and an older theory that Pluto is an escaped satellite of Neptune has been largely abandoned. Rather, both Pluto and Charon may be related to other icy worlds such as Triton, Neptune's largest moon. Recently, astronomers have begun discussing the possibility that Pluto, Charon, and Triton are merely the remains of a large number of small, icy bodies that once populated the outer solar system. Tentatively called Plutons or ice dwarfs, these bodies could explain a number of peculiarities.

The tilted axes of rotation of Uranus and Pluto and the disturbed orbits of Neptune's moons suggest collisions with other objects. Such collisions seem highly improbable given the present rarity of bodies in the outer solar system. If, however, a thousand or so ice dwarfs existed in the outer solar system during or soon after planet formation, then such collisions would be more likely. Most of those bodies would have been incorporated into planets or ejected into the far fringes of the solar system, but we might see the last remaining ice dwarfs as a few of the larger icy satellites, such as Triton, and as Pluto and its companion Charon.

In fact, new searches have found some of these icy worlds. For example, 1992 QB1, only about 200 km in diameter, follows a circular orbit about 41 AU from the sun, even more distant than the orbit of Pluto. Other small, icy asteroids, such as Chiron (not the same as Charon), follow orbits between the orbits of Saturn and Uranus. Hubble Space Telescope images confirm that there are thousands of these objects in the outermost reaches of our solar system. Astronomers are excited about these objects, not only because they suggest an origin for Pluto and Charon, but also because their icy composition and large size suggest that the outer solar system was once crowded with vast numbers of such objects.

If this is true, then Pluto and Charon are not sisters but more like distant relatives that joined as planet and moon some time after they formed. In that case, Charon's inclined orbit and Pluto's inclined axis of rotation might be the result of a collision in which Pluto captured Charon. This hypothesis is speculative, but it is being tested by the search for more small, icy worlds beyond Neptune's orbit.

CRITICAL INQUIRY

What evidence do we have that cataclysmic impacts have occurred in our solar system?

We would expect objects that formed together from the solar nebula to rotate and revolve in the plane of the solar system, so the high inclination of Pluto and the orbit of Charon suggest a past collision or, at least, a close gravitational interaction with a passing body. And the Pluto–Charon system is not the only evidence for major impacts in the past. The peculiar orbits of Neptune's moons Triton and Nereid, the fractured and cratered state of the Jovian satellites, and the peculiar rotation of Uranus all hint that impacts and encounters with large planetesimals or the nuclei of comets have been important in the history of these worlds. Furthermore, the existence of planetary rings suggests that impacts have scattered small particles and replenished the ring systems. Even in the inner solar system, the backward rotation of Venus, the density of Mercury, and the formation of Earth's moon are seen as possible consequences of major impacts when the solar system was young.

Certainly, we do not imagine that planets bounce around like billiard balls, but we must recognize that the planets are not totally isolated now and may have been victims (or beneficiaries) of large impacts as they formed long ago. For example, how do the size and composition of Pluto and Charon, compared with the moons of the outer Jovian worlds, suggest an unexpected population of icy worlds in the outer solar system? ∎

Like clues at the scene of a crime, the evidence we find in the outer solar system alerts us to the reality that all of the planets in our solar system have been and still are sitting ducks for impacts by the smaller objects that hurtle along their orbits around the sun. What are these objects? What can they tell us about the origin and evolution of our solar system? We explore these questions in the next chapter. ∎

∎ Summary

Although Uranus was discovered unexpectedly in 1781, we should probably not refer to its discovery as accidental. In 1846, the position of Neptune was predicted from the orbital motion of Uranus, and Pluto was discovered in 1930 based on predictions derived from the motion of Neptune.

Uranus and Neptune are commonly classified as Jovian planets, but their interiors are different from the interiors of Jupiter and Saturn. The density, rotation periods, and oblateness of Uranus and Neptune tell us that they contain small cores of heavy materials, such as rock and metal, and deep mantles of partly frozen ices mixed with denser material, such as rock. Their atmospheres are rich in hydrogen and helium, and absorption of red light by small amounts of methane causes both planets to have a blue color.

The axis of rotation of Uranus is tipped 97.9° from the perpendicular to its orbit, and its south pole was facing the sun when Voyager 2 flew past. The spacecraft discovered that belts and zones lie deep in the hazy atmosphere. Thus, the rotation of the planet, rather than solar energy, dominates atmospheric circulation.

The rings of Uranus are very narrow and dark. Two shepherd satellites were discovered orbiting on either side of the ε ring, and other shepherd satellites are believed to bound the other rings. These satellites and the larger satellites appear to anchor the rings and prevent them from spreading and dissipating. The dark color of the ring particles appears to come from the methane-rich ices darkened by the radiation belts around the planet. The surfaces of the moons are also very dark.

The rings of Uranus are not very bright in forward scattering and thus must be very poor in small particles. Some dust was detected, but it seems to be falling into the planet due to the resistance of low-density hydrogen gas leaking out of the upper atmosphere.

Voyager 2 discovered ten new satellites orbiting Uranus, all of them small and dark. The five largest moons show signs of impact cratering and some surface activity in the past. The innermost moon, Miranda, is marked by large ovals called ovoids. Astronomers believe that Miranda was heated by tides when it was young and that great convection currents stirred its partially frozen mantle and formed the ovoids.

When Voyager 2 flew past Neptune, it revealed a world similar in size to Uranus but with visible atmospheric circulation. The Great Dark Spot is an anticyclonic storm much like the Great Red Spot on Jupiter, and high, bright clouds are composed of methane-ice crystals. Like Uranus, Neptune has a magnetic field that is highly inclined and off center, suggesting that these magnetic fields originate not far below the cloud belts. Also like Uranus, Neptune has narrow rings of dark particles.

Triton, Neptune's largest moon, is an icy world with a thin atmosphere of nitrogen and methane. A polar cap of methane ice was visible as Voyager 2 flew past, but few impact craters mark the surface, hinting that the moon has been active. Nitrogen geysers detected in the south polar cap appear to be powered by heat from the sun.

Thanks to its elliptical orbit, Pluto will be closer to the sun than Neptune is until 1999. Pluto is a small world about two-thirds the size of Earth's moon, and it has a satellite of its own called Charon. As that satellite crossed in front of the planet, the total light of the system varied, allowing us to find the diameter of the planet and its moon and to map the surface of Pluto with very low resolution. The density of Pluto suggests that it is made up of about 80 percent rock mixed with ices.

■ New Term

occultation

■ Questions

1. Describe the location of the equinoxes and solstices in the Uranian sky. What are seasons like on Uranus?

2. Why is belt–zone circulation difficult to detect on Uranus?

3. Discuss the origin of the rings of Uranus and Neptune. Cite evidence to support hypotheses.

4. How do the magnetic fields of Uranus and Neptune suggest that the mantles inside those planets are fluid?

5. If Neptune had no satellites at all, would you expect it to have rings? Why or why not?

6. Why might the surface brightness of ring particles and small moons orbiting Uranus and Neptune depend on whether those planets have magnetic fields?

7. Both Uranus and Neptune have a blue-green tint when observed through a telescope. What does that tell us about their composition?

8. How can small worlds like Triton and Pluto have atmospheres when a larger world such as Ganymede has none?

9. Why do we suspect that Triton has had an active past? What source of energy could power such activity?

10. If you visited Pluto and found Charon a full moon directly overhead, where would Charon be in the sky when it was new? when it was first quarter?

11. What evidence do we have that Pluto and Charon are made of mixtures of rock and ice?

■ Discussion Questions

1. Why might it be unfair to describe William Herschel's discovery of Uranus as accidental? Why might it be unfair to describe the discovery of the rings of Uranus as accidental?

2. Suggest a single phenomenon that could explain the inclination of the rotation axis of Uranus, the orbits of Neptune's satellites, and the existence of Pluto's moon.

■ Problems

1. What is the maximum angular diameter of Uranus as seen from Earth? of Neptune? of Pluto? (HINT: Use the small-angle formula.)

2. One way to recognize a distant planet is by its motion along its orbit. If Uranus circles the sun in 84 years, how many seconds of arc will it move in 24 hours? (This does not include the motion of Earth. Assume a circular orbit for Uranus.)

3. What is the orbital velocity of Miranda around Uranus?

4. What is the escape velocity from the surface of Miranda? (HINTS: Radius = $2\frac{1}{2}$ km. Assume that the density is 2 g/cm^3. See Chapter 5.)

5. The magnetosphere of Uranus rotates with the planetary interior in 17.24 hours. What is the velocity of the outer

portion of the magnetic field just beyond the orbit of Oberon? (HINT: The circumference of a circle is $2\pi r$.)

6. If the ϵ ring is 50 km wide and the orbital velocity of Uranus is 6.81 km/sec, how long a blink should we expect to see when the ring crosses in front of a star? Is this consistent with Figure 25-7a?

7. What is the escape velocity from the surface of an icy moon with a diameter of 20 km? (HINTS: The density of ice is 1 g/cm^3. The volume of a sphere is $\frac{4}{3}\pi r^3$. See Chapter 5.)

8. What is the difference in the orbital velocities of the two shepherd satellites Cordelia and Ophelia? (HINTS: Orbital radii = 49,800 km and 53,800 km. See Chapter 5.)

9. Repeat Problem 2 for Pluto. That is, ignoring the motion of Earth, how far across the sky would Pluto move in 24 hours? (Assume a circular orbit for Pluto.)

10. Given the size of Triton's orbit (R = 355,000 km) and its orbital period (P = 5.877 days), calculate the mass of Neptune. (HINT: See Chapter 5.)

■ *Critical Inquiries for the Web*

1. Occasionally news stories surface that raise the question "Is Pluto a planet?" Determine what properties characterize a planet and examine Pluto-related Web sites. Take a position on this "yes or no" issue based on your research and justify your stand.

2. If you lived on the surface of Pluto and looked into the sky to observe Charon, what phases would you see? HINT: Be sure to consider your location on the planet when answering this question.

3. What factors caused Voyager 2 to see a bland atmosphere when it encountered Uranus in 1986? Given these circumstances, would images of the Uranian atmosphere from a space probe arriving at Uranus in 2006 be similar to those from 1986, or would you expect significant differences?

Go to the Wadsworth Astronomy Resource Center (www.wadsworth.com/astronomy) for critical thinking exercises, articles, and additional readings from InfoTrac College Edition, Wadsworth's online student library.

Chapter Twenty-Six

METEORITES, ASTEROIDS, AND COMETS

When they shall cry "PEACE, PEACE"

then cometh sudden destruction!

COMET'S CHAOS?—What Terrible

events will the Comet bring?

From a religious pamphlet predicting the end of the world
because of the appearance of Comet Kohoutek, 1973

GUIDEPOST

In Chapter 20, we began our study of planetary astronomy by asking how our solar system formed. In the five chapters that followed, we surveyed the planets, but we gained only limited insight in-

to the origin of the solar system. The planets are big, and they have evolved as heat has flowed out of their interiors. In this chapter, we have our best look at unevolved matter left over from the solar nebula. These small bodies are, in fact, the last remains of the nebula that gave birth to the planets. This chapter is unique in that it covers small bodies. In past chapters, we have used the principles of comparative planetology to study large objects—the planets. In this chapter, we see that some of these principles apply to smaller bodies, but we also see that we need some new tools in order to think about the tiniest worlds in the solar system. ■

FIGURE 26-1
Comet Hyakutake swept through the inner solar system in 1996 and was dramatic in the northern sky. Seen here from Kitt Peak National Observatory, the comet passed close to the north celestial pole (behind the observatory dome in this photo). Notice the Big Dipper below and to the left of the head of the comet. (Courtesy Tod Lauer)

In 1910, Comet Halley was spectacular, and on the night of May 19, Earth actually passed through the tail of the comet—and millions of people panicked. The spectrographic discovery of cyanide gas in the tails of comets led many to believe that life on Earth would end. Householders in Chicago stuffed rags around doors and windows to keep out the gas, and bottled oxygen was sold out. Con artists in Texas sold comet pills and inhalers to ward off the noxious fumes. An Oklahoma newspaper reported (in what was apparently a hoax) that a religious sect tried to sacrifice a virgin to the comet.

Throughout history, bright comets have been seen as portents of doom. Even in our own time, the appearance of bright comets has generated predictions of the end of the world. Comet Kohoutek in 1973 and Comet Halley in 1986 caused concern among the superstitious. More recently Comet Hyakutake in 1996 (Figure 26-1) and Comet Hale–Bopp produced spectacular sights in the sky and superstitious concern on earth. A bright comet moving slowly through the night sky is so out of the ordinary that we should not be surprised if it generates some instinctive alarm.

In fact, comets are graceful and beautiful visitors to our skies. Astronomers think of comets as messengers from the age of planet building. By studying comets, we learn about the conditions in the solar nebula from which planets formed. But comets are only the icy remains of the solar nebula; the asteroids are the rocky debris left over from planet building. Because we cannot easily visit comets and asteroids, we begin our discussion with the fragments of those bodies that fall into our atmosphere—the meteorites.

26-1 METEORITES

On the afternoon of November 30, 1954, Mrs. E. Hulitt Hodges of Sylacauga, Alabama, lay napping on her living room couch.* An explosion and a sharp pain jolted her awake, and she found that a meteorite had smashed through the ceiling and bruised her left leg. Mrs. Hodges is the only person known to have been injured by a meteorite.

Meteorite impacts on homes are not common, but statistical calculations show that a meteorite should damage a building somewhere in the world about once every 16 months. About two meteorites large enough to produce visible impacts strike the earth each day, but most meteors are small particles ranging from a few centime-

*Across the street from the Comet Drive-in Theater.

a b

FIGURE 26-2

(a) Meteors in a meteor shower enter the earth's atmosphere along parallel paths, but perspective makes them appear to diverge from a radiant point. (b) Similarly, parallel railroad tracks appear to diverge from a point on the horizon.

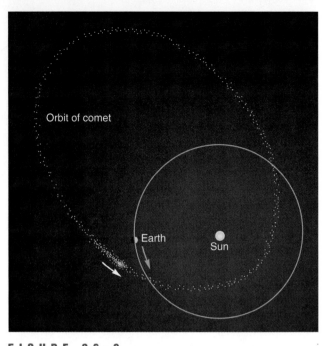

Orbit of comet

Earth   Sun

FIGURE 26-3

As a comet's ices evaporate, it releases rocky and metallic bits of material that spread along its orbit. If the earth passes through such material, it experiences a meteor shower.

ters down to microscopic dust. The earth gains about 40,000 tons of mass per year from meteorites of all sizes.

Recall from Chapter 20 that astronomers distinguish between the words *meteoroid*, *meteor*, and *meteorite*. A small body in space approaching the earth is a meteoroid, but once it begins to vaporize in the atmosphere it is called a meteor. If it survives to strike the ground, it is called a meteorite.

We have two questions to consider about meteorites: Where in the solar system do these objects originate? What can meteorites tell us about the origin of the solar system? To answer these questions, we must consider the orbits of meteoroids and the minerals found in meteorites.

Meteoroid Orbits

Meteoroids are much too small to be visible through even the largest telescopes, but once they enter the earth's atmosphere at speeds of 10–30 km/sec* and are

*For comparison, a rifle bullet travels at about 0.7 km/sec.

heated to incandescence by friction, we can determine their speed and direction of travel. Thus, we can backtrack the flaming trail of a meteor to discover the orbit of the meteoroid. These orbits hint at the origins of the meteorites.

One way to backtrack meteor trails is to observe meteor showers. On any clear night, you can see 3–15 meteors an hour, but on some nights there are meteor showers during which you can see as many as 60 meteors an hour.

a

The meteors in a shower are related to one another. To confirm this, try observing a meteor shower. Select a shower from the table printed inside the back covers of this book, and on the appropriate night stretch out in a lawn chair and watch the sky. When you see a meteor, plot its path on a star chart. You will discover that most of the meteors you see seem to come from a single area of the sky, the **radiant** of the shower (Figure 26-2a). In fact, meteor showers are named after the constellation from which they seem to radiate. Thus, the Perseids are a meteor shower in August that radiates from the constellation Perseus.

Meteors in a shower appear to come from a radiant because they are all traveling through space along parallel paths. When they encounter the earth and are vaporized at altitudes of 80 km or so, we see their fiery tracks in perspective; they appear to come from a single radiant point, just as railroad tracks seem to come from a single point on the horizon (Figure 26-2b).

Studies of meteor-shower radiants reveal that these meteors are produced by bits of matter orbiting the sun along the paths of comets. The vaporizing head of the comet releases bits of debris that become spread along the entire orbit (Figure 26-3). When the earth passes through this stream of material, we see a meteor shower.

In some cases the comet has wasted away and is no longer visible, but in other cases the comet is still prominent though somewhere else along its orbit. For example, in May the earth comes near the orbit of Comet Halley, and we see the Eta Aquarids shower. Around October 20 the earth passes near the other side of the orbit of Comet Halley, and we see the Orionids shower.

Even when there is no shower, we still see meteors, and these, too, may be related to comets. Such meteors are called **sporadic meteors** and are caused by random bits of matter entering the earth's atmosphere. Many of these specks of debris come from comets.

If we photographed a meteor from two locations on the earth a few miles apart, we could use triangulation to find the altitude, speed, and direction of the meteor as it moves through the atmosphere. Then we could work backward to find out what its orbit looked like before it entered the earth's atmosphere. These studies reveal that meteors belonging to showers and some sporadic meteors have orbits that are similar to the orbits of comets. In many cases, the comets can be identified, so the origin of these meteors is well known. But some sporadic meteors have orbits that lead back to the asteroid belt. Thus, we can conclude that meteors have a dual source: some come from comets, and some come from the asteroid belt.

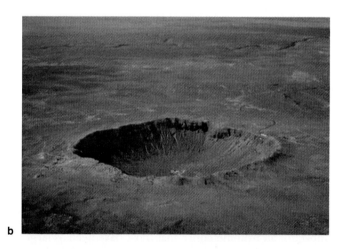

FIGURE 26-4

(a) The Barringer Meteorite Crater, nearly a mile in diameter, was formed about 50,000 years ago by the impact of an iron meteorite roughly 90 m in diameter. It hit with energy equivalent to that of a 3-megaton hydrogen bomb. Notice the raised and deformed rock strata all around the crater. For scale, locate the brick building on the far rim at right. (M. A. Seeds) (b) Like all larger impact features, the Barringer Meteorite Crater has a raised rim and scattered ejecta. (U. S. Geological Survey)

b

FIGURE 26-5

(a) W. A. Cassidy, leader of several expeditions to Antarctica, holds a numbering device near an achondritic meteorite exposed on the icy surface. (b) An iron meteorite shown as it was found. (NASA photographs, Ursula B. Marvin)

Meteorites

When a meteor is massive enough and mechanically strong enough to survive its passage through the earth's atmosphere, it finally reaches the earth's surface. A very large meteorite might hit with the energy of a very large nuclear weapon and dig a crater many kilometers in diameter. However, most are small, and, because their energy of motion is spent in the atmosphere, they fall to the earth with no more energy than a falling hailstone.

For an example of the impact of a large meteorite, we can study the Barringer Meteorite Crater near Flagstaff, Arizona. The crater is 1.2 km in diameter and 200 m deep. It was produced 50,000 years ago by the impact of an iron meteorite that must have been as large as a big building (80–100 m). The impact deformed the rock strata and scattered debris over a large area (Figure 26-4). (For a summary of the crater-forming process, see Figure 22-5.) Although the Barringer Meteorite Crater seems large and the hike around its rim is long, hot, and dry, the crater is actually small compared with some impact features on the earth. About 65 million years ago, an impact in the northern Yucatan of Central America produced a crater, now covered by sediment, that was between 180 and 300 km in diameter. (The 1994 impacts on Jupiter were even larger, but they formed no craters because Jupiter has no surface.) Thus, impact craters are a natural feature of planetary surfaces.

Large meteorites can produce craters, but most meteorites are small and don't create significant craters. Also, craters erode rapidly on the earth, so most mete-orites are found without associated craters. The best place to look for meteorites turns out to be certain areas of Antarctica—not because more meteorites fall there but because the meteorites are easy to recognize on the icy terrain. Most meteorites look like earth rocks, but on the ice of Antarctica there are few natural rocks. Also, the slow creep of the ice toward the sea concentrates the meteorites in certain areas where the ice runs into mountain ranges, slows down, and evaporates. Thus, teams of scientists can easily recover meteorites from the ice (Figure 26-5). Such unbiased collections of meteorites are important in helping define the frequency of different meteorite types.

Meteorites that are seen to fall are called **falls.** A fall is known to have occurred at a given time and place, and thus the meteorite is well-documented. A meteorite that is discovered but was not seen to fall is called a **find.** Such a meteorite could have fallen thousands of years ago.

Studies of meteorites reveal three major classes: irons, stones, and stony-irons. **Iron meteorites** make up 66 percent of finds (Table 26-1) but only 6 percent of falls. Why? Because an iron meteorite does not look like a rock. If you trip over one on a hike, you are more likely to carry it home and show it to your local museum. Also, some stone meteorites deteriorate rapidly when exposed to weather; irons survive longer. Thus, there is a **selection effect** that makes it more likely that iron meteorites will be found (Window on Science 26-1). That only 6 percent of falls are irons shows that iron meteorites are fairly rare.

Selection Effects in Science

Many different kinds of science depend on selecting objects to study. Scientists studying insects in the rain forest, for example, must choose which ones to catch. They can't catch every insect they see, so they might decide to catch and study any insect that is red. If they are not careful, a selection effect could bias their data and lead them to incorrect conclusions without their ever knowing it. The reason selection effects are dangerous in science is that they can have powerful influences without being evident. Only by carefully designing a research project can scientists avoid selection effects.

For example, suppose we decide to measure the speed of cars on a highway. There are too many cars to measure every one, so we reduce the work load and measure only the speed of red cars. It is quite possible that this selection criterion will mislead us because people who buy red cars may be more likely to be younger and drive faster. Should we measure

only brown cars? No, because we might suspect that only older, more sedate people would buy a brown car. Should we measure the speed of any car in front of a truck? Perhaps we should pick any car following a truck? Again, we may be selecting cars that are traveling a bit faster or a bit slower than normal. Only by very carefully designing our experiment can we be certain that the cars we measure are traveling at representative speeds.

Public-opinion companies that survey voters in shopping malls must think carefully about selection effects. If they are seeking opinions about property taxes and decide to interview only men with gray hair, they will get a biased result. Interviewing men wearing T-shirts would give them quite a different result. Thus, they must be very careful to avoid unnoticed selection effects.

Astronomers face the danger of selection effects quite often. Iron meteorites are easier to

find than stone meteorites. Very luminous stars are easier to see at great distances than faint stars. Spiral galaxies are brighter, bluer, and more noticeable than elliptical galaxies. What we see through a telescope depends on what we notice, and that is powerfully influenced by selection effects.

Scientists engaged in observation must spend considerable time designing their experiments. They must be careful to observe an unbiased sample if they expect to make logical deductions from their results. The scientists in the rain forest, for example, should not catch and study only the red insects. Often, the most brightly colored insects are poisonous. If they are good scientists, they will plan their work with great care and avoid any possible selection effects. ■

TABLE 26-1

Proportions of Meteorites

Type	Falls (%)	Finds (%)
Stony	92	26
Iron	6	66
Stony-iron	2	8

When iron meteorites are sliced open, polished, and etched with nitric acid, they reveal regular bands called **Widmanstätten patterns** (Figure 26-6). These patterns are caused by alloys of iron and nickel called kamacite and taenite that formed as the molten iron cooled and solidified. The size and shape of the bands indicate that the molten metal cooled very slowly, no faster than 20 K per million years.

A lump of molten metal floating in space would cool very quickly, so the presence of the Widmanstätten pattern tells us that the molten metal must have been well insulated to have cooled so slowly. Such slow cooling is typical of the interiors of bodies at least 30–50 km in diameter. On the other hand, the iron meteorites show no evidence of the very high pressures that would develop in larger bodies. Thus, the iron meteorites appear to have formed from the interiors of molten objects of planetesimal size.

FIGURE 26-6

An iron meteorite sliced open, polished, and etched with acid reveals a Widmanstätten pattern. Such patterns are evidence that the iron cooled from a molten state no faster than a few degrees per million years. (Jordan Marché, North Museum, Franklin and Marshall College)

In contrast to irons, **stony meteorites** are common. Among falls, 92 percent are stones. Although there are many different types of stony meteorites, we will classify them into two subgroups depending on their physical and chemical content—chondrites and achondrites.

a

b

FIGURE 26-7

(a) A stony meteorite sliced open and polished to show the small, spherical inclusions called chondrules. (Jordan Marché, North Museum, Franklin and Marshall College) (b) A sliced portion of the Allende meteorite showing round chondrules and irregularly shaped white inclusions called CAIs. (NASA)

Roughly 80 percent of all meteorite falls are stony meteorites called **chondrites.** These are dark gray, granular rocks containing **chondrules,** round bits of glassy rock no larger than 5 mm across (Figure 26-7a). Chondrules must have cooled very quickly (in minutes), but their origin is not clear. They may be bits of molten rock splashed into space by impacts between planetesimals. Another possibility is that they are bits of rock that condensed from the solar nebula and were later melted by a sudden flare-up of the young sun or by electrical discharges (lightning). Whatever their origin, they are very old, and, because melting of the meteoroid would have destroyed them, their presence in chondrites indicates that these meteorites have never been melted. Chondrites can never have been hotter than 1000–1300 K.

On the other hand, the chondrites have been heated to moderate temperatures for long periods sometime in their past, as shown by studies of the granulation and chemical composition of the rock. Also, the chondrites

contain no volatiles. These easily vaporized compounds evidently either were baked out of the chondrites or were never present because the chondrites formed at too high a temperature.

In contrast, **carbonaceous chondrites,** a subclass of chondrites, are rare, only about 5.7 percent of falls, but they are important because they contain volatiles. They are rich in water and compounds of carbon that would have been driven out if they had ever been heated above about 500 K. Thus, if the chondrites have never been hot, the carbonaceous chondrites have never even been warm. They are the least modified of all the meteorites.

One of the most important meteorites ever recovered was a carbonaceous chondrite that was seen falling on the night of February 8, 1969, near the Mexican village of Pueblito de Allende. The brilliant fireball was accompanied by tremendous sonic booms and showered an area about 50 km by 10 km with over 4 tons of fragments. About 2 tons were recovered.

Studies of the Allende meteorite disclosed that it contained—besides volatiles and chondrites—small, irregular inclusions rich in calcium, aluminum, and titanium. Now called **CAIs,** for calcium–aluminum-rich inclusions, these bits of matter are highly refractory (Figure 26-7b); that is, they can survive very high temperatures. If we could scoop out a ton of the sun's surface matter and cool it, the CAIs would be the first particles to form. As the temperature fell, other materials would condense in accord with the condensation sequence described in Chapter 20. When the material finally reached room temperature, we would find that all of the hydrogen, helium, and a few other gases like argon and neon had escaped and that the remaining lump, weighing about 18 kg (40 lb), had almost the same composition as the Allende meteorite, including CAIs. Thus, the Allende meteorite seems to be a very old sample of the solar nebula.

Unlike the carbonaceous chondrites, stony meteorites called **achondrites** (7.1 percent of falls) are highly modified. They contain no chondrules and no volatiles. This suggests that they have been hot enough to melt chondrules and drive off volatiles, leaving behind rock with compositions similar to the earth's lavas.

Iron meteorites and stony meteorites make up most falls, but 2 percent of falls are meteorites that are made up of mixed iron and stone. These **stony-iron meteorites** appear to have solidified from a region of molten iron and rock—the kind of environment we might expect deep inside a planetesimal with a molten iron core and a rock mantle.

The Origins of Meteorites

The properties of meteorites suggest that they formed in the solar nebula. Their radioactive ages are about 4.6 billion years (Chapter 20), and some appear not to have been modified since they condensed from the solar nebula. Others have been heated or melted sometime after formation.

Meteorites almost certainly do not come from comets. Most cometary particles are very small specks of low-density, almost fluffy, material. When these enter the atmosphere, we see them incinerated as meteors, and most meteors are produced by this cometary debris. But such meteors are small and weak and do not survive to reach the ground. Although most meteors come from comets, most meteorites are stronger chunks of matter—more like fragments of asteroids.

Thus, when we talk of the origin of the meteorites, we are concerned with the history of asteroidlike planetesimals rich in metals and silicates, not the cometlike planetesimals that were rich in ices. Indeed, many of the properties of the meteorites suggest that they originated in stony-iron planetesimals, which evolved in complicated ways and were eventually broken up by collisions. Here we will concentrate on the answers to two questions: How did these planetesimals evolve? When did these planetesimals break up?

The iron and achondritic meteorites show that at least some of the planetesimals melted, but what produced this heat? Planets the size of the earth can accumulate a great deal of heat from the decay of radioactive atoms such as uranium, thorium, and the radioactive isotope of potassium. The heat is trapped deep underground by many kilometers of insulating rock. But in a small planetesimal, the insulating layers are not as thick, and the heat leaks out into space as fast as the slowly decaying atoms can produce it. If a small planetesimal is to melt, it must have a more rapidly decaying heat source.

Recent studies have shown that some meteorites may have contained the radioactive element aluminum-26. These atoms are now gone, decayed to magnesium-26, but if they were once present their rapid decay could have melted the center of a planetesimal as small as 20 km in diameter. This is a sufficient heat source to melt the parent bodies of the meteorites.

Once the planetesimals were melted and the aluminum-26 had decayed, they cooled and slowly solidified. In the meantime, however, the planetesimals would have differentiated. The heavy metals would have sunk to the center to form a molten metal core, and the less dense silicates would have floated upward to form a stony mantle. In some cases, the surface may not have been melted. If such a planetesimal were broken up (Figure 26-8), fragments from the center would look much like iron meteorites with their Widmanstätten patterns, and fragments from the outer portions might look like stony-irons and achondrites.

The chondrites and carbonaceous chondrites might have formed as smaller bodies farther from the sun. Planetesimals in the outer asteroid belt would have been cooler and could have retained volatiles more easily. They could not have been large, or heating and differentiation would have altered them.

The evidence seems strong that the meteorites are the result of the breaking up of larger bodies. For example,

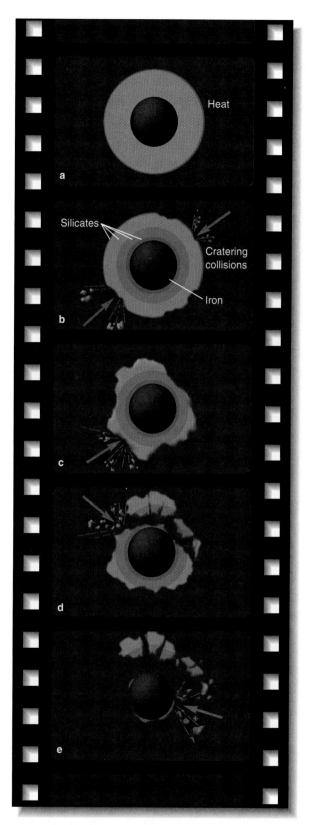

FIGURE 26-8

Production of various kinds of meteoroids from asteroids. If it melted, an asteroid (a) could differentiate into an iron core surrounded by layers of different silicate compositions (b). Impacts (c, d, e) could break up these layers and produce various kinds of meteorites. (Based on "The Nature of Asteroids" by C. R. Chapman. Copyright © 1975 by Scientific American, Inc. All rights reserved.)

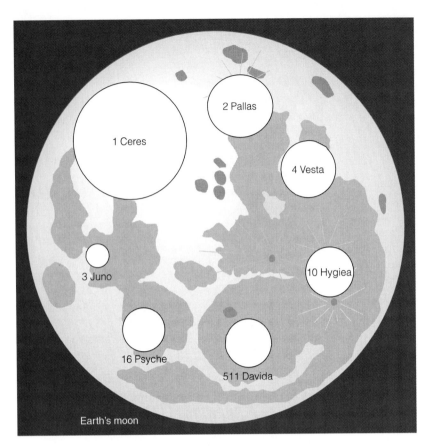

FIGURE 26-9

The relative sizes of some of the larger asteroids are shown here as circles compared with the size of the moon. Smaller asteroids may be highly irregular in shape, and even Pallas is slightly ellipsoidal.

some meteorites are breccias—collections of fragments cemented together. Breccias are found on the earth and are very common on the moon, but studies of the meteoric breccias show that they were produced by impacts. A collision between planetesimals produces fragments, and the slower-moving particles fall back to the surface of the planetesimal to form a regolith, a soil of broken rock fragments. Later impacts may add to this regolith and stir it. Still later, an impact may be so violent that the fragments are pressed together and momentarily melt where they touch. Almost instantly, the material cools, and the fragments weld themselves together to form a breccia. Most of the breccias were formed as the planetesimals formed. Much later, as the planetesimals were broken up by collisions, the layers of breccia were exposed, shattered, and became brecciated meteoroids.

The collisions that broke up the planetesimals could not have happened long ago, when the solar system was young. Small meteoroids would have been swept up by the planets in only a billion years or less. Also, cosmic rays striking meteoroids in space produce isotopes such as helium-3, neon-20, and argon-38. Studies of these atoms in meteorites show that most meteorites have not been exposed to cosmic rays for more than a few tens of millions of years. The meteorites must have been buried under protective layers until recently. Thus, the 3000 or so meteorites now in museums all over the world must have broken off from planetesimals somewhere in our solar system within the last billion years.

CRITICAL INQUIRY

How can we say that meteors come from comets, but meteorites come from asteroids?

A selection effect can affect what we notice when we observe nature, and a very strong selection effect prevents us from finding meteorites that originated in comets. Cometary particles are physically weak, and they vaporize in our atmosphere easily. Thus, very few ever reach the ground, and we are unlikely to find them. Furthermore, even if a cometary particle reached the ground, it would be so fragile that it would weather away rapidly, and, again, we would be unlikely to find it. Asteroidal particles, however, are made from rock and metal and so are stronger. They are more likely to survive their plunge through the atmosphere and more likely to survive erosion on the ground. Meteors from the asteroid belt are rare. Almost all of the meteors we see come from comets, but not a single meteorite is known to be cometary.

A similar selection effect makes it more likely that we will find iron meteorites than stony meteorites. Irons are easier to notice and survive longer. Thus, most *finds* are iron, but most *falls* are stone.

The meteorites are valuable because they give us hints about the process of planet building in the solar nebula. What evidence do we find in the meteorites that tells us they were once part of larger bodies broken up by impacts? ■

FIGURE 26-10

Kirkwood's gaps are regions in the asteroid belt where there are few asteroids. These gaps are caused by orbital resonances with Jupiter.

The vast majority of meteorites must have originated in the asteroids. Our next goal is to try to understand the nature and origin of these tiny rocky worlds.

26-2 ASTEROIDS

Until recently, few astronomers knew or cared much about asteroids. They were small chunks of rock drifting between the orbits of Mars and Jupiter that occasionally marred long-exposure photographs by drifting past more interesting objects. Asteroids were more irritation than fascination.

Now we know differently. The asteroids seem to be the last remains of the planetesimals that built the planets 4.6 billion years ago. Almost all of the original planetesimals were either incorporated into planets or ejected from the solar system, but out in the asteroid belt, trapped in an orbital dustbin, are a few small fragments. The study of the asteroids is now an exciting exploration of the ancient past of our planetary system.

The Asteroid Belt

The first asteroid was discovered on January 1, 1801 (the first night of the 19th century), by the Sicilian monk Giuseppe Piazzi. It was later named Ceres after the Roman goddess of the harvest (thus our word *cereal*). We know now that it is a spherical ball of rock about 900 km in diameter.

Astronomers were excited by Piazzi's discovery. There seemed to be a gap where a planet might exist between Mars and Jupiter at an average distance from the sun of 2.8 AU. Ceres fit. Its average distance from the sun is 2.766 AU. It was a bit small to be a planet, and because three more objects—Pallas, Juno, and Vesta—were discovered in the following years, astronomers realized that Ceres and the other asteroids were not true planets.

Today over 3500 asteroids have accurately known orbits, and about 6000 more have poorly known orbits.

There are many more as yet undiscovered, but they are all small bodies. All of the larger asteroids in the asteroid belt have been found.

When an asteroid is discovered, the discoverer is allowed to choose a name for it, and asteroids have been named for spouses, lovers, dogs, Greek gods, politicians, and others.* Once an orbit has been calculated, the asteroid is assigned a number listing its position in the Russian catalog known as the *Ephemerides of Minor Planets*. Thus, Ceres is known as 1 Ceres, Pallas as 2 Pallas, and so on. Only three are larger than 400 km in diameter, and most are much smaller (Figure 26-9).

Most of these objects orbit the sun in the asteroid belt between Mars and Jupiter, and we might suspect that massive Jupiter was responsible for their origin. Certainly the distribution of asteroids in the belt shows the importance of Jupiter's gravitation. Certain regions of the belt, called **Kirkwood's gaps** (Chapter 24) after their discoverer, are almost free of asteroids (Figure 26-10). These gaps lie at certain distances from the sun where an asteroid would find itself in resonance with Jupiter. For example, if an asteroid lay 3.28 AU from the sun, it would revolve twice around its orbit in the time it took Jupiter to revolve once around the sun. Thus, on alternate orbits, the asteroid would find Jupiter at the same place in space tugging outward. The cumulative perturbations would rapidly change the asteroid's orbit until it was no longer in resonance with Jupiter. Our example is in a 2:1 resonance, but gaps occur in the asteroid belt at many resonances, including 3:1, 5:2, and 7:3.

Modern research shows that the motion of asteroids in Kirkwood's gaps is described by a newly developed theory in mathematics that deals with chaotic behavior. The smooth motion of water sliding over the edge of a waterfall decays rapidly into a chaotic jumble. The same mathematical theory of chaos that describes the motion of the water shows how the slowly changing orbit of an asteroid within one of Kirkwood's gaps can suddenly

*Some sample asteroid names: Olga, Chicago, Vaticana, Noel, Ohio, Tea, Gaby, Fidelio, Hagar, Geisha, Dudu, Tata, Mimi, Dulu, Tito, Zulu, Beer, and Zappafrank (after the late musician Frank Zappa).

change into a long, elliptical orbit that carries the asteroid into the inner solar system, where it is likely to be removed by a collision with Mars, Earth, or Venus. This ejection process may also explain how Jupiter's gravity can throw so many stony meteoroids from the asteroid belt into the inner solar system.

Nonbelt Asteroids

Not all asteroids orbit within the asteroid belt. Some of the most interesting follow orbits that cross the orbits of the terrestrial planets or wander among the Jovian worlds. In fact, some asteroids even share the orbits of the larger planets.

The **Apollo–Amor objects** are asteroids whose orbits carry them into the inner solar system. The Amor objects follow orbits that cross the orbit of Mars but don't reach the orbit of Earth. The Apollo objects have Earth-crossing orbits. These Apollo–Amor objects are dangerous. Jupiter's influence makes their orbits precess, so they are bound to interact with a planet eventually. About one-third will be thrown into the sun, and a few will be ejected from the solar system, but many of these objects are doomed to collide with a planet—perhaps ours. Earth is hit by an Apollo object once every 250,000 years, on average. With a diameter of up to 2 km, they hit with the power of a 100,000-megaton bomb and can dig craters 20 km in diameter. About 30 Apollo objects are known, and none will hit Earth in the foreseeable future. The bad news is that objects as small as a few meters in diameter could be dangerous, and there may be as many as several hundred thousand such Apollo objects.

It is easy to assume that the Apollo–Amor objects are rocky asteroids that have been thrown into their extreme orbits by collisions in the main asteroid belt. At least some of these objects, however, may be comets that have exhausted their ices and become trapped in short orbits that keep them in the inner solar system. We will see later that the distinction between comets and asteroids is not totally clear.

Early estimates suggested that smaller Apollo objects were relatively rare, but the Spacewatch Telescope atop Kitt Peak in Arizona has been searching for near-earth asteroids, and it has found 10–100 times more than originally expected. For example, in January 1991, an undiscovered asteroid about 9 m in diameter raced past the earth at a distance only half that of the moon. Had it struck the earth, it would have hit with an energy ten times greater than that of the bomb that destroyed Hiroshima. Objects of this size, a few tens of meters or less, fragment and explode in the atmosphere, and recently declassified data from military satellites show that the earth is hit about once a week by these house-size asteroids. The largest impact ever detected was the atmospheric explosion of a rocky meteorite over the South Pacific on February 1, 1994. It was seen by fishermen as a blast as bright as the sun. It produced about five times the energy of the Hiroshima bomb and was probably a rocky body about 15 m in diameter—probably a small Apollo object. We will see later in this chapter what a slightly larger impactor can do.

While the Apollo–Amor objects rush through the inner solar system, there are also nonbelt asteroids in the outer solar system. These objects, being farther from the sun, move more slowly. The object Chiron, found in 1977, appears to be a body about 170 km in diameter. Its orbit carries it from just inside the orbit of Uranus to just inside the orbit of Saturn. Although it was first classified as an asteroid, its status is now less certain. Ten years after its discovery, Chiron surprised astronomers by suddenly brightening with a release of vapor and dust. Like a comet, Chiron was releasing a cloud of gas and dust in jets. Studies of older photographs showed that Chiron had been brighter even when it was farther from the sun, and astronomers now suspect that it may have a rocky crust covering deposits of ices such as solid nitrogen, methane, and carbon monoxide. Thus, Chiron may be more comet than asteroid, and it warns us that the distinction is not clear-cut.

Chiron's orbit is not stable. It cannot have remained in its present orbit for the entire history of the solar system but must have been perturbed into its present orbit within the last million years or so. Other, smaller bodies have been found in the outer solar system, and astronomers are now trying to understand how their origin and history are linked to the formation of the planets. We saw in the previous chapter that Pluto, Charon, and Triton might be large members of this family of icy planetesimals.

The icy asteroids of the outer solar system seem to be widely scattered, but Jupiter ushers two groups of asteroids within its own orbit. These nonbelt asteroids have become trapped in the Lagrangian points along Jupiter's orbit. These points lie 60° ahead of and 60° behind the planet and are regions where the gravitation of the sun and Jupiter combine to trap small bodies (Figure 26-11). Like cosmic sinkholes, the Lagrangian points have trapped chunks of debris now called **Trojan asteroids** and named after the heroes of the Trojan war (588 Achilles, 624 Hektor, 659 Nestor, and 1143 Odysseus, for example). Although about 15 Trojans have been found and named, astronomers estimate that 700 may be trapped in the Lagrangian points.

In 1990, observers found an asteroid at the trailing (L_5) Lagrangian point in the orbit of Mars, and growing evidence suggests that other planets may have Trojan asteroids trapped in their orbits. As we search for smaller objects, we are finding that our solar system still contains large numbers of these small bodies. The challenge is to explain their physical nature and their origin.

Composition and Origin

We now suspect that the asteroids are fragments of planetesimals left over from the formation of the planets. The most popular theory to account for the asteroids sup-

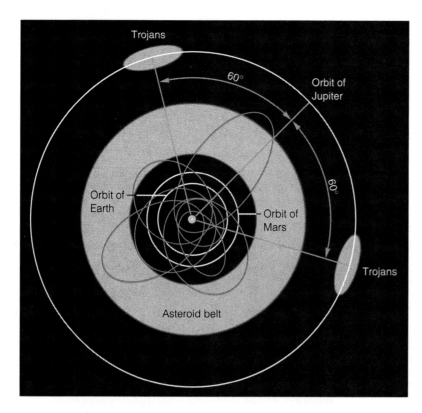

FIGURE 26-11

The Trojan asteroids are trapped in the Lagrangian points 60° ahead of and 60° behind Jupiter. The Apollo asteroids move along orbits that cross Earth's orbit. The orbits of only 7 of the 31 known Apollo asteroids are shown here.

poses that Jupiter's gravity stirred the planetesimals in the asteroid belt and caused collisions at unusually high velocities. These impacts tended to break up the planetesimals rather than assemble them into a planet. Much of the material has been lost—swept up by planets, captured as satellites, or ejected from the solar system.

Collisions are surely common among the asteroids, and we can see evidence of catastrophic impacts powerful enough to break up an asteroid. Early in this century, Japanese astronomer Kiyotsugu Hirayama discovered that some groups of asteroids share similar orbits. They have the same average distance from the sun, the same eccentricity, and the same inclination. These **Hirayama families,** as they are now called, appear to be the products of catastrophic collisions that broke a single asteroid into fragments. The fragments still travel along similar orbits around the sun.

In 1983, the Infrared Astronomy Satellite detected the infrared glow of sun-warmed dust scattered in bands throughout the asteroid belt. These dust bands appear to be the products of past collisions. The dust will eventually be blown away, but collisions occur constantly in the asteroid belt, so new dust bands will presumably be produced as the present bands dissipate.

The asteroids are too small and too far away to show any detail when viewed through earth-based telescopes, so until recently astronomers had to use their imaginations to visualize the asteroids as great, irregular, cratered fragments of rock tumbling ponderously through space. It was reasonable to suppose that they would be irregular in shape, because all but the largest have too little gravity to overcome the strength of rock. As the asteroids collide

and fragment, most of the fragments should remain irregular in shape. In addition, meteorite impacts would produce craters ranging from microscopic pits to vast scars only slightly smaller than the asteroid itself.

Beginning in the 1990s, astronomers have gotten close-up looks at a few asteroids. Radar observations of the asteroid 4769 Castalia (Figure 26-12) show that it is roughly peanut-shaped and tumbling end over end. More powerful radar probes reveal that the asteroid 4179 Toutatis (Figure 26-13a) is also peanut-shaped. Astronomers suspect that these shapes result from the gentle collision of two bodies. It is possible that the two bodies orbit a single center of mass, but it is more likely that they are in actual contact.

The Galileo spacecraft, on its way to Jupiter, was able to image two asteroids, and both show evidence of past fragmentation. Gaspra (Figure 26-13b) is $20 \times 12 \times 11$ km and is heavily battered by crater impacts. Planetary astronomers estimate from the number of craters that the surface has been exposed to space for about 200 million years. Ida (Figure 26-13c) is a larger object, 52 km long, but it, too, is scarred by craters. Both asteroids are covered by a layer of crushed-rock regolith. The photos reveal that Ida has a small moon, dubbed Dactyl, only about 1.5 km in diameter. Although amateur astronomers observing asteroids crossing in front of stars have claimed to see evidence of moons orbiting asteroids, most astronomers thought such moons would not be stable. Ida's moon must have originated as a fragment in a collision not too many million years ago. In fact, twin impact craters common on Earth, the moon, and Venus appear to have been produced by the impact of double asteroids.

F I G U R E   2 6 - 1 2

Asteroid Castalia passed only 5.6 million km from Earth in 1989, and radar studies were used to construct a 3-D model. This sequence shows the peanut-shaped asteroid rotating as it falls along its orbit. The asteroid is about 1.8 km in diameter as its widest point, and it appears to be two objects that have collided and fused. The radar could not detect features smaller than about 100 m. Thus, the smooth surface of the computer model is not realistic. (Courtesy R. Scott Hudson and Steven Ostro)

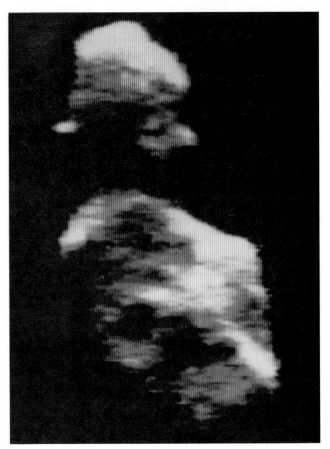

a

The colors in the Galileo images have been computer-enhanced to reveal subtle differences in surface composition. If we could view the two asteroids with our eyes, they would look gray. Nevertheless, the colors of asteroids can be helpful.

In the 1930s, astronomers discovered that some asteroids are redder than others, but only in the last few decades has infrared photometry made it possible to study the color of the light reflected from asteroids and compare it with the colors reflected from various kinds of minerals and meteorites. This has led to the discovery of different kinds of asteroidal worlds.

The S type asteroids are bright, with albedos of about 0.1–0.2, and tend to be reddish. In a diagram of albedo versus ultraviolet minus visual color, the S type asteroids lie toward the upper right (Figure 26-14). At present, the data cannot tell us whether they are composed of mixed silicates and metals or of chondritic minerals.

The C type asteroids are very dark, with albedos of less than 0.06. That is, they reflect less than 6 percent of the light that hits them and thus are darker than a lump of sooty coal. Their properties match those of carbonaceous chondrites.

The M type asteroids are moderately bright, with albedos about the same as those of the S types, but the M types are not as red. Thus, we can separate the two types in the albedo–color diagram (Figure 26-14). Their opti-

b

c

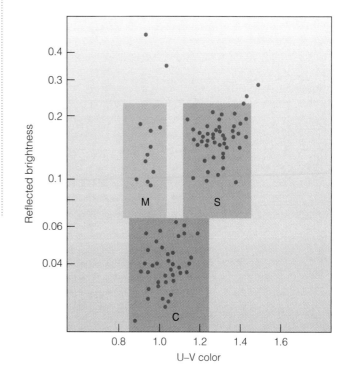

d

FIGURE 26-13
FIGURE 26-13

(a) A radar image of asteroid Toutatis shows two bodies in contact. (Steven J. Ostro) (b) Asteroid Gaspra photographed by the Galileo spacecraft is 20 km long. (NASA) (c) A Galileo image of Ida (52 km long) and its moon Dactyl (only 15 km in diameter) show a cratered surface. (NASA) (d) The NEAR spacecraft imaged asteroid Mathilda (52 km in diameter). It is scarred by giant craters; its low density suggests that it is a fractured ball of rubble and not solid rock. (Johns Hopkins University, Applied Physics Laboratory)

cal properties suggest that M types are composed of iron and nickel alloys and contain very few silicates.

A few other types of asteroids are known, and a number of individual asteroids have been found that are unique, but these three classes contain a majority of the known asteroids.

How did these three types originate? A clue lies in their distribution in the asteroid belt. The S type asteroids are much more common in the inner belt. In fact, there are almost none beyond a distance of about 3.45 AU. The C types are rather rare in the inner belt but are very common in the outer belt. Many astronomers believe that this distribution reflects the differences in temperature during the formation of the planetesimals. It was cooler in the outer belt, and the planetesimals there tended toward the composition of carbonaceous chondrites. In the warmer regions of the inner belt, the composition of the growing planetesimals was more like that of the chondrites.

FIGURE 26-14 ▶

The three principal types of asteroids. In a diagram of albedo versus ultraviolet minus visual color index, the most reflective asteroids lie near the top, and the reddest at the right. The S types, which are mixed silicates and metals or perhaps chondrites, are clearly redder than the M types, which seem to be metallic. The darkest asteroids, the C types, are believed to be similar to carbonaceous chondrites. (Diagram adapted from a figure by B. Zellner)

a

b

FIGURE 26-15

(a) A Hubble Space Telescope image of the asteroid Vesta shows a complex surface with a large crater near the south pole. From 78 such images, astronomers constructed a model of Vesta, showing the impact crater with its central peak. The crater is up to 13 km deep. (Ben Zellner, Georgia Southern University; Peter Thomas, Cornell University; NASA)
(b) A meteorite identified spectroscopically as a fragment of Vesta. The chemical and crystalline structure of the meteorite shows that it was once part of a molten lava flow. Presumably, it was chipped off Vesta by a large impact.
(R. Kempton, New England Meteoritical Services)

The large asteroid Vesta is reminding astronomers that these are complex worlds. With a diameter of 525 km, Vesta is too small to image from Earth's surface, but the Hubble Space Telescope reveals that is has a complex surface with light and dark patches. The spectra of these regions are like those of basalts typical of solidified lavas. Furthermore, a study of the best photos shows a giant crater 460 km in diameter (87 percent of Vesta's diameter) near Vesta's south pole (Figure 26-15a). A certain type of basaltic meteorite has been identified as matching Vesta, and astronomers feel confident that these meteorites are samples of Vesta. The impact that formed the large crater could have blasted loose as much as 1 percent of Vesta's mass as meteoroids, some of which later fell to earth. Thus, we have strong evidence that the asteroid Vesta has at some time in the past been at least partly resurfaced by lava flowing up from its interior.

How could a small asteroid be heated enough to produce lava flows? One source of heat in newly formed asteroids would be the decay of short-lived radioactive elements such as aluminum-26. Such radioactive elements could have been produced in a supernova explosion occurring shortly before the formation of our solar system. The smallest asteroids lose their heat too fast to melt, but the decay of aluminum-26 occurs fast enough to melt the interiors of bodies larger than a few dozen kilometers in diameter. Thus, it is reasonable to suppose that Vesta and some other asteroids larger than about 100 km in diameter might have been strongly heated and modified by lava flows.

In contrast, the largest asteroid, Ceres, about 900 km in diameter, does not seem to have been modified by internal heating. The light reflected from its surface suggests a claylike surface related to carbonaceous chondrites. This is surprising, since clays form when minerals are exposed to water. In fact, spectroscopic observations reveal what seem to be traces of water ice on Ceres. Thus, we must suspect that some asteroids may have had significant amounts of water bound into their crusts when they were young.

All of the evidence suggests that the asteroids are the broken remains of planetesimals that formed in the solar nebula as planet building began. The largest remaining asteroids, such as Ceres and Vesta, may be largely unbroken planetesimals, and some of these may have evolved due to internal heat or the presence of water. Nevertheless, the vast majority of the asteroids are fragments of larger bodies, and many may consist of bodies that have been shattered and then re-formed as gravity pulled the fragments back together. Some compositional differences seem to have originated when the asteroids formed in that the objects in the outer asteroid belt contain more carbon-rich compounds. The presence of massive Jupiter orbiting just beyond this region prevented the original planetesimals from accreting to build a planet. Instead, collisions fragmented the planetesimals, and nearly all of the material is now lost. The entire mass of the asteroid belt today adds up to only 0.08 percent of the mass of the earth.

Some experts discuss the mining of the asteroid belt for its mineral riches, but the asteroid belt is nearly empty. The small, fragmented asteroids drift through space far from one another and only rarely collide with each other or with planets. A voyage to explore or exploit the asteroid belt would find the distance between bodies forbiddingly large.

CRITICAL INQUIRY

What evidence do we have that the asteroids have been fragmented?

First, we might note that the solar nebula theory of the formation of the solar system predicts that planetesimals collided and either stuck together or fragmented. This is suggestive, but it is not evidence. A theory can never be used as evidence to support some other theory or hypothesis. Evidence means observations or the results of experiments, so we must turn to observations of asteroids. The Galileo spacecraft photographs of Ida and Gaspra show irregularly shaped little worlds heavily scarred by impact craters, and the radar images obtained so far of asteroids also show irregularly shaped bodies that appear to be the fragments of larger bodies. In fact, radar images of Castalia and Toutatis show what may be pairs of bodies in contact, and the Galileo image of Ida reveals its small satellite, Dactyl. Furthermore, some meteorites appear to come from the asteroid belt, and a few have been linked to specific asteroids such as Vesta. All of this evidence suggests that the asteroids have suffered violent collisions in their history.

The impact fragmentation of asteroids has been important, but it has not erased all traces of the original planetesimals from which the asteroids formed. What evidence do we have to tell us what those planetesimals were like? ■

The asteroids are only the last crumbs left behind by failed planet building near Jupiter. Farther out in the solar system are other remains—ice shards from the age of planet formation.

26-3 COMETS

Few things in astronomy are more beautiful than a bright comet hanging in the night sky (Figure 26-16). Whereas meteors shoot across the sky like demented fireflies, a comet moves with the stately grace of a great ship at sea, its motion hardly apparent. Night by night it shifts against the stars and may remain visible for months. Faint comets are common; a number are discovered every year. But a truly bright comet appears about once a decade. Comet Hyakutake in 1996 (Figure 26-1)

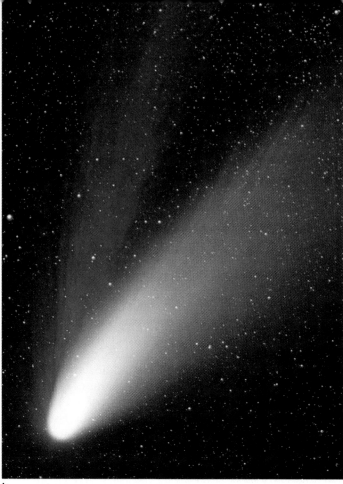

a

b

FIGURE 26-16

(a) Comet Ikeya–Seki was beautiful in the dawn sky in 1965. The long, graceful tail sprang from a nucleus of ice only a dozen kilometers in diameter. (Stephen M. Larson) (b) Comet Hale–Bopp passed through the inner solar system in 1997. Its large nucleus, estimated to be 40 km in diameter, produced large amounts of gas and dust, making it quite bright in the dawn and later in the evening sky. (Dean Ketelsen)

and Comet Hale–Bopp in 1997 (Figure 26-16b) were both dramatic, but the later comet was so bright that we might class it with the great comets such as Halley's comet in 1910. A patient person might see half a dozen bright comets in a lifetime.

While everyone enjoys the beauty of comets, astronomers study them for their cargo of clues to the origin of our solar system.

Observational Properties of Comets

We saw in Chapter 20 that the overall properties of comets could be explained by the dirty snowball hypothesis. It proposes that a comet is a lump of dirty ices, mostly water ice, orbiting the sun in a long elliptical orbit. Only when that nucleus of ice comes within a few astronomical units of the sun does the ice vaporize to form a tail of gas and dust pointing away from the sun.

Now we are ready to compare this hypothesis in detail with the observed properties of comets and especially with Comet Halley, which was studied exhaustively during its return in 1986. Two Soviet spacecraft, Vega 1

and Vega 2, flew past the comet at distances of 8900 km and 7900 km, respectively. Two Japanese spacecraft, Sakigake and Suisei, studied the comet from 20 times farther away, but the European Space Agency's Giotto spacecraft flew within about 600 km of the comet's heart. The data from those missions paint a detailed picture of what a comet is like.

We can think of a comet as comprising a head and a tail. The visible head, or **coma,** of a comet is a vast gas and dust cloud, which can be up to 15 times the diameter of the earth (Figure 26-17). The Vega spacecraft approaching Comet Halley encountered the first dust particles at a distance of 47 earth radii from the nucleus. Spectra made from the earth and measurements by spacecraft show that the gas is made of hydrogen, carbon, nitrogen, and oxygen atoms linked to form various ions and molecules. These atoms are what we would expect to find if ultraviolet radiation from the sun had broken down vaporized ices consisting of water (H_2O) and carbon dioxide (CO_2). The ices may also contain ammonia (NH_3) and methane (CH_4).

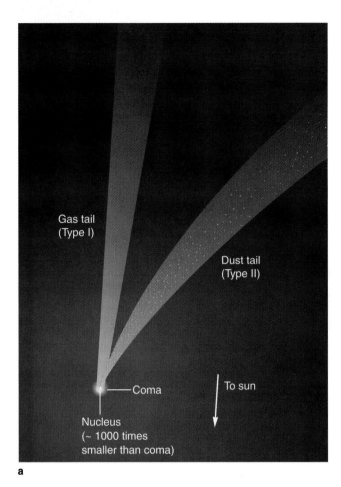

Gas tail
(Type I)

Dust tail
(Type II)

Coma

To sun

Nucleus
(~ 1000 times
smaller than coma)

a

b

FIGURE 26-17

(a) The parts of a comet include an extended coma, or head, of gas and dust surrounding an icy nucleus only a few kilometers in diameter. Gas is carried away by the solar wind in a wispy gas tail, but dust moves more slowly and forms a curving dust tail. (b) Comet Mrkos in 1957 had pronounced gas and dust tails, which changed dramatically from night to night. (Palomar Observatory/Caltech)

In 1970, observations made from earth orbit revealed a vast cloud of hydrogen surrounding the nucleus of bright comets. This sphere is larger than the sun and must be continually resupplied with fresh hydrogen coming from the head of the comet at the prodigious rate of 10^{29} atoms per second. Apparently, these atoms come from disrupted water molecules.

The tail of a comet springs from the coma and typically extends 10–100 million km through space. The longest tails reach a length of 1 AU, about 150 million km. Seen from the earth, the tail of a major comet can span 30° in the sky. This immense structure contains nothing more than very tenuous gas and scattered bits of microscopic dust. Comets have been described as "as close to nothing as something can be and still be something."

Comets have two kinds of tails. A **type I tail,** also known as a **gas tail,** is straight and streaked with wisps and irregularities, which make it look like a streamer of smoke caught in a smooth, fast wind (Figure 26-17). The spectra of type I tails show emission lines of ionized atoms excited by the ultraviolet radiation from the sun (Table 26-2). This ionized gas is carried away from the head by the solar wind and its embedded magnetic field pushing the gas away from the sun. In fact, type I tails provided some of the earliest clues to the existence of a solar wind.

Occasionally, a comet will disconnect its gas tail, allow it to blow away, and begin developing a new tail.

TABLE 26-2

Elements and Ions Detected in Comets

Coma	Gas Tail
H, OH, O	CO^+, CO_2^+
C, C_2, C_3, CH	H_2O^+, OH^+
CN, CO, CS, S	CH^+, CN^+, N_2^+
NH, NH_2, HCN, CH_3CN	C^+, Ca^+
Na, Fe, K, Ca, V	
Cr, Mn, Co, Ni, Cu	

This seems to happen when the comet moves across a boundary in the solar wind into a region where the magnetic field is reversed. This illustrates the importance of the magnetic field in the solar wind in controlling the structure of gas tails. In September 1985, a spacecraft called ICE (International Cometary Explorer) flew through the tail of the comet Giacobini–Zinner and confirmed the magnetic nature of the gas tail. The spacecraft found the magnetic field of the solar wind draped over the head of the comet like seaweed draped over a fishhook. The gas tail was composed of ionized gas trapped in this magnetic tail, streaming outward with the solar wind.

a

b

FIGURE 26-18

(a) Comet Halley is a typical comet in that much of the light we receive from its tail is reflected by microscopic dust particles released from the head of the comet. (National Optical Astronomy Observatory) (b) NASA aircraft flying at an altitude of 18 km (59,000 ft) can collect dust falling into the earth's atmosphere from space. This dust speck is only 5 by 6 μm. Comets continually release such dust throughout the inner solar system, and it can be detected by infrared telescopes in earth orbit. (NASA)

A **type II tail,** also known as a **dust tail,** is smooth and featureless. Its spectrum is merely the spectrum of the sun. This tells us that the type II tail is made up of dust particles that reflect sunlight and that the tail does not contain atoms that would produce emission lines. The dust particles are typically about 1 μm in diameter (Figure 26-18). (One micrometer or micron, abbreviated 1 μm, is a millionth of a meter.) The dust is pushed outward by the pressure of sunlight and forms the long, featureless type II tails. These tails are curved because the comet moves along its orbit with a velocity comparable to the velocity of the dust. Like a stream of water from the moving nozzle of a lawn sprinkler, the dust tail appears to curve.

The dust in a comet's coma poses a serious threat to spacecraft. The Vega probes were damaged by the dust they encountered some 8000 km from the nucleus of Comet Halley, but the Giotto probe, traveling at 70 km/sec (155,000 mph), passed only 600 km from the nucleus. It was struck by 12,000 dust specks and was finally knocked tumbling and damaged when it plunged through a denser dust jet only seconds before its closest approach. Astronauts may someday visit the nucleus of a comet, but they will have to approach slowly because of the dust.

The dust released by comets becomes scattered throughout the solar system and can be collected by spacecraft. The Infrared Astronomy Satellite found that this dust was much more abundant than expected. IRAS observed a number of comets and found that the comets themselves and their orbits were rich in dust. Spacecraft measurements of the dust grains around Comet Halley revealed that the dust is partially silicate material, but it has a high content of carbon and organic compounds and thus resembles the meteorites known as carbonaceous chondrites. Comet dust falls into the earth's atmosphere continuously. During the summer of 1986, a team of geologists reported vacuuming up about 10 kg of comet dust from the bottom of an ice lake in Greenland. Such dust must have collected on the ice over thousands of years.

Both dust tails and gas tails point generally away from the sun. Thus, a comet approaches the sun with its tail streaming out behind, but it recedes from the sun tail first (Figure 20-16b). Also, a single comet can have both a dust tail and a gas tail at the same time (Figure 26-17).

Many comets (IRAS data suggest most) have no visible tail. This condition may arise when a comet exhausts its supply of ices. A comet that passes close to a massive planet such as Jupiter can be deflected into a smaller orbit that keeps it in the inner solar system (Figure 26-19). Nearer the sun, its ices vaporize rapidly until little is left but a ball of dust and debris. A number of these "dried-out" comets are known, and IRAS found that, although such comets are poor in ice and have no visible tails, they are, indeed, rich in dust.

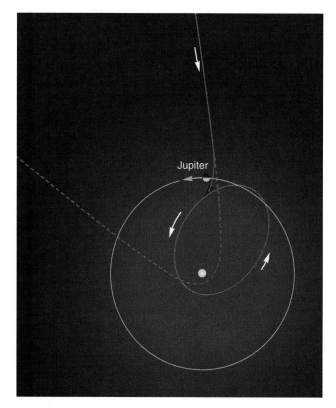

FIGURE 26-19
Massive Jupiter can alter the orbit of a comet (from the dashed path into the smaller ellipse) that keeps it in the inner solar system.

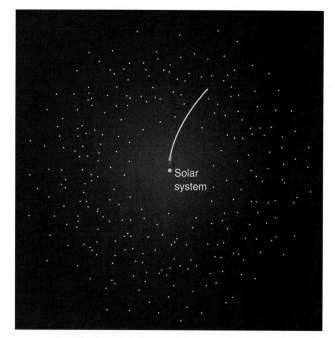

FIGURE 26-20
The long-period comets appear to originate in the Oort cloud, a spherical volume of icy bodies between 20,000 and 150,000 AU from the sun. Objects that fall into the solar system from this cloud arrive from all directions. The short-period comets appear to come from the Kuiper belt, a disk of icy bodies ranging from 30 or 40 to at least 100 AU from the sun in the plane of the solar system. At the scale of this diagram, the Kuiper belt is much smaller than the center dot that symbolizes the solar system.

The Oort Cloud and the Kuiper Belt

If we look at the orbits of comets instead of their comas and tails, we discover some family relations. Most comets have long, narrow orbits with periods much greater than 200 years. The orbits are random in shape and inclination, with comets falling into the inner solar system from all directions. As many long-period comets circle the sun clockwise as counterclockwise. Comet Kohoutek, which was visible in 1973–74, is a long-period comet; it has a period of 80,000 years and a maximum distance from the sun of 4000 AU—about 100 times the distance from the sun to Pluto.

In contrast, about 100 or so of the 600 well-studied comets have small orbits with periods of less than 200 years. These short-period comets have orbits that tend to lie within 30° of the plane of the solar system, and a majority revolve around the sun counterclockwise. Comet Halley, with a period of 76 years, is a short-period comet, but its orbit is retrograde (clockwise). It never travels farther from the sun than about 35 AU, just beyond the orbit of Neptune. It is presently falling outward along its orbit and will not return to the inner solar system until 2062. The shortest-period comet known, Encke's Comet, has a period of only 3.3 years. It is trapped in the inner solar system and has lost much of its ice, so it no longer produces a dramatic coma and tail.

Short-period comets cannot survive long before the heat of the sun drives away their ices. They may last only about 1000 orbits and can't have survived 4.6 billion years since the solar system formed. Thus, there must be a continuous supply of new short-period comets.

In the 1950s, Dutch astronomer Jan Oort proposed that the long-period comets are objects that fall in from the **Oort cloud** (Figure 26-20), a spherical cloud of icy bodies believed to extend from 20,000–150,000 AU from the sun. Because they fall from the spherical Oort cloud, these long period comets have randomly tipped orbits. The cloud may hold several trillion icy bodies without comas or tails. Occasional passing stars could perturb a few to fall into the inner solar system; Hipparcos data show that star Gliese 710 will pass within 1 ly of our solar system in about one million years. That event may shower the inner solar system with Oort cloud objects, which, warmed by the sun, will become comets.

The short-period comets, however, require a different explanation. In 1951, Dutch-American astronomer Gerard P. Kuiper proposed that the formation of our solar system should have left behind a belt of small icy bodies beyond the Jovian planets and in the plane of the solar system. This **Kuiper belt** could not be observed at the time, and it did not attract the attention of planetary astronomers for some decades. Instead, the short-period comets were thought to be former long-period comets

FIGURE 26-21

A comet strikes the sun. This sequence of photographs was taken by the Naval Research Laboratory's satellite Solwind, which carried telescopes to study the sun's corona and solar wind. The comet approaches the sun from the left but fails to reappear on the other side (shown by arrows). The comet must have collided with the sun or vaporized in the corona. (Naval Research Laboratory)

trapped in shortened orbits by gravitational encounters with the Jovian planets. In recent years, however, numerical calculations have shown that the effects of the Jovian planets cannot produce the orbits seen among the short-period comets, and astronomers now suppose that these comets originate in the Kuiper belt.

Observations give us further confidence that the Kuiper belt is real. Searches for small icy bodies in the outer solar system have so far identified over two dozen objects that may be larger members of the Kuiper belt. Also, the physical and chemical similarities among Chiron, Triton, Pluto, and Charon lead some astronomers to suspect that these are larger icy bodies related to Kuiper-belt objects. In addition, the appearance of the dust disks around nearby stars such as Beta Pictoris (Chapter 20) hints that these disks may be made up of debris released by comets in the equivalent of Kuiper belts around those stars. Finally, working at the extreme limit of its capability, the Hubble Space Telescope has produced statistical evidence of a population of Kuiper-belt objects roughly the size of the nucleus of Comet Halley. Although the evidence is not all in, many astronomers now conclude that the Kuiper belt really does exist.

The sky seems abuzz with comets, but no comet has ever been identified as coming from beyond our solar system. That is, all known comet orbits are consistent with origin in the Oort cloud or the Kuiper belt. Interstellar comets may occasionally fall through our solar system, but they are sufficiently rare that none have been recognized yet.

Between the long-period comets and the short-period comets, there is almost always a comet in the sky, but most are too faint to see without a good telescope. Each year, about half a dozen new ones are discovered, many by amateur astronomers, and another half a dozen known comets reappear. IRAS data and other infrared observations of known comets suggest that many comets pass near the sun and we never see them. They are too poor in ices to have bright comas and tails.

Sungrazers are comets that come very close to the sun, and some have slipped into the inner solar system unnoticed only to appear during total solar eclipses. Some have vaporized entirely in the sun's atmosphere (Figure 26-21). The Solwind satellite detected six sungrazers during 5 years in orbit, and the Solar Max satellite detected ten in only 13 months.

The orbits of sungrazing comets seem to make up a family, and some astronomers believe they were all formed by the breakup of a larger comet. As we will see in the next section, the icy heart of a comet can break.

The Nuclei of Comets

The nucleus of a comet is so small that it cannot be resolved by any earth-based telescope, and thus its size and composition must be inferred from the gases and dust that it releases. Until recently, astronomers thought

a

b

c

FIGURE 26-22

(a) This image of the nucleus of Comet Hale–Bopp shows curving jets venting from active regions on the spinning nucleus. (William Keel) (b) Hubble Space Telescope photo of the inner coma of Comet Hale–Bopp shows multiple jets springing from the nucleus, which is too small to resolve in this photo. (Harold Weaver, Johns Hopkins University; NASA) (c) The active regions and their jets are visible in this image of Comet Halley's nucleus recorded by the Giotto spacecraft. These active regions turn on as the spinning nucleus carries them into sunlight and turn off as they rotate into darkness. (© 1986 Max-Planck Institute)

that cometary nuclei were only a few kilometers in diameter. But when the Vega and Giotto spacecraft sent back pictures of Comet Halley's nucleus, it proved to be shaped like a peanut 16 km long, 8 km wide, and 7 km thick with a very black crust (Figure 26-22).

The nucleus of Comet Halley has a density of only 0.1–0.25 g/cm³, which is surprisingly low. The density of water ice is about 1 g/cm³. Thus, the icy nucleus must be less like a hard-packed snowball of ice and more like a fluffy snowbank containing large amounts of empty volume.

Unlike freshly formed snowbanks on the earth, the nucleus of Comet Halley is covered by a black crust that insulates the ices below. Vega spacecraft measured the temperature of the crust as 300–400 K (27°–127°C). This is far above the temperature of vaporizing ice and shows that the crust insulates the interior from the sun's heat. Judging by the albedo of the crust (0.04) and the composition of the dust found near the nucleus, the crust is apparently made up of carbon-rich material mixed with

silicates and may vary from a few centimeters to a kilometer thick. Vapor and dust spew primarily from faults in the crust, which make up no more than 15 percent of the surface area.

It now seems clear that the vaporization of the ice is only the simplest part of the process that releases gases into the coma. The composition of the ice is complex, and a number of chemical reactions, some aided by ultraviolet light from the sun, break down parent molecules into the simpler forms we detect in the coma and gas tail. For the first time, in Comet Halley, water vapor, carbon dioxide, and carbon monoxide have been positively identified as parent molecules. Also detected was the CN molecule, a possible fragment of hydrogen cyanide (HCN). Methane (CH_4) was not detected in Comet Halley, although many theorists expected methane molecules to be trapped in the spaces between the frozen

a

b

FIGURE 26-23
(a) The heart of Comet Shoemaker–Levy 9 came within the Roche limit of Jupiter in July 1992, and tidal forces pulled it apart. At least 21 fragments slowly spread apart as they looped out into space and eventually returned to impact on Jupiter 2 years later.
(b) This 40-km-long chain of craters on the moon may have been formed by fragments from the nucleus of a comet ripped to fragments by the earth's tidal forces. Unlike the impacts that occurred on Jupiter, the impacts that made these craters probably happened within a few seconds. (NASA)

water molecules. Apparently, methane either is not very abundant in cometary ice or is transformed into other molecules as it is released from the icy nucleus.

As the ices vaporize, gases vent through faults in the crust and produce jets that are visible in the coma before the solar wind blows them back to form the tail (Figure 26-22). These jets can act like small rocket motors and slightly modify the orbit of the comet. This effect was used to estimate the mass of Comet Halley's nucleus.

Studies of the motions of these jets have also made it possible to find the period of rotation of some cometary nuclei. The nucleus of Comet Halley, for example, rotates with a period of 52.5 h. This rotation was detected in photographic data from 1910 and spacecraft data from 1986. But earth-based observations suggest a period of 7.4 days. Apparently, the peanut-shaped nucleus can rotate with one period and precess with the other like a badly thrown football. An active site on the nucleus turns on rapidly when the rotation of the nucleus carries it into sunlight and turns off as soon as it is carried into darkness.

The violence of gases bursting through the crust plus tidal forces can break up a cometary nucleus. A number of comets have been observed to break into two or more parts while near the sun or near Jupiter. On July 8, 1992,

Comet Shoemaker–Levy 9 passed within 1.29 planetary radii of Jupiter's center, well within its Roche limit, and tidal forces ripped it into at least 21 pieces (Figure 26-23a). The fragmented pieces of the comet were as large as a few kilometers in diameter, and they released gas and dust and spread into a long string of objects that returned to Jupiter in July 1994 to strike the planet and produce massive impacts over a period of 6 days (Chapter 24). In fact, this may explain long chains of craters found on the earth-facing side of the moon. A comet passing near the earth could be pulled apart by the earth's tidal forces, and the fragments could pound out a chain of craters on the moon (Figure 26-23b). Similar crater chains are found on Jupiter's moon Calisto. The breakup of Comet Shoemaker–Levy 9 and similar comets illustrates how fragile the nucleus of a comet really is.

A comet near the sun is a snowball in hell. Each time it swings past the sun, the nucleus can lose ices equivalent to a surface layer 1 m thick. Even if it does not break up, it must eventually lose most of its ices and become a burnt-out comet after 1000 orbits or so. A number of such comets are known, and many resemble asteroids. Comet Schwassman–Wachmann-1 is 40 km in diameter, follows a nearly circular orbit beyond Jupiter, and usually looks like an asteroid. Occasionally, however, it has outbursts and develops a coma. Similarly, asteroids with a faintly cometary appearance, such as Chiron, are known. This suggests that comets and asteroids are not fundamentally different but are similar objects that differ in their content of ices. Dead comets may look much like asteroids, and some astronomers believe that the Apollo asteroids that cross the earth's orbit are exhausted comets.

The Origin of Comets

The most interesting question we can ask about comets concerns their origin. Since they gradually vaporize in the sun's heat, the comets we see now cannot be very old. There must be a continuous supply of new comets to make up for those that exhaust their ices.

Two parts of the explanation are the Oort cloud and the Kuiper belt. These reservoirs of icy bodies continuously resupply the inner solar system with long- and short-period comets. As comets exhaust their ices and die, new icy bodies fall inward from these reservoirs. But this only sidesteps the real question. How did these objects get into the reservoirs in the first place?

When Gerard Kuiper first suggested the existence of the Kuiper belt, he proposed that the icy bodies there had formed from the solar nebula. Closer to the sun, massive planets had accreted or ejected the planetesimals, but beyond Neptune there is no massive planet, and some of the ice-rich planetesimals could have survived. So the answer for the Kuiper-belt objects is that they are the original planetesimals that formed on the outer fringes of the solar system.

The bodies in the Oort cloud, however, could not have formed at their present location, because the solar nebula would not have been dense enough so far from the sun. Also, had they formed from the solar nebula, we would expect them to be distributed in a disk and not in a sphere. Rather, astronomers think the future comets formed as icy planetesimals in the outer solar system near the present orbits of Saturn and Uranus. As the Jovian planets grew more massive, they swept up some of these planetesimals and ejected others to form the Oort cloud. If this idea is true, long-period comets are also icy planetesimals.

CRITICAL INQUIRY

How do comets help us understand the formation of the planets?

According to the solar nebula hypothesis, the planets formed from planetesimals that accreted in a disk-shaped nebula around the forming sun. The planetesimals that formed in the inner solar nebula were warm and could not incorporate much ice. The asteroids may be the last remains of such bodies. In the outer solar nebula, it was colder, and the planetesimals would have contained large amounts of ices. Many of these planetesimals were destroyed when they fell together to make

the planets, but some may have survived. The icy bodies of the Oort cloud and the Kuiper belt may be the last surviving icy planetesimals in our solar system. When these bodies fall into the inner solar system, we see them as comets, and the gases they release tell us that many of them are rich in volatile materials such as water and carbon dioxide. These are the ices we would expect to find in the icy planetesimals. Furthermore, comets are rich in dust, and the planetesimals must have included large amounts of dust frozen into the ices when they formed. Thus, the nuclei of comets seem to be frozen samples of the ancient solar nebula.

Nearly all of the mass of a comet is in the nucleus, but the light we see comes from the coma and the tail. What kind of spectra do we get from comets, and what does that tell us about the process that converts a dirty iceberg into a comet? ▪

Our discussion of comets has revealed that there is a continuous supply of these objects, that they fragment easily when they encounter planets, and that entire comets or fragments of comets can hit planets. That makes us wonder what happens when a comet hits the earth.

26-4 IMPACTS ON THE EARTH

Asteroids and comets fall through our solar system like runaway trucks on a busy highway. These objects must hit the planets. In fact, it is these collisions that built the planets, and most of the original planetesimals have now been either incorporated into the planets or ejected from the solar system. Nevertheless, planets still get hit, as was dramatically demonstrated in 1994 when Comet Shoemaker–Levy hit Jupiter. Like any other planet, the earth must get hit now and then, and we should look for evidence that such impacts have occurred.

Impacts and Dinosaurs

Comet nuclei and asteroids may seem small compared to planets, but they hit the earth with tremendous impact. Many scientists suspect that such impacts throw vast quantities of dust into the atmosphere and cause widespread changes in climate. The extinction of entire species, such as the dinosaurs, may have been caused by comet impacts (Figure 26-24a). One of the most important clues is an element called iridium.

In 1980, Luis and Walter Alvarez announced the discovery of an excess of the element iridium in sediments laid down at the end of the Cretaceous period—just when the dinosaurs and many other species became extinct. Iridium is rare in the earth's crust but common in meteorites, which led the Alvarezes to suggest that a major meteorite impact at the end of the Cretaceous threw vast

a

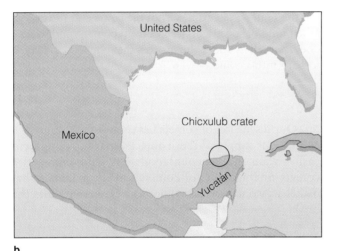

b

FIGURE 26-24

(a) The theory that the impact of one or more comets altered the earth's climate and drove dinosaurs to extinction has become so popular it appeared on this Hungarian stamp. The spacecraft shown (ICE) flew through the tail of Comet Giacobini–Zinner in 1985. Note the dead dinosaurs in the background. (b) The giant impact scar buried in the earth's crust near the village of Chicxulub in the northern Yucatán was formed about 65 million years ago. Such a large impact would have altered the earth's climate dramatically, and thus the Chicxulub impact is suspected of being the trigger that ended the Cretaceous period and destroyed the dinosaurs.

amounts of iridium-rich dust into the atmosphere. This dust might have plunged the earth into a winter that lasted many years, killing off many species of plants and animals, including the dinosaurs.

This theory was rejected at first, but soon scientists found the iridium anomaly in sediments of the same age all over the world. Others found related elements and mineral forms typical of meteorite impacts. At the same time, theorists studying nuclear weapons predicted that a nuclear war would throw so much dust into the atmosphere that our planet would be plunged into a "nuclear winter" that could last a number of years. Within a few years, scientists generally agreed that the Cretaceous extinctions were caused by one or more major impacts.

Mathematical models and observations of the impact of the comet fragments on Jupiter in 1994 have com-

bined to create a plausible scenario of the events following a major impact on the earth. Of course, creatures living near the site of the impact would probably die in the initial shock, and an impact at sea would create tsunamis (tidal waves) many hundreds of meters high that would devastate coastal regions for many kilometers inland halfway around the world. But the worst effects would begin after the initial explosion. On land or sea, a major impact would excavate large amounts of pulverized rock, heat it to high temperatures, and loft it high above the atmosphere. As this material fell back, the earth's atmosphere would be turned into a glowing oven of red-hot rock streaming through the air as a rain of meteors, and the heat would trigger massive forest fires around the world. Soot from such fires has been detected in the layers of clay laid down at the end of the Cretaceous. Once the fire storms cooled, the remaining dust in the atmosphere would block sunlight and produce deep darkness for a year or more. At the same time, large amounts of carbon dioxide locked in limestone deposits would be released into the atmosphere, where it would produce acid rain. All of these consequences make it surprising that any life could have survived such an impact.

Geologists have located a crater at least 150 km in diameter centered near the village of Chicxulub in the northern Yucatan (Figure 26-24b). Although the crater is totally covered by sediments, mineral samples show that it contains shocked quartz typical of impact sites and that it is the right age. A gigantic impact formed the crater about 65 million years ago, just when the dinosaurs and many other species died out, and many earth scientists now believe that this is the scar of the impact that ended the Cretaceous.

At first, these climate-changing impacts were blamed on large meteorites, but the nucleus of a comet would do as much damage. In fact, we have seen in this chapter that the distinction between an asteroid and a comet is not clear-cut.

Some studies of mass extinctions have suggested that they occur every 27–30 million years, and a few scientists have suggested that this might be linked to an astronomical cycle. If the sun had a low-mass companion star orbiting it with a period of 27–30 million years, it would be too distant to detect easily and would have no significant influence on the solar system. If its orbit were elliptical, however, it would come closer to the sun once on each orbit and might stir up the Oort cloud and cause a storm of comets to fall into the inner solar system. A few of these could hit the earth and trigger extinctions. This is an entertaining idea, and the proposed companion star has even been given a name—Nemesis, after the goddess of divine retribution. Nevertheless, the evidence for the periodic nature of extinctions is very shaky, and there is no evidence at all to support the existence of Nemesis.

The controversy continues, but it is an interesting example of how scientists work with new ideas. The iridium anomaly and its meaning for the Cretaceous extinctions are now widely accepted. The periodic nature of mass extinctions and impacts is viewed with caution and, in some cases, skepticism. Suggestions that showers of comets caused these extinctions are highly speculative, and most scientists are waiting for better evidence before they take such hypotheses seriously.

A solar system is a dangerous place to put a planet. Comets, asteroids, and meteoroids constantly rain down on the planets, and the earth gets hit quite often. The Chicxulub Crater isn't the only large impact scar on the earth. For example, a giant crater has been found buried under sediment in Iowa, and another giant impact crater may underlie most of the Chesapeake Bay. What would it be like to live on the earth when such an impact occurs? Humanity got a hint in 1908 when something hit Siberia.

The Tunguska Event

On the morning of June 30, 1908, scattered reindeer herders and homesteaders in central Siberia were startled to see a brilliant blue-white fireball brighter than the sun streak across the sky. Still descending, it exploded with a blinding flash and an intense pulse of heat.

The whole northern part of the sky appeared to be covered with fire. . . . I felt great heat as if my shirt had caught fire . . . there was a . . . mighty crash. . . . I was thrown on the ground about [7 m] from the porch. . . . A hot wind, as from a cannon, blew past the huts from the north.

The blast was heard up to 1000 km away, and the resulting pulse of air pressure circled the earth twice. For a number of nights following the blast, European astronomers, who knew nothing of the explosion, observed a glowing reddish haze high in the atmosphere.

Travel in the wilderness of Siberia was difficult early in this century; moreover, World War I, the Bolshevik Revolution, and the Russian Civil War prevented expeditions from reaching the site before 1927. When at last an expedition arrived, it found that the blast had occurred above the Stony Tunguska River valley and had flattened trees in an irregular pattern extending out 30 km (Figure 26-25). The trees were knocked down radially away from the center of the blast, and limbs and leaves had been stripped away. The trunks of trees at the very center of the area were still standing, although they had lost all their limbs. No expedition has ever found a crater, so it seems that the explosion, equaling a 12-megaton nuclear weapon, occurred at least a few kilometers above the ground.

All kinds of speculative theories have been proposed in tabloid newspapers and occult books. One popular idea was that a flying saucer with engine trouble tried to make an emergency landing and exploded. Slightly more responsible ideas have proposed that the earth was hit by a piece of antimatter or a miniature black hole. The evidence does not necessarily rule out either of these last two ideas, but they violate an important principle in science. They ask us to assume more than we need to. Science always prefers the simplest adequate explanation

FIGURE 26-25

The aftermath of the Tunguska blast. Trees were knocked down pointing radially away from the center of the explosion. (Sovfoto)

for any given phenomenon, and a small asteroid or comet would explain the event.

The comet hypothesis proposes that the head of a small comet struck the earth and exploded in the atmosphere. We might not see such a comet if it was coming from the direction of the sun. Witnesses reported that the fireball in the atmosphere came from the east, which in the morning means from the direction of the sun. If the object was a comet, perhaps a sungrazer, whatever tail it had would have been projected straight at the earth and would not have been visible. More likely, say some astronomers, the object was the exhausted core of a comet and had no appreciable tail.

A fragment of a partially exhausted comet should consist of silicate material resembling carbonaceous chondrites, mixed with some remaining ices. Modern studies of the area show that thousands of tons of powdered material resembling carbonaceous chondrites are scattered in the soil. A chunk of dirty ice 50 m in diameter would contain about the right mass and would be totally invisible approaching from the sun. Also, a comet fragment would be so fragile that it would explode in the atmosphere rather than dig a crater, as would a denser, stronger meteorite.

The falling-comet hypothesis has been very popular with astronomers. It does not ask us to make any fantastic assumptions, and it explains the observed phenomena, even the reddish glow over Europe. Some astronomers claim the glow was the faint traces of the comet's tail and scattered dust, trapped high in the atmosphere and illuminated by the summer sun shining over the North Pole.

As compelling as the comet hypothesis may seem, not all astronomers agree. In the early 1980s, a detailed analysis of all the evidence suggested that the object was not cometary. Its speed and direction were not consistent with the motion of a comet. The analysis suggested instead that the object was a small Apollo asteroid, 90–190 m in diameter.

In 1993, astronomers studied computer models of objects entering the earth's atmosphere at high speed and concluded that the fragile head of a comet would have exploded much too high in the atmosphere. An iron-rich asteroid, in contrast, would be so strong that it would survive to reach the ground and would form a large crater. The most likely candidate seems to be a stony asteroid about 30 m in diameter. It would explode at just about the right height to produce the Tunguska blast.

One natural question is, Why should a piece of rock explode at all? To understand, we must recall that a meteor enters the earth's atmosphere at roughly 15 km/sec. The air resistance is tremendous, and as the meteor descends lower and meets denser air, the air resistance increases rapidly. The resistance that the meteor experiences depends on its surface area. When the meteor breaks into smaller pieces, its surface area suddenly increases, so the air resistance suddenly increases and thus causes more fragmentation. Because air resistance depends on surface area, when the meteor begins to break up, it explodes and vaporizes almost instantly. That is most likely what created the blast that equaled a 12-megaton explosion at Tunguska.

Can such explosions be common? Recent searches for small asteroids passing near the earth have found that space is littered with such objects. The earth must be hit often, and large impacts like the one at Tunguska may occur every century or so. This has led to a controversy. Some experts say that major, climate-changing meteorite impacts, such as the one that may have killed off the dinosaurs, pose such a dreadful danger that we should develop massive nuclear-tipped missiles, ready to blast a meteoroid to pieces long before it can reach the earth. Some astronomers point out that lots of small fragments slamming into the earth may be even worse than one big impact. Other astronomers argue that the big objects are so rare we can ignore them. The real danger, they say, lies in the smaller, more common meteoroids, which are too small to detect with current telescopes.

Some people may prefer to believe that the Tunguska event was an alien spaceship; that is a titillating idea. But the truth about the Tunguska event is far more exciting. The earth is struck by meteorites every day. Some impacts are large enough to alter the climate, and the future of our civilization on the earth may depend on how much time we have until the next big one.

CRITICAL INQUIRY

How can a large meteorite impact alter the earth's climate?

Even a meteorite no bigger than a house strikes the earth with as much energy as a nuclear bomb and can produce a crater a mile in diameter. A large impact could produce a crater 100 km or more in diameter. Rock from the earth's crust would be pulverized and lifted high above the atmosphere. As it fell back in a rain of fiery meteors, it would be so hot it would trigger massive fires over much of the earth's surface. We can detect traces of such fires at about the time of the Chicxulub impact, 65 million years ago. Even after the fires cooled, the smoke and dust in the atmosphere would prevent sunlight from reaching the ground, and plants dependent on sunlight would be starved of energy. This world-wide darkness, called a nuclear winter, might last a year or more before the dust settled out of the atmosphere. By that time, many species would have died out, and the climate could have changed dramatically.

As scary as this is, we need not ask if it will happen. The probability of such an event is 100 percent. Our question can only be, How soon? In fact, such large impacts happen regularly on all of the terrestrial planets. Summarize the evidence we have to show that large impacts have occurred on the earth and on other planets and satellites in our solar system. ∎

Life on the earth seems fragile and exposed. Are we just lucky to have survived so long? Does life exist on other worlds? If so, has it, too, survived long enough to develop intelligence? That is the story of the next chapter. ∎

∎ *Summary*

Most of the meteors we see, and certainly those in showers, are caused by small particles released by comets. When the earth crosses the orbit of a comet, we see a meteor shower. Most meteorites, however, are more massive, stronger fragments that probably originated in the asteroid belt.

The different kinds of meteorites could have originated in a planetesimal-size body that was heated by the decay of short-lived radioactive nuclei like aluminum-26. Differentiation would form an iron–nickel core that would cool slowly. Fragments from this core would show the Widmanstätten pattern common in iron meteorites. The outer parts of such a body might resemble the stony-iron and achondritic meteorites.

Chondrites, with their beadlike chondrules, may have formed as smaller bodies farther from the sun. The carbonaceous chondrites, being rich in volatile compounds, also probably formed farther from the sun. The presence of volatiles, chondrules, and CAIs shows that the carbonaceous chondrites are the least altered of the meteorites.

Collisions must have been common among the planetesimals, just as they are still common in the asteroid belt. We find breccias in some meteorites that must have been formed in impacts long ago. In the asteroid belt today, we can find evidence that the asteroids are fragmented. The Hirayama families appear to be fragments with similar orbits. Photographs of two asteroids taken by the Galileo spacecraft and radar images of other asteroids made from the earth show irregular-shaped bodies heavily battered by impact craters.

Of the three kinds of asteroids, the S types are more common in the inner belt, whereas the C types are more common in the outer belt. This may represent a difference in temperature at the time of formation or different evolution. The M types appear to be the exposed iron–nickel cores of broken planetesimals.

Some of the asteroids show clear evidence of geological processes. The reflected light from Vesta, for instance, looks much like the light reflected from achondritic meteorites that resemble basalts. Thus, it seems that Vesta has melted and has had lava flows on its surface.

Jupiter clearly dominates the asteroid belt and may have prevented the formation of a planet there long ago. It now causes Kirkwood's gaps and traps the Trojan asteroids at the Lagrangian points along its orbit.

According to the dirty snowball theory, a comet is a ball of dirty ices that produces a coma as it approaches within 3 AU of the sun. The pressure of the solar wind pushes ionized gas back into a type I, or gas, tail; the pressure of sunlight pushes dust particles back into a type II, or dust, tail.

When Comet Halley passed through the inner solar system in 1986, five spacecraft flew past it and discovered that its nucleus was larger and blacker than expected. The nucleus was about 16 km × 8 km × 7 km and was covered with a black crust having an albedo of only 0.04. The crust is probably made of material resembling carbonaceous chondrites. Gas and dust jets vent from active regions, which cover only 15 percent of the surface.

According to the Oort cloud theory, comets occupy a hollow spherical cloud centered on the sun and extending out about 150,000 AU. Perturbations from the motions of nearby stars may cause a few of these icy bodies to fall into the inner solar system, where they become long-period comets. These dirty snowballs may have originated at their present distance or as icy planetesimals among the Jovian planets. In either case, comets are almost unaltered samples of the original solar nebula.

The short-period comets have small orbits and periods of less than 200 years. Comet Halley is an example. These comets appear to originate in the Kuiper belt, a disk-shaped cloud of icy bodies believed to extend from 30 or 40 AU to at least 100 AU from the sun. A few of these bodies have been found, and astronomers speculate that Pluto, Charon, Triton, and Chiron may be large icy worlds related to the Kuiper-belt objects.

The earth runs into space debris quite often. The Apollo asteroids must hit the earth with great regularity, and the heads of short-period comets must also intersect the earth now and then. One such collision, the Tunguska event, occurred as recently as 1908.

■ New Terms

radiant	achondrite
sporadic meteor	stony-iron meteorite
fall	Kirkwood's gaps
find	Apollo–Amor objects
iron meteorite	Trojan asteroids
selection effect	Hirayama families
Widmanstätten pattern	coma
stony meteorite	gas (type I) tail
chondrite	dust (type II) tail
chondrule	Oort cloud
carbonaceous chondrite	Kuiper belt
CAI	sungrazer

■ Questions

1. If most falls are stony meteorites, why are most finds iron?
2. How do observations of meteor showers reveal one of the sources of meteoroids?
3. How can most meteors be cometary if most, perhaps all, meteorites are asteroidal?
4. Why are meteorites easier to find in Antarctica? Why are stony meteorites better represented among these finds than among finds in the United States?
5. What evidence do we have that some meteorites have originated inside large bodies?
6. Why couldn't uranium, thorium, and radioactive potassium have melted the planetesimals? What evidence do we have that some asteroids have had molten interiors?
7. How do Kirkwood's gaps resemble Cassini's division? How do the Trojan asteroids resemble some of the satellites of Saturn?
8. The first asteroids discovered were much larger than the typical asteroid known today. How does this illustrate a selection effect?
9. Describe three types of asteroids (S, M, and C), and explain a theory to account for their differences.
10. How might the Apollo–Amor objects have originated?
11. Why doesn't a type II comet tail ever suffer a disconnection?
12. If comets are icy planetesimals left over from the formation of the solar system, why haven't they all vaporized by now?

■ Discussion Questions

1. Futurists suggest that we may someday mine the asteroids for materials to build and supply space colonies. What kinds of materials could we get from asteroids? (HINT: What are S, M, and C type asteroids made of?)
2. If cometary nuclei were heated by internal radioactive decay rather than by solar heat, how would comets differ from what we observe?
3. From what you know now, do you think the government should spend money to locate near-earth asteroids? How serious is the risk?

■ Problems

1. Large meteorites are hardly slowed by the earth's atmosphere. Assuming that the atmosphere is 100 km thick and that a large meteorite falls vertically toward the ground, how long does it spend in the atmosphere? (HINT: How fast do meteoroids travel?)
2. If a single asteroid 1 km in diameter were fragmented, how many meteoroids 1 m in diameter could it yield? (HINT: The volume of a sphere is $\frac{4}{3}\pi r^3$.)
3. What is the orbital velocity of a meteoroid orbiting the sun at the distance of the earth? (HINT: See Chapter 5.)
4. If a half-million asteroids, each 1 km in diameter, were assembled into one body, how large would it be? (HINT: The volume of a sphere is $\frac{4}{3}\pi r^3$.)

5. The asteroid Vesta has a mass of 2×10^{20} kg and a radius of about 250 km. What is its escape velocity? Could you jump off the asteroid? (HINT: See Chapter 5.)

6. The asteroid Pallas has a mass of 2.5×10^{20} kg. What is the orbital velocity of a small satellite orbiting 500 km from the center of Pallas? (HINT: See Chapter 5.)

7. What is the maximum angular diameter of Ceres as seen from the earth? Could earth-based telescopes detect surface features? Could the Hubble Space Telescope (Chapter 6)? (HINT: Use the small-angle formula.)

8. At what distances from the sun would you expect to find Kirkwood's gaps where the orbital period of asteroids is one-half of and one-third of the orbital period of Jupiter? Compare your results with Figure 26-10. (HINT: Use Kepler's third law.)

9. If the velocity of the solar wind is about 400 km/sec and the visible tail of a comet is 100 million km long, how long does it take an atom to travel from the nucleus to the end of the visible tail?

10. If you saw Comet Halley when it was 0.7 AU from the earth and it had a visible tail 5° long, how long was the tail in kilometers? Suppose that the tail was not perpendicular to the line of sight. Is your answer too large or too small? (HINT: Use the small-angle formula.)

11. Calculate the orbital velocity of a comet while it is in the Oort cloud. (HINT: See Chapter 5.)

12. The mass of a comet's nucleus is about 10^{14} kg. If the Oort cloud contains 2 trillion (2×10^{12}) cometary nuclei, what is the mass of the cloud in earth masses? (HINT: Mass of the earth = 6×10^{24} kg.)

■ Critical Inquiries for the Web

1. Some nights are better for looking for meteors than others (see the Table of Meteor Showers inside the back cover). We know that these showers are associated with comets, but how are these associations made? Several sites on the internet provide information on meteor showers, including historical data and information on parent comets. Pick a shower whose parent comet is known and summarize how we came to know that the meteors and the comet are related.

2. Chances are small that you will be killed by an asteroid impact, but if objects are out there that astronomers are not aware of whose orbits intersect Earth, we could be in for a surprise one day. Look for information on the Spacewatch project and other investigations into near-Earth asteroids. How many such objects have been discovered? What is the record for closest known passage of an asteroid to Earth?

Go to the Wadsworth Astronomy Resource Center (www.wadsworth.com/astronomy) for critical thinking exercises, articles, and additional readings from InfoTrac College Edition, Wadsworth's online student library.

Chapter Twenty-Seven

LIFE ON OTHER WORLDS

GUIDEPOST

This chapter is either unnecessary or critical depending on our point of view. If we believe that astronomy is the study of the physical universe above the clouds, then this chapter does not belong here. But if we believe that astronomy is the study of our position in the universe, not only our physical position but also our position as living beings in the origin and evolution of the universe, then everything else in this book is just preparation for this chapter.

Astronomy is the only science that truly acts as a mirror. In studying the universe up there, we learn what we are down here. Astronomy is not really about stars, galaxies and planets; it is about us. ■

Did I solicit thee from darkness to promote me?

JOHN MILTON

Paradise Lost

As living things, we have been promoted from darkness. We are made of heavy atoms that could not have formed at the beginning of the universe. Successive generations of stars fusing light elements into heavier elements have built the atoms so important to our existence. When a dark cloud of interstellar gas enriched in these heavy atoms fell together to form our sun, a small part of the cloud gave birth to the planet we inhabit.

Are there intelligent beings living on other planets? That is the last and perhaps the most challenging question in our study of astronomy. We will try to answer it in three steps, each dealing with a different aspect of life.

First, we must decide what we mean by life. A living thing is not so much a physical object as a process. We are not simply the matter that forms our bodies, but rather a tremendously complex system that has the ability to duplicate and protect itself. Thus, life is based on information that contains the directions for the processes of duplication and preservation.

Our second step is to study the origin of life. Direct investigation is limited to Earth, but if we can understand how life began here, we can better estimate the chances that it occurred elsewhere. We will find that life on Earth probably began with simple chemical reactions that happened naturally. If these gave rise to life on Earth, similar reactions may have provided the spark on other worlds.

Our third step is to study evolution, the process by which life improves its ability to survive. The survival of stable species has transformed the simple organisms that began in Earth's oceans into a wide variety of creatures with special adaptations. The rose's thorn, the deer's quickness, and the human's intelligence are protective adaptations. Evolution is a natural, physical process. If it can work on Earth, then it surely works on any planet where life begins, and if we assume that intelligence is a valuable trait, then intelligent beings may eventually emerge.

If life is common in the universe, where might we look for it, and how might we communicate with other intelligent beings? Certainly the prospects of finding life, intelligent or otherwise, on any of the other planets in our solar system are bleak. If we are to find extraterrestrial life, we must go beyond our solar system and search among any planets that may orbit other stars.

Communication with intelligent races on other worlds may be possible, but we cannot expect to travel between solar systems. Interstellar distances are so great that only in science fiction do spaceships flit from star to star. However, it may be possible to communicate via radio. If civilizations can survive for long periods of time at a technological level at which they can build large radio telescopes, then we may be able to send and receive messages. Such messages would mark a turning point in the history of humanity. If life is common in the universe, such signals may be detected during our lifetime.

Our goal in this chapter is to use our knowledge of astronomy, combined with the rules of evidence that guide all of science, to try to understand the range of possibilities for life in our universe (Window on Science 27-1). We begin with a simple question: Is life on other worlds possible?

27-1 THE NATURE OF LIFE

What is life? Philosophers have struggled with that question for thousands of years, so it is unlikely that we will answer it here. But we must agree on a working model of life before we can speculate on its occurrence on other worlds. To that end, we will identify in living things three properties: a process, a physical base, and a unit of controlling information.

The life process is aimed at survival. Living things extract energy from their environment and use that energy to modify their surroundings to make their own preservation more likely. For example, human beings obtain energy by eating and breathing, and they use that energy to build houses, cities, and stable societies to protect themselves. The same can be said of a bacterium absorbing food and reinforcing its cell structure.

This apparently selfish protection of the individual is aimed at the preservation of the race through safe reproduction. The ability to reproduce is one of the distinguishing characteristics of living organisms, and any organism that does not, in some way, ensure safe reproduction will not survive many generations. The entire life process is aimed at safe reproduction because any other target is self-destructive.

The physical basis of life on Earth is the chemistry of the carbon atom (Figure 27-1). Because of the way in which this atom bonds to other atoms, it can form long, complex, stable chains that are capable of extracting, storing, and utilizing energy. Other chemical bases may exist. Science fiction stories and movies abound with silicon creatures, living things whose body chemistry is based on silicon rather than carbon. However, silicon forms weaker bonds than carbon does, and it cannot form double bonds as easily. Consequently, it cannot form the long, complex, stable chains that carbon can. Silicon is 135 times more common on Earth than carbon, yet there are no silicon people among us. All Earth life is carbon-based. Thus, the likelihood that distant planets are inhabited by silicon people seems small, but we cannot rule out life based on noncarbon chemistry.

In fact, nonchemical life might be possible. All that nature requires is some mechanism capable of supporting the extraction and utilization of energy that we have identified as life. One could at least imagine life based on electromagnetic fields and ionized gas. No one has ever met such a creature, but science fiction writers conjure up all sorts.

Clearly, we could range far in space and time theorizing about alien life, but to make progress we must discuss what we know best—carbon-based life on Earth. We must try to understand how it works and how it came to exist. Only then can we consider life on other worlds.

The Nature of Scientific Explanation

Science is a way of understanding the world around us, and the heart of that understanding is the explanations that science gives us for natural phenomena. Whether we call these explanations stories, histories, theories, or hypotheses, they are attempts to describe how nature works based on fundamental rules of evidence and intellectual honesty. While we may take these explanations as factual truth, we should understand that they are not the only explanations that satisfy the rules of logic.

A separate class of explanations involves religion, and those explanations can be quite logical. The Old Testament description of the creation of the world, for instance, does not fit scientific observations, but if we accept the existence of an omnipotent being, then the biblical explanation is internally logical and acceptable. Of course, it is not a scientific explanation, but religion is a matter of faith and not subject to the rules of evidence. Religious explanations follow their own logic, and we would be wrong to demand that they follow the rules of evidence that govern scientific explanations.

If scientific explanations are not the only logical explanations, then why do we give them such weight? First, we must notice the tremendous success of scientific explanations in producing technological advances in our daily lives. Smallpox is a disease of the past thanks to the application of scientific explanations to modern medicine. The power of science to shape our world can lead us to think that its explanations are unique. Second, the process we call science depends on the use of evidence to test and perfect our explanations, and the logical rigor of this process gives us great confidence in our conclusions.

Scientific explanations have given us tremendous insight into the workings of nature, and consequently both scientists and nonscientists tend to forget that there can be other logical explanations. The so-called conflict between science and religion has been symbolized for centuries by the trial of Galileo. That conflict is easier to understand when we consider the nature of scientific explanations and the role of evidence in testing scientific understanding. ■

a

b

FIGURE 27-1

All living things on Earth are based on carbon chemistry. (a) Katie, a complex mammal containing about 30 astronomical units of DNA. (Michael Seeds) (b) Tobacco mosaic virus. Each rod is a single spiral strand of RNA about 0.01 mm long surrounded by a protein coat. (L. D. Simon)

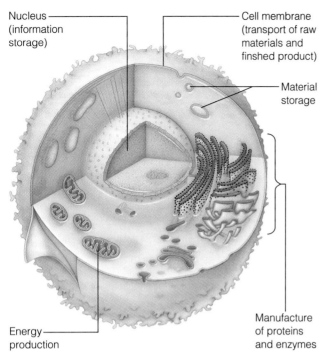

Nucleus
(information
storage)

Cell membrane
(transport of raw
materials and
finshed product)

Material
storage

Manufacture
of proteins
and enzymes

Energy
production

FIGURE 27-2
A living cell is a self-contained factory that absorbs raw materials from its surroundings and uses them to maintain itself and manufacture finished products for the use of the organism as a whole.

FIGURE 27-3
The DNA molecule consists of two vertical strands of sugars and phosphates (dark) and rungs made of bases: adenine (A), cytosine (C), guanine (G), and thymine (T).

The DNA Code

The key to understanding life is information—the information the organism uses to control its utilization of energy. We must discover how life stores and uses that information and how the information changes and thus preserves the species.

The unit of life on Earth is the cell (Figure 27-2), the self-contained factory capable of absorbing nourishment from its surroundings, maintaining its own existence, and performing its task within the larger organism. The foundation of the cell's activity is a set of patterns that describes how it is to function. This information must be stored in the cell in some safe location, yet it must be passed on easily to new cells and be used readily to guide the cell's activity. To understand how matter can be alive, we must understand how the cell stores, reproduces, and uses this information.

The information is stored in long, carbon-chain molecules called **DNA (deoxyribonucleic acid),** most of which reside in the cell nucleus. The structure of DNA resembles a long, twisted ladder. The vertical pieces of the ladder are made of alternating phosphates and sugars; the rungs are made of pairs of molecules called bases (Figure 27-3). Only four kinds of bases are present in DNA, and the order in which they appear on the DNA ladder represents the information the cell needs to function. One human cell stores about 1.5 m of DNA, containing about 4.5 billion pairs of bases. Thus, 4.5 billion pieces of information are available to run a human cell.

That is enough to record all the works of Shakespeare over 200 times. Because the human body contains about 60×10^{12} cells, the total DNA in a single human would stretch 9×10^{13} m, about 600 AU.

Storing all these data in each cell does the organism no good unless the data can be reproduced and passed on to new cells. The DNA molecule is specially adapted for duplicating itself by splitting its ladder down the center of the rungs, producing two vertical strands with protruding bases (Figure 27-4). These quickly bond with the proper bases, phosphates, and sugars to reconstruct the missing part of the molecule, and, presto, the cell has two complete copies of the critical information. One set goes to each of the newly forming cells. Thus, DNA is the genetic information passed from parent to offspring.

Segments of DNA molecules are patterns for the production of **proteins.** Many proteins are structural molecules; the cell might make protein to repair its cell wall, for example. **Enzymes** are special proteins that control other processes—growth, for example. Thus, the DNA molecule contains the recipes to make all the different molecules required in an organism.

Actually, the cell does not risk its precious DNA patterns by involving them directly in the manufacture of proteins. DNA stays safely in the cell nucleus, where it produces a copy of the patterns by assembling a long carbon-chain molecule called **RNA (ribonucleic acid).** RNA carries the information out of the nucleus and then assembles the proteins from simple molecules called **amino acids,** the basic building blocks of protein. Thus, RNA acts as a messenger, carrying copies of the necessary plans from the central office to the construction site.

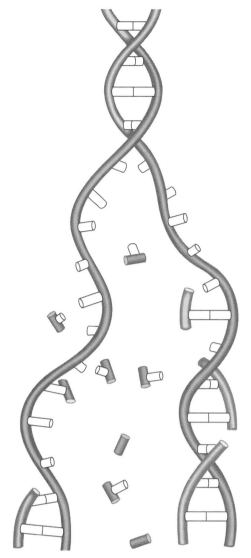

FIGURE 27-4
The DNA molecule can duplicate itself by splitting in half (top), assembling matching bases, sugars, and phosphates (center), and thus producing two DNA molecules (bottom). The actual duplication process is significantly more complex than this schematic diagram.

Notice that the information stored in DNA consists of units of information commonly called genes. A gene may consist of a long length of DNA. We commonly think of genes as units of information strung together to make up our chromosomes. When we blame our heredity on our genes and chromosomes, we are talking about the DNA that makes us who we are.

Clearly, it is important that an organism protect its DNA information from harm, but it is important to the species that the DNA be capable of change. If the information in DNA couldn't be modified, the species would become extinct. To see why, we must study evolution, the process that rewrites the data in DNA.

The Evolution of Life Forms

Every living thing on Earth is part of a web of interdependence. Not only now but ever since life began, life forms have depended on one another for food and shelter. This dependence exposes life to a serious danger in that gradual changes in climate may destroy one life form and endanger hundreds of others. A slight warming of the climate, for example, might kill a species of plant, starving the rabbits, deer, and other herbivores and leaving the hawks, wolves, and mountain lions with no prey. If a species is to survive in such a world, it must be able to adapt to changing conditions, and that means the data coded in DNA must change. The species must evolve.

Species evolve because of a process called **natural selection.** Each time an organism reproduces, its offspring receive the data stored in DNA, but some variation is possible. For example, most of the rabbits in a litter may be normal, but it is possible for one to get a DNA recipe that gives it stronger teeth. If it has stronger teeth, it may be able to eat something other than the plant the others depend on, and if that plant is becoming scarce, the rabbit with stronger teeth has a survival advantage. It can eat other plants and will be healthier than its litter-mates and have more offspring. Some of these offspring may also have stronger teeth as the altered DNA data are handed down to the new generation. Thus, nature selects and preserves those attributes that contribute to the survival of the species. Those that are unfit die. Natural selection is merciless to the individual, but it gives the species the best possible chance to survive.

The only way nature can obtain new DNA patterns from which to select the best is to alter actual DNA molecules. This can happen through chance mismatching of base pairs—errors—in the reproduction of the DNA molecule. Another way this could occur is through damage to reproductive cells from exposure to radioactivity. Cosmic rays or natural radioactivity in the soil might play this role. In any case, an offspring born with altered DNA is called a **mutant.** Most mutations are fatal, and the individual dies long before it can have offspring of its own. But rarely a mutation may give a species a new survival advantage. Then natural selection makes it likely that the new DNA message will survive and be handed down, making the species more capable of surviving.

CRITICAL INQUIRY

Why can't the information in DNA be permanent?

The information stored in a creature's DNA provides the recipes that make the creature what it is. For example, the DNA in a starfish must contain all the recipes for making the various kinds of proteins needed to consume and digest food. That information must be passed on to offspring starfish, or they will be unable to survive. But the information must be changeable because our

FIGURE 27-5

Trilobites made their first appearance in the Cambrian oceans about 600 million years ago. This example, about the size of a human hand, lived 400 million years ago in an ocean floor that is now a limestone deposit in Pennsylvania. (Grundy Observatory)

FIGURE 27-6

Among the oldest fossils known, this microscopic filament resembles modern bacterial forms (artist's reconstruction at right). This fossil was found in the 3.5-billion-year-old chert of the Pilbara Block in northwestern Australia. (J. William Schopf)

environment is changeable. Ice ages come and go, mountains rise, lakes dry up, and ocean currents shift. If the environment changes in some way, one or more of the recipes may no longer work. In our example, a change in the temperature of the ocean water may kill off the shellfish the starfish eat. If they can't digest other shellfish, the entire species will become extinct. Natural variation in DNA means that among all the infant starfish in any generation, some of the recipes are different; if the environment changes, all of the old-style starfish may die, but a few—those with the different DNA—can carry on.

The survival of life depends on this delicate balance between reliable reproduction and the introduction of small variations in DNA information. What are some of the ways these small changes in DNA can arise? ∎

Life is based not only on information, but on the duplication of information. Today, that process seems so complex that it is hard to imagine how it could have begun.

27-2 THE ORIGIN OF LIFE

Clearly the carbon chemistry of life on Earth is extremely complex. How could it have ever gotten started? Obviously, 4.5 billion chemical bases didn't just happen to drift together to form the DNA formula for a human. The key is evolution. Once a life form begins to reproduce itself, natural selection preserves the most advantageous traits. Over long periods spanning thousands, perhaps millions, of generations, the life form becomes more fit to survive. This nearly always means the life form becomes more complex. Thus, life could have begun as a very simple process that gradually became more sophisticated as it was modified by evolution.

We begin our study on Earth, where fossils and an intimate familiarity with carbon-based life give us a glimpse of the first living matter. Once we discover how earthly life could have begun, we can look for signs that life began on other planets in our solar system. Finally, we speculate on the chances that other planets, orbiting other stars, have conditions that give rise to life.

The Origin of Life on Earth

The oldest fossils hint that life began in the oceans. The oldest easily identified fossils appear in sedimentary rocks that formed 500–600 million years ago—the **Cambrian period.** Such Cambrian fossils were simple ocean crea-

a

FIGURE 27-7

The Miller experiment (a) circulated gases through water in the presence of an electric arc. This simulation of primitive conditions on Earth produced amino acids, the building blocks of proteins. (b) Stanley Miller with a Miller apparatus. (Courtesy Stanley Miller)

b

tures, the most complex of which were trilobites (Figure 27-5), but there are no Cambrian fossils of land plants or animals. Evidently, land surfaces were totally devoid of life until only 400 million years ago.

Precambrian deposits contain no obvious fossils, but microscopes reveal microfossils that were the ancestors of the Cambrian creatures. Fig-tree chert* in South Africa is 3.0–3.3 billion years old, and Onverwacht shale, also found in South Africa, may be as old as 3.6 billion years. These and similar deposits contain structures that appear to be microfossils of bacteria or simple algae such as those that live in water (Figure 27-6). Apparently, life was already active in Earth's oceans a billion years after the planet formed.

The key to the origin of this life may lie in an experiment performed by Stanley Miller and Harold Urey in 1952. This **Miller experiment** sought to reproduce the conditions on Earth under which life began (Figure 27-7). In a closed glass container, the experimenters placed water (to represent the oceans), the gases hydrogen, ammonia, and methane (to represent the primitive atmosphere), and an electric arc (to represent lightning bolts). The apparatus was sterilized, sealed, and set in operation.

After a week, Miller and Urey stopped the experiment and analyzed the material in the flask. Among the

many compounds the experiment produced, they found four amino acids that are common building blocks in protein, various fatty acids, and urea, a molecule common to many life processes. Evidently, the energy from the electric arc had molded the atmospheric gases into some of the basic components of living matter. Other energy sources such as hot silica (to simulate hot lava spilling into the sea) and ultraviolet radiation (to simulate sunlight) give similar results.

Recent studies of the composition of meteorites and models of planet formation suggest that Earth's first atmosphere did not resemble the gases used in the Miller experiment. Earth's first atmosphere was probably composed of carbon dioxide, nitrogen, and water vapor. This change, however, does not invalidate the Miller experiment. The point of the experiment is to show how easily organic molecules occur in natural settings.

The Miller experiment did not create life, nor did it necessarily imitate the exact conditions on the young Earth. Rather, it is important because it shows that complex organic molecules form naturally in a wide variety of circumstances. The chemical deck is stacked to deal nature a hand of complex molecules. If we could travel back in time, we would probably find Earth's oceans filled with a rich mixture of organic compounds in what some have called the **primordial soup.**

The next step on the journey toward life is for the compounds dissolved in the oceans to link up and form larger molecules. Amino acids, for example, can link to

*Chert is a rock form that resembles flint.

FIGURE 27-8
Amino acids can link through the release of a water molecule to form long carbon-chain molecules. The amino acid in this hypothetical example is alanine, one of the simplest.

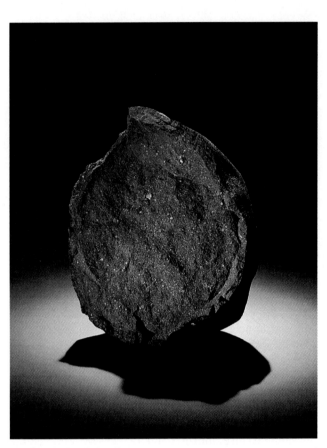

FIGURE 27-9
A sample of the Murchison meteorite, a carbonaceous chondrite that fell in 1969 near Murchison, Australia. Analysis of the interior of the meteorite revealed evidence of amino acids. Whether the first building blocks of life originated in space is unknown, but the amino acids found in meteorites illustrate how commonly amino acids and other complex molecules occur in nonorganic settings. (Chip Clark, National Museum of Natural History)

FIGURE 27-10
Single amino acids can be assembled into long protein-like molecules. When such material cools in water, it can form microspheres, microscopic spheres with double-layered boundries similar to cell membranes. Microspheres may have been an intermediate stage in the evolution of life between complex molecules and cells holding molecules reproducing genetic information. (Courtesy Sidney Fox and Randall Grubbs)

form proteins. This linkage occurs when amino acids join end to end and release a water molecule (Figure 27-8). For many years, scientists have assumed that this process must have happened in sun-warmed tidal pools where evaporation concentrated the broth. But recent studies suggest that the young Earth was subject to extensive volcanism and large meteorite impacts that periodically modified the climate enough to destroy any life forms exposed on the surface. It seems most likely that the early growth of complex molecules took place among the hot springs along the midocean ridges. The heat from such springs could have powered the growth of long protein chains. Deep in the oceans, they would have been safe from climate changes.

Although these proteins might have contained hundreds of amino acids, they were not alive. Not yet. Such molecules did not reproduce but merely linked and broke apart at random. However, because some molecules are more stable than others and because some molecules bond more readily than others, this **chemical evolution** led to the concentration of the varied smaller molecules into the most stable larger forms. Eventually, somewhere in the oceans, a molecule took shape that could reproduce a copy of itself. At that point, the chemical evolution of molecules became the biological evolution of living things.

FIGURE 27-11
A 3.5-billion-year-old fossil stromatolite from western Australia is one of the oldest known fossils (below). Stromatolites were formed, layer by layer, by mats of blue-green algae or bacteria living in shallow water. Such algae may have been common in shallow seas when Earth was young (right). Stromatolites are still being formed today in similar environments. (Mural by Peter Sawyer; photo courtesy Chip Clark, National Museum of Natural History)

It is natural for us to assume that the first reproducing molecule was some form of DNA, but that is not necessary. DNA is a very complex molecule. Some experts believe that the first reproducing molecule was a primitive form of RNA, a simpler molecule. The first life may have lived in an RNA world, and DNA may have developed later. Still others argue that the first replicating molecules were some form of protein that then formed the first RNA. The details are far from clear, but there seem to be a number of ways that reproducing molecules could have formed.

The building blocks that created these reproducing molecules—amino acids and even simpler organic molecules—are very common. They seem to form easily, and some of these building blocks have even been found in space by radio astronomers. Meteorites have been found to harbor such organic materials (Figure 27-9), and the interplanetary dust sifting into our atmosphere from space is contaminated with these organic molecules. It is possible that meteorite and comet impacts brought the building blocks of life to Earth, but given how readily they form, Earth probably had an adequate supply of its own.

Which came first, reproducing molecules or the cell? Because we think of the cell as the basic unit of life, this question seems to make no sense, but in fact the cell may have originated during chemical evolution. If a dry mixture of amino acids is heated, the acids form long, proteinlike molecules that, when poured into water, collect to form microscopic spheres that function in ways similar to cells (Figure 27-10). They have a thin membrane surface, absorb material from their surroundings, grow in size, and divide and bud just as cells do. However, they contain no large molecule that copies itself. Thus, the structure of the cell may have originated first and the reproducing molecules later.

An alternative theory supposes that the replicating molecule developed first. Such a molecule would be exposed to damage if it were bare, so the first to manufacture or attract a protective coating of protein would have a significant survival advantage. If this is the case, the protective cell membrane is a later development of biological evolution.

The first living things must have been single-celled organisms much like modern bacteria and simple algae. Some of the oldest fossils known are **stromatolites,** structures produced by communities of blue-green algae or bacteria, which grew in mats and year by year deposited layers of minerals, which were later fossilized. One of the oldest such fossils known is 3.5 billion years old (Figure 27-11). If such algae were common when Earth was young, they might have been able to produce a small amount of oxygen in the early atmosphere. Recent studies suggest that only 0.1 percent oxygen would have been sufficient to provide an ozone screen, which would protect organisms from the sun's ultraviolet radiation.

How evolution shaped creatures to live in the ancient oceans, molded them into multicellular organisms, and developed sexual reproduction, photosynthesis, and respiration is a fascinating story, but we cannot explore it in detail here. We can see that life could begin through simple chemical reactions building complex molecules and that once some DNA-like molecule formed, it protected its own survival with selfish determination. Over billions of years, the genetic information stored in living things kept those qualities that favored survival and discarded the rest. As Samuel Butler said, "The chicken is the egg's way of making another egg." In that sense, all living matter on Earth is merely the physical expression of DNA's mindless determination to continue its existence.

Perhaps this seems harsh. Human experience goes far beyond mere reproduction. *Homo sapiens* has art, poetry, music, philosophy, religion, science—all of the great, sensitive accomplishments of our intelligence. Perhaps that is more than mere reproduction of DNA, but intelligence, the ability to analyze complex situations and respond with appropriate action, must have begun as a survival mechanism. For example, a fixed escape strategy stored in DNA is a disadvantage for a creature that frequently moves from one environment to another. A rodent that always escapes from predators by automatically climbing the nearest tree would be in serious jeopardy if it met a hungry fox in a treeless clearing. Even a faint glimmer of intelligence might allow the rodent to analyze the situation and, finding no trees, to choose running over climbing. Thus, intelligence, of which *Homo sapiens* is so proud, may have developed in ancient creatures as a way of making them more versatile.

Any discussion of the evolution of life seems to involve highly improbable coincidences until we consider how many years have passed in the history of Earth. We read the words—4.6 billion years—so easily, but in truth it is hard to grasp the meaning of such a long period of time.

Geologic Time

Humanity is a very new experiment on planet Earth. We can fit all of the evolution that led from the primitive life forms in the oceans of the Cambrian period 600 million years ago to fishes, amphibians, reptiles, and mammals into a single chart such as Figure 27-12. We can take comfort in thinking that humanlike creatures have walked on Earth for at least 3 million years, but when we add our history to the chart, we discover that the entire history of the human race makes up no more than a thin line at the top. In fact, if we try to represent the entire 4.6-billion-year history of Earth on the chart, the portion describing the rise of life on land is an unreadably small segment.

One way to represent the evolution of life is to compress the 4.6-billion-year history of Earth into a 1-year-long videotape. In such a program, Earth forms as the video begins on January 1, and through all of January and February it cools and is cratered and the first oceans form. Search as we might, we will find no trace of life in these oceans until sometime in March or early April, when the first living things develop. The slow development of these simplest of living forms grinds on slowly

through the spring and summer of our videotape. The entire 4-billion-year history of Precambrian evolution lasts until the video reaches mid-November, when the primitive ocean life begins to evolve into more complex organisms such as trilobites.

While our year-long videotape plays on and on, we might amuse ourselves by looking at the land instead of the oceans, but we would be disappointed. The land is a lifeless waste, with no plants or animals of any kind. Not until November 28 in our video does life appear on the land, but once it does, it evolves rapidly into a wide range of plants and animals. Dinosaurs, for example, appear about December 12 and vanish by Christmas evening as mammals and birds flourish.

Throughout the 1-year run of our video, there have been no humans, and even during the last days of the year, as the mammals rise to dominate the landscape, there are no people. In the early evening of December 31, vaguely human forms move through the grasslands, and by late evening they begin making stone tools. The Stone Age lasts until about 11:45 PM, and the first signs of civilization, towns and cities, do not appear until 11:54 PM. The Christian era begins only 14 seconds before the New Year, and the Declaration of Independence is signed with but 1 second to spare.

By converting the history of Earth into a year-long videotape, we have placed the rise of life in perspective. Tremendous amounts of time were needed to create the first simple living things in the oceans, and even more time was needed to develop complex creatures that could colonize the land. As life became more complex, it evolved and diversified faster and faster, as if evolution were drawing on a growing library of solutions that had previously been invented with great effort to solve earlier problems. The burst of diversity on land led slowly to the rise of intelligent creatures like us, a process that has taken 4.6 billion years.

If life could originate on Earth and develop into intelligent creatures, perhaps the same thing could have happened on other planets. This raises three questions. First, could life originate if conditions were suitable? Second, if life begins on a planet, will it evolve toward intelligence? The answer to both these questions seems to be yes. The direction of chemical and biological evolution is directed toward survival, which should lead to versatility and intelligence. But what of the third question: Are suitable conditions so rare that life almost never gets started? The only way to answer this question is to search for life on other planets. We begin with Mars.

Evidence of Past Life on Mars?

Moderate temperatures and evidence of water make Mars a promising candidate, but the two Viking landers found no evidence of life in 1976 when they scooped up soil samples and tested them for living organisms. Those tests, however, could not detect fossil evidence.

In 1996, a team of scientists announced the discovery of what appear to be fossil traces of bacterial life in a

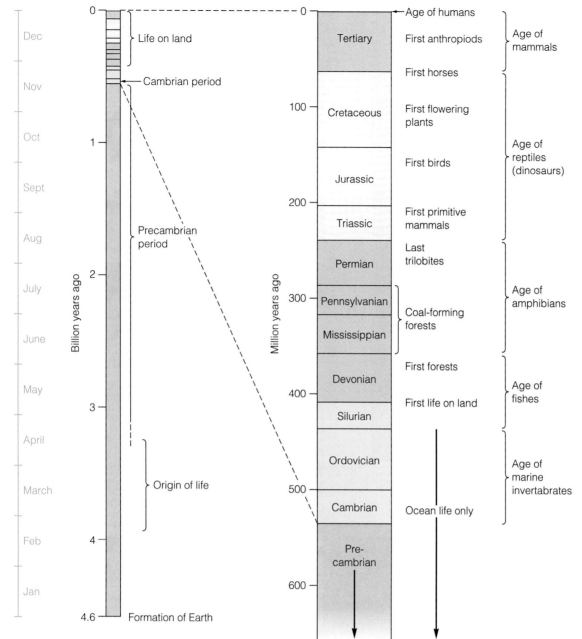

FIGURE 27-12

Complex life has developed on Earth only recently. If the entire history of Earth were represented in a time line (left), we would have to magnify the end of the line to see details such as life leaving the oceans and dinosaurs appearing. The age of humans would still be only a thin line at the top of our diagram. If the history of the earth were a year-long videotape, humans would not appear until the last hours of December 31st.

meteorite that originated on Mars. Meteorite ALH84001 (see Figure 23-24b) was found on the Antarctic ice in 1984 and only recently recognized as one of 12 meteorites believed to have come from Mars. How do we know it is from Mars? An even dozen of these SNC-class meteorites are known, and one of them contains gases trapped in small glassy particles. Analysis of these gases shows that they have the same abundance as the gases in the atmosphere of Mars as measured by the two Viking spacecraft that landed there in 1976. The abundances are like fingerprints and they match exactly, so it seems very likely that the SNC meteorites are rocks that were

blasted off of the surface of Mars at some point in the past when a comet or an asteroid struck the planet. From traces in ALH84001, scientists conclude it left the surface of Mars about 16 million years ago, orbited the sun for a long time, and then fell to Earth in Antarctica about 13,000 years ago.

Chemical, mineral, and fossil evidence suggest that the Mars meteorite once contained life. The chemical evidence was found in the form of PAHs (polycyclic aromatic hydrocarbons), complex carbon-based molecules that are characteristic of living things. Had there been living bacteria in the sample long ago, the decay of the

a

b

c

FIGURE 27-13

(a) Microscopic orange-colored carbonate mineral globules were found inside meteorite ALH84001, which is believed to have originated on Mars. The chemical and geometrical structure of the globules resembles mineral deposits created by earthly bacteria. (b) An electron micrograph of a section from carbonate globules in the Mars meteorite reveals egg shapes and elongated tube shapes that resemble small bacteria on Earth. (c) Concentrations of possible fossil bacteria are located in the carbonate globules, which is also where chemical analysis reveals carbon compounds (PAHs) that may be the molecular remains of living tissue. (NASA)

bacteria would have released many different kinds of molecules, including PAHs. Of course, PAHs have been found in many settings, including other meteorites and interplanetary dust, but those were simpler molecules and had a chemical structure typical of nonbiological origins. The PAHs found in the Mars meteorite are quite different and resemble the carbon compounds produced on Earth by living things.

The mineral evidence consists of small globules of carbonates in fine cracks deep inside the meteorite (Figure 27-13a). Evidently deposited by moderate-temperature water (0°C to 80°C) seeping through the rock, these mineral deposits are very similar to minerals released by some of Earth's bacteria. Yet some of the globules are found along fractures in the rock that scientists can date to billions of years before the meteorite landed on Earth. Furthermore, some of the globules themselves are frac-

tured, and so the globules can't be contamination from water on Earth and probably were produced by water on Mars. The structure of these carbonate globules includes small amounts of magnetite and iron-rich sulfides near the center, with a layered rim of sulfides, and that is typical of some bacterial colonies known on Earth. Dramatically, the traces of PAHs are found in the regions rich in carbonates.

Finally, scanning electron microscopes were able to detect at the centers of some of the globules structures that look like fossilized bacteria (Figure 27-13b and c). The tiny egg-shaped objects or elongated tube shapes resemble certain earthly bacteria that are known to precipitate minerals as they grow. These objects are about 100 to 1000 times smaller than the diameter of a human hair, so they are difficult to study; but their location in the carbonate globules and their primitive forms are consistent with the idea that they are the fossilized remains of early Martian life.

Chemical, mineral, and fossil evidence are individually suggestive but not conclusive. That all three are found very close together lends strength to the belief that they are traces of life.

Nevertheless, scientists, being professionally skeptical (see Window on Science 20-2), are testing these results in every possible way, and some tests contradict the life explanation. For example, some experts argue that the carbonate globules were deposited by water as hot as 650°C, much too hot for biological processes. But other studies suggest temperatures below 80°C, comfortable for bacteria. Some scientists contend that the microscopic fossils are natural whiskers of the mineral magnetite, but others claim that they are fossils of living organisms similar to the smallest bacteria found on earth. Some suspect that the meteorite was contaminated while in the Antarctic ice, but others point to evidence that earthly water never soaked into the meteorite's interior.

The debate continues. Long after the first announcement, no conclusive evidence has been found. A final verdict may not be found until humans visit Mars in the next century.

Life on Jovian Moons

Once we leave Mars, our solar system harbors only a few places where life might exist. Mercury, the moon, and most of the satellites are airless, and most other bodies such as Venus, Pluto, and the Jovian worlds have no liquid water. The moons of the Jovian worlds, however, may have conditions more favorable to life.

Methane could be important. Saturn's moon, Titan, has a nitrogen atmosphere containing methane gas and may have oceans of ethane on its surface (Chapter 24). Sunlight acting on the methane can form complex organic molecules (molecules with a carbon "backbone"), and this material produces the thick smog that makes the atmosphere opaque. These smog particles should settle out of the atmosphere to form deposits on the surface, where the organic goo could give rise to living things. The low temperature (−178°C), however, may forbid the chemical reactions that are the basis of life. Neptune's moon Triton also has methane in its atmosphere, but it is even colder than Titan (−236°C).

Methane can make organic molecules, but liquid water seems to be the key to life. Jupiter's moon Europa may be our best candidate after Mars. Cracks in its icy crust suggest that the ice floats on an ocean of liquid water (Figure 27-14), and the clean, fresh surface of the ice suggests that heat continues to flow out of the interior. On Earth, most living things depend on energy from sunlight, but creatures have been found in Earth's oceans that extract energy from hot water seeping out of the ocean floor. Thus, heat flowing up into Europa's water ocean might support living things. It may be a very long time before humans can test that hypothesis in person.

FIGURE 27-14

The icy surface of Europa, digitally enhanced and colored here, is covered by a network of cracks whose shape suggests that the crust floats on an ocean of water. Heat flowing from the interior might provide the energy to support living things in the dark ocean. (NASA)

Except for Earth, our solar system seems inhospitable to life, so our search turns to other planetary systems.

Life in Other Planetary Systems

Although we cannot visit other stars, we can draw some conclusions about the chances of finding life on other planets. Part of the problem is astronomical: Are planets common? But part is biological: Can life begin on other worlds? Our ability to discuss life on other worlds is severely limited by our lack of experience. We know of only one planet with life—Earth.

In Chapter 20, we found strong evidence that planets are common. Observational evidence shows that star formation creates disks around forming protostars, and we can detect such disks around young stars. We also considered the theory that planets form in such disks, and we saw evidence that some stars are orbited by planets.

If a planet is to become a suitable home for life, it must have a stable orbit around its sun. This is simple in a solar system like our own, but in a binary system most planetary orbits are unstable. Most planets in such systems would not last long before they were swallowed up by one of the stars or ejected from the system.

Thus, it seems that single stars are the most likely to have planets suitable for life. Because our galaxy contains about 10^{11} stars, half of which are single, there should be roughly 5×10^{10} planetary systems in which we might look for life.

A few million years of suitable conditions does not seem to be enough time for life to originate. Our planet required at least 0.5–1 billion years to create the first cells and 4.6 billion years to create intelligence. Clearly, conditions on a planet must remain acceptable over a long

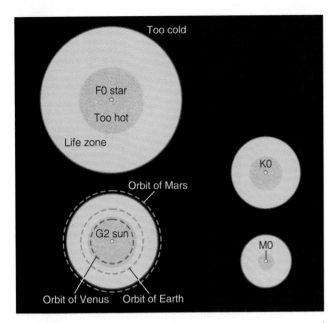

FIGURE 27-15

The life zone around a star is the region where a planet would have a moderate temperature. Too close to the star, the planet will be too hot; too far from the star, the planet will be too cool. The size of a life zone depends on the temperature of the star.

time. This eliminates giant stars that change their luminosity rapidly as they evolve. It also eliminates massive stars that remain stable on the main sequence for only a few million years. If life requires a few billion years to originate and evolve to intelligence, no star hotter than about F5 will do. This is not really a serious restriction, because upper main-sequence stars are rare anyway.

In previous sections, we decided that life, as we know it at least, requires liquid water. That requirement defines a **life zone** (or ecosphere) around each star, a region within which a planet has temperatures that permit the existence of liquid water.

The size of the life zone depends on the temperature of the star (Figure 27-15). Hot stars have larger life zones because the planets must be more distant to remain cool. But the short main-sequence lives of these stars make them unacceptable. M stars have small life zones because they are extremely cool; only planets very near the star receive sufficient warmth. However, planets that are close to a star would probably become tidally coupled, keeping the same side toward the star. This might allow the water and atmosphere to freeze in the perpetual darkness of the planet's night side and end all chance of life. Also, M stars are subject to sudden flares that might destroy life on a planet close to the star. Thus, the life zone restricts our search for life to main-sequence G and K stars. Some of the cooler F stars and warmer M stars might also be good candidates.

Even a star on the main sequence is not perfectly stable. Main-sequence stars gradually grow more luminous as they convert their hydrogen to helium, and thus the life zone around a star gradually moves outward. A planet might form in the life zone, and life might begin and evolve for billions of years only to be destroyed as the slowly increasing luminosity of its star moved the life zone outward, evaporated the planet's oceans, and drove off its atmosphere. If a planet is to remain in the life zone for 4–5 billion years, it must form on the outer edge of the zone. This may be the most serious restriction we have yet discussed.

If all of these requirements are met, will life begin? Early in this chapter, we decided that life could begin through simple chemical reactions, so perhaps we should change our question and ask, What could prevent life from beginning? Given what we know about life, it should arise whenever conditions permit, and our galaxy should be filled with planets that are inhabited with living creatures.

CRITICAL INQUIRY

What evidence do we have that life is at least possible on other worlds?

Our evidence is limited almost entirely to Earth, but it is promising. The fossils we find show that life originated in Earth's oceans almost 4 billion years ago, and biologists have proposed relatively simple chemical processes that could have created these first reproducing molecules. The fossils show that life developed very slowly at first into more and more complex creatures that filled the oceans. The pace of evolution quickened dramatically about 600 million years ago, the beginning of the Cambrian period, when life began taking on complex forms. Later, when life emerged onto the surface of the land, it again evolved rapidly to produce the tremendous diversity we see around us. Human intelligence has been a very recent development, only a few million years old.

If this process occurred on Earth, then it seems reasonable that it could have occurred on other worlds as well. We have only limited evidence from the Viking landers on Mars, which searched for and did not find recognizable signs of life, but we can expect that life might begin and evolve to intelligent forms on any world where conditions are right. What are the conditions we should expect of other worlds that host life? ■

It is both easy and fun to speculate about life on other worlds, but it leads to a simple question: Is there really life beyond Earth? If we can't leave Earth and visit other worlds, then the only life we can detect will be beings intelligent enough to communicate with us. Why haven't we heard from them?

UFOs and Space Aliens

When we discuss life on other worlds, we might be tempted to use UFO sightings and supposed visits by aliens from outer space as evidence to test our hypotheses. We don't do so for two reasons, both related to the reliability of these observations.

First, the reputation of the sources of UFO sightings and alien encounters does not give us confidence that these data are reliable. Most people hear of such events via grocery store tabloids, daytime talk shows, or sensational "specials" on viewer-hungry cable networks.

We must take note of the low reputation of the media that report UFOs and space aliens. Most of these reports are simply made up for the sake of sensation, and we cannot use them as reliable evidence.

Second, the remaining UFO sightings, those not simply made up, do not survive careful examination. Most are mistakes or unconscious misinterpretations of natural events made by honest people. A number of unbiased studies have found no grounds for believing in UFOs. In short, there is no conclusive evidence

that Earth has been visited by aliens from space.

That's too bad. A confirmed visit by intelligent creatures from beyond our solar system would answer many of our questions. It would be exciting, enlightening, and, like any real adventure, a bit scary. But none of the UFO sightings are dependable, and we are left with no direct evidence of intelligent life on other worlds. ■

27-3 COMMUNICATION WITH DISTANT CIVILIZATIONS

If other civilizations exist, perhaps we can communicate with them in some way. Sadly, travel between the stars appears more difficult in real life than in science fiction. It may in fact be almost impossible. If we can't physically visit, perhaps we can communicate by radio. Again, nature places restrictions on such conversations, but the restrictions are not too severe. The real problem lies with the nature of civilizations.

Travel Between the Stars

Roaming among the stars is, in practice, tremendously difficult because of three limitations: distance, speed, and fuel. The distances between stars are almost beyond comprehension. It does little good to explain that if we represent the sun by a golf ball in New York City, the nearest star would be another golf ball in Chicago. It is only slightly better to note that the fastest commercial jet would take about 4 million years to reach the nearest star.

The second limitation is a speed limit—we cannot travel faster than the speed of light. Although science fiction writers invent hyperspace drives so their heroes can zip from star to star, the speed of light is a natural and unavoidable limit that we cannot exceed. This, combined with the large distances between stars, makes interstellar travel very time consuming.

The third limitation says that we can't even approach the speed of light without using a fantastic amount of fuel. Even if we ignore the problem of escaping from Earth's gravity, we must still use energy stored in fuel to accelerate to high speed and to decelerate to a stop when we reach our destination. To return to Earth, assuming

we wish to, we have to repeat the process. These changes in velocity require a tremendous amount of fuel. If we flew a spaceship as big as a large yacht to a star 5 ly (1.5 pc) away and wanted to get there in only 10 years, we would use 40,000 times as much energy as the United States consumes in a year.

Travel for a few individuals might be possible if we accepted very long travel times. That would require some form of suspended animation (currently unknown) or colony ships that carry a complete, though small, society in which people are born, live, and die generation after generation. Whether the occupants of such a ship would retain the social characteristics of humans over a long voyage is questionable.

These three limitations not only make it difficult for us to leave our solar system but also make it difficult for aliens to visit Earth. Reputable scientists have studied UFOs and related phenomena and have never found any evidence that Earth is being visited or has ever been visited by aliens from other worlds (Window on Science 27-2). Thus, it seems unlikely that humans will ever meet an alien face to face. The only way we can communicate with other civilizations is via radio.

Radio Communication

Nature places two restrictions on our ability to communicate with distant societies by radio. One has to do with simple physics, is well understood, and merely makes the communication difficult. The second has to do with the fate of technological civilizations, is still unresolved, and may severely limit the number of societies we can detect by radio.

Radio signals are electromagnetic waves that travel at the speed of light. Because even the nearest civilizations must be a few light-years away, this limits our ability to carry on a conversation with distant beings. If we ask a question of a creature 4 ly away, we will have to wait

a

b

FIGURE 27-16

(a) The Arecibo message of pulses transmitted toward the globular cluster M 13 is shown as a series of 0s and 1s.
(b and c) Arranged in 73 rows of 23 pulses each and represented as light and dark squares, the message would tell aliens about human life (color added for clarity). (a and c from "The Search for Extraterrestrial Intelligence" by Carl Sagan and Frank Drake. Copyright © 1975 by Scientific American, Inc. All rights reserved. b: The Arecibo Message of November 1974 was prepared by the staff of the National Astronomy and Ionosphere Center, which is operated by Cornell University under contract with the National Science Foundation.)

8 years for a reply. Clearly, the give-and-take of normal conversation will be impossible.

Instead, we could simply broadcast a radio beacon of friendship to announce our presence. Such a beacon would have to consist of a pattern of pulses obviously designed by intelligent beings to distinguish it from natural radio signals emitted by nebulae, pulsars, and so on. For example, pulses counting off the first dozen prime numbers would do. In fact, we are already broadcasting a recognizable beacon. Short-wavelength radio signals, such as TV and FM, have been leaking into space for the last 40 years or so. Any civilization within 40 ly might already have detected us.

If we intentionally broadcast such a signal, we could give listening aliens a good idea of what humanity is like by including coded data in the signal. For example, in 1974, at the dedication of the new reflecting surface of the 1000-ft radio telescope at Arecibo, Puerto Rico, radio astronomers transmitted a series of pulses toward the globular cluster M 13 in Hercules (Figure 27-16). The number of data points in the message was 1679, a number selected because it can be factored only into 23 and 73. When the signal arrives at the globular cluster 26,000 years from now, any aliens who detect it will be able to arrange the data in only two ways—23 rows of 73 data points each, or 73 rows of 23 points each. The first way yields nonsense, but the second produces a picture that describes our solar system, the chemical basis of our life form, the general shape and size of the human body, and the number of humans on Earth. Whether there will still be humans on earth in 52,000 years when any reply to our message returns cannot be predicted.

It took only minutes to transmit the Arecibo message. If more time were taken, a more detailed picture could be sent, and if we were sure our radio telescope was pointed at listeners, we could send a long series of pictures. With pictures we could teach them our language and tell them all about our life, our difficulties, and our accomplishments.

If we can think of sending such signals, aliens can think of it, too. If we point our radio telescopes in the right direction and listen at the right wavelength, we might hear other intelligent races calling out. This raises two questions: Which stars are the best candidates, and what wavelengths are most likely? We have already answered the first question. Main-sequence G and K stars have the most favorable characteristics. But the second question is more complex.

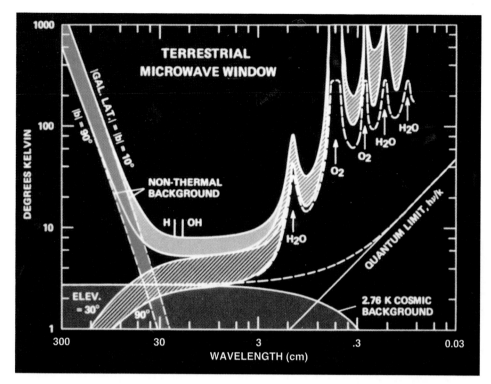

FIGURE 27-17
The microwave-radio wavelength range between 30 cm and 3 cm is plausible for communication between civilizations. Absorption by water and oxygen in any earthlike planet's atmosphere limits communication at short wavelengths. Radio noise from the galaxy (nonthermal background) limits communication at long wavelengths. The "water hole" between the radio emissions of H and OH is an especially likely wavelength range. (NASA–SETI program)

Only certain wavelengths would be useful for communication. We cannot use wavelengths longer than about 30 cm because the signal would be lost in the background radio noise from our galaxy. Nor can we go to wavelengths much shorter than 1 cm because of absorption within our atmosphere. Thus, only a certain range of wavelengths, a radio window, is open for communication (Figure 27-17).

This communications window is very wide, so a radio telescope would take a long time to tune over all the wavelengths searching for intelligent signals. However, nature may have given us a way to narrow the search. Within the communications window lie the 21-cm line of neutral hydrogen and the 18-cm line of OH. The interval between these two lines has been dubbed the **water hole** because the combination of H and OH yields water (H_2O). Water is the fundamental solvent in our life form, so it might seem natural for similar water creatures to call out to each other at wavelengths in the water hole. But even silicon creatures would be familiar with the 21-cm line of hydrogen. Thus, they too might select wavelengths near the water hole.

This is not idle speculation. A number of searches for extra-terrestrial radio signals have been made, and some major searches are now under way. The field has become known as **SETI,** Search for Extra-Terrestrial Intelligence, and it has generated heated debate. Some scientists and philosophers argue that life on other worlds can't possibly exist, so it is a waste of money to search. Others argue that life on other worlds is common and could be detected with present technology. Congress funded a NASA search for a short time but then ended support because political leaders feared public reaction.

In fact, the annual cost of a major search is only about as much as a single Air Force attack helicopter. The controversy may spring in part from the theological and philosophical controversy that would result from the discovery of life on another world.

In spite of the controversy, some searches have been made, and others are continuing. The first was Project Ozma in 1960. It listened to two stars at a wavelength of 21 cm. Since then, numerous small-scale searches have been conducted, but in order to have some assurance of success a search must scan the entire sky at millions of closely spaced frequency bands. Only recently has technology made that possible, and as technology improves, the searches can scan more sky and more frequencies.

Since 1985, META, Megachannel Extra-Terrestrial Assay, has been searching the entire sky at millions of frequency bands in the water hole. Funded by the Planetary Society, the search uses radio antennas at Harvard and near Buenos Aires, Argentina. The project is now instituting a second, even more efficient search called BETA. Since it began, META has detected over three dozen candidate signals. That is, it has found dozens of radio signals that satisfy its most basic criteria. Unfortunately, none of these signals has proved to be continuous, and no signal has been detected when the radio telescopes were redirected at any of the candidate stars. Many of these signals are probably unusual noise from sources on Earth, but the candidate signals are still under study.

One ingenious search is called SERENDIP, Search for Extraterrestrial Radio Emission from Nearby Developed Intelligent Populations. Rather than monopolize an entire radio telescope, SERENDIP rides piggyback on the 305-m Arecibo Telescope. Wherever the radio astronomers point the telescope, the receiver samples the signal, looking for intelligent signals over millions of frequency bands. If candidate signals are found, the receivers note the position of the radio telescope for future investigation.

The NASA search originally called SETI was renamed High Resolution Microwave Survey to reduce criticism, but Congress cut off all of its funds in the fall of 1993. The search continues, however, with private funding as Project Phoenix. It uses an 8.4-million-channel receiver on major radio telescopes and plans to survey the entire sky in 10 years. In addition, it will listen carefully to 800 nearby solar-type stars. The search officially began on October 12, 1992, 500 years after the landing of Columbus in the New World.

What should we expect if signals are found? By international agreement, those searching for signals have agreed to share candidate signals and obtain reliable confirmation before making public announcements. False alarms would be embarrassing. The chances of success depend on the number of inhabited worlds in our galaxy, and that number is difficult to estimate.

How Many Inhabited Worlds?

The technology exists, and given enough time the searches will find other inhabited worlds, assuming there are at least a few out there. If intelligence is common, then we should find the signals soon—in the next few decades—but if intelligence is rare in the universe, it may be a very long time before we confirm that we are not alone.

Simple arithmetic can give us an estimate of the number of technological civilizations with which we might communicate. The formula for the number of communicative civilizations in a galaxy, N_c, is:

$$N_c = N^* \cdot f_P \cdot n_{LZ} \cdot f_L \cdot f_I \cdot F_S$$

N^* is the number of stars in a galaxy, and f_P represents the probability that a star has planets. If all single stars have planets, f_P is about 0.5. The factor n_{LZ} is the average number of planets in a solar system suitably placed in the life zone, f_L is the probability that life will originate if conditions are suitable, and f_I is the probability that the life form will evolve to intelligence. These factors can be roughly estimated, but the remaining factor is much more uncertain.

F_S is the fraction of a star's life during which the life form is communicative. Here we assume that a star lives about 10 billion years. If a society survives at a technological level for only 100 years, our chances of communicating with it are small. But a society that stabilizes and remains technological for a long time is much more likely to be in the communicative phase at the proper time to signal to us. If we assume that technological societies destroy themselves in about 100 years, F_S is 100 divided by 10 billion, or 10^{-8}. But if societies can remain technological for a million years, then F_S is 10^{-4}. The influence of the factors in the formula is shown in Table 27-1.

TABLE 27-1

The Number of Technological Civilizations per Galaxy

		Estimates	
Variables		Pessimistic	Optimistic
N^*	Number of stars per galaxy	2×10^{11}	2×10^{11}
f_P	Fraction of stars with planets	0.01	0.5
n_{LZ}	Number of planets per star that lie in life zone for longer than 4 billion years	0.01	1
f_L	Fraction of suitable planets on which life begins	0.01	1
f_I	Fraction of life forms that evolve to intelligence	0.01	1
F_S	Fraction of star's life during which a technological society survives	10^{-8}	10^{-4}
N_c	Number of communicative civilizations per galaxy	2×10^{-5}	10×10^6

If the optimistic estimates are true, there may be a communicative civilization within a few dozen light-years of us, and we could locate it by searching through only a few thousand stars. On the other hand, if the pessimistic estimates are correct, we may be the only planet in our galaxy capable of communication. We may never know until we understand how technological societies function.

CRITICAL INQUIRY

Why does the number of inhabited worlds we might hear from depend on how long civilizations survive at a technological level?

When we turn radio telescopes toward the sky and scan millions of frequency bands, we take a snapshot of the universe at the particular time when we are living and able to build radio telescopes. To detect other civilizations, they must be in a similar technological stage so they will be broadcasting either intentionally or accidentally. If we search for decades and detect no other signal, it may mean not that life is rare but rather that civilizations do not survive at a technological level for very long. Only a few decades ago, our civilization was threatened by nuclear war, and now we are threatened by the pollution of our environment. If nearly all civilizations in our galaxy are either on the long road up from primitive life forms in oceans or on the long road down from nuclear war or environmental collapse, there may be no one transmitting during the short interval when we are capable of building radio telescopes with which to listen.

Radio communication between inhabited worlds is limited because it requires very fast computers to search many frequency intervals. Why must we search so many frequencies when we suspect that the water hole would be a good place to listen? ■

Are we the only thinking race? If we are, we bear the sole responsibility to understand and admire the universe. Then we are the sole representatives of that state of matter called intelligence. The mere detection of signals from another civilization would demonstrate that we share the universe with others. Although we might never leave our solar system, such communication would end the self-centered isolation of humanity and stimulate a reevaluation of the meaning of our existence. We may never realize our full potential as humans until we communicate with nonhuman intelligent life. ■

■ Summary

To discuss life on other worlds, we must first understand something about life in general, life on Earth, and the origin of life. In general, we can identify three properties in living things—a process, a physical basis, and a controlling unit of information. The process must extract energy from the surroundings, maintain the organism, and modify the surroundings to promote the organism's survival. The physical basis is the arrangement of matter and energy that implements the life process. On Earth all life is based on carbon chemistry. The controlling information is the data necessary to maintain the organism's function. Data for Earth life are stored in long carbon-chain molecules called DNA.

The DNA molecule stores information in the form of chemical bases linked together like the rungs of a ladder. When these patterns are copied by RNA molecules, they can direct the manufacture of proteins and enzymes. Thus, DNA information is the chemical formulas the cell needs to function. When a cell divides, the DNA molecule splits lengthwise and duplicates itself so that each of the new cells has a copy of the information. Errors in the duplication or damage to the DNA molecule can produce mutants, organisms that contain new DNA information and have new properties. Natural selection determines which of these new organisms are most suited to survive, and the species evolves to fit its environment.

The Miller experiment duplicated conditions in Earth's primitive environment and suggests that energy sources such as lightning could have formed amino acids and other complex molecules. Chemical evolution would have connected these together in larger and more complex, but not yet living, molecules. When a molecule acquired the ability to produce copies of itself, natural selection perfected the organism through biological evolution. Although this may have happened in the first billion years, life did not become diverse and complex until the Cambrian period, about 600 million years ago. Life emerged from the oceans about 400 million years ago, and humanity developed only a few million years ago.

The existence of life seems unlikely on other planets in our solar system. Most of the planets are too hot or too cold. Mars once had a thicker atmosphere and liquid water, and analysis of a meteorite known to come from Mars has produced controversial evidence of primitive life there long

ago. Titan and Triton, moons of Saturn and Neptune, contain methane and have been discussed as places where life might arise, but the water ocean beneath Europa's icy crust may be the best place to look after Mars.

To find life, we must look beyond our solar system. Because we suspect that planets form from the leftover debris of star formation, we suspect that most stars have planets. The rise of intelligence may take billions of years, however, so short-lived massive stars and binary stars with unstable planetary orbits must be discounted. The best candidates are G and K main-sequence stars.

The distances between stars are too large to permit travel, but communication by radio could be possible. A certain wavelength range called a radio window is suitable, and a small range between the radio signals of H and OH, the so-called water hole, is especially likely.

■ New Terms

DNA (deoxyribonucleic acid)	Miller experiment
protein	primordial soup
enzyme	chemical evolution
RNA (ribonucleic acid)	stromatolite
amino acid	life zone
natural selection	water hole
mutant	SETI
Cambrian period	

■ Questions

1. If life is based on information, what is that information?
2. What would happen to a life form if the information handed down to offspring was always the same? How would that endanger the future of the life form?
3. How does the DNA molecule produce a copy of itself?
4. Why do we believe that life on Earth began in the sea?
5. What is the difference between chemical evolution and biological evolution?
6. How does intelligence make a creature more likely to survive?
7. In your opinion, where in our solar system is the most likely place to find life beyond Earth?
8. Why do scientists feel confident that the meteorite ALH84001 came from Mars? What evidence does it contain?
9. Why are upper main-sequence stars unlikely sites for intelligent civilizations?
10. How does the stability of technological civilizations affect the probability that we can detect them?
11. Make as strong an argument as you can that we are alone in our galaxy.

■ Discussion Questions

1. What would you change in the Arecibo message if humanity lived on Mars instead of Earth?
2. What do you think it would mean if decades of careful searches for radio signals from extraterrestrial intelligence turned up nothing?

3. How do you suppose various political, religious, and social leaders would react to the unambiguous reception of a simple greeting from another race beyond our solar system?

▪ Problems

1. A single human cell encloses about 1.5 m of DNA containing 4.5 billion base pairs. What is the spacing between these base pairs, in nanometers? That is, how far apart are the rungs on the DNA ladder?

2. If we represent the history of Earth by a line 1 m long, how long a segment would represent the 400 million years since life moved onto the land? How long a segment would represent the 3-million-year history of humanity?

3. If a human generation (the time from birth to childbearing) is 20 years, how many generations have passed in the last 1 million years?

4. If a star must remain on the main sequence for at least 5 billion years for life to evolve to intelligence, how massive could a star be and still harbor intelligent life on one of its planets?

5. If there are about 1.4×10^{-4} stars like the sun per cubic light-year, how many lie within 100 ly of Earth? (HINT: The volume of a sphere is $\frac{4}{3}\pi r^3$.)

6. The mathematician Karl Gauss suggested planting forests and fields in a gigantic geometric proof to signal to possible Martians that intelligent life existed on Earth. If Martians had telescopes that like ours could resolve details no smaller than 1 second of arc, how large would the smallest element of Gauss's proof have to be? (HINT: Use the small-angle formula.)

7. If we detected radio signals with an average wavelength of 20 cm and suspected that they came from a civilization on a distant planet, roughly how much of a change in wavelength should we expect to see because of the orbital motion of the distant planet? (HINT: Use the Doppler formula in Chapter 7.)

8. Calculate the number of communicative civilizations per galaxy from your own estimates of the factors in Table 27-1.

▪ Critical Inquiries for the Web

1. The popular movie *Contact* focused interest in the SETI program by profiling the work of a radio astronomer dedicated to the search for extraterrestrial intelligence. Visit Web sites that give information about the movie, SETI programs, and radio astronomy. Discuss whether or not the movie realistically depicted how such research is done.

2. How many inhabited worlds are out there? Search for a Web site that allows interactive manipulation of our equation for the number of technological civilizations (sometimes called the Drake Equation) and consider a range of possibilities in the variables.

3. Where outside the solar system would you look for habitable planets? NASA has increasingly focused its interests on this question. Look for information online about programs dedicated to detect planets on which the existence of life as we know it might be possible. What criteria are used to choose targets for the planned searches? What methods will be used to carry out the searches.

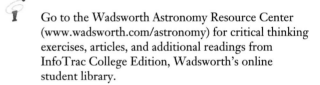

Go to the Wadsworth Astronomy Resource Center (www.wadsworth.com/astronomy) for critical thinking exercises, articles, and additional readings from InfoTrac College Edition, Wadsworth's online student library.

AFTERWORD

Our journey is over, but before we part company, there is one last thing to discuss—the place of humanity in the universe. Astronomy gives us some comprehension of the workings of stars, galaxies, and planets, but its greatest value lies in what it teaches us about ourselves. Now that we have surveyed astronomical knowledge, we can better understand our own position in nature.

To some, the word *nature* conjures up visions of furry rabbits hopping about in a forest glade dotted with pastel wildflowers. To others, nature is the blue-green ocean depths filled with creatures swirling in a mad struggle for survival. Still others think of nature as windswept mountaintops of gray stone and glittering ice. As diverse as these images are, they are all earthbound. Having studied astronomy, we can view nature as a beautiful mechanism composed of matter and energy interacting according to simple rules to form galaxies, stars, planets, mountaintops, ocean depths, and forest glades.

Supernatural is a null word.

ROBERT A. HEINLEIN

The Notebooks of Lazarus Long

Perhaps the most important astronomical lesson is that we are a small but important part of the universe. Most of the universe is lifeless. The vast reaches between the galaxies appear to be empty of all but the thinnest gas, and the stars, which contain most of the mass, are much too hot to preserve the chemical bonds that seem necessary for life to survive and develop. Only on the surfaces of a few planets, where temperatures are moderate, could atoms link together to form living matter.

If life is special, then intelligence is precious. The universe must contain many planets devoid of life, planets where the wind has blown unfelt for billions of years. There may also exist planets where life has developed but has not become complex, planets on which the wind stirs wide plains of grass and rustles dark forests. On some planets, insects, fish, birds, and animals may watch the passing days unaware of their own existence. It is intelligence, human or alien, that gives meaning to the landscape.

Science is the process by which intelligence tries to understand the universe. Science is not the invention of new devices or processes. It does not create home computers, cure the mumps, or manufacture plastic spoons—that is engineering and technology, the adaptation of scientific understanding for practical purposes. Science is understanding nature, and astronomy is understanding on the grandest scale. Astronomy is the science by which the universe, through its intelligent lumps of matter, tries to understand its own existence.

As the primary intelligent species on this planet, we are the custodians of a priceless gift—a planet filled with living things. This is especially true if life is rare in the universe. In fact, if Earth is the only inhabited planet, our responsibility is overwhelming. In any case, we are the only creatures who can take action to preserve the existence of life on Earth, and, ironically, our own actions are the most serious hazards.

The future of humanity is not secure. We are trapped on a tiny planet with limited resources and a population growing faster than our ability to produce food. In our efforts to survive, we have already driven some creatures to extinction and now threaten others. If our civilization collapses because of starvation, or if our race destroys itself somehow, the only bright spot is that the rest of the creatures on Earth will be better off for our absence.

But even if we control our population and conserve and recycle our resources, life on Earth is doomed. In 5 billion years, the sun will leave the main sequence and swell into a red giant, incinerating Earth. However, Earth will be lifeless long before that. Within the next few billion years, the growing luminosity of the sun will first alter Earth's climate and then boil its atmosphere and oceans. Earth, like everything else in the universe, is only temporary.

To survive, humanity must eventually leave Earth and search for other planets. Colonizing the moon and other planets of our solar system will not save us, since they will face the same fate as Earth when the sun dies. But travel to other stars is tremendously difficult and may be impossible with the limited resources we have in our small solar system. We and all of the living things on Earth may be trapped.

This is a depressing prospect, but a few factors are comforting. First, everything in the universe is temporary. Stars die, galaxies die, perhaps the entire universe will fall back in a "big crunch" and die. That our distant future is limited only assures us that we are a part of a much larger whole. Second, we have a few billion years to prepare, and a billion years is a very long time. Only a few million years ago, our ancestors were learning to walk erect and communicate. A billion years ago, our ancestors were microscopic organisms living in the primeval oceans. To suppose that a billion years hence we humans will still be human, that we will still be the dominant species on Earth, or that we will still be the sole intelligence on Earth is the ultimate conceit.

Our responsibility is not to save our race for all eternity but to behave as dependable custodians of our planet, preserving it, admiring it, and trying to understand it. That calls for drastic changes in our behavior toward other living things and a revolution in our attitude toward our planet's resources. Whether we can change our ways is debatable—humanity is far from perfect in its understanding, abilities, or intentions. However, we must not imagine that we and our civilization are less than precious. We have the gift of intelligence, and that is the finest thing this planet has ever produced.

■

We must not cease from exploration and the end of all our exploring will be to arrive where we began and to know the place for the first time.

—T. S. Eliot

UNITS AND ASTRONOMICAL DATA

Introduction

The metric system is used worldwide as the system of units not only in science but also in engineering, business, sports, and daily life. Developed in 18th-century France, the metric system has gained acceptance in almost every country in the world because it simplifies computations.

A system of units is based on the three fundamental units for length, mass, and time. Other quantities such as density and force are derived from these fundamental units. In the English (or British) system of units (commonly used in the United States) the fundamental unit of length is the foot, composed of 12 inches. The metric system is based on the decimal system of numbers, and the fundamental unit of length is the meter, composed of 100 centimeters.

To see the advantage of having a decimal-based system, try computing the volume of a bathtub that is 5′9″ long, 1′2″ deep, and 2′10″ wide. In the metric system the length of the tub is 1 m and 75 cm or 1.75 m. The other dimensions are 0.35 m and 0.86 m, and the volume is just $1.75 \times 0.35 \times 0.86$ cm^3. To make the computation in English units, we must first convert inches to feet by dividing by 12. We can convert centimeters to meters by the simpler process of moving the decimal point. Thus, the computation is much easier if we measure the tub in meters and centimeters instead of feet and inches.

Because the metric system is a decimal system, it is easy to express quantities in larger or smaller units as is convenient. We can express distances in centimeters, meters, kilometers, and so on. The prefixes specify the relation of the unit to the meter. Just as a cent is $\frac{1}{100}$ of a dollar, a centimeter is $\frac{1}{100}$ of a meter. A kilometer is 1000 m, and a kilogram is 1000 g. The meanings of the commonly used prefixes are given in Table A-1.

The SI Units

Any system of units based on the decimal system would be easy to use, but by international agreement, the preferred set of units, known as the *Système International d'Unités* (SI units) is based on the meter, kilogram, and second. These three fundamental units define the rest of the units as given in Table A-2.

The SI unit of force is the newton (N), named after Isaac Newton. It is the force needed to accelerate a 1 kg mass by 1 m/sec^2, or the force roughly equivalent to the weight of an apple at the earth's surface. The SI unit of energy is the joule (J), the energy produced by a force of

TABLE A-1

Metric Prefixes

Prefix	Symbol	Factor
Mega	M	10^6
Kilo	k	10^3
Centi	c	10^{-2}
Milli	m	10^{-3}
Micro	μ	10^{-6}
Nano	n	10^{-9}

TABLE A-2

SI Metric Units

Quantity	SI Unit	English Unit
Length	Meter (m)	Foot
Mass	Kilogram (kg)	Slug (sl)
Time	Second (sec)	Second (sec)
Force	Newton (N)	Pound (lb)
Energy	Joule (J)	Foot-pound (fp)

1 N acting through a distance of 1 m. A joule is roughly the energy in the impact of an apple falling off a table.

Exceptions   Units help us in two ways. They make it possible to make calculations, and they help us to conceive of certain quantities. For calculations the metric system is far superior, and we will use it for our calculations throughout this book.

But Americans commonly use the English system of units, so for conceptual purposes we can express quantities in English units. Instead of saying the average person would weigh 133 N on the moon, we could express the weight as 30 lb. Thus, the text will commonly give quantities in metric form followed by the English form in parentheses: the radius of the moon is 1738 km (1080 miles).

In SI units, density should be expressed as kilograms per cubic meter, but no human can enclose a cubic meter in his or her hand, so this unit does not help us grasp the significance of a given density. This book will refer to density in grams per cubic centimeter. A gram is roughly the mass of a paperclip, and a cubic centimeter is the size of a small sugar cube, so we can conceive of a density of 1 g/cm^3, roughly the density of water. This is not a bothersome departure from SI units because we will not make complex calculations using density.

Conversions

To convert from one metric unit to another (from meters to kilometers, for example), we have only to look at the prefix. However, converting from metric to English or English to metric is more complicated. The conversion factors are given in Table A-3.

Example: The radius of the moon is 1738 km. What is this in miles? Table A-3 indicates that 1 mile equals 1.609 km, so:

$$1738 \text{ km} \times \frac{1 \text{ mile}}{1.609 \text{ km}} = 1080 \text{ miles}$$

Temperature Scales

In astronomy, as in most other sciences, temperatures are expressed on the Kelvin scale, although the centigrade (or Celsius) scale is also used. The Fahrenheit scale commonly used in the United States is not used in scientific work.

Temperatures on the Kelvin scale are measured from absolute zero, the temperature of an object that contains no extractable heat. In practice, no object can be as cold as absolute zero, although laboratory apparatuses have reached temperatures less than 10^{-6} K. The scale is named after the Scottish mathematical physicist William Thomson, Lord Kelvin (1824–1907).

The centigrade scale refers temperatures to the freezing point of water (0°C) and to the boiling point of water (100°C). One degree centigrade is 1/100th the temperature difference between the freezing and boiling points of water. Thus the prefix *centi*. The centigrade scale is also called the Celsius scale after its inventor, the Swedish astronomer Anders Celsius (1701–1744).

The Fahrenheit scale fixes the freezing point of water at 32°F and the boiling point at 212°F. Named after the German physicist Gabriel Daniel Fahrenheit (1686–1736), who made the first successful mercury thermometer in 1720, the Fahrenheit scale is used only in the United States.

It is easy to convert temperatures from one scale to another using the information given in Table A-4.

Powers of 10 Notation

Powers of 10 make writing very large numbers much simpler. For example, the nearest star is about 43,000,000,000,000 km from the sun. Writing this number as 4.3×10^{13} km is much easier.

Very small numbers can also be written with powers of 10. For example, the wavelength of visible light is about 0.0000005 m. In power of 10 this becomes 5×10^{-7} m.

The powers of 10 used in this notation appear in the table on the next page. The exponent tells us how to move the decimal point. If the exponent is positive, we move the decimal point to the right. If the exponent

---

TABLE   A-3

Conversion Factors

1 inch = 2.54 centimeters	1 centimeter = 0.394 inch
1 foot = 0.3048 meter	1 meter = 39.36 inches = 3.28 feet
1 mile = 1.6093 kilometers	1 kilometer = 0.6214 mile
1 slug = 14.594 kilograms	1 kilogram = 0.0685 slug
1 pound = 4.4482 newtons	1 newton = 0.2248 pound
1 foot-pound = 1.35582 joules	1 joule = 0.7376 foot-pound
1 horsepower = 745.7 joules/sec	1 joule/sec = 1 watt

TABLE   A-4

Temperature Scales

	Kelvin (K)	Centigrade (°C)	Fahrenheit (°F)
Absolute zero	0 K	−273°C	−459°F
Freezing point of water	273 K	0°C	32°F
Boiling point of water	373 K	100°C	212°F
Conversions	$K = °C + 273$	$°C = \frac{5}{9}(°F - 32)$	$°F = \frac{9}{5}°C + 32$

is negative, we move the decimal point to the left. Thus, 2×10^3 equals 2000.0, and 2×10^{-3} equals 0.002.

$$
\begin{aligned}
&\vdots \\
10^5 &= 100,000 \\
10^4 &= 10,000 \\
10^3 &= 1,000 \\
10^2 &= 100 \\
10^1 &= 10 \\
10^0 &= 1 \\
10^{-1} &= 0.1 \\
10^{-2} &= 0.01 \\
10^{-3} &= 0.001 \\
10^{-4} &= 0.0001 \\
&\vdots
\end{aligned}
$$

Astronomy, and science in general, is a way of learning about nature and understanding the universe we live in. To test hypotheses about how nature works, scientists use observations of nature. The tables that follow contain some of the basic observations that support our best understanding of the astronomical universe. Of course, these data are expressed in the form of numbers, not because science reduces all understanding to mere numbers, but because the struggle to understand nature is so demanding we must use every tool available. Quantitative thinking, reasoning mathematically, is one of the most powerful tools ever invented by the human brain. Thus the tables that follow are not nature reduced to mere number, but number supporting humanity's growing understanding of the natural world around us.

TABLE A-5

Units Used in Astronomy

1 Ångstrom (Å)	$= 10^{-8}$ cm
	$= 10^{-10}$ m
1 astronomical unit (AU)	$= 1.495979 \times 10^{11}$ m
	$= 92.95582 \times 10^6$ miles
1 light-year (ly)	$= 6.3240 \times 10^4$ AU
	$= 9.46053 \times 10^{15}$ m
	$= 5.9 \times 10^{12}$ miles
1 parsec (pc)	$= 206265$ AU
	$= 3.085678 \times 10^{16}$ m
	$= 3.261633$ ly
1 kiloparsec (kpc)	$= 1000$ pc
1 megaparsec (Mpc)	$= 1,000,000$ pc

TABLE A-6

Constants

astronomical unit (AU)	$= 1.495979 \times 10^{11}$ m
parsec (pc)	$= 206265$ AU
	$= 3.085678 \times 10^{16}$ m
	$= 3.261633$ ly
light-year (ly)	$= 9.46053 \times 10^{15}$ m
velocity of light (c)	$= 2.997925 \times 10^8$ m/sec
gravitational constant (G)	$= 6.67 \times 10^{-11}$ N·m²/kg²
mass of earth ($M_\oplus$)	$= 5.976 \times 10^{24}$ kg
equatorial radius of earth ($R_\oplus$)	$= 6378.164$ km
mass of sun ($M_\odot$)	$= 1.989 \times 10^{30}$ kg
radius of sun ($R_\odot$)	$= 6.9599 \times 10^8$ m
solar luminosity ($L_\odot$)	$= 3.826 \times 10^{26}$ J/sec
mass of moon	$= 7.350 \times 10^{22}$ kg
radius of moon	$= 1738$ km
mass of H atom	$= 1.67352 \times 10^{-27}$ kg

If you use scientific notation in calculations, be sure you correctly enter numbers into your calculator. Not all calculators can accept scientific notation, but those that can have a key labeled EXP, EEX, or perhaps EE that allows you to enter the exponent of ten. To enter a number such as 3×10^8 we press the keys 3 EXP 8. To enter a number with a negative exponent, we must use the change-sign key, usually labeled $+/-$ or CHS. To enter the number 5.2×10^{-3} we press the keys 5.2 EXP $+/-$ 3. Try a few examples.

To read a number in scientific notation from a calculator we must read the exponent separately. The number 3.1×10^{25} may appear in a calculator display as 3.1 25 or on some calculators as 3.1 10^{25}. Examine your calculator to determine how such numbers are displayed.

TABLE A-7

The Nearest Stars

Name	Absolute Magnitude (M_v)	Distance (ly)	Spectral Type	Apparent Visual Magnitude (m_v)
Sun	4.83		G2	−26.8
α Cen A	4.38	4.3	G2	0.1
B	5.76	4.3	K5	1.5
Barnard's Star	13.21	5.9	M5	9.5
Wolf 359	16.80	7.6	M6	13.5
Lalande 21185	10.42	8.1	M2	7.5
Sirius A	1.41	8.6	A1	−1.5
B	11.54	8.6	white dwarf	7.2
Luyten 726–8A	15.27	8.9	M5	12.5
B (UV Cet)	15.8	8.9	M6	13.0
Ross 154	13.3	9.4	M5	10.6
Ross 248	14.8	10.3	M6	12.2
ε Eri	6.13	10.7	K2	3.7
Luyten 789–6	14.6	10.8	M7	12.2
Ross 128	13.5	10.8	M5	11.1
61 CYG A	7.58	11.2	K5	5.2
B	8.39	11.2	K7	6.0
ε Ind	7.0	11.2	K5	4.7
Procyon A	2.64	11.4	F5	0.3
B	13.1	11.4	white dwarf	10.8
Σ 2398 A	11.15	11.5	M4	8.9
B	11.94	11.5	M5	9.7
Groombridge 34 A	10.32	11.6	M1	8.1
B	13.29	11.6	M6	11.0
Lacaille 9352	9.59	11.7	M2	7.4
τ Ceti	5.72	11.9	G8	3.5
BD + 5° 1668	11.98	12.2	M5	9.8
L 725–32	15.27	12.4	M5	11.5
Lacaille 8760	8.75	12.5	M0	6.7
Kapteyn's Star	10.85	12.7	M0	8.8
Kruger 60 A	11.87	12.8	M3	9.7
B	13.3	12.8	M4	11.2

Properties of Main-Sequence Stars

Spectral Type	Absolute Visual Magnitude *(M_v)*	Luminosity*	Temp. (K)	λ max (nm)	Mass*	Radius*	Average Density (g/cm³)
O5	−5.8	501,000	40,000	72.4	40	17.8	0.01
B0	−4.1	20,000	28,000	100	18	7.4	0.1
B5	−1.1	790	15,000	190	6.4	3.8	0.2
A0	+0.7	79	9900	290	3.2	2.5	0.3
A5	+2.0	20	8500	340	2.1	1.7	0.6
F0	+2.6	6.3	7400	390	1.7	1.4	1.0
F5	+3.4	2.5	6600	440	1.3	1.2	1.1
G0	+4.4	1.3	6000	480	1.1	1.0	1.4
G5	+5.1	0.8	5500	520	0.9	0.9	1.6
K0	+5.9	0.4	4900	590	0.8	0.8	1.8
K5	+7.3	0.2	4100	700	0.7	0.7	2.4
M0	+9.0	0.1	3500	830	0.5	0.6	2.5
M5	+11.8	0.01	2800	1000	0.2	0.3	10.0
M8	+16	0.001	2400	1200	0.1	0.1	63

*Luminosity, mass, and radius are given in terms of the sun's luminosity, mass, and radius.

The Brightest Stars

Star	Name	Apparent Visual Magnitude *(m_v)*	Spectral Type	Absolute Visual Magnitude *(M_v)*	Distance (ly)
α CMa A	Sirius	−1.47	A1	1.4	8.7
α Car	Canopus	−0.72	F0	−3.1	98
α Cen	Rigil Kentaurus	−0.01	G2	4.4	4.3
α Boo	Arcturus	−0.06	K2	−0.3	36
α Lyr	Vega	0.04	A0	0.5	26.5
α Aur	Capella	0.05	G8	−0.6	45
β Ori A	Rigel	0.14	B8	−7.1	900
α CMi A	Procyon	0.37	F5	2.7	11.3
α Ori	Betelgeuse	0.41	M2	−5.6	520
α Eri	Achernar	0.51	B3	−2.3	118
β Cen AB	Hadar	0.63	B1	−5.2	490
α Aql	Altair	0.77	A7	2.2	16.5
α Tau A	Aldebaran	0.86	K5	−0.7	68
α Cru	Acrux	0.90	B2	−3.5	260
α Vir	Spica	0.91	B1	−3.3	220
α Sco A	Antares	0.92	M1	−5.1	520
α PsA	Fomalhaut	1.15	A3	2.0	22.6
β Gem	Pollux	1.16	K0	1.0	35
α Cyg	Deneb	1.26	A2	−7.1	1600
β Cru	Beta Crucis	1.28	B0.5	−4.6	490

OBSERVING THE SKY

Observing the sky with the naked eye is of no more importance to modern astronomy than picking up pretty pebbles is to modern geology. But the sky is a natural wonder unimaginably bigger than the Grand Canyon, the Rocky Mountains, or any other natural wonder that tourists visit every year. To neglect the beauty of the sky is equivalent to geologists neglecting the beauty of the minerals they study. This supplement is meant to act as a tourist's guide to the sky. We analyzed the universe in the regular chapters, but here we will admire it.

The brighter stars in the sky are visible even from the centers of cities with their air and light pollution. But in the countryside only a few miles beyond the cities, the night sky is a velvety blackness strewn with thousands of glittering stars. From a wilderness location, far from the city's glare, and especially from high mountains, the night sky is spectacular.

Using Star Charts

The constellations are a fascinating cultural heritage of our planet, but they are sometimes a bit difficult to learn because of the earth's motion. The constellations above the horizon change with the time of night and the seasons.

Because the earth rotates eastward, the sky appears to rotate around us westward. A constellation visible in the southern sky soon after sunset will appear to move westward, and in a few hours it will disappear below the horizon. Other constellations will rise in the east, so the sky changes gradually through the night.

In addition, the orbital motion of the earth makes the sun appear to move eastward among the stars. Each day the sun moves about twice its own diameter eastward along the ecliptic, and each night at sunset, the constellations are about 1° farther toward the west.

Orion, for instance, is visible in the southern sky in January, but as the days pass, the sun moves closer to Orion. By March, Orion is difficult to see in the southwest sky soon after sunset. By June, the sun is so close to Orion it sets with the sun and is invisible. Not until late July is the sun far enough past Orion for the constellation to become visible rising in the eastern sky just before dawn.

FIGURE B-1

To use the star charts in this book, select the appropriate chart for the date and time. Hold it overhead, and turn it until the direction at the bottom of the chart is the same as the direction you are facing.

Clearly the rotation and orbital motion of the earth require that we use more than one star chart to map the sky. Which chart we select depends on the time of night and the time of year. The charts given at the end of this book show the evening sky for each month.

To use the charts, select the chart for the appropriate month and hold it overhead as shown in Figure B-1. If you face south, turn the chart until the words *Southern Horizon* are at the bottom. If you face other directions, turn the chart appropriately.

JANUARY

Early in Month	9 P.M.
Midmonth	8 P.M.
End of Month	7 P.M.

Months along the ecliptic show the location of the sun during the year.

Numbers along the celestial equator show right ascension.

FEBRUARY

Early in Month	9 P.M.
Midmonth	8 P.M.
End of Month	7 P.M.

Months along the ecliptic show the location of the sun during the year.

Numbers along the celestial equator show right ascension.

MARCH

Early in Month	9 P.M.
Midmonth	8 P.M.
End of Month	7 P.M.

Months along the ecliptic show the location of the sun during the year.

Numbers along the celestial equator show right ascension.

APRIL

Early in Month	9 P.M.
Midmonth	8 P.M.
End of Month	7 P.M.

Months along the ecliptic show the location of the sun during the year.

Numbers along the celestial equator show right ascension.

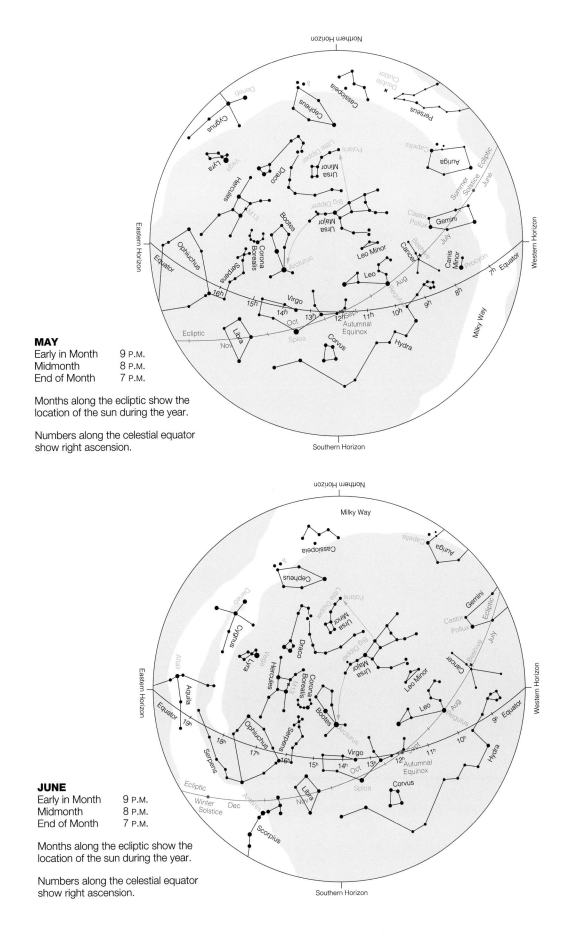

MAY

Early in Month	9 P.M.
Midmonth	8 P.M.
End of Month	7 P.M.

Months along the ecliptic show the location of the sun during the year.

Numbers along the celestial equator show right ascension.

JUNE

Early in Month	9 P.M.
Midmonth	8 P.M.
End of Month	7 P.M.

Months along the ecliptic show the location of the sun during the year.

Numbers along the celestial equator show right ascension.

JULY

Early in Month	9 P.M.
Midmonth	8 P.M.
End of Month	7 P.M.

Months along the ecliptic show the location of the sun during the year.

Numbers along the celestial equator show right ascension.

AUGUST

Early in Month	9 P.M.
Midmonth	8 P.M.
End of Month	7 P.M.

Months along the ecliptic show the location of the sun during the year.

Numbers along the celestial equator show right ascension.

SEPTEMBER

Early in Month	9 P.M.
Midmonth	8 P.M.
End of Month	7 P.M.

Months along the ecliptic show the location of the sun during the year.

Numbers along the celestial equator show right ascension.

OCTOBER

Early in Month	9 P.M.
Midmonth	8 P.M.
End of Month	7 P.M.

Months along the ecliptic show the location of the sun during the year.

Numbers along the celestial equator show right ascension.

NOVEMBER

Early in Month	9 P.M.
Midmonth	8 P.M.
End of Month	7 P.M.

Months along the ecliptic show the location of the sun during the year.

Numbers along the celestial equator show right ascension.

DECEMBER

Early in Month	9 P.M.
Midmonth	8 P.M.
End of Month	7 P.M.

Months along the ecliptic show the location of the sun during the year.

Numbers along the celestial equator show right ascension.

absolute bolometric magnitude   The absolute magnitude we would observe if we could detect all wavelengths.

absolute visual magnitude (M_v)   Intrinsic brightness of a star; the apparent visual magnitude the star would have if it were 10 pc away.

absolute zero   The lowest possible temperature. The temperature at which the particles in a material, atoms or molecules, contain no energy of motion that can be extracted from the body.

absorption line   A dark line in a spectrum. Produced by the absence of photons absorbed by atoms or molecules.

absorption spectrum (dark line spectrum)   A spectrum that contains absorption lines.

acceleration   A change in a velocity; a change in either speed or direction. (See **velocity**.)

acceleration of gravity   A measure of the strength of gravity at a planet's surface.

accretion   The sticking together of solid particles to produce a larger particle.

accretion disk   The whirling disk of gas that forms around a compact object such as a white dwarf, neutron star, or black hole as matter is drawn in.

achondrites   Stony meteorites containing no chondrules or volatiles.

achromatic lens   A telescope lens composed of two lenses ground from different kinds of glass and designed to bring two selected colors to the same focus and correct for chromatic aberration.

active galactic nucleus   The central energy source of an active galaxy.

active galaxy   A galaxy that is a source of excess radiation, usually radio waves, X rays, gamma rays, or some combination.

active optics   Optical elements whose position or shape is continuously controlled by computers.

adaptive optics   Computer-controlled telescope mirrors that can at least partially compensate for seeing.

alt-azimuth mounting   A telescope mounting capable of motion parallel to and perpendicular to the horizon.

amino acids   Carbon-chain molecules that are the building blocks of protein.

Angstrom (Å)   A unit of distance 1 Å $= 10^{-10}$ m; often used to measure the wavelength of light.

angular momentum   The tendency of a rotating body to continue rotating. Mathematically it is the product of mass, velocity, and radius.

annular eclipse   A solar eclipse in which the solar photosphere appears around the edge of the moon in a bright ring, or annulus. The corona, chromosphere, and prominences cannot be seen.

anorthosite   Rock of aluminum and calcium silicates found in the lunar highlands.

aphelion   The orbital point of greatest distance from the sun.

apogee   The orbital point of greatest distance from earth.

Apollo–Amor objects   Asteroids whose orbits cross that of earth (Apollo) and Mars (Amor).

apparent relative orbit   The orbit of one star in a visual binary with respect to the other star as seen from earth. (See **true relative orbit**.)

apparent visual magnitude (m_v)   The brightness of a star as seen by human eyes on earth.

arachnoid   On Venus, one of a number of round networks of fractures in the crust.

archaeoastronomy   The study of the astronomy of ancient cultures.

associations   Groups of widely scattered stars (10 to 1000) moving together through space; not gravitationally bound into clusters.

asterism   A named group of stars not identified as a constellation, e.g., the Big Dipper.

asteroids   Small rocky worlds, most of which lie between Mars and Jupiter in the asteroid belt.

astrometric binary   A binary star identified by its irregular proper motion.

astronomical unit (AU)   Average distance from the earth to the sun; 1.5×10^8 km, or 93×10^6 miles.

atmospheric window   Wavelength regions in which our atmosphere is transparent—at visual wavelengths, infrared, and at radio wavelengths.

aurora   The glowing light display that results when a planet's magnetic field guides charged particles toward the north and south magnetic poles, where they strike the upper atmosphere and excite atoms to emit photons.

autumnal equinox   The point on the celestial sphere where the sun crosses the celestial equator going southward. Also, the time when the sun reaches this point and autumn begins in the Northern Hemisphere—about September 22.

Bailey's beads   Bright spots of the solar surface visible at the edge of the moon during a total solar eclipse.

Balmer series   Spectral lines in the visible and near-ultraviolet spectrum of hydrogen produced by transitions whose lowest orbit is the second.

barred spiral galaxy   A spiral galaxy with an elongated nucleus resembling a bar from which the arms originate.

basalt   Dark, igneous rock characteristic of solidified lava.

belts Dark bands of clouds that circle Jupiter parallel to its equator; generally red, brown, or blue-green; believed to be regions of descending gas.

β Canis Majoris variables Short-period variable stars that do not lie in the instability strip.

big bang theory The theory that the universe began with a violent explosion from which the expanding universe of galaxies eventually formed.

binary stars Pairs of stars that orbit around their common center of mass.

binding energy The energy needed to pull an electron away from its atom.

bipolar flows Oppositely directed jets of gas ejected by some protostellar objects.

black body radiation Radiation emitted by a hypothetical perfect radiator. The spectrum is continuous, and the wavelength of maximum emission depends only on the body's temperature.

black dwarf The end state of a white dwarf that has cooled to low temperature.

black hole A mass that has collapsed to such a small volume that its gravity prevents the escape of all radiation; also, the volume of space from which radiation may not escape.

BL Lac objects Objects that resemble quasars; thought to be highly luminous cores of distant galaxies.

blue shift The shortening of the wavelengths of light observed when the source and observer are approaching each other.

Bok globules Small, dark clouds only about 1 ly in diameter that contain 10–1000 $M_\odot$ of gas and dust. Believed related to star formation.

bolometric magnitude The magnitude we would measure if we could detect electromagnetic radiation of all wavelengths.

bow shock The boundary between the undisturbed solar wind and the region being deflected around a planet or comet.

breccia A rock composed of fragments of earlier rocks bonded together.

bright line spectrum See **emission spectrum.**

brown dwarf A very cool, low-luminosity star whose mass is not sufficient to ignite nuclear fusion.

burster A source of bursts of X rays or, in some cases, gamma rays; believed associated with neutron stars.

butterfly diagram See **Maunder butterfly diagram.**

CAI Calcium–aluminum–rich inclusions found in some meteorites. Believed to be very old.

Cambrian period A geological period 0.6–0.5 billion years ago during which life on earth became diverse and complex. Cambrian rocks contain the oldest easily identifiable fossils.

capture hypothesis The theory that the moon formed elsewhere in the solar system and was later captured by the earth.

carbonaceous chondrites Stony meteorites that contain both chondrules and volatiles. They may be the least altered remains of the solar nebula still present in the solar system.

carbon detonation The explosive ignition of carbon burning in some giant stars. A possible cause of some supernova explosions.

carbon–nitrogen–oxygen (CNO) cycle A series of nuclear reactions that use carbon as a catalyst to combine four hydrogen atoms to make one helium atom plus energy; effective in stars more massive than the sun.

Cassegrain telescope A reflecting telescope in which the secondary mirror reflects light back down the tube through a hole in the center of the objective mirror.

celestial equator The imaginary line around the sky directly above the earth's equator.

celestial sphere An imaginary sphere of very large radius surrounding the earth and to which the planets, stars, sun, and moon seem to be attached.

center of mass The balance point of a body or system of bodies.

Cepheid variable stars Variable stars with a period of 1–60 days. Their period of variation is related to luminosity.

Chandrasekhar limit The maximum mass of a white dwarf, about 1.4 $M_\odot$. A white dwarf of greater mass cannot support itself and will collapse.

charge-coupled device (CCD) An electronic device consisting of a large array of light-sensitive elements used to record very faint images.

chemical evolution The chemical process that led to the growth of complex molecules on the primitive earth. This did not involve the reproduction of molecules.

chondrite A stony meteorite that contains chondrules.

chondrules Round, glassy bodies in some stony meteorites; believed to have solidified very quickly from molten drops of silicate material.

chromatic aberration A distortion found in refracting telescopes because lenses focus different colors at slightly different distances. Images are consequently surrounded by color fringes.

chromosphere Bright gases just above the photosphere of the sun; responsible for the emission lines in the flash spectrum.

circular velocity The velocity required to remain in a circular orbit about a body.

circumpolar constellation Any of the constellations so close to the celestial pole that they never set (or never rise) as seen from a given latitude.

closed orbit An orbit that returns to its starting point; a circular or elliptical orbit. (See **open orbit.**)

closed universe A model universe in which the average density is great enough to stop the expansion and make the universe contract.

cluster method The method of determining the masses of galaxies based on the motions of galaxies in a cluster.

CNO cycle See **carbon–nitrogen–oxygen cycle.**

co-accretion hypothesis   The theory that the earth and moon formed together.

cocoon   The cloud of gas and dust around a contracting protostar that conceals it at visible wavelengths.

collisional broadening   The smearing out of a spectral line because of collisions among the atoms of the gas.

color index   A numerical measure of the color of a star.

coma   The glowing head of a comet.

comet   One of the small, icy bodies that orbit the sun and produce tails of gas and dust when they near the sun.

compact object   A star that has collapsed to form a white dwarf, neutron star, or black hole.

comparative planetology   The study of planets by comparing the characteristics of different examples.

comparison spectrum   A spectrum of known spectral lines used to identify unknown wavelengths in an object's spectrum.

condensation   The growth of a particle by addition of material from surrounding gas, atom by atom.

condensation sequence   The sequence in which different materials condense from the solar nebula as we move outward from the sun.

conic sections   The family of curves generated by slicing a cone—the circle, ellipse, hyperbola, and parabola.

constellation   One of the stellar patterns identified by name, usually of mythological gods, people, animals, or objects; also, the region of the sky containing that star pattern.

continuity of energy law   One of the basic laws of stellar structure. The amount of energy flowing out of the top of a shell must equal the amount coming in at the bottom plus whatever energy is generated within the shell.

continuity of mass law   One of the basic laws of stellar structure. The total mass of the star must equal the sum of the masses of the shells, and the mass must be distributed smoothly throughout the star.

continuous spectrum   A spectrum in which there are no absorption or emission lines.

Copernican principle   The belief that the earth is not in a special place in the universe.

corona   The faint outer atmosphere of the sun; composed of low-density, very hot, ionized gas. On Venus, round networks of fractures and ridges up to 1000 km in diameter.

coronagraph   A telescope designed to photograph the inner corona of the sun.

coronal hole   An area of the solar surface that is dark at X-ray wavelengths; thought to be associated with divergent magnetic fields and the source of the solar wind.

cosmic rays   Atomic nuclei that enter earth's atmosphere at nearly the speed of light. Some originate in solar flares, and some may come from supernova explosions, but their true nature is not well understood.

cosmological principle   The assumption that any observer in any galaxy sees the same general features of the universe.

cosmological test   A measurement or observation whose result can help us choose between different cosmological theories.

cosmology   The study of the nature, origin, and evolution of the universe.

coudé focus   The focal arrangement of a reflecting telescope in which mirrors direct the light to a fixed focus beyond the bounds of the telescope's movement, typically in a separate room. Usually used for spectroscopy.

Coulomb barrier   The electrostatic force of repulsion between bodies of like charge; commonly applied to atomic nuclei.

critical density   The average density of the universe needed to make its curvature flat.

critical point   The temperature and pressure at which the vapor and liquid phases of a material have the same density.

cultural shock   The bewildering impact of an advanced society upon a less sophisticated one.

dark line spectrum   See **absorption spectrum**.

dark matter   The matter believed to exist that would make up the missing mass in the universe. (See **missing mass**.)

dark nebula   A nebula consisting of dust and gas blocking our view of more distant stars.

decameter radiation   Radio signals from Jupiter with wavelengths of about 10 m.

decimeter radiation   Radio signals from Jupiter with wavelengths of about 0.1 m.

declination   A coordinate used on the celestial sphere just as latitude is used on earth. An object's declination is measured from the celestial equator—positive to the north and negative to the south.

declination axis   The pivot in a telescope mounting that allows the telescope to move north and south.

deferent   In the Ptolemaic theory, the large circle around the earth along which the center of the epicycle moved.

degenerate matter   Extremely high-density matter in which pressure no longer depends on temperature, due to quantum mechanical effects.

density wave theory   Theory proposed to account for spiral arms as compressions of the interstellar medium in the disk of the galaxy.

diamond ring effect   A momentary phenomenon seen during some total solar eclipses when the ring of the corona and a bright spot of photosphere resemble a large diamond set in a silvery ring.

differential rotation   The rotation of a body in which different parts of the body have different periods of rotation. This is true of the sun, the Jovian planets, and the disk of the galaxy.

differentiation   The separation of planetary material according to density.

diffraction fringe   Blurred fringe surrounding any image caused by the wave properties of light. Because of this, no image detail smaller than the fringe can be seen.

dilation of time   The slowing of time as recorded by moving clocks.

direct orbit   An orbit that carries a moon or planet in the same direction as most other motion in the solar system—counterclockwise as seen from the north.

dirty snowball theory   The hypothesis that comets are kilometer-size balls of ices with embedded impurities.

disk component   All material confined to the plane of the galaxy.

distance indicators   Objects whose luminosities or diameters are known; used to find the distance to a star cluster or galaxy.

distance modulus ($m - M_v$)   The difference between the apparent and absoute magnitude of a star. A measure of how far away the star is.

diurnal motion   Daily motion, as in the rising and setting of the sun.

DNA (deoxyribonucleic acid)   The long carbon-chain molecule that records information to govern the biological activity of the organism. DNA carries the genetic data passed to offspring.

Doppler broadening   The smearing of spectral lines because of the motion of the atoms in the gas.

Doppler effect   The change in the wavelength of radiation due to relative radial motion of source and observer.

double exhaust model   The theory that double radio lobes are produced by pairs of jets emitted in opposite directions from the centers of active galaxies.

double galaxy method   A method of finding the masses of galaxies from orbiting pairs of galaxies.

double-line spectroscopic binary   A spectroscopic binary star in which spectral lines from both stars are visible in the spectrum.

double-lobed radio source   A galaxy that emits radio energy from two regions (lobes) located on opposite sides of the galaxy.

double stars   A pair of stars close together in the sky. Not all double stars are necessarily in orbit around each other.

dust tail (type II)   The tail of a comet formed of dust blown outward by the pressure of sunlight. (See **gas tail**.)

dwarf nova   A star that undergoes novalike explosions every few days or weeks; believed to be associated with mass transfer onto a white dwarf in a binary system.

dynamo effect   The theory that the earth's magnetic field is generated in the conducting material of its molten core.

eccentric   In astronomy, an off-center circular path.

eclipse season   That period when the sun is near a node of the moon's orbit and eclipses are possible.

eclipse year   The time the sun takes to circle the sky and return to a node of the moon's orbit; 346.62 days.

eclipsing binary   A binary star system in which the stars eclipse each other.

ecliptic   The apparent path of the sun around the sky.

ejecta   Pulverized rock scattered by meteorite impacts on a planetary surface.

electroglow   The ultraviolet radiation produced in the upper atmospheres of Jupiter, Saturn, and Uranus by high-energy particles in the planets' magnetospheres.

electromagnetic radiation   Changing electric and magnetic fields that travel through space and transfer energy from one place to another—for example, light, radio waves, etc.

electrons   Low-mass atomic particles carrying negative charges.

electron volt (eV)   A unit of energy equal to the energy produced by an electron accelerated through a voltage difference of 1 volt.

ellipse   A closed curve enclosing two points (foci) such that the total distance from one focus to any point on the curve back to the other focus equals a constant.

elliptical galaxy   A galaxy that is round or elliptical in outline. It contains little gas and dust, no disk or spiral arms, and few hot, bright stars.

emission line   A bright line in a spectrum caused by the emission of photons from atoms.

emission nebula   A cloud of glowing gas excited by ultraviolet radiation from hot stars.

emission spectrum (bright line spectrum)   A spectrum containing emission lines.

energy level   One of a number of states an electron may occupy in an atom, depending on its binding energy.

energy machine   An object that releases energy. Commonly used to refer to the source of energy in active galactic nuclei.

energy transport   Energy must flow from hot regions to cooler regions by conduction, convection, or radiation.

enzymes   Special proteins that control processes in an organism.

epicycle   The small circle followed by a planet in the Ptolemaic theory. The center of the epicycle follows a larger circle (deferent) around earth.

equant   The point off center in the deferent from which the center of the epicycle appears to move uniformly.

equation of time   The difference between apparent solar time and mean solar time.

equatorial mounting   A telescope mounting that allows motion parallel to and perpendicular to the celestial equator.

ergosphere   The region surrounding a rotating black hole within which one could not resist being dragged around the black hole. It is possible for a particle to escape from the ergosphere and extract energy from the black hole.

escape orbit   An orbit that does not return to its starting point. (See **open orbit**.)

escape velocity   The initial velocity an object needs to escape from the surface of a celestial body.

ether The medium through which light traveled, according to late-19th-century physics; rejected by modern physics.

evening star Any planet visible in the sky just after sunset.

event horizon The boundary of the region of a black hole from which no radiation may escape. No event that occurs within the event horizon is visible to a distant observer.

excited atom An atom in which an electron has moved from a lower to a higher orbit.

extinction The dimming of light by intervening material; commonly, dimming by the interstellar medium.

eyepiece A short-focal-length lens used to enlarge the image in a telescope; the lens nearest the eye.

faculae Bright areas of the solar surface associated with sunspots and prominences.

fall A meteorite seen to fall. (See **find**.)

false color A representation of graphical data in which the colors are altered or added to reveal details.

field (gravitational, electric, or magnetic) A way of explaining action at a distance. A particle produces a field of influence to which another particle in the field responds.

filar micrometer An instrument that permits precise measurements at the telescope of the position of visual binary stars and similar objects.

filtergram A photograph (usually of the sun) taken in the light of a specific region of the spectrum—e.g., an H-alpha filtergram.

find A meteorite that is found but was not seen to fall. (See **fall**.)

fission hypothesis The theory that the moon formed by breaking away from the earth.

flare A violent eruption on the sun's surface.

flash spectrum The emission spectrum of the chromosphere that is visible for the few seconds during a total solar eclipse when the moon has covered the photosphere but has not yet covered the chromosphere.

flatness problem In cosmology the circumstance that the early universe must have contained almost exactly the right amount of matter to close space-time (to make space-time flat).

flat universe A model of the universe in which space-time is not curved.

flocculent Woolly, fluffy; used to refer to certain galaxies that have a woolly appearance.

focal length The distance from a lens to the point where it focuses parallel rays of light.

focus (of an ellipse) One of two points inside an ellipse that satisfy the condition that the distance from one focus to any point on the ellipse to the other focus is a constant.

forward scattering The optical property of finely divided particles to preferentially direct light in the original direction of the light's travel.

frequency The number of times a given event occurs in a given time; for a wave, the number of cycles that pass the observer in 1 second.

galactic cannibalism The theory that large galaxies absorb smaller galaxies.

galactic corona The low-density extension of the halo of a galaxy; now suspected to extend many times the visible diameter of the galaxy.

galaxy seeds Small centers of gravitational attraction in the early universe that caused the concentration of matter into clusters of galaxies, filaments, and walls.

Galilean satellites The four largest satellites of Jupiter, named after their discoverer, Galileo.

gas tail (type I) The tail of a comet produced by gas blown outward by the solar wind. (See **dust tail**.)

Gauss (G) A unit used to measure the strength of a magnetic field.

geocentric universe A model universe with the earth at the center, such as the Ptolemaic universe.

geosynchronous orbit An eastward orbit whose period is 24 hours. A satellite in such an orbit remains above the same spot on the earth's surface.

giant molecular clouds Very large, cool clouds of dense gas in which stars form.

giant stars Large, cool, highly luminous stars in the upper right of the H–R diagram. Typically 10 to 100 times the diameter of the sun.

glacial period An interval when ice sheets cover large areas of the land.

glitch A sudden change in the period of a pulsar.

globular cluster A star cluster containing 50,000 to 1 million stars in a sphere about 75 ly in diameter; generally old, metal-poor, and found in the spherical component of the galaxy.

graben rille A linear feature on a planetary surface caused by faulting and sinking of portions of the crust.

grand unified theories (GUTs) Theories that attempt to unify (describe in a similar way) the electromagnetic, weak, and strong forces of nature.

granulation The fine structure visible on the solar surface caused by rising currents of hot gas and sinking currents of cool gas below the surface.

grating A piece of material in which numerous microscopic parallel lines are scribed. Light encountering a grating is dispersed to form a spectrum.

gravitation lens effect The focusing of light from a distant galaxy or quasar by an intervening galaxy to produce multiple images of the distant body.

gravitational red shift The lengthening of the wavelength of a photon due to its escape from a gravitational field.

gravitational wave The transport of energy by the motion of waves in a gravitational field; predicted by general relativity.

grazing incidence optics Reflecting mirrors using very shallow angles of incidence to focus X rays.

greatest elongation The maximum angular separation between an object and the sun; typically said of Mercury or Venus.

greenhouse effect The process by which a carbon dioxide atmosphere traps heat and raises the temperature of a planetary surface.

Gregorian calendar The calendar now in use, instituted by Pope Gregory XIII in 1582.

grooved terrain Regions of the surface of Ganymede consisting of parallel grooves; believed to have formed by repeated fracture and refreezing of the icy crust.

ground state The lowest permitted electron orbit in an atom.

half-life The time required for half of the atoms in a radioactive sample to decay.

halo The spherical region of a spiral galaxy containing a thin scattering of stars, star clusters, and small amounts of gas.

head–tail radio galaxy A radio galaxy with a contour consisting of a head and a tail; believed caused by the motion of an active galaxy through the intergalactic medium.

heat Thermal energy present in a body as agitation (motion) among its particles (atoms or molecules).

heat of formation In planetology, the heat released by the infall of matter during the formation of a planetary body.

heliocentric universe A model of the universe with the sun at the center, such as the Copernican universe.

helioseismology The study of the interior of the sun by the analysis of its modes of vibration.

helium flash The explosive ignition of helium burning that takes place in some giant stars.

Herbig–Haro objects Small nebulae that vary irregularly in brightness; believed associated with star formation.

Hertzsprung–Russell diagram A plot of the intrinsic brightness versus the surface temperature of stars. It separates the effects of temperature and surface area on stellar luminosity; commonly absolute magnitude versus spectral type, but also luminosity versus surface temperature or color.

heterogeneous accretion The formation of a planet by the accumulation of planetesimals of different composition—e.g., first iron particles, then silicates. (See **homogeneous accretion**.)

high-velocity star A star with a large space velocity. Such stars are halo stars passing through the disk of the galaxy at steep angles.

Hirayama families Families of asteroids with orbits of similar size, shape, and orientation; believed to be fragments of larger bodies.

homogeneity The assumption that, on the large scale, matter is uniformly spread through the universe.

homogeneous accretion The formation of a planet by the accumulation of planetesimals of the same composition. (See **heterogeneous accretion**.)

horizon problem In cosmology the circumstance that the primordial background radiation seems much more isotropic than could be explained by the standard big bang theory.

horizontal branch In the H–R diagram of a globular cluster, the sequence of stars extending from the red giants toward the blue side of the diagram; includes RR Lyrae stars.

horoscope A chart showing the positions of the sun, moon, planets, and constellations at the time of a person's birth; used in astrology to attempt to read character or foretell the future.

horseshoe orbit The path followed by a small moon when it occupies the same orbit as a larger moon. The small moon moves along a horseshoe-shaped orbit with respect to the larger moon.

hot spot In geology a place on the earth's crust where volcanism is caused by a rising convection cell in the mantle below. In radio astronomy, a bright spot in a radio lobe.

H–R diagram (See **Hertzsprung–Russell diagram**.)

H II region A region of ionized hydrogen around a hot star.

Hubble constant (H) A measure of the rate of expansion of the universe; the average value of velocity of recession divided by distance; presently believed to be about 50 km/sec/megaparsec.

Hubble law The linear relation between the distance to a galaxy and its radial velocity.

hydrostatic equilibrium The balance between the weight of the material pressing downward on a layer in a star and the pressure in that layer.

image intensifier An electronic device that increases the brightness of telescope images.

inflationary universe A version of the big bang theory that includes a rapid expansion when the universe was very young.

infrared cirrus A fine network of filaments covering the sky detected in the far infrared by the IRAS satellite; believed associated with dust in the interstellar medium.

infrared outburst A sudden brightening of an object at infrared wavelengths.

infrared radiation Electromagnetic radiation with wavelengths intermediate between visible light and radio waves.

instability strip The region of the H–R diagram in which stars are unstable to pulsation. A star passing through this strip becomes a variable star.

intercrater plains The relatively smooth terrain on Mercury.

interglacial period A period when ice sheets melt back and the climate is warmer.

interstellar absorption lines Dark lines in some stellar spectra that are formed by interstellar gas.

interstellar medium The gas and dust distributed between the stars.

interstellar reddening The process in which dust scatters blue light out of starlight and makes the stars look redder.

inverse square law The rule that the strength of an effect (such as gravity) decreases in proportion as the distance squared increases.

Io flux tube A tube of magnetic lines and electric currents connecting Io and Jupiter.

ion An atom that has lost or gained one or more electrons.

ionization The process in which atoms lose or gain electrons.

iron meteorite A meteorite composed mainly of iron–nickel alloy.

irregular galaxy A galaxy with a chaotic appearance, large clouds of gas and dust, and both population I and population II stars, but without spiral arms.

isotopes Atoms that have the same number of protons but a different number of neutrons.

isotropy The assumption that in its general properties the universe looks the same in every direction.

joule (J) A unit of energy equivalent to a force of 1 newton acting over a distance of 1 meter; 1 joule per second equals 1 watt of power.

Jovian planets Jupiterlike planets with large diameters and low densities.

Julian calendar The calendar established in 46 BC by Julius Caesar. It included a leap day every 4 years.

jumbled terrain Strangely disturbed regions of the moon opposite the locations of the Imbrium Basin and Mare Oriental.

Kelvin temperature scale The temperature, in Celsius (Centigrade) degrees, measured above absolute zero.

Keplerian motion Orbital motion in accord with Kepler's laws of planetary motion.

Kerr black hole A solution to the equations of general relativity that describes the properties of a rotating black hole.

kiloparsec (kpc) A unit of distance equal to 1000 pc, or 3260 ly.

Kirchoff's laws A set of laws that describes the absorption and emission of light by matter.

Lagrangian points Points of stability in the orbital plane of a binary star system, planet, or moon. One is located 60° ahead and one 60° behind the orbiting bodies. Another is located between the orbiting bodies.

large-impact hypothesis The theory that the moon formed from debris ejected during a collision between the earth and a large planetesimal.

life zone A region around a star within which a planet can have temperatures that permit the existence of liquid water.

light curve A graph of brightness versus time commonly used in analyzing variable stars and eclipsing binaries.

light-gathering power The ability of a telescope to collect light. Proportional to the area of the telescope objective lens or mirror.

lighthouse theory The theory that a neutron star produces pulses of radiation by sweeping radio beams around the sky as it rotates.

light-year (ly) The distance light travels in 1 year.

limb The edge of the apparent disk of a body, as in "the limb of the moon."

limb darkening The decrease in brightness of the sun or other body from its center to its limb.

line of nodes The line across an orbit connecting the nodes; commonly applied to the orbit of the moon.

line profile A graph of light intensity versus wavelength showing the shape of an absorption line.

liquid metallic hydrogen A form of hydrogen under high pressure that is a good electrical conductor.

lobate scarp A curved cliff such as those found on Mercury.

local apparent time The time defined by the true location of the sun in the sky; the time kept by a sundial.

local celestial meridian A north–south line around the sky passing through the zenith and nadir.

local hypothesis The theory that quasars were not at great distances but relatively nearby.

local mean time The time defined by the location the sun would have if it moved at a constant rate along the ecliptic.

long-period variable A variable star with a period ranging from 100 days to over 400 days.

look-back time The amount by which we look into the past when we look at a distant galaxy; a time equal to the distance to the galaxy in light-years.

luminosity The total amount of energy a star radiates in 1 second.

luminosity class A category of stars of similar luminosity; determined by the widths of lines in their spectra.

lunar eclipse The darkening of the moon when it moves through the earth's shadow.

Lyman series Spectral lines in the ultraviolet spectrum of hydrogen produced by transitions whose lowest orbit is the ground state.

Magellanic Clouds Small, irregular galaxies that are companions to the Milky Way; visible in the southern sky.

magnetosphere The volume of space around a planet within which the motion of charged particles is dominated by the planetary magnetic field rather than the solar wind.

magnifying power The ability of a telescope to make an image larger.

magnitude scale The astronomical brightness scale. The larger the number, the fainter the star.

main sequence The region of the H–R diagram running from upper left to lower right, which includes roughly 90 percent of all stars.

mantle The layer of dense rock and metal oxides that lies between the molten core and the surface of the earth; also, similar layers in other planets.

mare (sea) One of the lunar lowlands filled by successive flows of dark lava.

mass A measure of the amount of matter making up an object.

mass function A measure of the ratio of the masses in a single-line spectroscopic binary. Also includes the inclination, which is unknown for such systems.

mass–luminosity relation The more massive a star is, the more luminous it is.

Maunder butterfly diagram A graph showing the latitude of sunspots versus time; first plotted by W. W. Maunder in 1904.

Maunder minimum A period of less numerous sunspots and other solar activity from 1645–1715.

mean solar day The average time between two passages of the sun across the local celestial meridian.

megalith A very large stone used in a prehistoric structure such as Stonehenge.

megaparsec (Mpc) A unit of distance equal to 1 million pc.

metals In astronomical usage, all atoms heavier than helium.

meteor A small bit of matter heated by friction to incandescent vapor as it falls into earth's atmosphere.

meteorite A meteor that has survived its passage through the atmosphere and strikes the ground.

meteoroid A meteor in space before it enters the earth's atmosphere.

midocean rift Chasms that split the midocean rises where crustal plates move apart.

midocean rise One of the undersea mountain ranges that push up from the seafloor in the center of the oceans.

Milankovitch theory The theory that small changes in the orbital and rotational motions of the earth cause the ice ages.

Miller experiment An experiment that reproduced the conditions under which life began on earth and manufactured amino acids and other organic compounds.

minute of arc An angular measure. Each degree is divided into 60 minutes of arc.

missing mass Unobserved mass in clusters of galaxies believed to provide sufficient gravity to bind the cluster together.

model In science a mental conception of how a specific aspect of nature works; could be expressed mathematically.

molecular cloud An interstellar gas cloud that is dense enough for the formation of molecules; discovered and studied through the radio emissions of such molecules.

molecule Two or more atoms bonded together.

momentum The tendency of a moving object to continue moving; mathematically, the product of mass and velocity.

morning star Any planet visible in the sky just before sunrise.

mutant Offspring born with altered DNA.

nadir The point on the celestial sphere directly opposite the zenith.

nanometer A unit of length equal to 10^{-9} m.

natural motion In Aristotelian physics, the motion of objects toward their natural places—fire and air upward and earth and water downward.

natural selection The process by which the best traits are passed on, allowing the most able to survive.

nebula A cloud of gas and dust in space.

neap tides Ocean tides of low amplitude occurring at first- and third-quarter moon.

neutrino A neutral, massless atomic particle that travels at the speed of light.

neutron An atomic particle with no charge and about the same mass as a proton.

neutron star A small, highly dense star composed almost entirely of tightly packed neutrons; radius about 10 km.

newton (N) A unit of force. One newton is the force needed to accelerate a mass of 1 kilogram by 1 meter per second in 1 second.

Newtonian focus The focal arrangement of a reflecting telescope in which a diagonal mirror reflects light out the side of the telescope tube for easier access.

node The points where an object's orbit passes through the plane of the earth's orbit.

noncosmological red shift A galaxy red shift caused by something other than the expansion of the universe. No conclusive examples are known.

north celestial pole The point on the celestial sphere directly above the earth's North Pole.

north circumpolar constellation One of the constellations near the north celestial pole. Such constellations never set as seen from moderate northern latitudes.

nova From the Latin "new," a sudden brightening of a star, making it appear as a "new" star in the sky; believed associated with eruptions on white dwarfs in binary systems.

nuclear bulge The spherical cloud of stars that lies at the center of spiral galaxies.

nucleosynthesis The production of elements heavier than helium by the fusion of atomic nuclei in stars and during supernovae explosions.

nucleus (of an atom) The central core of an atom containing protons and neutrons; carries a net positive charge.

objective lens In a refracting telescope, the long-focal-length lens that forms an image of the object viewed; the lens closest to the object.

objective mirror In a reflecting telescope, the principal mirror (reflecting surface) that forms an image of the object viewed.

oblateness The flattening of a spherical body; usually caused by rotation.

oblate spheroid A sphere flattened such that its polar diameter is smaller than its equatorial diameter.

occultation The passage of a larger body in front of a smaller body.

Olbers' paradox The conflict between observation and theory as to why the night sky should or should not be dark.

135 km/sec arm A receding cloud of neutral hydrogen lying on the far side of the galactic center.

Oort cloud theory The hypothesis that the source of comets is a swarm of icy bodies believed to lie in a spherical shell 50,000 AU from the sun.

opacity The resistance of a gas to the passage of radiation.

open orbit An orbit that does not return to its starting point; an escape orbit. (See **closed orbit**.)

open star cluster A cluster of 10 to 10,000 stars with an open, transparent appearance. The stars are not tightly grouped. Usually relatively young and located in the disk of the galaxy.

open universe A model universe in which the average density is less than the critical density needed to halt the expansion.

optical binary A binary star in which the stars are only apparently associated. One star is nearby and one is more distant.

oscillating universe theory The theory that the universe begins with a big bang, expands, is slowed by its own gravity, and then falls back to create another big bang.

outgassing The release of gases from a planet's interior.

ozone layer In earth's atmosphere, a layer of oxygen ions (O_3) lying 15 to 30 km high that protects the surface by absorbing ultraviolet radiation.

parallax (p) The apparent change in the position of an object due to a change in the location of the observer. Astronomical parallax is measured in seconds of arc.

parsec (pc) The distance to a hypothetical star whose parallax is one second of arc; 1 pc = 206,265 AU = 3.26 ly.

partial eclipse A lunar eclipse in which the moon does not completely enter the earth's shadow; a solar eclipse in which the moon does not completely cover the sun.

Paschen series Spectral lines in the infrared spectrum of hydrogen produced by transitions whose lowest orbit is the third.

path of totality The track of the moon's umbral shadow over the earth's surface. The sun is totally eclipsed as seen from within this path.

penumbra The portion of a shadow that is only partially shaded.

penumbral eclipse A lunar eclipse in which the moon enters the penumbra of the earth's shadow but does not reach the umbra.

perfect cosmological principle The belief that, in general properties, the universe looks the same from every location in space at any time.

perigee The orbital point of closest approach to the earth.

perihelion The orbital point of closest approach to the sun.

period–luminosity diagram A graph showing the relation between period of pulsation and intrinsic brightness among Cepheid variable stars.

permitted orbit One of the energy levels in an atom that an electron may occupy.

photometer An instrument used to measure the intensity and color of starlight.

photon A quantum of electromagnetic energy. Carries an amount of energy that depends inversely on its wavelength.

photosphere The bright visible surface of the sun.

planetary nebula An expanding shell of gas ejected from a star during the latter stages of its evolution.

planetesimal One of the small bodies that formed from the solar nebula and eventually grew into protoplanets.

plastic A material with the properties of a solid but capable of flowing under pressure.

plate tectonics The constant destruction and renewal of earth's surface by the motion of sections of crust.

polar axis The axis around which a celestial body rotates.

poor galaxy cluster An irregularly shaped cluster that contains fewer than 1000 galaxies, many spiral, and no giant ellipticals.

population I Stars rich in atoms heavier than helium; nearly always relatively young stars found in the disk of the galaxy.

population II Stars poor in atoms heavier than helium; nearly always relatively old stars found in the halo, globular clusters, or the nuclear bulge.

position angle The angular direction of one body with respect to another; measured from north toward the east; typically used in the study of visual binaries.

precession The slow change in the direction of the earth's axis of rotation. One cycle takes nearly 26,000 years.

pressure broadening The blurring of spectral lines due to the gas pressure in a star's atmosphere.

pressure (P) waves In geophysics, mechanical waves of compression and rarefaction that travel through the earth's interior.

primary minimum In the light curve of an eclipsing binary, the deeper eclipse.

prime focus The point at which the objective mirror forms an image in a reflecting telescope.

primeval atmosphere Earth's first air, composed of gases from the solar nebula.

primordial background radiation Radiation from the hot clouds of the big bang explosion. Because of its large red shift it appears to come from a body whose temperature is only 2.7K.

primordial black holes Low-mass black holes that may have formed during the big bang explosion.

primordial soup The rich solution of organic molecules in the earth's first oceans.

prolate spheroid A sphere stretched along its polar axis so its polar diameter is greater than its equatorial diameter.

prominences Eruptions on the solar surface. Visible during total solar eclipses.

proper motion The rate at which a star moves across the sky. Measured in seconds of arc per year.

proteins Complex molecules composed of amino acid units.

proton A positively charged atomic particle contained in the nucleus of atoms. The nucleus of a hydrogen atom.

proton–proton chain A series of three nuclear reactions that build a helium atom by adding together protons. The main energy source in the sun.

protoplanet Massive object resulting from the coalescence of planetesimals in the solar nebula and destined to become a planet.

protostar A collapsing cloud of gas and dust destined to become a star.

pulsar A source of short, precisely timed radio bursts; believed to be spinning neutron stars.

quantum mechanics The study of the behavior of atoms and atomic particles.

quasar (quasi-stellar object, or QSO) Small, powerful source of energy believed to be the active core of very distant galaxies.

quasi-periodic object Certain X-ray sources that "flicker" rapidly for short intervals.

radial velocity (V_r) That component of an object's velocity directed away from or toward the earth.

radial velocity curve A graph of the velocity of recession or approach of the stars in a spectroscopic binary.

radiant The point in the sky from which meteors in a shower seem to come.

radiation pressure The force exerted on the surface of a body by its absorption of light. Small particles floating in the solar system can be blown outward by the pressure of the sunlight.

radio galaxy A galaxy that is a strong source of radio signals.

radio interferometer Two or more radio telescopes that combine their signals to achieve the resolving power of a larger telescope.

rays Ejecta from meteorite impacts forming white streamers radiating from some lunar craters.

recombination The stage within 10^6 years of the big bang when the gas became transparent to radiation.

recurrent novae Stars that erupt as novae every few dozen years.

red dwarf Cool, low-mass stars on the lower main sequence.

red shift The lengthening of the wavelengths of light seen when the source and observer are receding from each other.

reflecting telescope A telescope that uses a concave mirror to focus light into an image.

reflection nebula A nebula produced by starlight reflecting off of dust particles in the interstellar medium.

refracting telescope A telescope that forms images by bending (refracting) light with a lens.

regolith A soil made up of crushed rock fragments.

relative age The age of a geological feature referred to other features. For example, relative ages tell us the lunar maria are younger than the highlands.

relativistic jet model An explanation of superluminal expansion based on a high-velocity jet from a quasar directed approximately toward the earth.

relativistic red shift The red shift due to the Doppler effect for objects traveling near the speed of light.

resolving power The ability of a telescope to reveal fine detail; depends on the diameter of the telescope objective.

resonance The coincidental agreement between two periodic phenomena; commonly applied to agreements between orbital periods, which can make orbits more or less stable.

rest mass The mass of a particle as measured by an observer not moving with respect to the particle.

retrograde loop The apparent backward (westward) motion of planets as seen against the background of stars.

retrograde motion Backward motion; in the sky, westward motion. In the solar system, clockwise as seen from the north.

retrograde orbit An orbit that carries a moon or planet in the opposite direction from most other motion in the solar system; that is, retrograde orbits are clockwise as seen from the north.

rich galaxy cluster A cluster containing over 1000 galaxies, mostly elliptical, scattered over a volume about 3 Mpc in diameter.

rift valley A long, straight, deep valley produced by the separation of crustal plates.

right ascension (R.A.) A coordinate used on the celestial sphere just as longitude is used on earth. An object's right ascension is measured eastward from the vernal equinox.

ring galaxy A galaxy that resembles a ring around a bright nucleus; believed to be the result of a head-on collision of two galaxies.

RNA (ribonucleic acid) Long carbon-chain molecules that use the information stored in DNA to manufacture complex molecules necessary to the organism.

Roche limit The minimum distance between a planet and a satellite that holds itself together by its own gravity. If a satellite's orbit brings it within its planet's Roche limit, tidal forces will pull the satellite apart.

Roche surface The outer boundary of the volume of space that a star's gravity can control within a binary system.

rolling plains The most common type of terrain on Venus.

rotation curve A graph of orbital velocity versus radius in the disk of a galaxy.

rotation curve method A method of determining a galaxy's mass by observing the orbital velocity and orbital radius of stars in the galaxy.

RR Lyrae variable stars   Variable stars with periods of 12–24 hours; common in some globular clusters.

Sagittarius A   The powerful radio source located at the core of the Milky Way galaxy.

saros cycle   An 18-year 11-day period after which the pattern of lunar and solar eclipses repeats.

Schmidt camera   A photographic telescope that takes wide-angle photographs.

Schmidt–Cassegrain telescope   A Cassegrain telescope that uses a thin correcting lens as in a Schmidt camera.

Schwarzschild radius (R_s)   The radius of the event horizon around a black hole.

scientific notation   The system of recording very large or very small numbers by using powers of 10.

secondary atmosphere   The gases outgassed from a planet's interior; rich in carbon dioxide.

secondary minimum   In the light curve of an eclipsing binary, the shallower eclipse.

secondary mirror   In a reflecting telescope, the mirror that reflects the light to a point of easy observation.

second of arc   An angular measure. Each minute of arc is divided into 60 seconds of arc.

seeing   Atmospheric conditions on a given night. When the atmosphere is unsteady, producing blurred images, the seeing is said to be poor.

seismic wave   A mechanical vibration that travels through the earth. Usually caused by an earthquake.

seismograph   An instrument that records seismic waves.

selection effect   An influence on the probability that certain phenomena will be detected or selected, which can alter the outcome of a survey.

self-sustaining star formation   The process by which the birth of stars compresses the surrounding gas clouds and triggers the formation of more stars; proposed to explain spiral arms.

semimajor axis   Half of the longest axis of an ellipse.

setting circle   One of two circular scales on a telescope used for setting right ascension and declination.

Seyfert galaxy   An otherwise normal spiral galaxy with an unusually bright, small core that fluctuates in brightness; believed to indicate the core is erupting.

shear (S) waves   Mechanical waves that travel through earth's interior by the vibration of particles perpendicular to the direction of wave travel.

shepherd satellite   A satellite that, by its gravitational field, confines particles to a planetary ring.

shield vocanoes   Wide, low-profile volanic cones produced by highly liquid lava.

shock wave   A sudden change in pressure that travels as an intense sound wave.

sidereal day   The period of rotation of the earth with respect to the stars.

sidereal drive   The motor and gears on a telescope that turn it westward to keep it pointed at a star.

sidereal period   The period of rotation or revolution of an astronomical body referred to the stars.

sidereal time   Time based on the rotation of the earth with respect to the stars. The sidereal time at any moment equals the right ascension of objects on the upper half of the local celestial meridian.

sidereal year   The period of the earth's revolution around the sun with respect to the stars.

single-line spectroscopic binary   A spectroscopic binary in which lines of one star are visible in the spectrum.

singularity   The object of zero radius into which the matter in a black hole is believed to fall.

sinuous rille   A narrow, winding valley on the moon caused by ancient lava flows along narrow channels.

smooth plains   Apparently young plains on Mercury formed by lava flows at or soon after the formation of the Caloris Basin.

solar constant   A measure of the energy output of the sun. The total solar energy striking 1 m^2 just above earth's atmosphere in 1 second.

solar eclipse   The event that occurs when the moon passes directly between the earth and sun, blocking our view of the sun.

solar granulation   The patchwork pattern of bright areas with dark borders observed on the sun. The tops of rising currents of hot gas in the convective zone.

solar nebula theory   The theory that the planets formed from the same cloud of gas and dust that formed the sun.

solar wind   Rapidly moving atoms and ions that escape from the solar corona and blow outward through the solar system.

south celestial pole   The point of the celestial sphere directly above the earth's South Pole.

special relativity   The first of Einstein's theories of relativity, which dealt with uniform motion.

spectral class or type   A star's position in the temperature classification system O, B, A F, G, K, and M. Based on the appearance of the star's spectrum.

spectral sequence   The arrangement of spectral classes (O, B, A, F, G, K, M) ranging from hot to cool.

spectrograph   A device that separates light by wavelength to produce a spectrum.

spectroscopic binary   A star system in which the stars are too close together to be visible separately. We see a single point of light, and only by taking a spectrum can we determine that there are two stars.

spectroscopic parallax   The method of determining a star's distance by comparing its apparent magnitude with its absolute magnitude as estimated from its spectrum.

spherical aberration   An image distortion caused by a telescope mirror failing to be exactly shaped (parabolic).

spherical component   The part of the galaxy including all matter in a spherical distribution around the center (the halo and nuclear bulge).

spicules   Small, flamelike projections in the chromosphere of the sun.

spiral arms   Long, spiral patterns of bright stars, star clusters, gas, and dust that extend from the center to the edge of the disk of spiral galaxies.

spiral galaxy   A galaxy with an obvious disk component containing gas; dust; hot, bright stars; and spiral arms.

spiral tracers   Objects used to map the spiral arms (e.g. O and B associations, open clusters, clouds of ionized hydrogen, and some types of variable stars).

spoke   A radial feature in the rings of Saturn.

sporadic meteor   A meteor not part of a meteor shower.

spring tides   Ocean tides of high amplitude that occur at full and new moon.

standard time   The local mean time on the central meridian of the time zone.

starburst galaxy   A bright blue galaxy in which many new stars are forming, believed caused by collisions between galaxies.

steady state theory   The theory (now generally abandoned) that the universe does not evolve.

stellar density function   A description of the abundance of stars of different types in space.

stellar model   A table of numbers representing the conditions in various layers within a star.

stellar parallax (p)   A measure of stellar distance. (See **parallax**.)

stony-iron meteorite   A meteorite that is a mixture of stone and iron.

stony meteorite   A meteorite composed of silicate (rocky) material.

stromatolite   A layered fossil formation caused by ancient mats of algae or bacteria that build up mineral deposits season after season.

subsolar point   The point on a planet that is directly below the sun.

summer solstice   The point on the celestial sphere where the sun is at its most northerly point; also, the time when the sun passes this point about June 22 and summer begins in the Northern Hemisphere.

sungrazer   A comet that comes very close to the sun.

sunspots   Relatively dark spots on the sun that contain intense magnetic fields.

supercluster   A cluster of galaxy clusters.

superconductor   A material that can conduct electricity with essentially zero resistance.

supergiant stars   Exceptionally luminous stars 10 to 1000 times the sun's diameter.

supergranule   A large granule on the sun's surface including many smaller granules.

superluminal expansion   The apparent expansion of parts of a quasar at speeds greater than the speed of light.

supernova remnant   The expanding shell of gas marking the site of a supernova explosion.

supernova (Type I)   The explosion of a star, believed caused by the transfer of matter to a white dwarf.

supernova (Type II)   The explosion of a star, believed caused by the collapse of a massive star.

synchrotron radiation   Radiation emitted when high-speed electrons move through a magnetic field.

synodic period   The period of rotation or revolution of a celestial body with respect to the sun.

temperature   A measure of the velocity of random motions among the atoms or molecules in a material.

terminator   The dividing line between daylight and darkness on a planet or moon.

terrestrial planets   Earthlike planets—small, dense, rocky.

third contact   During an eclipse, the moment when the edge of the sun reappears from behind the moon or when the leading edge of the moon reaches the edge of the umbra.

3-kpc arm   A cloud of neutral hydrogen moving outward from the nucleus of our galaxy at about 53 km/sec. It lies 3 kpc from the center of the galaxy.

tidal coupling   The locking of the rotation of a body to its revolution around another body.

tidal heating   The heating of a planet or satellite because of friction caused by tides.

Titius–Bode rule   A simple series of steps that produces numbers approximately matching the sizes of the planetary orbits.

total eclipse   A solar eclipse in which the moon completely covers the bright surface of the sun; a lunar eclipse in which the moon completely enters the dark shadow of the earth.

transition   The movement of an electron from one atomic orbit to another.

transverse velocity   The velocity of a star perpendicular to the line of sight.

triaxial ellipsoid   A geometrical solid whose three axes are unequal.

trilithon   Literally, "three stones"; any of the five large arches at Stonehenge composed of two uprights and a horizontal top piece.

triple alpha process   The nuclear fusion process that combines three helium nuclei (alpha particles) to make one carbon nucleus.

Trojan asteroids   Small, rocky bodies caught in Jupiter's orbit at the Lagrangian points, 60° ahead of and behind the planet.

tropical year   The time from one vernal equinox to the next.

true relative orbit   The orbit of one star in a visual binary with respect to the other star after correction for orbital inclination. (See **apparent relative orbit**.)

T Tauri stars   Young stars surrounded by gas and dust. Believed to be contracting toward the main sequence.

tuning fork diagram   A system of classification for elliptical, spiral, and irregular galaxies.

turnoff point   The point in an H–R diagram where a cluster's stars turn off of the main sequence and move toward the red giant region, revealing the approximate age of the cluster.

21-cm radiation   Radio emission produced by cold, low-density hydrogen in interstellar space.

twin paradox   The seeming contradiction when one twin travels near the speed of light and returns younger than the twin who stayed behind. (See **dilation of time.**)

ultraviolet radiation   Electromagnetic radiation with wavelengths shorter than visible light but longer than X rays.

umbra   The region of a shadow that is totally shaded.

uncompressed density   The density a planet would have if its gravity did not compress it.

uniform circular motion   The classical belief that the perfect heavens could move only by the combination of uniform motion along circular orbits.

universality   The assumption that the physical laws observed on earth apply everywhere in the universe.

Van Allen belts   Radiation belts of high-energy particles trapped in the earth's magnetosphere.

variable star   A star whose brightness changes periodically.

velocity   A rate of travel that specifies both speed and direction.

velocity dispersion method   A method of finding a galaxy's mass by observing the range of velocities within the galaxy.

vernal equinox   The place on the celestial sphere where the sun crosses the celestial equator moving northward; also, the time of year when the sun crosses this point, about March 21, and spring begins in the Northern Hemisphere.

very long baseline interferometry (VLBI)   The use of radio telescopes located thousands of miles apart to resolve detail in radio sources.

vesicular basalt   A porous rock formed by solidified lava with trapped bubbles.

vidicon   A type of vacuum tube used to record television images.

violent motion   In Aristotelian physics, motion other than natural motion. (See **natural motion.**)

visual binary   A binary star system in which the two stars are separately visible in the telescope.

VLBI   See **very long baseline interferometry.**

water hole   The interval of the radio spectrum between the 21-cm hydrogen radiation and the 18-cm OH radiation. Likely wavelengths to use in the search for extraterrestrial life.

wavelength   The distance between successive peaks or troughs of a wave; usually represented by λ.

wavelength of maximum (λ_{max})   The wavelength at which a perfect radiator emits the maximum amount of energy; depends only on the object's temperature.

white dwarf stars   Dying stars that have collapsed to the size of the earth and are slowly cooling off; at the lower left of the H–R diagram.

Widmanstätten patterns   Bands in iron meteorites due to large crystals of nickel–iron alloys.

winter solstice   The point on the celestial sphere where the sun is farthest south; also, the time of year when the sun passes this point, about December 22, and winter begins in the Northern Hemisphere.

Zeeman effect   The splitting of spectral lines into multiple components when the atoms are in a magnetic field.

zero-age main sequence (ZAMS)   The locus in the H–R diagram where stars first reach stability as hydrogen burning stars.

zodiac   The band around the sky centered on the ecliptic within which the planets move.

zone of avoidance   The region around the Milky Way where almost no galaxies are visible because our view is blocked by dust in our galaxy.

zones   Yellow-white regions that circle Jupiter parallel to its equator; believed to be areas of rising gas.

Answers to Even-Numbered Problems

CHAPTER 1 **2.** 1.3 sec **4.** 1.8×10^7 km

CHAPTER 2 **2.** 4 **4.** 2800 **6.** A is brighter than B by a factor of 170. **8.** 66.5°; 113.5° **10.** The sun would follow a path that coincides with the location of the celestial equator rising directly east and setting directly west.

CHAPTER 3 **2.** (a) full, (b) first quarter, (c) waxing gibbous, (d) waxing crescent **4.** 12, 12.4. The moon orbits the earth and moves eastward about 13° per day. **6.** About 32 arc seconds **8.** (a) The moon won't be full until Oct. 17. (b) The moon will no longer be near the node of its orbit. **10.** August 12, 2026 [July 10, 1972 + 3 × (6585$\frac{1}{3}$ days). To get August 12 instead of August 11, you must take into account the number of leap days in the interval.]

CHAPTER 4 **2.** Retrograde loops: Jupiter, Saturn, Uranus, Neptune, and Pluto. Never seen as crescents: Jupiter, Saturn, Neptune, and Pluto **4.** Mars, about 18 seconds of arc; Saturn's rings, about 44 seconds of arc **6.** $\sqrt{27}$ = 5.2 years

CHAPTER 5 **2.** 19.6 m/s; 39.2 m/s **4.** 3070 m/s; 86,600 s (24 hours) **6.** Circular velocity would be 29.8 m/s. Yes, a 90 mph fast ball travels at 40 m/s.

CHAPTER 6 **2.** Short radio waves **4.** The 10 m Keck telescope has a light-gathering power that is 1.56 million times greater than the human eye. **6.** No, his resolving power should have been about 5.8 seconds of arc at best. **8.** 0.013 m (1.3 cm or about 0.5 inch) **10.** About 50 cm (From 400 km above, a human is about 0.25 seconds of arc from shoulder to shoulder.)

CHAPTER 7 **2.** 150 nm **4.** By a factor of 16 **6.** 250 nm **8.** a. B; b. F; c. M; d, K **10.** About 0.58 nm

CHAPTER 8 **2.** 730 km **4.** About 3.6 times **6.** 400,000 years

CHAPTER 9 **2.** 63 pc; absolute magnitude is 2 **4.** About B7 **6.** About 1580 solar luminosities **8.** 160 pc **10.** a, c, c, d (Use Figure 9-13 to determine the absolute magnitudes.)

CHAPTER 10 **2.** 1.28 solar masses **4.** 6.48 solar masses; mass ratio about 1.8; 4.17 and 2.31 solar masses **6.** a = 24.8 sec of arc or 84.5 AU; the total mass is about 1.4 solar masses. **8.** 3.69 days or 0.010 years; about 1.2 solar masses; about 0.67 and 0.53 solar masses **10.** 1.38×10^6 km, about 0.002 solar radius

CHAPTER 11 **2.** 60,000 nm **4.** 0.0001 **6.** 1.5×10^6 years **8.** 4.2×10^{35} kg or 210,000 solar masses (Note: each hydrogen molecule contains two H atoms.)

CHAPTER 12 **2.** 0.35 solar luminosity **4.** 3.2 solar radii **6.** 9×10^{16} J **8.** 0.22 kg

CHAPTER 13 **2.** About 9.8×10^6 years for a 16 solar mass star; about 5.7×10^5 years for a 50 solar mass star **4.** About 1×10^6 times smaller than present or about 1.4×10^{-6} g/cm^3 **6.** 2.4×10^{-9} or 1/420,000,000th **8.** About 3 pc **10.** 3.04 minutes early after 1 year; 30.4 minutes early after 10 years

CHAPTER 14 **2.** About 1.75 ly **4.** About 16,000 years ago **6.** About 940 years ago (approximately 1055 AD) **8.** 22 km/sec **10.** About 300 years ago

CHAPTER 15 **2.** 7.1×10^{25} J/sec or about 0.19 solar luminosity **4.** 820 km/sec (assuming mass is one solar mass) **6.** About 11 sec of arc **8.** About 490 sec

CHAPTER 16 **2.** About 11% **4.** 3.8×10^6 years **6.** Over estimate by a *factor* of 1.58 **8.** About 21 kpc **10.** 7.8×10^{10} solar masses **12.** 1500 K

CHAPTER 17 **2.** 4 Mpc **4.** 2.7×10^{10} solar masses **6.** 2.2×10^{11} solar masses

CHAPTER 18 **2.** 7.8×10^6 years **4.** 0.024 pc **6.** −28.5 **8.** About 91,000 km/sec **10.** 1.5 sec of arc; 4100 Mpc

CHAPTER 19 **2.** 1.72×10^{10} years; 11.5×10^{10} years **4.** 1.6×10^{-30} gm/cm^3 (assuming R = 1/2 distance between galaxies = 1 Mpc) **6.** 76 km/sec/Mpc **8.** 52.6 Mpc

CHAPTER 20 **2.** It will look $206,265^2 = 4.3 \times 10^{10}$ times fainter, which is about 26.6 magnitudes fainter; 22.6 mag **4.** About 3.3 times the half-life or 4.3 billion years **6.** Large amounts of methane and water ices **8.** About 1400

CHAPTER 21 **2.** About 17 per cent

CHAPTER 22 **2.** The rate at which an object radiates energy is proportional to its surface area (For example, for a blackbody the rate of emission of energy is given by $4\pi R^2 \sigma T^4$). However, the energy an object has stored as heat is proportional to its mass and hence to its volume [i.e., mass = (4/3)πR^3 (density)]. The cooling time will therefore be proportional to the mass divided by the surface area:

$$\text{cooling time} = \frac{\text{mass}}{\text{surface area}} = \frac{\frac{4}{3}\pi R^3 \times \text{density}}{4\pi R^2} = \frac{1}{3} R \times \text{density}$$

and we find that for planets of nearly the same density the one with the largest radius will take the longest time to cool. **4.** No. Their angular diameter would be only 0.5 sec of arc. They are visible in photos from orbit around the moon. **6.** 0.5 sec of arc (assuming an astronaut seen from above is 0.5 meters in diameter) **8.** 10.0016 cm

CHAPTER 23 **2.** 33,400 km (39,500 km from the center of Venus) **4.** 0.074 seconds of arc; none **6.** 1.8×10^{15} kg; 6.3 m/sec (about 14 mph)

CHAPTER 24 **2.** 35 Earth days **4.** 4.4° **6.** About 0.056 nm **8.** 5.2 m/sec

CHAPTER 25 **2.** 42 sec of arc **4.** 0.25 km/sec (570 mph) **6.** About 7 seconds; yes **8.** 0.43 km/sec **10.** 1.04×10^{26} kg (17.2 Earth masses)

CHAPTER 26 **2.** One billion **4.** 80 km in diameter **6.** 0.18 km/sec **8.** 3.28 AU and 2.5 AU **10.** 9 million km (0.06 AU) **12.** 33 earth masses

CHAPTER 27 **2.** 8.9 cm; 0.67 mm **4.** About 1.3 solar masses **6.** 380 km **8.** Pessimistic: 2×10^{-5}. Optimistic: 10^7

NOTE: Page numbers in bold typeface indicate definitions.

eccentrics (off-center circles), **58**
eclipse seasons, **47, 49**
eclipse year, **48**
eclipses, 32
   predicting, 46–50
   saros cycle of, **49**–50
   See also lunar eclipses; solar eclipses
eclipsing binaries, **195**–199
ecliptic, 20–22, **21,** 24
Eddington, Sir Arthur, 241
Eddy, John, 53, 164
Egg Nebula, 268
EGGs (evaporating gaseous globules), 225, 226
Einstein, Albert, 77, 90–95, 291, 372
   general theory of relativity, 93–95, 382–383
   special theory of relativity, 90–93
Einstein Cross, 397
Einstein Observatory, 118
ejecta (impact crater debris), **448**
ejection, from the solar system, 422
electromagnetic radiation, 99–100
   21-cm radiation, **210**–211
   black body radiation, **126**
   cosmic rays, **119**–120
   energy transport by, 231–232
   gamma rays, 100, 101, 118
   infrared radiation, **100,** 101, 212–213
   synchrotron radiation, **276**–277, 288, 359
   ultraviolet radiation, **100,** 101, 116, 214
   See also light; radio waves; X rays
electromagnetic spectrum, 100–101
electrons, **125,** 129
   energy levels (shells), **129,** 130, 205
elements
   and big bang theory, 390, 391, 392–393
   in comets, 567
   fusion of, 250–253
   metals in stars, **314**–316, 317, 318
   origin, 315–316, 405
   and radioactive dating, 416–417
   in the sun, 139
Eliot, T. S., 600
ellipses, **71**–72
elliptical galaxies, **332**–334, 341, 350–351, 352,
   360, 398, 399
   formation, 350–351
   giant, 347–348, 362–365
Elysium region (Mars), 482, 484, 485
emission lines, **131**–132, 150, 373
emission nebulae, **204**–205
emission spectrum, **131**–132, 133, 149–150
Enceladus (moon of Saturn), 514, 517,
   519–520, 521
Encke's Gap moons (in Saturn's rings), 515
energy
   binding, **129,** 272
   chemical, 227
   from hydrogen fusion, 229–231
   jets from black holes, 301–303, 326, 364,
      365–366
   law of conservation of, 87, 88
   law of continuity of, **242**
   solar energy generation, 227–228, 235
   of stars, 226–231
energy flow, as a key to understanding, 452
energy levels (of electrons), **129,** 130
   metastable, **205**
energy transport, in stars, 231–232, 242
engineering, vs. science and technology, 501
enzymes, **582**
epicycles (planetary motion), **59,** 60

Epsilon Aurigae, 198
Epsilon Eridani, 407
equant, **59,** 61
equatorial mountings (telescopes), **105**
equinoxes, **22,** 23–24
equivalence principle, 93, 95
Eratosthenes, 58
ergosphere, **298**
erosion
   on Earth, 429
   on Mars, 485, 486
escape orbits, **86**
escape velocity, **86,** 296
   of Jovian planets, 480, 481
Eta Aquarids meteor shower, 553
Eta Carinae (supergiant star), 302
Eudoxus of Cnidus, 56, 57
Europa (moon of Jupiter), 504, 506, 509, 591
European Southern Observatory, 106
evening star, **25**
event horizon (of black holes), **296**
evidence
   in scientific enquiry, 27
   statistical, 357
evolution
   chemical, **586**
   of life, 429, 580, 583, 588, 589
   surface, **428**
evolutionary hypothesis, **403,** 404, 405
Exosat X-ray observatory, 118, 295
exponential scales, 177
Extreme Ultraviolet Explorer (EUVE), 117,
   214
eyepieces (of telescopes), **102,** 109

F
Fahrenheit temperature scale, 602
falls (meteorites), **554**
false-color images, **110,** 275, 472
Faraday, Michael, 501
field, gravitational, **82**–83
filtergrams, **150,** 160
finds (meteorites), **554**
fission, nuclear, 272
fission hypothesis, **456**
flares, solar, **160**–161
flash spectrum, **149**–150
flatness problem, big bang theory, **394**
Fleming, Alexander, 543
flocculent galaxies, **322**
flooding (lava). *See* lava flows
flooding (water), 428
   on Mars, 485–486
focal length (of lenses), **102**
foci (of ellipses), **71**
folded mountain chains, **435**
forbidden lines (emission spectra), **205**
forces of nature, four, 227
forward scattering (of light), **500**–501
fossils
   Cambrian period, 584–585
   Martian bacteria, 590–591
Frail, Dale A., 294
fraud, publication and, 300
Frederik II (King of Denmark), 68, 69
free-fall contraction, **222**
frequency, **99**–100
Frost, Robert, 98
funding, for research, 512
fusion, 230–231, 272
   of carbon, 252, 253, 271, 272

in comets, 570–573
of heavy elements, 252–253
of helium, 250–252, 271, 272, 316
of hydrogen, 229–231, 246, 248–249, 271,
   272
in the sun, 227–228

G
G stars, 324, 592
galactic cannibalism, **346**–347
galactic corona, **314**
Galatea (moon of Neptune), 540
galaxies, **305,** 330–352
   active, **356**–367
   active galactic nuclei (AGN), **356,** 361–362,
      365–367
   black holes in, 7, 364–367, 374–375
   classification, 332
   clusters, 6–7, 348–350, 388, 396
   collisions between, 345–348, 350–351
   color, 337
   dark matter, 342–344
   diameters, 340, 341
   discovery, 331–332
   distances to, 337–340, 344–345
   distribution, 348–350, 388, 389
   double-lobed radio sources, **358**–361
   dwarfs, 348, 349, 350
   elliptical. *See* elliptical galaxies
   eruptions, 361, 366
   evolution, 350–352
   flocculent, **322**
   grand design, 322, 323
   and Hubble law, 344
   irregular, 332, **336**–337, 341, 350
   life in, 596
   luminosity, 340, 341
   Magellanic Clouds, 290, 291, **306,** 336–337,
      348
   mass, 340–342, 344–345
   mergers between, 346–347, 350–351, 360
   properties, 337–345
   and quasars, 370–373
   radio, **356,** 361, 362
   ring, **346**
   Seyfert, **357**–358, 367
   SO, 335, 351
   spiral. *See* spiral galaxies
   starburst, **350**–351, 364
   structure, 350, 351–352
   tides between, 345
   tuning fork diagram, **332,** 350
   unified model, **366,** 368, 374
   See also specific galaxies
galaxy clusters, 6–7, 348–350, 388, 396
   poor, **348**–350
   rich, **348**
   superclusters, 7, **350**
galaxy seeds, **397**
Galilean moons (Jupiter), 65, 67, **414,** 504–509
Galileo Galilei, 52, 73, 74, 77, 308
   and the Copernican model, 64, 65–67
   discoveries with telescopes, 64–65, 67
   and Jupiter's moons, 65, 67
   and motion, 78–80, 81
   and Neptune, 537
   and Saturn's rings, 512
Galileo spacecraft
   asteroid images, 415, 561, 562, 563, 565
   Jupiter, 494, 498, 499, 500, 505, 507, 508
Galle, Johann, 537

Periodic Table of the Elements

Group

Noble Gases

IA(1)

Atomic number →	11
Symbol →	Na
Atomic mass →	22.99

Atomic masses are based on carbon-12. Numbers in parentheses are mass numbers of most stable or best known isotopes of radioactive elements.

Transition Elements

Inner Transition Elements

Period

	IA(1)	IIA(2)	IIIB(3)	IVB(4)	VB(5)	VIB(6)	VIIB(7)	(8) VIII	(9)	(10)	IB(11)	IIB(12)	IIIA(13)	IVA(14)	VA(15)	VIA(16)	VIIA(17)	(18)
1	1 H 1.008																	2 He 4.003
2	3 Li 6.941	4 Be 9.012											5 B 10.81	6 C 12.01	7 N 14.01	8 O 16.00	9 F 19.00	10 Ne 20.18
3	11 Na 22.99	12 Mg 24.31											13 Al 26.98	14 Si 28.09	15 P 30.97	16 S 32.06	17 Cl 35.45	18 Ar 39.95
4	19 K 39.10	20 Ca 40.08	21 Sc 44.96	22 Ti 47.90	23 V 50.94	24 Cr 52.00	25 Mn 54.94	26 Fe 55.85	27 Co 58.93	28 Ni 58.7	29 Cu 63.55	30 Zn 65.38	31 Ga 69.72	32 Ge 72.59	33 As 74.92	34 Se 78.96	35 Br 79.90	36 Kr 83.80
5	37 Rb 85.47	38 Sr 87.62	39 Y 88.91	40 Zr 91.22	41 Nb 92.91	42 Mo 95.94	43 Tc 98.91	44 Ru 101.1	45 Rh 102.9	46 Pd 106.4	47 Ag 107.9	48 Cd 112.4	49 In 114.8	50 Sn 118.7	51 Sb 121.8	52 Te 127.6	53 I 126.9	54 Xe 131.3
6	55 Cs 132.9	56 Ba 137.3	57* La 138.9	72 Hf 178.5	73 Ta 180.9	74 W 183.9	75 Re 186.2	76 Os 190.2	77 Ir 192.2	78 Pt 195.1	79 Au 197.0	80 Hg 200.6	81 Tl 204.4	82 Pb 207.2	83 Bi 209.0	84 Po (210)	85 At (210)	86 Rn (222)
7	87 Fr (223)	88 Ra 226.0	89** Ac (227)	104 Unq (261)	105 Unp (262)	106 Unh (263)	107 Uns (262)	108 Uno (265)	109 Une (266)									

	58 Ce 140.1	59 Pr 140.9	60 Nd 144.2	61 Pm (145)	62 Sm 150.4	63 Eu 152.0	64 Gd 157.3	65 Tb 158.9	66 Dy 162.5	67 Ho 164.9	68 Er 167.3	69 Tm 168.9	70 Yb 173.0	71 Lu 175.0
Lanthanide Series *6														
Actinide Series **7	90 Th 232.0	91 Pa 231.0	92 U 238.0	93 Np 237.0	94 Pu (244)	95 Am (243)	96 Cm (247)	97 Bk (247)	98 Cf (251)	99 Es (252)	100 Fm (257)	101 Md (258)	102 No (259)	103 Lr (260)

Greatest Elongations of Mercury

Evening Sky	Morning Sky
Jan. 2, 1996	Feb. 11, 1996
April 23, 1996*	June 10, 1996
Aug. 21, 1996	Oct. 3, 1996*
Dec. 15, 1996	Jan. 24, 1997
April 6, 1997*	May 22, 1997
Aug. 4, 1997	Sept. 16, 1997*
Nov. 28, 1997	Jan. 6, 1998
March 20, 1998*	May 4, 1998
July 17, 1998	Aug. 31, 1998*
Nov. 11, 1998	Dec. 20, 1998
March 3, 1999*	April 16, 1999
June 28, 1999	Aug. 14, 1999
Oct. 24, 1999	Dec. 3, 1999
Feb. 15, 2000	March 28, 2000
June 9, 2000	July 27, 2000
Oct. 6, 2000	Nov. 15, 2000
Jan. 28, 2001	March 11, 2001
May 22, 2001	July 9, 2001
Sept. 18, 2001	Oct. 29, 2001*

*Most favorable elongations.

Greatest Elongations of Venus

Evening Sky	Morning Sky
April 1, 1996	Aug. 20, 1996
Nov. 6, 1997	March 27, 1998
June 11, 1999	Oct. 30, 1999
Jan. 17, 2001	June 8, 2001
Aug. 22, 2002	Jan. 11, 2003
March 29, 2004	Aug. 17, 2004
Nov. 3, 2005	March 12, .2007
May 27, 2007	Oct. 15, 2007

Meteor Showers

Shower	Dates	Hourly Rate	Radiant R.A.	Radiant Dec.	Associated Comet
Quadrantids	Jan. 2–4	30	15^h24^m	$50°$	
Lyrids	April 20–22	8	18^h4^m	$33°$	1861 I
η Aquarids	May 2–7	10	22^h24^m	$0°$	Halley?
δ Aquarids	July 26–31	15	22^h36^m	$-10°$	
Perseids	Aug. 10–14	40	3^h4^m	$58°$	1982 III
Orionids	Oct. 18–23	15	6^h20^m	$15°$	Halley?
Taurids	Nov. 1–7	8	3^h40^m	$17°$	Encke
Leonids	Nov. 14–19	6	10^h12^m	$22°$	1866 I Temp
Geminids	Dec. 10–13	50	7^h28^m	$32°$	